SEX, ECOLOGY, SPIRITUALITY

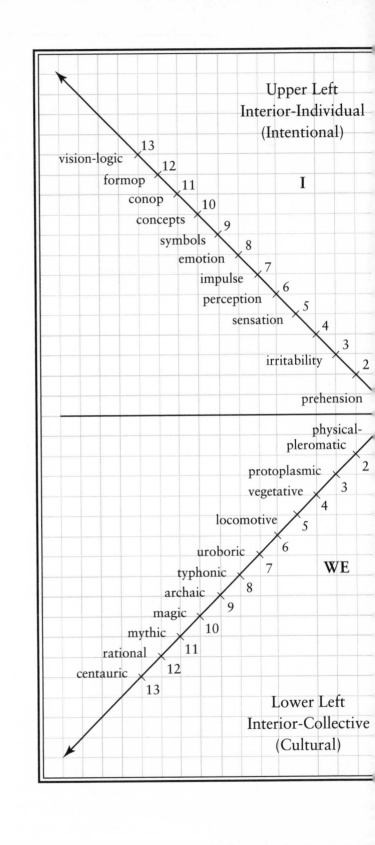

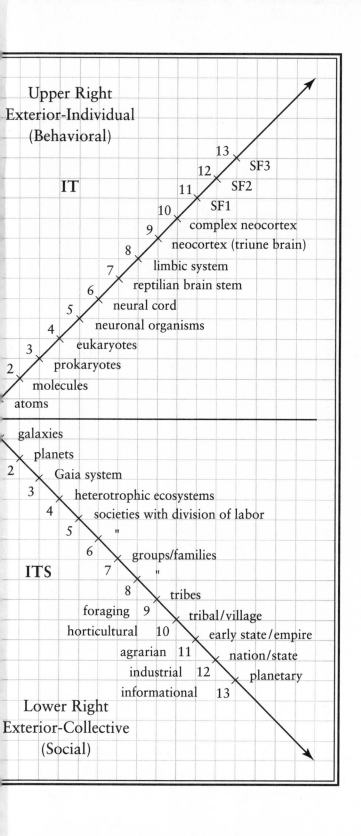

Upper Right
Exterior-Individual
(Behavioral)

IT

13
12   SF3
11   SF2
10   SF1
9   complex neocortex
8   neocortex (triune brain)
7   limbic system
6   reptilian brain stem
5   neural cord
4   neuronal organisms
3   eukaryotes
2   prokaryotes
        molecules
atoms

galaxies
planets
2   Gaia system
3   heterotrophic ecosystems
4   societies with division of labor
5   "
6   groups/families
ITS   7   "
        8   tribes
foraging   9   tribal/village
horticultural   10   early state/empire
    agrarian   11   nation/state
    industrial 12   planetary
    informational   13

Lower Right
Exterior-Collective
(Social)

SEX, ECOLOGY, SPIRITUALITY

The Spirit of Evolution

SECOND EDITION, REVISED

Ken Wilber

SHAMBHALA

Boston & London

2000

SHAMBHALA PUBLICATIONS, INC.
Horticultural Hall
300 Massachusetts Avenue
Boston, Massachusetts 02115
www.shambhala.com

© 1995, 2000 by Ken Wilber

9 8 7 6 5 4 3 2

SECOND EDITION, REVISED
Printed in the United States of America
⊗This edition is printed on acid-free paper that meets the
American National Standards Institute z39.48 Standard.
Distributed in the United States by Random House, Inc.,
and in Canada by Random House of Canada Ltd

Library of Congress Cataloging-in-Publication Data
Wilber, Ken.
Sex, ecology, spirituality: the spirit of evolution/
Ken Wilber.—2nd ed.
p.   cm.
Includes bibliographical references and index.
ISBN 1-57062-744-4 (alk. paper)
1. Cosmology. 2. Consciousness. 3. Evolution—Philosophy.
4. Whole and parts (Philosophy) 5. God. I. Title.
BD511 .W54 2000
110–dc21
00-044003

Contents

LIST OF FIGURES

Preface to the Second Edition, Revised

The Genesis of *Sex, Ecology, Spirituality*

SEX, ECOLOGY, SPIRITUALITY was the first theoretical book I had written in almost ten years, following the events described in *Grace and Grit*. The previous book, *Transformations of Consciousness* (with Jack Engler and Daniel P. Brown), was completed in 1984; I wrote *Grace and Grit* in 1991; and then I settled down to finally write a textbook of psychology that I had been planning on doing for several years. I was calling that textbook *System, Self, and Structure*, but somehow it never seemed to get written. Determined to complete it, I sat down and began transcribing the two-volume work, whereupon I realized, with a shock, that four of the words I used in the very first paragraph were no longer allowed in academic discourse (development, hierarchy, transcendental, universal). This, needless to say, put a considerable cramp in my attempt to write this book, and poor *System, Self, and Structure* was, yet again, shelved. (I recently brought out an abridged version of it with the title *Integral Psychology*.)

What had happened in my ten-year writing hiatus, and to which I had paid insufficient attention, is that extreme postmodernism had rather completely invaded academia in general and cultural studies in particular—even the alternative colleges and institutes were speaking postmodernese with an authoritarian thunder. The politically correct were policing the types of serious discourse that could, and could not, be uttered in academe. Pluralistic relativism was the only acceptable worldview. It claimed that all truth is culturally situated (except its own truth, which is true for all cultures); it claimed there are no transcendental truths (except its own pronouncements, which transcend specific contexts); it claimed that all hierarchies or value rankings are oppressive and marginalizing (except its own value ranking, which is superior to the alternatives); it claimed that there are no universal truths (except its own pluralism, which is universally true for all peoples).

The downsides of extreme postmodernism and pluralistic relativism are now well known and widely acknowledged, but at the time I was trying to write *System, Self, and Structure,* they were thought to be gospel and were as religiously embraced, making any sort of developmental and transcendental studies anathema. I therefore set *System, Self, and Structure* aside and began to ponder the best way to proceed, feeling rather like a salmon who had first to swim upstream in order to have any fun at all.

But I have been dwelling merely on the downsides of postmodernism and pluralistic relativism. Their positive benefits are equally numerous and far-reaching, and deserve a hearing as well. As I have tried to suggest in several places (e.g., *The Marriage of Sense and Soul, Integral Psychology,* and *A Theory of Everything*), pluralistic relativism is actually a very high developmental achievement, stemming from the postformal levels of consciousness, which disclose a series of very important truths. ("Postformal" means the cognitive stages lying immediately beyond linear rationality or formal operational thinking. Thus, cognitive development proceeds from sensorimotor to preoperational to concrete operational to formal operational to postformal cognition, to possibly higher modes [see below]. I also refer to postformal cognition as *network-logic* or *vision-logic*—Gebser called it *integral-aperspectival*—and it is vision-logic that drives the best of postmodernism.)

As I suggested in those publications, the truths of postmodernism include constructivism (the world is not just a perception but an interpretation); contextualism (all truths are context-dependent, and contexts are boundless); and integral-aperspectivism (no context is finally privileged, so an integral view should include multiple perspectives; pluralism; multiculturalism). All of those important truths can be derived from the beginning stages of postformal vision-logic, and postmodernism at its best is an elucidation of their profound importance.

In particular, the previous stages of concrete operational (which supports a worldview called "mythic-membership") and formal operational (which supports a worldview called "universal formalism") have inherent limitations and weaknesses in them, and these limitations, when pressed into social action, produce various types of rigid social hierarchies, mechanistic worldviews that ignore local color, and universalistic pronouncements about human beings that violate the rich differences between cultures, peoples, and places. But once consciousness evolves from formal to postformal—and thus evolves from universal formalism to pluralistic relativism—these multiple contexts and pluralistic tapestries come jumping to the fore, and postmodernism has spent much of the last two decades attempting to *deconstruct* the rigid hierarchies, formalisms, and oppressive schemes that are inherent in the preformal-to-formal stages of consciousness evolution.

But pluralistic relativism is not itself the highest wave of development, as numerous studies have consistently shown (see *Integral Psychology*). When vision-logic matures into its middle and late phases, pluralistic relativism increasingly gives way to more holistic modes of awareness, which begin to weave the pluralistic voices together into beautiful tapestries of integral intent. Pluralistic relativism gives way to *universal integralism*. Where pluralism frees the many different voices and multiple contexts, universal integralism begins to bring them together into a harmonized chorus. (Universal integralism thus stands on the brink of even higher developments, which directly disclose the transpersonal and spiritual realms—developments wherein the postformal mental gives way to the postmental or supramental altogether.)

But this leaves pluralistic relativism in a difficult position. Having heroically developed beyond a rigid universal formalism, it became suspicious of any universals at all, and thus it tended to fight the emergence of universal integralism with the same ferocity that it deconstructed all previous systems. It turned its critical guns not just on pre-pluralistic stages (which was appropriate), but also on post-pluralistic stages (which was disastrous). Deconstructive postmodernism thus began to actively fight any higher stages of growth, often turning academia into a charnel ground of deconstructive fury. Little new was created; past glories were simply torn down. Little novel was constructed; previous constructions were merely deconstructed. Few new buildings were erected; old ones were simply blown up. Postmodernism often degenerated into the nihilism and narcissism for which it is now so well known, and the vacant, haunted, hollow eyes of professional academia, peering through the smoking ruins, told the tale most sadly.

One thing was very clear to me as I struggled with how best to proceed in an intellectual climate dedicated to deconstructing anything that crossed its path: I would have to back up and start at the beginning, and try to create a vocabulary for a more constructive philosophy. Beyond pluralistic relativism is universal integralism; I therefore sought to outline a philosophy of universal integralism.

Put differently, I sought a world philosophy. I sought an *integral* philosophy, one that would believably weave together the many pluralistic contexts of science, morals, aesthetics, Eastern as well as Western philosophy, and the world's great wisdom traditions. Not on the level of details—that is finitely impossible; but on the level of orienting generalizations: a way to suggest that the world really is one, undivided, whole, and related to itself in every way: a holistic philosophy for a holistic Kosmos: a world philosophy, an integral philosophy.

Three years later, *Sex, Ecology, Spirituality* was the result. During that period I lived the hermit's life; I saw exactly four people in three years (Roger Walsh, who is an M.D., stopped by once a year to make sure I was alive); it was very much a typical three-year silent retreat (this period is described in *One Taste,* June 12 entry). I was locked into this thing, and it would not let go.

The hard part had to do with *hierarchies.* Granted, rigid social hierarchies are deplorable and oppressive social rankings are pernicious. Postmodernism has fortunately made us all more sensitive to those injustices. But even the antihierarchy critics have their own strong hierarchies (or value rankings). The postmodernists value pluralism over absolutism—and that is their value hierarchy. Even the ecophilosophers, who abhor hierarchies that place humans on the top of the evolutionary scale, have their own very strong hierarchy, which is: subatomic elements are parts of atoms, which are parts of molecules, which are parts of cells, which are parts of organisms, which are parts of ecosystems, which are parts of the biosphere. They thus value the biosphere above particular organisms, such as the human being, and they deplore our using the biosphere for our own selfish and ruinous purposes. All of that comes from their particular value hierarchy.

Feminists have several hierarchies (e.g., partnership societies are better than power societies; linking is better than ranking; liberation is better than oppression); systems theorists have hundreds of hierarchies (all natural systems are arranged hierarchically); biologists and linguists and developmental psychologists all have hierarchies. *Everybody* seemed to have some sort of hierarchy, even those who claimed they didn't. The problem is, none of them matched with the others. None of the hierarchies seemed to agree with each other. And that was the basic problem that kept me locked in my room for three years.

At one point, I had over two hundred hierarchies written out on legal pads lying all over the floor, trying to figure out how to fit them together. There were the "natural science" hierarchies, which were the easy ones, since everybody agreed with them: atoms to molecules to cells to organisms, for example. They were easy to understand because they were so graphic: organisms actually contain cells, which actually contain molecules, which actually contain atoms. You can even see this directly with a microscope. That hierarchy is one of actual embrace: cells literally embrace or enfold molecules.

The other fairly easy series of hierarchies were those discovered by the developmental psychologists. They all told variations on the cognitive hierarchy that goes from sensation to perception to impulse to image to symbol to concept to rule. The names varied, and the schemes were slightly different, but the hierarchical story was the same—each succeeding stage incorporated its predecessors and then added some new capacity. This seemed very similar to the natural science hierarchies, except they still did not match up in any obvious way. Moreover, you can actually see organisms and cells in the empirical world, but you can't see interior states of consciousness in the same way. It is not at all obvious how these hierarchies would—or even could—be related.

And those were the easy ones. There were linguistic hierarchies, contextual hierarchies, spiritual hierarchies. There were stages of development in phonetics, stellar systems, cultural worldviews, autopoietic systems, technological modes, economic structures, phylogenetic unfoldings, superconscious realizations. . . . And they simply refused to agree with each other.

G. Spencer Brown, in his remarkable book *Laws of Form,* said that new knowledge comes when you simply bear in mind what you need to know. Keep holding the problem in mind, and it will yield. The history of human beings is certainly testament to that fact. An individual runs into a problem and simply obsesses about that problem until he or she solves it. And the funny thing is: the problem is *always* solved. Sooner or later, it yields. It might take a week, a month, a year, a decade, a century, or a millennium, but the Kosmos is such that solutions are always forthcoming. For a million years, people looked at the moon and wanted to walk on it. . . .

I believe any competent person is capable of bearing problems in mind until they yield their secrets; what not everybody possesses is the requisite will, passion, or insane obsession that will let them hold the problem long enough or fiercely enough. I, at any rate, was insane enough for this particular problem, and toward the end of that three-year period, the whole thing started to become clear to me. It soon became obvious that the various hierarchies fall into four major classes (what I would eventually call the four quadrants); that some of the hierarchies are referring to individuals, some to collectives; some are about exterior realities, some are about interior ones, but they all fit together seamlessly; the ingredients of these hierarchies are *holons,* wholes that are parts of other wholes (e.g., a whole atom is part of a whole molecule, which is part of a whole cell, which is part of a whole organism, and so on); and therefore the correct word for hierarchy is actually *holarchy.* The Kosmos is a series of nests within nests within nests indefinitely, expressing greater and greater holistic embrace—holarchies of holons everywhere!—which is why *everybody* had their own value holarchy, and why, in the end, all of these holarchies intermesh and fit perfectly with all the others.

The universe is composed of holons, all the way up, all the way down. And with that, much of *Sex, Ecology, Spirituality* began to write itself. The book is divided into two parts (three actually, counting the endnotes, a separate book in themselves). Part one describes this holonic Kosmos—nests within nests within nests indefinitely—and the worldview of universal integralism that can most authentically express it. This part of the book covers a great deal of ground, and one of my regrets is that I could not include the voluminous research material and explanations that would flesh out the details with much more persuasion. As those who have seen some of the research notes will attest, many of the paragraphs in *Sex, Ecology and Spirituality* (SES) are summaries of short books. (One reviewer actually spotted this, and began the review, "No summary of this book is possible. The book, all 524 pages of text and 239 pages of notes, *is* a summary, which should reveal the depth and breadth of its scope." Other reviewers found this very irritating, but I really had no choice. I hope to be able to get these research notes into print at some point, not so much to show the material itself as to make it available for criticism and scrutiny. But that reviewer is right: SES is a summary.)

If the first part of the book attempts to outline a universal integralism—a view of the holonic Kosmos from subconscious to self-conscious to superconscious—the second part of the book attempts to explain why this holistic Kosmos is so often ignored or denied. If the universe really is a pattern of mutually interrelated patterns and processes—holarchies of holons—why do so few disciplines acknowledge this fact (apart from their own narrow specialties)? If the Kosmos is *not* holistic, not integral, not holonic—if it is a fragmented and jumbled affair, with no common contexts or linkings or joinings or communions—then fine, the world is the jumbled mess the various specialties take it to be. But if the world is holistic and holonic, then why do not more people see this? And why do many academic specialties actively deny it? If the world is whole, why do so many people see it as broken? And why, in a sense, *is* the world broken, fragmented, alienated, divided?

The second part of the book therefore looks at that which prevents us from seeing the holistic Kosmos. It looks at what I call *flatland.*

(At one point I had named part one and part two, before deciding not to narrow their content with a name; but part one was "Spirit-in-Action," and part two was "Flatland." Part two, in any event, attempts to explain why part one isn't more often seen and understood.)

When I went over this book for its inclusion in the *Collected Works,* I decided to do a second, revised edition (which is also the book you now hold in your hands), mostly because I wanted to clarify a few sections in light of the constructive criticism of the first edition. In particular, I wanted to explain more clearly the historical rise of scientific materialism (a version of flatland), and thus I added several new sections in several chapters (especially 12 and 13), along with six new diagrams, which I believe help the narrative considerably. I have also carefully gone over the endnotes, including new material where appropriate.

Speaking of the endnotes, they really were written as a book in themselves. Many of the most important ideas in SES are mentioned and developed only in the notes (such as the Basic Moral Intuition), as is much of the dialogue with other scholars (Heidegger, Foucault, Derrida, Habermas, Parmenides, Fichte, Hegel, Whitehead, Husserl) and with alternative present-day theorists (Grof, Tarnas, Berman, Spretnak, Roszak). The notes also contain a handful of polemical bursts, which I will explain in a moment. All of these have been retouched for the second edition.

Once the book was conceived, the actual writing went fairly quickly. It was published in 1995 and, I'm told, was the largest-selling academic tome in any category for that year, at one point going into three printings in four months alone. The reactions to it were extreme, from incredibly positive statements to infuriated rants. But the specific criticisms were straightforward, and they deserve a respectful hearing.

THE MAJOR CRITICISMS OF
SEX, ECOLOGY, SPIRITUALITY

Some critics of the book claimed that it too rigidly categorized various approaches and thus marginalized important differences. They therefore charged the book with various "isms" of one sort or another (sexism, anthropocentrism, speciesism, logocentrism, and invidious monism). Those defending the book claimed that most of those criticisms came from individuals whose worldviews were shown to be narrow and partial by comparison, and they were reacting in a negative fashion for that reason. Both sides refused to budge, generally.

In my opinion, there are a handful of serious criticisms that need to be addressed. Although I believe the bulk of these criticisms are based on an unfamiliarity with my work as a whole, some are more serious. Here are the major criticisms.

Piaget

One of the most common charges was that I used Piaget as the basis for my entire view of psychological development. This is very inaccurate, but I understand how the book gave that impression. One of the most difficult problems I face in writing about my ideas is that I always assume the audience has no prior knowledge of my work. With each new book I therefore must start from scratch and explain my "system" from the beginning. Usually, around the first third of a book is taken up introducing the system, and then the new material is presented in the last part of the book. This gives readers familiar with my work the impression that I am repeating myself; but this is for the benefit of those new to the game.

With SES, I did this introducing using a few shortcuts, which was perhaps a bad idea. For the higher or transpersonal stages of development, instead of explaining the stages themselves, I simply used examples of each (Emerson, Saint Teresa, Eckhart, and Sri Ramana Maharshi), and for the ontogenetic development of worldviews, I simply used the work of Jean Piaget. Many reviewers—especially the postmodern pluralists—jumped on Piaget as an example of the fact that I was using old-paradigm, hierarchical, Eurocentric, sexist schemes, and therefore the entire book was suspect.

Of course, those who were familiar with my work knew that Piaget was only one of dozens of theorists that I had attempted to integrate into a more holistic overview of development, and that, even then, I was not a strict Piagetian by any means. But before I briefly state my view, let us not rush over the attacks on Piaget too quickly, because the unfairness of those attacks applies equally to those aimed at SES. For the fact is, if we focus on the aspects of cognition that Piaget studied, his general scheme has held up to intense cross-cultural investigation. Those who attack Piaget often seem uninformed of the evidence.

After almost three decades of intense cross-cultural research, the evidence is virtually unanimous: Piaget's stages up to formal operational are universal and cross-cultural. As only one example, *Lives Across Cultures: Cross-Cultural Human Development* is a highly respected textbook written from an openly liberal perspective (which is often suspicious of "universal" stages). The authors (Harry Gardiner, Jay Mutter, and Corinne Kosmitzki) carefully review the evidence for Piaget's stages of sensorimotor, preoperational, concrete operational, and formal operational. They found that cultural settings sometimes alter the *rate* of development, or an *emphasis* on certain aspects of the stages—but not the stages themselves or their cross-cultural validity.

Thus, for sensorimotor: "In fact, the qualitative characteristics of sensorimotor development remain nearly identical in all infants studied so far, despite vast differences in their cultural environments." For preoperational and concrete operational, based on an enormous number of studies, including of Nigerians, Zambians, Iranians, Algerians, Nepalese, Asians, Senegalese, Amazon Indians, and Australian Aborigines: "What can we conclude from this vast amount of cross-cultural data? First, support for the universality of the structures or operations underlying the preoperational period is highly convincing. Second, . . . the qualitative characteristics of concrete operational development (e.g., stage sequences and reasoning styles) appear to be universal [although] the rate of cognitive development . . . is not uniform but depends on ecocultural factors." Although the authors do not use exactly these terms, they conclude that the deep features of the stages are universal but the surface features depend strongly on cultural, environmental, and ecological factors (as I would put it, all four quadrants are involved in individual development). "Finally, it appears that although the rate and level of performance at which children move through Piaget's concrete operational period depend on cultural experience, children in diverse societies still proceed in the same sequence he predicted."

Fewer individuals in any culture (Asian, African, American, or otherwise) reach formal operational cognition, and the reasons given for this vary. It might be that formal operational is a genuinely higher stage that fewer therefore reach, as I believe. It might be that formal operational is a genuine capacity but not a genuine stage, as the authors believe (i.e., only some cultures emphasize formal operational and therefore teach it). Evidence for the existence of Piaget's formal stage is therefore strong but not conclusive. Yet this one item is often used to dismiss *all* of Piaget's stages, whereas the correct conclusion, backed by enormous evidence, is that all of the stages up to formal operational have now been adequately demonstrated to be universal and cross-cultural.

I believe the stages at and beyond formop are also universal, including vision-logic and the general transrational stages, and my various books have presented substantial evidence for that. But the point is that any model that does not include Piaget's stages up to formop is an inadequate model.

Waves, Streams, and States

Although I include the Piagetian cognitive line in my model, as demanded by the cross-cultural evidence, his scheme is, as I suggested, only a small part of an overall view. In my model, there are the various levels or waves of consciousness (stretching from matter to body to mind to soul to spirit), through which pass various developmental lines or streams (including cognitive, affective, moral, interpersonal, spiritual, self-identity, needs, motivations, and so on). A person can be at a very high level in one line (say, cognitive), at a medium level of development in others (e.g., emotional intelligence), and at a low level in still others (e.g., morals). Thus, *a person's overall development follows no linear sequence whatsoever.* Development is far from a sequential, ladder-like, clunk-and-grind series of steps, but rather involves a fluid flowing of many waves and streams in the great River of Life.

Moreover, a person at virtually any wave or stage of development can experience an altered *state* of consciousness or a peak experience of any of the transpersonal realms (psychic, subtle, causal, or nondual). Thus, transpersonal peak experiences and altered states *are available to virtually anybody at virtually any stage of development;* the notion that transpersonal states are available only at higher levels of development is quite incorrect. My overall model, then, consists of waves, streams, and states, and thus there is precious little about it that is linear.

And yet that was by far the most common criticism of SES: it represented a model of merely linear development. Since I had not subscribed to a linear model since 1981 (see the Introduction to Volume Three of the *Collected Works*)—and since, in fact, I had written at length criticizing such a view (the rejection of which marked the transition from phase-2 to phase-3 in my own work)—I must confess I was astonished to see critics ascribe this view to me and then criticize it at length. A book purporting to be a dialogue with my work contained these or similar errors throughout, and it has taken several years to dig out from under those unfortunate distortions. Still, it is finally the case that, due to vigorous support by scholars of my work, one hears less and less the charge that my model is linear (it is multidimensional), or that it is Eurocentric (it is based on much cross-cultural evidence), or that it is marginalizing (holarchies transcend and include in multiple contexts), or that transpersonal experiences occur only at higher levels (they are available as states at virtually any level).

At the same time, I repeat that I understand how critics could have gotten the wrong impression if they only read SES. I should have made my overall model much clearer, which would have helped to ward off these misunderstandings. I have attempted to do so in the second edition and, obviously, in this Introduction.

Spirituality in Children and Dawn Humans

Closely related to the previous criticism was the charge that I denied any sort of spirituality to both children and early humans. This, too, is an unfortunate misrepresentation of my work, based exactly on the notion that my model is merely linear. A few critics went apoplectic at my "linear" model and accused me of things slightly worse than well-poisoning. Since my model is one of waves, streams, and states—and since spiritual states can occur at virtually any wave of unfolding—that particular criticism is considerably off the mark. I can, as I said, understand how a critic who had only read SES might get that impression, but the impression is false. (For a specific discussion of spirituality in children and in early humans, see *Integral Psychology,* chaps. 10, 11, and 12.)

As is perhaps obvious, a good deal of the major criticisms of SES were based on simple misrepresentations of my work, with blame to be shared on both sides: I did not clearly outline my overall model, and the critics were not well informed about my other works. My responses began to sound like a broken record: "That is not my view, that is not my view, that is not. . . ." Nobody got more tired of this than I.

The Treatment of Ecophilosophies

One quite accurate criticism was that I lumped together the many various ecophilosophies and treated them indiscriminately. This is true, and the criticism is well taken. In my defense, I can only say that I explained, in several endnotes, that volume 2 of the Kosmos trilogy (tentatively entitled *Sex, God, and Gender: The Ecology of Men and Women*) treats the various ecophilosophies separately and deals with each on its own terms. I was simply stating certain broad conclusions from those studies. At the same time, SES levels a very powerful critique at many, I would say most, of the current ecophilosophies, pointing out that they are, in fact, representatives of a very flatland view. One reviewer of SES concluded that "this presentation, which I believe is generally true, is fatal to most forms of ecotheory," and Michael Zimmerman (author of *Radical Ecology*) pointed out that most (not all) forms of ecophilosophy do indeed appear to be caught in flatland as described.

SES went on to suggest a type of ecophilosophy that is profoundly ecological but not in flatland terms, and, in my own opinion, this *holonic ecology* is one of the book's most important contributions. However, because SES does not subscribe to the flatland version of ecology that most (not all) ecophilosophies adopt, SES was not well received by ecophilosophers generally. It is still not. And yet, as SES carefully explains, most ecophilosophies do indeed contain the major problems, inherent in flatland, that will very likely continue to hobble them (both theoretically and practically) until a more holonic ecology is embraced.

Emerson and Plotinus

A few ecophilosophers objected to my treatment of Emerson and Plotinus. I made two minor factual errors in reporting their views. One, I failed to use ellipses correctly in several Emerson quotes. Two, I reported the final words of Plotinus according to the translation given by Karl Jaspers, not William Inge as indicated. Both errors were corrected in subsequent printings. But those minor infractions became the starting point for an agitated onslaught as to my interpretations of Emerson and Plotinus altogether. (See *The Eye of Spirit,* chapter 11, endnotes 1, 2, and 3). Unfortunately, in my opinion, this attack simply allowed some of the ecophilosophers to draw attention away from my substantial criticisms of their views, and also to ignore the major criticisms that both Emerson and Plotinus themselves leveled against nature mysticism (and would therefore level against most forms of present-day ecopsychology, deep ecology, ecofeminism, and neopaganism).

Here, from *The Eye of Spirit,* is a summary of the widely accepted interpretation of Emerson's view: (1) nature is not Spirit but a symbol of Spirit (or a manifestation of Spirit); (2) sensory awareness in itself does not reveal Spirit but obscures it; (3) an Ascending (or transcendental) current is required to disclose Spirit; (4) Spirit is understood only as nature is transcended (i.e., Spirit is immanent in nature, but fully discloses itself only in a transcendence of nature—in short, Spirit transcends but includes nature). Those points are uncontested by Emerson scholars.

As far as those points go, Plotinus would have completely agreed. Thus, both Emerson and Plotinus would condemn—as *true* but *partial*—most (not all) forms of ecopsychology, Gaia worship, neopaganism, deep ecology, and ecofeminism. This is why it became important for these particular ecophilosophers to assert that the common and widely accepted interpretations of Emerson and Plotinus (which I presented) were in fact massive distortions, because otherwise they could not claim support for their theories from these two towering figures. Of course, one is free to try to bring fresh and novel interpretations to the classics, which is always worthwhile. But to try to get these new interpretations across by simply asserting that I had massively distorted these theorists was one of the most ham-handed of the criticisms aimed at SES (not to mention the fact that, even if it were true, it wouldn't affect the conclusions of SES one way or the other). But I must say, it made for some fun and lively fireworks.

Minor Points

Chapter 2 outlines "twenty tenets" that are common to evolving or growing systems wherever we find them. Many people counted them up and didn't get twenty, and they wanted to know if they had missed something. This simply depends on what you count as a tenet. I give twelve numbered tenets. Number 2 contains four tenets, and number 12 contains five. That's nineteen altogether. Throughout the book, I give three additions. That's twenty-two. But one or two of the tenets are not really characteristics, just simple word definitions (e.g., tenet 7 and possibly 9). So that leaves around twenty actual tenets, or actual characteristics of evolution. But there is nothing sacred about the number twenty; these are just some of the more noticeable trends, tropisms, or tendencies of evolution.

Chapter 9, "The Way Up Is the Way Down," discusses evolution and involution. Evolution is the unfolding from matter to body to mind to soul to spirit, with each higher dimension transcending and including its juniors, resulting in the Great Nest of Being. Involution is the reverse process, or the higher dimensions "enfolding" and "involving" themselves in the lower, depositing themselves in the lower as great potentials, ready to unfold into actuality with evolution. Some readers felt that this made the universe completely deterministic and fated. But involution, in my opinion, simply creates a vast field of potentials, which are not determined as to their surface features at all. Those are co-created during evolution, depending on an almost infinite number of variables, from individual initiative to random chance. (I deal specifically with this topic in the Introduction to Volume Two of the *Collected Works* and in *Integral Psychology*.) Within very broad spaces, evolution is playfully creative at every point!

A few Jungians wished that I had expanded my discussion of archetypes. Further material can be found in *The Eye of Spirit*, chapter 11 (which also answers common criticisms from Jungians), and *Integral Psychology* (especially chapter 8).

One critic wondered why I had relied so much on Habermas for my account of phylogenetic evolution. Actually, I relied on dozens of major anthropological researchers—many of which are listed in the bibliography (and hundreds of which are listed in volume 2 of the trilogy)—but because I was using Habermas as an example of a theorist who recognizes all three domains of art, morals, and science (the "Big Three"), I simply presented his extensive anthropological research as long as it did not conflict with generally accepted conclusions in the field.

Smile When You Say That, Mister

There is, finally, the tone of the book. *Sex, Ecology, Spirituality* is in some ways an angry book. Anger, or perhaps anguish, it's hard to say which. After three years immersed in postmodern cultural studies, where the common tone of discourse is rancorous, mean-spirited, arrogant, and aggressive; after surveying countless "new paradigm" treatises, many of which boasted, without irony, that they possessed the new paradigm that was the greatest transformation in history and that would save the planet and save the world; after being exposed to a relentless onslaught of anti-Western, anti-male, anti-culture, anti-almost-anything rhetoric that was some of the most toxic and venomous writing I have ever seen, and which reduced cultural studies to this or that pet theory and narcissistic display of self—after all of that, in anger and anguish, I wrote SES, and the tone of the book indelibly reflects that.

In many cases it is very specific: I often mimicked the tone of the critic I was criticizing, matching toxic with toxic and snide with snide. Of course, in doing so I failed to turn the other cheek. But then, there are times to turn the other cheek, and there are times not to. If you happen to *agree* with the holistic vision presented in SES, you, too, might get angry at the narrowness of what passes for cultural studies nowadays. You might also share a sense of sadness, of melancholy, at the shallowness that pervades postmodernism. Between anger and anguish you might oscillate, as did I when writing the book. And, to be honest, I think all of that is appropriate. But SES definitely was, for me, a cry of anger and anguish.

Still, I could have toned the book down. I chose not to. I sincerely believed, as I still do, that the occasional polemical burst was necessary to get the conversation moving in an integral direction. For over two decades I had seen numerous excellent books with an integral intent completely ignored by "new paradigm" theorists who claimed to be integral and holistic. I chose to rattle the cage and see what happened.

Did it work? What was its effect? Several critics took the polemic to be evidence of my nasty character: I just couldn't help myself, I had to attack. This overlooked the fact that in all of my first twelve books, stretching over two decades, there is not a single polemical sentence.

Other critics maintained that the tone prevented its message from getting out. I truly understand what they mean, but I claim exactly the contrary. These ideas had been studiously ignored for decades, until a little polemical rattling, whereupon they took center stage, for better or worse.

One critic inadvertently demonstrated what was involved by calling for a "dialogue" in the wake of SES, wherein all parties would care for each other in a dance of mutual respect, and not conduct theoretical discourse as if it were a war. This critic then proceeded to do exactly what he professed he despised, and instead of presenting both sides of the argument fairly and respectfully, simply condemned my tone from start to finish.

The fact is, the pro and con stances on the tone of the book lined up almost exactly with whether or not one agreed with it. Those who agreed with the holistic vision of SES shared my anger and anguish, and applauded the polemic. As one critic put it, "Let us not forget: many of us really liked the polemical notes in SES, for their refreshing critiques and liberating humor."

On the other side of the aisle, those who were themselves criticized in the book, or found the vision deficient, lashed out at the tone. As one put it, "Worse than ignorant, Wilber is also unmannered, rude, and offensive."

No doubt, both sides were right.

THE KOSMOS TRILOGY

But by far the most common overall reaction to SES was one of what I suppose we might call joy. I was flooded with mail from readers who told of the liberating influence that SES had on their view of the world, on their view of reality, on their consciousness itself. SES is, after all, a story of the feats of your very own Self, and many readers rejoiced at that remembrance. Women forgave me any patriarchal obnoxiousness, men told me of weeping throughout the last chapter. Apart from *Grace and Grit,* I have never received such heartfelt and deeply moving letters as I received from SES, letters that made those difficult three years seem more than worth it.

I am often asked when volume 2 of the trilogy will be published. My original plan was to release one volume a decade, which means volume 2 would be ready around the year 2005. But now I have no idea exactly when the other two volumes will be ready. Volume 2 is more or less fully written. Volume 3 exists in outline. But I want each to have the chance to absorb the constructive criticisms of its predecessor. In the previous section on Objections, I only focused on the major criticisms, each of which, in my opinion, can be satisfactorily answered. What I didn't mention are all the dozens of minor criticisms that I found valid and well taken, and which I have attempted to incorporate in subsequent writing. I would like the Kosmos trilogy to stand as a solid version of a truly integral philosophy, a believable, if initial, world philosophy, and thus I would like all the many cogent criticisms to have plenty of time to sink in.

There is one other reason I am in no rush to bring out the other volumes. SES itself was begun in part due to a lament at the state of postmodern cultural studies. In the time since SES was conceived, postmodernism's stance has weakened perceptibly. We are truly entering a post-postmodern, post-pluralistic world—by any other name, *integral*. Genuinely integral philosophies will become, and are becoming, more and more acceptable, even eagerly embraced. With every passing year, there is one less chapter of criticism I have to write. With every passing year, a universal integralism becomes more and more welcome.

One critic wrote of SES that "it honors and incorporates more truth than any approach in history." I obviously would like to believe that is the case, but I also know that every tomorrow brings new truths, opens new vistas, and creates the demand for even more encompassing views. SES is simply the latest in a long line of holistic visions, and will itself pass into a greater tomorrow where it is merely a footnote to more glorious views.

In the meantime, it is quite a ride.

SEX, ECOLOGY, SPIRITUALITY

Introduction

I T IS FLAT-OUT strange that something—that *anything*—is happening at all. There was nothing, then a Big Bang, then here we all are. This is extremely weird.

To Schelling's burning question, "Why is there something rather than nothing?," there have always been two general answers. The first might be called the philosophy of "oops." The universe just occurs, there is nothing behind it, it's all ultimately accidental or random, it just is, it just happens—oops! The philosophy of oops, no matter how sophisticated and adult it may on occasion appear—its modern names and numbers are legion, from positivism to scientific materialism, from linguistic analysis to historical materialism, from naturalism to empiricism—always comes down to the same basic answer, namely, "Don't ask."

The question itself (Why is anything at all happening? Why am I here?)—the *question itself* is said to be confused, pathological, nonsensical, or infantile. To stop asking such silly or confused questions is, they all maintain, the mark of maturity, the sign of growing up in this cosmos.

I don't think so. I think the "answer" these "modern and mature" disciplines give—namely, oops! (and therefore, "Don't ask!")—is about as infantile a response as the human condition could possibly offer.

The other broad answer that has been tendered is that *something else is going on*: behind the happenstance drama is a deeper or higher or wider pattern, or order, or intelligence. There are, of course, many varieties of this "Deeper Order": the Tao, God, Geist, Maat, Archetypal Forms, Reason, Li, Mahamaya, Brahman, Rigpa. And although these different varieties of the Deeper Order certainly disagree with each other at many points, they all agree on this: the universe is not what it appears. *Something else* is going on, something quite other than oops. . . .

* * *

This book is about all of that "something other than oops." It is about a possible Deeper Order. It is about evolution, and about religion, and, in a sense, about everything in between. It is a brief history of cosmos, bios, psyche, theos—a tale told by an idiot, it goes without saying, but a tale that, precisely in signifying Nothing, signifies the All, and there is the sound and the fury.

This is a book about holons—about wholes that are parts of other wholes, indefinitely. Whole atoms are parts of molecules; whole molecules are parts of cells; whole cells are parts of organisms, and so on. Each *whole* is simultaneously a *part*, a whole/part, a holon. And reality is composed, not of things nor processes nor wholes nor parts, but of whole/parts, of holons. We will be looking at holons in the cosmos, in the bios, in the psyche, and in theos; and at the evolutionary thread that connects them all, unfolds them all, embraces them all, endlessly.

The first chapters deal with holons in the physical cosmos (matter) and in the biosphere (life). This is the general area of the natural and ecological sciences, the life sciences, the systems sciences, and we will explore each of them carefully. This is particularly important, given not only the ecological crisis now descending on this planet with a vengeance, but also the large number of movements, from deep ecology to ecofeminism, that have arisen in an attempt to find spirituality and ecology connected, not divorced; and we will look at the meaning of all of that.

The middle chapters explore the emergence of the mind or the psyche or the noosphere, and at the holons that compose the psyche itself (the mind is composed of units that have meaning only in contexts: wholes that are parts of other wholes, endlessly). These psychic holons, like all holons, emerged and evolved—in time and history—and we will look briefly at the historical evolution of the mind and consciousness, and at how these psychic holons relate to the holons in the cosmos and in the bios.

The last chapters deal with theos, with the Divine Domain, with a Deeper Order, and how it might indeed be related to the cosmos, the biosphere, and the noosphere. And here, I think, some surprises await us.

This book is the first of three volumes (the series itself is simply called *Kosmos*, or *The Kosmos Trilogy*; brief summaries of the other two volumes are given throughout this book). Many of the questions raised in this volume are more carefully examined in the other two; and, in any event, this volume stands more as a broad overview and introduction, rather than a finished conclusion.

As such, the book is built upon what I would call *orienting generalizations*. For example, in the sphere of moral development, not everybody agrees with the details of Lawrence Kohlberg's seven moral stages, nor with the details of Carol Gilligan's reworking of Kohlberg's scheme. But there is general and ample agreement that human moral development goes through at least three broad stages: the human at birth is not yet socialized into any sort of moral system (it is "preconventional"); the human then learns, from itself and from others, a general moral scheme that represents the basic values of the society it is raised in (it becomes "conventional"); and with even further growth, the individual may come to reflect on society and thus gain some modest distance from it and gain a capacity to criticize or reform it (the individual is to some degree "post-conventional").

Thus, although the actual details and the precise meanings of that developmental sequence are still hotly debated, everybody pretty much agrees that something like those three broad stages do indeed occur, and occur universally. These are *orienting generalizations*: they show us, with a great deal of agreement, where the important forests are located, even if we can't agree on how many trees they contain.

My point is that if we take these types of largely-agreed-upon orienting generalizations from the various branches of knowledge (from physics to biology to psychology to theology), and if we string these orienting generalizations together, we will arrive at some astonishing and often profound conclusions, conclusions that, as extraordinary as they might be, nonetheless embody nothing more than our already-agreed-upon knowledge. The beads of knowledge are already accepted: it is only necessary to provide the thread to string them together into a necklace.

These three volumes are one attempt to string together such a necklace; whether it succeeds or not remains to be seen. But if nothing else, I think it is at least a good example of how this type of work can be done in today's postmodern world. In working with broad orienting generalizations, the trilogy delivers up a broad orienting map of the place of men and women in relation to Universe, Life, and Spirit, the details of which we can all fill in as we like, but the broad outlines of which really have an awful lot of supporting evidence, culled from the orienting generalizations, simple but sturdy, from the various branches of human knowledge.

Nonetheless, this broad orienting map is nowhere near fixed and final. In addition to being composed of broad orienting generalizations, I would say this is a book of a thousand hypotheses. I will be telling the story as if it were simply the case (because telling it that way makes for much better reading), but not a sentence that follows is not open to confirmation or rejection by a community of the adequate. I suppose many readers will insist on calling what I am doing "metaphysics," but if "metaphysics" means thought without evidence, there is not a metaphysical sentence in this entire book.

Because this book (or this trilogy) offers a broad orienting map of men and women's place in the larger Kosmos (of matter, life, mind, and spirit), it naturally touches on a great number of topics that have recently become "hot," from the ecological crisis to feminism, from the meaning of modernity and postmodernity to the nature of "liberation" in relation to sex, gender, race, class, creed; to the nature of techno-economic developments and their relation to various worldviews; to the various spiritual and wisdom traditions the world over that have offered telling suggestions as to our place in a larger scheme of things.

How can we become more fully human and at the same time be saved from the fate of being merely human? Where is Spirit in this God-forsaken, Goddess-forsaken world of modernity? Why are we destroying Gaia in the very attempt to improve our own condition? Why are so many attempts at salvation suicidal? How do we actually fit into this larger Kosmos? How are we *whole* individuals who are also *parts* of something Larger?

In other words, since human beings, like absolutely everything else in the Kosmos, are *holons*, what does that mean? How do we fit into that which is forever moving beyond us? Does liberation mean being whole ourselves, or being a part of something Larger—or something else altogether? If history is a nightmare from which I am trying to awaken, then what exactly is it that I am supposed to awaken *to*?

And, most important, can we not stare into that vast and stunning Kosmos and respond with something more mature than oops?

From those who have already read this book in manuscript come two suggestions for the reader:

First, skip the endnotes on the first reading, and save them for (and if) a second look. This book was intentionally written on two levels: the main text, which makes every attempt to be as accessible as possible, and the notes (a small book in themselves), meant for serious students. But in both cases, the notes are, for the most part, best reserved for a second reading, as they greatly disrupt the narrative flow. (Alternatively, some have simply read the notes by themselves, as a type of appendix, just for the information, which is fine.)

Second, read the book a sentence at a time. People who try to skip around get competely lost. But pretty much everybody reports that if you simply read each sentence, the text will carry you along, and any problems encountered are usually cleared up down the road. This is a long book, obviously, but apparently it comes in nice, small, bite-sized chunks, and its readers all seem to have a great good time—a bite at a time.

It is often said that in today's modern and postmodern world, the forces of darkness are upon us. But I think not; in the Dark and the Deep there are truths that can always heal. It is not the forces of darkness but of shallowness that everywhere threaten the true, and the good, and the beautiful, and that ironically announce themselves as deep and profound. It is an exuberant and fearless shallowness that everywhere is the modern danger, the modern threat, and that everywhere nonetheless calls to us as savior.

We might have lost the Light and the Height; but more frightening, we have lost the Mystery and the Deep, the Emptiness and the Abyss, and lost it in a world dedicated to surfaces and shadows, exteriors and shells, whose prophets lovingly exhort us to dive into the shallow end of the pool head first.

"History," said Emerson, "is an impertinence and an injury if it be anything more than a cheerful apologue or parable of my being and becoming." What follows, then, is a cheerful parable of your being and your becoming, an apologue of that Emptiness which forever issues forth, unfolding and enfolding, evolving and involving, creating worlds and dissolving them, with each and every breath you take. This is a chronicle of what you have done, a tale of what you have seen, a measure of what we all might yet become.

BOOK ONE

What is it that has called you so suddenly out of nothingness to enjoy for a brief while a spectacle which remains quite indifferent to you? The conditions for your existence are as old as the rocks. For thousands of years men have striven and suffered and begotten and women have brought forth in pain. A hundred years ago, perhaps, another man—or woman—sat on this spot; like you he gazed with awe and yearning in his heart at the dying light on the glaciers. Like you he was begotten of man and born of woman. He felt pain and brief joy as you do. Was he someone else? Was it not you yourself? What is this Self of yours?

—ERWIN SCHRÖDINGER

1

The Web of Life

So the world, grounded in a timeless movement by the Soul which suffuses it with intelligence, becomes a living and blessed being.
—PLOTINUS

IT'S A STRANGE WORLD. It seems that around fifteen billion years ago there was, precisely, absolute nothingness, and then within less than a nanosecond the material universe blew into existence.

Stranger still, the physical matter so produced was not merely a random and chaotic mess, but seemed to organize itself into ever more complex and intricate forms. So complex were these forms that, many billions of years later, some of them found ways to reproduce themselves, and thus out of matter arose life.

Even stranger, these life forms were apparently not content to merely *reproduce* themselves, but instead began a long evolution that would eventually allow them to *represent* themselves, to produce signs and symbols and concepts, and thus out of life arose mind.

Whatever this process of evolution was, it seems to have been incredibly driven—from matter to life to mind.

But stranger still, a mere few hundred years ago, on a small and indifferent planet around an insignificant star, evolution became conscious of itself.

And at precisely the same time, the very mechanisms that allowed evolution to become conscious of itself were simultaneously working to engineer its own extinction.

And that was the strangest of all.

THE ECOLOGICAL CRISIS

I will not belabor the point by bringing out all the ghastly statistics, from the fact that we are at present exterminating approximately one hundred species a day to the fact that we are destroying the world's tropical forests at the rate of one football field per second. The planet, indeed, is headed for disaster, and it is now possible, for the first time in human history, that owing *entirely* to manmade circumstances,[1] not one of us will survive to tell the tale. If the Earth is indeed our body and blood, then in destroying it we are committing a slow and gruesome suicide.

Arising in response to the alarming dimensions of this ecological catastrophe (the nature and extent of which I will simply assume is evident to every intelligent person) have been a variety of popularly based responses generally referred to as the environmental movement (usually dated from the 1962 publication of Rachel Carson's *Silent Spring*).[2] Starting in part within the environmental movement, but going quite beyond it, are two "ecophilosophies" that will particularly interest us: ecofeminism and deep ecology (and we will see that they almost perfectly embody, respectively, the female and male value sphere approaches to the same topic).

Central to these ecological approaches is the notion that our present environmental crisis is due primarily to a *fractured worldview*, a worldview that drastically separates mind and body, subject and object, culture and nature, thoughts and things, values and facts, spirit and matter, human and nonhuman; a worldview that is dualistic, mechanistic, atomistic, anthropocentric, and pathologically hierarchical—a worldview that, in short, erroneously separates humans from, and often unnecessarily elevates humans above, the rest of the fabric of reality, a broken worldview that alienates men and women from the intricate web of patterns and relationships that constitute the very nature of life and Earth and cosmos.

These approaches further maintain that the only way we can heal the planet, and heal ourselves, is by replacing this fractured worldview with a worldview that is more holistic, more relational, more integrative, more Earth-honoring, and less arrogantly human-centered. A worldview, in short, that honors the entire web of life, a web that has intrinsic value in and of itself, but a web that, not incidentally, is the bone and marrow of our own existence as well.

Fritjof Capra, for example, has argued that the world's present social, economic, and environmental crises all stem from the same fractured worldview:

> Our society as a whole finds itself in an [unprecedented] crisis. We can read about its numerous manifestations every day in the newspapers. We have high unemployment, we have an energy crisis, a crisis in health care, pollution and other environmental disasters, a rising wave of violence and crime, and so on. The basic thesis of this book [*The Turning Point*] is that these are all different facets of one and the same crisis, and that this crisis is essentially a crisis of perception. It derives from the fact that we are trying to apply the concepts of an outdated worldview—the mechanistic worldview . . . —to a reality that can no longer be understood in terms of those concepts. We live today in a globally interconnected world, in which biological, psychological, social, and environmental phenomena are all interdependent. To describe this world appropriately we need an ecological perspective. . . .[3]

As many scholars have pointed out, because of the perceived closeness of "woman" and "nature," the despoliation of the Earth and the subjugation of women have historically gone hand in hand. *Ecofeminism* is a powerful response to the denigration of both of these "others." As Judith Plant explains:

> Historically, women have had no real power in the outside world, no place in decision making and intellectual life. Today, however, ecology speaks for the Earth, for the "other" in human/environmental relationships; and feminism speaks for the "other" in female/male relations. And ecofeminism, by speaking for *both* the original others, seeks to understand the interconnected roots of all domination as well as ways to resist and change. The ecofeminist's task is one of developing the ability to take the place of the other when considering the consequences of possible actions, and ensuring that we do not forget that we are all part of one another . . . , as we mend our relations with each other and with the Earth.[4]

Bill Devall and George Sessions, representing the movement known as *deep ecology*, point out that "this is the work we call cultivating ecological consciousness, the insight that everything is connected," which Jack

Forbes explains as perceiving ourselves as "being deeply bound together with other people and with the surrounding nonhuman forms of life in a complex interconnected web of life, that is to say, a true community. All creatures and things [are] brothers and sisters. From this idea comes the basic principle of non-exploitation, of respect and reverence for all creatures."[5]

The extraordinary thing about all these quotes is that, although they might sound rather romantic and poetic and, some would say, even mushy-minded, they are in fact rooted in the hardest of scientific evidence. After all, holistic theories of the "web of life" are as old as civilization itself, forming the very core of the world's great religions and wisdom traditions (as we will see). But it is one thing to merely have God on your side; quite another to have science on your side.

And the ecological sciences are just that: hard sciences. What I would like to do, then, is briefly review the systems (holistic or ecological) sciences, and demonstrate just exactly what they mean when they refer to the interconnectedness or "weblike" nature of all life. This will give us some necessary background information, and also serve as a platform for our discussion of ecofeminism and deep ecology.

Finally, as important as the ecological or holistic systems approach is, we will have to make a few very important revisions in it (for reasons that will become obvious), and, most important, we will have to set it in its *own* larger context, a context almost always ignored (with rather dire consequences). And we will find that the same strengths—and the same weaknesses—beset ecofeminism and deep ecology as well.

We will find, in other words, that those who talk of the "web of life" are basically half right and half wrong (or seriously incomplete), and the "half-wrong" part has caused almost more problems than the "half-right" part has solved.

But first, the half that seems definitely right.

TWO ARROWS OF TIME

The new systems sciences are, in a sense, the sciences of wholeness and connectedness. If we now add the notion of *development* or *evolution*—the idea that wholes grow and evolve—we have the essence of the modern systems sciences. As Ervin Laszlo puts it, "A new system, scientific in origin and philosophic in depth and scope, is now on the rise. It encompasses the great realms of the material universe, of the world of the living, and of the world of history. This is the evolutionary paradigm. . . ." As he explains it:

> The old adage "everything is connected with everything else" describes a true state of affairs. The results achieved [by the evolutionary sciences] furnish adequate proof that the physical, the biological, and the social realms in which evolution unfolds are by no means disconnected. At the very least, one kind of evolution prepares the ground for the next. Out of the conditions created by evolution in the physical realm emerge the conditions that permit biological evolution to take off. And out of the conditions created by biological evolution come the conditions that allow human beings—and many other species—to evolve certain social forms of organization.[6]

Laszlo then offers this important conclusion:

> Scientific evidence of the patterns traced by evolution in the physical universe, in the living world, and even in the world of history is growing rapidly. It is coalescing into the image of basic regularities that repeat and recur. It is now possible to search out these regularities and obtain a glimpse of the fundamental nature of evolution—of the evolution of the cosmos as a whole, including the living world and the world of human social history.
>
> To search out and systematically state these regularities is to engage in the creation of the "grand synthesis" that unites physical, biological, and social evolution into a consistent framework with its own laws and logic.[7]

Exactly what those laws and logic are, we will explore later in this and the next chapter. For the moment, notice that Laszlo refers to the three "great realms" of evolution: material, biological, and historical. Erich Jantsch refers to them as cosmic, biosocial, and sociocultural. Michael Murphy summarizes them as physical, biological, and psychological. In popular terms: matter, life, and mind. I will refer to these three general domains as the physiosphere (matter), the biosphere (life), and the noosphere (mind).[8]

The central claim made by the evolutionary systems sciences is that, whatever the actual nature of these three great domains, they are all united, not necessarily by similar contents, but because they all express the same general laws or *dynamic patterns*. As Ludwig von Bertalanffy, founder of General System Theory, put it, the "Unity of Science is granted, not by a utopian reduction of all sciences to physics and chemistry, but by the structural uniformities [regularities of dynamic patterns] of the different levels of reality."[9]

Now historically it had been maintained, from the time of Plato and Aristotle until around the end of the nineteenth century, that all of these great domains—physiosphere, biosphere, noosphere—were one continuous and interrelated manifestation of Spirit, one Great Chain of Being, that reached in a perfectly unbroken or uninterrupted fashion from matter to life to mind to soul to spirit.

As Arthur Lovejoy[10] demonstrated, the various Great Chain theorists maintained three essential points: (1) all phenomena—all things and events, people, animals, minerals, plants—are manifestations of the superabundance and plenitude of Spirit, so that Spirit is woven intrinsically into each and all, and thus even the entire material and natural world was, as Plato put it, "a visible, sensible God"; (2) therefore, there are "no gaps" in nature, no missing links, no unbridgeable dualisms, for each and every thing is interwoven with each and every other (the "continuum of being"); and (3) the continuum of being nonetheless shows gradation, for various emergents appear in some dimensions that do not appear in others (e.g., wolves can run, rocks can't, so there are "gaps" in the special sense of emergents).

Now whatever we moderns might think of the Great Chain as a theory, it nonetheless "has been the official philosophy of the larger part of civilized humankind through most of its history"; and further, it was the worldview that "the greater number of the subtler speculative minds and of the great religious teachers [both East and West], in their various fashions, have been engaged in."[11]

Thus, whether or not we accept some version of the Great Chain (and we will be examining this in later chapters), scholars agree that its worldview saw matter and bodies and minds as a vast network of mutually interweaving orders subsisting in Spirit, with each node in the continuum of being, each link in the chain, being absolutely necessary and intrinsically valuable. To give only one example now, Plotinus would soon explain that since each link was a manifestation of the goodness of Spirit, each had intrinsic value—it was valuable in and of itself—and no link whatsoever, no matter how "lowly," existed merely or even primarily for the instrumental use of others: destroy any precious strand and the whole fabric would come unravelled.

But with the rise of modern science—associated particularly with the names of Copernicus, Kepler, Galileo, Bacon, Newton, Kelvin, Clausius— this great unified and holistic worldview began to fall apart, and fall apart in ways, it is clear, that none of these pioneering scientists themselves either foresaw or intended.

And it fell apart in a very peculiar way. These early scientists began their experimental studies in that realm which is apparently the least complicated: the physiosphere, the material universe, the world of inanimate matter. Kepler focused on planetary motion and Galileo on terrestrial mechanics; Newton synthesized their results in his universal law of gravitation and laws of motion; and Descartes worked all the results into a most influential philosophy. In all of these endeavors, the physiosphere began to look like a vast mechanism, a universal machine governed by strict causality. And worse, a machine that was running down.

Here was the problem: in the material world, science soon discovered, there are at least two very different types of phenomena, one described by the laws of classical mechanics and the other by the laws of thermodynamics. In the former, in classical Newtonian mechanics, time plays no fundamental role, because the processes described are reversible. For example, if a planet is going one way around the sun or the reverse way around the sun, the laws describing the motion are the same, because in these types of "classical mechanics" time changes nothing essential; you can as easily turn your watch forward as you can turn it backward—the mechanism and its laws don't care which way you turn it.

But in thermodynamic processes, "time's arrow" is absolutely central. If you put a drop of ink in a glass of water, in a day or so the ink will have evenly dispersed throughout the water. But you will *never* see the reverse process happen—you will never see the dispersed ink gather itself together into a small drop. Hence, time's arrow is a crucial part of these types of physical processes, because these processes always proceed in one direction only. They are irreversible.

And the infamous Second Law of Thermodynamics added a dismal conclusion: the direction of time's arrow is downward. Physical processes, like the inkdrop, always go from *more ordered* (the inkdrop) to *less ordered* (dispersed throughout the water). The universe may be a giant clockwork, but the clock is winding down . . . and will eventually run out.

The problem was not that these early conceptions were simply wrong. Aspects of the physiosphere do indeed act in a deterministic and mechanistic-like fashion, and some of them are definitely running down. Rather, it was that these conceptions were *partial*. They covered some of the most obvious aspects of the physiosphere, but because of the primitive means and instruments available at the time, the subtler (and more significant) aspects of the physiosphere were overlooked.

And yet, as we will see, it was precisely in these subtler aspects that the physiosphere's *connection* to the biosphere could be established. At the time, however, lacking these connections, the physiosphere and the biosphere simply fell apart—in the sciences, in religion, and in philosophy. Thus, it was the *partialness* of the early natural sciences, and not any glaring errors, that would inadvertently contribute to the subsequent and rather horrendous fracturing of the Western worldview.

Against this early (and partial) scientific understanding of the physiosphere, which was now seen as a reversible mechanism irreversibly running down, came the work of Alfred Wallace and Charles Darwin on evolution through natural selection in the *biosphere*. Although the notion of evolution, or irreversible development through time, had an old and honorable history (from the Ionian philosophers to Heraclitus to Aristotle to Schelling), it was of course Wallace and Darwin who set it in a scientific framework backed by meticulous empirical observations, and it was Darwin especially who lit the world's imagination with his ideas on the evolutionary nature of the various species, including humans.

Apart from the specifics of natural selection itself (which most theorists now agree can account for microchanges in evolution but not macrochanges), there were two things that jumped out in the Darwinian worldview, one of which was not novel at all, and one of which was very novel. The first was the continuity of life; the second, speciation by natural selection.

The idea of the continuity of life—the web of life, the tree of life, the "no gaps in nature" view—was at least as old as Plato and Aristotle, and, as I briefly mentioned, it formed an essential ingredient in the notion of the Great Chain of Being. Spirit manifests itself in the world in such a complete and full fashion that it leaves no gaps in nature, no missing links in the Great Chain. And, as Lovejoy noted, it was the philosophical belief that there are no gaps in creation that *directly* led to the scientific attempts to find not only the missing links in nature (which is where that phrase originated) but also evidence of life on other planets. All of these "gaps" needed to be filled in, in order to round out the Great Chain, and there was precisely nothing new or unusual in Darwin's presentation of the continuous tree of life.

What was rather novel was his thesis that the various links in the Great Chain, the various species themselves, had in fact unfolded or evolved over vast stretches of geological time and were not simply put there, all at once, at the creation. There were precedents for this thesis, particularly in Aristotle's version of the Great Chain, which, he maintained, showed a progressive and unbroken development of nature through what he called metamorphosis, from inorganic (matter) to nutritive (plant) to sensorimotor (animal) to symbol-utilizing animals (humans), all displaying progressive organization and increasing complexity of form. Leibniz had taken profound steps to "temporalize" the Great Chain, and with Schelling and Hegel we see the full-blown conception of a process or developmental philosophy applied to literally all aspects and all spheres of existence.

But it was Darwin's meticulous descriptions of natural species and his unusual clarity of presentation, combined with his hypothesis of natural selection, that propelled the concept of development or evolution to the scientific forefront. Of the *biosphere*.

And in the biosphere Darwin (and many others) noticed that there is also a crucial time's arrow. Evolution is irreversible. We may see amoebas eventually evolve into apes, but we never see apes turn into amoebas. That is, evolution proceeds irreversibly in the direction of increasing differentiation/integration, increasing structural organization, and increasing complexity. It goes from less ordered to more ordered. But obviously, the direction of *this* time's arrow was diametrically opposed to time's arrow in the (known) physiosphere: the former is winding up, the latter is winding down.

It was at this point, historically, that the physiosphere and the biosphere fell apart. It was an extremely difficult situation. For one thing, both physics and biology were *supposed* to be part of the new natural sciences, relying on empirical observation, measurement, theory formation, and rigorous testing (this overall procedure was indeed novel, dating from 1605 with Kepler and Galileo). But although the *methods* of physics and biology were similar, their *results* were fundamentally incompatible, saddled, as Laszlo put it, with "the persistent contradiction between a mechanistic world slated to run down and an organic world seeming to wind up."[12]

A further complication was the relation of physics and biology to the noosphere itself, to mind and values and history. In the earlier conception of the Great Chain of Being, matter and body and mind were seen as perfectly continuous aspects of the superabundant overflowing of Spirit. They were *all* organically related as manifestations or emanations of the Divine, with no gaps and no holes (we find this from Plato to Plotinus to Pascal). But with the separation of the physiosphere and the biosphere (due to their two different arrows of time), the links in the entire Chain began to fall into alienated and seemingly unrelated spheres—dead matter versus vital body versus disembodied mind.

There were immediate and rather desperate attempts to repair the damage, to return the universe to a unified conception. The first and by far the most influential attempt to resurrect a coherent worldview was material reductionism, the attempt to reduce all mind and all body to various combinations of matter and mechanism (Hobbes, La Mettrie, Holbach). Equally alluring was the reverse agenda: elevate all matter and bodies to the status of mental events (as in the phenomenalism of Mach or Berkeley). In between the two extremes of reductionism and elevationism were a whole host of uneasy compromises: the dualism of Descartes, a noble and, at the time, wholly understandable attempt to salvage the status of the mind from its reduction to mere material mechanism, by unfortunately throwing the whole biosphere over to the mechanists and snatching only the noosphere from the jaws of the sharks; the pantheism of Spinoza (who regarded himself as a good Cartesian), seeing mind and matter as two parallel attributes of God that *never* interacted (which he assumed took care of *that* problem); the epiphenomenalism of T. H. Huxley, who saw the mind as an "epiphenomenon," real enough in itself, but purely the byproduct of physiological causes and having no causal power itself— the "ghost in the machine."

All of these attempts were sabotaged from the beginning, not so much by the split between mind and body (which was at least as old as civilization and had never bothered anybody before), but by the more primitive and radical split between body and matter—that is, life and matter (the particular form of which was indeed very new and very disturbing). As Henri Bergson put it, the universe shows two tendencies, a "reality which is making itself in a reality which is unmaking itself."

The upshot of all this was that, precisely because the physiosphere and the biosphere had parted ways, the world was, indeed, *fractured*. The immediate effect was that physics and biology went their separate ways. Much more disturbing, natural philosophy was split from moral philosophy, and the natural sciences were split from the human sciences. The physiosphere was seen as the realm of *facts* unaffected by history, and the noosphere as the realm of *values* and *morals* created primarily *by* history, and this gap was felt to be absolutely unbridgeable. And poor biology, caught in the middle between the "hard sciences" of the physiosphere and "soft sciences" of the noosphere, became positively schizophrenic, trying now to ape physics and reduce all life to mechanism, now to ape the noosphere and see all life as fundamentally embodying élan vital and values and history.

As several researchers have noted, "not until the puzzle of the two and opposing arrows of time was resolved in the late twentieth century was there a sound basis for bridging the gap between matter and mind, the natural world and the human world, and thus between the 'two cultures' of modern Western civilization."[13] And, as I said, it was not so much the gap between mind and body as the gap between body and matter that needed to be closed.

THE MODERN EVOLUTIONARY SYNTHESIS

The closing of the gap between the physiosphere and the biosphere came precisely in the rather recent discovery of those subtler and originally hidden aspects of the material realm that, under certain circumstances, *propel themselves* into states of higher order, higher complexity, and higher organization. In other words, under certain circumstances matter will "wind itself up" into states of higher order, as when the water running down a drain suddenly ceases to be chaotic and forms a perfect funnel or whirlpool. Whenever material processes become very chaotic and "far from equilibrium," they tend under their own power to escape chaos by transforming it into a higher and more structured order—commonly called "order out of chaos."

Notice that these types of purely material systems also have an arrow of time, but this arrow is pointed in the *same direction* as the arrow of time *in living systems*, namely, to higher order and higher structural organization. In other words, aspects of the physiosphere *are headed in the same direction* as the biosphere, and that, put roughly, closes the gap between them. The material world is perfectly capable of winding itself up, long before the appearance of life, and thus the "self-winding" nature of matter itself sets the stage, or prepares the conditions, for the complex self-organization known as life. The two arrows have joined forces.

The nature of these chaotic transitions and transformations are still being explored. But the point is that, where there once appeared a single and absolutely unbridgeable gap between the world of matter and the world of life—a gap that posed a completely unsolvable problem—there now appeared only a series of minigaps. And whatever the exact nature of these minigaps, they looked unavoidably like a series of bridges inherently *relating* matter to life, and not like a moat forever separating them. Thus the old continuity between the physiosphere and the biosphere—a continuity that was a hallmark of the Great Chain of Being—was once again established.

As I said, there are still certain types of very important gaps or "leaps" in nature (expressed in emergents), but these now make sense, and seem somehow inevitable, in ways that early science found incomprehensible.

The new sciences dealing with these "self-winding" or "self-organizing" systems are known collectively as the sciences of complexity—including General System Theory (Bertalanffy, Weiss), cybernetics (Wiener), nonequilibrium thermodynamics (Prigogine), cellular automata theory (von Neumann), catastrophe theory (Thom), autopoietic system theory (Maturana and Varela), dynamic systems theory (Shaw, Abraham), and chaos theories, among others.

I do not mean to minimize the very real differences between these various sciences, or the great advances that the more recent sciences of complexity (especially self-organizing systems and chaos theories) have made over their predecessors. But since my aim is very general, I will refer to them collectively as systems theory, dynamic systems theory, or evolutionary systems theory. For the general claim, remember, of evolutionary systems theory is that there have now been discovered basic regularities, patterns, or laws that apply in a broad fashion to all three great realms of evolution, the physiosphere and the biosphere and the noosphere, and that a "unity of science"—a coherent and unified worldview—is now possible.[14] They claim, in other words, that "everything is connected to everything else"—the web of life as a scientific and not just religious conclusion.

THE PROBLEM WITH HIERARCHY

Before we discuss some of the finer points and conclusions of the evolutionary systems sciences, the first thing that we can't help noticing is that these sciences are shot through with the notion of *hierarchy*—a word that has fallen on very hard times. All sorts of theorists, from deep ecologists to social critics, from ecofeminists to postmodern poststructuralists, have found the notion of hierarchy not only undesirable but a bona fide cause of much social domination, oppression, and injustice.

And yet here are the actual system sciences talking openly and glowingly of hierarchy. I will present the evidence for this later, but we find it everywhere: from the founder of General System Theory, Ludwig von Bertalanffy ("Reality, in the modern conception, appears as a tremendous hierarchical order of organized entities") to Rupert Sheldrake and his "nested hierarchy of morphogenetic fields"; from the great systems linguist Roman Jakobson ("Hierarchy, then, is the fundamental structural principle of language") to Charles Birch and John Cobb's ecological model of reality based on "hierarchical value"; from Francisco Varela's groundbreaking work on autopoietic systems ("It seems to be a general reflection of the richness of natural systems . . . to produce a hierarchy of levels") to the brain research of Roger Sperry and Sir John Eccles and Wilder Penfield ("a hierarchy of nonreducible emergents"), and even the social critical theory of Jurgen Habermas ("a hierarchy of communicative competence")—hierarchy seems to be everywhere.

Now, the opponents of hierarchy—their names are legion—basically maintain that all hierarchies involve a *ranking* or dominating judgment that oppresses other values and the individuals who hold them (hierarchies are a "hegemonic domination that marginalizes differential values"), and that a linking or *nonranking* model of reality is not only more accurate but, we might say, kinder and gentler and more just.

So they propose instead, in their various fashions, the notion of *heterarchy*. In a heterarchy, rule or governance is established by a pluralistic and egalitarian interplay of all parties; in a *hierarchy*, rule or governance is established by a set of priorities that establish those things that are more important and those that are less.

Nowhere in the literature of modern social theory is there more acrimony expressed than over the topic of hierarchy/heterarchy. On the one side we have the champions of egalitarian and "equalitarian" views (heterarchy), who see all creatures as equal nodes in the web of life, and who, with good reasons, inveigh against harsh social ranking and domination, arguing instead for a pluralistic wholeness that intrinsically values each strand in the web, with "higher" and "lower" being forms not of organization but domination and exploitation. Even the notions of "higher" and "lower" are said to be part of "old-paradigm" thinking, and not part of the "new-paradigm" or "network" or "web-of-life" thinking.

And yet, when it comes to the actual sciences of this web of life, the sciences of wholeness and connectedness, we find them speaking unmistakably of hierarchy as the basic organizing principle of wholeness. They maintain that you cannot have *wholeness* without *hierarchy*, because unless you organize the parts into a larger whole whose glue is a principle higher or deeper than the parts possess alone—unless you do that, then you have heaps, not wholes. You have strands, but never a web. Even if the whole is a mutual interaction of parts, the wholeness cannot be on the same level as the partness or it would itself be merely another part, not a whole capable of embracing and integrating each and every part. "Hierarchy" and "wholeness," in other words, are two names for the same thing, and if you destroy one, you completely destroy the other.

It is ironic, to say the least, that the social champions of the web of life deny hierarchy in any form while the sciences of the web of life insist upon it. And it is doubly ironic that the former often point to the latter for support (e.g., "The new physics supports the equalitarian web of life").

What's going on here? In part, I will try to show, this is a colossal semantic confusion—the two parties are actually much closer than either supposes. The real world does indeed contain some natural or normal hierarchies (as we'll see), and it definitely contains some pathological or dominator hierarchies. And just as important, it contains some normal heterarchies and some pathological heterarchies (I'll give some examples of all four in a moment). The semantic confusions surrounding these topics are an absolute nightmare, confusions that have bred an inordinate amount of ideological fury on both sides, and unless we attempt to clear up some of this confusion, the discussion simply cannot go forward. So let us try. . . .

HOLONS

Hiero- means sacred or holy, and *-arch* means governance or rule. Introduced by the great sixth-century Christian mystic Saint Dionysius the Areopagite, the "Hierarchies" referred to nine celestial orders, with Seraphim and Cherubim at the top and archangels and angels at the bottom. Among other things, these celestial orders represented higher knowledge and virtue and illuminations that were made more accessible in contemplative awareness. These orders were *ranked* because each successive order was more inclusive and more encompassing and in that sense "higher." "Hierarchy" thus meant, in the final analysis, "sacred governance," or "governing one's life by spiritual powers."

In the course of Catholic Church history, however, these celestial orders of contemplative awareness were translated into *political* orders of power, with the Hierarchies supposedly being represented in the pope, then archbishops, then bishops (and then priests and deacons). As Martineau put it in 1851, "A scheme of a hierarchy which might easily become a despotism." And already we can start to see how a normal developmental sequence of increasing wholes might pathologically degenerate into a system of oppression and repression.

As used in modern psychology, evolutionary theory, and systems theory, a hierarchy is simply a ranking of orders of events *according to their holistic capacity.* In any developmental sequence, what is whole at one stage becomes a part of a larger whole at the next stage. A letter is part of a whole word, which is part of a whole sentence, which is part of a whole paragraph, and so on. As Howard Gardner explains it for biology, "Any change in an organism will affect all the parts; no aspect of a structure can be altered without affecting the entire structure; each whole contains parts and is itself part of a larger whole."[15] Or Roman Jakobson for language: "The phoneme is a combination of distinctive features; it is composed of diverse primitive signaling units and can itself be incorporated into larger units such as syllables and words. It is simultaneously a whole composed of parts and is itself a part that is included in larger wholes."[16]

Arthur Koestler coined the term *holon* to refer to that which, being a *whole* in one context, is simultaneously a *part* in another. With reference to the phrase "the bark of a dog," for example, the word *bark* is a whole with reference to its individual letters, but a part with reference to the phrase itself. And the whole (or the context) can *determine* the meaning and function of a part—the meaning of *bark* is different in the phrases "the bark of a dog" and "the bark of a tree." The whole, in other words, is more than the sum of its parts, and that whole can influence and determine, in many cases, the function of its parts (and that whole itself is, of course, simultaneously a part of some other whole; I will return to this in a moment).

Normal hierarchy, then, is simply an order of increasing holons, representing an increase in wholeness and integrative capacity—atoms to molecules to cells, for example. This is why hierarchy is indeed so central to systems theory, the theory of wholeness or holism ("wholism"). To be a part of a larger whole means that the whole supplies a principle (or some sort of glue) not found in the isolated parts alone, and this principle allows the parts to join, to link together, to have something in common, to be connected, in ways that they simply could not be on their own.

Hierarchy, then, converts heaps into wholes, disjointed fragments into networks of mutual interaction. When it is said that "the whole is greater than the sum of its parts," the "greater" means "hierarchy." It doesn't mean fascist domination; it means a higher (or deeper) commonality that joins isolated strands into an actual web, that joins molecules into a cell, or cells into an organism.

This is why "hierarchy" and "wholeness" are often uttered in the same sentence, as when Gardner says that "a biological organism is viewed as a totality whose parts are integrated into a hierarchical whole."[17] Or why, as soon as Jakobson explains language as "simultaneously a whole composed of parts and itself a part included in a larger whole," he concludes, "Hierarchy, then, is the fundamental structural principle." This is also why normal hierarchies are often drawn as a series of concentric circles or spheres or "nests within nests." As Goudge explains:

> The general scheme of levels is not to be envisaged as akin to a succession of geological strata or to a series of rungs in a ladder. Such images

fail to do justice to the complex interrelations that exist in the real world. These interrelations are much more like the ones found in a nest of Chinese boxes or in a set of concentric spheres, for according to emergent evolutionists, a given level can contain other levels within it [i.e., holons].[18]

Thus, the common charge that all hierarchies are "linear" completely misses the point. Stages of growth in any system can, of course, be written down in a "linear" order, just as we can write down: acorn, seedling, oak; but to accuse the oak of therefore being linear is silly. As we will see, the stages of growth are not haphazard or random, but occur in some sort of pattern, but to call this pattern "linear" does not at all imply that the processes themselves are a rigidly one-way street; they are interdependent and complexly interactive. So we can use the metaphors of "levels" or "ladders" or "strata" only if we exercise a little imagination in understanding the complexity that is actually involved.

And finally, hierarchy *is* asymmetrical (or a "higher"-archy) because the process does not occur in the reverse. Acorns grow into oaks, but not vice versa. There are first letters, then words, then sentences, then paragraphs, but not vice versa. Atoms join into molecules, but not vice versa. And that "not vice versa" constitutes an unavoidable hierarchy or ranking or asymmetrical order of increasing wholeness.

All developmental and evolutionary sequences that we are aware of proceed in part by hierarchization, or by orders of increasing holism— molecules to cells to organs to organ systems to organisms to societies of organisms, for example. In cognitive development, we find awareness expanding from simple images, which represent only one thing or event, to symbols and concepts, which represent whole groups or classes of things and events, to rules which organize and integrate numerous classes and groups into entire networks. In moral development (male or female), we find a reasoning that moves from the isolated subject to a group or tribe of related subjects, to an entire network of groups beyond any isolated element. And so on.

(It is sometimes said that Carol Gilligan denied, not just the specific nature of the stages of Kohlberg's scheme, but his entire hierarchical approach. This is simply not true. Gilligan, in fact, accepts Kohlberg's general three-stage or three-tiered hierarchical scheme, from preconventional to conventional to postconventional—"meta-ethical"—

development; she simply denies that the logic of justice alone accounts for the sequence; men seem to emphasize rights and justice, she says, and that needs to be supplemented with the logic of care and responsibility with which females progress through the *same* hierarchy—points we will return to later.)

These hierarchical networks necessarily unfold in a sequential or stage-like fashion, as I earlier mentioned, because you first have to have molecules, *then* cells, *then* organs, *then* complex organisms—they don't all burst on the scene simultaneously. In other words, growth occurs in *stages*, and stages, of course, are *ranked* in both a logical and chronological order. The *more holistic* patterns appear *later* in development because they have to await the emergence of the parts that they will then integrate or unify, just as whole sentences emerge only *after* whole words.

And some hierarchies do involve a type of control network. As Roger Sperry points out, the lower levels (which means, less holistic levels) can influence the upper (or more holistic) levels, through what he calls "upward causation." But just as important, he reminds us, the higher levels can exert a powerful influence or control on the lower levels—so-called "downward causation." For example, when you decide to move your arm, all the atoms and molecules and cells in your arm move with it—an instance of downward causation.

Now, *within* a given level of any hierarchical pattern, the elements of that level operate by *heterarchy*. That is, no one element seems to be especially more important or more dominant, and each contributes more or less equally to the health of the whole level (so-called "bootstrapping"). But a higher-order whole, of which this lower-order whole is a part, can exert an overriding influence on each of its components. Again, when you decide to move your arm, your mind—a higher-order holistic organization—exerts influence over all the cells in your arm, which are lower-order wholes, but *not vice versa*: a cell in your arm cannot decide to move the whole arm—the tail does not wag the dog.

And so systems theorists tend to say: *within* each level, heterarchy; *between* each level, hierarchy.

In any developmental or growth sequence, as a more encompassing stage or holon emerges, it *includes* the capacities and patterns and functions of the previous stage (i.e., of the previous holons), and then adds its own unique (and more encompassing) capacities. In that sense, and that sense only, can the new and more encompassing holon be said to be "higher" or "deeper." ("Higher" and "deeper" both imply a vertical dimension of integration not found in a merely horizontal expansion, a point we will return to in a moment.) Organisms *include* cells, which *include* molecules, which *include* atoms (but not vice versa).

Thus, whatever the important value of the previous stage, the new stage has that enfolded in its own makeup, plus something extra (more integrative capacity, for example), and that "something extra" means "extra value" *relative* to the previous (and less encompassing) stage. This crucial definition of a "higher stage" was first introduced in the West by Aristotle and in the East by Shankara and Lieh-tzu; it has been central to developmental studies ever since.

A quick example: in cognitive and moral development, in both the boy and the girl, the stage of preoperational or preconventional thought is concerned largely with the individual's own point of view ("narcissistic"). The next stage, the operational or conventional stage, still takes account of the individual's own point of view, but *adds* the capacity to take the *view of others* into account. Nothing fundamental is lost; rather, something new is added. And so in this sense it is properly said that this stage is higher or deeper, meaning more valuable and useful for a wider range of interactions. Conventional thought is *more valuable* than preconventional thought in establishing a balanced moral response (and postconventional is even more valuable, and so on).

As Hegel first put it, and as developmentalists have echoed ever since, each stage is adequate and valuable, but each deeper or higher stage is more adequate and, in that sense only, more valuable (which always means more holistic, or capable of a wider response).

It is for all these reasons that Koestler, after noting that all such hierarchies are composed of holons, or increasing orders of wholeness, pointed out that the correct word for "hierarchy" is actually *holarchy*.[19]

He is absolutely correct, and so from now on I will often use "hierarchy" and "holarchy" interchangeably.

Thus heterarchists, who claim that "heterarchy" and "holism" are the same thing (and that both are contrasted to the divisive and nasty "hierarchy"), have got it exactly backward: The only way to get a holism is via a holarchy. Heterarchy, in and by itself, is merely differentiation without integration, disjointed parts recognizing no common and deeper purpose or organization: heaps, not wholes.

PATHOLOGY

That is normal or natural holarchy, the sequential or stagelike unfolding of larger networks of increasing wholeness, with the larger or wider wholes being able to exert influence over the lower-order wholes. And as natural, desirable, and unavoidable as that is, you can already start to see how holarchies can go pathological. If the higher levels can exert influence over the lower levels, they can also overdominate or even repress and alienate the lower levels. And that leads to a host of pathological difficulties, in both the individual and society at large.

It is precisely *because* the world is arranged holarchically, precisely because it contains fields within fields within fields, that things can go so profoundly wrong, that a disruption or pathology in one field can reverberate throughout an entire system. And the cure for this pathology, in all systems, is essentially the same: rooting out the pathological holons so that the holarchy itself can return to harmony. The cure does not consist in getting rid of holarchy per se, since, even if that were possible, it would simply result in a uniform, one-dimensional flatland of no value distinctions at all (which is why those critics who toss out hierarchy in general immediately replace it with a new scale of values of their own, i.e., with their own particular hierarchy).

Rather, the cure of any diseased system consists in rooting out any holons that have usurped their position in the overall system by abusing their power of upward or downward causation. This is exactly the cure we see at work in psychoanalysis (shadow holons refuse integration), critical social theory (ideological holons distort open communication), democratic revolutions (monarchical or fascist holons oppress the body politic), medical science interventions (cancerous holons invade a benign system), radical feminist critiques (patriarchal holons dominate the public sphere), and so on. It is not getting rid of holarchy per se, but arresting (and integrating) the arrogant holons.

In short, the existence of pathological hierarchies does not damn the existence of hierarchies in general. That distinction is crucial and, for the most part, very easy to spot. Thus Riane Eisler, herself a rather staunch champion of heterarchy, nonetheless emphatically notes that "an important distinction should be made between domination and actualization hierarchies. The term *domination hierarchies* describes hierarchies based on force or the express or implied threat of force. Such hierarchies are very different from the types of hierarchies found in progressions from lower to higher orderings of functioning—such as the progression from cells to organs in living organisms, for example. These types of hierarchies may be characterized by the term *actualization hierarchies* because their function is to maximize the organism's potentials. By contrast, human hierarchies based on force or the threat of force not only inhibit personal creativity but also result in social systems in which the lowest (basest) human qualities are reinforced and humanity's higher aspirations (traits such as compassion and empathy as well as the striving for truth and justice) are systematically suppressed."[20]

Let us further note that, by Eisler's own definitions, what dominator hierarchies are suppressing is in fact the individual's own actualization hierarchies!—what she calls "humanity's higher aspirations" instead of its "lowest (basest) qualities." In other words, the cure for pathological hierarchy is actualization hierarchy, not heterarchy (which would produce more heaps and fragments, not wholes and cures).

These types of distinctions are crucial, because not only are there pathological or dominator hierarchies, there are pathological or *dominator heterarchies* (which is a topic that heterarchists studiously avoid). I just suggested that normal hierarchy, or the *holism between levels*, goes pathological when there is a breakdown between levels and a particular holon assumes a repressive, oppressive, arrogant role of dominance over other holons (whether in individual or social development). On the other hand, *normal* heterarchy, which is holism *within* any level, goes pathological when there is a blurring or fusion of that level with its environment: a particular holon doesn't stand out too much, it blends in too much; it doesn't arrogate itself above others, it loses itself in others—and all distinctions, of value or identity, are lost (the individual holon finds its value and identity only through others).

In other words, in pathological *hierarchy*, one holon assumes agentic dominance to the detriment of all. This holon doesn't assume it is *both* a whole and a part, it assumes it is the whole, period. On the other hand, in pathological *heterarchy*, individual holons lose their distinctive value and identity in a communal fusion and meltdown. This holon doesn't assume it is *both* a whole and a part, it assumes it is a part, period. It becomes only *instrumental* to some other use; it is merely a strand in the web; it has no intrinsic value.

Thus, pathological heterarchy means not union but fusion; not integration but indissociation; not relating but dissolving. All values become equalized and homogenized in a flatland devoid of individual values or identities; nothing can be said to be deeper or higher or better in any meaningful sense; all values vanish into a herd mentality of the bland leading the bland.

Whereas pathological hierarchy is a type of ontological fascism (with the one dominating the many), pathological heterarchy is a type of ontological totalitarianism (with the many dominating the one)—all of which we will discuss in detail in later chapters (where we will see that pathological hierarchy and pathological heterarchy are, respectively, types of pathological *agency* and pathological *communion*; and we will further see that these two pathologies are often associated, respectively, with the *male* and *female* value spheres—as with the work of Gilligan, Eisler, et al.—the males "ranking" and the females "linking," with the possible *respective pathologies* of dominance and fusion; feminists center on the male pathologies of dominance and miss the equally catastrophic pathologies of fusion).

In the meantime, beware any theorist who pushes solely hierarchy or solely heterarchy, or attempts to give greater value to one or the other in an ontological sense. When I use the term "holarchy," I will especially mean the balance of normal hierarchy and normal heterarchy (as the context will make clear). "Holarchy" undercuts both extreme hierarchy and extreme heterarchy, and allows the discussion to move forward with, I believe, the best of both worlds kept firmly in mind.

Finally, I would say that in trying to redress the severe imbalances of pathological hierarchy (which, as I indicated, we will explore under the category of pathological masculinity and pathological agency, sometimes called "the patriarchy"), we are allowed, even enjoined, to give normal femininity and normal heterarchy an *exaggerated* emphasis and a *greater* value, simply because we are trying to balance the scales. We are not allowed, I believe, to go to the other extreme and replace pathological masculinity with pathological femininity—or, to say virtually the same thing, we don't cure pathological hierarchy with pathological heterarchy.

QUALITATIVE DISTINCTIONS

The fact that actualization hierarchies involve a *ranking* of increasing holistic capacity—or even a ranking of *value*—is deeply disturbing to believers in extreme heterarchy, who categorically reject any sort of actual ranking or judgments whatsoever. With very good and often noble reasons (many of which I heartily support), they point out that value ranking is a hierarchical judgment that all too often translates into social oppression and inequality, and that in today's world the more compassionate and just response is a radically egalitarian or pluralistic system—a heterarchy of equal values. And while some of these critics are, as I said, quite nobly inspired, some of them have become quite rancorous, even vicious, in their vocal condemnation of any sort of value hierarchies. "Higher" has become their all-purpose dirty word.

What they don't seem to realize is that their valued embrace of heterarchy is itself a hierarchical judgment. They value heterarchy; they feel it embodies more justice, and compassion, and decency; they contrast it with hierarchical views, which they feel are dominating and denigrating. In other words, they *rank* these two views, and they feel one is definitely *better* than the other. That is, they have their own hierarchy, their own value ranking.

But since they consciously deny hierarchy altogether, they must obscure and hide their own. They must pretend that their own hierarchy is not a hierarchy. Their ranking becomes unacknowledged, hidden, covert. Further, not only is their own hierarchy hidden, it is self-contradictory: it is a hierarchy that denies hierarchy. They are presupposing that which they deny; they are consciously disavowing what their actual stance assumes.

By refusing even to look at hierarchy, even while making massively hierarchical judgments anyway, they are saddled with a rather crude and very poorly-thought-out hierarchy of values. This all too often, and unfortunately, lends an unmistakable air of hypocrisy to their stance. With much righteous indignation, they hierarchically denounce hierarchy. With their left hand they are doing what their right despises in everybody else. By *hating* judgments, and by *hiding* their own, they convert self-loathing into righteous condemnation of others.

In essence, their stance amounts to: "I have my ranking, but you shall not have yours. And further, by pretending that my ranking is not a ranking"—that move is done unconsciously—"I will say that I am without ranking altogether; and I shall then, in the name of compassion and equality, despise and attack ranking wherever I find it, because ranking is very bad."

By making these hierarchical judgments in an unacknowledged fashion.

they avoid and suppress the really difficult issues of just how we go about making our value judgments in the first place. They are articulate on the lamentable hierarchical value judgments of others, but strangely inarticulate—totally silent, actually—as to how and why they arrived at their own. Their self-ethic of inarticulacy and their other-ethic of vocal condemnation combine to form a large club with which they simply bash others in the name of kindness. This does little to help articulate the nature of human value systems, the nature of how men and women go about choosing the good, and the true, and the beautiful, choices that involve ranking, and choices that these critics make and then deny they have made.

Their heterarchy is a stealth hierarchy. They bury their tracks, then claim they have no tracks, and thus avoid and repress the truly profound and difficult topic: why do human beings *always* leave tracks? Why is the finding of *value* in the world inherent in the human situation? And since, even if we decide to value everything equally, that involves *rejecting* value systems that do not, why is some sort of ranking *unavoidable*? Why are *qualitative distinctions* built into the fabric of the human orientation? Why is trying to deny value itself a value? Why is denying ranking itself a ranking? And given that, how can we sanely and *consciously* choose our unavoidable hierarchies, and not merely fall into the ethics of unacknowledgment and suppression and inarticulacy?

Charles Taylor, whose book *Sources of the Self* will be one of our constant companions later in this volume, has done a masterful job of tracing the modern rise of the worldview that claims it is not a worldview. That is, the rise of certain value judgments that deny they are value judgments, the rise of certain hierarchies that deny the existence of hierarchies. It is an altogether fascinating story, and one we will later follow in detail, but for the moment we might observe the following.

Taylor begins by noting that the making of what he calls "qualitative distinctions" is an unavoidable aspect of the human situation. We simply find ourselves existing in various *contexts*, in various *frameworks* (as I would put it, we are holons within holons, contexts within contexts), and these contexts unavoidably constitute various values and meanings that are embedded in our situation. "What I have been calling a *framework*," says Taylor, "incorporates a crucial set of qualitative distinctions [a value hierarchy]. To think, feel, judge within such a framework is to function with the sense that some action. or mode of life. or mode of feeling is incomparably higher than the others which are more readily available to us. I am using 'higher' here in a generic sense. The sense of what the difference consists in may take many forms. One form of life may be seen as fuller, another way of feeling and acting as purer, a mode of feeling or living as deeper, a style of life as more admirable, and so on."[21]

Thus, even those who embrace heterarchy or radical pluralism are making deep and profound qualitative distinctions, even though they denounce qualitative distinctions as brutal and vicious, even though they deny the notion of frameworks altogether. "But this person doesn't lack a framework. On the contrary, he has a strong commitment to a certain ideal of benevolence. He admires people who live up to this ideal, condemns those who fail or who are too confused even to accept it, feels wrong when he himself falls below it. He lives within a moral horizon which *cannot be explicated by his own moral theory*."[22]

The point is that, even though this individual espouses diversity and *equality* of values, the idea is *never*, as Taylor puts it, that "*whatever* we do is acceptable."

> I want to defend the strong thesis that doing without frameworks is utterly impossible for us; otherwise put, that the horizons within which we live our lives and which make sense of them have to include these strong qualitative discriminations [value hierarchies]. Moreover, this is not meant just as a contingently true psychological fact about human beings, which could perhaps turn out one day not to hold for some exceptional individual or new type, some superman of disengaged objectification. Rather, the claim is that living within such strongly qualified horizons is constitutive of human agency . . . and not some optional extra we might just as well do without.[23]

Yet there is a modern outlook, says Taylor, "which is tempted to deny these frameworks altogether. My thesis here is that this idea is deeply mistaken . . . and deeply confused. It reads the affirmation of life and freedom as involving a repudiation of qualitative distinctions, a rejection of constitutive goods as such, while these are themselves reflections of qualitative distinctions and presuppose some conception of qualitative goods."[24]

In the course of historically tracing the curious rise of this curious stance, Taylor notes that "the more one examines the motives—what

Nietzsche would call the 'genealogy'—of these theories, the stranger they appear. It seems that they are motivated by the strongest moral ideals, such as freedom, altruism, and universalism [i.e., universal pluralism]. These are among the central moral aspirations of modern culture, the hypergoods [strong hierarchies] which are distinctive to it. And yet what these ideals drive the theorists toward is a denial of all such goods. They are caught in a strange pragmatic contradiction, whereby the very goods which move them push them to deny or denature all such goods. They are constitutionally incapable of coming clean about the deeper sources of their own thinking. Their thought is inescapably cramped."[25] They are, Taylor says, morally superior in a universe where nothing is supposed to be superior.

The resultant "frameworkless agent," says Taylor, "is a monster," motivated by the "deep incoherence and self-illusion which this denial involves." This hierarchical denial of hierarchy involves an *ethics of suppression*, according to Taylor, because "layers of suppression" are required to so thoroughly conceal from oneself the sources of one's own judgments.

And this further explains why these theorists are, as Taylor puts it, "parasitic." Since they can't "come clean about the deeper sources of their own thinking," they necessarily live off nothing but the vocal denouncements of those views that do manage to consciously acknowledge their own qualitative distinctions. "Because their moral sources are unavowable, they are mainly invoked in polemic. Their principal words of power are denunciatory. Much of what they [actually] live by has to be inferred from the rage with which their enemies are attacked and refuted. This self-concealing kind of philosophy is also thereby parasitic. . . ."

Thus, even the radical pluralists (the heterarchists) are motivated by values of freedom, altruism (universal benevolence), and universal pluralism. These are *deeply* hierarchical judgments, and judgments that— rightly, I believe!—vigorously *reject* other types of value judgments and other types of hierarchies that have flourished throughout history. They *deeply reject* the warrior ethic, the ethic of elite aristocracy, the male-only ethic, and the master-slave ethic, to name a few.

In other words, their heterarchical values are held in place by hierarchical judgments (most of which I thoroughly agree with), and they might as well come clean and join the rest of us in trying to *consciously* understand all that, and not simply bury their tracks in parasitic and denunciatory and suppressive rhetoric.

The same problems, of course, beset the "cultural relativists," who maintain that all diverse cultural values are equally valid (in a functional sense), and that no *universal* value judgments are possible. But that judgment is itself a universal judgment. It claims to be universally true that no judgments are universally true. It makes its own universal judgment and then simultaneously denies all others, because universal judgments are very, very bad. It thus ignores the crucial issue of how we go about making valid universal judgments in the first place. It *exempts* its own universal claims from any scrutiny by simply claiming they aren't claims.

The extreme cultural relativists thus maintain that "truth" is basically what any culture can come to agree on, and thus no "truth" is inherently better than any other. There was a certain vogue for this type of stance during the sixties and seventies, but its self-contradictory nature became apparent with, to give the most notorious example, Michel Foucault's book *The Order of Things*. In this work Foucault maintained, in essence, that what humans come to call "truth" is simply an arbitrary play of power and convention, and he outlined several epochs where the "truth" seemed to depend entirely on shifting and conventional epistemes, or discursive formations governed not by "truth" but by exclusionary transformation principles. All truth, in other words, was ultimately arbitrary.

The argument seemed quite persuasive, and even caused a bit of an international sensation. Until his brighter critics simply asked him: "You say all truth is arbitrary. Is your presentation itself true?"

Foucault, like all relativists, had exempted himself from the very criteria he aggressively applied to others. He was making an extensive series of truth claims that denied all truth claims (except his own privileged stance), and thus his position, as critics from Habermas to Taylor pointed out, was profoundly incoherent. Foucault himself abandoned the extreme relativism of this "archaeological" endeavor and subsumed it in a more balanced approach (that would include continuities as well as abrupt discontinuities; he called the merely archaeological approach "arrogant").

Nobody is denying that many aspects of culture are indeed different and equally valuable. The point is that that stance itself is universal and *rejects* theories that merely and arbitrarily rank cultures on an ethnocentric bias (a rejection I share). But because it claims that all ranking is either bad or arbitrary, it cannot explain its own stance and the process of its own (unacknowledged) ranking system. And if nothing else, unconscious ranking is bad ranking, by any other name.

And the relativists are very bad rankers.[26] Jürgen Habermas and numerous others (Charles Taylor, Karl-Otto Apel, Quentin Skinner, John Searle, etc.) have launched devastating critiques of these positions, pointing out that they all involve a "performative contradiction": another way of saying that they are implicitly presupposing universal validity claims that they deny can even exist.

In short, extreme cultural relativity and merely heterarchical value systems are no longer enjoying the vogue they once did. The word is out that qualitative distinctions are inescapable in the human condition, and further, that there are *better* and *worse* ways to make our qualitative distinctions.

In many ways, we want to agree with the broad *conclusions* of the cultural diversity movements: we do want to cherish all cultures in an equal light. But that *universal* pluralism is not a stance that all cultures agree with; that universal pluralism is a very special type of ranking that most ethnocentric and sociocentric cultures do not even acknowledge; that universal pluralism is the result of a very long history hard-fought against dominator hierarchies of one sort or another.[27]

Why is universal pluralism better than dominator hierarchies? And *how* did we develop or evolve to a stance of universal pluralism, when most of history despised that view? These are some of the many developmental and evolutionary themes of this volume. *How* we arrive at that universal pluralism, and how we can defend it against those who would, in a dominating fashion, elevate their culture or their beliefs or their values above all others—these are the crucial questions whose answers are aborted by merely denying ranking and denying qualitative distinctions in the first place.

CONCLUSION

But here is my point: if frameworks are inescapable (we are contexts within contexts, holons within holons), and if frameworks involve qualitative distinctions—in other words, if we are inextricably involved in judgments that are hierarchical—then we can begin to consciously join these judgments with the *sciences* of hierarchy, that is, the sciences of holarchy, of frameworks within frameworks, of contexts within contexts, of holons within holons—with the result that values and facts are no longer automatically divorced.

This unifying and integrative move was *blocked* as long as the heterarchists were calling their view "holistic" (when it was really "heapistic"). *Blocked*, because the heterarchists insisted that reality was nonhierarchical, whereas the sciences of wholeness insisted altogether otherwise. But with the understanding that *the only way you get a holism is via a holarchy*, we are now in a position to realign facts and values in a gentler embrace, with science working with us, not against us, in constructing a truly holistic, not heapistic, worldview.

Further, let us simply note that the Great Chain of Being was in fact a Great Holarchy of Being—with each link being an intrinsic whole that was simultaneously a part of a larger whole—and the entire series nested in Spirit.[28]

If these various holarchies—in the sciences, in value judgments, in the great wisdom traditions—could in fact be sympathetically aligned with one another, a truly significant synthesis might indeed lie in our collective future.

2

The Pattern
That Connects

Matter, which appears to be merely passive and without form and arrangement, has even in its simplest state an urge to fashion itself by a natural evolution into a more perfect constitution.
—IMMANUEL KANT

God does not remain petrified and dead; the very stones cry out and raise themselves to Spirit. —GEORG HEGEL

THE NATURE OF THE PATTERN

WE BEGIN WITH the sciences of wholeness, or dynamic systems theory. The rest of this chapter and all of chapters 3 and 4 will be devoted to exploring some of the basic conclusions of the modern evolutionary sciences, with a view toward their possible integration in a larger scheme of things.

What follows are twenty basic tenets (or conclusions) that represent what we might call "patterns of existence" or "tendencies of evolution" or "laws of form" or "propensities of manifestation." These are the common patterns or tendencies, recall, that modern systems sciences have concluded are operative in all three domains of evolution—the physiosphere, the biosphere, and the noosphere—and tendencies that therefore make this universe a genuine *uni-versum* ("one turn"), or an emergent pluralism undergirded by common patterns—the "patterns that connect."

(At this point, I don't want to get involved in intricate arguments over whether these are "eternal laws" or simply "relatively stable habits" of the universe, and so I will be satisfied with the latter.)[1]

These patterns (listed as the Twenty Tenets below) are drawn from the modern evolutionary and systems sciences, but I would like to emphasize that they are not confined to those sciences. As I mentioned earlier, we are now looking at the "half" of those sciences that seem accurate, and we have yet to examine the half that is extremely questionable (this will begin in chapter 3). As we will see in great detail, the problem with virtually every attempt to outline the common patterns found in all three domains of evolution is simply that the patterns are presented in the *language of objective naturalism* ("it"-language), and thus they fail miserably when applied to domains described only in I-language (aesthetics) and we-language (ethics). Every "unified systems attempt" that I have seen suffers from this crippling inadequacy.

I have been very careful, therefore, to cut these tenets at a level and type of abstraction that is, I believe, fully compatible with it-, we-, and I-languages (or the true, the good, and the beautiful), so that the synthesis can proceed nonviolently into domains where previously systems theory was intent upon subtle reductionism to its own naturalistic and objectifying terms. (All of this will, as I said, be discussed in detail, beginning in chapter 3.)

Finally, a small warning. Many readers have found these tenets to be the most interesting part of the book; others have found them too abstract and rather boring. If you are of the latter, I should mention that they will be fleshed out and made obvious, I trust, in the succeeding chapters. In the meantime, a *Reader's Digest* version of them might go as follows:

Reality is not composed of things or processes; it is not composed of atoms or quarks; it is not composed of wholes nor does it have any parts. Rather, it is composed of whole/parts, or holons.

This is true of atoms, cells, symbols, ideas. They can be understood neither as things nor processes, neither as wholes nor parts, but only as simultaneous whole/parts, so that standard "atomistic" and "wholistic" attempts are both off the mark. There is nothing that isn't a holon (upwardly and downwardly forever).

Before an atom is an atom, it is a holon. Before a cell is a cell, it is a holon. Before an idea is an idea, it is a holon. All of them are wholes that exist in other wholes, and thus they are all whole/parts, or holons, first and foremost (long before any "particular characteristics" are singled out by us).

Likewise, reality might indeed be composed of processes and not things, but all processes are only processes within other processes—that is, they are first and foremost holons. Trying to decide whether the fundamental units of reality are things or processes is utterly beside the point, because either way, they are all holons, and centering on one or the other misses the central issue. Clearly some things exist, and some processes exist, but they are each and all holons.

Therefore we can examine what *holons* have in common, and this releases us from the utterly futile attempt to find common processes or common entities on all levels and domains of existence, because that will never work; it leads always to reductionism, not true synthesis.

For example, to say that the universe is composed primarily of quarks is already to privilege a particular domain. Likewise, at the other end of the spectrum, to say that the universe is really composed primarily of our symbols, since these are all we really know—that, too, is to privilege a particular domain. But to say that the universe is composed of holons neither privileges a domain nor implies special fundamentalness for any level. Literature, for example, is *not* composed of subatomic particles; but both literature and subatomic particles are composed of holons.

Starting with the notion of holons, and proceeding by a combination of *a priori* reasoning and *a posteriori* evidence, we can attempt to discern what all known holons seem to have in common. These conclusions are refined and checked by examining any and all domains (from cellular biology to physical dissipative structures, from stellar evolution to psychological growth, from autopoietic systems to spiritual experiences, from the structure of language to DNA replication).

Since all of those domains operate with holons, we can attempt to discern what all these holons have in common when they interact—what their "laws" or "patterns" or "tendencies" or "habits" are. And this gives us a list of some twenty tenets, which I have grouped into twelve categories (some of these are simple definitions, but for convenience I will always refer to the entire list as "twenty tenets." There is nothing special about twenty; some of these might not hold up, others can be added, and I have not tried to be exhaustive).

TWENTY TENETS

1. *Reality as a whole is not composed of things or processes, but of holons*. Composed, that is, of wholes that are simultaneously parts of other wholes, with no upward or downward limit. To say that holons are processes instead of things is in some ways true, but misses the essential point that processes themselves exist only within other processes. There are no things or processes, only holons.

Since reality is not composed of wholes, and since it has no parts—since there are only whole/parts—then this approach undercuts the traditional argument between atomism (all things are fundamentally isolated and individual wholes that interact only by chance) and wholism (all things are merely strands or parts of the larger web or whole). Both of those views are absolutely incorrect. There are no wholes, and there are no parts. There are only whole/parts.

This approach also undercuts the argument between the materialist and idealist camps. Reality isn't composed of quarks, or bootstrapping hadrons, or subatomic exchange; but neither is it composed of ideas, symbols, or thoughts. It is composed of holons.

There is an old joke about a King who goes to a Wiseperson and asks how is it that the Earth doesn't fall down? The Wiseperson replies, "The Earth is resting on a lion." "On what, then, is the lion resting?" "The lion is resting on an elephant." "On what is the elephant resting?" "The elephant is resting on a turtle." "On what is the . . ." "You can stop right there, your Majesty. It's turtles all the way down."

Holons all the way down. "Subatomic particles are—in a certain sense which can only be defined rigorously in relativistic quantum mechanics—nested inside each other. The point is that a physical particle—a renormalized particle—involves (1) a bare particle and (2) a huge tangle of virtual particles, inextricably wound together in a recursive mess. Every real particle's existence therefore involves the existence of infinitely many other particles, contained in a virtual 'cloud' which surrounds it as it propagates. And each of the virtual particles in the cloud, of course, also drags along its own virtual cloud, bubbles within bubbles [holons within holons], and so on ad infinitum. . . ."[2]

But it's also turtles all the way up. Take mathematics, for example. The notorious "paradoxes" in set theory (Cantor's, Burali-Forti's, Russell's), which, among other things, led to Tarski's Theorem and Gödel's Incompleteness Theorem, placed mathematics in an *irreversible, ever-expanding, no-upper-limit universe*: "The totality of sets cannot be the terminus of a well-defined generating process, for if it were we could take all of what we had generated so far as a set and continue to generate still larger universes. The totality of sets [mathematical holons] is an 'unconditioned' or absolute totality which for just that reason cannot be adequately conceived by the human mind, since the object of a normal conception can always be incorporated in a more inclusive totality. Moreover, the sets are arranged in a transfinite hierarchy"—a holarchy that continues upwardly forever, and must continue upwardly forever ("transfinitely"), or mathematics comes to a screeching self-contradictory halt.[3] Even mathematics is set in time's arrow, and time's arrow is indefinitely—"transfinitely"—holarchical.

This is important for philosophy as well, and particularly for many of the "new age" paradigms that now trumpet "Wholism." "Transfinite" (turtles all the way up) means that the sum total of all the whole/parts in the universe is *not itself a Whole*, because the moment it comes to be (as a "whole"), that totality is merely a *part* of the very next moment's whole, which in turn is merely a part of the next . . . and so ad infinitum.

This means that there is no place where we can rest and say, "The universe's basic principle is Wholeness" (nor, of course, can we say, "The basic principle is Partness"). This prevents us from ever saying that the principle of the Whole rules the world, for it does not; any whole is a part, indefinitely.

Thus, holons within holons within holons means that the world is without foundation in either wholes or parts (and as for any sort of "absolute reality" in the spiritual sense, we will see that it is neither whole nor part, neither one nor many, but pure groundless Emptiness, or radically *nondual* Spirit).

This is important because it prevents a totalizing and dominating Wholeness. "Wholeness"—this is a very dangerous concept (a point that will accompany us throughout this book)—dangerous for many reasons, not the least of which is that it is always available to be pushed into ideological ends. Whenever anybody talks of wholeness being the ultimate, then we must be very wary, in my opinion, because they are often telling us that we are merely "parts" of their particular version of "wholeness," and so we should be subservient to their vision—we are merely strands in their wonderful web.

And then, since we are all defined as strands in the web, a totalizing social agenda seems eminently reasonable. It is not beside the point that theorists as diverse as Habermas and Foucault have seen such totalizing agendas as the main modern enemy of the life-world (a point we will return to in chapter 12).

I belabor this issue because it is extremely important to emphasize the indefiniteness of holarchy, its openness, its dizzifyingly nesting nature—an actualization holarchy, not a dominator holarchy. And a dominator holarchy, recall, occurs precisely whenever any holon is established, not as a whole/part, but as the whole, period. "Ultimate Wholeness": this is the essence of dominator holarchies, pathological holarchies. "Pure wholeness": this is the totalizing lie.

For all these reasons, I will usually refer to the sum total of events in the universe not as the "Whole" (which implies the ultimate priority of wholeness over partness) but as "the All" (which is the sum total of whole/parts). And this sum total is not itself a whole but a whole/part: as soon as you think "the All," your own thought has *added* yet another holon to the All (so that the first All is no longer *the* All but merely *part* of the new All), and so off we go indefinitely, never arriving at that which we symbolize as the "All," which is why it is *never* a whole, but an unending series of whole/parts (with the series itself a whole/part—and so on "transfinitely").

The Pythagoreans introduced the term "Kosmos," which we usually translate as "cosmos." But the original meaning of Kosmos was the patterned nature or process of all domains of existence, from matter to math to theos, and not merely the physical universe, which is usually what both "cosmos" and "universe" mean today.

So I would like to reintroduce this term, *Kosmos*. The Kosmos contains the cosmos (or the physiosphere), the bios (or biosphere), nous (the noosphere), and theos (the theosphere or divine domain)—none of them being foundational (even spirit shades into Emptiness).

So we can say in short: The Kosmos is composed of holons, all the way up, all the way down.

Let me end this section with one last example, from perhaps an unexpected source. The "postmodern poststructuralists"—usually associated with such names as Derrida, Foucault, Jean-François Lyotard, and stretching back to George Bataille and Nietzsche—have been the great foes of any sort of systematic theory or "grand narrative," and thus they might be expected to raise stern objections to any overall theory of "holarchy." But a close look at their work shows that it is driven precisely by a conception of holons within holons within holons, of texts within texts within texts (or contexts within contexts within contexts), and it is this sliding play of texts within texts that forms the "foundationless" platform from which they launch their attacks.

George Bataille, for instance. "In the most general way"—and these are his italics—"*every isolable element of the universe always appears as a particle that can enter into composition with a whole that transcends it. Being is only found as a whole composed of particles whose relative autonomy is maintained* [a part that is also a whole]. These two principles [simultaneous wholeness and partness] dominate the uncertain presence of an *ipse* being across a distance that never ceases to put *everything* in question."[4]

Everything is put into question because everything is a context within a context forever. And *putting everything in question* is precisely what the postmodern poststructuralists are known for. And so in a language that would soon become quite typical (and is by now almost comical), Bataille goes on to point out that "putting everything into question" counters the human need to violently arrange things in terms of a pat wholeness and smug universality: "With extreme dread imperatively becoming the demand for universality, carried away to vertigo by the movement that composes it, the *ipse* being that presents itself as a universal is only a challenge to the diffuse immensity that escapes its precarious violence, the tragic negation of all that is not its own bewildered phantom's chance. But, as a man, this being falls into the meanders of the knowledge of his fellowmen, which absorbs his substance in order to reduce it to a component of what goes beyond the virulent madness of his autonomy in the total night of the world."[5] Um, and so forth.

The point is *not* that Bataille himself was without any sort of system, but simply that the *system is sliding*—holons within holons. So the claim to simply have "no system" is a little disingenuous. Which is why André Breton, the leader of the surrealists at the time, began a counterattack on this part of Bataille, also in terms that are echoed by today's critics of postmodernists: "M. Bataille's misfortune is to reason: admittedly, he reasons like someone who 'has a fly on his nose,' which allies him more closely with the dead than with the living, but *he does reason*. He is trying, with the help of the tiny mechanism in him which is not completely out of order, to share his obsessions: this very fact proves that he cannot claim, no matter what he may say, to be opposed to any system, like an unthinking brute."[6]

Both sides are correct, in a sense. There is system, but the system is sliding. It is unendingly, dizzifyingly holonic. This is why Jonathan Culler, perhaps the foremost interpreter of Jacques Derrida's deconstruction, can point out that Derrida does *not* deny truth per se, but only insists that truth and meaning are *context-bound* (each context being a whole that is also part of another whole context, which itself . . .). "One could therefore," says Culler, "identify deconstruction with the twin principles of the *contextual determination of meaning* and the *infinite extendability of context*."[7]

Turtles all the way up, all the way down. What deconstruction puts into question is the desire to find a final resting place, in either wholeness or partness or anything in between. Every time somebody finds a final interpretation or a foundational interpretation of a text (or life or history or Kosmos), deconstruction is on hand to say that the total context—or Wholistic interpretation—does not exist, because it is also unendingly a part of yet another text forever. As Culler puts it, "Total context [final Wholism] is unmasterable, both in principle and in practice. *Meaning is context bound, but context is boundless*."[8] Transfinite turtles.

Even Habermas, who generally takes Breton's position to Derrida's Bataille, agrees with that particular point. As he puts it, "These variations of context that change meaning cannot in principle be arrested or controlled, because contexts cannot be exhausted, that is, they cannot be theoretically mastered once and for all."[9]

That the system is sliding does *not* mean that meaning can't be established, that truth doesn't exist, or that contexts won't hold still long enough to make a simple point. Many postmodern poststructuralists have not simply discovered holonic space, they have become thoroughly lost in it (a point we will return to in chapter 13; George Bataille, for example, took a good, long, hard look at holonic space and went properly insane, though which is cause and which effect is hard to say).

As for our journey, we need only note that there is system, but the system is sliding: The Kosmos is the unending All, and the All is composed of holons—all the way up, all the way down.

2. *Holons display four fundamental capacities*: *self-preservation, self-adaptation, self-transcendence, and self-dissolution*. These are all very important, and we'll take them one at a time.

a. *Self-preservation*. All holons display some capacity to preserve their individuality, to preserve their own particular wholeness or autonomy. A hydrogen atom, in a suitable context, can remain a hydrogen atom. It doesn't necessarily display intentionality in any developed sense, but it does *preserve its agency over time*: it manages to remain itself across time's fluctuations—it displays self-preservation in the simple sense of maintaining identity across time (that itself is a remarkable accomplishment!).

A holon in a living context (e.g., a cell) displays an even more sophisticated capacity for self-preservation: the capacity for self-renewal (autopoiesis); it retains its own recognizable pattern (or structure) even as its material components are exchanged; it *assimilates* the environment to itself (it is a "meta-stable dissipative structure").

In other words, although holons exist *by virtue* of their interlinking relationships or context, they are *not defined* solely by their context but also by their own individual form, pattern, or structure (even particles that are bootstrappings of other particles retain their *individual* perspective, as Leibniz put it).

This intrinsic form or pattern is known by various names: entelechy (Aristotle), morphic unit/field (Sheldrake), regime or code or canon (Koestler), deep structure (Wilber). In the physiosphere, the form of a holon is relatively simple (although even that is bewilderingly complex); in the biosphere and noosphere, it takes on elaborate rules of relational exchange with its environment and self-organizing rules for its own structures, all geared to preserve the stable (or teleologically recognizable), coherent, and relatively autonomous pattern that is the essence of any holon. Francisco Varela, for example, explains that the "old biology" was based on "heteronomous units operating by a logic of correspondence," whereas the entire essence of the new biology is "autonomous units operating by a logic of coherence."[10]

In short, holons are defined, not by the stuff of which they are made (there is no stuff), nor merely by the context in which they live (though they are inseparable from that), but by the relatively autonomous and coherent pattern they display, and the capacity to preserve that pattern is one characteristic of a holon: the *wholeness* aspect of a holon is displayed in its pattern-preservation.

b. *Self-adaptation.* A holon functions not only as a self-preserving *whole* but also as a *part* of a larger whole, and in its capacity as a *part* it must adapt or accommodate itself to other holons (not autopoiesis but allopoiesis; not assimilation but accommodation). The *partness* aspect of a holon is displayed in its capacity to accommodate, to register other holons, to fit into its existing environment. Even electrons accommodate themselves, for example, to the number of other electrons in an orbital shell; they register, and react to, their environment. This doesn't imply intentionality on the part of the electron, just a capacity to react to surrounding actions. As a *whole*, it remains itself; as a *part*, it must fit in—and those are tenets 2a and 2b.

We can just as well think of these two opposed tendencies as a holon's *agency* and *communion*. Its agency—its self-asserting, self-preserving, assimilating tendencies—expresses its *wholeness*, its relative autonomy; whereas its communion—its participatory, bonding, joining tendencies—expresses its *partness*, its relationship to something larger.

Both of these capacities or tendencies are absolutely crucial and equally important; an excess of either will kill a holon immediately (i.e., destroy its identifying pattern); even a moderate imbalance will lead to structural deformity (whether we're talking about the growth of a plant or the growth of the patriarchy). And we have already suggested (in chapter 1) that an imbalance of these two tendencies in any system expresses itself as *pathological agency* (alienation and repression) or *pathological communion* (fusion and indissociation).

This primordial polarity runs through all domains of manifest existence, and was archetypally expressed in the Taoist principles of yin (communion) and yang (agency). Koestler: "On different levels of the inorganic and organic hierarchies, the polarisation of 'particularistic' [agency] and 'holistic' [communion] forces takes different forms, but it is *observable on every level*."[11]

(I will have much more to say about these two tendencies when we reach the psychological and political levels of self-organization, and particularly with reference to the male and female value spheres, and to political theories of rights [agency] and responsibilities [communion]).

c. *Self-transcendence* (or self-transformation). When an oxygen atom and two hydrogen atoms are brought together under suitable circumstances, a new and in some ways unprecedented holon emerges, that of a water molecule. This is not just a communion, self-adaptation, or association of three atoms; it is a transformation that results in something novel and emergent—different wholes have come together to form a new and different whole. There is some sort of creative twist on what has gone before. This is what Whitehead referred to as *creativity* (which he called "the ultimate category"—the category necessary to understand *any* other category),[12] and what Jantsch and Waddington call *self-transcendence.*

Some writers, such as Koestler, lump together self-adaptation and self-transcendence and refer to them interchangeably, because both embody a type of "going beyond." But apart from that similarity, the two are different in degree and in kind. In self-adaptation or communion, one finds oneself to be *part* of a *larger* whole; in self-transformation one *becomes* a *new whole*, which has its own new forms of agency *and* communion.[13] As Jantsch puts it:

> It is not sufficient to characterize these systems simply as open, adaptive, nonequilibrium, or learning systems [communion]; they are all that and more: they are *self-transcendent* [his italics], which means that they are capable . . . of transforming themselves. Self-transcendent systems are evolution's vehicle for qualitative change and thus ensure its continuity; evolution, in turn, maintains self-transcendent systems which can only exist in a world of interdependence. For self-transcendent systems, Being falls together with Becoming. . . .[14]

As Ilya Prigogine puts it, the various levels and stages of evolution are irreducible to each other because the transitions between them are characterized by *symmetry breaks*, which simply means that they are not equivalent rearrangements of the same stuff (whatever that "stuff" might be), but are in part a significant transcendence, a novel and creative twist. Jantsch explains:

> Symmetry breaks introduce new dynamic possibilities for morphogenesis and signal an act of self-transcendence. Complexity becomes possible only through symmetry breaks. The world which emerges from them becomes increasingly irreducible to a single level of basic [properties]. The reality which emerges is coordinated at many levels.

All of which he summarizes this way: "In the self-organization paradigm, *evolution is the result of self-transcendence at all levels.*"[15] He also calls this "self-realization through self-transcendence."

In other words, this introduces a *vertical* dimension that cuts at right angles, so to speak, to the horizontal agency and communion. In self-transcendence, agency and communion do not just interact; rather, *new forms* of agency and communion emerge through symmetry breaks, through the introduction of new and creative twists in the evolutionary stream. There is not only a *continuity* in evolution, there are important *discontinuities* as well. "Nature progresses by sudden leaps and deep-seated transformations rather than through piecemeal adjustments. The diagram of the branching tree of life no longer resembles the continuous Y-shaped joints of the synthetic theory; it is now pictured in terms of abrupt switches. . . . There is significant evidence accumulated in many fields of empirical science that dynamic systems do not evolve smoothly and continuously over time, but do so in comparatively sudden leaps and bursts."[16]

Paleontologist George Simpson referred to this as "quantum evolution," because these bursts "involved relatively abrupt alterations of adaptive capacity or bodily structure and left little or no evidence in the fossil record of the transitions between them."[17] This "quantum evolution" led to Niles Eldredge and Stephen Jay Gould's "punctuational model," all of which Michael Murphy summarizes as "evolutionary transcendence."

Murphy points out that "evolutionary theorists Theodosius Dobzhansky and Francisco Ayala have called these . . . events instances of 'evolutionary transcendence' because in each of them there arose a new order of existence. G. Ledyard Stebbins, a principal architect of modern evolutionary theory, described certain differences between large and small steps in organic evolution, distinguishing minor from major advances in grade of plants and animals. The term *grade* is used among biologists to denote a set of characteristics or abilities that clearly give descendent species certain advantages over their ancestors. According to Stebbins, the development of pollination mechanisms in milkweeds and orchids is an example of minor advances in grade, while the appearance of the digestive tube, central nervous system, elaborate sense organs, vertebrate limbs, and elaborate social behavior represent major advances. There have been about 640,000 of the former, he estimated, and from 20 to 100 of the latter during the several hundred million years of eukaryote evolution."[18]

The point is that there is nothing particularly metaphysical or occult about this. Self-transcendence is simply a system's capacity to reach beyond the given and introduce some measure of novelty, a capacity without which, it is quite certain, evolution would never, and could never, have even gotten started. Self-transcendence, which leaves no corner of the universe untouched (or evolution would have *no point* of departure), means nothing more—and nothing less—than that the universe has an intrinsic capacity to go beyond what went before.

d. *Self-dissolution*—Holons that are built up (through vertical self-transformation) can also break down. Not surprisingly, when holons "dissolve" or "come unglued," they tend to do so along the same vertical sequence in which they were built up (only, of course, in the reverse direction).

> If a structure is forced to retreat in its evolution (e.g., by a change in the non-equilibrium), as long as there are no strong perturbations it does so along the same path which it has come. . . . This implies a primitive, holistic *system memory* which appears already at the level of chemical reaction systems.[19]

This is true for psychological and linguistic (noospheric) holons as well. Roman Jakobson speaks of "those stratified phenomena which modern psychology uncovers in the different areas of the realm of the mind. New additions are superimposed on earlier ones and dissolution begins with the higher strata—the amazingly exact agreement between the chronological succession of these acquisitions and the general laws of irreversible solidarity which govern the synchrony of all the languages of the world."[20]

In other words, that which is vertically built up can vertically break down, and the pathways in both cases are essentially the same.

Taken together, these four capacities—agency or self-preservation, communion or self-accommodation, self-transcendence, and self-dissolution—can be pictured as a cross, with two horizontal "opposites" (agency and communion) and two vertical "opposites" (self-transcendence and self-dissolution).

These four "forces" are in constant tension (as we will see throughout this volume). *Horizontally*: the more agency, the less communion, and vice versa. That is, the more intensely a holon preserves its own individuality, preserves its *wholeness*, the less it serves its communions or its *partness* in larger or wider wholes (and vice versa: the more it is a part, the less it is its own whole). When we say, for example, that an element (such as helium) is "inert," we mean it intensely resists joining with other elements to form compounds; it retains its agency and resists communion—it is relatively inert.

This is a constant tension, as it were, across all domains, and shows up in everything from the battle between self-preservation and species-preservation, to the conflict between rights (agency) and responsibilities (communions), individuality and membership, personhood and community, coherence and correspondence, self-directed and other-directed, autonomy and heteronomy. . . . In short, how can I be both my own wholeness and a part of something larger, without sacrificing one or the other?

(Part of the answer, we will see, at all stages of evolution, including the human, involves self-transcendence to new forms of agency and communion that integrate and incorporate both partners in a supersession: not just a *wider* whole—a horizontal expansion—but a *deeper* or *higher* whole—a vertical emergence—which is indeed why "evolution is the result of self-transcendence at all levels," and why it is "self-realization through self-transcendence"—but that is somewhat ahead of the story.)

This constant horizontal battle between agency and communion extends even to the forms of *pathology* on any given level, where *too much agency*, too much individuality, leads to a severing (repression and alienation) of the rich networks of communion that sustain individuality in the first place; and *too much communion* leads to a loss of individual integrity, leads to fusion with others, to indissociation, to a blurring of boundaries and a meltdown and loss of autonomy.

(And we will see that the typical "male pathology" tends to be hyperagency, or fear of relationship, like the inertness of helium!, and the typical "female pathology" tends to be hypercommunion, or fear of autonomy—the one leading to domination, the other to fusion; and we will see how this played itself out in the "patriarchy" and the "matriarchy.")

If the horizontal battle is between agency and communion, the constant vertical battle is between self-transcendence and self-dissolution, the tendency to build up or to break down (and, of course, these forces complexly interact with agency and communion on any given level; for example, either too much agency or too much communion can both lead to breakdown—as we will see, this is a constant problem in human affairs, where the desire to find a "larger meaning" often leads to too much communion or fusion with a "greater cause," and this fusion is mistaken for transcendence, whereas it is simply loss of autonomy and release from responsibility, which apparently is the attraction).

Now I have been giving mostly examples from the human domain, but the point is that, in simpler forms, these four forces are operative in even the simplest of holons, because it's turtles all the way down. Every holon is actually a holon within other holons transfinitely—that is, every holon is *simultaneously* both a *subholon* (a part of some other holon) and a *superholon* (itself containing holons). As a holon, it must preserve its own pattern (agency over time), and it must register and react to its environment (its communions in space). If it does not appropriately respond, it is erased: too much agency, or too much communion, will destroy its identifying pattern.

Because each holon is also a superholon, then when it is erased—when it undergoes self-dissolution into its subholons—it tends to follow the same path down as its subholons followed up: cells break down into molecules, which break down into atoms, which break into particles, which disappear into transfinite clouds of probabilistic "bubbles within bubbles." . . .

Preserve or accommodate, transcend or dissolve—the four very different pulls on each and every holon in the Kosmos.

3. *Holons emerge.* Owing to the self-transcendent capacity of holons, new holons emerge. First subatomic particles, then atoms, then molecules, then polymers, then cells, and so on. The emergent holons are in some sense novel; they posses properties and qualities that cannot be strictly and totally deduced from their components; and therefore they, and their descriptions, cannot be reduced without remainder to their component parts. "Levels of organization involve ontologically new entities beyond the elements from which the self-organization process proceeds. There is no absolute deterministic level of description."[21] As Hofstadter puts it, "It is important to realize that the high-level law cannot be stated in the vocabulary of the low-level description"—and, he points out, this is true from gas particles to biological speciation, from computer programs to DNA replication, from musical scores to linguistic rules.[22]

Emergence also means that *indeterminacy* (and one of its correlates, degrees of freedom) is sewn into the very fabric of the universe, since *unprecedented* emergence means *undetermined* by the past (although pockets of the universe can regularly collapse in a deterministic fashion, as in classical mechanics). Holons, that is, are fundamentally indeterminate in some aspects (precisely because they are fundamentally self-transcending). As Laszlo summarizes the available evidence, "The selection from among the set of dynamically functional alternative steady states is not predetermined. The new state is decided neither by initial conditions in the system nor by changes in the critical values of environmental parameters; when a dynamic system is fundamentally destabilized it acts indeterminately."[23]

It now seems virtually certain that determinism arises only as a limiting case where a holon's capacity for self-transcendence approaches zero, or when its own self-transcendence hands the locus of indeterminacy to a higher holon (we will return to this important topic below).

Emergence is neither a rare nor an isolated phenomenon. As Varela, Thompson, and Rosch summarize the available evidence: "It is clear that emergent properties have been found across all domains—vortices and lasers, chemical oscillations, genetic networks, developmental patterns, population genetics, immune networks, ecology, and geophysics. What all these diverse phenomena have in common is that in each case a network gives rise to new properties. . . . The emergence of global patterns or configurations in systems of interacting elements is neither an oddity of isolated cases nor unique to [special] systems. In fact, it seems difficult for any densely connected aggregate to escape emergent properties."[24] As Ernst Mayr put it in his exhaustive work *The Growth of Biological Thought*:

> Systems almost always have the peculiarity that the characteristics of the whole cannot (not even in theory) be deduced from the most complete knowledge of the components, taken separately or in other partial combinations. This appearance of new characteristics in wholes has been designated as *emergence*. Emergence has often been invoked in attempts to explain such difficult phenomena as life, mind, and consciousness. Actually, emergence is equally characteristic of inorganic systems. . . . Such emergence is quite universal and, as Popper said, "We live in a universe of emergent novelty."[25]

Let me note in passing that this means *all* sciences are fundamentally *reconstructive sciences*. That is, we never know, and never can know, exactly what any holon will do tomorrow (we might know broad outlines and probabilities, based on *past* observations, but self-transcendent emergence always means, to some degree: surprise!). We have to wait and see, and from that, after the fact, we *reconstruct* a knowledge system.[26]

However, when a holon's self-transcendence approaches zero (when its creativity is utterly minimal), then the *reconstructive sciences* collapse into the *predictive sciences*. Historically, the empirical sciences got their start by studying precisely those holons that show minimal creativity. In fact, they basically studied nothing but a bunch of *rocks* in motion (mass moving through space over time), and thus they mistook the nature of science to be essentially predictive.

I mean no offense to rocks, but by taking some of the dumbest holons in existence and making their study the study of "really real reality," these physical sciences, we have seen, were largely responsible for the collapse of the Kosmos into the cosmos, for the reduction of the Great Holarchy of Being to the dumbest creatures on God's green Earth, and for the leveling of a multidimensional reality to a flat and faded landscape *defined* by a *minimum* of creativity (and thus *maximum* of predictive power).[27]

It would take such a turn of events as Heisenberg's uncertainty principle to remind us that even the constituents of rocks are neither as predictable nor as dumb as these silly reductionisms. In the meantime, the "ideal" of knowledge as predictive power would ruin virtually every field it was applied to (including rocks), because its very methods would erase any creativity it would find, thus erasing precisely what was novel, significant, valuable, meaningful. . . .

4. *Holons emerge holarchically.* That is, as a series of increasing whole/parts. Organisms contain cells, but not vice versa; cells contain molecules, but not vice versa; molecules contain atoms, but not vice versa. And it is that *not vice versa*, at *each* stage, that constitutes unavoidable asymmetry and nested hierarchy (holarchy). Each deeper or higher holon embraces its junior predecessors and then *adds* its own new and more encompassing pattern or wholeness—the new code or canon or morphic field or agency that will define this as a whole and not merely a heap (as Aristotle clearly spotted). This is Whitehead's famous dictum: "The many become one and are increased by one." Laszlo comments:

> The empirical evidence for this process is indisputable. Diverse atomic elements converge in molecular aggregates; specific molecules converge in crystals and organic macromolecules; the latter converge in

cells and the subcellular building blocks of life; single-celled organisms converge in multi-cellular species; and species of the widest variety converge in ecologies.[28]

Bertalanffy put it very bluntly: "Reality, in the modern conception, appears as a tremendous hierarchical order of organized entities, leading, in a superposition of many levels, from physical and chemical to biological and sociological systems. Such hierarchical structure and combination into systems of ever higher order, is characteristic of reality as a whole and is of fundamental importance especially in biology, psychology and sociology."[29] Likewise, Edward Goldsmith's *The Way: An Ecological Worldview*, an explanation of the "new paradigm," summarizes it thus: "Ecology explains events in terms of their role within the spatio-temporal Gaian hierarchy."[30] And even Rupert Sheldrake's theory of morphic fields—which is considered extremely innovative and very daring—does not evade the obvious; as Sheldrake concludes: "Thus the morphogenetic fields, like morphic units themselves, are essentially hierarchical in their organization."[31]

Francisco Varela, whose research in the autonomous nature of holonic cognition is a cornerstone of the "new biology" (see tenet 2a), points out that "it seems to be a general reflection of the richness of natural systems that indication can be iterated to produce a hierarchy of levels. The choice of considering the level above or below corresponds to a choice of treating the given system as autonomous or constrained"[32]—that is, every holon is both a whole and a part, and can be considered in terms of its autonomy (agency) or in terms of its being constrained by other holons (in communion), both views being correct (but partial).

Virtually all deep ecologists and ecofeminists reject the notion of holarchy, for rather confused reasons, it seems to me. From what I can tell, they seem to think that hierarchy and atomism are "bad," and that their "wholism" is the opposite of both. But no less than the patron saint of deep ecology, Arne Naess, clearly points out that "wholism" and "atomism" are actually two sides of the same problem, and that the *cure for both* is hierarchy. All reality, he points out, consists of what he calls "subordinate wholes" or "subordinate gestalts"—that is, holons. "We have therefore," he says, "a complex realm of gestalts, in a vast hierarchy. We can therefore speak of lower- and higher-order gestalts."[33]

Further, Naess points out, this hierarchic conception is necessary to counteract *both* wholism (meaning an emphasis on just wholes) *and* atomism (just parts), because gestalts are holons; they are *both* wholes and parts, arranged in a hierarchic order of higher and lower. As he puts it, "This terminology—a vast hierarchy of lower- and higher-order gestalts—is more useful than speaking about wholes and holism, because it induces people to think more strenuously about the relations between wholes and parts. It [also] facilitates the emancipation from strong atomistic or mechanistic trends in analytical thought."[34]

Hierarchy, in short, is for Naess the antidote both to atomism and wholism (extreme heterarchy). I don't know why his followers have such a difficult time grasping his notion. Perhaps it is, as I suggested earlier, that they are in such reaction to pathological hierarchies that they toss the baby out with the bathwater. I think this is very understandable, because in some early versions of hierarchy, the conception was rather rigid and "fascist," and resulted, no doubt, from what we have called pathological agency.

The point, of course, is to tease apart pathological hierarchies—where one holon usurps its position in the totality—from normal holarchies in general, which express the natural interrelations between holons that are always both parts and wholes in horizontal and vertical relationships. As we saw earlier, both pathological hierarchies and pathological heterarchies do exist and need to be addressed *as pathologies*, but that does not damn the existence of normal heterarchies and normal hierarchies themselves, both of which are necessary for wholes and parts to coexist.

Jantsch expresses the more recent and balanced understanding:

> Evolution [appears] as a multilevel reality in which the evolutionary chain of autopoietic levels of existence appear in hierarchic order. Each level includes all lower levels—there are systems within systems within systems . . . within the total system in question. However, it is essential that this hierarchy is not a control hierarchy in which information streams upward and orders are handed from the top down. Each level maintains a certain [relative] autonomy and lives its proper existence in horizontal relations [heterarchy] with its specific environment. [For example:] The organelles within our cells go about their business of energy exchange in an autonomous way and maintain their horizontal relationships within the framework of the world-wide Gaia system.[35]

If the deep ecologists and ecofeminists would follow Naess's lead, the whole discussion could move forward more smoothly. As it is now, they are often the defenders of flatland "wholism" and extreme heterarchy, which indeed *is* the opposite of atomism: two sides of the same problem.

5. *Each emergent holon transcends but includes its predecessor(s).* Each newly emergent holon, as we have seen, includes its preceding holons and then adds its own new and defining pattern or form or wholeness (its new canon or code or morphic field). In other words, it *preserves* the previous holons themselves but *negates* their separateness or isolatedness or aloneness. It preserves their being but negates their partiality or exclusiveness. "To supersede," said Hegel, "is at once to preserve and to negate."[36]

Another way to express this is: all of the lower is in the higher, but not all of the higher is in the lower. For example, hydrogen atoms are in a water molecule, but the water molecule is not in the atoms (we might say the water molecule "pervades" or "permeates" the atoms but is not actually in them, just as all of a word is in a sentence but not all of the sentence is in a word).

Varela points out that "at a given level of the hierarchy, a particular system can be seen as an *outside* to systems below it, and as an *inside* to systems above it; thus the status of a given system changes as one passes through its level in either the upward or the downward direction."[37] This is simply another example of asymmetry, and it means that all holons are not equally internal to all other holons (some have a relation of partial externalness marked by the emergence or the symmetry break): all of the lower is in the higher, but not all of the higher is in the lower.

Nobel laureate Roger Sperry puts it very straightforwardly: "It is important to remember in this connection that all of the simpler, more primitive, elemental forces remain present and operative; none has been canceled. These lower-level forces and properties, however, have been superseded in successive steps, encompassed or enveloped as it were, by those forces of increasingly complex organizational entities."[38] The higher embraces the lower, as it were, so that all development is envelopment.

But there is an important distinction that needs to be emphasized, which I indicated earlier by saying that supersession "preserves the being but negates the partiality" (or "preserves and negates"). A simple example will suffice: before Hawaii became a state, it was its own nation, and as a nation it had all the prerogatives of sovereignty: it could declare war, print its own money, conscript an army, and so forth. These were all parts of its regime or code or cannon—its self-preservation or *agency* as a whole, as an individual nation.

When Hawaii became an American state, all of its basic property, land, and fundamental features became part of the U.S.A. All of these basic structures were preserved in the new Union; none of them were destroyed or harmed in the least. What was not preserved, however, what was lost and actually negated, was Hawaii's capacity to be its own nation, to declare war, print its own money, and so on. Hawaii's isolated and autonomous regime was subsumed in the higher regime of the United States (where it retained some relatively autonomous rights as a state).

This is what I mean when I say "transcends but includes" or "negates and preserves"—all the *basic structures* and functions are preserved and taken up in a larger identity, but all the *exclusivity* structures and functions that existed because of isolation, set-apartness, partialness, exclusiveness, separative agency—these are simply dropped and replaced with a deeper agency that reaches a wider communion.[39]

This leads to a general phenomenon observed at all levels of holistic organization: in normal holarchies, the new and senior pattern or wholeness can to some degree *limit the indeterminacy* (organize the freedom) of its junior holons (precisely because it transcends but includes them; i.e., via "downward causation," or more generally, "downward influence"). Rupert Sheldrake is worth quoting at length, to show exactly what is involved, and how widespread and crucial the phenomenon is:

> In living organisms, as in the chemical realm, the morphogenetic fields [holons] are hierarchically organized: those of organelles—for example the cell nucleus, the mitochondria and chloroplasts—act by ordering physico-chemical processes within them; these fields are subject to the higher-level fields of cells; the fields of cells to those of tissues; those of tissues to those of organs; and of organs to the morphogenetic field of the organism as a whole. *At each level the fields work by ordering processes which would otherwise be indeterminate.*
>
> For example, in the case of free atoms, electronic events take place with the probabilities given by the unmodified probability structures of the atomic morphogenetic fields. But when the atoms come under the influence of the higher-level morphogenetic field of a molecule,

these probabilities are modified in such a way that the probability of events leading toward the actualization of the final form are enhanced, while the probability of other events is diminished. Thus the morphogenetic fields of molecules *restrict* the possible number of atomic configurations which would be expected on the basis of calculations which start from the probability structures of free atoms. And this is what is found in fact; in the case of protein folding, for example, the rapidity of the process indicates that the system does not "explore" the countless configurations in which the atoms could conceivably be arranged. Similarly, the morphogenetic fields of crystals restrict the large number of possible arrangements which would be permitted by the probability structures of their constituent molecules.

At the cellular level the morphogenetic field orders the crystallization of microtubules and other processes which are necessary for the coordination of cell division. But the planes in which the cells divide may be indeterminate in the absence of a higher-level field: for instance, in plant wound-calluses the cells proliferate more or less randomly to produce a chaotic mass. Within organized tissue, on the other hand, one of the functions of the tissue's morphogenetic field may be to impose a pattern on the planes of cell division, and thus control the way in which the tissue as a whole grows. Then the development of tissues itself is inherently indeterminate in many respects, as revealed when they are artificially isolated and grown in tissue culture; under normal conditions this indeterminacy is restricted by the higher-level field of the organ. Indeed at each level in biological systems, as in chemical systems, the morphic units in isolation behave more indeterminately than they do when they are part of a higher-level morphic unit. The higher-level morphogenetic field restricts and patterns their intrinsic indeterminism.[40]

This brings us directly to our sixth major tenet.

6. *The lower sets the possibilities of the higher; the higher sets the probabilities of the lower.* We just saw that as a higher level of creative novelty emerges, it in many ways goes beyond (but includes) the givens of the previous level. However, even though a higher level "goes beyond" a lower level, it does not *violate* the laws or the patterns of the lower level. It cannot be reduced to the lower level; it cannot be determined by the lower level; but neither can it *ignore* the lower level. My body follows

the laws of gravity; my mind follows other laws, such as those of symbolic communication and linguistic syntax; but if my body falls off a cliff, my mind goes with it.

This is what is meant by saying that a lower sets the possibilities, or the large framework, within which the higher will have to operate, but to which it is not confined. To use Polanyi's example, nothing in the laws governing physical particles can predict the emergence of a wristwatch, but nothing in the wristwatch violates the laws of physics. As Laszlo nicely summarizes it for any level:

> The processes of cosmic evolution bring forth a variety of matter-energy systems; these systems define the range of constraints and the scope of possibilities within which systems on higher levels can evolve. Systems on the lower level clusters can permit the evolution, but can never determine the nature, of systems on higher-level clusters. The evolution of physical matter-energy systems sets the stage and specifies the rules of the game for the evolution of biological species, and biological evolution sets the stage and specifies the rules of the game for the evolution of sociocultural systems.[41]

As for the higher restricting the probabilities of the lower, we already heard Sheldrake's summary of the extensive evidence, which he concludes in this way:

> At every level, the fields of the holons are probabilistic, and the material processes within the holon are somewhat random or indeterminate. Higher-level fields may act upon the fields of lower-level holons in such a way that their probability structures are modified. This can be thought of in terms of a restriction of their indeterminism: out of the many possible patterns of events that could have happened, some now become much more likely to happen as a result of the order imposed by the higher-level field. This field organizes and patterns the indeterminism that would be shown by the lower-level holons in isolation.[42]

Now, a "level" in a holarchy is established by several objective criteria: by a qualitative emergence (as explained by Popper); by asymmetry (or "symmetry breaks," as explained by Prigogine and Jantsch); by an inclusionary principle (the higher includes the lower, but not vice versa, as explained by Aristotle); by a developmental logic (the higher negates and preserves a lower, but not vice versa, as explained by Hegel); by a chronological indicator (the higher chronologically comes after the lower, but all that is later is not higher, as explained by Saint Gregory).

Despite the clarity of the concept of levels (and the meaning of "higher" and "lower"), some critics have maintained that the concept itself cannot be maintained, because of the purely arbitrary nature of the number of levels in any given holon. It is definitely true that the number of levels in any holon has an element of arbitrariness to it, simply because there is no upper or lower limit to a manifest holarchy and therefore no absolute referent.

For example, when it was thought that atoms were the ultimate holons, then a simple atom had one level, a water molecule had two levels, an ice crystal had three levels, and so on. When it was then thought that protons, neutrons, and electrons were the ultimate holons, then they had one level, a nucleus had two levels, an atom had three levels, and so forth. But that doesn't change the *relative* placement of the holons themselves and thus doesn't change in the least the meaning of higher and lower.

Also notice that we can count as one level any truly emergent quality, which means we can divide and subdivide a series in any number of ways. To give a crude analogy, take a three-story house with a stairway consisting of twenty steps between each floor. We usually say the house has three main levels (just as we refer to the three general realms of evolution), but we could also use the stairs as the scale and say the house has sixty levels or steps. Obviously this is somewhat arbitrary.

The level claim is simply that (1) the existence of floors and steps is not itself purely arbitrary—there are quanta in the universe with or without the presence of humans—and (2) in making any comparison, use the same ruler, thus factoring out that particular arbitrariness (just as we can measure the temperature of water with any number of arbitrary scales, Celsius or Fahrenheit or whatever, as long as we agree which to use). Whenever we refer to the "number of levels" in a holon, then, we are using a relative scale consistently applied within the particular comparison.

With that in mind, we can introduce two very important definitions, first clearly suggested by Arthur Koestler. In his words:

7. *"The number of levels which a hierarchy comprises determines whether it is 'shallow' or 'deep'; and the number of holons on any given level we shall call its 'span.'"*[43]

For this example, let us arbitrarily assign atoms a depth of three (they contain as components at least two other levels). We can imagine a time, early in the universe, when there were only atoms and not yet molecules. Atoms had a small depth (3) but an enormous span, stretching, we presume, throughout the existent universe and numbering in the megazillions (thus, depth = 3, span = zillions). When molecules first emerged, they had a greater depth, a depth of four, but initially a very small span (presumably in the hundreds, or whatever, and then growing rapidly).

This again recognizes a crucial distinction found in any holarchy—the vertical dimension and the horizontal dimension. The greater the vertical dimension of a holon (the more levels it contains), then the *greater* the *depth* of that holon; and the more holons on that level, the *wider* its *span*.

This is important, because it points out that it is not merely population *size* that establishes order of richness (or order of qualitative emergence), but rather *depth*. And we will see that one of the greatest confusions in general ecological or new-paradigm theories (whether "pop" or "serious") is that they often mistake great span for great depth. For what we will find is that:

8. *Each successive level of evolution produces GREATER depth and LESS span.* The greater the depth of a holon, the more precarious is its existence, since its existence depends also on the existence of a whole series of other holons internal to it. And since the lower holons are *components of* the higher, there physically cannot be more numbers of the higher than there are numbers of components.

Thus, for example, the number of molecules in the universe will always be less than the number of atoms in the universe. The number of cells in the universe will always be less than the number of molecules in the universe, and so on. It simply means the number of wholes will always be less than the number of parts, indefinitely.

Thus, greater depth always means less span, in relation to a holon's predecessor(s). Of course, what we might call "total span" increases—if there is more of anything, the sum total of everything increases—but simple span, the span of any given type of holon, becomes less and less vis-à-vis its predecessor(s). (This is why the span of mental holons is much less than the span of living holons, which is much less than the span of material holons—the so-called pyramid of development.) We will return to this momentarily.

Throughout this book I will be adding what might be called "fundamental additions" to the twenty tenets. These additions are tenets that seem crucial to an understanding of evolution and the Kosmos, but tenets that, for reasons we will be investigating in detail, simply cannot be established in the it-language of instrumental and objectifying naturalism (i.e., systems theory). *Why* they cannot be established in systems theory will form a large part of the discussion about the "wrong half" of modern systems sciences.

These additions do not violate systems theory; they simply can find no place whatsoever in its tenets (the subjective world of "I," for example, cannot be captured without remainder in the objectivistic language of "its" and "objects" and "processes"; nor can the collective worlds of "we" be captured in systems theory without a reduction to totalitarian principles). As I said, we will later return to all of this in much detail, and so for the moment we need only note our first addition:

Addition 1: The greater the depth of a holon, the greater its degree of consciousness. The spectrum of evolution is a spectrum of consciousness. And one can perhaps begin to see that a spiritual dimension is built into the very fabric, the very *depth*, of the Kosmos.

But this addition takes us a bit ahead of our story, and so let me return to the more orthodox account and pick up the story at tenet 7, the distinction between depth and span, or the distinction between vertical richness and horizontal reach. We have two different scales here: a vertical scale of deep versus shallow, and a horizontal scale of wide versus narrow.

One of the difficulties in keeping these distinctions in mind is that, once we disavow reductionism, once we disavow the flatland approach to the Kosmos, then theoretically we are no longer just playing chess, which is hard enough; we are now playing three-dimensional chess, which is extremely complicated! And I won't even mention *n*-dimensional chess! For wherever development proceeds, we must, at the very least, distinguish between depth and span, not only because there are objective (real) correlates to the distinction, but because to fail to do so, even under the guise of being "holistic," simply reintroduces a subtle form of reductionism, for everything is then compared merely in terms of population *span* (bigger or smaller, wider or narrower), and thus all *depth* is erased from existence.

As we will see in the next chapter, many such theorists then attempt to construct holistic sequences based only on *size*. They completely mistake great span for great depth, because now the only *qualitative* distinction they possess is actually *quantitative*: they only have one way to go— namely, that "bigger is better." And thus their "holistic sequences," be- cause they are based on *span* alone, are actually in themselves *regressive*, because evolution is *not* bigger and better, but smaller and better (greater depth, less span). These theorists, as we will see in detail, end up unknow- ingly recommending pure regression as our salvation.[44]

One last point in this regard: other critics accept the term "depth," but strenuously object to the word "height," again feeling, apparently, that "height" is the height of human arrogance. And yet they accept "depth," this being, I suppose, the depth of human daffiness. Any way we slice it, depth without height is just another name for shallow.

With reference to the vertical scale (which just as well could be any direction we agree upon; I will use vertical), we can speak of height and depth synonymously. The height of a holon and the depth of a holon refer to the same thing, the (relative) number of levels of other holons internal to it. Huston Smith, however, makes the interesting observation that tradi- tionally *height* was used when referring to the Kosmos (as in "high Heaven") and *depth* in regard to individuals (as in "the depth of her soul").

But I will use the two words interchangeably, so that the vertical dimen- sion is a scale of depth/height versus shallow/flat, and the horizontal scale is one of width versus narrowness.

We have, of course, on several occasions already met these two dimen- sions. Agency and communion (or self-preservation and self-accommoda- tion) refer to changes in the horizontal dimension; self-transcendence and self-dissolution refer to changes in the vertical dimension. So we can in- troduce a few more simple definitions:

Changes in the horizontal dimension I will call *translation*, and changes in the vertical dimension I will call *transformation*.

The agency (or regime or code) of any given holon *translates* the world according to the terms of its code or regime—it will recognize, or register, or respond to, only those items that fit its code. An electron, for example, will register numerous other physical forces, but will not register or respond to the meaning of literature, nor will it respond to the sexual advances of a rabbit.

Holons, in other words, do not simply reflect a pregiven world. Rather, according to their capacity, they select, organize, give form to, the multitude of stimuli cascading around them. Their responses never simply "correspond" to something "out there"; they *register* (and thus *respond to*) only that which fits the *coherency* of their *regime* (or code or agency or deep structure). This is what Varela means, for example, when he says holons are not "heteronomous units operating by a logic of correspondence," but rather relatively "autonomous units operating by a logic of coherence." Holons *translate* their reality according to the patterns of their agency, their relatively autonomous and coherent deep structures; and stimuli that don't fit the deep structure or regime are simply not registered and might as well not exist (in fact, do not exist, do not disclose themselves, for that holon).

In *transformation*, however, new forms of agency emerge, and this means a *whole new world* of available stimuli becomes accessible to the new and emergent holon. The new holon can respond to deeper or higher worlds, because its translation processes transcend and include those of its subholons. A deer, for example, will register and respond not only to physical forces that continue to impact it, but also to a whole range of biological forces, from hunger to pain to sexual drive, that simply make no impression on its constituent atoms. To the individual atom, these new forces are all "otherworldly"—literally out of this world, or not in *its* world.

Thus, in *transformation* (or self-transcendence), whole new worlds of translation disclose themselves. These "new worlds" are not physically located someplace else; they exist simply as a *deeper perception* (or deeper registration) of the available stimuli in *this world*. They appear to be—and might as well be—"other worlds" to the junior holons, but these "other worlds" disclose themselves—they become *this worldly*—via transformation and self-transcendence. We will see, in later chapters, that development is a constant conversion of "otherworldly" into "this worldly" via a deepening of perception brought about by emergent evolution and transformation. Greater depth brings other worlds into this world, constantly. . . .

A holon's regime or deep structure, then, governs the range of its possible translations, governs the *types* of worlds that it *can* respond to.

This is what actually allows a holon's stability, its relative autonomy or coherence. We noted that any stable holon *is* stable by virtue of its identifying pattern—its relatively autonomous code, canon, regime, entelechy, morphic unit—what I have been calling its *deep structure*. The human body, for example, has 208 bones, 1 heart, 2 lungs, and so forth, wherever it appears. But what we do with the human body—its various modes of play, work, sex, and so on—varies from culture to culture. These variations within the deep structure I call *surface structures*. Koestler refers to this difference as "fixed codes and flexible strategies"—that is, relatively stable deep structures and changing surface structures within the basic guidelines (or basic limiting principles) of the deep structure.

And so we can say: *translation* is a change in *surface structures* ("horizontal"), whereas *transformation* is a change in *deep structures* ("vertical").

(The relation between deep structures and surface structures I call "transcription," which we will discuss later, and for now, in a note.)[45]

This is yet another reason why evolution cannot be explained merely in terms of horizontal additions—that is, cannot be explained merely in terms of translation, of horizontal changes in agency and communion. When the regimes or deep structures of hydrogen and oxygen are brought together in a certain fashion, a water molecule emerges with a new regime or deep structure, a regime that incorporates the separate regimes of the atoms into a single "morphic unit," a holon with a new deep structure that subsumes its predecessors. This is a vertical *transformation*, or *change in deep structure*; it is *not* the mere addition of more and more atoms, which would simply produce a mess, not a molecule; a heap, not a whole. Translation shuffles parts; transformation produces wholes.[46]

We can use our simple analogy of the three-story building to summarize all of these definitions. Each of the three main floors is a deep structure; the furniture, chairs, and tables on each floor are the surface structures. Rearranging the furniture on any given floor is translation; changing floors is transformation. (And the relation of the furniture to each floor is transcription.)

The point, then, is that evolution is first and foremost a series of transformations ("self-realization through self-transcendence"). Transformation is how you get *levels* in the first place. And each major transformation produces greater depth, and less span, in relation to its previous level(s), in relation to its predecessors.

9. *Destroy any type of holon, and you will destroy all of the holons above it and none of the holons below it.* There is an enormous amount of confusion in the literature as to how to determine whether a given holon is "higher" or "lower" in the developmental sequence—not to mention the large numbers of critics who deny higher and lower altogether.

But we can actually locate the level of a holon in any evolutionary or holistic sequence very simply, by asking ourselves, as a type of thought experiment, "What other types of holons would be destroyed if we destroyed this type of holon?"

For example, take the holistic sequence: subatomic particles, atoms, molecules, cells. Everybody agrees that that is indeed a holistic sequence—each member includes its predecessor(s) but not vice versa, and thus each successive member is indeed more encompassing (or more holistic).

But notice what that also unavoidably means: If we destroyed, for example, all of the molecules in the universe, we would also destroy all of the cells in the universe (all of the holons *above* molecules in the sequence), but atoms and subatomic particles would or could still exist (none of the *lower* holons would have to cease existence).

In other words, destroy any type of holon, and that also destroys all the *higher* holons in the sequence, because those higher wholes depend upon the lower as constituent parts.

Molecules cannot exist without atoms; atoms can exist without molecules. There is nothing arbitrary in this arrangement; it is not the product of a particular "human value judgment." It is true physically, and logically, and chronologically as well. (Atoms and molecules existed perfectly well on their own before the emergence of any cells. Destroy every cell in the universe, and everything that came after it—plants, animals, societies—will likewise be destroyed, but nothing that came before it— subatomic particles, atoms, molecules, polymers.)

This allows us to easily determine what is lower, and what is higher, in any holistic sequence: destroy the particular holon, and everything else that is also destroyed is higher; those holons not destroyed are lower.[47]

This is true for any developmental sequence, and is partially responsible for the "retreat" we see when holons dissolve: they regress to the *next* lower level, whereas if any lower level is destroyed, all the higher levels automatically go with it. This also points up very clearly why there is a vertical and not just a horizontal dimension to evolution, and shows why "higher" and "lower" are *not* arbitrary.[48]

In the next chapter, we will apply this principle to the work of several theorists who are attempting to construct "holistic sequences." These theorists, in order to help men and women more accurately position themselves in the larger universe, are attempting to find a "more holistic" reality in which humans can situate their communions—an altogether commendable task. But because so many of these theorists confuse great span with great depth, they end up creating "holistic sequences" that are in fact regressive, that slide down the pyramid of greater depth / less span and land us in the shallower end of the Kosmos.

As we will see, the careful application of this thought experiment allows us to determine if their sequences are really holistic (or merely regressive), and allows us to offer an alternative "big picture" in which men and women can situate themselves in a truly holistic fashion.

In the meantime, this thought experiment allows us to introduce another important definition, the difference between *fundamental* and *significant*. For what we will find is that the more fundamental a holon is, the less significant it is, and vice versa. That is:

The *less depth* a holon has, the *more fundamental* it is to the Kosmos, because it is a *component* of so many other holons. Atoms, for example, are very fundamental, relatively speaking, because molecules and cells and organisms all depend upon them. The more fundamental a holon is, the more of the universe *contains* that holon as a *necessary* part or constituent, without which these other holons could not function (or even exist). Less depth means more fundamental, means the particular holon is a "building block" of so many other holons.

At the same time, the *less depth* a holon has, the *less significant* it is to the Kosmos, because it embraces (as its own components) so little of the Kosmos. There is, relatively speaking, less of the Kosmos that is actually internal to this holon, that is actually embraced within the being of the holon itself. (Put differently, it is less significant because more of the Kosmos is external to it.)

On the other hand, the *greater the depth*, or the greater the particular wholeness of a holon, then the *less fundamental* it is, because fewer other holons depend on it for their own existence. Primates, for example, are not very fundamental holons, because neither atoms nor molecules nor cells depend upon them. But by the same token, the less fundamental, the more significant: the *more significant* that holon is for the universe, because more of the universe is reflected or embraced in that particular wholeness (more of the Kosmos is internal to it, as part of its own being). Primates are very significant, relatively speaking, because they represent and contain atoms and molecules and cells: they *signify* more of the Kosmos.

Thus an atom, for example, is more fundamental, and less significant, than a cell. *More fundamental*, because everything above it (including the cell) depends upon it for its existence. *Less significant*, because less of the Kosmos is embraced in it (or is actually internal to it). A cell, on the other hand, is more significant because it *embraces* atoms and therefore reflects or *signifies* more of the Kosmos in its own being. It is more significant, but less fundamental, than an atom. And so on.

(We will return throughout the book to this topic, and see that it is related to the issue of intrinsic and extrinsic values; and see also that theorists who mistake great span for great depth are always confusing more fundamental with more significant, and thus, once again, end up recommending regression as a direction for further growth; with a flat-land ontology, the crucial depth dimension is missing.)

10. *Holarchies coevolve.* Holons do not evolve alone, because there are no alone holons (there are only fields within fields within fields). This principle is often referred to as *coevolution*, which simply means that the "unit" of evolution is not an isolated holon (individual molecule or plant or animal) but a holon plus its inseparable environment. Evolution, that is, is ecological in the broadest sense.

Jantsch refers to this as the interdependence of microevolution and macroevolution, by which he simply means the coevolution of the individual (micro) and its larger environment (macro). This is just another way of saying that *all* agency is *always* agency-in-communion.[49] Jantsch:

> The immediate consequence is . . . the simultaneity of macro- and microevolution in the universe. Macroscopic structures become the environment for microscopic structures and influence their evolution in decisive ways, or make it possible at all. Vice versa, the evolution of microscopic structures becomes a decisive factor in the formation and evolution of macroscopic structures. This interdependence consti-tutes nothing but an aspect of *co-evolution* [operative, he says, "in all three realms"]. This principle implies that every system is linked with

its environment by circular processes which establish a feedback link between the evolution of both sides. This holds not only for systems at the same hierarchical level; the entire complex system plus environment evolves as a whole.[50]

In other words, according to Jantsch (and others), micro and macro—individual and social/environmental—evolve heterarchically to new holarchical levels of each.

This distinction between an individual holon and its social holon (environment in the broadest sense) is not as easy to draw as it may first appear, however, because it's almost impossible to define what we mean by an individual in the first place. The word itself, from the Latin *individualis*, means not divisible or not separable; by that definition, there are no individuals anywhere in the Kosmos. There are only holons, or dividuals.[51]

On the other hand, we do recognize that enduring holons possess a specific form or pattern, and this pattern is to some degree autonomous, or resistant to environmental obliteration. And this is usually what we mean by calling a holon an "individual"—we mean an *enduring compound individual*, compounded of its junior holons and adding its own defining form or wholeness or canon or deep structure (which is the novel holon in its own compound individuality). And we further mean, usually (although we might not use these terms), that the overall wholeness or morphic field of the individual holon organizes the indeterminateness of its junior partners or subholons.

In other words, even though an "individual holon" exists inseparably from its social environment, its defining factor is its own particular form or pattern. To the degree that we can reasonably recognize that pattern, we will refer to an *individual holon*.

This is still somewhat arbitrary, of course, because there are some social holons that seem to act as individual holons or "superorganisms"—an ant colony, for example. But in human affairs, to give a counterexample, most of us resist the temptation to describe a social holon, such as the State, as being literally a superorganism, because all organisms have priority over all of their components, and yet with the rise of democratic structures, we like to think that the State is subservient to the people, and to the degree that that is true, then the social system is not a true organism (it is a social or environmental holon, not an individual holon).

Further, the State, unlike a concrete individual, does not have a locus of self-prehension, a unitary feeling *as* a oneness. In more general terms, it lacks a locus of individual self-being (one of Whitehead's main conclusions; and Habermas: the State is not a macrosubject). And finally, the parts in this social system are conscious, but the "whole" is not.

We will devote much of the next chapter to this topic. For the time being, let us simply recognize that there are important distinctions between micro- and macroevolution, between an individual holon and a social or environmental holon (even though they are inseparably interactive, which is the meaning of "coevolution"). A social holon is still a holon—and not a mere heap or aggregate—because it displays a whole/part pattern, it is rule-bound, it in a sense develops (we speak coherently of stellar evolution, ecosystem evolution, social evolution, etc.), and it can function with various degrees of upward and downward causation (depending on its depth). But it is not a true individual holon, as Whitehead and Habermas and others have noted. Jantsch refers to this as the difference between "vertical organismic and horizontal ecosystemic (symbiotic) organization"—but the point, again, is that they coevolve.

11. *The micro is in relational exchange with the macro at all levels of its depth.* This tenet is extremely important, particularly when it comes to holons of greater depth and the types of ecosystems (in the broad sense) that they must co-create and upon which their existence depends.

Take, for example, a human being, using just the three levels of matter, life, and mind: all of these levels maintain their own existence through an incredibly rich network of relational exchange *with holons of the same depth* in the environment. The *physical* body exists in a system of relational exchange with other *physical* bodies—in terms of gravitation, material forces and energies, light, water, environmental weather, and so on—and the physical body itself depends for its existence on these physical relationships. Further, the human race *reproduces itself physically* through food production and food consumption, through social labor organized in an economy for basic material exchanges in the physiosphere.

Likewise, humanity *reproduces itself biologically* through emotional-sexual relations organized in a family and an appropriate social environment, and depends for its biological existence on a whole network of other biological systems (and ecosystems)—it depends upon harmonious relational exchanges with the biosphere.

Finally, human beings *reproduce themselves mentally* through exchanges with cultural and symbolic environments, the very essence of which is the relational exchange of symbols with other symbol exchangers. These relational exchanges are embedded in the traditions and institutions of a particular society in such a way that that society can reproduce itself on a cultural level, can reproduce itself in the noosphere.

In short, as holons evolve, each layer of depth continues to exist in (and depend upon) a network of relationships with other holons at the *same level of structural organization.* I usually refer to this, for short, as "same-level relational exchange."[52] The point is that all holons are compound individuals, compounded of their previous holons and adding their own distinctively emergent pattern; and each level of these holons (i.e., *every* holon) maintains its existence through relational exchanges with same-depth holons in the social (or macro-) environment.

12. *Evolution has directionality.* This is the famous arrow of evolutionary time first recognized in the biosphere, but now understood, in the sciences of complexity, to be present in all three of the great domains of evolution. This directionality is usually stated as being one of increasing differentiation, variety, complexity, and organization. But I think it would be helpful to gather together, from different sources, the various indicators of this evolutionary directionality and briefly examine them one at a time.

We have already seen that evolution is marked by creative emergence (novelty), symmetry breaks, self-transcendence, increasing depth (and greater consciousness, which we will discuss later as an addition). Those are already indicators of evolution's directionality. Here are some others. Aside from regressions, dissolutions, arrest, etc., evolution tends in the direction of:

a. *Increasing complexity.* The German biologist Woltereck coined the term *anamorphosis*—literally, "not being formless"—to describe what he saw as the central and universal feature of nature: the emergence of ever-increasing complexity. As Jantsch puts it,

> The evolution of the universe is the history of an unfolding of differentiated order or complexity. Unfolding is not the same as building up. The latter emphasizes structure and describes the emergence of hierarchical levels by the joining of systems "from the bottom up." Unfolding, in contrast, implies the interweaving of processes which lead si-

multaneously to phenomena of structuration at different hierarchical levels. Evolution acts in the sense of simultaneous and interdependent structuration of the macro- and the micro-world. Complexity thus emerges from the interpenetration of processes of differentiation and integration. . . .[53]

Ballmer and von Weizsacker, in fact, refer to this maximization of complexity as the "general statement of evolution," and L. L. Whyte called it "the fundamental principle of the development of pattern."

We need one qualification here. As Laszlo points out, the emergence of a new level of complexity also brings with it, from a slightly different angle, a new *simplicity*, precisely because the new whole, as a single whole, is simpler than its many parts. Thus, says Laszlo, "the emergence of a higher-level system is a . . . *simplification* of system function." He continues:

> However, once a new hierarchical level has emerged, systems on the new level tend to become progressively more complex. For example, on the atomic level of organization, hydrogen, the first element to be synthesized in the processes of cosmic evolution, is structurally simpler than the subsequently synthesized, heavier elements. On a higher level of organization, a molecule of water is simpler than a protein molecule; on a still higher organizational level, a unicellular organism is less complex than a multicellular one. . . . Thus, while a new level of organization means a simplification of system function, and of the corresponding system structure, it also means the initiation of a process of progressive structural and functional complexification.[54]

b. *Increasing differentiation/integration.* This principle was given its first modern statement by Herbert Spencer (in *First Principles*, 1862): evolution is "a change from an indefinite, incoherent homogeneity, to a definite, coherent heterogeneity, through continuous differentiations and integrations" (this definition of the term *evolution* allowed biologists to begin using it instead of Darwin's phrase "descent with modification"). And we already quoted Jantsch: "Complexity thus emerges from the interpenetration of processes of differentiation and integration. . . ."

Differentiation produces partness, or a new "manyness"; integration produces wholeness, or a new "oneness." And since holons are whole/

parts, they are formed by the joint action of differentiation and integration.

The differentiating processes are obviously necessary for the undeniable novelty and diversity created by evolution, but integration is just as crucial, converting manyness into oneness (the regime or canon or pattern of a holon *is* its integrative coherence). Thus Whitehead's view that "the ultimate character pervading the universe is a drive toward the endless production of new syntheses [integrations]."[55] We saw that Whitehead called this drive "creativity"; he says it is "the eternal activity," "the underlying energy of realization"—"nothing escapes it."[56]

Thus Whitehead's all-important dictum: "The many [differentiation] become one [integration] and are increased by one [the new holon]."

These two processes are very obvious in the physiosphere (atoms integrating differentiated particles, molecules integrating differentiated atoms, etc.) and in the biosphere (e.g., the progressive differentiation of the zygote and the progressive integration of the resultant parts into tissues, organ systems, organism), but they are also rampant in the noosphere. Even psychoanalysis is on board. Gertrude Blanck and Rubin Blanck, for example, pioneers in psychoanalytic developmental psychology, have persuasively argued that the aggressive drive is the *drive to differentiation*, and Eros is the *drive to integration*, and disruption of either one results in serious pathology (we will return to this in chapter 9).

And for all the commotion surrounding Derrida's notion of *différance* (to differ and defer)—some critics have used the notion to deconstruct anything they didn't happen to like—Derrida himself defines the notion very simply as "the process of differentiation."[57] Before communication can begin, entities have to be differentiated in the first place, and in one sense these entities do not exist prior to the differentiating process per se. Différance is thus part of that "eternal activity" of creativity, a dynamic force of bringing into being—"it has impulsive force, the force of articulation or differentiation."[58] In other words, as one of his interpreters points out, "Derrida sees the dynamic difference that characterizes reality as also composing the nature of language itself. This enables language, through its inherent process of difference, to function as a means of realization . . . language participates in the reality it manifests . . . the dynamic becoming of reality itself."[59] But differentiating also demands integrations and syntheses—as Derrida puts it, "The play of differences supposes, in effect, syntheses and referrals. . . ."[60]

This play of differentiating and integrating forces, or *différance* and synthesis and referral, is behind Derrida's notorious critique of "presence." As Derrida words it:

> The play of differences involves syntheses and referrals which prevent there from being at any moment or in any way a simple element which is *present* in and of itself and refers only to itself. Whether in written or spoken discourse, no element can function as a sign without relating to another element which itself is not simply present. This linkage means that each "element" is constituted with reference to the trace in it of the other elements of the system. Nothing, in either the elements or the system, is anywhere ever simply present or absent.[61]

In other words, there is nothing that isn't a holon, a context within a context forever. You cannot point to any thing, to any holon, and say it's just *that* and nothing else, because every holon is simultaneously a superholon and a subholon: it is composed of holons and composes others (across both space and time); nothing is ever simply present.[62]

As for the conjoint operation of differentiation and integration in the noosphere, Habermas notes that "the different lifeworlds that collide with one another do not stand *next to each other* without any mutual understanding. As totalities [holons], they follow the pull of their claims to universality and work out their differences until their horizons of understanding 'fuse' [integrate] with one another."[63] Across the board we see this dual operation of extension and condensation, he says: "The increasing reflexivity of culture, the generalization of values and norms, and the heightened individuation of socialized subjects, the enhancement of critical consciousness, autonomous will formation, and individuation, takes place under conditions of an ever more extensive and ever more finely woven net of linguistically generated intersubjectivity." And his point is that all of this "means differentiation and condensation [integration] at once—a thickening of the floating web of intersubjective threads that simultaneously holds together the ever more sharply differentiated components of culture, society, and person."[64]

That phrase is significant: it is not just an *extension* of the web, but a *thickening*: not just span, but also depth.

Finally, to touch bases with Foucault, we might note that his "*archaeological holism* asserts that the whole determines what can count even as a possible element. The whole verbal content is more fundamental and thus is more than the sum of its parts. Indeed, there are no parts except within the field which identifies and individuates them"—that is, differentiates and integrates them.[65] And let us immediately note that Foucault eventually abandoned archaeology as an exclusive methodology precisely because archaeology was itself merely a *part* of the larger holon of social practices: not wholes, but whole/parts.

In sum, evolution requires both differentiation and integration operating together—"the many become one and are increased by one"—and, indeed, these two normally (in healthy holarchies) occur conjointly, which is why I usually write it as "differentiation/integration." These appear as opposites (or utterly opposed tendencies) only in a flatland ontology, where more of the one means less of the other. But in the multidimensional Kosmos, more of one means more of the other. They join hands endlessly to produce new whole/parts or many/ones or holons. The dialectics of depth . . .

c. *Increasing organization/structuration.* "The further evolution of the suprasystem leads to the progressive complexification of its defining system level—and ultimately to the creation of hypercycles that shift it to the next organizational level. Thus evolution moves from the simpler to the more complex type of system, and from the lower to the higher level of organization."[66] This is, for example, behind the standard distinction, in evolutionary biology, between clades and grades: "A group of species with a recent common ancestor forms a clade; a group with the *same level of structural organization* forms a grade."[67] *Grade*, of course, is another term for depth.

d. *Increasing relative autonomy.* This is a much misunderstood concept. It simply refers to a holon's capacity for self-preservation in the midst of environmental fluctuations (*relative autonomy* is another term for agency). And according to the sciences of complexity, the greater the depth of a holon, the greater its relative autonomy. This does not mean greater permanence or greater concrete stubbornness. Worms are less durable than rocks. Relative autonomy simply refers to a certain flexibility in the face of changing environmental conditions. A fox can maintain its internal temperature relatively independently of changing weather, whereas a rock's temperature fluctuates immediately with every passing circumstance.

By the time we reach the noosphere, in humans, relative autonomy is of such a high degree that it can produce not just *differentiation* from the environment, which is necessary, but *dissociation* from the environment, which is disastrous—an expression of *pathological agency* that, among many other things, lands it squarely in ecological hell (we will, of course, be returning to this topic).

The reason that autonomy is always *relative* is that there are no wholes, only whole/parts. As *a whole*, a holon possesses a degree of autonomy, expressed in its identifying and enduring pattern, its self-preservation. It possesses coherence and identity across space and time (or else it simply ceases to exist), which is why autonomy is virtually synonymous with agency, regime, code, canon, deep structure. As *a part*, however, every holon—the autonomy of every holon—is subjected to larger forces and systems of which it is merely a component. This doesn't change its fundamental pattern or identifying regime, but it does subject it to a host of limiting circumstances and conditions that can alter its expression, and it often moves the source of the initiation of action to other regimes (e.g., if my country declares war, I am included, like it or not).

In other words, as we have seen time and again, all agency is agency-in-communion. (This is nicely captured in Varela's notion of "structural coupling": the agency of a biological system is relatively autonomous, but the form of the autonomy evolved in structural coupling with its environs; i.e., its present agency is the result of evolutionary communions.)

Thus, autonomy, like all aspects of a holon, is *sliding*: a holon is relatively autonomous vis-à-vis its juniors and relatively subservient vis-à-vis its seniors.[68]

Much of the fun, if one can call it that, of postmodern poststructuralist games consists in the upsetting of *established autonomies* by pointing to *larger contexts* which actually "determine" the supposed "autonomy" of the isolated unit, whereupon the isolated unit itself is promptly declared "dead" (death of the writer, death of the subject, death of the patriarchy, death of the mythic god, death of the ego, death of rationality, death of logocentrism, etc.), with the "autonomy" or systemic structure of the larger context being in turn merely a *part* of. . . . The game simply continues until the critic gets tired (or settles on an ideology), because there is nothing in reality that can stop the sliding game any sooner, for contexts are boundless.

But the "decentering" of previously "autonomous" units is indeed part of the important truths of postmodern critiques, and we will return to them throughout this volume. Thus, to give a few examples now: the autonomous ego of the Enlightenment is not *that* autonomous because it is actually set in the context of its own organic drives (the psychoanalytic critique of the Enlightenment), and these previously unconscious drives must be integrated for true autonomy to emerge. But even the entire integrated and autonomous person of psychoanalysis is not *really* autonomous, because that individual is actually set in contexts of *linguistic structures* that autonomously determine meaning without the individual even knowing about it (the critique launched by structuralism, archaeology). But linguistic structures aren't really *that* autonomous, because they exist only in the context of pre-articulate worldviews that use language without language ever registering that fact (the critique by Heidegger, Gebser). But further, worldviews themselves are merely a small component of massive networks and contexts of social practices (in various ways, Marx, Habermas, the later Foucault). And further yet, theorists from Kierkegaard to Schelling to Hegel would insist that those social practices only exist in, and because of, the larger context of Spirit.

In every one of those cases, the theorist (Freud, Marx, Heidegger, Foucault, Schelling, etc.) tells us something important about the *meaning* of our existence by situating our existence in a larger *context*—since meaning and context are in important ways synonymous, as we earlier noted. And each successive theorist gives a deeper or bigger or wider meaning to existence by finding *previously hidden contexts* that suddenly shift the autonomy out from under our feet and point to larger communions in which we live and breathe and have our being.

And in a sense, each of them is quite right: the ego does exist in the context of the total organism and its drives, which does exist in the context of its linguistically disclosed world, which does exist in terms of overall networks of social practices, which themselves subsist in Spirit. That is the very nature of *holon*, contexts within contexts within contexts. And each time we spot one of these larger (deeper) contexts, we find a new *meaning* conferred on a given holon, because, as we earlier noted, the larger context confers a meaning on its holons that the holons themselves, alone and isolated, do not and cannot possess.[69]

Likewise, each discovery of a new and deeper context and meaning is a discovery of a new *therapia*, a new therapy, namely: we must *shift our perspectives*, *deepen our perception*, often against a great deal of *resistance*, to embrace the deeper and wider context. The self is situated in contexts within contexts within contexts, and each shift in context is an often painful process of growth, of death to a shallow context and rebirth to a deeper one.

But for just that reason, each time we identify a deeper context, our relative autonomy actually increases, because in identifying with a deeper perception, we have found a wider freedom.

We will be discussing all these notions as we proceed. The simpler point of this section is just that, even though all autonomy is relative, nonetheless *relative* autonomy *increases* with evolution (as we said, a fox is more autonomous than a rock). And the reason that relative autonomy *increases* with emergent evolution is that more *external* forces *impinging* on the autonomy of a holon become *internal* forces *cooperating* with it (due to supersession, or transcendence and inclusion). It identifies with deeper contexts and thus finds wider freedom. We will see numerous examples of this as we proceed.

e. *Increasing telos.* The regime, canon, code, or deep structure of a holon acts as a magnet, an attractor, a miniature omega point, for the *actualization* of that holon in space and time. That is, the end point of the system tends to "pull" the holon's actualization (or development) in that direction, whether the system is physical, biological, or mental.

Again, this was an item that was largely ignored as long as science was intent on studying a bunch of rocks in motion, but even in the physiosphere the "entelechy" (regime, canon, deep structure, morphic field) of a holon governs the final form of its actualization, whether from an electron cloud or a chaotic attractor of complex systems. "In the geometric orientation that dominates in contemporary dynamic systems theory, the principal features of dynamic systems are the attractors: they characterize the long-run behavior of the systems. Dynamic systems evolve from a given initial state along a unique trajectory of states in accordance with the laws of evolution [some version of the twenty tenets]. This leads eventually to some recognizable pattern where the trajectory remains trapped. The pattern defines the *attractors* of the system. . . ."

> If the series of system states come to a rest, their evolution is governed by a static attractor [studied, for example, by René Thom's topological "catastrophe theory"]. If the states consist of a repeated cycle of states with a definite periodicity, the system is under the sway of a periodic attractor. And if the trajectory of system states neither comes

to rest nor exhibits periodicity but is highly erratic, it is under the influence of a so-called chaotic attractor.[70]

"In recent years chaotic behavior has been discovered in a wide variety of natural systems, and their mathematical modeling has made rapid progress. An entire discipline has sprung up within dynamic systems theory devoted to the study of the properties of chaotic attractors and of the systems governed by them; it became popularly known as chaos theory. Despite its name, the theory seeks to eliminate rather than discover or create chaos—it studies processes that appear chaotic on the surface but on detailed analysis prove to manifest subtle strands of order. Chaotic attractors are complex and subtly ordered structures that constrain the behavior of seemingly random and unpredictable systems."[71]

These attractors, in other words, are examples of the regimes or organizing forces of social holons and their inherent teleological pull to pattern (without which a holon would simply not exist). Of particular interest are "bifurcations," which consist of the shift from one type of attractor to another. "Models with catastrophic bifurcations (conducing from turbulent to newly ordered states through the reconfiguration of the attractors) simulate rapid *evolutionary leaps* with the greatest fidelity. The significant simulations occur when dynamic systems are destabilized and pass through a chaotic phase on the way toward essentially new—and in practice unpredictable—steady states."[72]

The conclusion for chaos theory is that these are "the kind of transformations that underlie the evolution of all third-state [far from equilibrium] systems in the real world, from atoms of the elements to societies of human beings."[73] These types of *transformations* (not merely translations) generate "a statistically significant tendency toward greater complexity and a higher level of organization. The system leaps to a new plateau and thus becomes more dynamic and more autonomous in its milieu."[74] So that, finally, these factors converge to "push bifurcating systems up the ladder of the evolutionary hierarchy"—that is, holarchy.[75]

> When growing fluctuations upset the dynamic stability of a system, its stable point of periodic attractors can no longer maintain it in its established state; chaotic attractors appear and with them an interval of transition hallmarked by transitory chaos. When the system achieves a new state of dynamic stability, the chaotic attractors of the

bifurcation epoch give way to a new set of point or periodic attractors. These attractors maintain the system in a condition far from thermodynamic equilibrium, with more effective use of information, greater efficiency in the use of free energies, greater flexibility [relative autonomy], as well as greater structural complexity on a higher level of organization.[76]

Telos—the miniature omega-point pull of the end state of a holon's regime—is, of course, rampant not only in physical systems but in the biosphere and noosphere as well. An acorn's code (its DNA) has oak written all over it. Through processes of *translation, transcription,* and *transformation,* the seed unfolds into a tree, holarchically. These biological processes have been studied so much, and are so familiar to most readers, that I will not dwell on them (we will see many examples later); we need only note that biologists "recognize the existence of direction toward future functions," and that "the purposeful aspect of organisms is incontrovertible."[77]

In the noosphere, once the mental and linguistic sciences stopped trying to ape the study of moving rocks and concentrated on mental processes, the mind's telos inexorably came to the fore. Freud's psychoanalysis, for example, was profoundly *developmental* in nature, and the whole point of development is that it is not a random hopping about but is indeed going somewhere, and therefore it can be derailed and sabotaged, with resultant pathology.

An old joke from Vermont: A city fellow, driving through the Vermont countryside, sees a man in a truck on the side of the road. The truck is axle-deep in the mud, and the wheels are spinning. "Are you stuck?" asks the city fellow. "I would be, if'n I was goin' somewhere."

Well, the psyche, for better or worse, *is* going somewhere, and that is why the process can get stuck, why it is fraught with frustration, arrest, fixation, stick points, logjams. If the mind weren't going somewhere, it could never get stuck, never get "sick." And these "sick points," these "stick points," can only be understood in terms of the mind's omega points, of where it wants to go. And not just Freud. "In an oft-cited formula, Piaget declares that there is no structure which lacks a development, and that the process of development can only be understood in view of the structure which exists at the beginning and the structures into which it will evolve."[78] The basin and their attractors, their miniature omega points.

Or America's undeniable philosophical genius, Charles Peirce: "The being governed by a purpose or other final cause is the very essence of the psychical phenomenon." He adds that "to say that the future does not influence the present is untenable doctrine."[79] Or Roman Jakobson, who was one of the first to introduce, as he put it, "a consistent application of a means-ends [telos] model to language design, to its self-regulating maintenance of integrity and dynamic equilibrium, as well as to its mutations [transformations]," so that the "adoption of teleology was a central concern."[80] All the way to Habermas: "There is a *relentless pressure* built into the use of language oriented toward mutual understanding."[81] And even Derrida: "This arche-trace contains within itself all the possibilities of manifestation as the primordial 'difference.' This 'difference' is the inherent *teleological force* within us that leads to self-manifestation."[82]

Now, I haven't actually mentioned the nature or the actual content of the mind's omega point(s), its basic attractor, its end state toward which earlier stages are struggling to reach, because here the story becomes truly fascinating. The mind's omega point, *for each theorist*, is the *context* that they believe *cannot* be outcontexted, the context beyond which growth or expansion cannot or does not or should not proceed.

For Freud, the omega point, the end of development, was genital organization and integrated ego: all stages and all roads lead to that Rome. For Piaget, the omega point is formal operational thinking, which alone reaches "equilibration" and thus alone marks the end of development. For Habermas, rational intersubjective exchange of uncoerced mutual understanding: when this fully unfolds, the relentless pressure subsides.

And so the omega points would proceed, with each theorist postulating, in some sense, the "end of history," the omega point that, when reached, would answer all the really difficult questions and usher in some sort of relatively paradisiacal condition. For Hegel, a rational State in which individuals could realize absolute Spirit acting in and through them in a mutual community. For Marx, a classless society in which alienation of labor and produce would be healed in shared mutual care. And most people are familiar with Teilhard de Chardin's ultimate omega point, the resurrection of Christ consciousness in each and all, which, like all omega points, is maintained to be the purpose of history and evolution itself.

Once again, there are important moments of truth in all of those omega points. Each stage of growth, because it is a holon, faces a dual tension.

As a *whole*, it is relatively autonomous and relatively "healthy, happy, whole." But as a *part*, it is in some sense alienated, set apart, or disconnected from those contexts that are beyond its own perception. And until it takes the larger and deeper context into account, the limitations of its own shallower position will torment it, inflict it with the agony of incompleteness, tear at its boundaries with hints of something deeper, higher, more meaningful. . . .

And I don't mean that poetically. Piaget, for example, found that the earlier/lower stages of cognition, such as preoperational and concrete operational thinking, had inherent limitations or conflicts built into them, and that thought stuck in those levels wobbles endlessly without resolution around certain very important tasks. Only with the emergence of formal operational thinking could these conflicts come to some sort of peace or resolution, which Piaget called equilibration, a type of balanced but dynamic harmony. And further, it was *only* from the view of formal operational that these earlier conflicts made any sense at all: they were struggling to their resolution, struggling to their omega point, struggling to equilibration in the larger context of formal operational cognition, and, put rather simply, they could not be happy until they found that larger horizon.[83]

Limited contexts find resolution, not by anything that can be done on the same level, but only by transcending that level, by finding its deeper and wider context. Deeper and wider contexts *exert a pull*, a telos, on present limited contexts.

And that is the truth common to all omega-point theorists (and any decent theorist is an omega-point theorist). Assuming that their contexts are well founded, and assuming that they do not engage in reductionism, then there is much we can learn from each of these types of theorists. They always point to *ways beyond* our present perception, and assuming their contexts are genuine, they are right: we will never be happy until we, too, can live with a larger horizon. Until we, too, can accept the therapia of embracing gently a greater depth. . . .

And a final Omega Point? That would imply a final Whole, and there is no such holon anywhere in manifest existence. But perhaps we can interpret it differently. Who knows, perhaps telos, perhaps Eros, moves the entire Kosmos, and God may indeed be an all-embracing chaotic Attractor, acting, as Whitehead said, throughout the world by gentle persuasion toward love.

But that, to say the least, is quite ahead of the story.

3

Individual and Social

If I am what I am because you are what you are, and you are what you are because I am what I am, then I am not I and you are not you. —HILLEL

W E SAW, in chapter 1, that various theorists with a "holistic" orientation—systems theorists, ecological theorists, "new paradigm" thinkers, deep ecologists, ecofeminists, and so on—maintain that the various crises facing the modern world are all fundamentally due to a "fractured" worldview: we lack a unifying vision, and thus our world is fragmented, broken, alienated. And so, they maintain, what is required is some sort of systems theory orientation, some way for us to see and feel that we are all interwoven into the single Web of Life. We need, they would all maintain in their various fashions, a profoundly *ecocentric* worldview.

But no sooner do such theorists make that claim than they run into theoretical problems. First of all, they maintain that, as John Seed summarizes this view, "the world is seen not as a pyramid but as a web. Humans are but one strand of that web." In other words, human beings are a part of a deeper (or greater) web of life, the biosphere. But we could destroy all humans and the biosphere would still exist (but not vice versa), showing that the biosphere is a lower and shallower, not deeper or higher, reality. This confuses most ecotheorists; they seem to have no coherent way to honor the biosphere without absolutizing it.

Second, as Fritjof Capra summarizes the "new paradigm" thought, "In the new paradigm, the properties of the parts can be understood only from the dynamics of the whole. Ultimately, there are no parts at all. What we call a part is merely a pattern in an inseparable web of relationships."[1] It's true that there are no parts, but equally true that there are no wholes—only whole/parts forever, which therefore resist all totalizing and dominating agendas. But even if there were only wholes, how would this translate into, say, political theory? Ultimately, there are no free citizens, only the whole State? If the "whole" is primary, and we are parts of the State, then clearly we exist to serve the State, and clearly a totalitarian regime would be a shining example of the new paradigm, where all those nasty parts will disappear in the glorious web.

It is not, as I will repeat a dozen times throughout this volume, that systems theory is wrong; it is that, ironically, it is incredibly partial and lopsided. And so we will begin, in this chapter and the next, to redress some of these partialities and imbalances. And we will see, as this book progresses, that systems theory, in and by itself, is not a force of healing and wholing the planet; rather, we will see, and again ironically, that systems theory (in all its many variants) is part of the flatland paradigm that is still contributing to the despoliation and devastation of Gaia.

MICRO AND MACRO

Virtually all of the popular theorists of systems theory, deep ecology, and ecofeminism present some version of a great holarchy of being, with fields within fields within fields. The ecofeminist Julia Russell:

> At the same time, a body, a nation, exists in a larger context in which it functions as a part. The nation exists in the context of all nations that make up the whole political body of humanity. And all of humanity exists in a biosphere that is the body of the whole Earth. We exist as part of a seamless whole in which everything is connected to everything else.[2]

Figure 3-1 is very typical of the various hierarchies (holarchies) that have been presented, either explicitly or implicitly, by different holistic or ecological or systems-oriented theorists (similar holarchies can be found in Popper, Laszlo, Sessions, Miller, Engels, and others). Many of these holarchies are represented by the theorists as a series of levels or "strati-

Biosphere

Society / Nation

Culture / Subculture

Community

Family

Person

Nervous System

Organs / Organ Systems

Tissues

Cells

Organelles

Molecules

Atoms

Subatomic Particles

Figure 3-1. Typical Holarchy

fied order" (Capra), like a "ladder," which is perfectly acceptable as long as we remember the circular complexity that these authors maintain is actually involved and don't make the unfair and unwarranted charge that these theorists are being "linear." That, as we will see, is not the problem.

If we look at figure 3-1—or at almost any of the other and similar holarchies—we notice immediately that there is a confusion and conflation of individual and social holons. That is, micro- and macroworlds are confused. The social holon is assumed to be of the same type and same nature as the compound individual holon, so that they can be arranged "above" or "below" each other. Look, for example, at Karl Popper's biospheric holarchy (fig. 3-2). Popper (and he is far from alone) counts individuals and populations as being of the same logical (and existential) type, and thus arrangeable one above the other, as different "levels" on the same scale (such as, in his scheme, level 7 and level 8, or level 10 and level 11). This is quite incorrect.

We saw in the last chapter (with tenet 9), that in any holistic sequence, individual or social, if we destroy any level, we destroy *all* the levels above it and *none* of the levels below it, by the simple definition of wholeness/partness. A higher-level holon is in part *composed of* its lower-level holons, and thus if we destroy any lower level, we will also destroy any levels above it because we have taken away some of their component

12) Level of ecosystems (total biosphere)
11) Level of populations of metazoa and plants
10) Level of metazoa and multicellular plants
 9) Level of tissues and organs
 8) Level of populations of unicellular organisms
 7) Level of cells and of unicellular organisms
 6) Level of organelles (and perhaps viruses)
 5) Liquids and solids (crystals)
 4) Molecules
 3) Atoms
 2) Elementary particles
 1) Sub-elementary particles
 0) Unknown: sub-sub-elementary particles?

Figure 3-2. Popper's Holarchy[3]

parts. But these parts themselves existed *before* the whole emerged and therefore can, in general terms, exist without it.

Look again at figure 3-1. The claim of these holists is that nation-states exist as *parts of* a larger biosphere, a higher (or deeper) wholeness. But if the biosphere were really a higher level of organization than nation-states, if it really were a higher-level whole organizing or containing nation-states as *parts*, then (1) you couldn't have a biosphere *until* you had nation-states, since the latter are claimed to be ingredients of the biosphere; and (2) if you destroyed all nation-states you would destroy the biosphere (since the whole cannot exist without its parts).

Both of those assumptions are obviously not true. In fact, it is just the opposite: destroy the biosphere and you destroy *all* nation-states, but theoretically remove all nation-states and the biosphere would and could continue (just as it did long before nations showed up). This means that the biosphere is a *lower* and shallower level (which does *not* mean less important for existence, but *more* important for existence, because the *lower* a level you destroy, the *more* higher levels it takes with it; it is, as we put it in tenet 9, *more fundamental* to existence).

So let us locate the "biosphere" or "total ecosystem" by asking what would or would not be destroyed if we theoretically "disappeared" it. Looking at Popper's diagram, we find that if we destroyed the biosphere (if a thermonuclear holocaust destroyed *all* life forms on earth), we would destroy all the levels down to and including his level 6, *but no lower.* That means the biological ecosystem *starts* at his level 6, and not, as he lists it, at level 12. (This, as we will see, is a crucial error in the ontologies of many ecotheorists as well.)

In other words, we obviously have ecosystems, in the broadest sense, *as soon as* we have different life forms interacting with each other and with a physical environment, and this begins to occur in its most fundamental fashion at the level of prokaryotes (organelles). Putting the ecosystem at level 12 would imply, for example, that poor level 10—plants and metazoa—don't have ecosystems.

Thus, if we consider "the" ecosystem as a single level, then it is obvious that both diagrams completely misplace it. But it's worse than that. Ecosystem (or "total population") isn't a *particular* level among other levels of individual holarchy, but the *social* environment of *each* and every level of individuality in the biosphere. And neither of these diagrams distinguishes between micro and macro (or individual and social) on *any* level; both treat them as *separate* levels on the *same* scale. (This is a second crucial error in most eco-ontologies.)

With reference to Popper's diagram, for example, level 8 and level 7 are not two different levels, as he imagines, but the individual and social aspects of the *same* level. (We know this to be true because if we destroy *either* "level," the other is also destroyed, which means *neither* is higher or lower relative to the other.) If one were really *lower*—if level 7 were really lower than level 8, as he states—then it *could* exist without the higher. But you cannot have cells without a society (population) of other cells, and you cannot have a population of cells with no cells at all—nothing exists alone without an environment of the similar.

In other words, the individual and social are not two different coins, one being of a higher currency than the other, but rather the heads and tails of the *same* coin at *every* currency. They are two aspects of the same thing, not two fundamentally different things (or levels).

What is necessary, then, is to construct a series of true holarchies of compound *individuals* and then indicate, at the *same* level of organization, the *type of environment* (or social holon) in which the individual holon is a participant (and on whose existence the individual holon depends). And this needs to be done in all three of the great realms of evolution—physiosphere, biosphere, and noosphere.

Fortunately, much of this work has already been done by Erich Jantsch, the only theorist, to my knowledge, to have investigated extensively (and often brilliantly) the relationships between micro and macro (individual and social) patterns of coevolution in all three domains (although, as we will see, his argument becomes somewhat garbled as he approaches the noosphere).

Figures 3-3 and 3-4 are Jantsch's schematic diagrams of the coevolution of micro and macro patterns in, respectively, the physiosphere and the biosphere. Figure 3-4 picks up where figure 3-3 leaves off (namely, the passing of matter into life). The explanations under each diagram are his own. (With the exceptions that I will mention, I am in substantial agreement with everything in these two diagrams and with his explanations of them.)

The lower portion of each diagram represents microevolution—the evolution of individual holons. The upper portion represents macroevolution—the *correlative* level of social or environmental relationships (relational exchanges) in which the individual holons are inextricably embed-

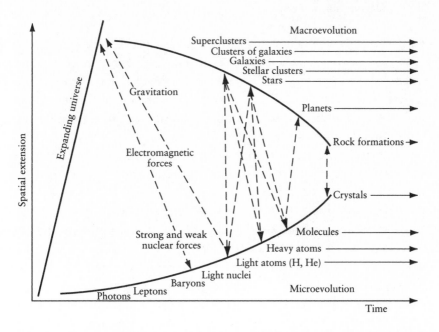

Figure 3-3. Cosmic Coevolution of Macro- and Microstructures. The asymmetrical unfurling of the four physical forces calls into play step by step new structural levels, from the macroscopic side as well as from the microscopic. These levels mutually stimulate their evolutions. (Jantsch, The Self-Organizing Universe, *p. 94)*

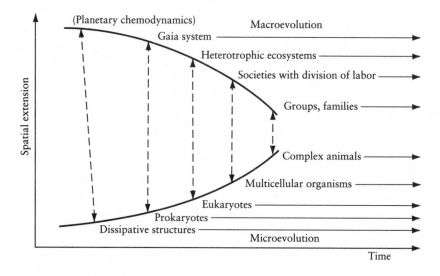

Figure 3-4. The history of life on earth expresses the coevolution of self-organizing macro- and microsystems in ever-higher degrees of differentiation. (Jantsch, The Self-Organizing Universe, p. 132)

ded. The idea is fairly simple: when you have a specific type of *individual* holon (say, heavy atoms), what kind of *social* or collective environment results? (in this case, stars). When molecules emerge, planets emerge (i.e., they coemerge, coevolve). When eukaryotes emerge, ecosystems emerge; when complex animals emerge, families emerge; and so on. All of these correspondences are listed on figures 3-3 and 3-4.

And notice: since evolution produces greater depth, less span, then the individual holons tend to get *bigger* (i.e., molecules are bigger than atoms, because they embrace and contain them), but social holons tend to get *smaller.* Because there are fewer holons at the greater depth (there are always fewer molecules than atoms), then when you put them together into collectives, the collectives are smaller. Thus families are smaller than ecosystems, which are smaller than planets, which are smaller than stars. (We will return to this increasing depth/decreasing span in a moment.)

The earlier figures (3-1 and 3-2) suffer badly by comparison because they treat micro and macro as *different levels* of the overall evolutionary process, instead of seeing them as two different aspects of each level of the overall evolutionary process, and they leave out altogether the macro component of the lower levels, a grave omission.

Notice in particular that both the micro and macro holarchies given by

Jantsch (in both the physiosphere and the biosphere) constitute a genuine higher/lower relationship: destroy any lower (whether individual or social), and the levels above it are also destroyed, but not vice versa. Jantsch is presenting genuine holarchies of both individual and social holons.

GAIA

Notice, in figure 3-4, where Jantsch places "Gaia"—it is the social holon composed primarily of the individual holons of prokaryotes. Jantsch is here using the term *Gaia* in its technically correct sense, as originally proposed by Lynn Margulis and James Lovelock (and named by William Golding, author of *Lord of the Flies*). As Jantsch explains, "The autocatalytic units in this system [Gaia] which make possible the formation of a dissipative structure far from equilibrium and maintain the through-flow of the various gases [in the atmosphere] are none else but the prokaryotes. It seems that after the profound transformations of the earth's surface by the oxidation of sediments and the accumulation of free oxygen, they have been instrumental in bringing the overall system bio- plus atmosphere into global, autopoietic stability, reigning now for 1500 million years."[4]

The prokaryotes mediate exchanges with the atmosphere (the physiosphere), but also form a global and interconnected network with all other prokaryotes—the overall Gaia system. This is precisely tenet 11, "The micro is in relational exchange with the macro at all levels of its depth." Further, and this is the fascinating part, since some prokaryotes were taken up and incorporated into higher (eukaryotic) cells and these in turn in complex organisms, the prokaryotic Gaia system is still serving its original global function. As Jantsch puts it, "The descendents of the prokaryotes, in the form of parts of more highly developed cells, are still at their old job." And this, of course, is precisely tenet 5, "Each emergent holon transcends but includes its predecessor(s)."

Jantsch points out that the Gaia system is the *largest* living social holon on the planet (has the most units, or the *greatest span*), precisely because it is the *shallowest* (most primitive). Again, evolution produces greater depth, less span. At the same time, precisely because Gaia is the shallowest of all living social holons, it is exactly the most fundamental. As we earlier put it, the *less depth* a holon has, the *more fundamental* it is to the

Kosmos, because it is a *component* of so many other holons. And thus Gaia is the most fundamental (and least significant) of all living social holons: all higher life forms depend upon it (destroy higher life forms, and Gaia is basically untouched; destroy Gaia, and everything else goes with it). Gaia is indeed our roots and our foundation.

But aside from this technical usage, Gaia has come to mean, for many people, the total biosphere; for some, it means the total planet and all life forms; and for a few, it means Spirit itself, or the Goddess, as the total Life Force of the planet. There is in all these approaches a tendency to confuse fundamental with significant. Nonetheless, although none of these usages is technically correct, using *Gaia* to mean the overall biosphere or overall life dimension of the planet is acceptable enough, as long as we are clear about what exactly is meant. I will often use *Gaia* in both the narrow technical sense (prokaryotic network) and the broader "mythic" sense (total biosphere or living planet as a whole), and the context will make clear which is intended.

SIZE, SPAN, EMBRACE

Notice that in both figures 3-3 and 3-4 Jantsch plots spatial extension (or physical size) against evolutionary time, and his conclusion in both figures is that, with increasing evolution, *individual* holons tend to become *larger* (in the sense that organisms are larger than molecules) and *social* holons tend to become *smaller* (planets are smaller than galaxies)—and he explicitly indicates this on his diagrams. Both of these are, we have seen, examples of the fact that evolution produces greater depth, less span.

On the one hand, individual holons transcent but include their predecessors—development is envelopment—and so the total depth or embrace of individual holons increases: cells embrace molecules which embrace atoms which embrace particles. Likewise, the greater the depth of a holon, the fewer of them that are produced and maintained, relative to the number of their predecessors (fewer molecules than atoms, fewer cells than molecules), and therefore the *population* of greater-depth holons will always be *less* than the population of its predecessors. Thus, greater depth, less span.

These correlations are important, we will see, because many ecotheorists and systems theorists confuse larger span with greater depth, and thus often recommend regressive ontologies and regressive directions for salvation.

But before we get to that, a simple point about Jantsch's important correlations. In equating these changes with actual size (or physical extension), Jantsch goes too far (or becomes too simplistic). Depth or embrace always increases, but embrace is not necessarily translatable into simple sizes. In the same way, increasing evolution means less relative span (or decreasing simple span), but this does not necessarily show up as decreasing physical size (see below).[5]

Of course, once a *given* level has emerged, then its simple span can begin to increase, and this will often show up as an increase in the spatial extension of the population—i.e., for a given depth, span can increase up to its carrying limits; thus human span has increased from small villages to global villages.[6] When this starts happening in human evolution, Jantsch gets confused, because, he supposes, macro-evolution is *supposed* to get physically smaller, and yet human sociocultural evolution keeps getting larger and *more* global, which leads Jantsch to saying that human evolution completely "turns on its head" the previous trend seen everywhere else in evolution; but, of course, it does nothing of the sort—all that gets turned on its head is his incorrect assumption about size. Because the fact is, although actual size or spatial extension is often a good indicator of these yardsticks, neither necessarily and always translates into simple physical dimensions.

The reason Jantsch's "size rule" gets into so much trouble with human or sociocultural evolution is that the human mind, as Descartes knew, is characterized more by intention than extension. That is why it is so important to state the "rule" strictly in terms of embrace (depth of holon, regardless of size of holon) and span (number of individuals at that depth, regardless of their actual size). A concept, we will see, always embraces a symbol (it is a symbol *plus* another cognitive function; it has greater depth), but we can't say a concept is physically *bigger* than a symbol. One value might be *better* than another, but it doesn't occupy a bigger space.

THE PROBLEM WITH SIZE AND SPAN

All of these seemingly hair-splitting distinctions are actually very important when it comes to the prevalent "holistic theories," because many of them try to construct their holarchies based on *increasing size* (or, secondarily, *increasing span*)—just as in figure 3-1—and this leads to some very unpleasant and confusing results. Whether they base their theories on size or span, they confuse those with depth, and this often leads to *regressive* ontologies and soteriologies. Understanding the "errors" in figure 3-1 is important for unraveling these common confusions found in many holistic, ecological, and systems thinkers.

I will mention only two problems at this point. For the evolution of individual holons, *greater embrace* means that more of the universe is being taken into the holon (is actually internal to that holon). As we will see, an individual holon can eventually *embrace the entire Kosmos*—its depth can go to infinity—but the *actual number* of holons (the span) that can realize this total embrace might be very, very few. Kosmic consciousness means Kosmic embrace, it doesn't mean Kosmic span.

Many holistic theorists are admirably trying to construct a series of "wider and wider wholes," and this series is of absolutely crucial significance, they maintain, because it is meant to help us humans find our place, our context, in the larger scheme of things, so that we may more accurately orient ourselves—morally, emotionally, cognitively, spiritually—to our fellows and to our world. I am in complete sympathy with that approach.

But *if depth is confused with span* (or with size), then the resultant "holistic sequence of wider wholes" ends up being perfectly regressive and anti-holistic! Since, with evolution, simple span always gets *smaller*, then by constructing a holistic sequence of *bigger* span, we are headed in precisely the opposite and altogether wrong direction.

For example, many holistic theorists create their "holistic" sequences using simple span, and end up with this: the noosphere is *part* of the larger whole called the biosphere, which itself is *part* of the larger whole called the cosmos (or the entire physiosphere).

Absolutely not true. Absolutely the other way around. As we have seen, if the noosphere were really a *part* of the biosphere, then if we destroyed the noosphere the biosphere would disappear, and that is clearly not the case. An atom is genuinely a part of a molecule, and thus if we destroy the atoms we also destroy the molecule: the whole needs its parts. Just so, if the noosphere were really a part of the biosphere, then destroying the noosphere would eliminate the biosphere, and yet it is just the other way around: destroy the biosphere, and the noosphere is gone, precisely because the biosphere is a part of the noosphere, and not vice versa.

In other words, because these theorists look mostly at simple span (and its size), and because they think that *bigger* span is a "higher or deeper whole," and since in fact greater evolution produces *less* span, then these theorists get the holistic sequence *exactly* backward.

What these holists mean to say, correctly, is that the noosphere *depends* on the biosphere, which *depends* on the physiosphere—and that is true *precisely* because the physiosphere is a *lower* component of the biosphere, which is a *lower* component of the noosphere, and not the other way around.

Because this is such an important topic, I want to go over it using some very simple examples.

SAME-LEVEL RELATIONAL EXCHANGE

We said that with emergent evolution we are, in effect, playing three-dimensional checkers or chess, not flatland chess. So imagine this: imagine one chessboard (or checkerboard) with forty black checkers placed on it. The depth is one, the span is forty. Place another chessboard over it, but leave it empty for the moment. The depth is two, the span zero.

Now, in evolution, the only way to get to level 2 is through level 1, and, in fact, the checkers on level 2 are composed in part of checkers from level 1—they are all holons or compound individuals. Represent this by taking one black checker from level 1, placing it on level 2, then adding a red checker on top of it. The new and "total holon" on level 2 thus incorporates its predecessor (the black checker) and adds its own distinctive properties (the red checker). If we did this, say, three times, the holons on level 2 would have a depth of 2 and a span of 3.

Now, the individual holons or checkers on level 1 (the physiosphere) depend for their existence on intricate networks of interrelationships with all the other black checkers in their environment—depend, that is, on networks of their own social holons (the coevolution of micro and macro). They exist in intricate networks of relational exchange with holons at the same level of structural organization (tenet 11).

But the situation on level 2 (the biosphere) is much more complicated, because the new total holon (the black-and-red compound checker) depends for its existence on intricate relationships on *both* levels. The red-and-black checkers on level 2 depend in part on their relationships with other red-and-black checkers—depend, that is, on ecological or macro relations with other *living* holons at the same level of structural organization. In other words, the "red" component of the red-and-black checkers depends on interrelations with the "red" component of other red-and-black holons—depends on *relational exchanges* such as sexual reproduction (which are *not* found on the black level and cannot be sustained by that level).

However, because the red-and-black checkers also have a black component, they *also* depend on the intricate relationships that sustain black holons themselves—depend, that is, on all of the mutually sustaining relationships and processes that constitute level 1 holons. So level 2 holons depend not only on the new and "red" relationships or social holons found *only* on level 2, they also depend upon the prior "black" relationships and sustainable patterns established on level 1 (but not vice versa: destroy level 1, and level 2 is destroyed; destroy level 2, and level 1 black checkers will still exist).

So any holon, or compound individual, depends on a whole series of intricate relational exchanges with social environments of the *same level* of structural organization for *each level* in the individual holon. And that means that a holon of depth three, for example, has to exist in an environment that also possesses holons of at least the same depth. So any holon is fundamentally a compound individual and its same-level relational exchanges at *all* of its levels—a compound individual in a compound environment, exchanging black with black and red with red and so forth.

We need one last point. The red-and-black checkers are not *in* the black universe. The only things that are *in* level 1 are more black checkers. The red-and-black checkers are to some degree *beyond* the level 1 universe (that is the meaning of emergence). In their "redness" they are beyond the mere blackness of level 1. They are a perfect expression of the self-transcending thrust of evolution, of the creative emergence of a "redness" that cannot be reduced to, or *found in*, the black universe.

And, in a sense, it is just the opposite. The red-and-black checkers contain "redness" as well as "blackness." The level 2 holons *embrace* level 1 holons and then go beyond them with their own defining emergents (tenet 5). Since the red-and-black checker depends for its existence on its own component black checkers, and since the black checkers themselves depend ultimately for their particular type of existence on all the other black checkers in *their* universe, then any level 2 holon in essence embraces *all* of its level 1 world by simple virtue of its own compound individuality. A single living cell embraces its entire physical universe. The many has become one and is increased by one—another meaning of Whitehead's crucial dictum. And according to Whitehead, in this "prehensive unification" *all* antecedent actual occasions participate to some degree (in a holarchy of embrace, as he carefully points out).

And that is what I mean when I say that a level 2 holon is not *in* level 1, but level 1 is *in* level 2, utterly and totally embraced by it—one of the meanings, we will see, of love, a loving embrace which, like consciousness and creativity and self-transcendence, is built into the very *depth* of the Kosmos.

All of this means, in short, that the biosphere is not in the physiosphere. The biosphere is *not* a *component* or a *part* of the greater whole called the physiosphere, because the only thing "greater" about the physiosphere is its span, not its depth or wholeness. The bios is not a part of the cosmos, but just the opposite: the cosmos is a part of, a component of, the bios.

Thus, a *part* of the biosphere (namely, its physical component) is indeed a *part* of the larger physiosphere; but its defining, emergent qualities are not *in* the set of physically definable determinants governing nonliving forms (as Maturana and Varela point out, autopoiesis is found nowhere in the nonliving world; autopoiesis is a red-checker quality found nowhere in black checkers). The biosphere is not in the physiosphere, but the physiosphere is indeed in the biosphere, contained in it as a *part*.

Thus, just as an atom is in a molecule but a molecule is not in an atom—even though the span of atoms is much, much larger than molecules—so the cosmos is in the bios but the bios is not in the cosmos, even though the span of the cosmos is astronomically larger.

And note: the bios is a part of the Kosmos, but not a part of the cosmos, and in that simple move we have forever disavowed reductionism: physics is the most fundamental, and least significant, of the sciences (the reason physics can't explain biology is precisely because the bios is not in the cosmos).[7]

This is just the opposite of what a naive faith in simple span or size would tell us. Just because the cosmos is bigger than the bios, we assume that the cosmos must be more significant. But it is only more fundamental; the bios is much more significant than the cosmos, because it contains much more reality internal to it, embraces a much deeper and greater wholeness, has more depth, and in fact subsumes the entire cosmos in its own being: transcends and includes.

Likewise, as we will see, the noosphere is not a part of the biosphere, but just the opposite: the bios is a lower component, a part of, the noosphere. So of course the biosphere has a greater span (is "bigger"), just as the cosmos is bigger than the bios. But from what we have now seen, that, of course, just proves my point.

Precisely because the biosphere is a component part of the noosphere, the destruction of the biosphere guarantees the destruction of the noosphere—and that is one of the points we will be returning to throughout this volume. This is a profoundly ecological orientation *without* regressively absolutizing the biosphere (or the noosphere, as we will see).

But for now, let us note that the *number* (the span) of holons at a higher level of development will always be *less* than the number at a preceding level (tenet 8). The number of red-and-black checkers will *always* be less than the number of black checkers. There are no exceptions, because the number of wholes will always be less than the number of parts contained in them. There will always be fewer fifth-graders than fourth-graders, because you have to go through the fourth to get to the fifth. There are always fewer oaks than acorns—the so-called pyramid of development that necessarily gets smaller and smaller toward the top. Laszlo, for example, represents this pyramid as in figure 3-5 (which also indicates accurately that, for example, the biosphere is not *in* the physiosphere). Where

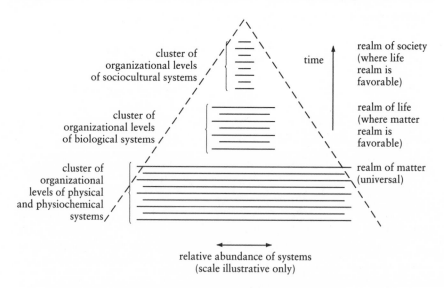

Figure 3-5. The Realms of Evolution (Laszlo, Evolution, p. 55)

matter is *favorable*, life emerges. Where life is *favorable*, mind emerges. Greater depth, less span.

Accordingly, when holistic theorists construct their holarchies in terms of increasing span ("wider and wider"), they take us right down the pyramid, scraping off more and more levels of depth in search of a bigger (and therefore shallower) span. They should be taking us to a greater depth that involves *both* a *deeper* embrace and a *wider* identity—an expansion vertically and horizontally, not just horizontally.

Thus, we can start to see why diagrams like figure 3-1 are such an unfortunate map: they confuse micro and macro, they confuse depth and span.

THE BRAIN OF A HUMAN HOLON

Jantsch's diagrams (figs. 3-3 and 3-4) cover the physiosphere and the biosphere. It is now time to look at the emergence of the noosphere.

It is with the brain that we usually associate the emergence of the noosphere from (or rather through) the biosphere. Jantsch, following Gunther Stent, divides communication in the biosphere into genetic, metabolic, and neural (and the highest part of the latter is said to correlate with noospheric evolution). Jantsch:

> Genetic communication acts in time intervals which are long compared to the lifetime of an individual. It makes phylogeny and coherent evolution across many generations possible. Metabolic communication which in the organism is transmitted by special messenger molecules, the hormones, fulfils two tasks within an organism. One task is the regulation of the development of multicellular organisms, in plants as well as in animals. The other task concerns the damping of the consequences of environmental fluctuations for the organism, or in other words, the enhancement of the organism's autonomy [tenet 12d]. Metabolic communication based on hormones acts relatively slowly, from seconds to minutes. The third type of biological communication is transmitted by the nervous system. This type acts in organisms typically within a hundredth to a tenth of a second and is thus about a thousand times faster than metabolic communication.[8]

Jantsch points out that these three systems emerge holarchically, and he then proceeds to give a holarchical breakdown of neural structures themselves, based primarily on Paul MacLean's highly influential notion of the triune brain (see fig. 3-6). In MacLean's own words:

> Man finds himself in the predicament that Nature has endowed him essentially with three brains which, despite great differences in structure, must function together and communicate with one another. The oldest of these brains is basically reptilian. The second has been inherited from lower mammals, and the third is a later mammalian development, which, in its culmination in primates, has made man peculiarly man.
>
> Speaking allegorically of these three brains within a brain [i.e., holons], we might imagine that when the psychiatrist bids the patient to lie on the couch, he is asking him to stretch out alongside a horse and a crocodile. . . . The reptilian brain is filled with ancestral lore and ancestral memories and is faithful in doing what its ancestors say, but it is not very good for facing up to new situations [relatively low autonomy expressed in reflex and instinctive behavior].
>
> In evolution one first sees the beginning of emancipation from the ancestral [inflexibility] with the appearance of the lower mammalian brain, which Nature builds on top of the reptilian brain. . . . Investigations of the last twenty years have shown that the lower mammalian

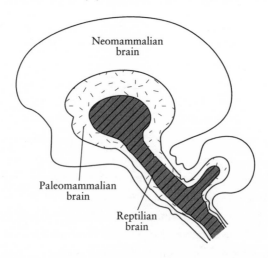

Figure 3-6. The "Triune Brain" (After MacLean)

brain plays a fundamental role in emotional behavior. It has a greater capacity than the reptilian brain for learning new approaches and solutions to problems on the basis of immediate experience. But like the reptilian brain, it does not have the ability to put its feelings into words.

MacLean is very specific about the holarchical nature of these three brains.

> In its evolution, the brain of [humans] retains the hierarchical organization of the three basic types which can be conveniently labelled as reptilian, paleo-mammalian and neo-mammalian. [The brain stem represents the reptilian brain, inherited from reptilian ancestors.] The limbic system represents the paleo-mammalian brain, which is an inheritance from lower mammals. Man's limbic system is much more highly structured than that of lower mammals, but its basic organization, chemistry, etc., are very similar. The same may be said of the other two basic types. And there is ample evidence that all three types have their own special subjective, cognitive (problem-solving) memory and other parallel functions.[9]

In other words, each of the brains is a relatively autonomous holon. And because each is a holon, we cannot say that any specific function is simply located in one of the holons; they all mutually act and interact with upward and downward influence. But in general, the three brains have the following basic functions:

Reptilian brain (or brain stem): "This is the phylogenetically oldest part of the brain, its core or chassis, roughly corresponding to the basic structures of the reptile's brain. It contains the essential apparatus for internal (visceral and glandular) regulations, for primitive activities based on instincts and reflexes, and also the centers for arousing the animal's vigilance or putting it to sleep."[10] We may refer to it as the overall level of rudimentary sensorimotor intelligence and instinctive drives or impulses.

Paleomammalian brain (or limbic system): "The limbic system is intimately connected by two-way neural pathways with the hypothalamus and other centers in the brain-stem concerned with visceral sensations and emotional reactions—including sex, hunger, fear and aggression; so much so that the limbic system once bore the name 'the visceral brain.'"[11]

In short, "The limbic system processes information in such a way that it becomes experienced as feelings and emotions, which become guiding forces for behavior."[12]

Neomammalian brain (or neocortex): "The explosive growth of the neocortex," says Jantsch, "in a later phase of evolution is one of the most dramatic events in the history of life on earth," and presumably connected with the emergence of the noosphere. It acts as "an immense neural screen on which the symbolic images of language and logics (including mathematics) appear. The neocortex is the location at which information is processed in the ways characteristic of the self-reflexive mind."[13]

MacLean underscores the holoarchic nature of these three brains by comparing them with the components of literature, which Jantsch elegantly summarizes in this fashion:

> The reptilian brain stands for the [basic] figures and roles which underlie all literatures. The limbic system brings emotional preferences, selection and development of the scenarios into play. And the neocortex, finally, produces on this substrate as many different poems, tales, novels and plays as there are authors.[14]

Again we see the lower setting the possibilities and the higher setting the probabilities (or actualizing potentials).

THE HUMAN SOCIAL HOLON

Jantsch gives several diagrams, similar to figures 3-3 and 3-4, which he believes represent developments in the noosphere (or what he calls sociocultural evolution). I will not reproduce them here, because I believe they are fundamentally confused. Jantsch says that once evolution reaches the noosphere, "things become completely inverted." But all that is inverted is his own prior confusion between span and size, and his insistence on finding extension where there is largely intention. Without that confusion, the same principles and the same evolutionary process can be seen at work—greater depth, less span—there is no "inversion." As I said, Jantsch often attempts to make spatial extension (physical size) a fundamental correlate of mental (sociocultural) growth, and as even Descartes was well aware, in the noosphere, extension gives way to intention, upon which rulers do not well fit.

I will return to that point in a moment and go over it carefully, but for now let us look instead to the work of those systems theorists who have plotted out the growth of human *social holons* during humanity's million-or-so-year history.

Keep in mind that we have already outlined (however briefly) the growth of *individual holons*, from atoms to cells to multicellular organisms to complex animals (the micro components as suggested roughly in the bottom halves of figs. 3-3 and 3-4), and we also indicated the correlative environmental or *social holons* at each stage (the macro components suggested in the upper halves in figs. 3-3 and 3-4). We then briefly sketched the further evolution of the *individual* holon of complex animals up to and including those with the triune brain (from reptile to mammal to primate). We want to look now at the *social* environment in which the triune-brained organisms existed, and the types of *social holons* upon which the triune-brained organism existed for its own relational exchanges.

And this moves us directly into the noosphere, or the realm of sociocultural (and not just biosocial) evolution. We are adding a third chessboard to the game—matter, life, mind.

If we look at Jantsch's figure 3-4, we will be reminded that individual complex animals, up to and including primates (which includes humans), needed *social* holons at the level of groups/families, and if given those appropriate social holons, the individual holons could be sustained quite adequately (assuming, of course, that *all* junior levels in the compound individual are also existing in a balanced and sustainable set of relationships with their own environments—assuming, that is, that the whole multilevel arrangement is ecologically sound in the broadest sense, which we have every reason to believe was the case at this point in evolution).

Put more simply, the social holon of the family/group could have sustained the human triune brain indefinitely, just as it still sustains the roughly similar triune brain of other primates (who continue to exist in kinship social holons). But the human holon pushed beyond an environment of kinship (or biospherically based) social holons such as the family and began also producing villages, towns, cities, states. . . . As I said, a third chessboard had been added to the evolutionary game.

Of course, this raises all the old and thorny problems, from the relation of mind and body to the whole issue of whether the noosphere itself was a good idea in the first place. All I can say is that I will try to address these issues as best I can, but we must proceed one step at a time, and each step will have its own true meaning disclosed only as the whole picture begins to emerge.

So, for the moment, let us simply note that there was no compelling *biological* reason to produce villages and cities and states. The social holon of the family/group could have sustained the human triune brain, just as it does other primates to this day. But just as matter had pushed forth life, the self-transcendent drive *within* biology pushed forth something *beyond* biology, pushed forth symbols and tools that both created and depended upon new levels of social holons in which the users of symbols and tools could exist and reproduce themselves, but the reproduction was now the reproduction of culture through symbolic communication and not just the reproduction of bodies through sexuality. Kinship gave way to "cultureship": add a blue checker on top of the red checker on top of the black checker.

And only a reductionist will be bold enough to claim that culture was really just a nifty new way to gain food, that the blue checker is just a sneaky rearrangement of red checkers, that the noosphere is nothing but a new twist on the biosphere. But just as the biosphere is not *in* the physiosphere (you can't find red checkers anywhere on level 1), so the noosphere is not *in* the biosphere (there are no blue checkers on level 2, no self-reflexive linguistic concepts, for example; no calculus, poetry, or formal logic).

As far as we can tell, the human brain has remained virtually unchanged in the past fifty thousand years. And yet during that fifty-thousand-year period, the same triune brain produced an extraordinary range of cultural achievements and cultural disasters. In other words, nothing truly novel happened to the triune brain during that period, no new major biospheric evolution occurred. And yet the entire majesty and catastrophe of culture paraded across the scene, all on the same biological base, but a majesty and catastrophe that could not be reduced to, explained by, or contained in that base.

Once again the lower had set the foundation and prepared the possibilities for (but did not determine) evolution in the higher, and those new possibilities were now played out in the noosphere, in the realm of culture and symbols and toys and tools.

We will be looking at the noospheric changes in the *individual* holons in the next chapter. For the moment let us center on the evolution of the *social* holons in the newly emerging noosphere. Figure 3-7 is adapted from Alastair Taylor's "sociocultural nonequilibrium systems model," a social approach based on dynamic systems theory (and at this point we are centering on the "true" aspects of systems theory). Taylor maintains that each succeeding social holon "builds upon the properties and societal experiences of the level(s) below and in turn contributes its own 'emergent qualities,' which take the form of new technologies and societal structures, accompanied by new apperceptions of the human-environment relationship. We can discern progressive developments in complexity and heterogeneity (although in any one historical situation a different, or even contrary, experience may occur)"[15]—all familiar concepts (all found in the twenty tenets).

We could of course include an enormous number of components in the evolving noosphere of social humans, from types of tools to various worldviews, from political institutions to styles of art, from modes of production to legal codes. And, indeed, we will be returning to these various aspects as we proceed with our story. But for the moment, figure 3-7 may serve as one example of the growth of social holons in the newly emergent noosphere.[16] Serve enough, that is, to allow us a few important observations.

DIFFERENTIATION AND DISSOCIATION, TRANSCENDENCE AND REPRESSION

The evolution of holons is not all sweetness and light, as some proponents of "evolutionary progress" seem to maintain. For the grim fact is that greater structural complexity—whether individual or social—means that more things *can* go horribly wrong. Atoms don't get cancer; animals do. Nonetheless, the opposite conclusion, that most evolution is really devolution, is not warranted either. The existence of cancer does not damn the existence of animals per se. And we will want to tease apart the cancers of the noosphere without damning the noosphere in toto.

As I indicated in the last chapter, the fact that evolution always produces greater transcendence and greater differentiation means that a factor of *possible* pathology is built into every evolutionary step, because

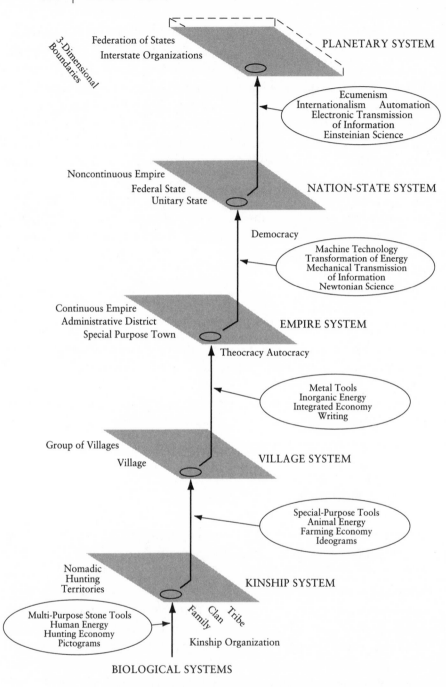

Figure 3-7. Emergent Geopolitical Systems Levels (Taylor)

transcendence *can* go too far and become *repression*—the higher does not negate and preserve the lower, it tries only to negate (or repress or deny) the lower, which works about as well as denying our feet.

Likewise, *differentiation* can go too far and become *dissociation*—a failure to adequately integrate the newly emergent differences into a coherent whole that is both internally cohesive and externally in harmony with other correlative holons and with all junior components. Whenever a new differentiation is not matched by a new and equal integration, whenever there is negation without preservation, the result is pathology of one sort or another, a pathology that, if severe enough, evolution sets about to erase in earnest.

This becomes particularly acute in the noosphere, in cultural evolution, simply because the human holon contains so many levels of depth—physiosphere and biosphere and noosphere—and something can go wrong at every level.

When the biosphere was first emerging on Earth, we can imagine all the fits and starts, all the dead-ends, that the first populations of cells tried out in their attempt to fit with the preconditions set by the physiosphere, whose territory the living cells were now invading. A false start—a start not consonant with the physiosphere—was simply erased. And throughout biological evolution, as the biosphere itself began adding layers and layers of new depth, each of these levels had to be brought into an adequate harmony with *both* its predecessors *and* its peers—no easy or trivial task, as the awesome example of the dinosaurs reminds us.

All of that was still true with humans, but with an extra and rather gruesome burden: fit the noosphere not only with its peers (holons at the same level) but also with *all* of its predecessors—mineral to plant to reptile to paleomammal—that not only continue to inhabit the same environmental space but also *exist as components in the human beings' own compound individuality.* When MacLean said that when humans lie on the couch for psychoanalysis, they lie down with a crocodile and a horse, that wasn't the half of it: we lie down with the planets and the stars, the lakes and the rivers, the plankton and the oaks, the lizards and the birds, the rabbits and the apes—and, to repeat, not simply because they are neighbors in our own universe, but because they are components in our own being, they are literally our bones and blood and marrow and guts and feelings and fears.

And just as the biosphere had to find its acceptable niche in (and beyond) the physiosphere, so too the noosphere had to find its allowable or harmonious place in (and beyond) the biosphere. And to borrow Shakespeare's phrase: "Aye, there's the rub."

The noosphere evolved. And as various stages of political, linguistic, and technical development emerged—incorporating and transcending their predecessors—not only could these higher stages of cultural development repress and alienate their own previous connections in the noosphere (as we will see), they could also come perilously close to severing their connections with the biosphere as well, and to such an alarming degree that, today, humans have earned, through enormously hard work and labor, the privilege and the possibility of being the first cultural dinosaurs in the fragile noosphere.

Instead of transcendence, repression; instead of differentiation, dissociation; instead of depth, disease. Because of the very nature of evolution, that type of dissociation can occur at any and all stages of growth and development. The noosphere is not privileged or unique in this regard. It is simply more alarming now because of its global dimensions. If we violate the *possibilities* given us by the biosphere, then the biosphere will flick us off its back like a bunch of fleas, and be none the worse for it.

But just as the existence of cancer does not damn the existence of animals per se, so the existence of cultural diseases and repressions does not damn cultural evolution itself. This particularly applies to successive stages of cultural evolution, from the hunt to the farm to the engine to the computer. Each successive stage brought new information, new potentials, new hopes and new fears; brought a greater complexity, a greater differentiation, a greater relative autonomy—*and* the capacity for a new and greater pathology if a corresponding integration and embrace did not ensue. And we will see that the history of cultural evolution is the history of new achievement, the history of new disease.

But I have no sympathy for those theorists who simply *confuse* differentiation and dissociation, transcendence and repression, depth and disease.

Whenever evolution produces a new differentiation, and that differentiation is not integrated, a pathology results, and there are two fundamental ways to approach that pathology.

One is exemplified by the Freudian notion (introduced by Ernst Kris) of "regression in service of ego." That is, the higher structure relaxes it grip on consciousness, regresses to a previous level where the failed integration first occurred, repairs the damage on that level by reliving it in a benign and healing context, and then integrates that level—embraces that level, embraces the former "shadow"—in the new and higher holon of the ego (or total self-system). For the ego's problem was that during its formative growth, where it should have *transcended and included* its lower-level drives (such as sex and aggression), it transcended and repressed them, split them off, alienated them—one of the prerogatives of a higher-level structure with its greater relative autonomy, but a prerogative, we have seen, that is bought only and always at the price of pathology. Thus the cure: regression in service of a higher reintegration—a regression that allows evolution to move forward more harmoniously by healing and wholing a previously alienated holon.

The other general approach is the retro-Romantic, which often recommends regression, period. This approach, in my opinion, simply confuses differentiation and dissociation, confuses transcendence and repression. Thus, whenever evolution produces a new differentiation, and that differentiation happens to go into pathological dissociation, then this approach seeks to permanently turn back the pages of emergent history to a time *prior* to the differentiation. Not prior to the dissociation—we all agree on that!—but prior to the differentiation itself!

That will indeed get rid of the new pathology, at the cost of getting rid of the new depth, the new creativity, the new consciousness. By that retro-Romantic logic, the only way to really get rid of pathology is to get rid of differentiation altogether, which means everything after the Big Bang was a Big Mistake.

All the more disturbing in the retro-Romantic approach is the whole problem of where these theorists rather arbitrarily *decide to stop* their retrogression. For example, many theorists, looking with justifiable alarm at the repressions and alienations that often accompanied the Machine Age, maintain that we should never have gone past farming, and there then follows a wonderful eulogizing of the glories of the "nonmechanized" and "nondehumanizing" farming societies, where few humans were alienated from the products of their own labor and the "Great Mother" ruled in peaceful and holistic happiness. Never mind that many of these societies introduced deliberate human sacrifice, multiplied the extent and means of war, put gender stratification at its peak, and made vast numbers of its populations into slaves.

Spotting those obvious difficulties, other theorists then go further and maintain that most of humankind's problems came with the invention of farming itself, because with farming the human animal began to deliberately alter the biosphere for its own gratification, produced a written language that ensconced power in the dogmatic text, produced an agricultural surplus that allowed some individuals to begin to economically control and enslave others, and began the wholesale subjugation of women. And, indeed, most of that did begin with agrarian farming.

So, these theorists maintain, we really should never have gone past hunting-and-gathering societies. The delightful things then said about these societies—some of which were peace-loving and rather egalitarian, and some of which most definitely were not—are, at the least, astonishingly one-sided. Until other theorists carefully point out that precious few of these societies were actually egalitarian, that warfare most definitely existed, that the very seeds of sexist subjugation were planted here, that slavery was not unheard of. . . .

We really should never have gone past gorillas, who at least don't deliberately sacrifice their own or engage in renegade warfare, where slavery is nonexistent and no animal is alienated from its own labors. Until you realize . . .

And so it goes, scraping layers and layers of depth off the Kosmos looking for a Garden of Eden that ever recedes into a shallower darkness.

My point is that it is one thing to remember and embrace and honor our roots; quite another to hack off our leaves and branches and celebrate that as a solution to leaf rot. So we will celebrate the new possibilities of evolution even as we gasp in horror—and try to redress—the multiplicity of new pathologies.

But to reiterate the one point of strong agreement with the Romantics in general: we have added such depth (height) to the noosphere itself that it is in danger of sliding off the biosphere altogether. And while that will merely detour and in no way stop the biosphere, which will go its merry way with or without us, it does spell catastrophe for the animal that not only transcends but represses.

INTERIORITY

If we look now at all of the diagrams presented in this chapter, we notice a startling fact. All of these diagrams claim to be holistic, to cover all of reality in an encompassing fashion. That is, if we take out the errors in their presentation (like confusing micro and macro), and put all of these diagrams together, the systems theorist would claim that the resultant "big picture" covers the whole of reality, from atoms to cells to animals, from stars to planets to Gaia, from villages to towns to planetary federations (all of which are on our diagrams). The noosphere transcends but embraces the biosphere, which transcends but embraces the physiosphere—one huge holistic and all-embracing system, stretching from here to eternity.

And yet, and yet. Something is terribly wrong. Or rather, terribly partial. All of these diagrams represent things that can be seen with the physical senses or their extensions (microscopes, telescopes). They are all, all of them, how the universe looks from the *outside*. They are all the *outward forms* of evolution, and not one of them represents how evolution looks from the *inside*, how the individual holons feel and perceive and cognize the world at various stages.

For example, take the progression: irritability, sensation, perception, impulse, image, symbol, concept. . . . We might believe that cells show protoplasmic irritability, that plants show rudimentary sensation, that reptiles show perception, paleomammals show images, primates show symbols, and humans show concepts. That may be true (and is true, I think), but the point is that *none* of those appear on any of our diagrams. Our diagrams (thus far) show only the outward forms of evolution, and none of the corresponding "interior prehensions" of the forms themselves (sensation, feelings, ideas, etc.).

So the diagrams themselves are not wrong (once we have revised a few errors), but they are terribly partial. They leave out the insides of the universe.

And there is a reason for this. The general systems sciences seek to be empirical, or based on sensory evidence (or its extensions). And thus they are interested in how cells are taken up into complex organisms, and how organisms are parts of ecological environments, and so on—all of which you can *see*, and thus all of which you can investigate empirically. And all of which is true enough.

But they are not interested in—because their empirical methods do not cover—how sensations are taken up into perceptions, and perceptions give way to impulses and emotions, and emotions break forth into images, and images expand to symbols. . . . The empirical systems sciences cover all of the outward forms of all that, and cover it very well; they simply miss, and leave out entirely, the *inside* of all of that.

Take, for example, the mind and the brain. Whatever else we may decide about the brain and the mind, this much seems certain: the brain looks something like figure 3-6 (or some anatomically correct figure), but my mind does not look like figure 3-6. I know my mind from the inside, where it seems to be seething with sensations and feelings and images and ideas. It looks nothing like figure 3-6, which is simply how my brain looks.

In other words, my mind is known interiorly "by acquaintance," but my brain is known exteriorly "by description" (William James, Bertrand Russell). That is why I can always to some degree see my own mind, but I can never see my own brain (without cutting open my skull and getting a mirror). I can know a dead person's brain by simply cutting open the skull and looking at it—but then I am *not* knowing or sharing that person's mind, am I? or how he felt and perceived and thought about the world.

The brain is the outside, the mind is the inside—and, as we will see, *a similar type of exterior/interior holds for every holon* in evolution. And the empirical systems sciences or ecological sciences, even though they claim to be holistic, in fact cover exactly and only one half of the Kosmos. And that is especially what is so partial about the web-of-life theories: they indeed see fields within fields within fields, but they are really only surfaces within surfaces within yet still other surfaces—they see only the exterior half of reality.

It is time now to look at the other half.

4

A View from Within

Things have their within. I am convinced that the two points of
view require to be brought into union, and that they soon will
unite in a kind of phenomenology or generalised physic in which
the internal aspect of things as well as the external aspect of the
world will be taken into account. Otherwise, so it seems to me, it
is impossible to cover the totality of the cosmic phenomenon by
one coherent explanation.
—PIERRE TEILHARD DE CHARDIN

ALONGSIDE THE without of things, the without of individual
and social holons, which we presented in the last chapter, we need
to present the within of things, the within of those same holons.

This part of our approach might seem at first a somewhat strange notion. How, for example, can we know the insides of a cell, the "interiority" of a cell? The answer, I believe, rests on the fact that cells are a *part*
of us, we *embrace* cells in our own compound individuality. That is, *nothing* in the preceding stages of evolution can be ultimately foreign to us
since they are all, in various degrees and ways, *in* us, as part of our very
being.

Thus, it is through an interior feeling of the shades of myself that I
might reasonably know the shades of other holons—which is how they
know me, too; for we are all ultimately in each other, in various degrees,
and right now. Gravity pulls at the minerals in my bones just as surely as
it pulls on the distant planets; hunger churns my belly as it does in every
starving wolf; the terror in the eyes of the gazelle being eaten by the lion

is not alien to me, or to you; and is that not joy in the song of the robin at the rising of the morning sun?

Nobody is more wary than I of the dangers of what Lovejoy called "retrotension"—reading "higher" thoughts and feelings into "lower" forms simply because we humans feel them—the anthropomorphic fallacy. But we are aided in this quest by precisely those evolutionary sciences that have already mapped the outward forms of the various holons at each stage of development, for we can reasonably make correlations between outward form and interior perception.

For example, it seems quite likely that, at this point in evolution anyway, intentional symbolic logic is found only where the outward form known as the complex neocortex has developed, and thus any other living holon on Earth whose exterior lacks the form of that neocortex—such as a plant, a lizard, a horse—will probably lack an interior that contains intentional symbolic logic. Likewise, animal holons that lack the exterior form known as the limbic system will probably also lack an interior that contains differentiated emotions, and so on. This puts a sharp curb to our "retrotensive tendencies," and helps us dig out, in the depths of our own feelings, just how far (and which of) those feelings extend into the depth of other beings.

Thus, we do indeed want to try to avoid retrotension, but this still leaves us in a far different place from the empiricists, who stare blankly at the rose and wonder how the epistemological gap shall ever be bridged, as if they were staring at an alien creature materialized from a wholly different dimension. They actually refer to it, with a puzzled expression, as "knowledge of the outer world." But I can know the outer world because the outer world is *already* in me, and I can know me. All knowledge of other is simply a different degree of self-knowledge, since self and other are of the same fabric, and speak softly to each other at any moment that one listens.

And, of course, this approach is not new. We find something like it from Aristotle to Spinoza, from Leibniz to Whitehead, from Aurobindo to Radhakrishnan. We don't find it in empiricism or positivism, and we don't find it in the "holistic" systems theories, which want ever so much to be empirical. That imbalance we wish to redress, drawing on, among others, the theorists I just mentioned.

Finally, we will want to look also at the higher stages of growth in the human holon, the higher stages of consciousness that, in mystical experi-

ence, for example, are said to issue forth in illuminations of the superconscious, illuminations of the very Divine (if such indeed exists). We will
want to look, that is, at the field known as transpersonal psychology. For
we are now entering the domain of our first Addition: the greater the
depth of evolution, the greater the degree of consciousness.

INTERIORITY AND CONSCIOUSNESS

Spinoza, Leibniz, Schopenhauer, Whitehead, Aurobindo, Schelling, and
Radhakrishnan are just a few of the major theorists who have explicitly
recognized that the within of things, the interiority of individual holons,
is in essence the same as *consciousness*, though of course they use different names with slightly different meanings.

Whitehead uses "prehension" to describe the contact and thus "feeling" of an object by any subject, no matter how "primitive," including
atoms (thus his famous statement, "Biology is the study of big organisms,
physics the study of little organisms"). Spinoza uses "cognition" for
knowing an event "from the inside" and "extension" (or matter) for
knowing the same event "from the outside." Leibniz uses "perception"
for the interior of his monads (holons) and "matter" for the exterior, with
the added proviso that only the interior is actually real and can be known
directly—matter (or extension) is only an appearance devoid of any substantial reality; it is simply what mind looks like from the outside. As for
Teilhard de Chardin, he put it very simply: "The within, consciousness,
spontaneity—three expressions for the same thing."[1]

I will not, at this point, get involved in the philosophical nuances of
those various positions, which are inextricably bound up with the problems of panpsychism and historical solutions to the mind-body problem
(we will return to this later). Rather, I will, for the time being, take a more
generalized position and simply say that, for me, the within of things is
consciousness, the without of things is *form*.

Or, as we put it earlier, the within of things is *depth*, the without is
surface. But all surfaces are surfaces of depth, which means, all forms are
forms of consciousness.

Further, I don't want to haggle over whether the very lowest holons are
totally or only mostly devoid of rudimentary forms of consciousness or
prehension. First, there is no lower limit to holons, so there is no rock

bottom to serve as a standard. Second, they are all forms *of* depth, so the actual amount of consciousness *in* them is a completely relative affair. Thus, whatever we take at present as the lowest or most primitive holons (quarks, for example), I will simply say that *they* have the least depth, the least consciousness, relatively speaking, and I will, with Whitehead, call that form "prehension." You are free to call the lowest levels "totally inert" if you wish, and pick up the argument from there.

Let me emphasize that it really does not matter, as far as I am concerned, how far down (or not) you wish to push consciousness. Whitehead, as we said, saw prehension as the irreducible "atom" of existence. Mahayana Buddhism maintains that literally all sentient beings possess Buddha Mind, and liberation involves a realization of that all-pervading consciousness. Lynn Margulis, the noted biologist, believes that cells possess consciousness. A handful of scientists think that plants show proto-sensation. Animal rights activists insist that most animal forms show rudimentary feelings. And I suppose most orthodox theorists don't really see consciousness emerging until primates and usually humans.

But my main point is not where precisely to draw this line—draw it wherever you feel comfortable—but that the line itself involves preeminently the distinction between interiority and exteriority.[2] We will be returning to this point throughout the book, and in the meantime, I will simply assume that consciousness is synonymous with depth, and depth goes all the way down, but progressively less and less and less . . . into those dark shades of night. As Teilhard eloquently put it, "Refracted rearwards along the course of evolution, consciousness displays itself qualitatively as a spectrum of shifting shades whose lower terms are lost in the night."[3]

Now for some simple correlations. Figure 4-1 lists some of the milestones in the evolution of the outward forms of individual holons, and alongside them some of the corresponding milestones in the emergent forms of consciousness that I suggest are correlated with them.

Each new interior holon, of course, transcends but includes its predecessor(s)—incorporates the essentials of what went before and then adds its own distinctive and emergent patterns[4] (as we will see in detail in a moment). And notice: these interior holons have nothing to do with size or spatial extension; a symbol is not bigger than an image, an image isn't bigger than an impulse—this is where the application of physicalist sciences becomes very distorting.

atoms	prehension
cells (genetic)	irritability
metabolic organisms	rudimentary sensation
(e.g., plants)	
proto-neuronal organisms	sensation
(e.g., Coelenterata)	
neuronal organisms	perception
(e.g., annelids)	
neural cord (fish/amphibians)	perception/impulse
brain stem (reptiles)	impulse/emotion
limbic system (paleomammals)	emotion/image
neocortex (primates)	symbols
complex neocortex (humans)	concepts

Figure 4-1. The Without and the Within

The important point, for now, is simply that each new and emergent interior holon transcends but includes, and thus *operates upon*, the information presented by its junior holons, and thus it fashions something *novel* in the ongoing cognitive or interior stream. Hence, each new growth in consciousness is not just the "discovery" of more of a pregiven world, but the co-creation of new worlds themselves, what Popper calls a "making and matching" of new epistemological domains, a discovery/ creation of higher and wider worlds.

There can be much discussion over the actual details of that list (fig. 4-1), the choice of words and the precise placement. But I think most people, even if they disagree with the details as given, would agree that *something* like that is indeed occurring. Greater depth, greater interiority, greater consciousness. Teilhard expressed this in his "law of complexity and consciousness"—namely, the more of the former, the more of the latter. Since, as we already saw, evolution tends in the direction of greater complexity, it amounts to the same thing to say that it tends in the direction of greater consciousness (again, depth = consciousness).

Most of the words in figure 4-1 are self-explanatory, but I should mention that an *image* is a mental construct that represents a thing by resemblance (the image of a dog "looks like" the actual dog); a *symbol* represents a thing by correspondence, not resemblance (the word *Fido* represents my dog but the word itself does not look like my dog at all—a more difficult cognitive task); and a *concept* represents an entire *class* of resemblance (the word *dog* represents the class of *all* dogs, an even more complicated cognitive task).

When a fox spots a rabbit on the other side of a fence and then runs hundreds of yards around the fence to get to the rabbit, presumably the fox carries an *image* or proto-image of the rabbit in its mind. Moving up, there is abundant evidence that apes and chimpanzees are capable of forming *symbols* (or at least paleosymbols) and indeed can be taught to recognize and use a simple denotative language. As far as we can tell, only humans create and consciously utilize fully formed *concepts* (or universals), and these concepts, among many other things, reach down and differentiate and color all previous levels in the human compound individual (a paleomammal may feel rage, but only humans conceptually elaborate that into anger and then hatred, a long slow burn maintained conceptually).

In other words, concepts transcend and include symbols, which transcend and include images (which transcend and include impulses, etc.)—and none of that has much to do with physical extension.

THE LIMITS OF THE EXTERIOR APPROACH

In the previous chapter I said that holistic systems theories leave out the interiority of the holons they describe. Let me be more precise. Some of the theories do, indeed, attempt to take into account the insides of the universe, for they at any rate mention things like feelings and symbols and ideas. But they then subject these interiors to the same analysis as they apply to the surfaces, because they are trying to be empirical, and this produces some very unpleasant results.

We can explain this with reference to figure 4-2. Here I have indicated the three general domains of evolution from the angle of a compound individual holon on level 3. The physiosphere is labeled A, the biosphere B, and the noosphere C, so that the holon on level 3 can be represented as A + B + C, transcending but including its predecessors.

Now the claim of the evolutionary systems theories (from Bertalanffy to Laszlo to Jantsch) is that, although *none* of the levels can be reduced to any other, general laws or regularities of dynamic patterns can be

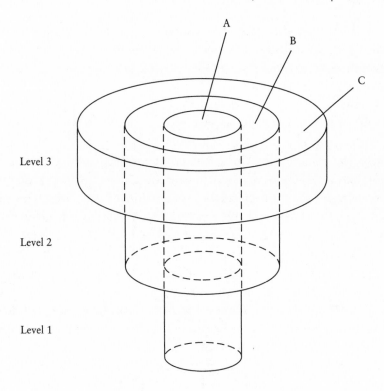

Figure 4-2. Levels of Evolution

found that are the *same* in all three realms. These are called "homolog laws" and not "analog laws," which means they are basically the same laws across domains.

I agree with that position (as far as it goes), and in chapter 2 we outlined twenty tenets, or "homolog laws," that are characteristic of holons *wherever* they appear. So far, so good.

But these tenets have to be of such a *general nature* that they will apply to all three general realms, and that means, in essence, that they apply basically to the realm labeled A, since A is the only thing that all three realms have in common. A runs from A through the core of B and the core of C, and thus what holds for A will hold for (but not totally cover) whatever happens on all three levels. Thus, on level 1, or A, we already find dissipative or self-organizing structures, holons with depth and span, creative emergence, increasing complexity, evolutionary development, differentiation, self-transcendence, teleological attractors, and so forth. When I presented the twenty tenets, I used examples from all three realms, but all of the tenets can be found, to some degree or another, in the physiosphere itself (thus acknowledging some degree of *continuity* in the overall evolutionary process—the "no gaps in nature" side of the equation). And all of that, too, is just fine.

But none of the twenty tenets—as crucially important as they are—describes specifically what is happening in B and C. To the extent that B and C are composed of holons—and they are—*they will follow all of the twenty tenets*. But holons with Life injected into them also do *other* things that A holons do not—such as sexual reproduction, metabolic communication, autopoietic self-preservation, and so forth; and holons with Mind injected into them do still other things that B holons do not—such as verbal communication, conceptual self-expression, artistic endeavors, and so on. The B and C qualities and functions and cognitions will follow the same tenets that also apply to A, but they will further follow other laws and patterns and actions not specifically derivable from the twenty tenets (which state, for example, that evolution gets more complex, but do not state that evolution will produce poetry). There is nothing in the twenty tenets that will tell us how to resolve an Oedipus complex, or why pride can be wounded, or what honor means, or whether life is worth living.

In other words, the twenty tenets—by which I mean dynamic systems theory in general—are the *most fundamental* tenets of *all* of development, and therefore the *least* interesting, least significant, least telling tenets when it comes to B and especially C (and super-especially anything higher). Systems theory—precisely in its claim and desire to cover *all* systems—necessarily covers the least common denominator, and thus nothing gets into systems theory that, to borrow a line from Swift, does not also cover the weakest noodle.

And the weakest noodles, the lowest holons, have the *least* depth, the *least* interiority, the *least* consciousness—so that a science of *that* is correspondingly a weakest noodle science. It is a science of surfaces. And this is invariably why, in such "systems books" as Laszlo's, one finds a very rich and impressive setting forth of something like the twenty tenets, and a rich application of them to the physiosphere and somewhat to the biosphere, but then the presentation, as it really explores the biosphere and then gets into the noosphere, becomes pale and anemic and thins out very quickly.

When it comes to social or historical evolution, for example, Laszlo correctly points out that noospheric evolution proceeds from less ordered to more ordered, from less complex to more complex, that it is irreversible, and a few other such items. That's about all he says, and that's about all he *can* say as a systems scientist. But we already know that; that's what happens to *any* complex system far from equilibrium. About the really interesting and utterly unique things that make history history and not just a dissipative structure, systems science can tell us precious little. Of the system A + B + C, it will tell us about the lowest common-denominator patterns, which, I repeat, are the most fundamental—and least significant—of patterns for the higher levels.

(Of course, many system theorists actually use the dynamics of B, or living systems, and make them paradigmatic for *all* levels, with equally unsatisfactory results but now in both directions: B underexplains C and overexplains or is retrotensive with A, trying, for example, to read autopoiesis into the physiosphere—unconvincingly. All of these difficulties come from the lowest common denominator approach and/or the attempt to privilege any single domain).[5]

This is why presentations that are otherwise often so brilliant, such as Jantsch's, will say things like, "In the same way, these mathematical relationships may be applied to the modelling of dynamics at other than the chemical levels"—by which he means (and explicitly states) that the system patterns holding for A are significantly inclusive for levels B and C. He then adds, "A certain imagination is required, of course. . . ."[6]

It takes more than imagination; it takes hallucination. This is exactly what Sheldrake is objecting to when he says, "In a similar way, to return to one of Prigogine's examples, a mathematical model of urbanization may shed light on the factors affecting the rate of urban growth, but it cannot account for the different architectural styles, cultures, and religions found in, say, Indian and Brazilian cities."[7] Cannot account, that is, for a single thing that is specifically cultural.

Michael Murphy makes the same point in a strong fashion. He first points out the similarities (or continuities) in the three major domains: "Though the kinds of development that occur in the physical, biological, and psychosocial domains are shaped by different processes and have different patterns, they proceed in sustained, irreversible sequences that are called evolutionary—these three domains have many features in common." These common features I have distilled in the twenty tenets. Yet Murphy then emphasizes, "But the three kinds of evolution—inorganic, biological, and psychosocial—though having many features in common, operate according to separate principles. Helpful scientific reductions do not erase the fact that these [domains] proceed according to their own distinctive patterns."[8]

So the twenty tenets are the backbone of our system, holding true for holons anywhere, or so I maintain. But for the meat and flesh and feelings and perceptions—for these we will have to look elsewhere, look instead to an empathic feel from within of those degrees of the All that are degrees of us as well. And we do not honor the richness of these feelings—in us, in others—by reducing or narrowing them to the lowest common denominator. The robin, the deer, and the amoeba, I presume, are as insulted by this as I. Our answer, as always, is never to be found in flatland, in the world of black checkers scurrying endlessly, meaninglessly, dimly, and disappearing finally into those dark shades of the night that are ever so fundamental, ever so insignificant.

THE EVOLUTION OF THE WITHIN OF HUMAN HOLONS

We have sketched the evolution of the without of individual holons up to the complex triune brain (embracing or enfolding all previous exteriors) and the correlative evolution of the interior of the same holons up to that of concepts (embracing or enfolding all previous interiors). It is time now to pick up the story with the emergence of the first human animals—with a complex triune brain producing concepts or proto-concepts and living in a social holon of the group/family—and briefly follow the succeeding evolution up to the present day.

Once again, most of our work has already been done for us, and we have two major sources to draw on in this regard. One is my own previous work in this area, published in *Up from Eden* and *Eye to Eye*. The other is the work of Jürgen Habermas, whom many (myself included) consider the world's foremost living philosopher and social theorist. The conclusions of these two sources are in strong agreement, even though they were arrived at independently and from very different angles. When I was working on *Up from Eden*, I was only beginning to come into contact with Habermas, and so I unfortunately did not have the chance to include his incisive commentaries. I drew instead on the works of such pioneers as Jean Gebser, Erich Neumann, L. L. Whyte, Georg Hegel, and

Joseph Campbell. In a sense this was fortunate; the fact that Habermas, working from a very different direction, arrived at the same general conclusions lends added strength to the overall thesis.

What I would like to do, then, in this and the next chapter, is very briefly outline Habermas's observations on the evolution of human consciousness and social communication (and parenthetically indicate the points of agreement with my own work; those familiar with *Eden* and *Eye to Eye* will immediately spot the similarities). Habermas's overall views on communication and the evolution of society are set in the context of his theory of communicative action (or action geared toward *mutual understanding* as an omega point); I will be cutting many corners and giving only the briefest skeleton of a small part of his rich theory; readers are urged to consult his original works.[9]

In *Up from Eden* (and *Eye to Eye*) I followed the groundbreaking work of Jean Gebser in recognizing four major epochs of human evolution, each anchored by a particular structure (or level) of *individual* consciousness that correspondingly produced (and was produced by) a particular *social worldview*. These general stages Gebser called the archaic, the magic, the mythic, and the mental.

Habermas, we will see, is in general agreement with this conclusion.

I (and to a lesser extent Gebser) further suggested that each of these structures of consciousness generated a different sense of space-time, law and morality, cognitive style, self-identity, mode of technology (or productive forces), drives or motivation, types of personal pathology (and defenses), types of social oppression/repression, degrees of death-seizure and death-denial, and types of religious experience.

In this chapter, and especially in the next, we will carefully examine the archaic, the magic, the mythic, and the mental worldviews (and suggest possible higher developments as well).

But first, let me be very clear about what is being attempted here. I have suggested that there are individual and social holons, each of which has an interior and an exterior. Thus, in evolution in general, and human evolution in particular, we are tracing *four different strands*, each of which is intimately related and indeed dependent upon all the others, but none of which can be reduced to the others.

The four strands are the interior and the exterior of the individual and the social, or the inside and the outside of the micro and the macro.

We have already looked at (1) the development of the *exterior* forms of *individual* holons, ranging from atoms to molecules to cells to organisms to neural organisms to triune-brained neural organisms. And we have already looked at (2) the development of the *exterior* forms of the *social* holon, from superclusters to galaxies to planets to Gaia to ecosystems to groups/families (and briefly suggested its continuation into villages, nation-states, and planetary systems). Further, we have already suggested (3) the *interior* development of the *individual* holon, from prehension to sensation to impulse to image to symbol to concept (and briefly suggested its continuation into concrete and formal operational thought, with higher stages yet to come). And finally, we have just suggested that, in human evolution anyway, (4) the *interior* development of the *social* holon evidences itself in a series of *shared worldviews* (from magic to mythic to mental, and possibly higher). (We'll go over all these correlations in a moment.)

The lower levels possess a "worldview"—by which I mean, a "common worldspace"—to precisely the same degree that you believe they possess a degree of consciousness or prehension. If depth is consciousness, which I believe it is, and if lower holons possess depth, which I believe they do, and if any holon exists only in a system of relational exchange with other same-level holons, which it does, then any holon possesses a shared depth with its peers, and that is a "worldview" or "common worldspace" in the broadest sense. We have already agreed, for example, that if a holon possesses a reptilian brain stem, then it possesses an interiority of impulse, and we have already agreed that that interiority cannot be captured without remainder in objective it-language, and thus it must possess a subjective (or proto-subjective) space, a space *shared* with similar-depth holons. It is not just shared surfaces, but shared depths, or a common *worldspace*.

Or again: if holons share common exteriors, which they do, then they share common interiors (or worldspaces). How far down we push this is a matter of reasonable disagreement, but confining it to humans alone is preposterous.

Nonetheless, as before, feel free to pick up the argument at the point in evolution where you feel that some form of rudimentary consciousness or prehension enters the scene; presumably, by the time we reach humans, we can all agree that shared worldviews exist, and these shared worldviews are simply the inside feel of a social holon, the inside space of collective awareness at a particular level of development; it is not just how "I" feel, it is how "we" feel.

THE FOUR QUADRANTS

We will be going into these distinctions in much detail as we proceed. For the moment, a quick summary is given in figure 4-3. (It might also help to glance at the more detailed diagram given in the next chapter, fig. 5-1, page 198.) The upper half of the diagram represents individual holons; the lower half, social or communal holons. The right half represents the exterior forms of holons—what they look like from the outside; and the left half represents the interiors—what they look like from within.

Thus we have four major quadrants or four major aspects to each and every holon. The Upper-Right (UR) is the *exterior* form or structure of an *individual* holon. This quadrant runs from the center—which is simply the Big Bang—to subatomic particles to atoms to molecules to cells to neural organisms to triune-brained organisms. With reference to human

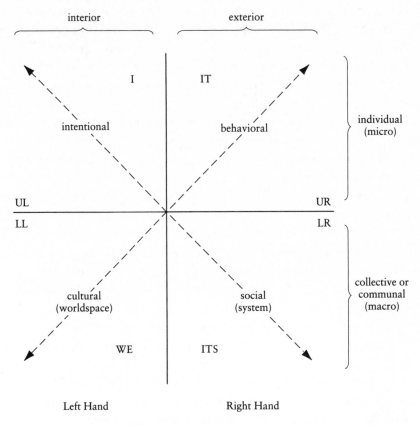

Figure 4-3. The Four Quadrants

beings, this quadrant is the one emphasized by *behaviorism*. Behavior can be *seen*, it is empirical—which is precisely why empirical science is always concerned only with the *behavior* of holons (the behavior of atoms, the behavior of gases, the behavior of fish, the behavior of humans) and wants nothing to do with nasty ol' introspection, which involves, of course, the interiors of individuals.

Which would be the Upper-Left (UL) quadrant. This quadrant—the *interior* form of an *individual* holon—runs from the center to prehension, sensation, impulse, image, symbol, concept (and so on). These interiors (UL) are correlated, we saw, with specific exteriors (UR), so that emotions "go with" limbic systems and concepts "go with" the neocortex of complex triune brains, and so forth (that is, every point on the right side has a correlate on the left side: every exterior has an interior). With reference to human beings, this quadrant contains all the "interior" individual sciences (among other things), from psychoanalysis to phenomenology to mathematics (nobody ever saw the square root of a negative one running around in the external world; that is apprehended only interiorly).

But individuals only exist in relational exchanges with other holons of similar depth (micro and macro, individual and social). In other words, every point on the upper half of the diagram has a corresponding point on the lower half (so that all four quadrants have corresponding points with each other). Taking the two lower quadrants one at a time:

The Lower-Right (LR) quadrant runs, as we saw, from the Big Bang to superclusters to galaxies to stars to planets to (on Earth) the Gaia system to ecosystems to societies with division of labor to groups/families (each getting "smaller" owing to less simple span). With reference to humans, this quadrant then runs from kinship tribes to villages to nation-states to global world-system (getting "bigger" on its own level). But this quadrant also refers to any of the concrete, material, embedded social forms of communities (the exterior forms of social systems), including modes of tools and technology, architectural styles, forces of production, concrete institutions, even written (material) forms, and so on.

The Lower-Right quadrant, in other words, represents all the *exterior* forms of *social* systems, forms that also can be *seen*, forms that are empirical and behavioral (everything on the Right half of the diagram is empirical, because it involves the exterior forms of holons; in this case, the social holon). This is why the study of human "sociology" (especially in Anglo-Saxon countries) has usually been the study of the observable *behavior* of social systems (or "social action systems"). Something is a "really real" science if its data can be seen empirically, and since all social holons do have an exterior form that can be seen empirically, sociology has all too often confined its studies to this one component (the LR quadrant) and been very distrustful of the study of anything other than monological, observable variables in a social action system.

This is why it has been so hard for sociologists to buck the positivistic trend of studying only behavior-oriented action systems, and to study not just society but also *culture*, or the *shared values* that constitute the common *worldviews* of various social systems—that is, the *interiors* of the social systems, the Lower Left (LL) quadrant. Thus, for example, a recent anthology entitled *Cultural Analysis* (as opposed to "social analysis") could find only four major theorists working this side of the street: Peter Berger, Mary Douglas, Michel Foucault, and Jürgen Habermas (we could certainly add Charles Taylor and Clifford Geertz, among a few others; the influence of these six theorists can be heavily felt throughout this presentation).

But the editors' point is well taken: "While theories, methods, and research investigations in other areas of the social sciences have accumulated at an impressive pace over the past several decades, the study of culture appears to have made little headway." The reason is that the orthodox and positivistically oriented researchers "turned from the ephemeral realm of attitudes and feelings . . . intersubjective realities . . . beliefs and values—the stuff of which culture is comprised—to the more obdurate [empirical] facts of social life—income inequality, unemployment, fertility rates, group dynamics, crime, and the like. On the whole, it may be only slightly presumptuous to suggest that the social sciences are in danger of abandoning culture entirely as a field of enquiry."[10] Abandoning, that is, the Lower Left in favor of the Lower Right.

At the same time, the editors note, these four theorists are nonetheless at the forefront of a revolution in the approach to the study of culture. However "ephemeral" aspects of culture might be, various phenomenological and analytical and structural tools can nonetheless be brought to bear on the issue. To say something is "subjective" is not to say it doesn't exist or can't be carefully studied. Each of these theorists, the editors point out, "have attempted to identify *systematic patterns among the elements of culture itself*, or patterns within culture"—as I would put it, cultural holons, structures or patterns from within, not just without.

"Largely outside the mainstream of social sciences, these approaches have been oriented primarily toward the realms of meaning, symbolism, language, and discourse. Each is rooted in deeper philosophical traditions themselves quite distinct and in significant ways alien to the so-called 'positivist' tradition of contemporary social science. The first, and perhaps most familiar of these, is phenomenology [Berger]; the second, cultural anthropology [Douglas]; the third, structuralism [Foucault]; and the fourth, critical theory [Habermas]."[11] These approaches have already had a profound impact and are helping to reverse the positivistic trend where, as the editors put it, "nuts and bolts have replaced hearts and minds."

But this distinction between *social* (in the sense of social action system, the empirical nuts and bolts) and *cultural* (meaning shared worldviews and values, hearts and minds) is very much the distinction between the exterior and the interior of a social (or communal) holon. These two dimensions are in intimate interaction and correlation, but neither can simply, without further ado, be reduced to the other (and thus not explained, but explained away).

Notice that, because we have only so many words to go around, I am forced to use *social* in two senses: in the *narrow* sense (as the social system or exterior observable patterns in a society [LR], contrasted to "cultural," or the interior values and meanings that cannot be captured empirically [LL]), and *social* in the *broad* or general sense (where it means both social, in the narrow sense, and cultural; where it means the entire lower half of the diagram). I trust that in the following discussion the context will make clear which meaning is intended. By and large, however, from now on I will use "cultural" to mean the Lower Left and will reserve "social" for the Lower Right.

As for the difference between social and cultural, here's a simple example. Imagine you go to a foreign country where you do not speak the language. As soon as you arrive in that country, you are *in* the social system, or the actual material components of the country. You are *in* that country. People around you are speaking a foreign language, which you do not understand, but the physically spoken words hit your ears just like they do anybody else's; you and the natives are both immersed in the *identical* physical vibrations of the social system.

But you don't understand a word. You are in the social system, but you are *not* in the worldview, you are not in the culture. You hear only the exteriors, you do not understand the interior meaning. All the social signifiers impinge on you, but none of the cultural signifieds come up. You are an insider to the social system but an outsider to the culture.

The study of the interior cultural meanings cannot be reduced to a study of exterior action systems (even though they have various correlations), or else you could know everything there is to know about a community without ever learning the language: you would just report the "behavior" of the inhabitants (like the behavior of gas particles), and to hell with that tricky "meaning" stuff. (Foucault, in his reductionistic archaeology phase, actually proceeded in just this manner, bracketing *both* the truth *and* the meaning of linguistic statements, and just reported their systematic behavior, which caused quite an uproar. He later recanted and labeled the *exclusive* use of that approach "arrogant."[12] But it is an arrogance that all such "happy positivism" is prone to.) Every holon has a component in the Lower-Right quadrant; it's just not the whole story.

The Lower-Left quadrant is the study of shared interior meanings that constitute the worldview (or common worldspace) of collective or communal holons. With regard to humans, we have seen that these run from archaic to magic to mythic to mental (with all sorts of variations, and holding open the possibility of still-further developments).

And as for the worldspace of lower holons, I simply mean a shared space of what they *can* respond to: quarks do not respond to *all* stimuli in the environment, because they *register* a very narrow range of what will have meaning to them, what will *affect* them. Quarks (and all holons) respond only to that which fits their worldspace: everything else is a foreign language, and they are outsiders. The study of what holons *can* respond to is the study of shared worldspaces.[13]

THE RIGHT- AND LEFT-HAND PATHS

There are an enormous number of correlations between this scheme of the four quadrants and the work of other theorists, and we will be exploring these in more detail as we proceed. For the moment, just a few preliminary notes.

The entire Right half of figure 4-3, the exterior half, can be described in "it" language (or "object" language) and can be studied *empirically* (in behavioristic or positivistic or monological terms). The entire Right half, as we said, is something you can *see* "out there," something you can register with the senses or their extensions (telescopes, microscopes, photographic plates, etc., and you tie your theorizing to these empirical observables, which is what "monological" or "empiric-analytic" means).[14] The components of the Right half—both the Upper-Right and Lower-Right quadrants—are, in themselves, neutral surfaces, neutral exteriors, neutral forms, all of which can fairly be described in "it" language. You don't ever have to engage the interiors of any of those holons: you don't have to engage in introspection or interpretation or meaning or values. You just describe the exterior form and its behavior. Nothing is better or worse, good or bad, desirable or undesirable, good or evil, noble or debased. The surface forms simply are, and you simply observe and describe them.

And you *can* do that; the surfaces forms *are* there; it is a *legitimate* and altogether *necessary* story. It is just not the whole story.

I will often refer to the study of both the Upper-Right and Lower-Right quadrants as the Right-Hand path: the path of that which can be seen with the eye of flesh or its extensions. In short, the path of "it"-language (objectivist, monological, observable, empirical, behavioral variables).

The Right-Hand path has two major and warring camps: the atomists, who study the surfaces of only individuals, and the wholists, who insist that whole systems, and not individuals, are the primary object of study. But both are equally exterior approaches confined to surfaces only. They are the two camps of flatland ontology: that which can be seen, detected with the senses, empirical through and through.

The entire Left half of the diagram, on the other hand, *cannot* be seen with the eye of flesh (except as those Left-side aspects become embedded in material or exterior forms, which we will discuss later). In other words, the Left half cannot be described in "it" language. Rather, the Upper-Left is described in "I"-language, and the Lower-Left is described in "we"-language (as I will explain in a moment). I will refer to both of them as the Left-Hand path.

Whereas the Right half can be *seen*, the Left half must be *interpreted*. The reason for this is that *surfaces can be seen*—there they are, anybody can look at them; but depth cannot be directly perceived—*depth must be interpreted*. The Right-Hand path always asks, "What does it *do*?" or "How does it *work*?" The Left-Hand path asks, "What does it *mean*?"

A deer sees me approach. It *sees* my exterior form, my shape, and registers all the appropriate physical stimuli coming from my form to the deer.

But what do they *mean*? Am I the friendly fellow with the food, or the hunter with the rifle? The deer must *interpret* its stimuli in the *context* of its own *worldspace* and how I might *affect* it. And this is not just a matter of seeing: the deer sees just fine. But it might be *mistaken* in its *interpretation*; I might actually have the rifle, not the food. All the physical stimuli are hitting the deer fully (that's not the problem); the problem is, what do they actually *mean*? The surfaces are *given*, but what is lurking in the depths? What are the intentions lying behind the surfaces? What is transmitted empirically but not merely given empirically?

Almost from its inception, and down to today, social theory has divided into two often sharply disagreeing camps: hermeneutics and structural-functionalism (or systems theory). Hermeneutics (the art and science of interpretation) attempted to reconstruct and empathically enter the shared cultural worldspace of human beings, and thus bring forth an *understanding* of the values contained therein. Structural-functionalism, on the other hand, dispensed with meaning (in any participatory sense) and looked instead at the external social structures and social systems that governed the behavior of the action system.

Both were holistic, in the sense that both situated individual existence in a larger network of communal practices and insisted that the individual could not be understood without reference to the holistic background of shared practices. But they were, almost exactly, representatives of the Left-Hand and Right-Hand paths, with hermeneutics asking always, "What does it *mean*?" and structural-functionalism asking instead, "What does it *do*?"

To reconstruct meaning (the Left-Hand path) I must engage in *interpretation* (hermeneutics); I must try to enter the shared depths, shared values, shared worldviews of the inhabitants; I must try to understand and describe the culture *from within* (while maintaining a delicate distance so as to be able to report at all). I cannot simply *see* meaning; meaning does not sit on the surface waiting, like a patch of color, to hit my senses. Rather, to the extent that I can, I must *resonate* with the interior depth of the inhabitants. The depth in me ("lived experience") must empathically align itself, intuitively feel into, the corresponding depth (or lived experience) that I seek to understand in others, and not simply blankly register an empirical patch. *Mutual understanding* is a type of interior harmonic resonance of depth: "I know what you mean!"

What, for example, does the Hopi Rain Dance *mean*? I might find, as many "participant observers" have, that it means, that it expresses, a connection with nature felt to be sacred, so that the dancing is both an expression *of* sacred nature and a request directed *toward* that same nature. To understand and articulate this in a sociological fashion (and not simply become a member of the tribe), I must look into the whole network of shared social practices and the "background unconscious" of linguistically (and prelinguistically) structured meanings and intersubjective exchanges that constitute the "pre-understanding" or "background" or "foreknowledge" of the particular worldspace or worldview—all of the contexts within contexts within contexts that structure the interior values and meanings of a culture, some of them explicit (requiring understanding), and some of them implicit (requiring excavation). (That is a short summary of the hermeneutic program from Wilhelm Dilthey and Weber and Heidegger down to today with Paul Ricoeur and Hans Gadamer and Geertz and Taylor.)

On the other hand, to reconstruct *function* (the Right-Hand path), I must carefully, and in largely *detached* fashion, observe not what the natives *say* they are doing, but what *function* the dance actually serves in the overall social action system (a function unknown to the natives). The detached observer concludes (as Talcott Parsons did) that the dance performs the function of securing social solidarity and social cohesion. Whether it is really going to make rain or not, or trying to make rain or not, is quite secondary to our concerns; because what it is really doing is providing an occasion that binds individuals together into the social fabric of the tribe (i.e., the self-organizing and autopoietic regime of the social action system). In order to determine this, the tribe is viewed as a holistic system, with its overall pattern (structure) and overall function (behavior) carefully observed, and then the "meaning" of any individual event is simply its place (or function) in the overall system. (And that is a summary of the structural/functional and "system theory" program in various forms, from Comte to Parsons to Niklas Luhmann, and even in its structuralistic and archaeological variants, Claude Lévi-Strauss to early Foucault).[15]

The one path asking always, "What does it *do*?" and thus seeking to offer *explanations* based on naturalistic, empirical, observable variables; the other path asking always, "What does it *mean*?" and thus attempting to arrive at *understanding* that would be mutual.[16]

Surfaces extend; interiors intend—it's still almost as simple as that.[17]

SUBTLE REDUCTIONISM

There are important truths in both the Left- and Right-Hand approaches, and both are required for a balanced or "all-quadrant" view. My position is that every holon has (at least) these four aspects or four dimensions (or four "quadrants") of its existence, and thus it can (and should) be studied in its intentional, behavioral, cultural, and social settings. No holon simply exists *in* one of the four quadrants; each holon *has* four quadrants.

Of course, the more primitive the holon, the less we seem to care about its intrinsic value or its intentions or its culture, which is sad. The point now is that, regardless of what we think of lower holons, when it comes at least to human beings, none of the four quadrants can be privileged. (If you don't believe that consciousness or prehension in any form extends below humans, a study of the four quadrants will at least show you in which directions and in which ways interiors would extend if you thought otherwise, which itself can be a very useful exercise.)

Which brings us to the subject of gross reductionism versus subtle reductionism. Gross reductionists, first of all, do not believe any interiors exist anywhere, so the issues of meaning, value, consciousness, depth, culture, and intentionality—these never come up for them; and, in fact, they hope to die with the boast on their lips that they never saw a value they couldn't reduce to atoms.

Gross reductionism *first* reduces *all* quadrants to the Upper-Right quadrant, and *second*—this is the gross part—then reduces all the higher-order structures of the Upper-Right quadrant to atomic or subatomic particles. The result is purely materialistic, mechanistic (usually), atomistic (always). Historically, only a very few, highly gifted nuts have actually taken this path with any sort of brilliance, starting with the Epicureans and running to Holbach, La Mettrie, and the like (what Polanyi referred to as "prejudice backed by genius" and Lovejoy summarized as, "There is no human stupidity that has not found its champion").

Opposed to these *flatland atomists* are the *flatland holists*. They do not reduce all holons to atoms. Rather, they reduce everything on the Left half of the diagram to a corresponding reality on the Right half. That is, they are the systems theorists and the structural/functionalists, in all their forms, from General System Theory to modern dynamic systems theory to many of the "new paradigm" and "ecological/holistic" theories.[18]

As I said, in many cases their hearts are in the right place, but their theories—because they are empirical and monological, because they are weakest-noodle sciences, because they deal with exteriors than can be seen and not interiors that must be arduously interpreted—because of all that, they end up with a rather insidious form of reductionism, insidious because they are almost completely unaware of what they have done. They claim to embrace the whole of reality, when in fact they have just devastated half of it.

(As only one example now, we can look at Habermas's withering attack on systems theory, and his demonstration that, although systems theory has an important but limited place, it is now, by virtue of its reductionism, one of the great modern enemies of the lifeworld—what he refers to as "the colonization of the lifeworld by the imperatives of functional systems that externalize their costs on the other. . . . a blind compulsion to system maintenance and system expansion."[19] And what is Foucault's biopower—the "modern danger"—if not, in large part, the systems/instrumental mentality biologized and applied to humans, converting each and all to strands and means in the great interlocking bio-web—the chief form of coercive power in the modern world?)

What these systems and "holistic" theorists don't seem to understand is that while they have indeed avoided *gross reductionism*—and for that are to be highly praised—they are nonetheless (and apparently unknown to them) the exemplars of *subtle reductionism*. They don't reduce everything to atoms; they reduce everything in the Left-Hand to a Right-Hand description in the "system." They reduce a four-quadrant holism (or Kosmic holism) to merely a Right-Hand holism, a flatland holism. And they are so understandably proud that they are holists, they overlook the flatland part.

Gross reductionism still rears its peculiar head every now and then, usually when somebody discovers a recursive law (a simple procedure that, when repeated, generates complex procedures), but by and large it has *not* been historically that influential, and certainly not as influential as many new-paradigm critics seem to think.

The truly devastating contributor to modern flatland ontology, to an erasure of the Kosmos, has been the Right-Hand path of systems theory or structural/functionalism in its many forms. The systems theorists like to claim that the reductionistic villains are the atomists, and that in emphasizing the wholistic nature of systems within systems, they themselves have overcome reductionism, and that they are therefore in a position to help "heal the planet." Whereas all they have actually done is use a subtle reduction to overcome a gross one.

These flatland holists claim, for example, that the great "negative legacy" of the Enlightenment was its atomistic and divisive ontology. But atomism was *not* the dominant theme of the Enlightenment. As we will see in great detail (in chapters 12 and 13)—and as virtually every historian of the period has made abundantly clear—the dominant theme of the Enlightenment was the "harmony of an interlocking order of being," a systems harmony that was behind everything from Adam Smith's great "invisible hand" to John Locke's "interlocking orders" to the Reformers' and the Deists' "vast harmonious whole of mutually interrelated beings."

To give only a few examples now, Charles Taylor represents the virtually uncontested conclusion of scholars that "for the mainstream of the Enlightenment, nature as the whole interlocking system of objective reality, in which all beings, including man, had a natural mode of existence which dovetailed with that of all others, provided the basic model, the blueprint for happiness and hence good. The Enlightenment developed a model of nature, including human nature, as a harmonious whole whose parts meshed perfectly," and the "unity of the order was seen as an interlocking set calling for actions which formed a harmonious whole."[20] As Alexander Pope would have it, speaking for an entire generation: "Such is the World's great harmony, that springs from Order, Union, full Consent of things; Where small and great, where weak and mighty, made to serve [each other], not suffer; strengthen, not invade; Parts relate to Whole; All served, all serving; nothing stands alone."[21]

Already the *Encyclopédie*, bastion of Enlightenment thought, had announced that "everything in nature is linked together," and Lovejoy points out that "they were wont to discourse with eloquence on the perfection of the Universal System as a whole."[22]

No, the downside of the Enlightenment was that it took a Kosmos of *both* Left and Right dimensions and reduced it to a cosmos that could be empirically (or monologically) described: it *collapsed the Left half to its correlates on the Right half.* Its great crime was not gross reductionism but *subtle reductionism.* The Great Holarchy of Being was collapsed into a "harmonious whole of interlocking orders," as John Locke put it, but orders that now had no within, no interiors, no qualitative distinctions, but instead could be approached through an objectifying and empiricist gaze (all being equal strands in the flatland web).

In short, the Left was reduced to the Right, and thus *interiors* tended to get lost and flattened into mere *exteriors* (and we will see more precisely what that entailed as we proceed). This indeed ended up being a *divisive* and *dualistic* ontology, precisely because, in describing the whole of reality in *objective* terms as a harmonious interlocking order or system (the great interlocking web), it left no room for the *subject* that was doing the describing.[23] Interpretation, consciousness, and interior depth were converted (reduced) to exterior, objective, systems interaction—"I" and "we" were reduced to holistic "its"—the precise essence of subtle reductionism.

In short, Nature was a harmonious whole known by a subject that could not fit into it. And once this *holistic/instrumental* or "just-parts-of-the-whole" approach was applied to the subject itself—what Foucault called "knowledge closing in on itself"—the subject inadvertently began to erase itself (and not in the sense of a genuine transcendence, but in the sense of shooting itself in the foot, or rather, in the head). Precisely because the subject's worldview was *empirically completely holistic,* the subject aced itself out of the picture altogether, and hovered above the holistic world, dangling and disengaged, now helplessly, now aggressively (thus the famous, or infamous, self-defining and disengaged subject cut off from the holistic/instrumental world, the "hyper-autonomous self" that so defined the Enlightenment, as we will see later in more detail).

In other words, the profound dualism inherent in flatland holism (namely, the isolated subject confronting an objectively holistic world), although hidden (or not immediately obvious), was nonetheless already at work, and contributed directly to the Enlightenment's "dehumanizing humanism" that has been so severely criticized, from Foucault's "Age of Man" to Habermas's philosophy of the subject to Taylor's disengaged subject of instrumentally interlocking orders. They all tell a similar tale of the objectifying holistic/instrumental mode of knowing turned back onto the subject, thus denying or destroying the very subjectivity it sought to understand.

As we will see later in this chapter (and again in chapter 12), the flatland holistic paradigm of the Enlightenment collapsed a Kosmos that was both interiorly and exteriorly holarchic (and thus included both Left and Right), into a flatland cosmos that was only exteriorly holarchic (or Right-Hand only): a flatland web that replaced interior depth with the great universal system of interlocking empirical surfaces, and thus reduced all interiors to exteriorly perceived strands in the great functional web. This "holistic" paradigm sounded like it covered "all" of reality (as against the atomists), but in fact it had just violently torn the Kosmos in half, tossed out depth and interiority, and settled aggressively on exteriors, surfaces, and the great interlocking web—interwoven "I" and "we" reduced to interwoven "its," which ironically left both "I" and "we" isolated, stranded, alienated, and alone, since none could fit into the harmonious web of holistic its.

To say that the subject is "an inseparable part of the great holistic web"—a web described in objective, process, dynamical, systems, third-person, it-language—is to destroy the subject in its own terms and devastate its authentic dimensions, and this, if anything, was the "crime of the Enlightenment."

A TOUR OF THE FOUR QUADRANTS

Once this flatland reduction occurred (which we will examine in great detail in chapters 12 and 13)—the Left-Hand aspects of reality were all reduced to their *corresponding* aspects on the Right Hand. Precisely because there *are* correlations between all four quadrants, this type of reductionism can be aggressively carried out. This is what makes subtle reductionism so hard to spot and so hard to redress.

For example, I have a thought; a thought occurs to me. That's the given holon, which we will use as an example. For this holon, in the Upper-Right quadrant, there is a change in brain physiology, a change that can be described in completely objective terms (it-language): there was a release of norepinephrine between the neural synapses in the frontal cortex, accompanied by high-amplitude beta waves . . . and so on. All of which is true enough, and all of which is very important.

But that is not how I experienced the thought, and I will never actually *experience* my thought in those terms. Instead, the thought had an interesting and important meaning to me, which I may or may not share with you. And even if you know what every single atom of my brain is doing, you will never know the actual details of my thought *unless I tell you.* That is the Upper-Left quadrant or aspect of this holon, this thought that occurred to me (and that is one of the many reasons why the Upper Left can never be reduced without remainder to the Upper Right; strong and general correlations and interactions, yes; detailed reduction, no).

Parenthetically, this is why the brain, even though it is "inside" my organism, is still only the *exterior* of my being—the brain is still Upper-Right quadrant, still *exterior*.[24] I can surgically cut open a human body and look at all the "insides," even down to the tissues and cells and molecules, but those are not the *within* or the *interior*, they are merely more surfaces, more exteriors that now can be *seen*. They are inside surfaces, not real interiors (which is why they are all listed on the Upper-Right quadrant; they are all the aspects of holons that can be empirically registered).[25]

But the crucial point is that you can look at the "insides" all you want and you will never see an *interior*, because whereas surfaces or exteriors can be seen, interiors must be *interpreted*. If I want to know what your brain looks like *from within*, what its actual lived *interior* is like (in other words, your *mind*), then *I must talk to you*. There is absolutely no other way. "What do you think? What's up? How do you feel?" And as we talk, I will have to *interpret* what you say, because all depth requires interpretation (whereas the brain physiologist doesn't have to talk to you at all; in fact, if a brain physiologist has got his hands on you, you are probably dead, which severely limits conversation; the brain physiologist can know all about your brain, and never know a single thought you had).

This is why the surfaces are all referred to as "monological"—they can be examined with a monologue, not a dialogue (which is why depth is dialogical or dialectical, or empathic in the broadest sense). You can study a bunch of moving rocks monologically; you can study brain physiology monologically; you can study suicide rates monologically. But you can only study interiors empathically, as a feel from within, and that means interpretation: both you and I might be mistaken as we try to assess each other.

In the Right-Hand path, all I basically have to do is *look*. I will need certain conceptual and material tools, it is true, but finally and fundamentally I just look: I look at a star with a telescope; I look at a cell with a microscope; I look at particle trails with a cloud chamber; I look at the behavior of rats in a maze; I look at a social system with empirical data and statistics. In each of these cases, I don't have to talk to the object of investigation. I don't try to reach mutual understanding. I don't try to interpret the depths or the interiors. I simply look at the surfaces, describe the exteriors, describe what I see. It can all be done in a monologue; it never needs a dialogue. And it's not that one is right and the other wrong; it's that both are extremely important.

This emphasizes the importance of the notion of *sincerity* in investigating the Upper-Left quadrant. When it comes to the developed forms of depth in humans, I only have access to that depth via interpreting what you tell me in a dialogue (and in body language or some such communication). *And you might be lying.* Further, you might be lying to yourself—which is where the numerous "sciences of depth," such as psychoanalysis, enter the picture. You and I can not only translate what we perceive, we can mistranslate. And all the various forms of "the hermeneutics of suspicion" (as Ricoeur called them) seek to dig beneath the stated surface content and assess the genuine or actual meaning of a communication: not what's sitting on the surface, but what hides in the depths.

The therapist, for example, based on formal training and on empathic (or sympathetic) intuition, might conclude that "the feeling of depression that you are experiencing is actually the disguised feelings of rage at your father for abandoning you." And if this makes sense to you, if you go, "Aha! that's exactly it!," then you have discovered a depth in you that you did not know was there, and that you consequently had been mistranslating (or misinterpreting) all those years. You and the therapist will then begin to retranslate those mistranslated feelings, correctly labeling them in the context of your own development: you are not "sad," you are actually "mad."

In this sense (which we will explore more later), my "unconscious," my "shadow," is the sum total of my past mistranslations carried into the present, where it distorts perception now ("transference"). My feelings of "sadness" were actually *insincere*; I was lying to myself in order to hide the worse pain of rage at a loved one; I deliberately (but "unconsciously") misinterpreted my feelings in order to protect myself ("defense mechanisms"). My shadow is the locus of my insincerity, my misinterpretation of my own depth (and this indeed is what all forms of "depth psychology," Freudian to Jungian to Gestalt, have in common).

Further, I will also mistranslate or misinterpret the depth in others ("What do you mean by that!") because I have mistranslated it in myself. And thus I have to reread the text of my own feelings, locate the source of my insincerity, and reinterpret my own depth more faithfully, with the help, usually, of somebody who has seen the mistranslation before and can help *interpret* me to myself. The issues are *meaning, interpretation,* and *sincerity* (or its lack).

Behaviorism, of course, wants nothing to do with any of this "black box" of interior meaning, and consigns the lot of it, at best, to "intervening variables" lying in that terra incognita between *observable* stimuli and *observable* response, internal variables that are defined merely as "tendencies to behavior," because behaviorism, being a Right-Hand path, does not trust anything it cannot *see* and monologically tinker with or reinforce.[26] (I wonder if any of the behaviorists ever explain to their spouses that their shared love is just an intervening variable.)

In other words, behaviorism, like all Right-Hand paths, is fundamentally concerned with *propositional truth.*[27] I, as a researcher, make a statement or a proposition about an objective state of affairs, and the research attempts to ascertain whether that statement is true or false (often by trying to disprove it). But in all cases it boils down to statements like, "Is it or is it not raining outside?" I go and *look,* and I find out what the objective situation is ("It is indeed raining outside"). Other researchers go and look, and if everybody agrees, we say it is propositionally true that it is raining outside.

All propositional truth is of that variety, although it can get quite complicated, and the monological *looking* often requires telescopes, microscopes, EEGs, and other complex instruments of one sort or another. But they all attempt to align propositions with an objective state of affairs. They are monological, propositional, empirical. The validity criterion is one of *truth,* of matching map and territory accurately enough.

But in the Upper-Left quadrant, the validity criterion is not so much truth as truthfulness, or sincerity. The question here is not "Is it raining outside?" The question here is: when I tell you it is raining outside, am I telling you the truth or am I lying? Or perhaps self-deceived? Am I being truthful and sincere, or deceitful and insincere? This is not so much a matter of objective *truth* as one of subjective *truthfulness.* It is not so much whether the map matches the territory, but whether the mapmaker can be trusted.

Truthfulness, then, is a matter of trust and sincerity. Precisely because *depth* does not sit on the surface for all to see, my reporting of depth may or may not be *trusted.* Truthfulness, sincerity, trustworthiness, integrity—

these are some of the crucial guideposts for navigating in the Upper-Left quadrant.

In short, it is not just a matter of true exteriors, but sincere interiors. And this sincerity cannot be determined empirically or objectively or monologically. We can give people empirical lie-detector tests, but if they have first lied to themselves, the monological machine will wrongly indicate they are telling the truth. The locus of sincerity is not objective, but subjective, and that can only be accessed in dialogical interpretation, not monological indication.

We will be returning to these points later, but those are some of the individual aspects of "this thought that occurred to me," this thought holon we are using as an example. The thought holon has brain correlates that can be determined objectively, monologically, propositionally (UR), guided by truth; and it has interior correlates that can be determined only dialogically and interpretively, guided by truthfulness. Not only are the two domains different in nature, they have *different criteria* for what is *valid* in each.

But, to continue the example, those individual aspects *also* have social or communal components (the lower half). Whether I tell you my specific thoughts or not, those thoughts do indeed have *meaning* to me (and would have meaning to you if I tell them to you), because that meaning is itself sustained by a whole network of background practices and norms and linguistic structures existing in our shared culture. If I tell them to you and you don't speak the language, then you won't get the meaning, even though all the physical words and stimuli freely bombard you. That is the Lower-Left quadrant, the *shared cultural worldspace* necessary for the communication of any meaning at all, and without which most (or even all) of my own private thoughts would be largely meaningless as well.

The question here is not one of truth so much, nor even truthfulness, but one of *cultural fit*, of the appropriateness or justness or "fitness" of my meanings and values with the culture that helps to produce them. My own individual meanings and values are not reducible to this cultural fitness (no quadrant is reducible to any other), but they do depend thoroughly on all the background contexts and cultural practices that allow me to form meanings in the first place. My thought-holon is inextricably situated in cultural contexts of relational exchange and intersubjective communications, without which my own subject (and its truthfulness) and the world of objects (with their truth) would not and could not disclose themselves in the first place.

If my thought-holon does not culturally fit, then I may be a genius rising above conventions, or I may be psychotic and totally out of touch with my fellows. But the criterion in any case is not so much truth or truthfulness, but justness, appropriateness: not whether my thought corresponds with a world of *objects*, nor whether I am being *subjectively* truthful, but whether I am *intersubjectively* in tune, appropriately meshed with the cultural worldspace that allows subjects and objects to arise in the first place. (Whether I agree or disagree with aspects of that culture, I have in all cases depended upon it to provide me with the capacity for intersubjective meaning in the first place.)

In other words, the criterion for validity in the Lower-Left quadrant is not just the *truth* of my statement, nor the *truthfulness* with which I put it, but whether you and I can come to *mutual understanding* with each other. Not objective, not subjective, but intersubjective.

But none of those meanings, whether individual or cultural, are simply or merely disembodied. My individual thoughts register a change in my brain physiology; likewise, cultural patterns are registered in exterior, material, observable social behaviors (even if they can't be reduced to them). And that is the Lower-Right quadrant.

That is, just as my thoughts have brain correlates, cultural meanings have correlates in objective social institutions and material social structures. Food production, transportation systems, written records, school buildings, geopolitical structures, behavioral actions of groups, written legal codes, architectural styles and the buildings themselves, types of technology, linguistic structures in their exterior aspects (written or spoken signifiers), techno-economic forces of production and distribution—all the physical components of a social action system, all the aspects of a social system that can be seen empirically or monologically.

Here the criterion is not the *truth* of objects, nor the *truthfulness* of subjects, nor the mesh of *intersubjective* understanding and meaning, but rather the *functional fit* or the *interobjective* mesh of social systems. A social action system, for example, can only produce so much butter or so many guns, and the more of one, the less of the other—they must *functionally fit* with what is physically possible (whereas, in intersubjective meaning, the more meaning I have does not mean the less meaning available to you: these are qualities, not quantities, so they don't have to physically add or subtract from each other, as quantities in functional fit must).

Functional fit ("What does it *do?*") is, of course, the major validity claim recognized by systems theory. And that is part of its *subtle reductionism.* Interpretive depth (sincerity and truthfulness) and cultural meaning (justness and moral appropriateness) are all reduced to functional fit in exterior surfaces: all are reduced to flatland holism.

(Even propositional truth, or subject and object corresponding fit, is reduced to functional fit, or *interobjective* fit, and "truth" becomes anything that furthers the autopoietic regime of the self-organizing social system; such theories dissolve their own truth value in the functional fit of that which they describe, so that their subjective theories become merely one among other interobjects, a very neat trick indeed; we will return to this later.)

Now the point of this overall example is simply that my "single" thought, the original holon, is not really a single thought as such, but rather a holon with four inseparable aspects (intentional, behavioral, cultural, and social), each with its own validity claims (subjective truthfulness, objective truth, intersubjective justness, and interobjective functional fit).[28] As I said, no holon whatsoever simply exists in one or another quadrant; all holons *possess* these four quadrants, and each quadrant is intimately correlated with, dependent upon, but not reducible to, the others.

But precisely because these quadrants are all so intimately correlated, I can indeed attempt an aggressive reductionism, and it can actually seem to make a lot of sense. For example, since every thought does indeed register *some* sort of change in brain physiology (even if I have an out-of-the-body experience!), I can *always* maintain that thoughts are just brain states, even though I will probably maintain (as most brain theorists do) that the brain states themselves are higher-order (or hierarchical) emergent patterns that cannot be reduced to *their* atomistic elements.[29]

This position is not yet gross reductionism. It is subtle reductionism. It has simply reduced interior depth, value, meaning, and consciousness (the Left Hand) to functional parts of a mutually interlocking order of holistic and empirical events (the Right Hand). It has reduced three-dimensional chess to flatland chess, the Kosmos to the cosmos, the interrelated pyramid of life to the interrelated web of life, interrelated I and we to interrelated its.

In short, this subtle reductionism *still recognizes holarchy* (hierarchy), but *only* the hierarchies on the Right Hand—only the hierarchies of size, extension, and surface. And therefore these hierarchies are all fundamentally defined by *physical inclusion* (a cell physically includes molecules, which physically include atoms, and so on; but these "depths" are all flatly *monovalent*: they are *not* holarchies of value, beauty, meaning, motivation, understanding, intention, consciousness, or anything else even vaguely Left Hand: all of those genuinely interior depths have been scrubbed clean and empirically whitewashed in the monochrome tones of exterior physical span, ex-span-sion, or extension/inclusion: no better or worse, only bigger or smaller).[30]

And this *subtle reductionism* is helped by the fact that *all* four strands of evolution, all four quadrants, follow the twenty tenets. They are all holarchical. But whereas the Right Hand is preeminently involved in physical extension (organisms are bigger than cells), the Left-Hand involves intentions (concepts per se are *not* bigger or smaller than symbols, even though they "include" them and are more intentional; these are gradations of value and beauty, not gradations of size).

Further, the *deeper* one proceeds, the less the twenty tenets tell us anything significant anyway (since they are the lowest-common-denominator factors, the weakest-noodle laws). Nonetheless, the twenty tenets do capture certain fundamentals in all four quadrants.

Thus, this subtle reductionism can still champion all the twenty tenets, the notion of development, the emergent characteristic of evolution, the mutually interrelated nature of all holons, the holarchical nature of all holons, and so on. *All of that can still be heartily embraced by the subtle reductionist*; it's just that all holarchies of quality have been lost entirely in holarchies of quantity, and all gradations of interior depth have been replaced by gradations of meaningless exteriors (compassion is better than murder, but quarks are not better than photons, and thus to explain the entire Kosmos in terms of exterior surfaces and empirical forms is to find only a cosmos with no value whatsoever: guaranteed).

And since for every event in the Left-Hand dimension there is always *something* registered in the Right Hand, it can appear that an exhaustive description of the Right-Hand path actually covers everything there is to be said. Put differently, since every event in the Kosmos does indeed have a Right-Hand component, it can mistakenly appear that the Right-Hand path alone exhausts the Kosmos, whereas it has merely registered and measured the footprints of the giant.

It's very subtle, this program of collapsing the Kosmos.

THE FUNDAMENTAL
ENLIGHTENMENT PARADIGM

As we will see in chapter 12, this subtle reductionism was the *fundamental Enlightenment paradigm*.[31] And within this subtle reductionism, within the fundamental Enlightenment paradigm, there were two warring camps: flatland atomists and flatland holists. And it is in an unbroken lineage that today's systems theorists carry on the latter pole of the Enlightenment paradigm: the task of *both* overcoming gross reductionism, which is altogether commendable, and *covertly* propagating subtle reductionism, the reduction of all intention to extension, all quality to quantity, all interpreted depths to unambiguously seen surfaces, all hierarchical values to monological mesh, all interiors to holistic strands, all truth and meaning to functional fit, all interwoven I's and we's to interwoven its.[32]

This is why Morris Berman can point out that the "contemporary quest for holism is just as formal, abstract, 'value-free,' and disembodied as the mechanistic paradigm it seeks to replace."[33] This confuses and angers the holists, because they sincerely think they are the good guys, since they managed to defeat *gross* reductionism. "But we have shown that everything is connected to everything else!" The flatland web of life. Although the intentions are often genuine, the gravitational pull of flatland makes it a planet very hard to escape.

Keep in mind that I am not saying systems theory and eco-holism is wrong; I would not have spent all of chapter 2 extolling it if I thought that. It is simply very partial—what Hegel called "a vanity of the understanding"—and that vanity tends to become dangerous, because when half the Kosmos claims to have the whole, certain types of aggression are about to ensue.

In chapter 1 we saw that, according to the holists themselves, we today suffer from a fractured worldview. That is true; but despite their best intentions, the holists have not solved it, they have merely cloned it. Reducing everything to functional fit destroys the integrity of each domain, and further renders true integration of each impossible. The world is indeed fractured; the flatland holists are some of the prime promoters of the fracture. It is from this flat and faded landscape, armed with good intentions and a weakest–noodle science, that they cry out to us as our saviors.

The holists, of course, tell the tale of the Enlightenment as being primarily an atomistic paradigm, which they maintain is the real cause of the world's fracture. This story allows the holists to hide the deeper crime of the Enlightenment that the holists themselves are still perpetuating. As we will see throughout the remainder of this volume, modernity's worldview is indeed fractured, not because everything isn't reduced to the Lower-Right quadrant of functional fit (the eco-holistic "solution"), but *because the four quadrants themselves have yet to be integrated.*

And it is in that integration that truth, truthfulness, meaning, and fit can be brought into a mutual harmony. That harmony—and not a reductionism that tears into the fabric of each in the name of a pretend wholeness—is one of the overall themes of this volume.

What we will find is that functional fit is indeed important, but it is only part of the harmony available. And it is not even the most important quadrant for today's world, because *before* we can even attempt an ecological healing, we must first reach a *mutual understanding* and mutual agreement among ourselves as to the best way to collectively proceed. In other words, the healing impulse comes not from championing functional fit (Lower Right) but mutual understanding (Lower Left). And that depends first and foremost, we will see, on individual growth and consciousness transformation (Upper Left). The Left-Hand path, not merely the Right-Hand path, must take the lead.

Anything short of that, no matter what the motives, perpetuates the fracture (or so I will try to demonstrate). In understandably emphasizing the importance and the urgency of eco-holistic fit, many holists have absolutized the Lower-Right quadrant, which, in thus sealing it off from any true integration, condemns it to the fate of all fragments. Reducing all domains to the Lower Right, to functional fit, doesn't just destroy the other domains, it destroys the Lower Right as well.

This is just another way of pointing out that absolutizing the biosphere destroys the biosphere. At this point of sorely needed integration, the flatland holists are ultimately no friends of Gaia. They might protect a patch of Gaia here or there, which is wonderful; but without an overall integration of all four quadrants, Gaia continues to wither in the winds of disregard. A genuinely Kosmic holism, not cosmic holism, is what is sorely needed.

THE BIG THREE

With reference to the four quadrants: because both of the Right-Hand quadrants are exteriors that can be described in it-language, I will sometimes count them as one major domain, the other two being the I-language of the Upper Left and the we-language of the Lower Left. And for simplicity's sake I will refer to these as the Big Three (I, we, it).

Numerous comparisons with other researchers could be pointed out here, and we will eventually discuss them in much detail. For the moment we might note that the Right half of the diagram is Karl Popper's World I (the objective world of it); the Upper Left, World II (the subjective world of I); and the Lower Left, World III (the cultural world of we, which can also, as Popper points out, be embodied or embedded in *material* social institutions, or the Lower Right).

Likewise, Habermas's three validity claims, for truth (objects), truthfulness or sincerity (subjects), and rightness or justice (intersubjectivity), refer respectively to the Right half, the Upper Left, and the Lower Left (I will return to Habermas in a moment and emphasize the importance of his formulations in this regard).[34] And in the broadest sense this is Plato's the True (or propositional truth referring to an objective state of affairs, it), the Good (or cultural justice and appropriateness, we), and the Beautiful (or the individual-aesthetic dimension, I). The Big Three are likewise Kant's three critiques: the Critique of Pure Reason (theoretical it-reason), of Practical Reason or intersubjective morality (we), and of personal Aesthetic Judgment (I). Thus, although other items are included as well, these three great domains—the Big Three—are especially the domains of empirical science, morality, and art.

But even when we move into a discussion of spirituality, we will see the Big Three appear in such all-important formulations as the Three Jewels of Buddhism: Buddha, Dharma, Sangha. Buddha is the ultimate I, Dharma the ultimate It, and Sangha the ultimate We. (I will be especially emphasizing this in subsequent chapters; it is absolutely crucial in spiritual concerns.)

Charles Taylor points out that "radically different senses of what the [cultural] good is go along with quite different conceptions of what a human agent is, different notions of the self. Our modern senses of the self not only are linked to and made possible by new understandings of good but also are accompanied by new forms of narrativity and new understandings of social bonds and relations. These all evolve together, in loose 'packages,' as it were."[35] We are always situated in relation to the I, the we, and the it—and they evolve together.

And, to return to Habermas, this is the essence of the Habermasian revolution. We are *inescapably situated* in relation to the Big Three, *each of which has its own validity claim* and its own standards, and none of which can be reduced to the others. "With any speech act," he says, "the speaker takes up a relation to something in the objective world [it], something in a common social world [we], and something in his own subjective world [I]." And the claims made with reference to each of those worlds have their own validity criteria, namely, *propositional truth* (referring to an objective state of affairs, or it), *normative rightness* (cultural justness or appropriateness, we), and *subjective truthfulness* (or sincerity, I). And this means that none of them can be reduced to the others:

> The "world" to which subjects can relate with their representations or propositions [propositional truth] was hitherto conceived of as the totality of objects or existing state of affairs [it]. The objective world is considered the correlative of all true assertoric [propositional] sentences. But if normative rightness [we] and subjective truthfulness [I] are introduced as validity claims analogous to truth, "worlds" [our "quadrants"] analogous to the world of facts have to be postulated for legitimately regulated interpersonal relationships [we] and for attributable subjective experiences [I]—a "world" [quadrant] not only for what is "objective," which appears to us in the attitude of the third person, but also one for what is normative, to which we feel obliged in the attitude of addresses, as well as one for what is subjective, which we either disclose or conceal [sincerity] in the attitude of the first person [the I].[36]

Likewise, each of these validity claims (for truth, truthfulness, and justness) can be exposed to its own different kinds of evidence, and thus its own kind of truth claims can be *exposed to evidence* and *checked* for their actual *validity*.

Correlative to the three fundamental functions of language [related to I, we, it], each elementary speech act can be contested under three different aspects of validity. The hearer can reject the utterance of a speaker by either disputing the *truth* of the proposition asserted in it (or of the existential presuppositions of its propositional content), or the *rightness* of the speech act in view of the normative context of the utterance (or the legitimacy of the presupposed context itself), or the *truthfulness* of the intention expressed by the speaker (that is, the agreement of what is meant with what is stated).[37]

And so we can now return to (and better understand) the notion of *subtle reductionism*, or the attempt to collapse the Kosmos to the Right-Hand path (or reduce the Left Hand to the Right, or reduce "I" and "we" to "it," or reduce interiors to exteriors, whether atomistic or functional):

Truth is reduced to mere representation (the "reflection paradigm").[38] That is, thoughts are no longer an integral part of the Kosmos, they are merely disengaged and hovering *propositions* that are supposed to *mirror* the cosmos, or simply "reflect accurately" a world of matter and facts out there (propositional thought as reflecting an objective state of affairs, or "its").[39] Reason is reduced from a *substantive* vision of the Kosmic order to a *procedural* process for mapping flatland.[40] Truth no longer means attunement with the Kosmos, but merely how to map the cosmos.

More subtly, truth comes to mean functional fit. Truth no longer refers to attunement with the Kosmic state of affairs (interiorly and exteriorly, vertically and horizontally), nor even reflection on a state of affairs; it comes to mean merely *instrumental means* in the autopoietic self-maintainance of the system of functional fit (at which point it bizarrely dissolves its own status; functional fitness can claim to be a true theory only by negating the meaning of truth: the theory, if it is true, must itself actually be an instrumental production of the system, at which point it can no longer be said to be true, only useful).[41]

Similarly, personal integrity and intentions (UL) are reduced to healthy brain functioning (UR), to the now rampant model of biological psychiatry (one is depressed, not if one's life becomes meaningless or *lacks values*, but if one's neurotransmitters lack serotonin). Or, likewise, one's personal integrity and meaning is reduced to behavioral modification (also UR). Not "Where am I situated in the Kosmos?" but "How can I function better in the cosmos?" Not "What does my existence mean?" but "How can I get it to work better?"

Personal meaning is thus reduced to behavioral fitness, and this is *unconsciously* judged by the prevailing *conventional* cultural reality (without acknowledging the implicit judgment involved). *"Adaptation to society"* is thus the standard yardstick against which behavioral (and brain chemistry) modifications are *judged* (explicitly or implicitly): not whether adaptation to a given society is a good idea or not, but how to function better in that society regardless (and if you can't adapt to being a good Serb while you rape and murder, then a little Prozac will help us over those rough spots).[42] Self-understanding is replaced by behavioral functioning, where one reinforces the desired response without a clue as to what responses are actually desirable and worth reinforcing in the first place.

Cultural meaning (LL) is reduced to *social integration* (LR), yet another version of functional fit. The whole *validity* of a cultural set of values is converted into a question of whether they promote social cohesion, functional fitness, and the integration of the social action system. By this criterion, the Nazis were altogether valid, because they certainly had one of the most coherent social orders ever devised: holism in action. Nothing beats fascism for a sturdy autopoietic regime. And nothing shows more clearly how limited social integration is as a criterion of the true and the good.

Put differently, since "What does it mean?" (the Left Hand) has been collapsed to "What does it do?" (the Right Hand), then the only criterion is "How well does it do it?" and not also "Is it worthy to pursue in the first place?"

Likewise, with subtle reductionism, morals no longer embody a statement of the good life, or what it means to lead a decent and worthy and noble life—a life worth being emulated, a life that empowers, a life plugged into moral sources that inspire worth and worship, awe and admiration—but rather what is merely required of us as a part, as a fraction, of a social action system. Intersubjective understanding is reduced to systems steering problems and technical tinkering. Morals and meanings become anemic markers of what procedures are required for systems maintenance and systems expansion. The True, and the Good, and the Beautiful are reduced to place settings at the dinner table of monological mesh.

And we are all reduced to whatever promotes the functional capacity of the autocratic system, reduced to what works, reduced to *fitting in*. Reduced to leading life by looking at a representational map of a flat and faded landscape, trying to fit into that landscape, and to trying to persuade others to embrace the same cheerful ontological suicide.

We will see that the great task of modernity and postmodernity, as theorists from Schelling to Hegel to Habermas to Taylor have pointed out, is not to replace gross reductionism with subtle reductionism (or atomism with flatland holism), but to *integrate the Big Three* (integrate I, we, and it; or art, morals, and science; or self, culture, and nature), not by reducing one to the others, but by finding a richly encompassing conception of the Kosmos that allows each to flourish in its own right.

In other words, if the great achievement of the Enlightenment (and "modernity") was the necessary *differentiation* of the Big Three, the great task of "postmodernity" is their *integration*, overcoming what Taylor called "a monster of arrested development" (both of these points will be discussed in later chapters).[43]

What we need to do, then, is look now to the evolution of the Left-Hand dimensions (I and we) as they appear in humans, so that these can be "added" to the accounts of the Right-Hand path for a more balanced overview, an overview that more hopefully might contribute to the integration of the Big Three domains.

We can start with Habermas.

MICRO AND MACRO, PHYLO AND ONTO

Habermas begins with the observation that the *same structures of consciousness* (his phrase) can be found in the individual self (UL) and its cultural setting (LL), that is, in the micro and macro branch of the evolution of human consciousness.[44] Habermas often uses "social" in the broad sense, but here we are centering mostly on what I have called cultural, as the context will make clear:

> If one examines social institutions and the action competences of socialized individuals for general characteristics, one encounters the same structures of consciousness. This can be shown in connection with [to give only one example] law and morality. One can see here

the identity of the conscious structures that are, on the one hand, embodied in the institutions of law and morality and that are, on the other hand, expressed in the moral judgements and actions of individuals. Cognitive developmental psychology has shown that in ontogenesis there are different stages of moral consciousness, stages that can be described in particular as preconventional, conventional, and postconventional patterns of problem-solving [which we will explain later]. The same patterns turn up again in the social evolution of moral and legal representations.[45]

As Habermas indicates, the individual and the cultural holon evidence the *same basic structures of consciousness,* and these same basic structures of consciousness show up in the development or evolution of both the *individual* and the *species*:

> The ontogenetic models are certainly better analyzed and better corroborated than their social-evolutionary counterparts. But it should not surprise us that there are homologous structures of consciousness in the history of the species, if we consider that linguistically established intersubjectivity of understanding marks that innovation in the history of the species which first made possible the level of sociocultural [noospheric] learning. At this level the reproduction of society and the socialization of its members are two aspects of the same process; *they are dependent on the same structures.*[46]

Let me repeat that Habermas is saying two things here: an individual human being and its sociocultural environment evidence the *same* basic structures of consciousness (correlation of micro and macro), and *further,* these same basic structures can be found in the evolution of the individual and the species (ontogenetic and phylogenetic parallels).

> The homologous structures of consciousness in the histories of the individual and the species [are not restricted to the domain of law and morality. They can also be found] in the domain of ego development and the evolution of worldviews on the one hand, and in the domain of ego and collective identities on the other [correlations between UL and LL].[47]

Here are a few of the correlations according to Habermas (and using only his terms): an individual at the level of preoperational thought participates in a natural or bodily identity, a shared worldview that is magic-animistic, and a preconventional morality; an individual at the level of concrete operational thought is open to a role identity, participates in a shared worldview of mythological thought, and in a conventional morality; and an individual at the level of formal operational thought possesses an ego identity, participates in a shared worldview that is rational, and evidences a postconventional morality (all of which we will be explaining).

Habermas's point is that, just as an infant of today develops from preconventional (magic) to conventional (mythic) to postconventional (rational)—which we will investigate in chapter 6—so the species itself evolved from magic to mythic to rational (in other words, quite similar to Gebser). It is the same basic structures of consciousness underlying both the micro and macro branch in both their ontogenetic and phylogenetic evolution, and it is the *same developmental logic* (holoarchic in nature, I would add) that governs their evolution.

I had taken virtually the identical approach in *Up from Eden* and *Eye to Eye*, and a few critics have objected to Habermas's and my use of phylogenetic and ontogenetic parallels, claiming that this is an old and outmoded way of thinking that nobody subscribes to anymore. And, indeed, the notion of onto/phylo parallels—first introduced by Ernst von Haeckel (1834–1919)—was originally used in such a rigid and simplistic way that the entire notion itself fell into disrepute. But in its modern versions it simply outlines a series of abstract patterns that indeed show similarities, and in this version it is even accepted by orthodox science itself, as witness that bastion of orthodoxy, Isaac Asimov, in his widely respected *New Guide to Science*:

> It is almost impossible to run down the roster of living things, as I have just done, without ending with a strong impression that there has been a slow development of life from the very simple to the complex [tenet 12a]. The phyla can be arranged so that each seems to add something to the one before [tenet 5]. Within each phylum, the various classes can be arranged likewise; and within each class, the orders.
>
> We can trace practically a re-enactment of the passage through the phyla, even in the development of a human being from the fertilized

egg. In the course of this development, the egg starts as a single cell (a kind of protozoon), then becomes a small colony of cells (as in a sponge), each of which at first is capable of separating and starting life on its own, as happens when identical twins develop. The developing embryo passes through a two-layered stage (like a coelenterate), then adds a third layer (like an echinoderm), and so continues to add complexities in roughly the order the progressively higher species do.[48]

As for cognitive development, the empirical conclusion of developmentalists from Silvano Arieti to Piaget is that there are indeed certain onto/phylo parallels in the evolution of deep structures (not surface structures), as Arieti explains: "What is of fundamental importance is that the [ontogenetic and phylogenetic] processes to a large extent follow similar developmental plans. This does not mean literally that in the psyche ontogeny recapitulates phylogeny, but that there are certain similarities in the [two] fields of development and that we are able to individualize schemes of highest forms of generality which involve all levels of the psyche in its [two] types of development. We also recognize concrete variants of the same overall structural plans in the types of development."[49]

That version of onto/phylo parallels is explicitly accepted by researchers and theorists from Erich Jantsch ("recapitulation does not act in a rigid and structurally oriented way, but in the sense of the flexible modification of a space-time structure")[50] to Karl Popper and John Eccles (see *The Self and Its Brain*). Massive evidence has been accumulated in this regard by the Jungians and given its most powerful statement in Erich Neumann's book *The Origins and History of Consciousness*.

But perhaps the most intriguing use of this notion is to be found in Rupert Sheldrake, who sees general recapitulation as a form of morphic resonance (or Kosmic memory) from all previous stages of evolution. It is as if, he says, nature develops habits—morphic units with morphic fields, which he also calls holons—and once these holons are developed or become set as habits of nature, then nature simply keeps reusing them in succeeding stages—another version of compound individuality.

Nobody is more aware than Habermas of the misuses and fallacies to which this notion has been put, and indeed he outlines eight ways that the notion has been erroneously used, objections which I strongly share (the critics of this approach might wish to consult Habermas, because he seems to be much clearer about its problems than they are).[51] Habermas nonetheless concludes, "All provisos notwithstanding, certain homologies can be found"—certain parallels between the development of individuals and the development of the species.[52]

We can now look directly to these homologies, remembering always that these are not just some archaeological digs from a distant past, interesting but irrelevant. On the contrary, this is also an archaeology of our own souls. These "digs" exist now in the present, enfolded *in us* as part of our own compound individuality, past interiors taken into our own present interiors, living on in the depths of our own being today, enriching us or destroying us, depending entirely on our own capacity to embrace and to transcend.

5

The Emergence of Human Nature

Today, makes yesterday mean. —EMILY DICKINSON

I think that there is a widespread and facile tendency, which one should combat, to designate that which has just occurred as the primary enemy, as if this were always the principal form of oppression from which one had to liberate oneself. Now this simple attitude entails a number of dangerous consequences: first, an inclination to seek out some cheap form of archaism or some imaginary past forms of happiness that people did not, in fact, have at all. There is in this hatred of the present a dangerous tendency to invoke a completely mythical past. —MICHEL FOUCAULT

A T THIS POINT in our discussion, we have sketched the evolution of the exterior of individual holons up to the complex triune-brained organisms (hominids) with an interior space of symbols and concepts. They lived in a social holon of a group/family (tribe) possessing a cultural archaic worldview. (Those are the four quadrants up to this point in our story.) But then something happened that transformed hominids (or "proto-humans") into true humans (*Homo sapiens*).

What was it?

THE EMERGENCE OF *HOMO SAPIENS*

Habermas begins with the Marxist idea that what separates humans—*Homo sapiens*—from hominids and other primates is *social labor* or the existence of an *economy*. But Habermas finds that it is very likely that hominids also possessed an economy, and thus the mere existence of economic exchange does not mark the specifically human form of life, does not separate hominids from *Homo sapiens*.[1]

> If we examine the concept of social labor [economy] in the light of more recent anthropological findings, it becomes evident that it cuts too deeply into the evolutionary scale; not only humans but hominids too were distinguished from the anthropoid apes in that they converted to reproduction through social labor and developed an economy. The [hominid] adult males formed hunting bands, which (a) made use of weapons and tools (technology), (b) cooperated through a division of labor, and (c) distributed the prey within the collective (rules of distribution). The making of the means of production and the social organization of labor, as well as of the distribution of its products, fulfilled the conditions for an economic form of reproducing life.[2]

Since part of my aim in this series is to follow the course of the male and female value spheres, it is instructive to note that, even in archaic hominid societies, these two spheres were already beginning to differentiate, and often sharply. Habermas:

> The society of hominids is more difficult to reconstruct than their mode of production. It is not clear how far beyond interactions mediated by gestures—already found among primates—their system of communication progressed. The conjecture is that they possessed a *language* of gestures and a system of *signal calls*. In any event, cooperative big-game hunting requires reaching understanding about experiences, so that we have to assume a protolanguage, which at least paved the way for the systematic connection of cognitive accomplishments, affective expressions, and interpersonal relations that was so important for hominization.
>
> The division of labor in the hominid groups presumably led to a development of two subsystems: on the one hand, the adult males, who were together in egalitarian hunting bands and occupied, on the whole, a dominant position; on the other hand, the females, who gathered fruit and lived together with their young, for whom they cared.

In comparison to primate societies, the strategic forms of cooperation and the rules of distribution were new [my italics; Habermas is searching for emergents]; both innovations were directly connected with the establishment of the first mode of production, *the cooperative hunt.*[3]

Thus, Habermas concludes, the emergence of an economy "is suitable for delimiting the mode of life of the hominids from that of the primates; but it does not capture the specifically human reproduction of life."[4] Rather, it *"appears now that the evolutionary novelty that distinguishes Homo sapiens is not the economy but the family."*[5]

Up to this point we have been using the term "family" in Jantsch's very loose sense, meaning any genetic-relative group of animals. But the specifically human family, according to Habermas, is marked by an emergent characteristic that is found nowhere else in evolution.

> Not hominids, but humans were the first to break up the social structure that arose with the vertebrates—the one-dimensional rank ordering in which every animal was transitively assigned one and only one status. Among chimpanzees and baboons this status system controlled the rather aggressive relations between adult males, sexual relations between male and female, and social relations between the old and the young. A familylike relationship existed only between the mother and her young, and between siblings. Incest between mothers and growing sons was not permitted; there was no corresponding incest barrier between fathers and daughters, because the father role did not exist. Even hominid societies converted to the basis of social labor did not yet know a family structure.[6]

The novel emergence of the human family, according to Habermas, occurred only as the male was also assigned the role of *father*, for it was only in this way that the two value spheres of the male and female could be linked. In ways (and for reasons) that we will examine in exhaustive detail in volume 2 of this series, the male and female value spheres had already been differentiated into social labor (hunting) and nurturance of the young. Lenski reports, for example, that of the known societies at this stage, an astonishing 97 percent show that pattern of male/female differentiation.[7] If evolution were to continue, *a new integration was also needed.* Since presumably the female could not be pregnant and simultaneously hunt, the integrating link was established by the novel emergence of the role of the father, with one foot in both spheres.

> The mode of production of the socially organized hunt created a system problem that was resolved by the familialization of the male, that is, by the introduction of a kinship system based on exogamy. The male society of the hunting band became independent of the plant-gathering females and the young, both of whom remained behind during the hunting expeditions. With this differentiation, linked to the division of labor, there arose a new need for integration, namely, the need for a controlled exchange between the two subsystems [what we called relational exchange with same-level holons]. But the hominids apparently had at their disposal only the pattern of status-dependent sexual relations. This pattern was not equal to the new need for integration, the less so, the more the status order of the primates was further undermined by forces pushing in the direction of egalitarian relations within the hunting band. Only a family system based on marriage and regulated descent permitted the adult male member to link— via the father role—a status in the male system of the hunting band with a status in the female and child system, and thus (1) integrate functions of social labor with functions of nurture of the young, and, moreover, (2) coordinate functions of male hunting with those of female gathering.[8]

The familialization of the male. Thus began the one, single, enduring, and nightmarish task of all subsequent civilization: the taming of testosterone.

MALE ADVANTAGE AND FEMALE ADVANTAGE

Notice that, unlike the role of the father, the role of the mother was not sufficient to link the two value spheres, because the mother could not, or at any rate did not, participate in the social labor of the hunt (with rare exceptions). And so we see that here, right at the very defining transformation from hominids to humans, a *sexual differentiation* emerged, a differentiation that could, as we will see, *dissociate* into extreme sexual polarization.

At this early point, this differentiation or asymmetry does not appear to have been strongly evaluative—that is, one sphere was not particularly more important or more valued than the other. It appears to have been mostly a simple differentiation of *function*, not a massive differentiation of *status*. There is no evidence that it was intensely ideological or exploitative, but rather seems to have been based largely on such simple biological factors as physical strength and mobility (male advantage) and procreation and biological nursing (female advantage).

These are intricate and emotionally charged issues, especially for the various schools of feminism, which is why the second volume of this series is devoted to an extensive review of just such issues.[9] Some liberal feminists feel that *any* sexual differentiation or asymmetry is due exclusively to male domination. Many radical feminists, on the other hand, strongly accept and embrace the differentiation between the male and female value spheres (which, we will see, they equate with *agency* and *communion*, respectively), and they urge society to embrace more of the latter and less of the former. Social feminists tend to see sexual differences arising from different modes of production and technology, and see these unfair differences "evening out" as capitalistic modes become more human and humanizing.

I will be examining all of those positions carefully in volume 2. But until then we need an agreed-upon truce in the gender wars. My claim is that the structures of human consciousness that I will be presenting in the rest of this volume are in fact gender neutral, that there is no fundamental gender bias in the deep structures themselves. *However,* in the course of historical development, these gender-neutral structures, for numerous reasons we will examine, became loaded with various factors (technological, economic, cultural, social, and intentional) that biased some of these structures in an often specifically subjugating and certainly polarizing (or dissociating) fashion.

Thus, in volume 2, I will be looking at each of the cultural stages or worldviews (archaic, magic, mythic, mental, centauric) and correlating them with their historical modes of material production and technology (foraging, horticultural, agrarian, industrial, informational)—as shown in Figure 5-1 (page 198). Further, I will then specifically examine the *status of men and women in each of those stages.* My claim is that a consensus of liberal, radical, and social feminist views already exists on many of these important issues, and I will present those at that point.

In the meantime, when we reach, for example, the mythic-agrarian structure in the present discussion, I will have little to say about the status of women in those societies, even though the strong consensus of orthodox and feminist researchers alike is that sexual polarization and male domination of the public sphere were at their *historical peak* (and almost reversed from some of the previous horticultural societies, where many feminists feel women were at their greatest power). I would just like to point out that the fact that I don't discuss that issue now does not mean I am ignoring it.

But I will indicate just a few of the ways that I believe this approach is novel. In studying gender and possible gender differences, what is required *initially* is a set of constants that acknowledge certain unambiguous differences in function (women give birth and lactate, and men have on average a moderate advantage in physical strength and mobility—these specific differences, for example, are emphasized by feminist researchers from Janet Chafetz to Joyce Nielsen, as we will see). These simple differences might not seem all that important today, but historically and prehistorically they were often some of the most important and determining factors in all of culture.

For example, when the animal-drawn plow was developed and eventually replaced the handheld hoe for farming (this is referred to as a shift from horticultural to agrarian modes), there was a massive shift from a largely female work force to a largely male work force, due almost entirely to the fact that the plow, unlike the hoe or digging stick, is a heavy piece of equipment. In horticultural societies, women produced about 80 percent of the foodstuffs (and consequently shared considerable public power with men); a pregnant woman could still easily use a handheld hoe, but not a plow. When the plow was invented, men took up virtually all of the productive work, the matrifocal modes of production gave way to patrifocal modes, and the reigning deity figures switched from Great Mother to Great Father focus.

(Peggy Sanday has demonstrated that predominant female deity figures appear almost exclusively in horticultural societies—about a third of them have female-only deities, and another third have male and female deities; whereas virtually *all* agrarian societies have *male-only* deities. Thus, one of the conclusions of volume 2 is that where women work the fields with a hoe, God is a Woman; where men work the fields with a plow, God is a Man. Exactly how that fits with any enduring masculine and feminine faces of Spirit is a major topic of volume 2. Are these deity figures just economic byproducts, or is something more enduring being reflected here?)

Feminist researchers such as Janet Chafetz have pointed out that women who participated in heavy plowing had a much higher rate of miscarriage, and thus it was not to *their* genetic advantage to do so. *This* aspect of the shift from matrifocal to patrifocal, in other words, cannot reasonably be ascribed to oppression or male domination, but to a *joint* decision on the part of men and women in the face of a set of natural givens.

That initial differentiation of *function*, however, could be (and in many cases was) parlayed into a difference in *status*, with males dominating the public/productive sphere and women relegated to the private/reproductive sphere. And what we will want to examine are these natural differentiations as they moved into *dissociations* (and extreme sexual polarizations) that disadvantaged one or the other sex (and usually both)—at *each* of the six or so major epochs/stages/structures we will be examining.

In other words, we will find that a set of constants (a handful of factors that, according to most feminist researchers, have not changed much from culture to culture, such as birthing/nursing), when applied to the various gender-neutral structures of consciousness and stages of technological unfolding, *will together generate the specific gender differences in the status of men and women at each stage* (and this will help us to decide what is, and what is not, oppression). These constants (such as physical strength/mobility and birthing/nursing) are presented in volume 2 ("Strong and Weak Universals in the Sexual Scheme"), and how they play out and unfold, for better and for worse, is examined in five or six chapters following.

What we are discussing now, however, is simply the gender-neutral structures of consciousness within which these sexual and gender polarizations would occur.

MALE AND FEMALE LIBERATION

Before we pick up the story of the evolution of these consciousness structures, let me give a final example of the ground we will be covering later in this chapter (and extensively in volume 2). Some of what I am going to say now will not make total sense until the end of this chapter (after I have introduced some much-needed evidence), but it would be useful to introduce the general topics at this point, because of both their importance and their intricacy. (We will return to these topics at the end of the chapter and review them in light of the material presented.)

Part of the disagreement between liberal feminism (men and women are fundamentally equivalent in capacities) and radical feminism (men and women represent two quite different value spheres, the former being hyperautonomous, individualistic, a bit power-crazed, but generally *agency-oriented*, the latter being more relational, nurturing, permeable, and *communion-oriented*) is that these two approaches are almost impossible to reconcile, and this has caused much dissension, sometimes quite bitter, within the ranks of the women's movement.

But what if the radical feminists are right when it comes to the biospheric component of human beings, and the liberal feminists are right when it comes to the noospheric component? If this integration could be effected in a genuine fashion, and not merely as a verbal flourish, we could simultaneously honor some of the important differences in sexual being—so emphasized by the radical feminists—and yet also insist on, and maintain, equality before the law (which means, in the noosphere)— the important demand of liberal feminists.

But it would mean something else as well, something even more important. In this chapter we will see the evidence (presented by numerous researchers) that the noosphere and the biosphere did not finally *differentiate* in the West until around the sixteenth-seventeenth century. (That this differentiation went too far into *dissociation*—that is one of our central topics, but for the moment, it is a separate issue.) The point now is that, with the differentiation of the noosphere and the biosphere, the roles of men and women were no longer necessarily or automatically determined (or largely influenced) by biological factors (such as physical strength/mobility and birthing/nursing), factors that, one way or another, had largely *dominated* not only the relation of men to women but also of men to men, up to that point in history.

In other words, with the differentiation of the noosphere and the biosphere, biology was no longer destiny. That is, no longer *necessarily* destiny: the relationships between men and women (and men and men) were no longer necessarily dominated by the heavy hand of biological differences and determinants, physical strength and reproduction.

Thus, as evolution continued to move into the noosphere and transcend its exclusive grounding in the biosphere, those two universals that we briefly mentioned—bodily strength and bodily reproduction—*which have value only in the biosphere, could be transcended as well* (i.e., preserved and negated). This, of course, is a matter of degree; biology is still preserved and its inclinations still taken up and reworked by culture; but the greater the differentiation, the greater the possible negation and reworking.

If, for the moment, we simply agree with the radical feminists that the "female value sphere" tends to emphasize *communion*, then this value sphere could be decoupled from *exclusive* biological reproduction and could begin to enter the world of agency or production or "male culture," while still retaining its roots in the richness of the reproductive sphere. The more evolution transcended the biosphere, the more women could participate in *both* worlds (noospheric-mental and biospheric-familial), because in the world of the mind, physical strength, for example, is *irrelevant*.

In other words, with the differentiation of the noosphere and the biosphere, women could also become *agents* in the noosphere (*historical agents*), as well as being grounded in the biosphere. *And this precisely reverses the roles that men and women had played up to this point in history.* Previously, the male and female spheres were integrated by placing the male in the family via the role of the father (as *biological agent*); now the two spheres could be integrated—demanded to be integrated—by placing the female in the public sphere (as *noospheric agent*).

Thus, at this present time in history—in *today's* world—it is especially the role of the *female as historical agent* that can *bridge the two value spheres*, whereas in all previous history that role rather necessarily fell to the male as father (this "familialization of the male" is the point that Habermas has just introduced in our narrative).[10]

In this way, and this way only, I believe, can we maintain the otherwise paradoxical and confusing truth—a paradoxical truth that has often crippled feminism vis-à-vis "female liberation"—in this way only can we maintain *both* that women *now* stand open to and in need of liberation yet *previously* were *not* acting in an unliberated (or duped) fashion. The widespread emergence of the women's movement in the seventeenth and eighteenth centuries occurred precisely because the noosphere and the biosphere were finally differentiated.[11] And this inescapably means that the widespread emergence of the women's movement was not primarily the *undoing* of a nasty state of affairs that easily could have been different, but rather it marked the *emergence* of an altogether *new* state of affairs that was in significant ways *unprecedented*.

What had grounded *both* men and women in narrowly based biological roles was an evolutionary process that was itself, until fairly recently, grounded in the biosphere, and that only recently is now in the process of liberating *both* men and women from confinement to those particular roles, which were necessary at the time but are now outmoded; and the lead in this evolutionary transformation can most easily come (as Ynestra King has pointed out) primarily in the form of women assuming cultural agency and not just family communion (just as in the previous integration, the familialization of the male, men were placed in family communion and not just cultural agency).

That is the new integration now demanded on a societal level. And viewing it in this evolutionary light allows us to bypass much of the standard and useless rhetoric that men have been oppressive pigs from day one, with the unavoidable implication that women have been herded sheep. It's not so much that that view is demeaning to men, although it is meant to be, but rather that one cannot claim women have been oppressed for five thousand years (some say for five hundred thousand years) without simultaneously implying that women are stupider and/or weaker than men—there are no other explanations.

If we see the biologically differentiated roles of men and women as something imposed by men on women, then we have to simultaneously assume the complete pigification of men and the total sheepification of women. To automatically assume that differentiation of roles is the result of domination is to automatically define a particular group as victim, which automatically and irrevocably disempowers that group in the very attempt to liberate it.[12]

But if men and women were largely operating in the face of certain biological givens, *and* if these givens are no longer primarily determinate, *then* both men and women stand in need of liberation from those previously confining roles. The "patriarchy" was not something that could have simply and easily been avoided or bypassed (wherever evolution moved beyond the hoe, the patriarchy accompanied it).[13] If it could have been easily bypassed but wasn't, then men indeed are pigs and women dolts. The patriarchy, however, is not something that needs to be reversed but rather something that needs to be outgrown: and that releases the male from blame and the female from sheepdom.

And it further allows us to see the real cause of the unfortunate resistance to today's feminism. It is not a resistance to undoing thousands of years of male obnoxiousness and female dupedom, but a resistance to the emergence of an *entirely novel structure of consciousness*, a structure that, having for the first time in history differentiated the noosphere and the biosphere (which, among many other things, *created* feminism), is now in the process of trying desperately to integrate both women and men into that utterly new worldspace. That is what is feared and resisted (by many men and many women). And that is the new integration of which I will speak at length (under the names "centaur" and "vision-logic," the stage beyond egoic-rational).

We will see that there have indeed been forms of oppression and subjugation, but these have to be judged, not against today's structures of consciousness, but against what could have been otherwise at a given previous structure. *Within the possibilities* of a given structure, we judge its degree of nastiness: the judgment needs to be stage-specific and stage-appropriate. And this means we have to carefully reconstruct the notion of oppression in this regard, because if something *could not occur* under any given circumstances at a particular time, then the "not occurring" cannot simply be ascribed to oppression without further ado. If, for example, widespread feminism could not have emerged until the differentiation of the noosphere and biosphere, then its nonexistence in earlier cultures was *not* the result of oppression, but the result of not-yet-enough evolution, an entirely different story: women were not being weak and ignorant and men were not being insensitive slobs (nothing, that is to say, out of the ordinary).

When the noosphere and the biosphere were finally differentiated, biology was no longer destiny. Women could begin to move out of the biosphere and into the noosphere, albeit against much resistance to this new emergence. (And indeed, every male today starts his own development as a biospheric beast, and he won't "get it" until he, too, differentiates noosphere and biosphere, and acts from the former, not the latter. The battle has to be fought anew with every birth, and the resistance, in different forms, is present in both women and men. Polls consistently showed, for example, that in the United States a majority of men favored the Equal Rights Amendment but a majority of women rejected it. The new emergence is difficult for all, precisely because it in many ways means leaving the predeterminatedness and comfort of biological givens.)

In the course of history and prehistory, it would take three or four major and profound cultural transformations to climb up and out of this biological destiny. (And each step would have its own and very new forms of pathology and possible dissociation, all of which would have to be negotiated—sometimes not very successfully.)

At this point in our narrative, we are still at the very first transformation. The first sexual-functional *differentiation* has just taken place, based largely on simple biospheric differences (physical strength/mobility versus birthing/nursing). And the first major *integration* of these two spheres has also occurred: placing the male in *both* worlds via the altogether novel emergence of the role of the father.

MAGICAL-ANIMISTIC

With the familialization of the male, and the emergence of human socio-cultural evolution, we move from the archaic to the magic. The archaic is for both Gebser and myself a loose "catchall" epoch that simply and globally represents *all* the structures of consciousness up to and including the first hominids. We could just as easily have broken the archaic down into its dozens of components and individual stages, such as the more detailed breakdown given in figure 4-1 (where everything up to concepts is "the archaic fund"). The archaic is simply a symbol of the totality of our rich evolutionary history, but it is not just a symbol, because this history lives on in each of us as part of our present compound individuality.

As for the magical structure itself, according to the extensive research of Habermas and his associates:

> Apparently the magical-animistic representational world of paleolithic societies was very particularistic and not very coherent. The ordering representations of mythology [i.e., early mythology, and not the complex mythology that is the next stage, the mythological stage proper] first made possible the construction of a complex of analogies in which

all natural and social phenomena were interwoven and could be transformed into one another ["magical displacement"]. In the egocentric world conception of the child at the preoperational level of thought, these phenomena are made relative to the center of the child's ego; similarly, in sociomorphic worldviews they are made relative to the center of the tribal group. This does not mean that the members of the group have formed a distinct consciousness of the normative reality of a society standing apart from objectivated nature—*these two regions have not yet been clearly separated* [i.e., the biosphere and the noosphere have not yet been clearly differentiated/integrated]. This magical-animistic stage is characterized by a conventional kinship organization, a preconventional stage of law, and an egocentric interpretive system. . . .[14]

Preoperational thinking, for cognitive anthropologists, is thinking that works with images, symbols, and concepts (but not yet complex rules and formal operations). It is also called "representational" because symbols and concepts essentially present and represent (make and match) sensory objects in the external world. For this reason, cognitive psychologists and anthropologists refer to representational thought as being *"close to the body."* That is, the noosphere is just emerging, the mind is just emerging, and as such it is still relatively undifferentiated from the biosphere, from the body and sensorimotor intelligence.[15]

This is why Habermas refers to the self-identity of this stage as a *natural* identity or a body-based identity (as Freud put it, "the ego is first and foremost a bodyego"). It is not yet an identity based upon extensive mental rules and roles—which develop only with the next stage, the concrete operational—and so the law and morality of this stage is likewise body-based, dependent upon physical pragmatic concerns and "naive instrumental hedonism." As McCarthy summarizes it, the individual at this *preconventional* moral stage "is responsive to cultural rules and labels of good and bad, right and wrong, but interprets these labels in terms of either the physical or the hedonistic consequences of action (punishment, reward, exchange of favor), or in terms of the *physical power* of those who enunciate the rules and labels."[16]

It is also called "magical" (by Habermas, Gebser, Piaget) because the mind and body are still relatively undifferentiated, and thus mental images and symbols are often confused or even identified with the physical events they represent, and consequently mental intentions are believed to be able to "magically" alter the physical world, as in voodoo, exoteric mantra, the fetish, magical ritual, "sympathetic magic," or magic in general (we will return to this topic in the next chapter).[17] Likewise, and from the other side of the indissociation, physical objects are "alive," possessing not just prehension but explicitly personal intentions (animism).

In other words, because the biosphere and the noosphere are not yet clearly differentiated/integrated, the subject has special power over the object (magic) and the object has special subjective qualities (animism). Whether that was or was not "spiritual" in any genuine sense we will investigate later.

Cognitive anthropologists often refer to this as "global syncretism" or "indissociation." Habermas explains:

> [In these societies] collective identity was secured through the fact that individuals traced their descent to the figure of a common ancestor [kinship] and thus, in the framework of their worldview, assured themselves of a common cosmogonic origin. On the other hand, the personal identity of the individual developed through identification with a tribal group, which was in turn perceived as part of a nature interpreted in interaction [magical] categories. As social reality was not yet clearly distinguished from natural reality, the boundaries of the social world merged into those of the world in general. Without clearly defined boundaries of the social system there was no natural or social environment in the strict sense; contacts with alien tribes were interpreted in accord with the familiar kinship connection.[18]

It is that indissociation that is so often eulogized by Romantics, because, I believe, they mistake indissociation for integration. The magical-animistic structure, as lovely as it might appear to us jaded moderns, was not an integration of the biosphere and the noosphere, because these had not yet been differentiated in the first place.

At the same time—and for precisely the same reason—that "lack of separation" might have embodied a type of ecological wisdom, a wisdom that many moderns are understandably trying to recapture. Namely, the tribal kinship consciousness, still lying "close to the body," close to the biosphere, was for just that reason sometimes more "ecologically sound," more in tune with natural wisdom, with the Earth and its many moods. And thus it is small wonder that, in these ecologically disastrous times, many moderns are attempting to resurrect the natural wisdom of tribal awareness more attuned with the biosphere.

I am in complete sympathy with that approach; I am not in sympathy with the attempt to turn back the clock and elevate this structure to a privileged status of integrative power that it simply did not possess. Further, whether "close to nature" automatically translates into "ecologically sound" is hotly debated. *Lack of capacity* to devastate the environment on a large scale does not automatically mean *presence of wisdom*, let alone reverence for the environment. And, in fact, many tribes, as Lenski points out, simply remained at one location until they had ecologically depleted the area, and then were forced to move on. Tribal awareness was in all cases close to nature, in the sense of indissociated; ecologically sound is another matter.

Riane Eisler is specific: "If we carefully examine both our past and present, we see that many peoples past and present living close to nature have all too often been blindly destructive of their environment. While many indigenous societies have a great reverence for nature, there are also both non-Western and Western peasant and nomadic cultures that have overgrazed and overcultivated land, decimated forests, and, where population pressures have been severe, killed off animals needlessly and indifferently. And while there is much we can learn today from tribal cultures, it is important not to indiscriminately idealize all non-Western cultures and/or blame all our troubles on our secular-scientific age. For clearly such tribal practices as cannibalism, torture, and female genital mutilation antedate modern times. And some indigenous . . . societies have been as barbarous as the most 'civilized' Roman emperors or the most 'spiritual' Christian inquisitors."[19]

As René Dubos summarizes the available evidence, "All over the globe and at all times in the past, men have pillaged nature and disturbed the ecological equilibrium, usually out of ignorance, but also because they have always been more concerned with immediate advantages than with long-range goals. Moreover, they could not foresee that they were preparing for ecological disasters, nor did they have a real choice of alternatives."[20]

Theodore Roszak, himself a staunch advocate of a certain type of "primal/tribal wisdom," nonetheless points out that in many instances "tribal societies have abused and even ruined their habitat. In prehistoric times, the tribal and nomadic people of the Mediterranean basin overcut and overgrazed the land so severely that the scars of the resulting erosion can still be seen. Their sacramental sense of nature did not offset their ignorance of the long-range damage they were doing to their habitat."

Likewise, he says, other indigenous or primal societies have, "in their ignorance, blighted portions of their habitat sufficiently to endanger their own survival. River valleys have been devastated, forests denuded, the topsoil worn away; but the damage was limited and temporary."[21]

The correct conclusion, in other words, is that the primal/tribal structure *in itself* did *not* necessarily possess ecological wisdom, it simply *lacked the means* to inflict its ignorance on larger portions of the global commons.

The main difference between tribal and modern eco-devastation is not presence or lack of wisdom, but presence of more dangerous means, where the *same* ignorance can now be played out on a devastating scale. As we will see, our massively increased means have led, for the first time in history, to an equally massive dissociation of the noosphere and the biosphere, and thus the *cure* is not to reactivate the tribal form of ecological ignorance (take away our means), nor to continue the modern form of that ignorance (the free market will save us), but rather to evolve and develop into an integrative mode of awareness that will—also for the first time in history—*integrate* the biosphere and noosphere in a higher and deeper union (all of which we will investigate later in this chapter).

Evolution moved beyond tribalism, and its limited capacity for social-planetary integration, for many reasons, it seems. Tribes may or may not have known how to remain reverential of nature; it was *other* tribes they could not integrate, and this because they lacked recourse to a binding *conventional* level of law and morality that very soon would start to build unified societies out of separate and conflicting tribal desires. The problem wasn't getting along with nature in the *biosphere*; it was getting along with other and conflicting interests in the *noosphere* that brought tribalism to its limits for evolutionary integrative power.

Habermas and his collaborators (particularly Klaus Eder and Rainer Dobert), who have been carefully studying the anthropological data, have mapped out some of the *inherent limitations* that brought tribal kinship systems to an evolutionary dead end.

In the evolutionarily promising neolithic societies, system problems arose which could not be managed with an adaptive capacity limited by the kinship principle of organization. For example, ecologically conditioned problems of land scarcity and population density or problems having to do with an unequal distribution of social wealth. These problems, *irresolvable within the given framework*, became more and more visible the more frequently they led to conflicts that overloaded the archaic [magic] legal institutions (courts of arbitration, feuding law).

A few societies under the pressure of evolutionary challenges from such problems made use of the cognitive potential in their worldviews [cognitive capacities of higher moral-pragmatic stages not yet socially embedded] and institutionalized—at first on a trial basis—an administration of justice at a conventional level [mythic-membership, concrete operational]. Thus, for example, the war chief was empowered to adjudicate cases of conflict, no longer only according to the distribution of power, but according to socially recognized norms grounded in tradition. Law was no longer only that on which the parties could agree.[22]

It was at this point that complex systems of mythology, whatever other functions they might have performed, began to serve first and foremost as a way to unify peoples *beyond mere blood lineage*. The human race was not only self-adapting, it was self-transcending.

The phenomenon to be explained is the emergence of a political order that organized a society *so that its members could belong to different lineages*. The function of social integration passed from kinship relations to political relations. Collective identity was no longer represented in the figure of a *common ancestor* but in that of a *common ruler* [or leader or governor, who sometimes, but not always, would abuse this newly emergent power; we will return to Habermas's thoughts on this later; he is at present focusing on the necessary and new integrative power brought by this transformation, and not on any of its particular abuses, which are legion and which he also discusses in great detail].

A ruling position gave the right to exercise legitimate [or agreed-upon] power. The legitimacy of power could not be based solely on authorization through kinship status; for claims based on family position, or on legitimate kinship relations in general, were limited precisely by the political power of the ruler. Legitimate power crystallized around the function of administering justice and around the position of the judge after the law was recognized in such a way that it possessed the characteristics of *conventional morality*. This was the case when the judge, instead of being bound as a mere referee to the contingent constellations of power of the involved parties, could judge according to intersubjectively recognized legal norms sanctified by tradition, when he took the intention of the agent into account as well as the concrete consequences of action, and when he was no longer guided by the ideas of reprisal for damages caused and restoration of a status quo ante [characteristic of preconventional morality], but punished the guilty party's violation of a rule. Legitimate power had in [this] instance the form of a power to dispose of the means of sanction in a conventional administration of justice. At the same time, mythological worldviews also took on—in addition to their explanatory function—justificatory functions. . . .[23]

In short, what was needed was not tribal but transtribal awareness, negating but preserving heretofore isolated tribal interests in a higher and wider communion; and mythology, not magic, provided the key for this new transcendence.

And so: some people today eulogize the primal tribal societies because of their "ecological wisdom" or their "reverence for nature" or their "nonaggressive ways." I don't think the evidence supports any of those views in a sweeping and general fashion. Rather, I eulogize the primal tribal societies for an entirely different reason: we are all the sons and daughters of tribes. The primal tribes are literally our roots, our foundations, the basis of all that was to follow, the structure upon which all subsequent human evolution would be built, the crucial ground floor upon which so much history would have to rest.

Today's existent tribes, and today's nations, and today's cultures, and today's accomplishments—all would trace their lineage in an unbroken fashion to the primal tribal holons upon which a human family tree was about to be built. And looking back on our ancestors in that light, I am struck with awe and admiration for the astonishing creativity—the *original* breakthrough creativity—that allowed humans to rise above a given nature and begin building a noosphere, the very process of which would bring Heaven down to Earth and exalt the Earth to Heaven, the very process of which would eventually bind all peoples of the world together in, if you will, one global tribe.

But in order for that to occur, the original, primal tribes had to find a way to transcend their *isolated* tribal kinship lineages: they had to find a way to go transtribal, and mythology, not magic, provided the key for this new transcendence.

MYTHOLOGICAL

> The transition to *societies organized through a state* required the relativization of tribal identities and the construction of a more abstract [meaning less body-bound] identity that no longer based the membership of individuals on common descent but on belonging in common to a territorial organization. This took place first through identification with the figure of a ruler who could claim close connection and privileged access to the mythological originary powers. In the framework of mythological worldviews the integration of different tribal traditions was accomplished through a large-scale, syncretic expansion of the world of the gods—a solution that proved to be rather unstable [and precipitated the next major transformation, which we will examine in a moment].[24]

It wasn't that magical-animistic societies had no mythologies, for they did. It was simply that, as Joseph Campbell explained, the rise of the first early states was marked by an explosion of codified mythologies—an enormous differentiation/integration of mythic motifs—and Habermas's point is that these mythologies became a large part of the integrating structures for society (i.e., providing both cultural meaning and social integration).

As we saw, in the previous or magical structure, personal identity was *natural* or body-based, and collective identity was likewise kinship or blood-bound, particularly through a common ancestor. Without a common ancestor (or kinship lineage), there was no way to socially integrate various interests. With the rise of the mythological structure, however, personal identity switched to a *role* identity in a society of a common *political* (not genetically related) ruler, and this ruler was given legitimacy not because of blood ties but because of his (or sometimes her) special relation to mythological gods/goddesses—"mythic-membership."

Under this arrangement, an enormous number of not-genetically-related tribes could be unified and integrated, something the preconventional magical structure—lying close to the body and close to nature—could never accomplish.

> Only with the transition to societies organized around a state do mythological worldviews also take on the legitimation of structures of [governance], which already presuppose the conventional stage of moralized law. Thus the naive attitude to myth [characteristic of the magical] must have changed by that time. Within a more strongly differentiated temporal horizon, myth is distantiated to a tradition that stands out from the normative reality of society and from a partially objectivated nature. With persisting sociomorphic traits, these developed myths establish a unity in the manifold of appearances; in formal respects, this unity resembles the sociocentric-objectivistic world conception of the stage of concrete operations [and conventional-stage morality].[25]

We will later address in more detail the topic of whether these myths actually contained any genuinely transcendental, mystical, or transformative spiritual capacities (as maintained by, e.g., Carl Jung and Joseph Campbell).

For the moment let us only note that these various "epochs," such as the magical or mythical, refer only to the *average* mode of consciousness achieved at that particular time in evolution—a certain "center of gravity" around which the society as a whole orbited. In any given epoch, some individuals will fall below the norm in their own development, and others will reach quite beyond it.

Habermas strongly believes, for example, that even in preoperational magical times, some individuals clearly developed cognitive capacities all the way up to formal operational cognition, not as fully formed structures but as potentials for understanding. This is an example of what I referred to in *Up from Eden* as the leading edge or *most advanced* form of consciousness evolution, in contrast to the *average* mode of the time. And I believe that the evidence clearly suggests that, to stay with the magical structure as an example, the most advanced mode of that time was not just formal operational, although I believe that was clearly present, but beyond that to the psychic level (evidenced, e.g., in shamans), as a potential for a *certain type* of understanding and awareness (which we will examine in chapter 8).

My point is that in each epoch, the *most advanced* mode of the time—in a very small number of individuals existing in relational exchange in microcommunities (lodges, academies, sanghas) *of the similarly depthed*—began to penetrate not only into higher modes of ordinary cognition (the Aristotles of the time) but also into genuinely transcendental, transpersonal, mystical realms of awareness (the Buddhas of the time).

Thus, in the magical, as I just mentioned, the most advanced mode seems to have been the psychic (embodied in a few genuine shamans or pioneers of yogic awareness); in mythological times the most advanced mode seems to have reached into what is known as the subtle realm (embodied in a few genuine saints); and in mental-egoic times the most advanced modes reached into the causal realm (embodied in a few genuine sages), all of which we will discuss in chapter 8.[26]

But the *average* mode of the mythological epoch did not reach into these subtler and transpersonal dimensions, but remained grounded in a concrete-literal interpretation of myth (e.g., Moses actually did part the Red Sea, as an empirical fact). There is precious little that is transpersonally inspired about such literal myths; rather, as Habermas suggests, they clearly represent the sociocentric integration offered by concrete operations and conventional morality, and for that reason were extremely important in moving social integration beyond the tribal and the preconventional—and with that their transcendental status, exceptionally important but limited, seems most definitely to end.

MYTHIC-RATIONAL

The word *rational* is an impossible label; it means a million things to a million people, and not all of the meanings are kind. Moreover, even if we decide what it means, there are still several different types of "it." Weber, for example, differentiated between purposive-rationality (such as scientific-technological knowledge), formal rationality (such as mathematics), and intersubjective or practical rationality (as displayed in morality and communication).

Cognitive psychologists and anthropologists tend to use *rationality* to mean "formal operational cognition," which simply means the capacity not just to think, but to think about thinking (and thus "operate upon" thinking: "formal operational"). Since you can operate upon or *reflect upon* your own thought processes, you are to some degree free of them; you can to some degree *transcend* them; you can take *perspectives* different from your own; you can entertain *hypothetical* possibilities; and you can become highly *introspective*. As we will see in the next chapter, all of these come into existence with the emergence of formal operational cognition, or "rationality."

Moreover, since you can now *reflect on* your own thoughts and patterns of behavior, you will now seek to justify your thoughts and actions, based not simply on what you were taught, or on what your society tells you (that is the rule-bound or conformist or conventional or sociocentric mode of the previous concrete operational cognition), but rather on a review of the *reasons* and the actual *evidence* for such beliefs. Did Moses *really* part the Red Sea?

Rationality, in this sense, means that you seek "reasonable reasons" for your beliefs. What is the evidence? Why should I believe this? Who says so? Where did you get that idea?

And finally, because we can reflect on our own thought processes, and thus to some degree remove ourselves from them, we become capable of imagining all sorts of other *possibilities*; we become *dreamers* in the true sense of the word. Other perspectives, other beliefs, other horizons open before the mind's eye, and the soul can take flight in the worlds of the not-yet-seen. Rationality is the great doorway to the invisible, through which, and then beyond which, lie so many secrets not given to the senses or to conventions (which is why all true mysticism is transrational and never antirational; "right thought" always precedes "right meditation").

Thus, the idea that rationality is somehow "dry and abstract," or that it has "no feelings," is way off the mark. Rationality creates a deeper *space* of possibilities through which deeper and wider feelings can run, feelings not bound to one's isolated desires or the narrow confines of official conventional reality. The fact that rationality is a relatively high level of development means, as always, that it *can* repress its lower holons, particularly emotions of sex and aggression, and *that*, as always, leads to pathology. And it is the *pathological* expression of rationality that has given rationality a bad name ("dry and abstract"), but that is definitely not characteristic of the structure as a whole. However, since *rational* is something of a loaded word, we might better use the word *reasonable*—they mean the same thing: what are your reasons, why are you doing that?

And this is why rationality or reasonableness tends to be *universal* in character, and is highly integrative. If my reasons are going to be valid, I want to know that they make sense, or that they hold true, not just for me or my tribe or my isolated culture (however important those might also be). If science, for example, is going to be true, then we are not going to have a Hindu chemistry that is different from a German chemistry that is different from a Greek chemistry. There is simply chemistry, and its truth is not forced or coerced or ideologically imposed, but is freely open to any who wish to look into its reasons.

This doesn't mean that we can't have special cultural differences that make each society unique and special; it means that only rationality will allow these differences to exist side by side by seeing them as different perspectives in a more universal space, something that cultural differences, left to their own conventional or sociocentric or ethnocentric devices, could never do. It is only rationality, in other words, that allows the beginning emergence of a truly global or planetary network, which, freed from any particular society, can allow all societies their own unique and special place.

But you can imagine what happened historically when the mythological structure (Moses parted the Red Sea) ran into the newly emerging rational structure (What does that mean, parted the Red Sea? Where is your evidence for that?). Was it really right for the gods to lie and rape, as Homer had maintained? Or could the myths simply be *wrong*? Plutarch records his father saying to him: "You seem to me to be handling a very large and dangerous question—or rather you are disturbing subjects which ought to be left alone, when you question the opinion we hold about the gods, and ask reason and proof for everything."

The indictment handed down by the city of Athens began, "Socrates is guilty of refusing to recognize the gods of the State." It ended with, "The penalty demanded is death." When asked, as was customary, if Socrates could suggest an alternative punishment, he suggested free meals for life.

But he refused a prison escape that had been arranged for him, and freely drank the hemlock. In choosing death instead of state mythology, Socrates died for the cause of emergent reason.

The rest of humanity, however, did not move quite as fast in that particular transformation. Rather, in this clash between myth and newly emerging reason, the traditional mythological structures were at first *rationalized*. That is, the old myths were propped up with rational reasons. And this structure, in general, I call mythic-rational.[27] Habermas does not specifically use the term *mythic-rational*; he speaks of the rationalization of mythological organization, an essentially similar notion.

Habermas makes several interesting observations about this mythic-rational space. First of all, we are dealing with the *universalizing* tendency in rationality—the desire of rationality to embrace, make room for, and integrate a *global* or *planetary* consciousness. But this planetary or global tendency first expressed itself in an attempt to extend a *particular mythology* to world-embracing dimensions. In other words, a military attempt to conquer as many peoples as possible.

Thus, according to Habermas, at this point historically we see the emergence, around the world, of the great Empires. From the Incas to the Aryans, from the Aztecs to the Alexanders, from the Khans to the Romans: the coming of the Empires was the first brutal step in a planetary expansion struggling to reach beyond isolated cultures by simple conquest.

Once conquered, the peoples were then usually given a genuinely equal access to *citizenship* (in many cases this included citizenship for women, slaves, and children). But these citizens could only be *global* citizens if the empire itself conquered the globe—that is, if everybody embraced the mythology that rationality was trying to prop up—if everybody became "true believers" in the chosen mythic god or goddess (we still see this type of mythic-imperialism operative in the world today), a rather twisted version of planetary citizenship.

Likewise, the old mythological structures had contained implicitly another type of partial move toward global or planetary citizenship, namely, *equal citizens of the faith*. Although this was also a faltering step in the right direction, it too suffered from the fact that the various mythologies themselves still *differed from each other*. All Christians anywhere, of whatever color or race or sex, were equally saved; but all Hindus would go to hell. The great mythologies, and the great empires that carried them to various corners of the globe, ran up against their own inherent limitations for integrative power, simply because they ran into each other; and the only way to overcome those differences was to shed particularistic and divisive mythologies and transform to a more global reasonableness (the *next* stage, the mental-rational).

But we are now at the point where the old mythologies, carried now by empires, are trying to be propped up and rationalized. But that means that there was already a profound break with pure mythological thought, which previously was simply accepted on its own terms. Habermas:

> For this reason, imperially developed civilizations [empires] had to secure their collective identity in a way that presupposed a break with mythological thought. The universalistic world interpretations of the great founders of religions and of the great philosophers grounded a commonality of conviction mediated through a teaching tradition and permitting only abstract objects of identification. As members of universal communities of faith, citizens could recognize their ruler and the order represented by him so long as it was possible to render political domination plausible in some sense as the legacy of an order of the world and of salvation that was believed in and posited absolutely [mythic dogma].
>
> The great empires had to demarcate themselves from a desocialized outer nature [biosphere] as well as from the social environment of those alien to the empire. But since collective identity could now be secured only by way of doctrines with a universal claim, the political order also had to be in accord with this claim—the empires were not universal in name alone [but in attempted global conquest as well]. . . .
>
> But the reality of *other empires* was incompatible with this definition of the boundaries and social environment of an empire. Despite the existence of trade relations, and despite the diffusion of innovations, the empires shielded themselves from this danger; between themselves they maintained no diplomatic relations in the sense of an institutionalized foreign policy. Their political existence was not dependent on a system of reciprocal recognition [they were not truly rational-planetary].[28]

Even though it was a mythological worldview that was trying to be propped up, the propping up was nevertheless being done with formal operational thought, and thus central parts of the mythic-rational world were centered around formal rationality and its universalizing or global outreach:

> There arise cosmological worldviews, philosophies, and the higher religions [so-called "rational religions"], which replace the narrative explanations of mythological accounts with argumentative foundations [they give reasons]. The traditions going back to the great founders are an explicitly teachable knowledge that can be dogmatized, that is, professionally rationalized. In their articulated forms rationalized worldviews are an expression of formal operational thought and of a moral consciousness guided by [postconventional] principles.[29]

But a propping-up is still a propping-up. And in one of the most difficult of all historical transformations, a universal or global reasonableness began slowly to supplement local and divisive mythologies, mythologies that, precisely because they could not be universally argued and supported by shared evidence, could only be supported militarily and imperialistically.

But slowly, in both the East and West, began to emerge rational philosophies, rational sciences, rational policies, rational religions—some of which indeed pointed *beyond* reason, but all of which depended on reason as a platform that could secure a common and mutual understanding for anyone, of any color, race, or creed, who cared to talk about it, share their evidence, discuss their reasons, and not simply shout dogmatisms and claim divine support.

RATIONAL

As the planetary reach of reasonableness began to shake off the divisive mythologies, empires gave way to modern states—which at least formally and mutually *recognized* one another, thus making room for each on the planet. The modern state thus separated from its embeddedness in particular mythologies—*the all-important separation of church and state*—and further removed itself from the economic sphere by allowing the emergence of a global market economy, all of which had to be grounded in universalistic reasons (but still tinged, initially, by remnants of imperialism, which indicated not an excess of reason but a lack of it).

The highest principles lost their unquestionable character; religious faith and the theoretical attitude became *reflective*. The advance of the modern sciences and the development of moral-practical will-formation were no longer prejudiced by an order that—although grounded—was posited absolutely [mythology]. For the first time, the universalistic [global] potential already contained in the rationalized worldviews could be set free. The *unity of the world* could no longer be secured objectively, through hypostasizing unifying principles (God, Being, or Nature); henceforth it could be asserted only reflectively, through the unity of reason. The unity of theoretical and practical reason then became the key problem for modern world interpretations. . . .[30]

In other words, it was no longer simply the case of a subject trying to look at and understand and operate on a world of objects (concrete operational); it was the subject looking at and trying to understand and operate on *itself*; it was the subject trying to understand the *subject* (formal operational). Suddenly, very suddenly, humanity had taken a new turn, a new transcendence, had discovered a new and deeper interior, with a new and higher consciousness, and that consciousness was to be found by *looking within*.

As we will see, Habermas, Gebser, and I (and many others) place the rough beginning of this new emergence (egoic-rational) in the middle of the first millennium BCE, but it reaches its fruition with the rise of the modern state, roughly the sixteenth century in Europe. We will be concentrating on this fruition, but let us at least note that what many of the central endeavors of this entire period have in common is a new type of *looking within*. In philosophy, for the first time, the theme is not "What is there to know?" but "How can I know it?" Not "What objects are out there?" but "What is the structure of the subject that wants to know?" From Socrates' Delphic "Know thyself" to the exhaustive introspections of Hume and Locke and Descartes and Kant (and Nagarjuna and Garab Dorje and Chih-i and Fa-tsang), the common theme heard over and over and over again: Look within.

We see precisely the same thing with the rise of the "rational religions," which does not mean that these religions were *only* rational—indeed, their aim was to go even beyond rationality in all cases. But these were the first great religions that took their stand, *not* in a mythological pantheon of gods and goddesses *out there* that had to be appeased through magic or mythic ritual, but rather took as their *starting* point: "The Kingdom of Heaven is *within*," as Jesus of Nazareth would put it.

Likewise, the essential message of Gautama the Buddha could be summarized in the statement "Don't worry about gods, goddesses, spirits, the afterlife, any of that—rather, look very carefully at the nature of your own subject, your own self, and try to penetrate to the bottom of that, for if enlightenment exists, it lies through an understanding of (and going beyond) the subject itself." All of that was radically, *radically* new.

For precisely the same reasons, individuals were no longer identified by the simple and unreflexive *roles* that they played in society, as was the case with ethnocentric and mythic-membership societies. Rather, in a postconventional society, individuals were identified in many ways by their own free choices, *as free subjects*, within the broad constraints of civil law. In other words, as Habermas puts it, an *ego* identity replaced a *role* identity.

> This domain of decentralized individual decisions was organized on universalistic principles in the framework of bourgeois civil law. It was thereby supposed that the private, autonomous, legal subjects pursued their interests in the morally neutralized domain of intercourse in a purposive-rational manner, in accord with general [universal] maxims. From this conversion of the productive sphere to universalistic [global] orientations there proceeded a strong structural compulsion for the development of personality structures that replaced conventional role identity with ego identity. [As I said, Habermas and I both place the beginning of this change to ego identity in the middle of the first millennium BCE, but it reaches its fruition here, with the rise of the modern state, beginning roughly in the sixteenth century in Europe]. In fact, emancipated members of bourgeois society, whose conventional [sociocentric] identity had been shattered, could know themselves as one with their fellow citizens in their character as (a) free and equal subjects of civil law, (b) morally free subjects, and (c) politically free subjects (the citizen as democratic citizen of the state).[31]

This is an extraordinary development, for many reasons (so much so that the third volume in this series is devoted to it). But, at this point, let us briefly return to the themes raised earlier in this chapter and at least note how this new egoic-rational emergence (with, among many other things, its differentiation of the noosphere and the biosphere) would affect the sexes, for it is here that we find some of the most surprising developments.

LIBERATION IN THE NOOSPHERE

According to most radical feminists, one of the special strengths of women is their ages-old connection with the Earth, with nature, with *embodiment* (in fact, a good summary of the female value sphere according to radical feminism would be "embodiment in communion"). The radical feminists maintain that woman's special association with embodied nature needs to be honored and cherished and celebrated. The connection of woman and nature is the source of female power and freedom and liberation.

According to most liberal feminists, it is precisely the opposite. The equation of woman and nature, they maintain, is *the* primary and overwhelming source of female oppression. The woman/nature "identity," they believe, has been the primary identification that men have, throughout the ages (literally from day one), used to keep women locked out of the noosphere (out of the public, productive, legal, cultural, power-wielding sphere). "Woman = nature" translates directly into "barefoot, pregnant, and in the kitchen," and the liberal feminists are alarmed that the radical feminists are even thinking such thoughts, let alone championing them.

But my point is that, once the noosphere and the biosphere had differentiated, it was—it is—an entirely different world, where all the old equations suddenly take on entirely new meanings. And although some of those meanings might indeed have been oppressive in the past, they are not necessarily so in the new world (in the new worldspace, literally).

First, even if the radical feminists are right, and woman's special power is her rootedness in nature and body (the biosphere), nonetheless the differentiation of the noosphere and the biosphere meant that women were no longer just that, or only that, or exclusively that. Since the noosphere and the biosphere had at this point been clearly differentiated on a collective scale, this is the first time that the *exclusive* identity "woman/nature" was no longer needed, or appropriate, or even relevant (as an *exclusive* identity, which does not deny the radical feminist claim that women have special connections with Earth and body and nature; it means only that that is not *all* they are or have).

It was here, in the clearly differentiated noosphere, that there was then no conceivable reason that women could not enter the world of *public agency* (bringing with them their own forms of communion). Biological roles were no longer primary; physical strength was no longer the major determinate of public power; the parameters of biospheric evolution were shading into those of noospheric evolution, such that *right* began to replace *might*. And it was *right* for the female to assume cultural agency in the newly differentiated world, a right that wasn't *prevented* before but was simply *meaningless* before (on any collective scale).

And indeed, it is precisely here, and *never* before, that we first start to see the beginning of a widespread and organized women's movement, a movement not just discontent about this or that particular mistreatment (which had happened since day one), but about women's place in society *on the whole*.

This was radically new—a new emergence, a new transcendence, and a new differentiation that *demanded a new integration*—woman as public and historical *agent* (where heretofore, as we saw, that role largely fell to the male as father).

Thus, we do not find the emergence of a widespread women's movement prior to this time, not because women were duped or brainwashed or submissive chattel, but because liberation in the feminist sense (woman as free agent) was virtually meaningless as long as the biosphere and the noosphere were not clearly differentiated, first, and as long as the state and the economic sphere were not clearly differentiated, second (both of which came to fruition, we have said, around the sixteenth century).[32] At that point, and not before, the rights of women as free agents would have profound meaning and profound desirability.

And indeed, the first great feminist treatise anywhere in history was Mary Wollstonecraft's *Vindication of the Rights of Woman*, written in 1792.[33] Women did not then suddenly become strong and intelligent and responsible after a million years of dupedom; but rather, using the *same* intelligence and strength and insight with which they had mined the natural wisdom of the biosphere for a million years, they now saw the first structural opening in the noosphere and they acted *immediately* on it. It would be a mere two hundred years later—a blink in evolutionary time—

that women would secure access to the public and political sphere of agency in the form of the right to vote and the right to own possessions. This was blazingly fast action on the part of women once the structural conditions of social life had switched its center of gravity from the biosphere to the noosphere.

And so henceforth, the role of the father (the "familialization of the male") *would no longer suffice to link the two value spheres*, but would have to be joined by the role of female as historical *agent*, equal noospheric agent (not just embodied in communion, but encultured in agency). And here is where the liberal feminist argument surely finds its strongest voice and its unshakable moral authority.

But since the noosphere transcends and includes the biosphere, there is no reason that the liberal cannot embrace the radical and then add its own extra demands. It's not like we're on a flatland playing field, where more of the one means less of the other. On a multidimensional field, in the real Kosmos, the more of both, the better.

In other words, special rootedness in the biosphere can indeed be reasonably claimed—there is literally *a million years* of rich tradition of the wise woman who feels the currents of embodiment in nature and communion, and celebrates it with healing rituals and knowing ways of connecting wisdom, a wisdom that does not worship merely the agentic sun and its glaring brightness, but finds in the depths and the organic dark the ways of being linked in relationship, that puts care above power and nurturance above self-righteousness, that reweaves the fragments with concern, and midwifes the communions and the unsung connections that sustain us each and all. And finds, above all, that being a self is always being a self-in-relationship.

The liberal position ought to be able gladly to embrace all that, as rootedness, and then add its own special and emergent ideas and demands—which are precisely those demands *made possible* for the first time in history by this extraordinary new emergence, by this differentiation of the noosphere and the biosphere, by the emergence of global rationality—and those demands are exactly the possibilities we just saw Habermas elaborate, applied now to women as well: to be (a) free and equal subjects of civil law, (b) morally free subjects, and (c) politically free subjects. And that is the enduring truth and strength of liberal feminism, which arose at this time, and not before this time, precisely to announce that emergent truth.

There is one final twist in the story. If the differentiation of the biosphere and noosphere (and the differentiation of the Big Three) began in earnest only with modernity, it is nonetheless the case, as we will see, that this differentiation has tended in many respects to slide into *dissociation*. And that gives the radical feminist component of the integration an important and urgent weight. We have somehow forgotten our embodiment.

And further: Since "woman, body, nature" have often been different aspects of the same dissociated fabric, this gives a certain cogency to the concerns of the ecofeminists as well (though we don't necessarily have to embrace their theoretical and historical explanations, which tend to the ideological).

This is all part of the new integration ("centauric," "vision-logic") that I have been referring to, and which involves, among other things, the integration of mind and body, the integration of noosphere and biosphere (*after* their differentiation). And since, for *whatever* reasons, throughout history and prehistory there was a special connection between woman, nature, and body,[34] then the integration of mind and body means virtually the same as the integration of male and female—in each of us.

At that point, *both* the male and the female would have one foot in the biosphere *and* one foot in the noosphere—an emergent integration utterly *unprecedented*, and the exact meaning of which we are all still trying to figure out.

Which is why the past is of no real help here. Calling on inherited archetypes and old mythological motifs is of little help in this new, emergent, and unprecedented endeavor. Likewise, many ecofeminists feel they have to be able to point to a past society (almost always horticultural) where women were "equal" in order to show that such equality is at least a possibility of human nature and human societies. But my point is that the possibility itself is an *emergent*—it *never* was, but is now coming to be—and so we don't *need* to wildly reinterpret the past in order to find hope for the future.

We are looking at an emergence, not an exhumation; at a birth, not a disinterment.

So I suppose it is all somehow deeply symbolic that Mary Wollstonecraft died giving birth to Mary Shelley, who wrote *Frankenstein*. A death, a birth, a monster.

The image of "that monster!" seems to fit how both men and women view each other as their new and unprecedented roles in the noosphere continue to be worked out, and why, I suppose, a certain kindness on both sides is so necessary. Throughout this new and difficult integration—this death and this new birth—each sex is the monster of the other's creation.

VISION-LOGIC/PLANETARY

Which brings us up to the present, and the new integration that is struggling to emerge, in all four quadrants. . . .

> We are concerned no longer with cultural inflections, but with a passage from one culture stage to another. In all previous ages, only restricted portions of the surface of the earth were known. Men looked out from the narrowest, upon a somewhat larger neighborhood, and beyond that, a great unknown. They were all, so to say, insular: bound in. Whereas our view is confined no longer to a spot of space on the surface of this earth. It surveys the whole of the planet. And this fact, this lack of horizon, is something new.

Those words, by the great Leo Frobenius in his *Monumenta Terrarum*, were written in 1929. The previous age, that of the early emergence of the rational and its first extraordinary discoveries, Frobenius called the Monumental Age, and the opening of the new, just now under way, he called the Global or World Culture.

As rationality continues its quest for a truly universal or global or planetary outlook, noncoercive in nature, it eventually gives way to a type of cognition I call vision-logic or network-logic. Where rationality gives all possible perspectives, vision-logic adds them up into a totality, which is simply the new and higher interior holon. Aurobindo gave the classic description of vision-logic, which "can freely express itself in single ideas, but its most characteristic movement is a mass ideation, a system or totality of truth-seeing at a single view; the relations of idea with idea, of truth with truth, self-seen in the integral whole."

What I am trying to do in this book, and what you are trying to do as you read it (or other similar books), is use vision-logic: not just reasonably decide the individual issues, but hold them together at once in mind, and judge how they all fit together as a truth-vision. In other words, vision-logic is a higher holon that *operates upon* (and thus transcends) its junior holons, such as simple rationality itself. As such, vision-logic can hold in mind contradictions, it can unify opposites, it is dialectical and nonlinear, and it weaves together what otherwise appear to be incompatible notions, as long as they relate together in the new and higher holon, *negated* in their partiality but *preserved* in their positive contributions.

This is, for example, what Hegel called "Reason" as opposed to "the understanding" (or the simpler, empiric-analytic rationality of propositions, or Aristotelian logic). And this is why Hegel maintained that the central defining characteristic of Reason (vision-logic) was its capacity to unify opposites and see identity-in-difference. (As such, Hegel was one of the first great philosophers of vision-logic, as were Schelling, Whitehead, and a few others we will explore later; and not just because of their systematic wholeness—that had been attempted before—but because of their explicit grasp of identity-in-difference or "nonbifurcated Reason" or vision-logic, beyond which lies the transrational altogether.)

And it is vision-logic that drives and underlies the possibility of a truly planetary culture (or, rather, the first true forms of planetary organization, which themselves will most likely evolve according to yet higher categories, as we will see).

Recall that we left off Habermas's account with the emergence (beginning on a collective scale around sixteenth-century Europe) of an *ego identity* (first secured by men, relatively swiftly followed by women) wherein *all* people were considered (1) free and equal subjects of civil law, (2) morally free subjects, and (3) politically free subjects as citizens of the democratic state (which also, of course, for the first time in history, would put a final end to slavery and back that emancipation by law).

But, Habermas continues, "these abstract determinations are best suited to the identity of *world citizens*, not to that of citizens of a *particular* state that has to maintain [and defend] itself against other states."[35] In other words, legally, morally, and politically free subjects: this is a state of affairs that actually applies to world citizens, not just to citizens of this or that state. The free subject, created by rationality, could not easily or for long be confined to the state, but belonged rather to what Habermas calls "global forms of intercourse" and "permanent variation of all reference systems"—that is, vision-logic.

Thus, true world citizenship belongs not to national organizations but to planetary organizations (as Habermas puts it, "The need for coordination at a supra-national level cannot easily be satisfied as long as governments have to legitimate themselves exclusively in terms of national decisions").[36] The modern nation-state, founded upon initial rationality, has run into its own internal contradictions or limitations, and can only be released by a vision-logic/planetary transformation.

THE CENTAUR IN VISION-LOGIC

The worldview or worldspace of vision-logic I also refer to as "existential" and "centauric." "Existential" we will examine in a moment. "Centaur" is the mythic beast, half human and half horse, which I (and others such as Hubert Benoit and Erik Erikson) have taken as a symbol of the integration of body and mind, or biosphere and noosphere. For if it is true that, a few hundred years ago, we finally succeeded in clearly differentiating these two great domains, it is equally true that we have not yet found a way to integrate them. On the contrary: the necessary differentiation of the biosphere and the noosphere has now moved clearly into the beginning stages of *dissociation*, and, indeed, some ecologists feel that the dissociation is fast becoming irreversible. Be that as it may, that some sort of dissociation is under way is all but undeniable.

It is the integrative power of vision-logic, I believe, and not the indissociation of tribal magic or the imperialism of mythic involvement that is desperately needed on a global scale. For it is vision-logic with its centauric/planetary worldview that, in my opinion, holds the only hope for the integration of the biosphere and noosphere, the supranational organization of planetary consciousness, the genuine recognition of ecological balance, the unrestrained and unforced forms of global discourse, the nondominating and noncoercive forms of federated states, the unrestrained flow of worldwide communicative exchange, the production of genuine world citizens, and the enculturation of female agency (i.e., the integration of male and female in both the noosphere and the biosphere)—all of which, in my opinion, is nevertheless simply the platform for the truly interesting forms of higher and transpersonal states of consciousness lying yet in our collective future—if there is one.

Jean Gebser refers to the emerging vision-logic as the "*integral-aperspectival*" mind, which is a particularly apt phrase. The previous structure (the egoic-rational), Gebser refers to as the "rational-perspectival," because rationality can indeed take different *perspectives*, as we saw. But vision-logic, or the integral-aperspectival mind, adds up all the perspectives *tout ensemble*, and therefore *privileges no perspective as final*: it is aperspectival.

The aperspectival mind, in other words, is holonic through and through: contexts within contexts within contexts forever. Of course, every structure of consciousness is actually holonic (there are only holons), but vision-logic consciously grasps this fact for the first time, and thus finds its own operation increasingly transparent to itself (this "transparency," according to Gebser, is a primary characteristic of the integral-aperspectival mind.) "Aperspectivity expresses itself in a structure of consciousness which is in process of emerging and is, therefore, new," he says. "We would like to designate this new structure the 'integral structure' of consciousness, and the modality of the world in the process of emerging the 'aperspective world.' "

Gebser's masterpiece, *The Ever-Present Origin*, was completed in 1953, the culmination of several decades of thought and research. He died in 1973 and thus he did not live to see the full extent of the explosion of aperspectivism in today's postmodern world. Not only has the aperspectival vision-logic defined postmodernity (in its best aspects), it has also defined many of postmodernity's self-conscious problems: individuals are not only aware of holonic and aperspectival space, but are often totally lost in it. The postmodern poststructuralists, for example, have gone from saying that no context, no perspective, is final, to saying that no perspective has any advantage over any other, at which point they careen uncontrollably in their own labyrinth of ever-receding holons, lost in aperspectival space.

That all perspectives interrelate, or that no perpsective is final (aperspectivism), does not mean that there are no relative merits among them. The postmodern poststructuralists go from saying "there is no final perspective" (or "perspectives are boundless") to saying "therefore there is no advantage in any perspective over another." This leveling of perspectives is not an interrelation of all perspectives but is itself merely one particular and covertly privileged perspective (and thus ends up, as we have seen, being perfectly self-contradictory: there is no advantaged perspective except mine, which maintains that all other perspectives are not so privileged). The postmodern poststructuralists confuse their own aperspectival madness with literary and political critique, following, I presume, the pioneering insanity of Bataille.

Nonetheless I believe, and I think Gebser would agree, that recent events have, if anything, strengthened his thesis: for better or for worse, the world is in the midst of the torturous birth throes of a collective emergence of an entirely new structure of consciousness, the centaur in vision-logic, the integral-aperspectival mind.

Gebser refers to such transformations as "mutations in consciousness," emphasizing their radically emergent nature. Nonetheless, each structure unfolds holarchically, transcending and including its predecessor(s). "None of these mutations of consciousness is responsible for the loss of previous possibilities and properties, but suddenly incorporates them into a new structure. Within the mutations of consciousness there takes place a process of re-arrangement beyond the reach of mere space-time-bound events, an [emergent] process, which manifests itself discontinuously, or by leaps and bounds. With every new mutation of awareness, consciousness unfolds more powerfully. . . ."[37]

This new integral-aperspectival structure is, of course, *integrative*. But integral vision-logic is not merely a sum of previous parts—that, again, would be mere horizontal expansionism, not vertical and creative emergence (and transcendence). As Gebser says, "The part is to a certain degree always a betrayal of the whole, for which reason the sum of the parts also only yields a fictitious but not an efficacious whole." Likewise, the dialectical nature of vision-logic—that is, the unity-of-opposites conceived mentally (as "mutual interpenetration")—is a hallmark of the integral structure, is "intrinsic to the emergent aperspectival consciousness."

But the emphasis, indeed, is on integration. As Georg Feuerstein, Gebser's ablest interpreter, says of the integral-aperspectival structure: "This nascent structure of consciousness, for the first time in human history, permits the conscious integration of all previous (but co-present) structures, and through this act of integration the human personality becomes, as it were, transparent to itself. . . ."[38]

Likewise, centauric-integral awareness integrates the body and mind in a new transparency; the biosphere and the noosphere, once finally differentiated, can now be integrated in a new embrace. Feuerstein therefore refers to this newly emerging structure as "psychosomatic," involving "the resurrection of the body," evidenced in such movements as holistic medicine and ecological sensitivity. "It is a whole-bodily event," he says, "feeling through the *lived* body. It does not take flight from bodily existence in any form. Rather it is grounded in one's unmitigated acceptance of, or primal trust in, corporeality. *It is the transparent body-mind.*"

Precisely because this centauric awareness transcends (but includes) so much of the verbal-mental-egoic dimension, that entire dimension itself becomes *increasingly objective*, increasingly transparent, to centauric consciousness. Where previously the verbal-mental-egoic self used those structures as something *with which* to view (and co-create) the world, now those structures themselves increasingly become an object of awareness and investigation by centauric consciousness (it is not just the mind looking objectively and "representationally" at external objects—the "reflection paradigm"—but the mind looking at the mind intersubjectively).

This is behind Gebser's claim that integral-aperspectival consciousness is especially a consciousness of language. As he puts it, "Language itself is treated as a primordial phenomenon by recognizing its originating-creative nature. Structurally, a new estimation of grammatical aspects and a novel use of syntactical freedom are evident" (and here we are on the trail of Foucault and Derrida, and the whole "linguistic turn" dating especially to de Saussure).

Likewise, Gebser relates the integral-aperspectival structure to the emergence of most forms of phenomenology, with their emphasis on a "postrational paradigm" that is also "embodied." He particularly focuses on Bergson, Husserl, and Heidegger, although he had some sharp criticisms of them as well (the last-named is also a "pioneer" of poststructuralism, and his notion of the *Destruktion* of rationalist ontology appears in Derrida as deconstruction).

Gebser's full genius comes to the fore in his brilliant descriptions, across virtually all disciplines, of the mutation, now globally in progress, from rational-perspectivism to integral-aperspectivism, and I cannot hope to do any sort of justice to his exhilarating presentation. (Readers are urged to consult *The Ever-Present Origin*—it is by far the pioneering book in "emergent worldviews"—and an excellent introduction, Georg Feuerstein's *Structures of Consciousness.*)

Gebser knows this aperspectival structure so intimately that he can spot even the most telltale signs of its emergence in virtually any endeavor. Here are just a few of the areas he covers in enormous detail (and remember, *The Ever-Present Origin* was published in 1953; the following list therefore includes only some of the very earliest signs of the emergence of vision-logic, of those movements that would be "postrational" or "post-structural" or "postmodern" in the best sense, those movements attempting to overcome the "self-defining subject" and the "Age of Man," those movements that would point not just to a differentiation of the Big Three, but to their integration as well): *biology* (e.g., Hans Driesch, Hugo de Vries), *mathematics* (Hilbert's axiomatics), *psychology* (both Freud and Jung in their own ways with an emphasis on depth psychology), *philosophy* (linguistics, phenomenology), *jurisprudence* (where care and "responsibility have entered the language of law in juxtaposition to justice and rights . . . this leads to 'open justice,' a hitherto impossible flexibility"—notice the integration of agency-rights and communion-responsibility, which we postulated as one of the centauric hallmarks), *economics* (Feuerstein takes E. F. Schumacher to be most recently representative), *history* (e.g., Arnold Toynbee and Swiss historian J. R. von Salis), *music* (Schoenberg, Stravinsky—"the new music is aperspectival inasmuch as it seeks to overcome the earlier fixities of meter or tonality as aspects of aesthetics"), *architecture* (Frank Lloyd Wright, Le Corbusier, Alvar Aalto), painting (Delacroix "revolted against the tyranny of the straight line, and Cézanne first seriously challenged the reign of perspectivity"), and, of course, the new *physics*, associated especially with Einstein and Planck (although this is not, as popularizers such as Fred Alan Wolf would have it, a "proof of mysticism"; as Feuerstein comments, "it would be imprudent to ignore the mythologizing trend within the new physics, which Gebser would readily have recognized as regressive rather than integrative").

As for the centauric-integral dimension being a fully worldcentric vision, Gebser refers to it as *"universal-integral"* and "world open," "world transparent"—which Feuerstein, correctly I believe, equates with the rising Global or Planetary Culture. In Gebser's words, the perspectives of the egoic-rational world are "replaced by the open expanse of the open world," the *"aperspectival world"*—the culmination of the worldcentric vision started by rationality and completed by vision-logic.

Neither Gebser nor I (nor Murphy nor Habermas, nor any of the evolutionary-oriented theorists) see the emergence of the "aperspectival worldcentric" structure as being a sure thing, as being somehow guaranteed. Not only does evolution, as Michael Murphy put it, meander more than progress; not only, when it does progress, is there always the "dialectic of progress"; there is also the ever-lurking possibility that the whole thing might simply blow up, that evolution will take a wrong turn (in the short run), but a wrong turn that includes us; that the stresses induced by the differentiation of the noosphere and the biosphere will make the whole system unsustainable. Evolution, we have amply seen, is not predictable, only reconstructable.

Nor does the fact that the integral structure is integral guarantee that the necessary integration will in fact occur. The claim is simply, to put it in the terms we have been using, that the integral structure *can* integrate the physiosphere, the biosphere, and the noosphere—it has the *potential* for that integration. Whether that potential becomes actual is up to you and me; it depends on the concrete actions that each of us takes.

As always, we have to make the future that is given to us.

THE FOUR QUADRANTS

Figure 5-1 is a summary of the four quadrants as we have followed them in the evolutionary curve thus far. Like all diagrams, it is very schematic and leaves out more than it includes. But it will serve to indicate the four quadrants and a handful of their milestones up to the present.

A few points about the diagram: in the Lower-Right quadrant, where evolution enters the human domain, I have only indicated the most concrete forms of geopolitical structures. But this quadrant includes the *exterior* of any of the *social* aspects of human interaction, including forces of production and techno-economic modes (bow and arrow, horticultural tools, agrarian implements, industrial machinery, computers, and so on), architectural structures, transportation systems, physical infrastructure, even written (material) forms of books, legal codes, linguistic structures, verbal signifiers, and so forth. Likewise, in the Lower Left, the cultural, I have included only worldviews; but this quadrant also includes interpretive understandings, cultural meanings in general, collective and group identities, intersubjective moral and ethical understanding, and so on.[39]

In the Upper-Right quadrant, as evolution moves into the human domain, I have indicated states marked by SF1, SF2, and SF3. These are the structure-functions of the human brain that correspond with concrete operational, formal operational, and vision-logic. These are currently

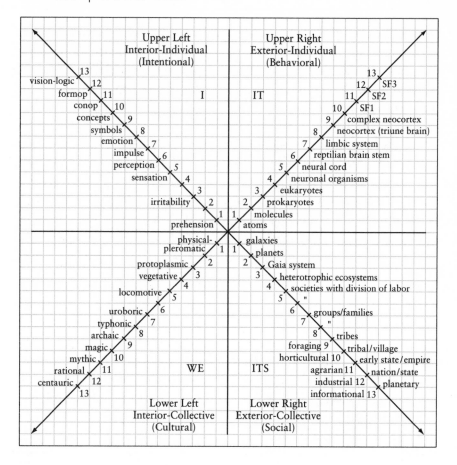

Figure 5-1. Some Details of the Four Quadrants

being mapped using PET and other sophisticated instruments, and I am simply indicating the correlation, whatever it turns out to be, with these symbols. Everybody agrees that mental states and structures have some sort of correlates in brain physiology, and this is all I mean by the symbols SF1, etc.[40]

Of course, we can also draw this diagram as a series of interfolding or nested pyramids, so that several of the multidimensional relationships within each quadrant could be better indicated. But then it would begin to lose the advantage of a certain type of simplicity. As long as its schematic and simplifying nature is kept firmly in mind, this figure will suffice. We will continue to draw on the implications of this diagram—and the four quadrants—as we continue to unfold our narrative.

THE WORLD IN TRANSFORMATION

About the centauric/planetary transformation, now haltingly but unmistakably in progress, I will make only a few passing comments now (all of volume 3 is devoted to this topic).

The previous, mythic-imperial empires (East and West, North and South), with their inherent dominator hierarchies, were deconstructed by the rise of egoic-rationality (wherever it did arise) and the switch from a role identity to an ego identity, and the correlative rise of the democratic state, where dominator hierarchies were replaced, in law, by (1) free and equal subjects of civil law, (2) morally free subjects, and (3) politically free subjects as citizens of the democratic state.

There still exist, of course, mythic-imperialisms that wish to dominate the world—that is, imperialisms that will accord you the status of equal world citizen *if* you embrace the particular mythology (with its particular dominator hierarchy). If you don't want to embrace the mythology, they will be glad to help you do so. We still hear the rumble of mythological fundamentalism arise from all corners of the globe, in secular and religious guises, none of which conceal their inherent drive to salvation via domination (and the women in these societies support the domination every bit as much as the men do).

But the rise of rationality—although it generated the worldspace in which all peoples *could* be recognized as free and equal subjects of the law and politically free subjects as citizens—did not produce, or has not yet produced, transformations of a *global* (and not just national) nature that would seek to socially empower the world in a noncoercive and non-dominating fashion, that would be driven by a recognition of what we all have in common as human beings and not what we have in common merely as believers in a particular and divisive mythology or as members of a particular ethnocentric tribe (however much those differences would also be cherished and honored in a *worldcentric* context).

With one major exception. The only serious *global* social movement, in all of history to date, has been the international labor movement (Marxism), which had one great, enduring, and legitimate strength—and one altogether fatal weakness. The strength was that it discovered a common trait that all humans possess, regardless of race, creed, nationality, mythology, or gender: we all have to secure our bodily survival through social labor of one sort or another. We all have to eat. And thus social labor puts us all in the same boat, makes us all world citizens. This movement was genuine enough, and serious enough, and *made such immediate good sense to so many people* that it set off the first modern *globally intent* revolutions from Russia to China to South America.

Such for its genuinely noble strengths. Its fatal weakness was that it did not just *ground* higher cultural endeavors in the economic or material realm (the physiosphere), it did not just *ground* them in social labor and material exchange—it *reduced* them to that exchange, reduced them to their lowest common denominator, reduced them to material productions and material values and material means, with all higher productions, especially spirituality, serving only as the opiate of the masses.[41]

In a nutshell, that movement did not just ground the noosphere in the physiosphere (which *is* vitally important because of compound individuality); it reduced the noosphere to the physiosphere, such an egregious reduction that it took evolution less than a mere century to begin to erase that mistake in earnest. This reductionistic thrust of Marxism, because it could find no support in the real Kosmos, had to be converted into a religious mythology, and thus had to press its vision in an imperialistic fashion.

The other major movement that has shown, or has claimed, some possibilities for underpinning a global citizenship is the Greens movement. While I have enormous sympathy for that movement as a *particular* endeavor, I believe it comes nowhere close to having the integrative potential for a planetary federation of world citizens freely embracing its tenets (forced to embrace its tenets, yes; but that is not unrestrained global discourse but a new coercion).

For the Greens are in essentially the same predicament as the Marxists—both of them are reducing higher levels to lower levels simply because of the undisputed fact that the lower is indeed *more fundamental* (and therefore necessary but not sufficient for the life of the higher and deeper). Where the Marxists tended to reduce all concerns to the material exchanges of the physiosphere, the Greens tend to reduce all concerns to the ecological exchanges of the biosphere. This is definitely a step up from the Marxists, but it does not escape the fact that it is still a type of lowest common denominator approach, which is true and important as far as it goes (as is Marxism), but catastrophic beyond that particular point, and utterly incapable of mobilizing world citizens *beyond* that particular point.

The Greens use, as their philosophical platform, two central notions: (1) the cultural noosphere is a part of the larger whole of the biosphere, and (2) the web-of-life systems theory. The first notion is incorrect, as we have seen, and the second notion is a form of subtle reductionism, as we have also seen. As such, any movement based on those tenets is unlikely to secure global integration and sustainable balance.

What is needed, rather, is a more integrative approach that works with our present historical actualities. A planetary culture will in effect have to deal with equitable material-economic distribution in the physiosphere (the enduring concern of Marx, even if we reject his particular solutions), and it will have to deal with sustainable ecological distribution in the biosphere (the enduring contribution of the Greens). But it will have to go much further and deal specifically and nonreductionistically with the noosphere and *its* distributions and distortions, and it will have to do so with something other than reductionistic web-of-life theories, *if* it is to freely engage the motivation of an entire globe. It will have to work toward specific theories of free noospheric exchange, including but transcending ecological concerns.[42]

Social labor could unite world citizens to the extent, but only to the extent, that we all share matter in common. The Greens can unite world citizens to the extent, but only to the extent, that we all share bodies in common. But it will take a vision-logic movement of tremendous integrative power (integral-aperspectival as universal-integral) in order to unite world citizens on the *centauric* basis that we all share matter *and* bodies *and* minds in common (not to mention a Spirit and a Self prior to *all* that). The Greens have produced a promising platform, but if it isn't any more than that (and it isn't so far, neglecting noospheric exchange), then it will be merely snapped up by egoic-rationality structures of capital production, and we will simply have McDonald's selling burgers in recyclable bags, which is nowhere near anything deserving to be called a Planetary Transformation.

In the meantime, the *platform* for an emerging world culture is being built by international markets of material-economic exchange, and by the increasingly free exchange of rationality structures, particularly empiric-analytic science and computer-transmitted information (the "Age of Information" is simply the "Age of Noosphere"), all of which are supranational in essential character (we will return to this in a moment).

As for the coming transformation itself, it is being built, as all past transformations have been, in the hearts and minds of those *individuals* who themselves evolve to centauric planetary vision. For these individuals create a "cognitive potential" in the form of *new worldviews* (in this case, centauric-planetary) that in turn feed back into the ongoing mainstream of *social institutions*, until the previously "marginalized" worldview becomes anchored in institutional forms which then catapult a collective consciousness to a new and higher order.[43] The revolution, as always, will come from the within and be embedded in the without.

And, at this point,[44] aside from the inner work that each of us individually can do, I personally see no obvious collective bearers of the new and deeper within.

THE DIALECTIC OF PROGRESS

Recall that one of the defining aspects of evolution is that it brings new and emergent possibilities and therefore new and potential pathologies. Habermas, working from a different angle, arrives at the same conclusion, and he refers to it as the "dialectic of progress":

> Evolutionarily important innovations mean not only a new level of learning but a new problem situation as well, that is, a new category of burdens that accompany the new social formation. The dialectic of progress can be seen in the fact that with the acquisition of problem-solving abilities new problem situations come to consciousness. A higher stage of development of productive forces and of social integration does bring relief from problems of the superseded [previous] social formation. But the problems that arise at the new stage of development can—insofar as they are at all comparable with the old ones—increase in intensity.
>
> Thus we can make an attempt to interpret social evolution taking as our guide those *problems* and *needs* that are first brought about by evolutionary advances. At every stage of development the social-evolutionary learning process itself generates new resources, which mean new dimensions of scarcity and thus new historical needs.[45]

Habermas's critique of Romantic regression is along the same lines as mine: of course there are new problems and new pathologies at each stage of development, but to take only the pathologies of the higher stage and compare them with only the achievements of the previous stage is perverse in the extreme. What is required, rather, is a balanced view that takes into account the limitations and failures of the previous stage that necessitated and propelled a new evolutionary transformation *beyond* them.

Habermas gives one very long paragraph where he outlines his view of what amounts to a phylogenetic needs hierarchy, including the achievements and the burdens of each of the major stages we have examined so far. It is worth studying (all italics are his):

> With the transition to the sociocultural form of life, that is, with the introduction of the family structure [during preoperational/magic], there arose *the problem of demarcating* [*beginning* to differentiate] *society from external nature*. In neolithic societies, at the latest, harmonizing society with the natural environment became thematic [problematic]. Power over nature came into consciousness as a scarce resource [which conflicts sharply with the eulogized accounts of this stage]. The experience of powerlessness in relation to the contingencies of external nature had to be interpreted away in myth and magic. With the introduction of a collective political order [mythic and mythic-rational], there arose *the problem of the self-regulation of the social system*. In developed civilizations, at the latest, the achievement of order by the state became a central need. Legal security came to consciousness as a scarce resource. The experience of social repression and arbitrariness had to be balanced with legitimations of domination. This was accomplished in the framework of rationalized world views [mythic-rational] (through which, moreover, the central problem of the previous stage—powerlessness—could be defused). In the modern age [egoic-rational], with the autonomization of the economy (and complementarization of the state), there arose *the problem of a self-regulated exchange of the social system with external nature* [biomaterial economic exchange]. In industrial capitalism, at the latest, society consciously placed itself under the imperatives of economic growth and increasing wealth. Value came into consciousness as a scarce resource. The experience of social inequality called into being social movements and corresponding strategies of appeasement. These seemed to lead to their goal in social welfare state mass democracies

(in which, moreover, the central problem of the preceding stage—legal insecurity—could be defused). Finally, if postmodern societies, as they are today envisioned from different angles [planetary/vision-logic], should be characterized by a primacy of the scientific and educational systems, one can speculate about the emergence of *the problem of a self-regulated exchange of society with internal nature* [by which he means, essentially, self-esteem and self-actualization and its pathologies; he specifically mentions existential anomie]. Again a scarce resource would become thematic—not the supply of power, security, or value, but the supply of motivation and meaning.[46]

The emergence of a new level solves or "defuses" some of the central problems and limitations of the previous stage (or else it wouldn't emerge), but it also introduces its own new problems and new *scarce resources*. One can't help but notice that this hierarchy of "deficiency needs" or "scarce resources" bears striking resemblance to Abraham Maslow's (and Jane Loevinger's and others') present-day ontogenetic needs hierarchy—from safety/power to conventional security/belonging to personal/egoic value to existential meaning. I believe it is no accident that I have long referred to the centauric level as the existential level of meaning, and I believe it is no accident that there would be Habermas/Maslow/Loevinger parallels (all of which we will return to in chapters 6 and 7).

But any way we slice the evolutionary pie, there is where we stand today: on the verge of a planetary transformation, struggling to be secured by rationality and completed by vision-logic, and embedded in global-planetary social institutions.

TRANSNATIONALISM

The *global* nature of this transformation is now being driven, particularly in its technological-economic base, by three interrelated factors: (1) the necessity to protect the "global commons," the common biosphere that belongs to no nation, no tribe, no creed, no race; (2) the necessity to regulate the world financial system, which no longer responds to national borders; (3) the necessity to maintain a modicum of international peace and security, which is now not so much a matter of major war between any two nations, but between a "new order" of loosely federated nations and renegade regimes threatening world peace.[47]

The point is that all three of those concerns no longer respond to actions taken merely on the part of individual nations. Not one of those problems can be solved on a national level. They are literally transnational crises demanding transnational, worldcentric responses. And exactly how to negotiate this difficult transition, with nations voluntarily surrendering some of their sovereignty for the global betterment—therein precisely is the extremely difficult nature of this "postnational" global transition.

But the situation is clear enough: these are transnational crises that have rendered national responses obsolete.

As I said, I will not at this time enter into a discussion of the intricacies of these crises and the requisite global transformation, except to indicate one factor crucial to our discussion at this point. Although each of those three factors have important material-economic components (the relational exchange of monetary-financial systems, the protection of the bio-material global commons, the physical security of nonaggression between nations)—none of those physical and material and economic components *can be secured* in the long run without a *corresponding change in consciousness* among the citizens of the nations *surrendering* some of their sovereignty for the transnational good.

Thus, solutions to the various global crises certainly demand efforts on the ecological-economic-financial front, which is where most efforts are now being concentrated. But it's a losing proposition without a corresponding shift in worldviews that will *allow* citizens and their governments to perceive the greater advantage in the lesser death (the surrender of some sovereignty for the greater good).

Tribal consciousness will not do; mythic-imperialism will surrender nothing; ethnocentric cleansing could care less about a worldcentric consensus. Mythic-membership is and always has been perfectly willing to sacrifice the lives of its true believers (and thus even more so the lives of nonbelievers) in order to advance the cause of the One True God, and if that means global suicide in exchange for life eternal, well, that's a small price, no? And magical blood cleansing would rather take its ethnic tribal self, and the world, into oblivion than mix with unpure tribes.

Thus, without in any way denying the crucial importance of the ecological and economic and financial factors in the world-demanding transfor-

mation, let us not forget that they all rest ultimately on a correlative transformation in human consciousness: the global embrace, and its pluralistic
world-federation, can only be *seen*, and *understood*, and *implemented*,
by individuals with a universal and global vision-logic, where the new
scarce resources involve not only material-economic shortages, but the
resources of a *meaning-in-life* that can *no longer be found* in self or tribe
or race or nation, but will find its context, its therapia, its omega, and its
release, in a worldcentric embrace through which runs the blood of a
common humanity and beats the single heart of a very small planet struggling for its own survival, and yearning for its own release into a deeper
and a truer tomorrow.

MULTICULTURALISM

In the meantime, it seems to be the case that the vast majority of the
world's population does not now need ways to get beyond rationality,
but ways to get up to it. The great mass of the world's social holons are
still caught in magic and warring tribalisms based on blood and ethnic
lineage, or in mythological empire-building: remnants of Marxism as a
mythic-rational "world religion"; Christian and Muslim fundamentalists
who would "convert" (coerce) the world; mythic-religious missionaries
with a global and proselytizing fury; a type of national-economic imperialism bordering on a mythology of the leading developed countries; and—
strangest of all—a dissolution of some of the great modern mythic-imperi-
alist States into their tribal subholons, bathed in blood and kinship
lineage and tribal warfare on a vicious scale: the retribalization of large
portions of the world.[48]

Thus, the single greatest *world transformation* would simply be the
embrace of global reasonableness and pluralistic tolerance—the global
embrace of egoic-rationality (on the way to centauric vision-logic).

The "multicultural movement," which claims a universal tolerance of
all cultures freed from the "logocentric, rational-centric, Eurocentric"
dominance and hegemony, is a step in the right direction, with all good
intentions, but ends up being self-contradictory and finally hypocritical.
It may claim to be "not rational-centric," but in fact cultural tolerance is
secured only by rationality as universal pluralism, by a capacity to mentally put yourself into the other person's shoes and then decide to honor

or at least tolerate that viewpoint even if you don't agree with it. You, operating from the pluralism of rational worldspace, might decide to tolerate the ideas of a mythic-believer; the problem is, they will not tolerate *you*—and, in fact, historically they would burn your tolerant tail at the stake in order to save your soul (whether your saviors be Christian, Marxist, Muslim, or Shinto).

In other words, multiculturalism is a noble, logocentric, and rational endeavor that simply *misidentifies its own stance* and claims to be not rational because some of the things *it tolerates* are not rational. But its own tolerance is rational through and through, and rightly so. Rationality is the only structure that will tolerate structures other than itself.

Nor can a genuine multiculturalism be established by "feelings" or by "coming from the heart," because my feelings are merely *mine*, not necessarily *yours* or *theirs*. Only in the space of rational pluralism can different feelings and thoughts and desires be given a fair play and an equal voice. It is from the platform of rational pluralism that the next stage, the truly aperspectival-integral (and universal-integral), can be reached.

Put differently, multiculturalism is a noble attempt to move to the integral-aperspectival structure, but, like many postmodern poststructuralists, it thoroughly confuses the fact that no perspective is final with the notion that all perspectives are therefore simply equal. It thus fails to notice that that stance itself actually (and appropriately) *rejects* all narrower perspectives (which clearly shows that all perspectives are *not* equal).

In other words, the "multicults" regress from "no stance is ultimate" to "every stance is equally acceptable," thus burying (and denying) their own otherwise accurate judgment that lesser stances and smaller perspectives are *unacceptable*. They correctly glimpse aperspectival space, but then get thoroughly disoriented in the dizzyingly holonic nature of eversliding contexts, failing to notice that sliding contexts do not in any way prevent some contexts from still being *relatively better* than less encompassing contexts.

Thus, that everything is relative does not mean nothing is better; it means some things are, indeed, relatively better than others, all the time. Neither atoms nor molecules are final or ultimate constituents of the universe; nonetheless, *wherever* they appear, molecules *always* contain atoms in a deeper embrace. (Indeed, the whole point of Einstein's somewhat mislabeled "relativity theory" was to find *invariant* transforms across relative spacetime.) The system is relative and sliding, for sure, but it slides in stable ways, and this stability-in-relativity allows the types of *correct* judgments that the multiculturalists do in fact make (namely, that pluralistic tolerance is *better* than narrow-minded intolerance)—even though their own theory (mistaken to be "we have no judgments since all judgments are relative")—even though their own theory cannot account for their actual stance (even though their own theory, in fact, denies their actual stance).

By failing to see the *definiteness* of *relative* judgments—and thus being totally disoriented and lost in aperspectival space—they miss the *integral* part, the *universal*-integral part, of their own stance, and thus they all too often regress into a riot of idiosyncratic differences that destroys the integrity of their own position. In a recent U.S. court case, a Chinese man found his wife, also Chinese, having an affair. He took a claw hammer and savagely beat her to death. He was *acquitted* on the "multicult defense": that's what they do in China, and we have to honor cultural differences, since no perspective is "better" than another. Aperspectival madness.

As for "Eurocentrism" being somehow despicable, the dividing line is not and should not be *between* cultures—some cultures are "good," some are "bad," some are "higher," some are "lower." Rather, the dividing line is *within* cultures—it is between those who, in *any* culture, wish to develop from egocentric to sociocentric to worldcentric, from magic to mythic to reason-planetary (first as formal rationality, then completed as centauric vision-logic).

For the most part, these rationality structures are *already* available for those individuals and social systems who can endure the transformation beyond their own provincial dogmatisms and embrace international recognition and mutual respect for each other's particular existence. This transformation—from the submergence of the individual in collective mythic-membership totalities to the emergence of the individual person or ego-subject, legally and morally and politically free within rational space secured by law—has already been embraced by Russia and the Eastern Bloc (which has temporarily thrown some of its junior holons back into vicious tribalistic warring simply because these tribal subholons were once held together by Marxism as a military mythic-rational empire: in holonic dissolution, the path of regression follows the reverse line of buildup, which now means: rampant retribalization).

Further, the transformation from mythic-membership to egoic-rationality (and its perils) is already open to China, Cuba, Libya, Iraq, North Korea, Serbia, and any other social holon that wishes to surrender its mythic "superiority" and join the community of nations governed by international law and mutual recognition, that wishes to cease dissociating and splitting off from the free exchange of planetary consciousness, that wishes to reintegrate into a common world spirit and collective sharing of reason and communication and vision. (This does not excuse those nation-states, particularly the United States, Japan, and an ominously emerging "fortress Europe"—which are still distorting supranational exchanges for their own particular interests and remnants of mythic-imperial hegemony—does not prevent or relieve them from continuing to search for more reasonable and equitable exchanges with the world community at large, and particularly with regard to the common biosphere).

But for those individuals who have already done so; for those who have begun to emerge as centauric vision-logic; for those who take their stand in a planetary awareness beyond this or that narrow parochialism; for those attempting to secure a centauric identity that integrates physiosphere and biosphere and noosphere (in male and female alike), and looking for that sense of meaning, existential meaning, global meaning—for those small few, the simple *living* of a planetary perspectivism (that is, an integral-aperspectivism) creates small pockets of leading-edge consciousness, creates small pockets of "cognitive potential" that slowly but surely feed back into collective worldviews and then into social institutions themselves (UL to LL to LR), and once materially embedded and institutionalized, those institutional structures automatically, as it were, act as pacers of transformation for all who follow.

And transformation into what, exactly? Is the centaur the end of the line? Put it another way: is there any conceivable reason that evolution, which has labored so mightily for fifteen billion years and produced so much undeniable wonderment, would just up and abruptly cease? Are there not higher spirals lying yet ahead? If we have discerned even the vaguest features of time's arrow, can we not stand on tiptoe and foresee dimly the arrow's arc into tomorrow?

6

Magic, Mythic, and Beyond

That little prigs and three-quarter-mad men may have the conceit that the laws of nature are constantly broken for their sakes—such an intensification of every kind of selfishness into the infinite, into the impertinent, *cannot be branded with too much contempt. And yet Christianity [any form of mythic-literalism] owes its triumph to this miserable flattery of personal vanity: it was precisely all the failures, all the rebellious-minded, all the less favored, the whole scum and refuse of humanity who were thus won over to it: The "salvation of the soul"—in plain language: "the world revolves around me."*
— FRIEDRICH NIETZSCHE

I N THE LAST chapter we looked at the formative stages in the collective evolution of the human species, up to the present—archaic to magic to mythic to mental (or egoic-rational on the verge of centauric vision-logic). In this chapter we will examine those same stages, those basic holons, as they appear in an individual's development today. And this will set the stage for a discussion of any future—and possibly higher—developments, in the individual and the species as a whole. And straining to see the arc of time's arrow into tomorrow, brings some surprises indeed.

THE PRE/TRANS FALLACY

Ever since I began writing on the distinctions between prerational (or prepersonal) states of awareness and transrational (or transpersonal) states—what I called the pre/trans fallacy—I have become more convinced than ever that this understanding is absolutely crucial for grasping the nature of higher (or deeper) or truly spiritual states of consciousness.

The essence of the pre/trans fallacy is itself fairly simple: since both prerational states and transrational states are, in their own ways, nonrational, they appear similar or even identical to the untutored eye. And once pre and trans are confused, then one of two fallacies occurs:

In the first, all higher and transrational states are *reduced* to lower and prerational states. Genuine mystical or contemplative experiences, for example, are seen as a regression or throwback to infantile states of narcissism, oceanic adualism, indissociation, and even primitive autism. This is, for example, precisely the route taken by Freud in *The Future of an Illusion*.

In these reductionistic accounts, rationality is the great and final omega point of individual and collective development, the high-water mark of all evolution. No deeper or wider or higher context is thought to exist. Thus, life is to be lived either rationally, or neurotically (Freud's concept of neurosis is basically anything that derails the emergence of rational perception—true enough as far as it goes, which is just not all that far). Since no higher context is thought to be real, or to actually exist, then whenever any genuinely transrational occasion occurs, it is immediately explained as a *regression* to prerational structures (since they are the only nonrational structures allowed, and thus the only ones to accept an explanatory hypothesis). The superconscious is reduced to the subconscious, the transpersonal is collapsed to the prepersonal, the emergence of the higher is reinterpreted as an irruption from the lower. All breathe a sigh of relief, and the rational worldspace is not fundamentally shaken (by "the black tide of the mud of occultism!" as Freud so quaintly explained it to Jung).

On the other hand, if one is sympathetic with higher or mystical states, but one still *confuses* pre and trans, then one will *elevate* all prerational states to some sort of transrational glory (the infantile primary narcissism, for example, is seen as an unconscious slumbering in the *mystico unio*). Jung and his followers, of course, often take this route, and are forced to read a deeply transpersonal and spiritual status into states that are merely indissociated and undifferentiated and actually lacking any sort of integration at all.

In the elevationist position, the transpersonal and transrational mystical union is seen as the ultimate omega point, and since egoic-rationality does indeed tend to deny this higher state, then egoic-rationality is pictured as the *low point* of human possibilities, as a debasement, as the cause of sin and separation and alienation. When rationality is seen as the anti-omega point, so to speak, as the great Anti-Christ, then *anything* nonrational gets swept up and indiscriminately glorified as a direct route to the Divine, including much that is infantile and regressive and prerational: *anything* to get rid of that nasty and skeptical rationality. "I believe *because* it is absurd" (Tertullian)—there is the battle cry of the elevationist (a strand that runs deeply through Romanticism of any sort).

Freud was a reductionist, Jung an elevationist—the two sides of the pre/trans fallacy. And the point is that they are *both* half right and half wrong. A good deal of neurosis is indeed a fixation/regression to prerational states, states that are not to be glorified. On the other hand, mystical states do indeed exist, beyond (not beneath) rationality, and those states are not to be reduced.

For most of the recent modern era, and certainly since Freud (and Marx and Ludwig Feuerbach), the reductionist stance toward spirituality has prevailed—*all* spiritual experiences, no matter how highly developed they might in fact be, were simply interpreted as regressions to primitive and infantile modes of thought. However, as if in overreaction to all that, we are now, and have been since the sixties, in the throes of various forms of elevationism (exemplified by, but by no means confined to, the New Age movement). All sorts of endeavors, of no matter what origin or of what authenticity, are simply elevated to transrational and spiritual glory, and the *only* qualification for this wonderful promotion is that *the endeavor be nonrational. Anything* rational is wrong; *anything* nonrational is spiritual.

Spirit is indeed nonrational; but it is trans, not pre. It transcends but includes reason; it does not regress and exclude it. Reason, like any particular stage of evolution, has its own (and often devastating) limitations, repressions, and distortions. But as we have seen, the inherent problems of one level are solved (or "defused") only at the next level of development; they are not solved by regressing to a previous level where the problem can be merely ignored. And so it is with the wonders and the terrors of reason: it brings enormous new capacities and new solutions, while introducing its own specific problems, problems solved only by a transcendence to the higher and transrational realms.[1]

Many of the elevationist movements, alas, are not beyond reason but beneath it. They think they are, and they announce themselves to be, climbing the Mountain of Truth; whereas, it seems to me, they have merely slipped and fallen and are sliding rapidly down it, and the exhilarating rush of skidding uncontrollably down evolution's slope they call "following your bliss." As the earth comes rushing up at them at terminal velocity, they are bold enough to offer this collision course with ground zero as a new paradigm for the coming world transformation, and they feel oh-so-sorry for those who watch their coming crash with the same fascination as one watches a twenty-car pileup on the highway, and they sadly nod as we decline to join in that particular adventure. True spiritual bliss, in infinite measure, lies up that hill, not down it.

We will want to look at the deeper, higher structures of awareness, the transrational structures. But in order to fully understand them, we need first to look carefully at the prerational structures, so we will know, if nothing else than by a process of subtraction, what the transpersonal states are like.

So as we pick up the story of ontogeny at the archaic and magic structures, and follow it into the realms of the superconscious, it is these two distortions—reductionism and elevationism—that we will most want to try to avoid.

COGNITIVE EVOLUTION

We are now following the development of the *interior* of an *individual* human holon in today's world, a recapitulation, in broad stroke, of the same basic holons we saw, in the last chapter, emerge over historical and prehistorical epochs in the various cultural worldspaces (worldspaces still available today owing to compound individuality, and spaces that unfold in the same sequence in individual development). As Sheldrake says, it is as if nature, once it has produced a basic holon and made sure it functions adequately, keeps using the same holon in subsequent development as building blocks (*preserving* the basic holon but *negating* its partialness or exclusivity by transcending it in a higher, deeper pattern).

Each interior development that we will be following is, of course, governed by the twenty tenets (though not by those alone), and thus each development involves a new and creative emergence, a new transcendence, a new depth, a new interiority, a new differentiation/integration, a greater degree of relative autonomy (greater capacity for both agency and for communion), a greater degree of consciousness, a greater total embrace—with new fears, new anxieties, new needs, new scarcities, new desires, new moral engagements in new shared worldviews, and the ever-present possibility of new and higher pathologies and distortions.

Nowhere have the vicissitudes of the mind's developmental emergence been chartered in greater detail than in the works of Jean Piaget; and although nobody imagines that Piaget's system is without its own inadequacies, nonetheless the wealth of research and data that he and his colleagues generated over a four-decade period stands as one of a handful of the truly great contributions to psychology (and philosophy and religion).

Without, therefore, endorsing all of the Piagetian system, I would like to draw on his data (and some of his conclusions) to point out very carefully the nature of the mind's development from archaic to magic to mythic to mental, as it appears in today's ontogeny. For if there can be legitimate questions about, for example, the "magical-animistic" times in the phylogenetic past, there can be much less doubt about that structure (and about the mythic structure) as it appears in development today.

Further, and just as important for our concern, a simple study of Piaget is the quickest cure for elevationism. For what we find as we carefully study Piaget is that the productions of, say, the preoperational mind initially look very holistic, very interconnected, very "religious" in a sense, until we even barely scratch the surface and find the whole production supported by egocentrism, artificialism, finalism, anthropocentrism, and indissociation. We then have two very clear choices: try to prop up those productions as being "really religious," or look elsewhere for a genuine God.

WAVES AND STREAMS

Before outlining Piaget's findings on cognitive development, a brief word about my overall psychological model, of which cognitive development is only a part. This model consists essentially of waves, streams, states, and self. *Waves* are the basic levels of consciousness (about a dozen major levels that range from subconscious to self-conscious to superconscious). I call them "waves" to indicate the fluid nature of these levels of consciousness, the fact that they interpenetrate and overlap (like colors in a rainbow) and are not rigid rungs in a ladder. *Streams* are the various developmental lines (such as cognitive, moral, psychosexual, affective, interpersonal, spiritual, and so on) that develop relatively independently through the basic levels or waves (so that a person can, for example, be at a fairly high level of development cognitively, a medium level emotionally, and a low level morally; development, in other words, is anything but a linear, step-by-step affair). *States* refers to altered states of consciousness (such as peak experiences), which are brief, temporary, but often powerful experiences, especially of the transpersonal realms (however, if development into the transpersonal is to become permanent and not merely passing, these *states* must be converted to *traits*: altered states must become permanent structures or levels of consciousness). And the *self* refers to the self-system that navigates the waves, streams, and states as it makes its way through the great River of Life.

Waves and streams—or levels and lines—are operative in all four quadrants. They are simply the grades and clades of evolution—the levels of holons and the lines of holons. That is, they are the levels of structural complexity, and the lines of development that move through those levels. In figure 5-1, the levels, waves, or grades are given as thirteen in number (a highly simplified summary). Each quadrant, although it shares the same basic levels or waves, has its own types of lines or streams. For example, UL: affective development, ego development, cognitive development (e.g., preop to conop to formop to postformal); UR: biological growth, neurophysiological development, behavioral evolution; LL: cultural worldviews, values, mutual understanding, group identities; LR: techno-economic forms, geopolitical structures, evolution of social systems, and so on. In each of the quadrants, the various lines can unfold in a relatively independent fashion, but the point is that any holon in any line has correlates at the same level in all the other quadrants ("same-level relationship"). And, of course, each line of development, in any quadrant, follows the twenty tenets.

What we have covered so far (in very general outline) is the phylogenetic evolution of all four quadrants up to today (as indicated in figure 5-1), using "phylogenetic" in the broad sense as the evolutionary history of any class of holons. Focusing on human holons, we also examined the phylogenetic evolution of cultural worldviews (archaic to magic to mythic to rational to vision-logic), which is only one line of development (but a very important line) in the Lower-Left quadrant (we also touched on a few developmental lines in the Lower Right, such as the techno-economic stream from foraging to horticultural to agrarian to industrial to informational). In this chapter, we will examine the development, in today's individual, of the cognitive line of development, which is only one line of development (but a very important line) in the Upper-Left quadrant. We have already seen that these lines are related (similar holons are "reused" ontogenetically and phylogenetically in both micro and macro), and now we will see ample evidence of this in today's ontogenetic development.

Finally, a word about Piaget's research itself. It has lately become fashionable and politically correct to dismiss Piaget's work entirely (it is "hierarchical," it is "rigidly linear," it is "patriarchal," etc.). Piaget's overall system, it is quite true, has several inadequacies, but these are rarely why it is criticized; while the reasons that it is commonly dismissed do not themselves hold up. The major inadequacy of Piaget's system, most scholars now agree, is that Piaget suggested that cognitive development (conceived as logico-mathematical competence) is the major axis of development, and that all other domains are subservient to the cognitive. In other words, Piaget made the cognitive stream the only major stream, and the implication was that all other streams fall within that one, whereas there is now abundant evidence that the various developmental streams can unfold in a relatively independent manner (so that a person can be at a high level of cognitive development while remaining at a medium level in other lines and at a low level in still others). In my model, for example, the cognitive line is merely one of some two-dozen developmental lines, none of which, as lines, can claim preeminence. (It should be remembered that in this chapter we are examining only the cognitive line—as a graphic example of evolutionary holons—and thus we are *not* presenting an overall or integral psychology [for which, see *Integral Psychology,* Volume Four of the *Collected Works*]).

But as for the cognitive line itself, Piaget's work is still very impressive; moreover, after almost three decades of intense cross-cultural research, the evidence is virtually unanimous: Piaget's stages up to formal operational are universal and cross-cultural. As only one example, *Lives Across Cultures—Cross-Cultural Human Development* is a highly respected textbook written from an openly liberal perspective (which is often suspicious of "universal" stages). The authors (H. Gardiner, J. Mutter, and Corinne Kosmitzki) carefully review the evidence for Piaget's stages of sensorimotor, preoperational, concrete operational, and formal operational. They found that cultural settings sometimes alter the *rate* of development, or an *emphasis* on certain aspects of the stages—but not the stages themselves or their cross-cultural validity.

Thus, for sensorimotor: "In fact, the qualitative characteristics of sensorimotor development remain nearly identical in all infants studied so far, despite vast differences in their cultural environments" (p. 88). For preoperational and concrete operational, based on an enormous number of studies that include Nigerians, Zambians, Iranians, Algerians, Nepalese, Asians, Senegalese, Amazon Indians, and Australian Aborigines: "What can we conclude from this vast amount of cross-cultural data? First, support for the universality of the structures or operations underlying the preoperational period is highly convincing. Second, . . . the qualitative characteristics of concrete operational development (e.g., stage sequences and reasoning styles) appear to be universal [although] the rate of cognitive development . . . is not uniform but depends on ecocultural factors." Although the authors do not use exactly these terms, they conclude that the deep structures of the stages are universal but the surface structures depend strongly on cultural, environmental, and ecological factors—in other words, all four quadrants are profoundly involved in individual development. "Finally, it appears that although the rate and level of performance at which children move through Piaget's concrete operational period depend on cultural experience [actually, on all four quadrants], children in diverse societies still proceed in the same sequence he predicted" (pp. 91–92).

Fewer individuals in any cultures (American, Asian, African, or otherwise) reach formal operational cognition, and the reasons given for this vary. It might be that formal operational is a genuinely higher stage that fewer therefore reach (greater depth, less span), as I believe. It might be that formal operational is a genuine capacity but not a genuine stage, as the authors believe (i.e., only some cultures emphasize formal operational and therefore teach it). Evidence for the existence of Piaget's formal stage is therefore strong but not conclusive. Yet the politically correct have used this one item to dismiss *all* of Piaget's stages, whereas the correct conclusion, backed by enormous evidence, is that all of the stages up to formal operational have now been adequately demonstrated to be universal and crosscultural.

We can now look at the contours of these universal stages. Although I

believe the stages at and beyond formop are also universal (including vision-logic and the general transrational stages), in a sense the most dis-
puted and controversial ground involves stages leading up to formal operational, for these are the stages—archaic, magic, and mythic—that are so
intensively debated as to their "spiritual" merit. Realizing that these are
exactly the stages of Piaget's studies that have consistently held up to
cross-cultural evidence allows us to see them in a more accurate light, I
believe.

As suggested, Piaget divided cognitive development into four broad
stages—sensorimotor (0–2 years), preoperational or "preop" (2–7 years),
concrete operational or "conop" (7–11 years), and formal operational or
"formop" (11 years onward).[2] Each of these broad stages he divided into
several substages. We will take them one at a time, with a few brief correlations with other major researchers.

Watching these many correlations between onto and phylo, micro and
macro, one can't help but marvel at the fact that once nature creates a
holon, nature uses it time and time again. The Kosmos creatively brings
forth a holon (tenet 2c: self-transcendence), whereupon it is subjected to
intense selection pressures (i.e., *it must fit with all four quadrants*). For as
long as it is selected (fits all four quadrants), it continues to exist, either
on its own, and/or taken up and included in a higher superholon, where
it lives on as a relatively independent subholon.

There is a Kosmic economy in all this that is genuinely awe-inspiring.
Who would have thought that the first atoms existing in the first far-flung
galaxies, stretching across billions of miles and existing billions of years
ago, would now be the ingredients, the actual subholons, in your very
own body, as you now sit here reading all this?

SENSORIMOTOR (ARCHAIC
AND ARCHAIC-MAGIC)

The human being at conception is a single-celled holon, embracing in
itself, as junior holons, organelles, molecules, atoms, and subatomic particles, reaching all the way back into those dark shadows that fade into the
evolutionary night.

By the time of birth, the human being has developed from protoplasmic
irritability to sensation to perception to impulse to proto-emotion, em-

bracing each as a successive holon in its own compound individuality. But none of these functions is yet clearly *differentiated* (or integrated), and the first years of life are a quick coming-to-terms with the physiosphere and the biosphere both within and without, in preparation for the emergence of the noosphere, which begins in earnest around age two with the emergence of language.

Thus Piaget, for example, in speaking of the first year of life, says that "the self is here *material*, so to speak." It is still, that is, embedded primarily in the physiosphere. In the first place, the infant cannot easily distinguish between subject and object or self and material environment, but instead lives in a state of "primary narcissism" (Freud) or "oceanic adualism" (Arieti) or "pleromatic fusion" (Jung) or primary "indissociation" (Piaget). The infant's self and the material environment (and especially the mother) are in a state of primitive nondifferentiation or indissociation. On the psychosexual side, this is the "oral phase" because the infant is coming to terms with food, physical nourishment, life in this physiosphere.

Sometime between the fourth and the ninth month, this archaic indissociation gives way to a *physical* bodyself *differentiated* from the *physical* environment—the "real birth" of the individual physical self. Margaret Mahler actually refers to it as "hatching." The infant bites its thumb and it hurts, bites the blanket and it doesn't. There is a difference, it learns, between the physical self and the physical other.

Another way to put this is to say that, with this first major differentiation (or first major "fulcrum of development," as researchers call it),[3] consciousness *seats* itself in the physical body, grounds itself in the physiosphere of the individual holon and not in the physiosphere of the social holon (its environment). Many researchers, from Kernberg to Mahler, have concluded that if, due to physiological/genetic factors or repeated trauma, consciousness fails to seat itself in the physical self, the result is *psychosis* of one sort or another.[4] Psychosis is many things (including, perhaps, an influx from higher dimensions), but it certainly includes a failure to establish a grounded physical self clearly differentiated from the environment. The psychotic, as R. D. Laing put it, is constantly "jumping out of the body"; he or she cannot easily differentiate where the body stops and the chair begins; subject and object collapse in a state of fusion and confusion, with hallucinatory blurring of boundaries, and so forth.

Psychosis, we may say, is a failure to differentiate and integrate the physi-osphere.

If all goes relatively well, then the infant *transcends* this archaic fusion state and emerges or hatches as a grounded physical self.

The *sensorimotor period* (0–2 years) is thus predominantly concerned with differentiating the *physical* self from the *physical* environment, and results, toward the end of the second year, in what Piaget calls physical "object permanence," the capacity of the infant to understand that physical objects exist independently of him or her (i.e., the physical world exists independently of one's egocentric wishes about it).

Thus, out of an initial state of *primary indissociation* ("protoplasmic," Piaget also calls it), the physical self and the physical other emerge:

> It is through a progressive differentiation that the internal world comes into being and is contrasted with the external. Neither of these two terms is given at the start. . . .
>
> Consequently, during the gradual and slow differentiation of the initial protoplasmic reality into objective and subjective reality, it is clear that each of the two terms [interior and exterior, Left Hand and Right Hand] in process of differentiation will evolve in accordance with its own structure. . . .
>
> This phenomenon is very general. During the early stages the [physical] world and the self are one; neither term is distinguished from the other. But when they become distinct, these two terms begin by remaining very close to each other: the world is still conscious and full of intentions, the self is still material, so to speak, and only slightly interiorized. At every stage there remain in the conception of nature what we might call "adherences," fragments of internal experience which still cling to the external world.[5]

At the end of the sensorimotor period, the physical self and physical other are clearly differentiated, but as the *mind begins to emerge* with preop, *the mental images and symbols themselves are initially fused and confused with the external world*, leading to what Piaget calls "adherences," which children themselves will eventually reject as being inadequate and misleading.

> We have distinguished [several] varieties of adherences defined in this way. There are, to begin with, during a very early stage, feelings of

participation accompanied sometimes by magical beliefs; the sun and moon follow us, and if we walk, it is enough to make them move along; things notice us and obey us, like the wind, the clouds, the night, etc.; the moon, the street lamps, etc., send us dreams "to annoy us," etc., etc. In short, the world is filled with tendencies and intentions which are [centered on] our own.

A second form of adherence, closely allied to the preceding, is that constituted by animism, which makes the child endow things with consciousness and life [oriented solely toward the child]. . . . In this magico-animistic order: on the one hand, we issue commands to things (the sun and the moon, the clouds and the sky follow us), on the other hand, these things acquiesce in our desires because they themselves wish to do so.[6]

A third form is artificialism [anthropocentrism]. The child begins by thinking of things in terms of his own "I": the things around him take notice of man and are made for man; everything about them is willed and intentional, everything is organized for the good of men. If we ask the child, or the child asks himself, how things began, he has recourse to man to explain them. Thus artificialism is based on feelings of participation which constitute a very special and very important class of adherences.[7]

As we will see, Piaget believes that the major and in many ways defining characteristic of all adherences is *egocentrism*, or an early and initial inability to transcend one's own perspective and understand that reality is not self-centered. Development thus proceeds slowly from egocentrism to perspectivism, from realism to reciprocity and mutuality, and from absolutism to relativity:

This formula means that the child, after having regarded his own point of view as absolute, comes to discover the possibility of other points of view and to conceive of reality as constituted, no longer by what is immediately given, but by what is common to all points of view taken together.

One of the first aspects of this process is the passage from realism of perception to interpretation properly so called. All the younger children take their immediate perceptions as true, and then proceed to interpret them according to their own egocentric relations. The most striking example is that of the clouds and the heavenly bodies, of which children believe that they follow us. The sun and moon are small globes traveling a little way above the level of the roofs of houses and following us about on our walks. Even the child of 6–8 years does not hesitate to take this perception as the expression of truth, and, curiously enough, he never thinks of asking himself whether these heavenly bodies do not also follow other people. When we ask the captious question as to which of two people walking in opposite directions the sun would prefer to follow, the child is taken aback and shows how new the question is to him. [Older children,] on the other hand, have discovered that the sun follows everybody. From this they conclude that the truth lies in the *reciprocity of the points of view*: that the sun is very high up, that it follows no one. . . .[8]

Piaget is at pains to indicate that the process of differentiation/integration between internal and external world is a long and slow one. It is not, for example, that magico-animistic beliefs are present at one stage and then completely disappear at the next, but rather that cognitions referred to as "magical" become progressively less and less as development proceeds, moving from a "pure magical autism" to mental egocentricity to reciprocal and mutual sharing. In a very important passage Piaget gets to the heart of the matter:

For the construction of the objective world and the elaboration of strict reasoning both consist in a gradual *reduction of egocentricity* in favor of . . . reciprocity of viewpoints. In both cases, the initial state is marked by the fact that the self is confused with the external world and with other people; the vision of the world is falsified by subjective adherences, and the vision of other people is falsified by the fact that the personal point of view predominates, almost to the exclusion of all others. Thus in both cases, truth is obscured by the ego.

Then, as the child discovers that others do not think as he does, he makes efforts to adapt himself to them, he bows to exigencies of control and verification which are implied by discussion and argument, and thus comes to replace egocentric logic by the logic created by social life. We saw that exactly the same process took place with regard to the idea of reality.

There is therefore an egocentric logic [LH] and an egocentric ontology [RH], of which the consequences are parallel: they both falsify the perspective of relations and of things, because they both start from the assumption that other people understand us and agree with us from the first, and that things revolve around us with the sole purpose of serving us and resembling us.[9]

A note on terminology: Piaget divides each of the major cognitive stages into at least two substages (early and late preop, early and late conop, early and late formop), and I have generally followed Piaget in this regard. Since we have also been using Gebser's general worldview terminology of archaic, magic, mythic, and mental (with the clear implication that they are referring to essentially similar stages), I will often hybridize Gebser's terminology to match Piaget's substages, so that we have a continuum of archaic, archaic-magic, magic, magic-mythic, mythic, mythic-rational, rational, rational-existential (and into vision-logic, psychic, etc.). These particular names are, of course, arbitrary; but the actual stages they refer to are based on extensive empirical/phenomenological research. I also believe these names (such as magic, mythic-rational, etc.) help to capture the essential "flavor" of each stage and substage.

The preponderance of indissociations and "adherences" at the sensorimotor and early preoperational (the archaic and the archaic-magic—roughly 0–3 years) have led Piaget himself to often refer to this general early period as one of "magical cognitions" or "magic proper." As he explains:

> The first [general stage] is that which precedes any clear consciousness of the self, and may be arbitrarily set down as lasting until the age of 2–3, that is, till the appearance of the first "whys," which symbolize in a way the first awareness of resistance in the external world.
>
> As far as we can conjecture, two phenomena characterize this first stage [the overall archaic-magic]. From the [internal] point of view, it is pure *autism*, or thought akin to dreams or daydreams, thought in which truth is confused with desire. To every desire corresponds immediately an image or illusion which transforms this desire into reality, thanks to a sort of pseudo-hallucination or play. No objective observation or reasoning is possible: there is only a perpetual play which transforms perceptions and creates situations in accordance with the

subject's pleasure [this is a stage that is often eulogized and "elevated" by the Romantics, such as Norman O. Brown, to a "spiritual non-dual" state, whereas it is actually, as we have seen, a very egocentric, narcissistic state: prerational, not transrational]. From the ontological viewpoint, what corresponds to this manner of thinking is primitive *psychological causality*, probably in a form that implies *magic* proper [his italics; I refer to it as archaic-magic]: the belief that any desire whatsoever can influence objects, the belief in the obedience of external things. Magic and autism are therefore two different sides of one and the same phenomenon—that confusion between the self and the world. . . .[10]

PREOPERATIONAL (MAGIC AND MAGIC-MYTHIC)

If all goes relatively well, the infant *transcends* the early archaic fusion state and emerges or hatches as a grounded physical self. But if the infant's physical body is now separated from the environment, its emotional body is not. The infant's *emotional self* still exists in a state of indissociation from other *emotional* objects, in particular the mothering one. But then, around eighteen months or so, the infant learns to *differentiate its feelings* from the *feelings of others* (this is the second major differentiation, or "second fulcrum"). Its own biosphere is differentiated from the biosphere of those around it—in other words, it transcends its embeddedness in the undifferentiated biosphere. Once again, as with every holon, it manifests not just a capacity for self-preservation and self-adaptation, but also for self-transcendence.

Mahler refers to this crucial transformation (the second fulcrum) as the "separation-individuation phase," or the differentiation-and-integration of a stable *emotional* self (whereas the previous fulcrum, as we saw, was the differentiation/integration of the *physical* self). Mahler actually calls this fulcrum "the psychological birth of the infant," because the infant emerges from its emotional fusion with the (m)other.

A developmental miscarriage at this critical fulcrum (according to Mahler, Kernberg, and others) results in the narcissistic and borderline pathologies, because if the infant does not differentiate-separate its feelings from the feelings of those around it, then it is open to being "flooded" and "swept away" by its emotional environment, on the one hand (the borderline syndromes), or it can treat the entire world as a mere extension of its own feelings (the narcissistic condition)—both of which result from a failure to transcend an embeddedness in the undifferentiated biosphere. One remains in indissociation with, or "merged" with, the biosphere, stuck in the biosphere, just as with the previous psychoses one remains merged with or stuck in the physiosphere.

By around age three, if all has gone relatively well, the young child has a stable and coherent physical self and emotional self; it has differentiated and integrated, transcended and preserved, its own physiosphere and bio-sphere. By this time language has begun to emerge, and development in the noosphere begins in earnest.

Thus, the intensity of the early archaic-magic declines with the differentiation of emotional self and emotional other (24–36 months)—but, according to Piaget, magical cognitions *continue to dominate the entire early preoperational period* (2–4 years), the period I simply call "magic."

In other words, the first major layer of the noosphere is magical. During this period, the newly emerging images and symbols do not merely represent objects; they are thought to be concretely *part of the things they represent*, and thus "word magic" abounds:

> Up to the age of 4–5, [the child] thinks that he is "forcing" or compelling the moon to move; the relation takes on an aspect of dynamic participation or of magic. From 4 to 5 he is more inclined to think that the moon is trying to follow him: the relation is animistic.
>
> Closely akin to this participation is *magical causality*, magic being in many respects simply participation: the subject regards his gestures, his thoughts, or the objects he handles, as charged with efficacy, thanks to the very participations which he establishes between those gestures, etc., and the things around him ["adherences"]. Thus a certain word acts upon a certain thing; a certain gesture will protect one from a certain danger; a certain white pebble will bring about the growth of water lilies, and so on. . . .[11]

Piaget refers to such magical cognitions as a form of "participation"— that is, the subject and the object, and various objects themselves, are "linked" by certain types of adherences, or felt connections, connections that nonetheless violate the rich fabric of relations actually constituting the object.

This is very much what Freud referred to as the primary process, which is governed by two general laws, that of displacement and that of condensation. In *displacement*, two different objects are equated or "linked" because they share similar parts or predicates (a relation of *similarity*: if one Asian person is bad, all Asians must be bad). In *condensation*, different objects are related because they exist in the same space (a relation of *contiguity*: a lock of hair of a great warrior "contains" in condensed form the power of the warrior).

(There is another way to categorize this. Each holon is an agency-in-communion. Displacement confuses different holons because they share similar agency. Condensation confuses different holons because they share similar communions. The former is metaphor, the latter is metonym. This leads to a host of interesting correlations, which I will reserve for a note).[12]

Put simply, such primary process or magical cognition is not yet capable of grasping the notion of a *holon*. It does not set whole and part in a rich network of *mutual* relationships, but short-circuits the process by merely collapsing or confusing various wholes and parts—what Piaget calls syncretism and juxtaposition (again, similarity and contiguity). Magical cognition, then, is of fused and confused wholes and parts, and not mutually related wholes and parts. These "fused networks" of "sycretic wholes" *appear* very holistic (or "holographic"), but are actually not very coherent and do not even match the already available sensorimotor evidence.

> [This] type of relation is *participation*. This type is more frequent than would at first appear to be the case, but it disappears after the age of 5–6. Its principle is the following: two things between which there subsist relations either of resemblance [similarity; metaphor] or of general affinity [contiguity; metonym], are conceived as having something in common which enables them to act upon one another at a distance, or more precisely, to be regarded one as a source of emanations, the other as the emanation of the first. Thus air or shadows in a room emanate from the air and shadows out of doors. Thus also dreams, which are sent to us by birds "who like the wind."
>
> [The child] begins, indeed, as we do, by feeling the analogy of the

shadow cast by the book with the shadows of trees, houses, etc. But this analogy does not lead him to any [mutual] relation: it simply leads him to identify the particular cases with one another. So that we have here, not analogy proper, but syncretism. The child argues as follows: "This book makes a shadow; trees, houses, etc., make shadows. The book's shadow (therefore) *comes from the trees* and the houses." Thus, from the point of view of the cause or of the structure of the object, there is participation, syncretistic schemas resulting from the fusion of singular terms. . . .[13]

THE SHIFT FROM MAGIC TO MYTHIC

As we move from early preoperational (2–4 years; "magic") to late preoperational (4–7 years; "magic-mythic"), similar types of adherences continue to dominate awareness. But one crucial difference comes to the fore: magic proper—the belief that the subject can magically alter the object—diminishes rapidly. Continued interaction with the world eventually leads the subject to realize that his or her thoughts do not egocentrically control, create, or govern the world. The "hidden linkages" don't hold up in reality.

Magic proper thus diminishes, or rather, the omnipotent magic of the individual subject—a magic that no longer "works"—is simply *transferred to other subjects*. Maybe I can't order the world around, but Daddy (or God or the volcano spirit) can.

And thus onto the scene come crashing a hundred gods and goddesses, all capable of doing what I can no longer do: miraculously alter the patterns of nature in order to cater to my wants. Whereas in the earlier magical stages proper, the secret of the universe was to learn the right type of word magic that would *directly* alter the world, the focus now is to learn the right rituals and prayers that will make the gods and goddesses intervene and alter the world for me. Piaget:

The possibility of miracles is, of course, admitted, or rather, miracles form part of the child's conception of the world, since law [at this stage] is a moral thing with the possibility of numerous exceptions ["suspended by God" or a *powerful other*]. Children have been quoted who asked their parents to stop the rain, to turn spinach into potatoes, etc.[14]

Thus the shift from magic to magic-mythic. Piaget: "The first stage is magical: we make the clouds move by walking. The clouds obey us at a distance. The average age of this stage is 5. The second stage [magic-mythic] is both artificialist and animistic. Clouds move because God or [other] men make them move. The average age of this stage is 6."[15]

It is from this magic-mythic structure that so many of the world's classical mythologies seem in large part to issue. As Philip Cowan points out, "During the [late preop or magic-mythic] stage, there is still a confusion between physical and personal causality; the physical world appears to operate much the way people do. All of these examples [show that late preop] children have already developed elaborate mythologies about cosmic questions such as the nature of life (and death) and the cause of wind [and so forth]. Further, these mythologies show many similarities from child to child across cultures and do not seem to have been directly taught by adults."[16]

MYTH AND ARCHETYPE

This directly brings us, of course, to the work of Carl Jung and his conclusion that the essential forms and motifs of the world's great mythologies—the "archaic forms" or "archetypes"—are collectively inherited in the individual psyche of each of us.

It is not often realized that Freud was in complete agreement with Jung about the existence of this archaic heritage. Freud was struck by the fact that individuals in therapy kept reproducing essentially similar "phantasies," phantasies that seemed therefore somehow to be collectively inherited. "Whence comes the necessity for these phantasies and the material for them?" he asks. "How is it to be explained that the same phantasies are always formed with the same content? I have an answer to this which I know will seem to you very daring. I believe that these primal phantasies are a phylogenetic possession. In them the individual stretches out to the experiences of past ages."[17]

This phylogenetic or "archaic heritage" includes, according to Freud, "abbreviated repetitions of the evolution undergone by the whole human race through long-drawn-out periods and from pre-historic ages." Although, as we will see, Freud and Jung differed profoundly over the actual nature of this archaic heritage, Freud nevertheless made it very clear that "I fully agree with Jung in recognizing the existence of this phylogenetic heritage."[18]

Piaget has also written extensively on his essential agreement with and appreciation of Jung's work. But he differs slightly with Jung in that he does not see the archetypes themselves as being directly inherited from past ages, but rather as being the secondary by-products of cognitive structures which themselves are similar wherever they develop and which, in interpreting a common physical world, generate common motifs.

But whether we follow Freud, Jung, or Piaget, the conclusion is essentially the same: all the world's great mythologies exist today in each of us, in me and in you. They are produced, and can at any time be produced, by the archaic, the magic, and the mythic structures of our own compound individuality (and classically by the magic-mythic structure).

The question then centers—and here Freud and Jung bitterly parted ways—on the nature and function of these mythic motifs, these archetypes. Are they *merely* infantile and regressive (Freud), or do they also contain a rich source of spiritual wisdom (Jung)? Piaget, needless to say, sided with Freud on this particular issue. I have already suggested that I do not see these particular "archetypes" as being quite the high source of transpersonal wisdom that Jung believed; but the situation is very subtle and complex, and we will return to it later in this chapter, in connection with Joseph Campbell, and discuss it more fully.

Campbell, we will see, believes that in certain circumstances (which we will explore), the early mythic archetypes can carry profound religious and spiritual meaning and power. But even Campbell clearly acknowledges (and indeed stresses) that the early and late preoperational stages themselves are both marked by a great deal of *egocentrism, anthropocentrism*, and *geocentrism*.

Put differently, still lying "close to the body," preoperational cognition does not easily take the role of other, nor does it still clearly differentiate the noosphere and the biosphere. Even in late preoperational thinking, the child firmly believes that names are a part of, or actually exist in, the objects that are named. "What are names for?" a child of five was asked. "They are what you see when you look at things." "Where is the name of the sun?" "Inside the sun." As one child summarized it: "If there weren't any words it would be very bad. You couldn't make anything. How could things have been made?" Joseph Campbell comments:

> In the cosmologies of archaic man, as in those of infancy, the main concern of the creator was in the weal and woe of man. Light was made so that we should see; night so that we might sleep; stars to foretell the weather; clouds to warn of rain. The child's view of the world is not only *geocentric*, but *egocentric*. And if we add to this simple structure the tendency recognized by Freud, to experience all things in association with the subjective formula of the family romance [Oedipus/Electra], we have a rather tight and very slight vocabulary of elementary ideas, which we may expect to see variously inflected and applied in the mythologies of the world.[19]

REPRESSION

The emergence of the noosphere: First images (at around 7 months), then symbols (the first full-fledged symbol probably being the word "no!"), then concepts (around 3–4 years), all aided immeasurably by the emergence of language.[20]

I mentioned the word "no!" as being the first symbol, and that word summarizes all the strengths—and all the weaknesses—of the newly emerging noosphere.

"No" is the first form of specifically mental transcendence. Images begin this mental transcendence, but images are tied to their sensory referents. With "no" I can for the first time decline to act on my bodily impulses or on your desires (which every parent discovers in the child during the "terrible twos"). For the first time in development, the child can begin to transcend its merely biological or biocentric or ecocentric embeddedness, begin to exert control over bodily desires and bodily discharges and bodily instincts, while also "separating-individuating" itself from the will of others. The Freudian fuss over "toilet training" and the "anal phase" simply refers to the fact that a mental-linguistic self is *beginning* to emerge and beginning to exert some type of conscious will and conscious control over its spontaneous biospheric productions, and over its being "controlled" by others as well.

In short, it is only with language that the child can differentiate its mind and body, differentiate its mental will and its bodily impulses, and then begin to *integrate* its mind and body. This is the third major differentiation, or the third fulcrum. The *failure* to differentiate mind and body—the failure to transcend at this stage—is another way to say "remains stuck in the body or the biosphere," which, we saw, is the primary developmental lesion underlying the narcissistic/borderline pathologies.

But "no!" can go too far, and therein lies all the horrors of the noosphere. For if it is indeed with language that the child can differentiate mind and body, differentiate the noosphere and the biosphere, that *differentiation* (as always) can go too far and result in *dissociation*. The mind does not just transcend and include the body, it represses the body, represses its sensuality, represses its sexuality, represses its rich roots in the biosphere. Repression, in the Freudian (and Jungian) sense, comes into existence only with the "language barrier," with a "no!" carried to extremes. And the result of this extreme "no!" is technically called "neurosis" or "psychoneurosis."

Every neurosis, in other words, is a miniature ecological crisis. It is a refusal to include in the compound individual some aspect of organic life, emotional-sexual life, reproductive life, sensuous life, libidinal life, biospheric life. It is a denial of our roots and our foundation. Neurosis, in this sense, is an assault on the biosphere by the noosphere, an attempt to render *extinct* some aspects of our own organic holons. But these biospheric holons do not thereby merely disappear (nor could they without killing the individual). Rather, the repressed holons return in disguised forms known as "neurotic symptoms"—anxieties and depressions and obsessions—painful symptoms of a biosphere ignored, a biosphere now forcing itself into consciousness in hidden forms, attempts to throw the noosphere off its back.[21]

And the neurotic symptoms disappear, or are healed, only as consciousness relaxes its repression, recontacts and befriends the biosphere that exists *in its own being*, and then integrates that biosphere with the newly emergent noosphere in its own case. This is called "uncovering the shadow," and the shadow is . . . the biosphere.

Thus, if remaining *stuck* in the biosphere results in the borderline/narcissistic conditions, going to the other extreme and *alienating* the biosphere results directly in the psychoneuroses. It follows that our present-day worldwide ecological crisis is, in the very strictest sense of the term, a worldwide collective neurosis—and is about to result in a worldwide nervous breakdown.

This crisis, I repeat, is in no way going to "destroy the biosphere"—the biosphere will survive, in some form or another (even if just viral and bacterial), no matter what we do to it. What we are doing, rather, is altering the biosphere in a way that will not support higher life forms and especially will not support the noosphere. That "alteration" is in fact a repression, an alienation, a denial of our common ancestry, a denial of our relational existence with all of life. It is not a destruction of the biosphere but a *denial* of the biosphere, and *that* is the precise definition of psychoneurosis.

What Freud found his patients doing on a couch in Vienna, we have now collectively managed to do to the world at large. And who shall be *our* doctor?

CONCRETE OPERATIONAL (MYTHIC AND MYTHIC-RATIONAL)

Assuming development goes relatively smoothly, then with the first significant differentiation of the mind and body (the third fulcrum), the mind can transcend its embeddedness in a merely bodily orientation—absorbed in itself (egocentric)—and begin to enter the world of other minds. But to do so it must learn to *take the role of other*—a new, emergent, and very difficult task.

In other words, the self has gone from a *physiocentric* identity (first fulcrum) to a *biocentric* identity (second fulcrum) to an early *noospheric* identity (third fulcrum), all of which are thoroughly *egocentric* and *anthropocentric* (magic and magic-mythic all centered on the self and oriented exclusively to the self, however "otherworldly" or "sacred" it might all appear).

If the sensorimotor and preoperational world is egocentric, the concrete operational world is *sociocentric* (centered not so much on a bodily identity as on a *role* identity, as we will see). It still contains "mythic" and "anthropocentric" elements because, as Cowan puts it, "there are still various colorings of the previous stages"[22] (which is why I call early and late conop, respectively, mythic and mythic-rational). A more differentiated causation by "five elements" (water, wind, earth, fire, ether) tends to replace more syncretic explanations, and there often emerges a belief in causation by "preformation" (the acorn contains a fully formed but miniature oak tree).

But by far the most significant transformation or transcendence occurs in the capacity to *take the role of other*—not just to realize that others have a different perspective, but to be able to mentally reconstruct that perspective, to put oneself in the other's shoes.

In what became known as the Three Mountains Task, Piaget and Bärbel Inhelder exposed children from four to twelve years old to a play set that contained three clay mountains, each of a different color, and a toy doll. The questions were simple: what do you see, and what does the doll see?

The typical response of the preoperational child is that the doll sees the same thing that the child is looking at, even if the doll is facing only, say, the green mountain. The child does not understand that there are different perspectives involved. At a later stage of preop, the child will correctly indicate that the doll has a different perspective, but the child cannot say exactly what that is.

But with the emergence of concrete operational, the child will easily and readily describe the true perspective of the doll (e.g., "I am looking at all three mountains, but the doll is only looking at the green mountain").

Investigation of these and similar tasks (by Robert Selman, John Flavell, and others) has confirmed the general conclusion: only with the emergence of concrete operational thought can the child transcend his or her egocentric perspective and take the *role of other*. As Habermas would put it, a *role* identity supplements a *natural* (or bodily) identity (the body cannot take the role of other). The child learns his or her *role* in a society of *other roles*, and must now learn to *differentiate* that role from the role of others and then *integrate* that role in the newly emergent worldspace (this is the fourth major fulcrum, the fourth major differentiation/integration of self-development). The fundamental locus of self-identity thus switches from *egocentric* to *sociocentric*.

Initially the child is indissociated from his or her role, is embedded or "stuck" in it (just as he or she was initially stuck in the physiosphere and then stuck in the biosphere). This unavoidable (and initially necessary) "sociocentric embeddedness" leads to what is variously known as the *conventional stages* of morality (Kohlberg/Gilligan), the *belongingness* needs (Maslow), the *conformist* mode (Loevinger).

Which is why pathology at this stage is known generally as "script pathology." One is having trouble, not with the physiosphere (psychoses), not with the biosphere (borderline and neuroses)—rather, one is *stuck* in the early roles and scripts given by one's parents, one's society, one's peer group: scripts that are not, and initially cannot be, checked against further evidence, and therefore scripts that are often outmoded, wrong, even cruel ("I'm no good, I'm rotten to the core, I can't do anything right," etc.; these do not so much concern *bodily impulses*, as in the psychoneuroses, but rather *social judgments* about one's *social standing*, one's role).

Therapy here involves digging up these scripts and exposing these *myths* to the light of more mature reason and more accurate information, thus "rewriting the script." (This is, for example, the primary approach of cognitive therapy and interpersonal therapy; not so much the digging up of buried and alienated bodily impulses, as important as that may be, but replacing false and distorting cognitive maps with more reasonable judgments).[23]

Equally important to the taking of *roles* is the capacity of conop to work with mental *rules*. We saw that preop works with images (pictorial representation), symbols (nonpictorial representation), and concepts (which represent an entire class of things). Rules go one step further and *operate upon* concrete classes, and thus these rules (like multiplication, class inclusion, hierarchization) begin to grasp the incredibly rich relationships between various wholes and parts.

That is, concrete operational is the first structure that can clearly grasp the nature of a holon, of that which in one relationship is a whole and *at the same time* in another relationship is merely a part (which is why *value holarchies* start to emerge spontaneously in children at this point; they switch from the rather strong "either-or" desires of preop to a *continuum of preferences*). All of this, of course, depends upon the capacity of conop to begin to take different perspectives and relate those perspectives to each other.

Because of its capacity to operate with both rules and roles, I also call this structure the rule/role mind. *In relation to the previous stage(s)*, it represents a greater transcendence, a greater autonomy, a greater interiority, a higher and wider identity, a greater consciousness, but one that, as in all previous stages, is initially "captured" by the self and the objects—now a social self and now social objects (roles)—that dominate this stage.

And thus a self now open to new and higher pathologies, which demand new and different therapies. No longer stuck in the physiosphere, stuck in the biosphere, or stuck in the early "egosphere," the pathological self is here stuck in the sociosphere, embedded in a particular society's rules and myths and dogmas, with no way to transcend that mythic-membership, and thus destined to play out the roles and rules of a particular and isolated society.

Mythic-membership is sociocentric and thus *ethnocentric*: one is *in* the culture (a member of the culture) if one accepts the prevailing mythology, and one is excommunicated from the culture if the belief system is not embraced. In this structure, there is no way a global or planetary culture can even be conceived unless it involves the imposing of one's particular mythology on all peoples: which is just what we saw with the mythic-imperialism of the great empires, from the Greek and Roman to the Khans and Sargons to the Incas and Aztecs. These great empires all overcame the egocentrism of local and warring tribes by subsuming their regimes into that of the empire (thus negating and preserving them in a larger reach or communion), and this was accomplished in part, as we saw, by the umbrella of a mythology that unified different tribes, not by blood or kinship (for that is impossible, since each tribe has a different lineage), but rather by a common mythological origin that could unite the various roles (as the twelve Tribes of Israel were united by a common Yahweh).

But as mature egoic-rationality begins to emerge, ethnocentric gives way to worldcentric.

The Ego

We saw that Habermas referred to the transcendence from conop to form-op as a transformation from a *role* identity to an *ego* identity. "Ego" here doesn't mean "egocentric"; on the contrary, it means moving from a *sociocentric* to a *worldcentric* capacity, a capacity to distance oneself from one's egocentric and ethnocentric embeddedness and consider what would be fair for all peoples and not merely one's own.

It would be helpful, then, to discuss the meaning of the word *ego*. Particularly in transpersonal circles, no word has caused more confusion. *Ego*, along with *rationality*, is generally *the* dirty word in mystical, trans-

236 | BOOK ONE

personal, and New Age circles, but few researchers seem even to define it, and those who do, do so differently.

We can, of course, define *ego* any way we like as long as we are consistent. Most New Age writers use the term very loosely to mean a separate-self sense, isolated from others and from a spiritual Ground. Unfortunately, these writers do not clearly distinguish states that are pre-egoic from those that are transegoic, and thus half of their recommendations for salvation are often recommendations for various ways to regress, and this rightly sends alarms through orthodox researchers. Nonetheless, their general conclusion is that all truly spiritual states are "beyond ego," which is true enough as far as it goes, but which terribly confuses the picture unless it is carefully qualified.

In most psychoanalytically oriented writers, the ego has come to mean "the process of organizing the psyche," and in this regard many researchers, such as Heinz Kohut, now prefer the more general term *self*. The ego (or self), as the principle that gives unity to the mind, is thus a crucial and fundamental organizing pattern, and to try to go "beyond ego" would mean not transcendence but disaster, and so these orthodox theorists are utterly perplexed by what "beyond ego" could possibly mean, and who could possibly desire it—and, as far as that definition goes, they too are quite right. (We'll return to this in a moment.)

In philosophy a general distinction is made between the empirical ego, which is the self insofar as it can be an object of awareness and introspection, and the Pure Ego or transcendental Ego (Kant, Fichte, Husserl), which is pure subjectivity (or the observing Self), which can never be seen as an object of any sort. In this regard, the pure Ego or pure Self is virtually identical with what the Hindus call Atman (or the pure Witness that itself is never witnessed—is never an object—but contains all objects in itself).

Furthermore, according to such philosophers as Fichte, this pure Ego is one with absolute Spirit, which is precisely the Hindu formula Atman = Brahman. To hear Spirit described as pure Ego often confuses New Agers, who generally want *ego* to mean only "the devil" (even though they heartily embrace the identical notion Atman = Brahman).

They are equally confused when someone like Jack Engler, a theorist studying the interface of psychiatry and meditation, states that "*meditation increases ego strength*," which it most certainly does, because "ego strength" in the psychiatric sense means "capacity for disinterested witnessing." But the New Agers think that meditation means "beyond ego," and thus anything that strengthens the ego is simply more of the devil. And so the confusions go.

Ego is simply Latin for "I." Freud, for example, never used the term *ego*; he used the German pronoun *das Ich*, or "the I," which Strachey unfortunately translated as the "ego." And contrasted to "the I" was what Freud called the *Es*, which is German for "it," and which, also unfortunately, was translated as the "id" (Latin for "it"), a term Freud never used. Thus Freud's great book *The Ego and the Id* was really called "The I and the It." Freud's point was that people have a sense of I-ness or selfness, but sometimes part of their own self appears foreign, alien, separate from them—appears, that is, as an "it" (we say, "The anxiety, it makes me uncomfortable," or "The desire to eat, it's stronger than me!" and so forth, thus relinquishing responsibility for our own states). When parts of the I are split off or repressed, they appear as symptoms or "its" over which we have no control.

Freud's basic aim in therapy was therefore to reunite the I and the it and thus heal the split between them. His most famous statement of the goal of therapy—"Where id was, there ego shall be"—actually reads, "Where it was, there I shall be." Whether one is a Freudian or not, this is still the most accurate and succinct summary of all forms of uncovering psychotherapy, and it simply points to an expansion of ego, an expansion of I-ness, into a higher and wider identity that integrates previously alienated processes.

The term *ego* can obviously be used in a large number of quite different ways, from the very broad to the very narrow, and it is altogether necessary to specify which usage one intends, or else interminable arguments arise that are generated only by an arbitrary semantic choice.

In the broadest sense, *ego* means "self" or "subject," and thus when Piaget speaks of the earliest stages being "egocentric," he does *not* mean that there is a clearly differentiated self or ego set apart from the world. *He means just the opposite*: the self is *not* differentiated from the world, there is *no* strong ego, and thus the world is treated as an extension of the self, "egocentrically." Only with the emergence of a strong and differentiated ego (which occurs from the third to the fifth fulcrums, culminating in formop, or rational perspectivism)—only with the emergence of the mature ego does egocentrism die down! The "pre-egoic" stages are the most egocentric!

Thus, it is only at the level of formal operational thought (as we will see in a moment) that a truly strong and differentiated self or ego emerges from its embeddedness in bodily impulses and pre-given social roles; and that, indeed, is what Habermas refers to as an *ego identity*, a fully separated-individuated sense of self.

To repeat: the "ego," as used by psychoanalysis, Piaget, and Habermas (and others), is thus less egocentric than its pre-egoic predecessors![24]

I will most often use the term *ego* in that specific sense, similar to Freud, Piaget, Habermas, and others—a rational, individuated sense of self, differentiated from the external world, from its social roles (and the super-ego), and from its internal nature (id).

In this usage, there are *pre-egoic* realms (particularly the archaic and the magic), where the self is poorly differentiated from the internal and external world (there are only "ego nuclei," as psychoanalysis puts it). These pre-egoic realms are, to repeat, the most egocentric (since the infant or child doesn't have a strong ego, it thinks that the world feels what it feels, wants what it wants, caters to its every desire: it does not clearly separate self and other, and thus treats the other as an extension of the self).

The ego begins more stably to emerge in the mythic stage (as a *persona* or *role*) and finally emerges, in the formal operational stage, as a self clearly differentiated from the external world and from its various roles (personae), which is the culmination of the overall *egoic* realms. Higher developments into more spiritual realms are then referred to as being *transegoic*, with the clear understanding that the ego is being negated but also preserved (as a functional self in conventional reality). The self in these higher stages I will refer to as the Self (and not the pure Ego, unless otherwise indicated, because this confuses everybody), and I will explain all of that in more detail as we proceed.[25]

These three large realms are also referred to, in very general terms, as the subconscious (pre-egoic), the self-conscious (egoic), and the superconscious (transegoic); or as the prepersonal, the personal, and the transpersonal; or as the prerational, the rational, and the transrational.

The point is that *each* of those stages is a *lessening of egocentrism* as one moves closer to the pure Self.[26] The *maximum* of egocentrism, as Piaget demonstrated, occurs in the primary or physical indissociation (the first fulcrum, where self-identity is physiocentric), because the entire material world is absorbed in the self-sense and cannot even be considered apart from the self-sense. This archaic-autistic stage is not "one with the entire world in bliss and joy," as many Romantics think, but a swallowing of the material world into the self: the child is all mouth, and everything else is merely food.

As identity switches from physiocentric to biocentric or ecocentric (fulcrum-2), there is a lessening of "pure autism" ("self-only!") but a blossoming of emotional narcissism or *emotional egocentrism* (fulcrum-2), which Mahler summarized as "narcissism at its peak!" (She also summarized it as "the world is the infant's oyster": grandiose-omnipotent fantasies). The emergence of the preop mind (fulcrum-3) is a lessening of that emotional egocentrism, but a blossoming of egocentric (and geocentric) magic—less primitive than the previous stage, but still shot through with egocentric adherences: the world exists centered on humans.

The emergence of the conop mind (fulcrum-4) is a lessening of that egocentric magic (where the self is central to the cosmos), but it is replaced with an *ethnocentrism*, where one's particular group, culture, or race is supreme. Nonetheless, at the same time this allows the beginning of what Piaget calls a *decentering*, where one can decenter or stand aside from the egocentrism of the early mind and instead take the role of other, and this comes to a fruition with a *further decentering*, a further lessening of egocentrism, in formal operational (where one can take the perspective, not just of others in one's group, but of others in other groups: *worldcentric* or non-ethnocentric).

In other words, each stage *transcends* its predecessor and is *therefore* less egocentric, less caught in the narrower and shallower perspective, but instead reaches out more and more to embrace deeper and wider occasions. The self becomes less and less egocentric, and thus embraces more and more holons as worthy of equal respect.

As we will see when we follow evolution into the transpersonal domain, these developments converge on an intuition of the very Divine as one's very Self, common in and to all peoples (in fact, all sentient beings), a Self that is the great omega point of this entire series of *decreasing egocentrism*, of decentering from the small self in order to find the big Self—a Self common in and to all beings and thus escaping the egocentrism (and ethnocentrism) of each. The completely decentered self is the all-embracing Self (as Zen would say, the Self that is no-self).

What else could *evolution* as *decreasing egocentrism* possibly mean? Embracing identities that include more and more beings, both deeper and wider—what is the chaotic Attractor here? Where else does it look like this series is going? Where else could it possibly go?

FORMAL OPERATIONAL

At this point, we are tracing the emergence of a strong rational ego out of its embeddedness in mythic-membership, and this brings us to Piaget's formal operational stage.

Formal operational awareness transcends but includes concrete operational thought, and thus formop can *operate upon* the holons that constitute conop—and that, in fact, is the primary definition of *formal operational*. Where concrete operational uses rules of thought to transcend and operate on the concrete world, formal operational uses a new interiority to transcend and operate on the rules of thought themselves. It is a new differentiation allowing a new integration (and a *deeper* and *wider* identity).

Again this sounds terribly dry and abstract, but the results are not. First and foremost, formal operational awareness brings with it a new world of feelings, of dreams, of wild passions and idealistic strivings. It is true that rationality introduces a new and more abstract understanding of mathematics, logic, and philosophy, but those are all quite secondary to the primary and defining mark of reason: *reason is a space of possibilities*, possibilities not tied to the obvious, the given, the mundane, the profane. Reason, we said earlier, is the great gateway to the unseen, the beginning of the invisible worlds, which is usually the last way people think of rationality.

But think of the great mystics such as Plato and Pythagoras, who saw rational Forms or Ideas as the grand patterns upon which all of manifestation was based, patterns that were utterly invisible to the eye of flesh and could only be seen interiorly, with the eye of mind. Or think of the great physicists such as Heisenberg and Jeans, who maintained that the ultimate building blocks of the universe are mathematical Forms, also seen only with the mind's eye. Or of the great Vedantin and Mahayana sages, who maintained that the entire visible world is just a precipitate of the mind's interior Forms or "seed-syllables." For all of these theorists (and many more like them), Reason was not an abstraction from the concrete physical world; rather, the concrete world was a reduction or condensation of the great mental Forms lying beyond the grasp of the senses, Forms which contained *en potentia* all *possible* manifest worlds.

Piaget approaches this whole topic by showing that, whereas the concrete operational child can indeed operate upon the concrete world, the child at that stage ultimately remains tied to the obvious and the given and the phenomenal, whereas the formal operational adolescent will *mentally see* various and different possible arrangements of the given. Again, a typical Piagetian experiment sounds very dry and abstract and far removed from ordinary events, but what it actually indicates is the power of the creative imagination. I'll simplify considerably:

A child is presented with five glasses which contain colorless liquids. Three of the glasses contain liquids that, if mixed together, will produce a yellow color. The child is asked to produce a yellow color.

The preop child will randomly combine a few glasses, then give up. If she accidentally hits upon the right solution, she will give a magical explanation ("The sun made it happen"; "It came from the clouds").

The conop child will eagerly begin by combining the various glasses, three at a time. She does this concretely; she will simply continue the concrete mixing until she hits upon the right solution or eventually gets tired and quits.

The formop adolescent will *begin* by telling you that you have to try all the possible combinations of three glasses. She has a mental plan or formal operation that lets her see, however vaguely, that *all possible* combinations have to be tried. She doesn't have to stumble through the actual concrete operations to understand this. Rather, she sees, with the mind's eye, that *all possibilities must be taken into account.*

In other words, this is a very *relational* type of awareness: all the possible relations that things can have with each other need to be held in awareness—and this is radically new. This is not the "wholeness" of syncretic fusion, where the integrity of wholes and parts is violated in a magical fusion, but rather a relationship of mutual interaction and mutual interpenetration, where wholes and parts, while remaining perfectly discrete and intact, are also seen to be what they are by virtue of their relationships to each other. The preop child, and to a lesser extent the conop child, thinks that the color yellow is a simple property of the liquids; the formop adolescent understands that the color is a *relationship* of various liquids.

Formal operational awareness, then, is the first truly *ecological mode of awareness*, in the sense of grasping mutual interrelationships. It is not *embedded* in ecology (that would be fulcrum-2); it *transcends* ecology (without denying it), and thus can reflect on the web of relationships constituting it. As various researchers have pointed out, to use Cowan's particular phrasing: "Again the emphasis in formal [operational] schemes is on the coordination of [various] systems. Not only can adolescents [at this stage] observe and reason about changes in the interior of [an individual], they can also be concerned with reciprocal changes in the surrounding environment. Only then, for example, will they be able to conceptualize an ecological system in which changes in one aspect may lead to a whole system of changes in the *balance* between other aspects of nature."[27]

So the first equation we need is: *formal operational = ecological.*

The fact that formop is also strong enough, as we have seen, to potentially repress the biosphere, resulting in ecological catastrophe, indicates merely that ecological catastrophe is an unfortunate possibility, but not an inherent component, of rationality. As always, we want to tease apart the pathological manifestations of any stage from its authentic achievements, and celebrate the latter even as we try to redress the former. The fact that ecological awareness becomes even greater at the next stage, the centauric, should not detract from the fact that it *begins* here, with the formal operational understanding of mutual relationships, and it does not begin prior to this stage at all (prior to this are syncretic wholes, which sound "ecological" but actually constitute a certain violence to the integrity of whole/parts).

The second equation we need is: *formal operational = understanding of relativity.*

The capacity to take different perspectives, we saw, begins in earnest with conop. But with the emergence of formop, all the various perspectives can be held in mind, however loosely, and thus all of them become *relative* to each other. "In a set of experiments, a snail moves along a board, which itself is moving along a table. Only children at the formal operations stage can understand the distance which the snail travels relative to the board *and* to the table. Here we find the intellectual equipment necessary for conceptions of relativity—that time taken or space travelled cannot be absolute, but must be measured relative to some arbitrary point."[28]

The third equation that we need is: *formal operational = non-anthropocentric*.

The egocentric, geocentric, anthropocentric notions of reality, so prevalent in the earlier magic and mythic stages, and so defining of those stages, finally begin to wind down and lose their grip on awareness, and humans take their rightful place in the great holarchy of being as one set of wholes that are parts in all sorts of other wholes, with no whole and no part being finally privileged.

And not merely egocentrism but sociocentrism or ethnocentrism begins to wind down. With the coming of formop, the rules and norms of any given society can themselves be reflected upon and judged by more universal principles, principles that apply not just to this or that culture, or this or that tribe, but to the multiculturalism of universal perspectivism. Not "My country right or wrong," but "Is my country actually right?" Not concrete moral rules such as the Ten Commandments ("Thou shalt have no other gods before me"—intertribal squabbling), but more universal statements, principles of justice and mercy and compassion, of reciprocity and equality, based on mutual respect for individuals and the dictates of conscience based on *rights* (as an autonomous *whole*) and *responsibilities* (as a *part* of a larger whole—agency and communion, rights and responsibilities).

Thus Kohlberg, Gilligan, and Habermas (to name a few) all refer to this general stage as *postconventional* (which doesn't mean postcultural or postsocial, but simply postconformist in some significant ways). Socrates versus Athens. Martin Luther King, Jr., versus segregation. Gandhi versus cultural imperialism.

Thus, we have seen moral development move from a *preconventional* orientation, which is strongly egocentric, geocentric, biocentric, narcissistic, bound to the body's separate feelings and nature's impulses (the first three fulcrums), to a *conventional* or sociocentric or ethnocentric orientation, bound to one's society, culture, tribe, or race, to a *postconventional* or worldcentric orientation, operating in the space of universal pluralism and global grasp (these stages have been elaborated by Piaget, Baldwin, Kohlberg, Gilligan, Habermas, Loevinger, Broughton, Selman, etc.).[29]

For all these reasons, the individual at this stage, who can no longer rely on society's given roles in order to establish an identity, is thus thrown back on his or her own inner resources. "Who am I?" becomes, for the first time, a burning question, and the self-esteem needs emerge from the belongingness needs (Maslow), or a "conscientious" self emerges from a "conformist" mode (Loevinger).

A failure to negotiate this painfully self-conscious phase ("fulcrum 5")—a *differentiation* from ethnocentrism and sociocentrism—results in the characteristic pathology of this stage, which Erikson called an "identity crisis." This is not a problem of merely finding an appropriate *role* in society (that would be script pathology); it is one of finding a self that may or may not fit with society at all (Thoreau on civil disobedience comes to mind).

In addition to formal operational awareness being ecological, relational, and nonanthropocentric, we have already mentioned several of its other properties: it is the first structure that is highly reflexive and highly introspective; it is experimental (or hypothetico-deductive) and relies on evidence to settle issues; it is universal as pluralism or perspectivism; and it is propositional (can understand "what if" and "as if" statements; the fact that formop is the first structure that can grasp "as if" statements turns out to be extremely important when it comes to interpreting mythology, as we will see in the following section on Joseph Campbell). But all of these are just variations on the central theme: reason is a space of possibilities.

No wonder adolescence and the emergence of formop is a time of wild passions and explosive idealisms, of fantastic dreaming and heroic urges, of utopian yells and revolutionary upsurge, of desires to change the entire world and idealistically straighten it all out, of feelings and emotions unleashed from the merely given and offered instead the space of all possibilities, a space through which they roam and rampage with love and passion and wildest terror. And all of this, all of it, comes from being able to see the possibilities of *what might be*, possibilities seen only with the mind's eye, possibilities that point toward worlds not yet in existence and worlds not yet really seen, the great, great doorway to the invisible and the beyond, as Plato, and Pythagoras, and Shankara, and every mystic worth his or her salt has always, always known.

The higher developments do indeed lie beyond reason, but never, never, beneath it.

JOSEPH CAMPBELL

There is no greater friend of mythology than Joseph Campbell, and I mean that in a good sense. In a series of articulate and extremely well-researched books, Campbell has done more than any other person, with the possible exceptions of Mircea Eliade and Carl Jung, to champion the position that mythological thought is the primary carrier of spiritual and mystical awareness. I and countless researchers have drawn on his works time and again, and his meticulous scholarship and detailed analyses never fail to inspire. And, among many other things, it is Joseph Campbell, via Robert Bly, who is responsible for much of the "men's movement" and the "mythopoetic" movement in general.

And yet his position, I believe, is finally untenable, and can be demonstrated to be so using his own assumptions and his own conclusions. For his position is, in the last analysis, a form of elevationism, and it is necessary to face this directly if we are ever to make sense of the truly deeper or higher developments of genuine spiritual and mystical experience. For, as we said earlier, one of the best ways to know what authentic mystical experience is, is to know what it is not.

To begin with, Campbell openly accepts the essentials of the Piagetian system. That is, he accepts the fact that the basic motifs of mythological thought are produced by the infantile and childhood structures of preop and early conop, and he explicitly says so using Piagetian terms. As just one of hundreds of instances:

> The two orders—the infantile and the religious—are at least analogous, and it may well be that the latter is simply a translation of the former to a sphere out of range of critical observation [reason]. Piaget has pointed out that although the little myths of genesis invented by children to explain the origins of themselves and of things may differ, the basic assumption underlying all is the same: namely, that things have been made by someone, and that they are alive and responsive to the commands of their creators. The origin myths of the world's mythological systems differ too; but in all [of them] the conviction is held (as in childhood), without proof, that the living universe is the handiwork . . . of some father-mother or mother-father God [artificialism/anthropocentrism].
>
> These three principles [magical participation, animism, and anthropocentrism] may be said to constitute the axiomatic, spontaneously supposed frame of reference of all childhood experience, no matter what the local details of this experience happen to be. And these three principles, it is no less apparent, are precisely those most generally represented in the mythologies and religious systems of the whole world.[30]

Campbell cheerfully and even enthusiastically acknowledges all of this, and he does so because he has a plan. He has a plan, that is, to salvage mythology, to prove that mythology is "really" religious and genuinely spiritual, and is not, in fact, merely a device of childhood.

The plan is this: The mythological productions of preop and conop, he says, are always taken very *literally* and *concretely*, a point I have also been at pains to emphasize. But, Campbell says, in a *very few* individuals, the myths are *not taken literally*, but are rather taken *in an "as if" fashion* (his terms), in a playful fashion that releases one from the concrete myth and ushers one into more transcendental realms.

And this, he says, is the *real* function of myth, and therefore this is how all myths have to be judged. For the masses it remains true that myth is an illusion, a distortion, an infantile and childish approach to reality (all his phrases), but for the very few who can see through them, myths become the gateway to the genuinely mystical. And he belabors the point that *myths, not reason, alone can do this*, and this is their wonderful function. And here he starts running into grave difficulties.

When myths are taken concretely and literally, Campbell says, they serve the mundane function of integrating individuals into the society and worldview of a given culture, and in that ordinary function, he says, they serve no spiritually transcendental or mystical purpose at all, which is true enough. I myself see that mundane integration as *the* central, enduring, and extremely important function of myths at that stage of development—simple cultural meaning and correlative social integration (at a preop and conop level).

Campbell acknowledges that function, but since he is looking for a way to elevate myths to a transpersonal status, those functions become quite secondary to him. In fact, he says, when people take myths literally—which, he says, 99.9 percent of mythic believers do—then *those myths become distorted*. He is very emphatic about this: "It must be conceded, as a basic principle of our natural history of the gods and heroes, that whenever a myth has been taken literally its sense has been perverted."[31]

Let us ignore, for the moment, that this implies that 99.9 percent of mythic believers are perverted (instead of stage-specifically quite adequate), and look instead to those very, very few individuals who do not take myth literally but rather in an "as if" fashion. By "as if" Campbell explicitly means the use given to it by Kant in his *Prolegomena to Every Future System of Metaphysics*, where Kant says that we can only hold our knowledge of the world in an "as if" or "possible realities" fashion. Campbell then drives to the heart of his argument:

> I am willing to accept the word of Kant, as representing the view of a considerable metaphysician. And applying it to the range of festival games and attitudes just reviewed [by which he means the attitude that does not take myth seriously or literally]—from the mask of the consecrated host and temple image, transubstantiated worshiper and transubstantiated world—I can see, or believe I can see, that a principle of *release* operates throughout the series by way of an "as if"; and that, through this, the impact of all so-called "reality" upon the psyche is transubstantiated.[32]

In other words, a myth is being a "real myth" when it is *not* being taken as true, when it is being held in an "as if" fashion. And Campbell knows perfectly well that an "as if" stance is *possible only with formal operational awareness*. Thus, according to his own conclusions, a myth offers its "release" *only* when it is transcended by, and held firmly in, the space of possibilities and as-ifs offered by rationality. It is *reason*, and reason alone, that can release myth from its concrete literalness and hold it in a playful, as-if, what-if fashion, using it as an *analogy* of what higher states might be like, which is something that myth, by itself, could *never* do (as Campbell bizarrely concedes).[33]

It is people such as Campbell and Jung and Eliade, *operating from a widespread access to rationality*—something the originators of myth did not have—who *then* read deeply symbolic "as ifs" into them, and who like to play with myths and use them as analogies and have great good fun with them, whereas the actual mythic-believers do not play with the myths at all, but take them deadly seriously and refuse in the least to open them to reasonable discourse or any sort of "as if" at all.

In short, a myth serves Campbell's main function *only* when it ceases to be a myth and is released into the space of reason, into the space of alternatives and possibilities and as-ifs. What structure does he think Kant is operating from?

Thus, in all of Campbell's presentations, he takes two tacks: he *first* lays out the concrete and literal way that 99.9 percent of believers take the myth. And here he is not often kind. He clearly despises concrete mythic-beliefs ("On the popular side, in their popular cults, the Indians [i.e., from India] are, of course, as positivistic in their readings of their myths as any farmer in Tennessee, rabbi in the Bronx, or pope in Rome. Krishna actually danced in manifold rapture with the gopis, and the Buddha walked on water").[34]

Instead of seeing the concrete myth as the *only* way that myth *can* be believed at that stage of development, and thus as being *perfectly adequate* and noble (if partial and limited) *for that stage*, he takes the concrete belief in magic and myth as a "perversion," as if this structure actually had a choice for which it could be condemned. He is in fact denigrating an entire series of developmental stages that represented extraordinary advances in their own ways, and were no more a perversion of spiritual development than an acorn is a perversion of an oak. But he *must* condemn these stages per se because he judges them against *his* elevated version of "real mythology," whereas I am not condemning these stages per se because they were themselves the real McCoy, the genuine item: they were doing *exactly* what is appropriate and definitive and stage-specific for mythology.

Second, Campbell then suggests the ways that those very few (who do not take the myth literally) have used to *transcend* the myth (and are therefore, I would like to point out, no longer doing anything that could remotely be called mythology). This involves first and foremost, for Campbell, holding the myth in the space of "as if"; that is, holding it in the space of reason (with the possibility of then going further and transcending reason as well).

And here Campbell commits the classic pre/trans fallacy. Since the prerational realms are definitely mythological, then Campbell wants to call the transrational realms "mythological" as well, since they too are nonrational (and since he wants to salvage mythology with a field promotion). So on the one side he lumps together *all* nonrational endeavors (from primitive mythology to highly developed contemplative encounters), and on the other side—the "bad" side—he dumps poor reason, even as he himself is in fact (and rather hypocritically) using the space of reason to salvage his myths. "Mythological symbols touch and exhilarate centers of life beyond the reach of vocabularies of reason."[35] There is indeed a "beyond reason," but *how much more so* is it "beyond mythology."

And, in fact, it is not "in praise of mythology," but rather "beyond mythology," to which the entire corpus of Campbell's work inexorably points. In surveying his truly magnificent, four-volume masterpiece, *The Masks of God*, Campbell leaves us with one final message:

> As any ethnologist, archaeologist, or historian would observe, the myths of the differing civilizations have sensibly varied throughout the centuries and broad reaches of mankind's residence in the world, indeed to such a degree that the "virtue" of one mythology has often been the "vice" of another, and the heaven of the one the other's hell. Moreover, with the old horizons now gone that formerly separated and protected the various culture worlds and their pantheons, a veritable *Götterdämmerung* has flung its flames across the universe. Communities that were once comfortable in the consciousness of their own mythologically guaranteed godliness find, abruptly, that they are devils in the eyes of their neighbors.[36]

The ethnocentric and divisive nature of mythology is fully conceded. Lamenting this state of affairs (even though it is inherent in mythology as mythic-imperialism), Campbell concludes that some more *global* understanding "of a broader, deeper kind *than anything envisioned anywhere in the past is now required.*" The hope, indeed, lies beyond parochial and provincial mythology. And beyond mythology is global and universal reason (and *then* beyond reason . . .).

Nowhere is Campbell's pre/trans confusion more painfully obvious than in his attempt to displace or deconstruct rational science (and thus simultaneously elevate mythology). And again, the embarrassment is established by his own premises and his own logical conclusions, which a mind as fine as Campbell's can ignore only by prejudice.

Since Campbell's aim is to prove that reason and science are in no sense "higher" than "real" mythology, he begins first by pointing out that even the worldview of science is actually a mythology. If he can do this successfully, he will have put science and mythology on the same level. He proceeds to outline four factors (or four functions) that all mythologies have in common.[37] The first he calls "metaphysical," whose function is "to reconcile waking consciousness to . . . this universe *as it is.*" The second function is to provide "an interpretative total image of the same," an interpretive cosmology. The third is sociocultural, "the validation and maintenance of a social order." And the fourth is psychological, or individual orientation and integration (in other words, his four functions are the Big Three with an integrating overview).

And, of course, defined in that way, science (or the scientific worldview) does indeed perform all four functions of mythology. But then, adds Campbell, science of course does some other things that mythology per se does not, such as its spectacular discoveries in evolution, medicine, engineering, and so forth.

In other words, rationality/science does everything myth does, plus something extra.

That, of course, *is* the definition of a higher stage. Campbell recognizes that mythology originates from a *particular stage* of human development (which he happily concedes is the childhood of men and women), and then he *also* defines it as what *all* stages have in common (namely, his four functions, which are simply a variation on the four quadrants and thus are present at all stages of human development). By this sneaky dual definition he hopes to be able *both* to concede mythology's childishness *and* run it through all higher stages, thus allowing him not only to salvage mythology (since it is now what all stages have in common) but also to push it all the way to infinity, all the way to transpersonal Spirit.

But the four functions (the four quadrants) are not a definition of mythology's functions; they are a definition of evolution's functions. They are a definition of human holons functioning at each and every stage of development, no more (and no less) present in mythology than in magic or in science or in centauric unfoldings. Mythology has no special claim whatsoever on the four quadrants; mythology is simply one particular instance of the four quadrants (mythology is the four quadrants as conceived by conop, just as science is the four quadrants as conceived by formop). Which only leaves Campbell's other definition: mythology can rightly lay claim only to the childhood of men and women.

And running *that* definition to infinity simply results in the infantilization of Spirit. Campbell's dual definitions actually undo each other, and point instead to the inexorable conclusion: beyond mythology is reason, and beyond both is Spirit.

The Romantic View of Mythic-Membership

Mythic-membership does indeed provide, or can provide, an *intensely cohesive social order*, principally because it can export disorder and excommunicate unbelievers. Nothing so wonderfully concentrates a community as the prospect of being burned at the stake for disagreeing with its worldview.

And, on the other hand, reason *can be disruptive* of societies, can lack the "social glue" of some mythic-membership cultures, precisely because the omega point of rationality is not sociocentric or ethnocentric but worldcentric (or global or universal): short of that it stops unhappily (as in a Gandhi or a Socrates or a Thoreau or a Martin Luther King, Jr.).

And if the ethnocentrically disruptive capacity of universal reason is, without further ado, compared with the happy ethnocentrism of mythology, it will appear that the emergence of rationality was somehow a massive loss of cultural meaning and social integration, whereas this is only true from an ethnocentric (or mythic-membership) bias.

New integrations are being sought by evolution on a *now-global level*, and this requires, as in all past transformations, the deconstruction of more limited and provincial perspectives. Mythology has tried to go global for the past ten thousand years and failed in every single attempt (attempts which, as we have seen, because they lacked the depth of genuinely universal reason, had to be pressed always in a military-imperialistic fashion). Reason has the global *capacity* and global *intent*, but is still developing and evolving the *means*, and in this endeavor it is resisted by every mythology which feels its god and goddess threatened.

Further, even within nation-states, reason provides a *deeper* form of integration (even if occasionally shaky)—namely, through the *relational exchange of self-esteem*, and not the relational exchange of conformist roles in the preestablished dominator hierarchy. Mythic members are happy if they fit snugly into the mythological hierarchy, for all are then sharing a similar depth.

But rational citizens are happy *only* if they coexist in relational exchange with other citizens at that new depth, that is, with other equally free citizens—*legally, morally,* and *politically* free subjects (as a worldcentric standard, regardless of race, color, creed, or gender). For somebody genuinely at the egoic-rational level, the very thought of other humans not having access to these freedoms is very painful (whereas for those at the mythic-membership level, the painful thought is the idea of other human beings not embracing their particular God and thus not being "saved": that others would not embrace their God sends agonies of proselytizing fury through their souls; infidels are *intolerable*, and can actually be killed in order to save them).

The rational structure can *tolerate* mythic-believers because it is a deeper and wider embrace (whereas mythic-believers cannot and will not tolerate rational attacks). And that rational transformation is more culturally *meaningful* and socially *integrating* on a *deeper* (and therefore *wider*) level, even if it appears less head-knockingly solid than that of the shallower and narrower engagement of mythic-membership.

That reason then introduces its own inherent problems and runs up against its own inherent limitations (and can be released only in the transrational domains)—that is no cause to board the Regress Express and set the Way Back Machine to medieval or horticultural or foraging or whatnot: they all had their chance. They all failed—each in their own special and wonderful and spectacular fashion.

THE BATTLE OF WORLDVIEWS

We have seen that each stage of development transcends and includes, negates and preserves, its predecessor(s), and this shows up in a particularly interesting fashion when it comes to the various worldviews themselves.

When we only have access to preoperational awareness, then the Kosmos appears as magical. When we have access to concrete operational awareness, then the Kosmos appears as mythical. When we have access to formal operational, the Kosmos appears rational-scientific (and it can, as a secondary issue, then be reduced to the flatland cosmos; this flatland then makes the previous magic and mythic structures look all that more appealing, a fact exploited by the Romantics—but that's another part of our story). With even higher structures, we will see, the Kosmos appears in other and quite different forms.

In each of these transformations or supersessions, notice what is *preserved* and what is *negated*. As magic gives way to mythic, the preoperational structures themselves are preserved. That is, the capacity to form images, symbols, and concepts (the components or holons of preop) are all taken up and enfolded in the concrete operational mind (as subholons or *necessary components*). None of these capacities are lost, but are now simply operated upon by the higher mind.

But the magical *worldview* is lost, or negated, or dropped altogether (at least as a dominant focus of awareness).[38] That is, the preop cognitive structures are taken up and *preserved* in the new cognitive structures of conop (in a very "friendly" manner as components or subholons of its own being), but the magical worldview is negated and *replaced* by the mythic worldview. Indeed, since they are now *mutually exclusive*, they become steadfast enemies (they are mutually exclusive because each was constituted by being *the* exclusive structure; I will return to this in a moment with some examples).

Thus, the mythic worldviews—which always situate miraculous *power* in a great Other (gods or goddesses), around which social cohesion can then be built *beyond* mere blood and kinship ties—these mythic worldviews are constantly at war with magic, because magic situates miraculous power in a human individual (sorcerer, witch, wizard), and this regression from sociocentric to egocentric power rightly alarms the mythic worldviews: it is taken, not altogether incorrectly, as a moral regression, for which mythic worldviews have historically devised many ingenious and unpleasant solutions.

Likewise, when formal operational awareness emerges, it incorporates and includes the cognitive structures of conop (its capacity to form rules, take roles, and so on), but it actively *negates* the mythic worldview that is generated when you *only* have concrete operations. Rationality *preserves*, as part of its own being, in a "friendly" way, the conop structures (as subholons of its new and higher regime), but it *replaces* the mythic worldview with the rational worldview, and once again, the two become steadfast enemies. This is why the Church, beginning around the sixteenth century, was involved in a war on two fronts: fighting regression to magic, and fighting supersession by science. The Galileos and the sorcerers were both introduced to the Inquisitor.

And this is why, conversely, science is not only always attempting to further its research, but also constantly battling myths, with their inherent tendencies to dominator-hierarchies, and always under the battle cry of Voltaire: "Remember the cruelties!"

(There are few or no dominator hierarchies in the magical structure or in magical cultures, because preop cannot construct a hierarchy in the first place, let alone a dominator hierarchy. With the emergence of conop, hierarchies are indeed created—not only mentally, but socially, politically, and religiously—a mania for caste systems and dominator hierarchies spring up wherever magic gives way to myth; and precisely because conop cannot reach into *postconventional* modes, these hierarchies are *always* rigidly and concretely set, as are the literal myths that prop them up: to change the social hierarchy is to *damage* the god/goddess/ruler and thus bring catastrophic retribution to the society itself. With rationality, these dominator hierarchies are deconstructed because they are not based on worldcentric or postconventional standards, and rationality is not happy short of that omega point. This is one of the main topics of volume 2, and I mention it now as simply another instance of the tensions inherent between the various worldviews.)

In each case, the basic structures of the previous stage are taken up and preserved in the new and wider holon, but the worldview, which was *generated* when there was *only* or *exclusively* the previous and lower stage—that is negated and replaced by a new worldview (which will in turn fade if development continues).[39]

Thus, in the same fashion, if development does proceed into the higher or transpersonal domains, the basic structures of rationality are all embraced and incorporated, in a friendly way, as subholons in the new awareness, but the *merely* scientific worldview, built by *exclusive* rationality—built when all you have is rationality—that worldview is replaced with a spiritual orientation (as we will see).

This is an example of what in chapter 2 was called basic structures and exclusivity structures, since development "includes the being but negates the partiality," and I gave Hawaii's statehood as an example. I want to give another example here, because a general grasp of this fundamental developmental process is crucial, I believe, to understanding any of the genuinely deeper or higher stages of growth and development.

If we use a crude ladder analogy the relationship can more easily be seen. The actual rungs in the ladder are the basic structures or levels in any developmental line (such as the cognitive). Each higher rung rests on the lower rungs, and no matter how high up the ladder you climb, all the rungs are necessary. Pull out a lower rung and the whole ladder collapses. Each rung is a *basic structure* (or basic holon—such as images, symbols, concepts, rules), and each is *preserved* in the overall ladder.

But as you climb the ladder, you get a different perspective on the world around you: each rung has a different worldview, and the higher rungs can see more of the world (they have a higher worldview). And although each of the *rungs* remains in existence as you climb (your climb actually depends upon them), the various worldviews are *replaced*. You cannot be at, say, rung 7 and simultaneously see the world from the shallower rung

1, even though you are still standing in part on rung 1 itself. The being, the rungs, are preserved, but the partiality or exclusivity, the partial worldviews, are *negated* and *replaced*. (Until, of course, you stop climbing, at which point you have "the" worldview that will attempt to subsume all else.)

Thus, each transformation upward is a "paradigm war"—a battle royale over how to exclusively view the world. And we will see, as we continue our account of the higher stages of growth, that as myth historically defeated magic, and reason defeated myth, the future paradigm wars will be fought around reason itself and its potential successors.

And if history is any guide, the new paradigm wars will be equally unpleasant.

THE VALUE OF THE MYTHOLOGICAL APPROACH

The value of including a properly interpreted mythological approach is that it can help us, of today's rational worldview, get in touch with aspects of our roots, our foundations, some of the archaeological layers of our own present awareness. From Freud to Jung to Piaget to Gebser, these roots are recognized. And that "getting in touch," that befriending of our roots, is empowering and enriching and energizing. There is a rush of energy released in reading Jung or Campbell or Eliade—we are watering our roots, and they help send forth new branches.

But all of that happens precisely because they are lower holons in a now-higher awareness—they are more fundamental (and less significant)—and the real power comes precisely from touching their basic structures ("archetypal") while *simultaneously robbing them of their exclusive worldview*. If we were simply, merely, actually reactivating the lower worldview itself, this would be wholesale regression to *de-differentiated* structures (borderline and psychotic, which indeed sometimes occurs). And this, I believe, is behind Jung's dual stance toward the archetypes: it is altogether necessary to contact and befriend them, but it is finally necessary to *differentiate* and *individuate* from them, break them of their power over us. In other words, befriend the images, rob them of their worldview.

But as for the images and symbols and early concepts themselves, as for these "archetypes," they lie in the direction of downward, not upward (shallower, not deeper). This is why, I believe, however Jung variously defined the archetypes, he always maintained they were "next to the instincts," or the "instincts' image of themselves." That is, as Liliane Frey-Rohn explains, "The connection between instinct and archetypal image appeared to [Jung] so close that he drew the conclusion that the two were coupled. He saw the primordial image ['archetype'] as the self-portrait of the instinct—in other words, the instinct's perception of itself."[40]

Instincts—the drives of the reptilian brain stem and paleomammalian limbic system—these are the structures "next to" the archetypes, which situates them perfectly on the spectrum of consciousness. Sensation, perception, impulse (instinct), image, symbol, concept, rule . . . the archetypes are, for the most part, collective, fundamental images lying next to impulse/instinct.[41] And proceeding in the *direction* of the archetypes we therefore eventually run into, not Spirit, but atoms. Jung: "The deeper layers of the psyche lose their individual uniqueness as they retreat farther and farther into darkness. 'Lower down' they become increasingly extinguished in the body's materiality, i.e., in chemical substances."[42]

In other words, the Jungian archetypes are *not* the transcendental archetypes or Forms found in Plato, Hegel, Shankara, or Asanga and Vasubandhu. These latter Forms—the true archetypes, the ideal Forms—are the creative patterns said to underlie all manifestation and give pattern to chaos and form to Kosmos (which we will investigate in chapter 9).

The Jungian archetypes, on the other hand, are for the most part the magico-mythic motifs and "archaic images"—they should really be called "prototypes"—collectively inherited by you and by me from past stages of development, archaic holons now forming part of our own compound individuality (they come from below up, not from above down). And coming to terms with these archaic holons—befriending and making conscious and differentiating/integrating these prototypes—is a useful endeavor, not because they are our transrational future, but because they are our prerational past. I agree *entirely* with Jung on the necessity of differentiating and integrating this archaic heritage; I do not believe that this has much to do with genuine mystical spirituality.[43]

Keep in mind also that for Jung, the archetypes are *collectively* inherited archaic images; they arise from the common, day-to-day, normal and *typical experiences* of men and women everywhere, so that there is an archetype of the Mother, the Father, the trickster, the shadow, the uro-boros (serpent), and so on. These situations are encountered by virtually everybody everywhere, and so the "imprints" of these encounters over the millennia became engraved in the brain, so to speak (or, less Lamarck-ianly, those who were born with the guiding imprint, which helps a person orient toward typical situations, would be more likely to survive and reproduce and pass the image along).

But in any given past era (and today as well), the number of individuals who *actually transcended* into profoundly mystical occasions was very, very small (as we saw Campbell point out). In other words, even under the best of circumstances, there simply weren't enough mystics to imprint the entire or collective human race: profound mystical experience is *not* a common, typical, everyday occurrence that could be swept up and in-cluded in past collective evolutionary imprints.[44] And in no way is today's profound mystical awareness an experience of yesterday's ordinary, col-lective, typical experiences. Excellence in any era is not an experience of yesterday's collective normality, made to sound "spiritual" by calling it "collective," as if a whole bunch more of average could equal extraordi-nary, as if multiplying mediocrity would equal excellence.

Thus, to give only one example, in Jean Bolen's books, such as *God-desses in Everywoman* and *Gods in Everyman*, there is a wonderful pre-sentation of all the "archetypal" gods and goddesses that are collectively inherited by men and women (from the steadiness and patience of Hestia to the sexuality and sensuality of Aphrodite to the strength and indepen-dence of Artemis). But these gods and goddesses are not transpersonal modes of awareness, or genuinely mystical luminosities, but simply a col-lection of typical and everyday self-images (and personae) available to men and women. Collective typical is not transpersonal.[45]

Now I believe that there are indeed mystical "archetypes" (in the tran-scendental sense I mentioned above, and to which we will return), but these archetypes cannot be explained as an inheritance from the past; they are strange Attractors lying in our future, omega points that have not been *collectively* manifested anywhere in the past, but are nonetheless available to each and every individual as *structural potentials*, as future structures attempting to come down, not past structures struggling to come up.

The Men's Movement

In retreating to the wilderness, and banging drums, and contacting the archaic images of man as warrior, hunter, king, wild man, and so forth, many modern men are helped to recontact these archaic holons and be thereby enriched and energized and given roots. Our modern culture has too often negated these holons while failing to also preserve them in an integrated and acceptable fashion, and thus a "return to roots" is altogether appropriate and benign.

But the point, notice, is that these are men who, in the plentiful space of reason, play out these myths in an as-if fashion, thus transcending them in the very process of playing with them (and those who don't transcend them, those who take these myths seriously, get into a great deal of emotional trouble—the "wild man" as a myth is one thing, as a literal role model, quite another).

But I have been struck by the fact that virtually every woman I know, and every woman who has written about the men's movement, is completely *alarmed* by it. They think it is either disgusting, or preposterous, or silly, or obscene. The liberal feminists are outraged by it, because they maintain that men, being essentially oppressors, have no right to complain about anything; and they see a "gathering of men" as nothing but a chance for jailors to fine-tune the art of oppression. The radical feminists, who otherwise so straightforwardly champion the differences between the sexes, appear never to have had exactly *this* difference in mind, and react with a very uneasy queasiness to the whole affair.

But under all these reactions there seems to be a pervasive fear. A gathering of wild men is *frightening* to most women, of whatever ideological persuasion, and I think with good reason. The archaic images, still lying "close to the body," close to the instincts, close to the biosphere, are under sway of those biological universals, strong universals, in which and by which biospheric men have always dominated women (whether or not intentionally). Only in the noosphere can equality be conceived and found and enforced. The wild man is the *merely* biospheric male, the instinctual male, where the physical body rules, where intersexual mutuality is suspended, where communication is a grunt and a groan and a "Ho!"

And in the sound of the drums pounding from the forest, today's woman hears, I believe, a million years of physical dominance rumbling from the depths, and not one of them likes what she hears.

REASON RELEASES MYTH

We already saw that in Campbell's opinion the "really real" (or most important) function of a myth occurs when it is taken in an as-if fashion. In other words, the "hidden power" of myth is *released* when it is taken up in the open space created by rationality. And now we can see why: when reason takes up myth, it befriends the being but negates its partialness, and thus *frees* the mythic structure from its imprisonment in a shallower occasion. It is reason, and reason alone, that frees the luminosity trapped in myth.

Thus reason *frees* whatever spirit manages to reside in myth—a spirit struggling from within, to get out. Whether rationality appears "spiritual" or not is utterly beside the point. It certainly does not appear "mythological," and if that is what one means by "spiritual," then reason is not that.

But genuine spirituality, we will see, is primarily a measure of depth and a disclosure of depth. There is *more* spirituality in reason's *denial* of God than there is in myth's *affirmation* of God, precisely because there is *more depth*. (And the transrational, in turn, discloses yet more depth, yet more Spirit, than either myth or reason).

But the very *depth* of reason, its capacity for universal-pluralism, its insistence on universal tolerance, its grasp of global-planetary perspectivism, its insistence on universal benevolence and compassion: these are the manifestations of its genuine depth, its *genuine* spirituality. These capacities are not *revealed* to reason from *without* (by a mythic source); they issue from *within* its own structure, its own *inherent* depth (which is why it does not need recourse to a mythic god to implement its agenda of universal benevolence, why even an "atheist" acting from rational-universal compassion is more spiritual than a fundamentalist acting to convert the universe in the name of a mythic-membership god). That the Spirit of reason does not fly through the sky hurling thunderbolts and otherwise spend its time turning spinach into potatoes speaks more, not less, on its behalf.

MYTHOLOGY TODAY

So perhaps we can begin to see that the uses of mythology in today's world are many and varied. For one thing, literal-fundamental mythological motifs are the main social cement in many cultures (including a very large segment of our own), and as divisive and imperialistic as those mythologies are, their particular ethnocentric and social-integrative power has to be reckoned with carefully. One cannot simply challenge or deconstruct the myths of such societies (or segments of societies) and expect them to survive (or expect them to acquiesce without a fight).

Moreover, even in societies organized around a rational-pluralistic worldview, each and every individual nonetheless *begins* life at the archaic, then the magic, then the mythic, before (and if) arriving at the rational. One of the great problems with the Enlightenment was that Reason was simply supposed to replace all of that right down to the roots, so that every child would be born, more or less, an enlightened and tolerant rationalist. Since the *tabula* was *rasa*, all we had to do was write R-E-A-S-O-N across the clean slate, and all would follow wonderfully.

Needless to say, it doesn't work like that. Every child still has to negotiate the archaic worldview, then deconstruct that with the magical worldview, then deconstruct that with the mythic, then deconstruct that with the rational (on the way to the transrational). Each of these transformations is a series of painful deaths and rebirths, and the typical individual attempts to stop the pain by stopping the transformation.

Gone is the time (in the developed East and West) when one could settle comfortably in the magic or even the mythic mode. The center of gravity of the "world soul," so to speak, has moved into rational modes of universal pluralism, and thus the earliest that an individual can abort transformation (and "stop" the pain), without social censure, is somewhere around the mythic-rational (that is, even if one clings to myth, it has to be propped up with rationalizations, because everywhere rationality impinges on it). And, indeed, the majority of individuals in rational societies still settle in somewhere around the mythic-rational, using all the formidable powers of rationality to prop up a particular, divisive, imperialistic mythology and an aggressively fundamentalistic program of systematic intolerance. As such, they are constantly at war with magic, with *other* myths, and with reason, all of which they view simply as the devil's work.

The modern solution to this developmental nightmare is that the rationality structure of the democratic state *tolerates* magic and mythic subholons, but it has, via the all-important separation of church and state, *removed the worldviews* of those subholons from the organizing regime of the society, which itself is defined by a rational tolerance of everything but intolerance. Because these subholons are robbed of their power to *govern exclusively*, they are robbed of pushing their mythic-imperialistic expansionism via national-military means, though this doesn't prevent them from always agitating to tilt the state toward their own fundamentalistic values. (And, of course, for those nations where the mythic holons *are* the governing regime, military expansionism is still the rule; and, in these cases, as usual, it isn't whether one can actually win and coerce others to the faith, but whether one can earn the right to die trying.)

But even for those who have developed beyond a concrete mythological approach to the world, the mythological structures themselves, as we just saw, are now junior holons in the person's own compound individuality, and should be honored as a rich source of one's own being and one's own roots. And therein, as I indicated, lies the real value of the Jungian/ Campbell/ mythopoetic approach (divested of its elevationism).

But if genuine spirituality is not, as I maintain, a product of the past (or of past archetypes), then what about the great spiritual figures, East and West, who definitely existed in the past? What about the Buddhas, the Christs, the Krishnas? Are they not expressing some past capacity, some past archetypal wisdom that we have lost touch with? And in contacting our own spirituality, aren't we contacting some past potential that we have lost or denied or forgotten?

Lost, perhaps; but not from the past. Let me repeat what I said above about the "true archetypes," the transcendental and transpersonal structures, as a reminder: they cannot be explained as an inheritance from the past; they are strange Attractors lying in our future, omega points that have not been *collectively* manifested anywhere in the past, but are nonetheless available to each and every individual as *structural potentials*, as future structures attempting to come down, not past structures struggling to come up.

The great and rare mystics of the past (from Buddha to Christ, from al-Hallaj to Lady Tsogyal, from Hui-neng to Hildegard) were, in fact, ahead of their time, and are still ahead of ours. In other words, they most definitely are *not* figures of the past. They are figures of the future.

In their spirituality, they did not tap into yesterday, they tapped into tomorrow. In their profound awareness, we do not see the setting sun, but the new dawn. They absolutely did not inherit the past, they inherited the future.

It is to that future we can now turn.

7

The Farther Reaches of Human Nature

This one being and consciousness is involved here in matter. Evolution is the process by which it liberates itself; consciousness appears in what seems to be inconscient [i.e., matter, the physiosphere], and once having appeared is self-impelled ["self-organizing"] to grow higher and higher and at the same time to enlarge and develop toward a greater and greater perfection. Life is the first step of this release of consciousness; mind is the second. But the evolution does not finish with mind; it awaits release into something greater, a consciousness which is spiritual and supramental. There is therefore no reason to put a limit to evolutionary possibility by taking our present organization or status of existence as final.
                                                              —SRI AUROBINDO

W E ARE YET the bastard sons and daughters of an evolution not yet done with us, caught always between the fragments of yesterday and the unions of tomorrow, unions apparently destined to carry us far beyond anything we can possibly recognize today, and unions that, like all such births, are exquisitely painful and unbearably ecstatic. And with yet just the slightest look—once again, within—new marriages unfold, and the drama carries on.

THE INTERIOR CASTLE

I have been constantly emphasizing that each stage of evolution, in whatever domain, involves a new emergence and therefore a new depth, or a new interiority, whether that applies to molecules or to birds or to dolphins; and that each new within is also a going beyond, a transcendence, a higher and wider identity with a greater total embrace. The formula is: going within = going beyond = greater embrace. And I want to make very clear exactly what that means.

This is extremely important, I think, because the higher stages of development, the transrational and transpersonal and mystical stages, all involve a new going within, a new interiorness. And the charge has been circulating, for quite some time now, that endeavors such as meditation are somehow narcissistic and withdrawn. Environmentalists, in particular, often claim that meditation is somehow "escapist" or "egocentric," and that this "going within" simply ignores the "real" problems in the "real" world "out there."

Precisely the opposite. Far from being some sort of narcissistic withdrawal or inward isolation, meditation (or transpersonal development in general) is a simple and natural continuation of the evolutionary process, where every going within is also a going beyond to a wider embrace.

Recall that two of our tenets (8 and 12d) stated that increasing evolution means increasing depth and increasing relative autonomy. In the realm of human development, this particularly shows up in the fact that, according to developmental psychology (as we will see), increasing growth and development always involve *increasing internalization* (or increasing interiorization). And as paradoxical as it initially sounds, the *more interiorized* a person is, the *less narcissistic* his or her awareness becomes. So we need to understand why, for all schools of developmental psychology, this equation is true: increasing development = *increasing* interiorization = *decreasing* narcissism (or decreasing egocentrism).

In short, we need to understand why the more interior a person is, the less egocentric he or she becomes.

Begin with interiorization. "Evolution, to Hartmann [founder of psychoanalytic developmental psychology], is a process of *progressive internalization*, for, in the development of the species, the organism achieves increased independence from its environment, the result of which is that 'reactions which originally occurred in relation to the external world are increasingly displaced into the interior of the organism.' The more independent the organism becomes, the greater its independence from the stimulation of the immediate environment."[1] This applies to the infant, for example, when it no longer dissolves in tears if food is not immediately forthcoming. By interiorizing its awareness, it is no longer merely buffeted by the immediate fluctuations in the environment: its relative autonomy—its capacity to remain stable in the midst of shifting circumstances—increases. This progressive internalization is a cornerstone of psychoanalytic developmental psychology (from Hartmann to Blanck and Blanck to Kernberg to Kohut). It is implicit in Jung's notion of individuation. Likewise, Piaget described thought as "internalized action," the capacity to internally plan an action and anticipate its course without being merely a reactive automaton—and so forth.

In other words, for developmental psychology, increasing development = increasing interiorization = increasing relative autonomy. This, of course, is simply tenet 12d as it shows up in humans.

The second link in the equation concerns narcissism, which is roughly synonymous with egocentrism, about which we have already said much. We need only recall that increasing development involves precisely the capacity to transcend one's isolated and subjective point of view, and thus to find higher and wider perspectives and identities. Piaget referred to the entire developmental process as one of *decreasing egocentrism*, or what he also called "decentering."

Putting these all together, we have: increasing development = increasing interiorization = increasing autonomy = decreasing narcissism (decentering).

In other words, the more one can go *within*, or the more one can introspect and reflect on one's self, then the more detached from that self one can become, the more one can rise above that self's limited perspective, and so the less narcissistic or less egocentric one becomes (or the more *decentered* one becomes). This is why Piaget is always saying things that *sound* paradoxical, such as: "Finally, as the child becomes conscious of his subjectivity, he rids himself of his egocentricity."[2]

The more he can subjectively reflect on his self, the more he can transcend it—his subjectivity, his *interiority*, rids him of his egocentricity. Howard Gardner does a masterful job of summarizing development as being the two processes of decreasing egocentrism and increasing interiority. "The first is the decline of egocentrism. The young child is, in Piaget's terms, totally egocentric—meaning not that he thinks selfishly only about himself, but to the contrary, that he is incapable of thinking about himself. The egocentric child is unable to differentiate himself from the rest of the world; he has not separated himself out from others or from objects. Thus he feels that others share his pain or his pleasure, that his mumblings will inevitably be understood, that his perspective is shared by all persons, that even animals and plants partake of his consciousness. In playing hide-and-seek he will 'hide' in broad view of other persons, because his egocentrism prevents him from recognizing that others are aware of his location. The whole course of human development can be viewed as a continuing decline in egocentrism. . . ."[3]

And this decreasing narcissism is directly connected with the "second trend in mental growth," namely, "the tendency toward internalization or interiorization. The infant either solves problems by his activity upon the world or he does not solve them at all. The older child, on the other hand, can achieve many intellectual breakthroughs without overt physical actions. He is able to realize these actions interiorly, through concrete and formal operations."[4] By acting on the self interiorly, that self is decentered, and this allows, among many other things, the continuing expansion (decentering) of moral response from egocentric to sociocentric to worldcentric (integral-aperspectival).

In short, the more one *goes within*, the more one *goes beyond*, and the more one can thus embrace a *deeper identity* with a *wider perspective*.

Meditation, then, as we will see in detail, involves yet a further *going within*, and thus a further *going beyond*, the discovery of a new and higher awareness with a new and wider identity—and thus meditation is one of the single strongest antidotes to egocentrism and narcissism (and geocentrism and anthropocentrism and sociocentrism).

And let us remember Piaget's central point about egocentrism, namely, that "egocentrism obscures the truth." It follows that meditation, as an antidote to egocentrism, would involve a substantial increase in capacity for *truth disclosure*, a clearing of the cobwebs of selfcentric perception and an opening in which the Kosmos could more clearly manifest, and be seen, and be appreciated—for what it is and not for what it can do for *me*.

In short, every *within* turns us *out* into more of the Kosmos. This is what seems to so confuse the flatland holists (and the ecological critics of contemplation), because in their flatland world of self and cosmos, the *more* attention you place on one, the *less* attention you have for the other (and they want everyone's eyes riveted on exterior nature), whereas in the pluridimensional and holoarchic Kosmos, the *more* the depths of the self are disclosed, the *more* the corresponding depths of the Kosmos reveal themselves. (We will see where all that leads in a moment.)

This general movement of *within-and-beyond* is nothing new with humans: it is a simple continuation of the Kosmic evolutionary process, which is "self-development through self-transcendence," the same process at work in atoms and molecules and cells, a process that, in the human domains, continues naturally into the superconscious, with precisely nothing occult or mysterious about it.

VISION-LOGIC

The capacity to go within and *look at* rationality results in a *going beyond* rationality, and the first stage of that going-beyond is vision-logic. If you are aware of being rational, what is the nature of that awareness, since it is now bigger than rationality? To be aware of rationality is no longer to have only rationality, yes?

Numerous psychologists (Bruner, Flavell, Arieti, Cowan, Kramer, Commons, Basseches, Arlin, etc.) have pointed out that there is much evidence for a stage beyond Piaget's formal operational. It has been called "dialectical," "integrative," "creative synthetic," "integral-aperspectival," "postformal," and so forth. I, of course, am using the terms *vision-logic* or *network-logic*. But the conclusions are all essentially the same: "Piaget's formal operational is considered to be a problem-solving stage. But beyond this stage are the truly creative scientists and thinkers who define important problems and ask important questions. While Piaget's formal model is adequate to describe the cognitive structures of adolescents and competent adults, it is not adequate to describe the towering intellect of Nobel laureates, great statesmen and stateswomen, poets, and so on."[5]

True enough. But I would like to give a different emphasis to this structure, for while very few people might actually gain the "towering status of a Nobel laureate," the *space* of vision-logic (its worldspace or worldview) is available for any who wish to continue their growth and development. In other words, to progress through the various stages of growth does not mean that one has to extraordinarily master each and every stage, and demonstrate a genius comprehension at that stage before one can progress beyond it. This would be like saying that no individuals can move beyond the oral stage until they become gourmet cooks.

It is not even necessary to be able to articulate the characteristics of a particular stage (children progress beyond preop without ever being able to define it). It is merely necessary to develop an *adequate competence* at that stage, in order for it to serve just fine as a platform for the transcendence to the next stage. In order to transcend the verbal, it is not necessary to first become Shakespeare.

Likewise, in order to develop formal rationality, it is not necessary to learn calculus and propositional logic. Every time you imagine different outcomes, every time you see a possible future different from today's, every time you dream the dream of what might be, you are using formal operational awareness. And from that platform you can enter vision-logic, which means not that you have to become a Hegel or a Whitehead in order to advance, but only that you have to think globally, which is not so hard at all. Those who will *master* this stage, or any stage for that matter, will always be relatively few; but all are invited to pass through.

Because vision-logic transcends but includes formal operational, it completes and brings to fruition many of the trends begun with universal rationality itself (which is why many writers refer to vision-logic as "mature reason" or "dialectical reason" or "synthetic reason," and so on). And some theorists simply subdivide formal operational awareness into several substages, with the highest of those stages being what we are calling vision-logic. James Fowler, for example, divides formop into early formop, dichotomizing formop, dialectical formop, and synthetic formop (the first two being what I am calling rationality, and the last two being vision-logic, although all four are "reason" in the very broadest sense). Incidentally, Fowler's extremely important work on the "stages of faith" (whose details I will reserve for a note)[6] is yet another clear account of the evolution from magic to mythic-literal to the universal "commonwealth of being."

In other words, rationality is global, vision-logic is more global. Take Habermas, for example (in *Communication and the Evolution of Society*). Formal operational rationality establishes the postconventional stages of, first, "civil liberties" or "legal freedom" for "all those bound by law," and then, in a more developed stage, it demands not just legal freedom but also "moral freedom" for "all humans as private persons." But even further, mature or communicative reason (our vision-logic) demands both "moral and political freedom" for "all human beings as members of a world society." Thus, where rationality began the *world-centric* orientation of universal pluralism, vision-logic brings it to a mature fruition by demanding not just legal and moral freedom, but legal and moral and political freedom (includes the previous stage and adds something crucial: transcends and includes).

In just the same way, ecological and relational awareness, which started to emerge with formal operational, comes to a major fruition with vision-logic and the centauric worldview. For, in beginning to *differentiate* from rationality (look at it, operate upon it), vision-logic can, for the first time, *integrate* reason with its predecessors, including life and matter, all as junior holons in its own compound individuality.

In other words, and I intend to emphasize this heavily, centauric vision-logic can integrate physiosphere, biosphere, and noosphere in its own compound individuality (and this is, as I suggested in chapter 5, the next major stage of leading-edge global transformation, even though most of the "work yet to be done" is still getting the globe up to decentered universal-rational pluralism in the first place).

This overall integration (physiosphere, biosphere, and noosphere, or matter, body, mind) is borne out, for example, by the researches of Broughton, Loevinger, Selman, Maslow, and others. As only one example, but an important one, we can take the work of John Broughton.

As usual, this new centauric stage possesses not just a new cognitive capacity (vision-logic)—it also involves a new sense of identity (centauric), with new desires, new drives, new needs, new perceptions, new terrors, and new pathologies: it is a new and higher self in a new and wider world of others. And Broughton has very carefully mapped out the developmental stages of self and knowing that lead up to this new centauric mode of being-in-the-world.[7]

To simplify considerably, Broughton asked individuals from preschool age to early adulthood: what or where is your *self*?

Since this was a verbal study, Broughton began with the late preop child (magic-mythic), which he calls level zero. At this stage, children uniformly reply that self is "inside" and reality is "outside." Thoughts are not distinguished from their objects (still magical adherences; the child has not completed fulcrum three).

At level one, still in the late preop stage, children believe that the self is identified with the physical body, but the mind controls the self and can tell it what to do, so it is the mind that moves the body. The relation of mind to body is one of authority: the mind is a big person and the body is a little person (i.e., mind and body are slowly differentiating). Likewise, thoughts are distinguished from objects, but there is no distinction between reality and appearance ("naive realism").

Level two occurs at about ages seven to twelve years (conop). Mind and body are initially differentiated at this level (completion of fulcrum three), and the child speaks of the self as being, not a body, but a *person* (a social *role* or *persona*, fulcrum four), and the person includes both mind and body. Although thoughts and things are distinguished, there is still a strong personalistic flavor to knowledge (remnants of egocentrism), so facts and personal opinions are not easily differentiated.

At level three, occurring around eleven to seventeen years (early formop), "the social personality or *role* is seen as a *false outer appearance*, different from the *true inner self*." Here we see very clearly the differentiation of the self (the rational ego) from its embeddedness in sociocentric roles—the emergence of a new interiority or relative autonomy which is *aware of, and thus transcends or disidentifies from*, overt social roles. "The self is what the person's nature normally is; it is a kind of essence and remains itself over changes in mental contents."

Likewise, and for precisely the same reasons, "reflective self-awareness appears at this level." (This is fulcrum five, the rational and reflexive ego differentiating from, and thus transcending, sociocentric or mythic-membership roles, with the correlative possible pathology of "identity crisis." Notice also that the new ego-self is *beginning* to remain as *witness* to the stream of mental events, and is not merely carried away by any passing thoughts; the adolescent at this stage reports that something "remains itself over changes in mental contents.")

At level four, or late formop, the person becomes capable of hypothetico-deductive awareness (what if, as if), and reality is conceived in terms of *relativity* and *interrelationships* (ecology and relativity, in the broadest sense, as we have seen). The self is viewed as a postulate "lending unity and integrity to personality, experience, and behavior" (this is the "mature ego").

But, and this is very telling, development can take a cynical turn at this stage. Instead of being the principle lending unity and integrity to experience and behavior, the self is simply *identified with experience and behavior*. In the cynical behavioristic turn of this stage, "the person is a cybernetic system guided to fulfillment of its material wants. At this level, radical emphasis on seeing everything within a relativistic or subjective frame of reference leaves the person close to a solipsistic position."

The world is seen as a great relativistic *cybernetic system*, so "holistic" that it leaves no room for the actual subject in the objective network. The self therefore hovers above reality, disengaged, disenchanted, disembodied. It is "close to a solipsistic position": hyperagency cut off from all communions. And this, as we have seen, is essentially the fundamental Enlightenment paradigm: a perfectly holistic world that leaves a perfectly atomistic self.[8]

A transcendental self can bond with other transcendental selves, whereas a merely empirical self disappears into the empirical web and interlocking order, never to be heard from again. (No strand in the web is ever or can ever be aware of the whole web; if it could, then it would cease to be merely a strand. This is not allowed by systems theory, which is why, as Habermas demonstrated, systems theory always ends up isolationist and egocentric, or "solipsistic.")

But for a more transcendental self to emerge, it has first to *differentiate* from the merely *empirical* self, and thus we find, with Broughton: "*At level five the self as observer is distinguished from the self-concept as known.*" In other words, something resembling a pure observing Self (a transcendental Witness or Atman, which we will investigate in a moment) is beginning to be clearly distinguished from the empirical ego or objective self—it is a new interiority, a new going within that goes beyond, a new emergence that transcends but includes the empirical ego. This beginning transcendence of the ego we are, of course, calling the centaur (the beginning of fulcrum six, or the sixth major differentiation that we have seen so far in the development of consciousness).[9] This is the realm of vision-logic leading to centauric integration, which is why at this stage, Broughton found that "reality is defined by the *coherence* of the *interpretive framework.*"

This integrative stage comes to fruition at Broughton's last major level (late centauric), where "*mind and body are both experiences of an integrated self,*" which is the phrase I have most often used to define the centauric or bodymind-integrated self. Precisely because awareness has *differentiated* from (or disidentified from, or transcended) an *exclusive* identification with body, persona, ego, and mind, it can now *integrate* them in a unified fashion, in a new and higher holon with each of them as junior partners. Physiosphere, biosphere, noosphere—exclusively identified with none of them, therefore capable of integrating all of them.

But everything is not sweetness and light with the centaur. As always, new and higher capacities bring with them the potential for new and higher pathologies. As vision-logic adds up all the possibilities given to the mind's eye, it eventually reaches a dismal conclusion: personal life is a brief spark in the cosmic void. No matter how wonderful it all might be now, we are still going to die: *dread*, as Heidegger said, is the authentic response of the existential (centauric) being, a dread that calls us back from self-forgetting to self-presence, a dread that seizes not this or that part of me (body or persona or ego or mind), but rather the totality of my being-in-the-world. When I authentically see my life, I see its ending, I see its death; and I see that my "other selves," my ego, my personas, were all sustained by inauthenticity, by an *avoidance* of the awareness of lonely death.

A profound existential malaise can set in—the characteristic pathology of this stage (fulcrum six). No longer protected by anthropocentric gods and goddesses, reason gone flat in its happy capacity to explain away the Mystery, not yet delivered into the hands of the superconscious—we stare out blankly into that dark and gloomy night, which will very shortly swallow us up as surely as it once spat us forth. Tolstoy:

> The question, which in my fiftieth year had brought me to the notion of suicide, was the simplest of all questions, lying in the soul of every man: "What will come from what I am doing now, and may do tomorrow? What will come from my whole life?" Otherwise expressed— "Why should I live? Why should I wish for anything?" Again, in other words, "Is there any meaning in my life which will not be destroyed by the inevitable death awaiting me?"

That question would *never* arise to the magical structure; that structure has abundant, even exorbitant meaning because the universe centers always on it, was made for it, caters to it daily: every raindrop soothes its soul because every confirming drop reassures it of its cosmocentricity: the great spirit wraps it in the wind and whispers to it always, I exist for you.

That question would *never* arise to a mythic-believer: this soul exists only for its God, a God that, by a happy coincidence, will save this soul eternally if it professes belief in this God: a mutual admiration society destined for a bad infinity. A crisis of faith and meaning is impossible from within this circle (a crisis occurs only when this soul suspects this God).

That question would never beset the happy rationalist, who long ago became a happy rationalist by deciding never to ask such questions again, and then forgetting, rendering unconscious, this question, and sustaining the unconsciousness by ridiculing those who ask it.

No, that question arises from a self that knows too much, sees too much, feels too much. The consolations are gone; the skull will grin in at the banquet; it can no longer tranquilize itself with the trivial. From the depths, it cries out to gods no longer there, searches for a meaning not yet disclosed, still to be incarnated. Its very agony is worth a million happy magics and a thousand believing myths, and yet its only consolation is its unrelenting pain—a pain, a dread, an emptiness that feels beyond the comforts and distractions of the body, the persona, the ego, looks bravely into the face of the Void, and can no longer explain away either the Mystery or the Terror. It is a soul that is much too awake. It is a soul on the brink of the transpersonal.

THE TRANSPERSONAL DOMAINS

We have repeatedly seen that the problems of one stage are only "defused" at the next stage, and thus the only cure for existential angst is the transcendence of the existential condition, that is, the transcendence of the centaur, negating and preserving it in a yet higher and wider awareness. For we are here beginning to pass out of the noosphere and into the theosphere, into the transpersonal domains, the domains not just of the self-conscious but of the superconscious.

A great number of issues need to be clarified as we follow evolution (and the twenty tenets) into the higher or deeper forms of transpersonal unfolding.

First and foremost, if this higher unfolding is to be called "religious" or "spiritual," it is a very far cry from what is ordinarily meant by those terms. We have spent several chapters painstakingly reviewing the earlier developments of the archaic, magic, and mythic structures (which are usually associated with the world's great religions), precisely because those structures are what transpersonal and contemplative development *is not*. And here we can definitely agree with Campbell: if 99.9 percent of people want to call magic and mythic "real religion," then so be it for them (that is a legitimate use);[10] but that is not what the world's greatest yogis, saints, and sages mean by mystical or "really religious" development, and in any event is not what I have in mind.

Campbell, however, is quite right that a very, very few individuals, during the magic and mythic and rational eras, were indeed able to go beyond magic, beyond mythic, and beyond rational—into the transrational and transpersonal domains. And even if their teachings (such as those of Buddha, Christ, Patanjali, Padmasambhava, Rumi, and Chih-i) were snapped up by the masses and translated downward into magic and mythic and egoic terms—"the salvation of the individual soul"—that is not what their teachings clearly and even blatantly stated, nor did they intentionally lend any support to such endeavors. Their teachings were about the *release* from individuality, and not about its everlasting perpetuation, a grotesque notion that was equated flat-out with hell or samsara.

Their teachings, and their contemplative endeavors, were (and are) transrational through and through. That is, although all of the contemplative traditions aim at going within and beyond reason, they all *start* with reason, start with the notion that truth is to be established by *evidence*, that truth is the result of *experimental* methods, that truth is to be *tested* in the laboratory of personal *experience*, that these truths are open to all those who wish to *try the experiment* and thus disclose *for themselves* the truth or falsity of the spiritual claims—and that dogmas or given beliefs are precisely what hinder the emergence of deeper truths and wider visions.

Thus, each of these spiritual or transpersonal endeavors (which we will carefully examine) claims that there exist higher domains of awareness, embrace, love, identity, reality, self, and truth. But these claims are not dogmatic; they are not believed in merely because an authority proclaimed them, or because sociocentric tradition hands them down, or because salvation depends upon being a "true believer." Rather, the claims about these higher domains are a *conclusion* based on hundreds of years of experimental introspection and communal verification. False claims are *rejected* on the basis of *consensual evidence*, and further evidence is used to adjust and fine-tune the experimental conclusions.

These spiritual endeavors, in other words, are scientific in any meaningful sense of the word, and the systematic presentations of these endeavors follow precisely those of any *reconstructive science*.

OBJECTIONS TO THE TRANSPERSONAL

The common objections to these contemplative sciences are not very compelling. The most typical objection is that these mystical states are private and interior and cannot be publicly validated; they are "merely subjective."

This is simply not true; or rather, if it is true, then it applies to any and all nonempirical endeavors, from mathematics to literature to linguistics to psychoanalysis to historical interpretation. Nobody has ever seen, "out there" in the "sensory world," the square root of a negative one. That is a mathematical symbol seen only *inwardly*, "privately," with the mind's eye. Yet a community of trained mathematicians know exactly what that symbol means, and they can share that symbol easily in intersubjective awareness, and they can confirm or reject the proper and consistent uses of that symbol. Just so, the "private" experiences of contemplative scientists can be shared with a community of trained contemplatives, grounded in a common and shared experience, and open to confirmation or rebuttal based on public evidence.

Recall that the Right-Hand path is open to empirical verification, which means that the Right-Hand dimension of holons, their form or exteriors, can indeed be "seen" with the senses or their extensions. But the Left-Hand dimension—the interior side—*cannot* be seen empirically "out there," although it can be internally *experienced* (and although it has empirical *correlates*: my interior thoughts register on an EEG but cannot be determined or interpreted or known from *that* evidence). Everything on the Left Hand, from sensations to impulses to images and concepts and so on, is an *interior experience* known to me *directly by acquaintance* (which can indeed be "objectively described," but only through an intersubjective community at the *same depth*, where it relies on *interpretation* from the *same* depth). *Direct spiritual experience* is simply the higher reaches of the Upper-Left quadrant, and those experiences are as real as any other direct experiences, and they can be as easily shared (or distorted) as any other experiential knowledge.[11]

(The only way to deny the validity of direct interior experiential knowledge—whether it be mathematical knowledge, introspective knowledge, or spiritual knowledge—is to take the behaviorist stance and identify interior experience with exterior behavior. Should somebody mention that this is the cynical twist or pathological agency of Broughton's level four?)

There is, of course, one proviso: the experimenter *must*, in his or her own case, *have developed the requisite cognitive tools*. If, for example, we want to investigate concrete operational thought, a community of those who have only developed to the preoperational level will not do. If you take a preop child, and in front of the child pour the water from a short fat glass into a tall thin glass, the child will tell you that the tall glass has more water. If you say, no, there is the same amount of water in both glasses, because you just saw me pour the same water from one glass to the other, the child will have no idea what you're talking about. "No, the tall glass has more water." No matter how many times you pour the water back and forth between the two glasses, the child will deny they have the same amount of water. (Interestingly, if you videotape the child at this stage, and then wait a few years until the child has developed conop—at which point it will seem utterly obvious to him that the glasses have the same amount of water—and then show the child the earlier videotape, he will deny that it's him. He thinks you've doctored the videotape; he cannot imagine anybody being that stupid.) The preop child is *immersed* in a world that includes conop realities, is *drenched* in those realities, and yet cannot "see" them: they are all "otherworldly."

At every stage of development, in fact, the next higher stage always appears to be a completely "other world," an "invisible world"—it has literally no existence for the individual, even though the individual is in fact *saturated* with a reality that contains the "other" world. The individual's "this-worldly" existence simply cannot comprehend the "otherworldly" characteristics lying all around it.

At the same time, these higher or deeper worldspaces (whether of conop or formop or anything higher) are not located *elsewhere* in physical space-time. They are located *here*, in *deeper* perceptions of *this* world. *Other worlds* become *this world* with increasing development and evolution. The worldspace of conop is a completely other world to the preop child (even though he or she is often *staring directly at it*), another world that nonetheless becomes perfectly obvious, present, seen, and "this-worldly" at the conop stage, and what was the fuss all about? But prior to the increased development and evolution, there is nothing from the present this-world that will allow the child to adequately grasp the other world (or else the other world would have already become a real this-world).

Just so with the higher or transpersonal developments. Explain them to someone at the rational level, and all you get, at best, is that deer-caught-in-the-headlights blank stare (at worst, you get something like, "And did we forget to take our Prozac today?").

So the first thing I would like to emphasize is that the higher stages of transpersonal development are stages that are taken from those who have actually developed into those stages and who display palpable, discernible, and repeatable characteristics of that development. The stages themselves can be *rationally reconstructed* (explained in a rational manner after the fact), but they cannot be rationally experienced. They can be experienced only by a transrational contemplative development, whose stages unfold in the same manner as any other developmental stages, and whose experiences are every bit as real as any others.

But one must be adequate to the experience, or it remains an invisible other world. When the yogis and sages and contemplatives make a statement like, "The entire world is a manifestation of one Self," that is *not* a merely rational statement that we are to think about and see if it makes logical sense. It is rather a description, often poetic, of a direct apprehension or a direct experience, and we are to test this direct experience, not by mulling it over philosophically, but by taking up the experimental method of contemplative awareness, developing the requisite cognitive tools, and then directly looking for ourselves.

As Emerson put it, "What we are, that only can we see."

LANGUAGE AND MYSTICISM

In this regard, another common objection is that mystical or contemplative experiences, because they cannot be put into plain language, or into any language for that matter, are therefore not epistemologically grounded, are not "real knowledge." But this simply bypasses the problem of what linguistically situated knowledge means in the first place.

Saussure, as I mentioned earlier, maintained that all linguistic signs have two components, the signifier and the signified, often represented as S/S. The *signifier* is the written or spoken symbol or sound, the *material component* of the sign (such as the physical ink forms written on this page, or the physical air vibrations as you speak). The *signified* is what comes to your mind when you see or hear the signifier.

Thus, I physically write the word *dog* on this page—that is the signifier.

You read the word, and you understand that I mean something like a furry animal with four legs that goes wuff-wuff—that is the signified, that is what comes to your mind. A *sign* is a combination of these two components, and these two components are, of course, the Right-Hand dimension of the sign (the physical exterior) and the Left-Hand dimension of the sign (the interior awareness or meaning).

And both of those are distinguished from the actual *referent*, or whatever it is that the sign is "pointing" to, whether interior or exterior. Thus, the signifier is the word *dog*, the referent is the real dog, and the signified is what comes to your mind when you read or hear the signifier *dog*. Saussure's genius was to point out that the *signified* is not merely or simply the same as the *referent*, because "what comes to mind" depends on a whole host of factors other than the real dog, and this is what makes linguistic reality so fascinating.

Saussure's point—and this is what actually ignited the whole movement of structuralism—is that the sign cannot be understood as an isolated entity, because in and by itself *the sign is meaningless* (which is why different words can represent the same thing in different languages, and why "meaning" is never a simple matter of a word pointing to a thing, because how could different words represent the same thing?). Rather, signs must be understood as part of a holarchy of differences integrated into meaningful structures. Both the signifiers and the signifieds exist as holons, or whole/parts in a chain of whole/parts, and, as Saussure made clear, it is their *relational standing* that *confers meaning* on each (language is a meaningful system of meaningless elements: *as always*, the regime or structure of the superholon confers meaning on the subholons, meaning which the subholons do not and cannot possess on their own).

In other words, the signifiers and the signifieds exist as a structure of contexts within contexts within contexts, and meaning itself is *context-bound*. Meaning is found not in the word but in the context: the bark of a dog is not the same as the bark of a tree, and the difference is not in the word, because the word *bark* is the same in both phrases—it is the relational context that determines its meaning: the *entire structure* of language is involved in the meaning of each and every term—this was Saussure's great insight.

And this, as usual, contributed to a split between Right- and Left-Hand theorists. The Right-Hand theorists, or the pure structuralists, wanted to study only the *exterior structure of the system of signifiers* in language and in culture (an approach which in turn gave way to the poststructuralists, who wanted to free the signifier from any grounding at all—as in Foucault's archaeology or Derrida's grammatology—and see it as free-floating or sliding, and anchored only by power or prejudice: meaning is indeed context-dependent, but contexts are boundless, and thus meaning is arbitrarily *imposed* by power or prejudice—the so-called "poststructural revolution" of "free-floating signifiers").

The Left-Hand theorists wanted to study the contexts within contexts of *interior meaning*, the *signifieds* that can only be *interpreted*, not seen, and interpreted only in a context of background cultural practices (the hermeneuticists, from Heidegger to Kuhn and Taylor and aspects of Wittgenstein).

But both the hermeneutical Left-Hand path and the structuralist Right-Hand path agreed that signs can only be understood contextually (whether in the context of shared cultural practices that provide the foreknowledge or background or context for common interpretation, or in the context of shared nonindividual linguistic structures. I argued in chapter 4 that both of these approaches are equally important—they represent the interior and the exterior of the linguistic holon—and indeed even Foucault came to this understanding).[12]

All of which relates to mysticism in this way: the word *dog* has a shared meaning to you and to me because that sign exists in a shared linguistic structure and a shared cultural background of social and interpretive practices. But what if you had never seen a real dog? What then?

I could of course describe one to you, but the word will be meaningless unless there are some points of shared experience that will allow you to "call up" in your mind the same *signified* that I mean with the *signifier* "dog." (Substitute the word *Buddha-nature* for *dog* and you can see the importance of this line of thought for mystical experience, which we will explore in a minute.) The hermeneuticists are quite right in that regard: the same linguistic structures that you and I share are *not enough*, in themselves, to give you the proper signified. You and I have to share a *common lived experience* in order to assume identical signification.

Further, the actual experience of seeing a dog is not itself a *merely* linguistic experience. The signifier, the word *dog*, is not the actual dog, not the actual referent. Obviously, the total experience of the real dog cannot itself be put into words, put into signifiers. But the fact that the real dog can't be fully captured in words does not mean that the real dog doesn't exist or isn't real. It means only that the signifier has sense only if you and I have had a similar experience, a common shared lifeworld experience, and then I will know what you mean when you say, "That dog scared me."

In short, no direct experience can be fully captured in words.[13] Sex can't be put into words; you've either had the experience or you haven't, and no amount of poetry will take its place. Sunsets, eating cake, listening to Bach, riding a bike, getting drunk and throwing up—believe me, none of those are captured in words.

And thus, so what if spiritual experiences can't be captured in words either? They are no more and no less handicapped in this regard than any other experience. If I say "dog" and you've had the experience, you know exactly what I mean. If a Zen master says "Emptiness," and you've had that experience, you will know exactly what is meant. If you haven't had the experience "dog" or the experience "Emptiness," merely adding more and more words will never, under any circumstances, convey it.

Thus, if we are going to level that charge at mysticism, then we must level it at dogginess and sunsetness and every other experience that happens to come our way. (This is really the cheapest of the cheap shots fired at mysticism.)

Conversely, words do just fine as signifiers for experience, whether mundane or spiritual, if we both, you and I, have had similar experiences in a context of shared background practices. Zen masters talk about Emptiness all the time! And they know exactly what they mean by the words, and the words are *perfectly adequate* to convey what they mean, *if* you have had the experience (for what they mean can only be disclosed in the shared praxis of zazen, or meditation practice).

Go one step further. If I say to a conop child, "It is as if I were elsewhere," the child might nod her head as if she actually understood all the meanings of that statement. The conop child already possesses the shared linguistic structure (and grammar) to decipher the words. But, as we have seen, since the conop child cannot fully grasp the implications of as-if statements, she doesn't really understand what is *signified* by my statement. Once the higher structure of formop emerges, however, this will usher the child into a *worldspace* where "as-if" is not just a signifier but a *signified* that has an existing *referent* in that formop worldspace: not just a word, but a direct understanding that more or less spontaneously jumps to mind whenever we hear or see the word, and which refers to a genuinely existing entity in the rational worldspace.

In other words, *all* signs exist in a continuum of *developmental referents* and *developmental signifieds*. The *referent* of a sign is not just lying around in "the" world waiting for any and all to simply look at it; the referent exists *only* in a *worldspace* that is itself *only* disclosed in the process of development, and the *signified* exists only in the *interior perception* of those who have developed to that worldspace (which structures the background interpretive meaning that allows the signified to emerge). No *amount of experience* by the conop child will ever show her the meaning of an "as-if" dog, because the as-if dog does not exist *anywhere* in the conop worldspace; it *exists* only in the formop worldspace, and thus it is a referent that demands a developmental signified to even be perceived in the first place.

To take it a point at a time: the *signifiers* of signs (such as the words on this page) are always physical, they are always material components, in which no meaning resides at all (Saussure's point); and because the signifiers are physical, even my dog can see them (and, of course, sees no meaning in them; or rather, sees them from a sensorimotor level, as something to eat, perhaps). That is because the actual *referent* of a sign exists only in a *worldspace* (sensorimotor, magical, mythical, mental, etc.) that is itself disclosed only at a particular level of depth (preop, conop, formop, etc.). And in the same way, the corresponding *signified* of the sign exists only in the *interior perception* of those who have developed the *requisite depth*. (All of this occurs in a context of cultural and social practices, or an intersubjective community of the same-depthed.)[14]

Both the conop child and my dog can see the physical words "as-if"; neither of them can understand the phrase. The empirical markings are meaningless. The child and the dog do not possess the developmental signified, and thus they cannot see the actual referent.

Several examples of referents and worldspaces: rocks exist in the sensorimotor worldspace; animistic clouds exist in the magic worldspace; Santa Claus exists in the mythic worldspace; the square root of a negative one exists in the rational worldspace; archetypes exist in the subtle worldspace, and so on—not as pregiven objects, but as the product of all four quadrants. And thus, in order to understand the referents represented by those signifiers (from "rocks" to "archetypes") one must possess the requisite depth through one's own interior development (so that those signifiers can evoke the appropriate signified: when you read "the square root of a negative one," you know what that means, what that signifies, but only if you have developed to formop).

Just so, the words *Buddha-nature* and *Godhead* and *Spirit* and *Dharmakaya* are signifiers whose *referents* exist only in the transpersonal or *spiritual worldspace*, and they therefore require, for their understanding, a *developmental signified*, an appropriately developed interior or Left-Hand dimension corresponding with the exterior word, or else they remain only words, like the unseen dog, this unseen Spirit. And without the developmental signified, words will capture neither the dog nor the Spirit.

And note: I can run around until I find a dog and show you the dog, because we both exist in the sensorimotor worldspace and there is no developmental reason why you can't spot a dog. Or, in the other example, there is no reason you can't understand an as-if dog, whose referent exists in the rational worldspace. We *already* share that worldspace. We have already *transformed* to that level of depth: an entire and shared world of referents are therefore lying around for us to apprehend (because we have *already* created the worldspace or the opening in which they can manifest).

But I can't run around and find Buddha and show you that, unless you have developed the requisite cognitions that will allow you to resonate with the *signifier* whose *referent* exists only in the spiritual worldspace and whose *signified* exists only in the consciousness of those who have awakened to that space.

VALIDITY CLAIMS OF MYSTICISM

If I want to know whether it is raining or not, I go to the window and look out, and sure enough, rain. But perhaps I am mistaken, or perhaps my eyesight is poor. Would you check? You go to the window and yes, rain.

That is a very simplified form of the three strands of any valid knowledge quest (whether of the Left- or Right-Hand path).[15] The first is *injunction*, which is always of the form, "If you want to know this, do this." If you want to know if a cell has a nucleus, then get a microscope, learn to take histological sections, stain the cell, put it under the microscope, and look. If you want to know the meaning of *Hamlet*, then learn English, get the book, and read. If you want to know whether 2 + 2 is really 4, then learn arithmetic theory, take the theorems, run them through your mind, and check the results.

The various *injunctions*, in other words, lead to or disclose or open up the possibility of an illumination, an apprehension, an intuition, or a direct experiencing of the domain addressed by the injunction. You "see" the meaning of *Hamlet*, or whether it is raining, or why 2 + 2 really is 4. This is the second strand, the illumination or *apprehension*. You see or apprehend, via a direct experience, the disclosed data of the domain.[16]

But you could be mistaken, and thus you check your results, your data, with others who have completed the first two strands, with others who have performed the injunctions and obtained the data. In this community of peers, you compare and confirm—or reject—your original data. And this is the third strand, *communal confirmation* (or refutation).

These three strands—injunction, illumination, confirmation—are the major components in any valid knowledge quest.[17] One of the great values of Thomas Kuhn's work (and that of the pragmatists before him, and in particular Heidegger's "analytic-pragmatic" side) was to draw attention to the importance of injunctions or *actual practices* in generating knowledge, and further, in generating the *type* of knowledge that *could* be articulated in a given worldspace.

That is, social *practices*, or social *injunctions* (and I mean "social" in the broad sense as "sociocultural"), are crucial in creating and disclosing the types of worldspace in which types of subjects and objects appear (and thus the types of knowledge that *can* unfold). The referents of knowledge, as we saw above, exist only in specific worldspaces, and those worldspaces are not simply given empirically, lying around for all and sundry to perceive.[18] Rather, these worldspaces are disclosed/created by cognitive transformations in the context of background injunctions or social practices.[19]

Put simply, the first strand of knowledge accumulation is never simply "Look"; it is "Do this, then look."

Kuhn, in one of the great misunderstood concepts of our era, pointed out that normal science proceeds by way of *exemplary injunctions*—that is, shared practices and methods that scientists agree disclose and address the important issues of their field. Kuhn called such an agreed-upon injunction an "exemplar" or a "paradigm"—an exemplary practice or technique or methodology that all agreed was central to furthering the knowledge quest. And it was the paradigm, the exemplary injunction, that disclosed a type of data, so that the paradigm itself was a matter of consensus, not merely correspondence.

In the academic world of the two cultures, many theorists in the underfunded humanities (and virtually everybody in the New Age movement) seized upon the notion of "paradigm" as a way to undercut the authority of normal science, bolster their own departments, reduce empirical facts to arbitrary social conventions—and then propose their own, new and improved "paradigm." In all of these, "paradigm" was mistaken as some sort of overall theory or concept or notion, the idea being that if you came up with a new and better theory, the factual evidence could be ignored because that was just "old paradigm."

Among other things, this meant that empirical science didn't really show any "progress," but was merely a shifting of opinions ("paradigms") that had no *referent* except in the arbitrary *conventions* of scientists (and these conventions were always charged with some sort of "ism" that the new paradigm would overcome).

All of this ignored Kuhn's repeated insistence that "later scientific theories are better than earlier ones for solving puzzles in the quite often different environments to which they are applied. This is not a relativist's position, and it displays the sense in which I am a convinced believer in scientific progress."[20]

But by collapsing "paradigm" into a mere theory (itself unanchored), the scientific enterprise could be collapsed into various forms of literary chitchat (and the new masters of the universe were therefore . . . the literary critics). And likewise, on the New Age front, a flurry of "new paradigms" could then step in and redress the ugliness of the old paradigm.

But paradigms are first and foremost *injunctions*, actual *practices* (all of which have nondiscursive components that never are entered in the theories they support)—they are methods for disclosing new data in an addressed domain, and the paradigms *work* because they are true in any meaningful sense of the word. Science makes real *progress*, as Kuhn said, because successive paradigms cumulatively disclose more and more interesting data. Even Foucault acknowledged that the natural sciences, even if they had started as structures of power, had separated from power (it was the pseudosciences of biopower that remained shot through with power masquerading as knowledge).

Neither the New Agers nor the "new paradigmers" had anything resembling a new paradigm, because all they offered was more talk-talk.

They had no new techniques, no new methodologies, no new exemplars, no new injunctions—and therefore no new data. All they possessed, through a misreading of Kuhn, was a pseudo-attempt to trump normal science and replace it with their ideologically favorite reading of the Kosmos.

The contemplative traditions, on the other hand, have always come first and foremost with a set of injunctions in hand. They are, above all else, a set of *practices*, practices that require years to master (much longer than the training of the average scientist). These injunctions (*zazen, shikan-taza, vipassana*, contemplative introspection, *satsang, darshan*—all of which we will discuss)—these are not things to think, they are things to do.

Once one masters the exemplar or the paradigmatic practice (strand one), then one is ushered into a worldspace in which new data disclose themselves (strand two). These are direct apprehensions or illuminations—in a word, *direct spiritual experiences* (*unio mystica, satori, kensho, shaktipat, nada, shabd*, etc.). These data are rigorously checked (strand three) in the community of those who have also completed the first two strands (injunction and illumination). Bad data are *rebuffed* by the community (the sangha) of those whose cognitive eyes are adequate to the addressed domain.

Thus, as I covered in more detail in *Eye to Eye*, authentic knowledge has a component that is similar to Kuhn's paradigm (namely, the injunction), a component that is similar to the broad empirical demand for evidence or data or experience (namely, the illumination or apprehension, whether that be from sensory experience, mental experience, or spiritual experience), and a component similar to Sir Karl Popper's fallibilistic criterion (namely, the potential confirmation or refutation by a community of the adequate).

Accordingly, contemplative knowledge is, or can be, genuine knowledge, because it follows all three strands of valid knowledge accumulation.

THE RECONSTRUCTION OF THE CONTEMPLATIVE PATH

Of course, this does not prevent the various contemplative traditions from possessing their own particular and culture-bound trappings, contexts, and interpretations. But to the extent that the contemplative endeavor discloses universal aspects of the Kosmos, then the deep structures of the contemplative traditions (but not their surface structures) would be expected to show cross-cultural similarities at the various levels of depth created/disclosed by the meditative injunctions and paradigms.

In other words, the deep structures of worldspaces (archaic, magic, mythic, rational, and transpersonal) show cross-cultural and largely invariant features at a deep level of abstraction, whereas the surface structures (the actual subjects and objects in the various worldspaces) are naturally and appropriately quite different from culture to culture. Just as the human mind universally grows images and symbols and concepts (even though the actual contents of those structures vary considerably), so the human spirit universally grows intuitions of the Divine, and those *developmental signifieds* unfold in an evolutionary and reconstructible fashion, just like any other holon in the Kosmos (and their referents are just as real as any other similarly disclosed data).

In the past few decades there has been a concerted effort on the part of many researchers (such as Stanislav Grof, Roger Walsh, Frances Vaughan, Daniel Brown, Jack Engler, Daniel Goleman, Charles Tart, Donald Rothberg, Michael Zimmerman, Seymour Boorstein, Mark Epstein, David Lukoff, Michael Washburn, Joel Funk, John Nelson, John Chirban, Robert Forman, Francis Lu, Michael Murphy, Mark Waldman, James Fadiman, myself, and others)[21] to rationally reconstruct the higher stages of transpersonal or contemplative development—stages that continue naturally or normally beyond the ego and centaur if arrest or fixation does not occur.

Much of this work has been summarized in *Transformations of Consciousness: Conventional and Contemplative Perspectives on Development* (Wilber, Engler, and Brown), and I will not repeat its contents. But the conclusion is straightforward. As Brown and Engler summarize it:

> The major [contemplative] traditions we have studied in their original languages present an unfolding of meditation experience in terms of a *stage model*: for example, the Mahamudra from the Tibetan Buddhist tradition; the Visuddhimagga from the Pali Theravada tradition; and the Yoga Sutras from the Sanskrit Hindu tradition. *The models are sufficiently similar to suggest an underlying common invariant sequence of stages*, despite vast cultural and linguistic differences as well as styles of practice.

This developmental model has also been found to be consistent with the stages of mystical or interior prayer found in the Jewish (Kabbalist), Islamic (Sufi), and Christian mystical traditions (see, for example, Chirban's chapter in *Transformations*), and Brown has also found it in the Chinese contemplative traditions. Theorists such as Da Avabhasha have given extensive hermeneutic and developmental readings from what now appears to be at least a representative sampling from every known and available contemplative tradition (see, for example, *The Basket of Tolerance*), and they are in fundamental and extensive agreement with this overall developmental model.

The evidence, though still preliminary, strongly suggests that, at a minimum, *there are four general stages of transpersonal development*, each with at least two substages (and some with many more). These four stages I call the *psychic*, the *subtle*, the *causal*, and the *nondual*.

Each of these stages follows the same patterns and shows the same developmental characteristics as all the other stages of consciousness evolution: each is a holon following the twenty tenets (a new differentiation/integration, a new emergence with a new depth, a new interiority, etc.); each possesses a new and higher sense of self existing in a new and wider world of others, with new drives, new cognitions, new moral stances, and so forth; each possesses a deep structure (basic defining pattern) that is culturally invariant but with surface structures (manifestations) that are culturally conditioned and molded; and each has a new and higher form of possible pathology (with the exception of the unmanifest "end" point, although even that is not without certain possible complications in its manifestation).

I have elsewhere given preliminary descriptions of the deep structures (and pathologies) of each of these four major stages.[22] Instead of repeating myself, I have for this presentation simply chosen four individuals who are especially representative of these stages, and will let them speak for us. They are (respectively) Ralph Waldo Emerson, Saint Teresa of Ávila, Meister Eckhart, and Sri Ramana Maharshi. Each also represents the *type of mysticism* typical at each stage: nature mysticism, deity mysticism, formless mysticism, and nondual mysticism (each of which we will discuss).

And each represents a form of tomorrow, a shape of our destiny yet to come. Each rode time's arrow ahead of us, as geniuses always do, and thus, even though looming out of our past, they call to us from our future.

8

The Depths of the Divine

Holy breath divinely streams through the luminous form when the feast comes to life, and floods of love are in motion, and, watered by heaven, the living stream roars when it resounds beneath, and the night renders her treasures, and up out of brooks the buried gold gleams.—

And, friendly spirit, just as from your serenely contemplative brow your ray descends, securely blessing, among mortals, so you witness to me, and tell me, that I might repeat it to others, for others too do not believe it.... —FRIEDRICH HÖLDERLIN

THE CONTINUING evolutionary process of within-and-beyond brings new withins . . . and new beyonds.

THE PSYCHIC LEVEL

We left off with the emergence of the centaur, which is, so to speak, on the border between the personal and the transpersonal. If the first three general domains were those of matter, life, and mind, the next general domain (that of the psychic and subtle) is the domain of the *soul*, as I will use the term. And the first rule of the soul is: it is transpersonal.

The word *transpersonal* is somewhat awkward and confuses many people. But the point is simply, as Emerson put it, *"The soul knows no persons.*" He explains (and note: Emerson throughout these quotes uses the masculine, as was the custom of the time; were he alive today he would use feminine and masculine, for the whole point of his notion of the Over-Soul was that it was neither male nor female, which is why it could anchor a true liberation from any and all restrictive roles: "The soul knows no persons"):

> Persons are supplementary to the primary teaching of the soul. In youth we are mad for persons. Childhood and youth see all the world in them. But the larger experience of man discovers the identical nature [the same self or soul] appearing through them all. In all conversation between two persons tacit reference is made, as to a third party, to a common nature. That third party or common nature is not social; it is impersonal; is God.[1]

The soul is without persons, and the soul is grounded in God. "Impersonal," however, is not quite right, because it tends to imply a complete negation of the personal, whereas in higher development the personal is negated and preserved, or transcended and included: hence, "transpersonal." So I think it's very important, in all subsequent discussion, for us to remember that *transpersonal* means "personal *plus*," not "personal minus."

But what could an actual "transpersonal" experience really mean? It's not nearly as mysterious as it sounds. Recall that at the centaur, according to the research of Broughton (and many others), the self is *already* beginning to transcend the empirical ego or the empirical person ("the observer is distinguished from the self-concept as known"). You yourself can, right now, be aware of your objective self, you can observe your individual ego or person, you are aware of yourself generally.

But who, then, is doing the observing? What is it that is observing or witnessing your individual self? That therefore *transcends* your individual self in some important ways? Who or what is *that*? The noble Emerson:

> All goes to show that the soul in man is not an organ, but animates and exercises all the organs; is not a function, like the power of memory, of calculation, of comparison, but uses these as hands and feet; is not a faculty, but a light; is not the intellect or the will, but the master of the intellect and the will; is the background of our being, in which

they lie,—an immensity not possessed and that cannot be possessed. From within or from behind, a light shines through us upon things and makes us aware that we are nothing, but the light is all.[2]

The observer in you, the Witness in you, transcends the isolated *person* in you and opens instead—from within or from behind, as Emerson said—onto a vast expanse of awareness no longer obsessed with the individual bodymind, no longer a respecter or abuser of persons, no longer fascinated by the passing joys and set-apart sorrows of the lonely self, but standing still in silence as an opening or clearing through which light shines, not from the world but into it—"a light shines *through us* upon things." *That which* observes or witnesses the self, the person, is precisely to that degree *free* of the self, the person, and *through that opening* comes pouring the light and power of a Self, a Soul, that, as Emerson puts it, "would make our knees bend."

> A man is the facade of a temple wherein all wisdom and all good abide. What we commonly call man [as an "individual person" or ego], the eating, drinking, counting man, does not, as we know him, represent himself, but misrepresents himself. Him we do not respect, but the soul, whose organ he is, if he would let it appear through his action, would make our knees bend. When it breathes through his intellect, it is genius; when it breathes through his will, it is virtue; when it flows through his affection, it is love. And the blindness of the intellect begins when it would be something of itself [be its "own person"]. The weakness of the will begins when the individual would be something of himself. All reform aims in some one particular to let the soul have its way through us. . . .[3]

And those persons *through whom* the soul shines, *through whom* the "soul has its way," are not therefore weak characters, timid personalities, meek presences among us. They are personal plus, not personal minus. Precisely because they are no longer exclusively identified with the individual personality, and yet because they still preserve the personality, then *through that* personality flows the force and fire of the soul. They may be soft-spoken and often remain in silence, but it is a thunderous silence that veritably drowns out the egos chattering loudly all around them. Or they may be animated and very outgoing, but their dynamism is magnetic, and people are drawn somehow to the presence, fascinated. Make no mistake: these are strong characters, these souls, sometimes wildly exaggerated characters, sometimes world-historical, precisely because their personalities are plugged into a universal source that rumbles through their veins and rudely rattles those around them.

I believe, for example, that it was precisely this fire and force that allowed Emerson, more than any other person in American history, to actually define the intellectual character of America itself. One of his essays, "The American Scholar," had, as one historian put it, "an influence greater than that of any single work in the nineteenth century." Oliver Wendell Holmes called it "our intellectual Declaration of Independence." James Russell Lowell explained: "The Puritan revolt had made us ecclesiastically, and the Revolution politically independent, but we were still socially and intellectually moored to English thought, till Emerson cut the cable and gave us a chance at the dangers and the glories of blue water. . . ."

And the message, this ringing Declaration of Independence? The Soul is tied to *no* individual, *no* culture, *no* tradition, but rises fresh in every person, beyond every person, and grounds itself in a truth and glory that bows to nothing in the world of time and place and history. We all must be, and can only be, "a light unto ourselves."[4]

And then in a phrase that, as Holmes indicated, rattled all of America, Emerson announced: "All that Adam had, all that Caesar could, you have and can do"—because it is the same Self in each of us. Why bow to past heroes, he asks, when all we are bowing to is our own Soul? "Suppose they were virtuous; did they wear out virtue?" The *magnetism* of the great heroes is only the call from our own Self, he says. Why this groveling to the past when the same Soul shines now and only now and always now? And then Emerson swiftly and irrevocably cut the cable and set us all— not just Americans—afloat on the dangers and the glories of blue water:

> Trust thyself: every heart vibrates to that iron string. . . . The magnetism which all original action exerts is explained when we inquire the reason of self-trust. Who is the Trustee? What is the aboriginal Self, on which a universal reliance may be grounded? . . . The inquiry leads us to that source, at once the essence of genius, of virtue, and of life. . . . In that deep force, the last fact behind which analysis cannot go, all things find their common origin.

For the sense of being which in calm hour arises, we know not how, in the Soul, is not diverse from things, from space, from light, from time, from man, but one with them and proceeds obviously from the same source whence their life and being also proceed. . . . Here is the fountain of action and of thought. Here are the lungs of that inspiration which giveth man wisdom. . . . We lie in the lap of immense intelligence, which makes us receivers of its truth and organs of its activity. When we discern justice, when we discern truth, we do nothing of ourselves, but allow a passage to its beams. . . .

The relations of the Soul to the divine spirit are so pure that it is profane to seek to interpose helps. It must be that when God speaketh he should communicate, not one thing, but all things; should fill the world with his voice; should scatter forth light, nature, time, souls, from the center of the present thought; and new date and new create the whole. Whenever a mind is simple and receives a divine wisdom, old things pass away—means, teachers, texts, temples fall; *it lives now* and absorbs past and future into the present hour. All things are made sacred by relation to it—one as much as another. All things are dissolved to their center by their cause, and in the universal miracle petty and particular miracles disappear.

If therefore a man claims to know and speak of God and carries you backward to the phraseology of some old mouldered nation in another country, in another world, believe him not. Is the acorn better than the oak which is its fullness and completion? Whence then this worship of the past? The centuries are conspirators against the sanity and authority of the Soul. Time and space are but physiological colors which the eye makes, but the Soul is light: where it is, is day; where it was, is night; and history is an impertinence and an injury if it be any thing more than a cheerful apologue or parable of my being and becoming.[5]

To emphasize that the Soul, the "aboriginal Self," is common in and to all beings, Emerson often refers to it as the "Over-Soul," one and the same in all of us, in all beings as such. The overall number of Souls is but one:

The only prophet of that which must be, is that great nature in which we rest as the earth lies in the soft arms of the atmosphere; that Unity, that Over-Soul, within which every man's particular being is contained and made one with all other; that common heart of which all sincere conversation is the worship, to which all right action is submission; that over-powering reality which confutes our tricks and talents, and constrains every one to pass for what he is, and to speak from his character [soul] and not from his tongue [ego], and which evermore tends to pass into our thought and hand and become wisdom and virtue and power and beauty. . . .

And this because the heart in thee is the heart of all; not a valve, not a wall, not an intersection is there anywhere in nature, but one blood rolls uninterruptedly an endless circulation through all men, as the water of the globe is all one sea, and, truly seen, its tide is one. . . .

It is one light which beams out of a thousand stars. It is one soul which animates all men. . . .

We live in succession, in division, in parts, in particles. Meantime within man is the Soul of the whole; the wise silence; the universal beauty, to which every part and particle is equally related; the eternal ONE.[6]

And so once again we see that a new and deeper *within* has brought us to a new and wider *beyond*, a beyond that "is not diverse from things, from space, from light, from time, from man, but one with them and proceeds obviously from the same source whence their life and being also proceed." This new within-and-beyond is not just beyond a sociocentric identity to a worldcentric identity with all human beings (which the rational-ego/centaur assumes in its global or universal postconventional awareness), but to an identity, a conscious union, with all of manifestation itself: not just with all humans, but with all nature, and with the physical cosmos, with all beings "great and small"—a union or identity that Bucke famously called "cosmic consciousness."

We could say: from the worldcentric centaur to direct cosmic consciousness. (When you consider the remarkable achievement that decentered worldcentric awareness is, it's not that far of a jump).[7]

Thus, the centaur could *integrate* the physiosphere and the biosphere and the noosphere, but the Over-Soul becomes, or is *directly one with*, the physiosphere and biosphere and noosphere. It is simple continuation of the deepening and widening of identity, grounded in an awareness very much within, and very much beyond, me.

And Emerson means this literally! According to Emerson, this cosmic consciousness is not poetry (though he often expresses it with unmatched poetic beauty)—rather, it is a direct realization, a direct apprehension, and "in that deep force, the last fact behind which analysis cannot go, all things find their common origin. It is one light which beams out of a thousand stars. It is one soul which animates all."

For the Over-Soul is also experienced as the World Soul, since self and world are here finding a "common fountain, common source."[8] The Over-Soul (or World Soul) is an initial apprehension of the pure Witness or aboriginal Self, which starts to emerge, however haltingly, as an experiential reality at this psychic stage.[9] (We will see how Emerson treats this Witness in a moment.)

With the Over-Soul, the World Soul, it is not that individuality disappears, but that—once again—it is negated and preserved in a deeper and wider ground, a ground that conspicuously includes all of nature and its glories. This cosmic consciousness is sometimes referred to as "nature mysticism," but that is a somewhat misleading term. For this psychic-level mysticism embraces not just *nature* but also *culture*, and calling it "nature mysticism" confuses it with a merely biocentric regression, an ecocentric indissociation, and this is not at all what Emerson has in mind (as we will see).

But since the Over-Soul is an experienced identity with all manifestation, it is an identity that most definitely and exuberantly *embraces nature*; and, to that degree, it begins to undercut the subject/object dualism.[10] Emerson explains:

> We see the world piece by piece, as the sun, the moon, the animal, the tree; but the whole, of which these are the shining parts, is the soul [the Over-Soul, the World Soul]. . . . And this deep power in which we exist and whose beatitude is all accessible to us, is not only self-sufficing and perfect in every hour, but the act of seeing and the thing seen, the seer and the spectacle, the subject and the object, are one.[11]

In his famous "transparent eyeball" section from the essay *Nature*, Emerson speaks movingly of the union of the Soul and nature, and of the capacity of nature, when rightly approached, to elicit this cosmic consciousness. The "transparent eyeball" is, of course, an intimation of the pure Witness in the form of the Over-Soul, wherein "all mean egotism vanishes; I am nothing; I see all":

> To speak truly, few adult persons can see nature. Most persons do not see the sun. At least they have a very superficial seeing. . . . Crossing a bare common, in snow puddles, at twilight, under a clouded sky, without having in my thoughts any occurrence of special good fortune, I have enjoyed a perfect exhilaration. I am glad to the brink of fear. . . . Within these plantations of God, a decorum and sanctity reign, a perennial festival is dressed, and the guest sees not how he should tire of them in a thousand years. In the woods, we return to reason and faith. There I feel that nothing can befall me in life,—no disgrace, no calamity, which nature cannot repair. Standing on the bare ground,—my head bathed by the blithe air, and uplifted into infinite space,—all mean egotism vanishes. I become a transparent eyeball; I am nothing; I see all; the currents of Universal Being circulate through me; I am part or parcel of God.[12]

What distinguishes this profound "nature mysticism" from a simple nature indissociation or ecocentric immersion or biospheric regression (which would be egocentric and anthropocentric, as we have seen) is the realization that nature is not Spirit but an *expression* of Spirit, radiant and glorious and perfect in its own way, but an expression nonetheless. Emerson says nature is not spirit but a *symbol* of spirit. Emerson is not regressing to fulcrum-2 (biocentric immersion and nondifferentiation)! Emerson is very clear on this distinction between nature *regression*, on the one hand, and a mysticism that also *embraces* nature, on the other—and this distinction rather upsets his environmentalist fans, who seem to want to *equate* a finite and temporal nature with an infinite and eternal Spirit:

> Beauty in nature is not ultimate. It is the herald of inward and eternal beauty, and is not alone a solid and satisfactory good. . . .
>
> Nature is a symbol of spirit. . . . Before the revelations of the Soul, time, space and nature shrink away. . . . In the hour of vision there is nothing that can be called gratitude, nor properly joy. The Soul raised

over passion beholds identity and eternal causation, perceives the self-existence of Truth and Right, and calms itself with knowing that all things go well. Vast spaces of nature, the Atlantic Ocean, the South Sea; long intervals of time, years, centuries, are of no account. . . .

Let us stun and astonish the intruding rabble of men and books and institutions by a simple declaration of the divine fact. Bid the intruders take the shoes from off their feet, for God is here within. Let our simplicity judge them, and our docility to our own law demonstrate the poverty of Nature beside our native riches.[13]

It is, in fact, according to Emerson, an allegiance to the senses and nature, in itself, that *blinds* us to the interior intuition of the Over-Soul and the God within and beyond:

To the senses and the unrenewed understanding, belongs a sort of instinctive belief in the absolute existence of nature. In their view man and nature are indissolubly joined. Things are ultimates, and they never look beyond their sphere [Piaget's egocentric "realism"]. . . . His mind is imbruted, and he is a selfish savage. . . .

The presence of intuition[14] mars this faith [in nature]. The first effort of thought tends to relax this despotism of the senses which binds us to nature as if we were a part of it, and shows us nature aloof, and, as it were, afloat. Until this higher agency intervened [intuition], the animal eye sees, with wonderful accuracy, sharp outlines and colored surfaces. When the eye of intuition opens, to outline and surface are at once added grace and expression. These proceed from imagination and affection, and abate somewhat of the angular distinctions of objects. If the intuition be stimulated to more earnest vision, outlines of surfaces become transparent, and are no longer seen; causes and spirits are seen through them. The best moments of life are these delicious awakenings of the higher powers, and the reverential withdrawing of nature before its God.[15]

At the same time, as Emerson points out, this does *not* mean that nature is apart from Spirit or divorced from Spirit or alien to Spirit—that is a common belief in the mythic structure (Campbell called it "mythic dissociation"), but it finds no place in genuine psychic mysticism. All of nature, every nook and cranny, is in Spirit, bathed by Spirit, awash in Spirit; there is no point in nature that is not totally permeated and enveloped by Spirit.

These distinctions are crucial, because they allow us to distinguish carefully and clearly between three quite different worldviews on the *relation between nature and spirit*:

- The first is *magical indissociation*, where spirit is simply equated with nature (nature = spirit); predifferentiated; very "this-worldly."
- The second is *mythic dissociation*, where nature and spirit are ontologically separate or divorced; very "otherworldly."
- The third is *psychic mysticism*: nature is a perfect expression of spirit (or as Spinoza put it, nature is a subset of spirit);[16] "otherworldly" and "this-worldly" are united and conjoined.

With reference to the third: One of the major and defining characteristics of psychic-level mysticism is that it is a conscious identity with physiosphere and biosphere and noosphere—it does not simply privilege the biosphere; it is no mere geocentric/egocentric indissociation and regression. Even though this mysticism often takes its glorious exultation in the wonders of nature, nonetheless, as Emerson constantly emphasizes, this is "the Self of *nation* and of *nature*"—that is, the mystical union of matter, life, *and* culture, not merely a biospheric immersion. Were it only the Self of nature and not also of nation (culture and morality), then it would be a perfectly regressive, dualistic, and amoralist stance, glorifying merely an egocentric joy in finding oneself vitally reflected in the biosphere (and the rain whispers in its ear, I am here for you).

Rather, it is the union of the human *moral endeavor* with the display of *nature as given* that so distinguishes "nation-nature" mysticism from the narcissistic "nature worship" of mere sentimentalism (the technical points of this argument are given in note 16). This is most definitely *not* an ecological self; it is an Eco-Noetic Self, "The Self of nation and of nature," the Over-Soul that is the World Soul.[17]

Indeed, if nature means the biosphere, and Nature (or Spirit) means the All, means the physiosphere and the biosphere and the noosphere and their Ground, then Emerson's point is very simple: the worshipers of nature are the destroyers of Nature.

Emerson, then, is singing songs to Nature, not nature. And that is why he maintains that nature immersion and nature worship *prevent* the realization of Nature, or the Spirit within and beyond, which transcends all, embraces all. And this is what he means by "nature-nation" mysticism: the biosphere and the noosphere united in the theosphere, or the Over-Soul that is simultaneously the World Soul.

And so he arrives at the very true conclusion: nature worshipers are the destroyers of Nature, the destroyers of Spirit; they would, he says, never *look within* long enough to find the true beyond, the Over-Soul *out of which* both culture and nature emerge (and which therefore lovingly embraces both); they would never look within for Nature, they only stare without at nature, and thus, as he puts it, their minds are imbruted, they remain a selfish savage—geocentric, egocentric.

Just as all of the lower is in the higher but not all the higher is in the lower (but rather "permeates" the lower), so all of nature is in Spirit but not all of Spirit is to be found in nature. Rather, Spirit permeates nature through and through, itself remaining behind nature, beyond nature, not confined to nature and not identified with nature, but *never*, at any point, divorced from nature or set apart from nature. Emerson is precise:

> But when, following the invisible steps of thought, we come to inquire, Whence is matter? and Whereto? many truths arise to us out of the recesses of consciousness. We learn that the highest is present to the soul of man; that the dread universal essence, which is not wisdom, or love, or beauty, or power, but all in one, and each entirely, is that for which all things exist, and that by which they are; that spirit creates; that behind nature, throughout nature, spirit is present; one and not compound it does not act upon us from without, that is, in space and time, but spiritually, or through ourselves: therefore, that spirit . . . does not build up nature *around* us, but *puts it forth through us*, as the life of the tree puts forth new branches and leaves through the pores of the old.[18]

That Spirit does not build up nature around us, but puts forth nature through us: there is the profound difference between nature/nation mysticism and mere biocentric immersion; there is the telling difference between the Eco-Noetic Self and the merely ecological self; there is the difference between transcendence and regression.

Here, then, is a summary of the widely accepted interpretation of Emerson's view: (1) nature is not Spirit but a symbol of Spirit (or a manifestation of Spirit); (2) sensory awareness in itself does not reveal Spirit but obscures it; (3) an ascending or transcendental current is required to disclose Spirit; (4) Spirit is understood only as nature is transcended (i.e., Spirit is immanent in nature, but fully discloses itself only in a transcendence of nature—in short, Spirit transcends but includes nature). Those points are largely uncontested by Emerson scholars [see the *Eye of Spirit*, chapter 11, note 2].

In concluding this brief survey of the psychic level and the cosmic consciousness of nature-nation mysticism, there are two points I would like to emphasize. The first is that, as I think is now obvious, this new *going within* has resulted in a new *going beyond*: a new and higher interior identity (Over-Soul) accompanied by a new and wider embrace of others (World Soul)—a single Soul embracing the physiosphere, biosphere, and noosphere in one loving caress.[19]

Once again we go within and fall without to find this time . . . an actual cosmic consciousness. But this movement itself is in no way any different from all the previous stages that we have examined, all of which were "self-development through self-transcendence," a new going within to a deeper and wider beyond.

And this Over-Soul recognition dawns precisely and only as the separate self, the ego or centaur, is transcended. Schopenhauer would agree entirely with Emerson (and so many others) on that crucial point. In Schopenhauer's words:

> When one is no longer concerned with the Where, the When, the Why and the What-for of things, but only and alone with the What, and lets go even of all abstract thoughts about them, intellectual concepts and consciousness, but instead of all that, gives over the whole force of one's spirit to the act of perceiving, becomes absorbed in it and lets every bit of one's consciousness be filled in the quiet contemplation of the natural object immediately present—be it a landscape, a tree, a rock, a building, or anything else at all; actually and fully *losing oneself* in the object: forgetting one's individuality, one's will, and remaining there only as a pure subject, a clear mirror to the object—so that it is as though the object alone were there, without anyone regarding it, and to such a degree that one might no longer distinguish the be-

holder from the act of beholding, [then] the two have become *one. . . .*[20]

Schopenhauer's "clear mirror to the object" is, of course, Emerson's "transparent eyeball," which is perfectly transpersonal, or no longer merely individual. Schopenhauer: "The person absorbed in this mode of seeing is no longer an individual—the individual has lost himself in the perception—but is a pure, will-less, painless, timeless, Subject of Apprehension." The Over-Soul as intimation of the pure and timeless Witness. . . .

The second thing I would like to emphasize is the relation of the global Self or Over-Soul to the whole notion of morality and moral development itself. This is a connection, it seems to me, overlooked by most of today's moral theorists, but utterly obvious to Emerson and Schopenhauer (and not them alone).

We have seen the development of the *moral sense* evolve from physiocentric to biocentric to egocentric to sociocentric to worldcentric ("worldcentric" being the global or planetary or universalizing reach of rationality and then vision-logic). And here, at the psychic level, the worldcentric *conception* gives way to a direct worldcentric *experience*, a direct experience of the global Self/World, the Eco-Noetic Self, where each individual is seen as an expression of the same Self or Over-Soul.

And what has that to do with morality? Everything, according to Emerson and Schopenhauer, for in seeing that all sentient beings are expressions of one Self, then all beings are treated *as* one's Self. And *that* realization—a profound fruition of the *decentering* thrust of evolution—is the only source of true *compassion*, a compassion that does not put self first (egocentric) or a particular society first (sociocentric) or humans first (anthropocentric), nor does it try merely in thought to act *as if* we are all united (worldcentric), but directly and immediately breathes the common air and beats the common blood of a Heart and Body that is one in all beings.

The whole point of the moral sequence, its very ground and its very goal, its omega point, its chaotic Attractor, is the drive toward the Over-Soul, where treating others as one's Self is not a moral imperative that has to be enforced as an ought or a should or a difficult imposition, but comes as easily and as naturally as the rising of the sun or the shining of the moon.[21]

This moral oneness intensifies in the subtle and causal (as we will see), but it first becomes directly obvious here, in the psychic, and issues naturally in the spontaneous compassion *inherent* in the Over-Soul, a compassion on which all previous moral endeavors depended, but a compassion of which all previous endeavors were but mere and partial glimmers.

In the light of the Over-Soul, it becomes perfectly obvious: all previous ethics were tried and found wanting, all previous struggles for the life Good and True were too partial and too limited and much too narrow to satisfy—all wanted to taste this, the universal compassion through universal identity with the commonwealth of all beings: that I would see in an Other my own Self, with love driving the embrace, and compassion issuing in the tenderest of mercies.

Schopenhauer:

> The sort of act that I am here discussing is . . . compassion, which is to say: immediate participation, released from all other considerations, first, in the pain of another, and then, in the alleviation or termination of that pain, which alone is the true ground of all autonomous righteousness and of all true human love. An act can be said to have genuine moral worth only in so far as it stems from this source [the common Self]; and conversely, an act from any other source has none. The weal and woe of another comes to lie directly in my heart in exactly the same way—though not always to the same degree—as otherwise only my own would lie, as soon as this sentiment of compassion is aroused, and therewith, the difference between him and me is no longer absolute. And this really is amazing—even mysterious.[22]

The mystery, of course, is the mystery of the Over-Soul allowing us to *recognize ourselves in each other*, beyond the illusions of separation and duality. Schopenhauer:

> For if plurality and distinction ["separate selves"] belong only to this world of *appearances*, and if one and the same Being is what is beheld in all these living things, well then, the experience that dissolves the distinction between the I and the Not-I cannot be false. On the contrary: its opposite must be false. The former experience underlies the mystery of compassion, and stands, in fact, for the reality of which compassion is the prime expression. *That* experience, therefore, must

be the metaphysical ground of ethics and consist simply in this: that *one* individual should recognize in *another*, himself in his own true being.[23]

THE SUBTLE LEVEL

At the psychic level, the universalizing and global tendencies of reason and vision-logic come to fruition in a direct experience—initial, preliminary, but unmistakable—of a truly universal Self, common in and to all beings; in a direct experience of the unity of the physiosphere, biosphere, and noosphere, as an expression and embrace of that Self or Soul; so much so that this Self is understood to be prior to, within, and beyond matter, life, and mind, so that, for all the glorious radiance of a Spirit embodied, nonetheless matter and nature and civilization all "withdraw before their God."

At the same time, this is no mere solipsism. It is a higher Self or I, most assuredly, but also a higher Truth (or It) and a wider Community (or We)—the Over-Soul as the World Soul in the commonwealth of all beings as an objective State of Affairs.[24] The Big Three in yet deeper unfoldings, higher reaches, wider communities, stronger affirmations. . . .

At the subtle level, this process of "interiorization" or "within-and-beyond" intensifies—a new transcendence with a new depth, a new embrace, a higher consciousness, a wider identity—and the soul and God enter an even deeper interior marriage, which discloses at its summit a divine union of Soul and Spirit, a union *prior* to any of its manifestations as matter or life or mind, a union that outshines any conceivable nature, here or anywhere else. Nature-nation mysticism gives way to Deity mysticism, and the God within announces itself in terms undreamt of in gross manifestation, with a Light that blinds the sun and a Song that thunders nature and culture into stunned and awestruck silence.

Nature lovers here scream "Foul!," as if beyond the glories of nature there should be no other glory, as if the visible and tangible scene exhausted the wonders of the Kosmos, as if in all the worlds and possible worlds through all eternity, their beloved nature alone should be allowed to shine.

But nature, dear sweet nature, is mortal and finite. It was born, it will remain a bit, and it will pass. It was created, it will be undone. And in all cases, it is bounded, and limited, and doomed to the decay that marks all manifest worlds. "I am somehow receptive of the great soul, and thereby I do overlook the sun and the stars and feel them to be the fair accidents and effects which change and pass," as Emerson said.

We are, of course, perfectly free to identify with nature, and to find a geocentric earth-religion that consoles us in our passing miseries. We are free to identify with a finite, limited, mortal Earth; we are not free to call it infinite, unlimited, immortal, eternal.

That Spirit which is within and beyond the Earth, which is prior to the Earth but not other to the Earth, that Spirit which is source and support and goal of all—that Spirit is intuited at the psychic and comes to the fore in the subtle stage of consciousness evolution, utterly including the previous stages, utterly outshining them. Let the Earth and Cosmos and Worlds dissolve, and see Spirit still shining in the Emptiness, never arising, never dissolving, never blinking once in the worlds of created time. "That joy," says Teresa, "is greater than all the joys of earth, and greater than all its delights, and all its satisfactions; and they are apprehended, too, very differently, as I have learned by experience."[25]

In the *Interior Castle*, one of the truly great texts of subtle-level development, Teresa describes very clearly the stages of evolution of the "little butterfly," as she calls her soul, to its union with the very Divine, and she does so in terms of "seven mansions," or seven stages of growth.

The first three stages deal with the ordinary mind or ego, "unregenerate" in the gross, manifest world of thought and sense. In the first Mansion, that of Humility, the ego is still in love with the creatures and comforts outside the Castle, and must begin a long and searching discipline in order to turn within. In the second Mansion (the Practice of Prayer), intellectual study, edification, and good company strengthen the desire and capacity to interiorize and not merely scatter and disperse the self in exterior distractions. In the Mansion of Exemplary Life, the third stage, discipline and ethics are firmly set as a foundation of all that is to follow (very similar to the Buddhist notion that *sila*, or moral discipline, is the foundation of *dhyana*, or meditation, and *prajna*, or spiritual insight). These are all *natural* (or personal) developments.

In the fourth mansion, a *supernatural* (or transpersonal) grace enters the scene with the Prayer of Recollection and the Prayer of Quiet (which Teresa differentiates by their bodily effects). In both, there is a calming and slowing of gross-oriented faculties (memory, thoughts, senses) and a consequent opening to deeper, more interior spaces with correlative "graces," which Teresa calls, at this stage, "spiritual consolations" (because they are consoling to the self, not yet transcending of the self). On the other hand, it is also as if the soul itself is actually beginning to emerge at this stage: "The senses and all external things seem gradually to lose their hold, while the soul, on the other hand, regains its lost control." And this carries a glimmer of the truth to come, "namely, that God is within us."[26]

In the fifth mansion, via the Prayer of Union, a Spiritual Betrothal occurs, where the soul first directly emerges and intuits Spirit residing in the deepest interior of its own heart (the psychic). I say "emerge," because even if the soul was previously present in the depths, it now comes to the fore.

And this occurs in one particularly significant transformation, according to Teresa. The individual experiences, for the first time, a complete *cessation*[27] of all faculties, and in that pure absorption, the self tastes its primordial union with God (or what Teresa also calls Uncreate Spirit). "For as long as such a soul is in this state, it can neither see nor hear nor understand: the period is always short [at this early stage]. God implants Himself in the interior of that soul in such a way that, when it returns to itself, it cannot possibly doubt that God has been in it and it has been in God."[28]

And here Teresa uses perhaps her most famous metaphor. Prior to this transformative absorption, the unregenerate self (or ego) is, says Teresa, like a silkworm. But one taste of union (literally, just a single experience of this, she says, however brief), and the self is changed forever. One taste of absorption in Uncreate Spirit, and the worm emerges a butterfly. As we might put it, the ego dies and the soul emerges. ("All mean egotism vanishes; the currents of Universal Being circulate through me; I am part or parcel of God.") Teresa:

> And now let us see what becomes of this silkworm. When it is in this state of [cessation/absorption], and quite dead to the world, it comes out a little white butterfly. Oh, greatness of God, that a soul should come out like this after being closely united for so short a time—never, I think, for as long as half an hour [in cessation]. For think of the difference between an ugly worm and a white butterfly; it is just the same here. The soul cannot think how it can have merited such a bless-

ing—whence such a blessing could have come to it, I meant to say, for
it knows quite well that it has not merited it at all.[29]

One taste, and the butterfly is born, the soul is born (or emerges from
its slumber in ego, its lostness in the exterior cocoon of form; and, of
course, the butterfly is the omega point of the silkworm). The rest of
Interior Castle describes the extraordinary journey of this little butterfly
toward that primordial Flame in which, at last, it will happily die (to be,
once again, reborn on yet a deeper level, that of union with Uncreate
Spirit).

In the sixth mansion, Lover and Beloved, butterfly and God, soul and
Uncreate Spirit, "see each other" for extended periods of time. Whereas
the absorption of the Fifth Mansion might last up to a half-hour, various
types of absorption here last a day or several days, she says (even if the
cessation itself is still shortlived). The soul is "so completely absorbed and
the understanding so completely transported—for as long as a day, or
even for several days—that the soul seems incapable of grasping anything
that does not awaken the will to love; to this it is fully awake, while asleep
as regards all attachment. . . ."[30]

But each new stage of growth, we have seen, introduces new types of
possible pathology, and so it is with the little butterfly. Many people think
that the famous "Dark Night of the Soul," a phrase introduced by Teresa's friend and collaborator Saint John of the Cross, is that terrible dark
period before one finds Uncreate Spirit. But not so; the Dark Night occurs
in that period *after* one has tasted Universal Being but before one is established in it, for one has now seen Paradise . . . and seen it fade.

The torment is now agonizing. The little butterfly suffers much, much
more "torture" (Teresa's term) than anything the ego suffers or even
could suffer. "This is a much greater trial," the little butterfly reports,
"especially if the pains are severe; in some ways, when they are very acute,
I think they are the greatest trial that exists. For they affect the soul both
outwardly and inwardly, till it becomes so much oppressed as not to
know what to do with itself. There are many things which assault her
soul with an interior oppression so keenly felt and so intolerable that I do
not know to what it can be compared. . . ."[31]

> She is conscious of a strange solitude, since there is not a creature on
> the whole earth who can be a companion to her—in fact, I do not

believe she would find any in Heaven, save Him Whom she loves: on the contrary, all earthly companionship is torment to her. She thinks of herself as of a person suspended aloft, unable either to come down and rest anywhere on earth or to ascend into Heaven. She is parched with thirst, yet cannot reach the water; and the thirst is not a tolerable one but a kind that nothing can quench. . . .[32]

The dialectic of progress, here unleashed in its most subtle yet agonizing form.

On the more positive side, it is here in the Sixth Mansion that all sorts of subtle-level phenomena begin to emerge in consciousness, and Teresa chronicles them with astonishing clarity: the interior illuminations, the raptures, the subtle sounds and visions, the types of tranquillity and recollection, "ecstasy, rapture, or trance (for I think these are all the same)." Most of these visions (late psychic and early subtle) are in themselves transverbal ("the revelations are communicated to it without words," "in a way that involves no clear utterance of speech").[33] But the central event remains, in each of them, the possibility of the absorption in Uncreate Spirit. "When the soul is thus cleansed, God unites it with Himself. The soul becomes one with God."[34]

All of which culminates in the Seventh Mansion, where actual Spiritual Marriage occurs, and vision gives way to direct apprehension or direct experience—"the union of the whole soul with God."[35] Once this union is apprehended, it seems so obvious, says Teresa, that we can't even find a "doorway" through which it occurred (which is very reminiscent of Zen's "gateless gate"), and, in trying to describe this secret yet perfectly obvious union, words, of course, miserably fail her (but only because she cannot assume that we have had the experience):

> This secret union takes place in the deepest center of the soul, which is where God dwells, and I do not think there is any need of a door by which to enter it. I say there is no need of a door because all that has so far been described [the earlier six stages or mansions] seems to have come through the medium of the senses and [mental] faculties. But what passes in the union of the Spiritual Marriage, in the center of the soul, is very different. This instantaneous [union] of God to the soul is so great a secret and sublime a favor, and such delight is felt by the soul, that I do not know with what to compare it. . . .[36]

In the sixth mansion, Teresa says, this divine union is indeed apprehended, but only briefly and sporadically. She likens this preliminary union to two candles joined at the ends: they then give one light, but the two candles can be broken apart again. Not so the true Spiritual Marriage:

> But here it is like rain falling from the heavens into a river or spring; there is nothing but water there and it is impossible to divide or separate the water belonging to the river from that which fell from the heavens. Or it is as if a tiny streamlet enters the sea, from which it will find no way of separating itself, or as if in a room there were two large windows through which the light streamed in: it enters in different places but it all becomes one.[37]

In this brief sketch, I have mentioned, but have not dwelled on the details of, the possible pathologies that beset the transpersonal stages (four different stages, four very different types of possible pathologies). Suffice it to say that they each involve (as always) problems of differentiation and integration at the new level, problems of agency and communion—too much of one or the other, and a failure of balance: problems of inflating the self at that stage or losing the self in the others of that stage (too much agency or too much communion).[38]

It is crucially important, then, to distinguish these new and higher pathologies—of fulcrums seven (psychic), eight (subtle), and nine (causal)—from the lower and primitive pathologies (particularly fulcrums one, two, and three). Teresa is positively brilliant in distinguishing the agonies of the soul in its higher mansions or stages from those emotional problems that characterize the lower faculties. She clearly distinguishes, for example, three types of "inner voices"—those of "the fancy" or "imagination," which can be hallucinatory and "diseased," she says; those that are verbal, and may or may not represent true wisdom (for they may also be deceptive and "diseased"); and those that are transverbal altogether, representing direct interior apprehension. She has an exquisite and precise discriminating awareness between "fancies" and "hallucinations" and direct intuitive apprehensions, and she explains the differences at length. She gives clear and classic phenomenological descriptions of so many of the subtle-level apprehensions: interior illumination, sound, bliss, and understanding beyond ordinary time and place; genuine archetypal Form as creative pattern (not mythic motif); and psychic vision giving way to pure nonverbal, transverbal, subtle intuition; all summating in the "union of the whole soul with uncreated Spirit."

As for the use of the term "supernatural" by certain contemplatives (both East and West), care should be taken to differentiate what they mean by that term and what, for example, the mythic or religious literalist means by it. Literal or mythic Christianity, for example, originating from the magic-mythic and mythic stages of development, and beset by "mythic dissociation," imagines God as a Cosmic Father set above and apart from nature (ontologically divorced), and thus any action on God's part is and must be "supernatural"—a "miraculous" suspension of the laws of nature on behalf of "His children," activities that are all nonetheless variations on turning spinach into potatoes.

This *dissociation* of "natural" and "supernatural," and a praying, a begging, for the latter to miraculously intervene in the former, Emerson calls "meanness and theft," a vicious craving for commodities:

> In what prayers do men allow themselves! Prayer looks abroad and asks for some foreign addition to come through some foreign virtue, and loses itself in endless mazes of natural and supernatural, and mediatorial and miraculous. Prayer that craves a particular commodity is vicious. [True] Prayer is the contemplation of the facts of life from the highest point of view. It is the soliloquy of a beholding and jubilant soul. It is the spirit of God pronouncing his works good. As soon as the man is at one with God, he will not beg. But prayer as a means to effect a private end is meanness and theft. It supposes dualism and not a unity in nature and consciousness.[39]

God's "supernatural" intervention in "nature": this bears no relation to the contemplative view of the psychic and subtle stages. God or Spirit is not set apart from nature, but rather is the Ground of nature, and indeed of all manifestation—as Teresa puts it, "God is *in all things by presence and power and essence.*" "Supernatural," in this usage, simply means that the *natural* union of Spirit *with all things* becomes a *conscious realization* in some, and that conscious realization is called supernatural, not because the union is present only in them and not in nature, but because they are directly realizing it. Teresa's spiritual friend and collaborator, the extraordinary John of the Cross, explains it thus:

This union between God and creatures always exists. By it He conserves their being so that if the union would end they would cease to exist [Spirit as Ground of Being]. Consequently, in discussing union with God, we are not discussing the substantial union which is always existing, but the union and transformation of the soul in God. This transformation is supernatural, the other natural.[40]

As for this inner transformation, this interior realization, Teresa speaks movingly to her sister nuns, and to us, from the direct experience of the contemplative heart:

It becomes evident that there is "someone" in the interior of the soul who sends forth these arrows and thus gives life to this life, and that there is a sun whence this great light proceeds, which is transmitted in the interior part of the soul. The soul, as I said, neither moves from that center nor loses its peace; [it] leaves the soul in a state of pure spirituality, so that it might be joined with Uncreated Spirit.

Oh, God help me! What a difference there is between hearing and believing these words and being led in this way to realize how true they are! Each day this soul wonders more, for she feels that they have never left her, and perceives quite clearly, in the way I have described, that They [the "true words"] are in the interior of her heart—in the most interior place of all and in its greatest depths.[41]

This new depth, this new within, which is a new beyond, utterly *transcends* nature, utterly *embraces* nature, and is thus *embodied* in nature, as perhaps Aurobindo explained most forcefully:

Its first effect has been the liberation of life and mind out of Matter; its last effect has been to assist the *emergence* of a spiritual consciousness, a spiritual will and spiritual sense of existence in the terrestrial being so that he is no longer solely preoccupied with his outermost life or with mental pursuits and interests, but has learned to look within, to discover his inner being, his spiritual self, to aspire to overpass [negate and preserve] earth and her limitations. As he grows more and more inward, his boundaries mental [noospheric], vital [biospheric], and spiritual begin to broaden, the bonds that held life, mind, soul to their first limitations loosen or snap, and man the mental being begins to have a glimpse of a larger kingdom of self and world closed to the first earth-life.

If he makes the inward movement which his own highest vision has held up before him as his greatest spiritual necessity, then he will find there in his inner being a larger consciousness, a larger life. An action from within and an action from above can overcome the predominance of the material formula, diminish and finally put an end to the power of the Inconscience, substitute Spirit for Matter as his conscious foundation of being and liberate its higher powers to their complete and characteristic expression in the life of the soul embodied in Nature.[42]

Which is why Teresa sings of *embracing all creatures* from the center of the love and joy that now overflows from her innermost being: "These are very unskillful comparisons to represent so precious a thing, but I am not clever enough to think out any more: the real truth is that this joy makes the soul so forgetful of itself, and of everything, that it is conscious of nothing, and able to speak of nothing, save of that which proceeds from its joy. . . . Let us join with this soul, my daughters all. Why should we want to be more sensible than she? What can give us greater pleasure than to do as she does? And may all the creatures join with us for ever and ever. Amen, amen, amen."[43]

And the little butterfly? What has become of her? As the ego (silkworm) died and was reborn as the soul (butterfly), so now the soul, after traversing the psychic and subtle domains and serving its purpose well, enters finally into its Spiritual Marriage, its own omega point, its deeper and greater context, and thus dies to its lesser being, dies as a separate self. "For it is here," she says softly, "that the little butterfly dies, and with the greatest joy, because Christ is now its life."

Which brings us to the causal.

THE CAUSAL

In the subtle level, the Soul and God unite; in the causal level, the Soul and God are both transcended in the prior identity of Godhead, or pure formless awareness, pure consciousness as such, the pure Self as pure Spirit (Atman = Brahman). No longer the "Supreme Union" of God and

Soul, but the "Supreme Identity" of Godhead. As Meister Eckhart put it, "I find in this breakthrough that God and I are one and the same."

As we will see, this pure formless Spirit is said to be the Goal and Summit and Source of all manifestation. And that is the *causal*.

Going *within* and *beyond* even this pure Source and pure Spirit—which is totally formless, boundless, unmanifest—the Self/Spirit awakens to an identity with, and as, *all* Form, all manifestation (gross, subtle, and causal), whether high or low, ascending or descending, sacred or profane, manifest or unmanifest, finite or infinite, temporal or eternal. This is not a particular stage among other stages—not their Goal, not their Source, not their Summit—but rather the Ground or Suchness or Isness of *all* stages, at all times, in all dimensions: the Being of all beings, the Condition of all conditions, the Nature of all natures. And that is the *Nondual*.

I have chosen Meister Eckhart and Sri Ramana Maharshi to illustrate both of these "stages" (causal and nondual), since we find in both of them not only a breakthrough *to* the causal, but also *through* the causal to the ultimate or Nondual, and as inadequate and misleading as words here invariably are, at least an indication of these two "movements" can be clearly and unmistakably discerned in both of these extraordinary sages.

Sri Ramana Maharshi (echoing Shankara) summarizes the "viewpoint" of the ultimate or Nondual realization:

> The world is illusory;
> Brahman alone is Real;
> Brahman is the world.

The first two lines represent pure causal-level awareness, or unmanifest absorption in pure or formless Spirit; line three represents the ultimate or nondual completion (the union of the Formless with the entire world of Form). The Godhead *completely transcends* all worlds and thus *completely includes* all worlds. It is the final within, leading to a final beyond—a beyond that, confined to absolutely *nothing*, embraces absolutely *everything*.

Eckhart begins by pointing to the need, first and foremost, for a transcendence or a "breakthrough" (a word he coined in German) from the finite and created realm to the infinite and uncreated source or origin (the causal), a direct and *formless awareness* that is without self, without other, and without God.

In the breakthrough, where I stand free of my own will and of the will of God and of all his works and of God himself, there I am above all creatures and am neither God nor creature. Rather, I am what I was and what I shall remain now and forever. Then I receive an impulse [awareness] which shall bring me above all the angels. In this impulse I receive wealth so vast that God cannot be enough for me in all that makes him God, and with all his divine works. For in this breakthrough I discover that I and God are one. There I am what I was, and I grow neither smaller nor bigger, for I am an immovable cause that moves all things.

Therefore also I am unborn, and following the way of my unborn [unmanifest] being I can never die. Following the way of my unborn being I have always been, I am now, and shall remain eternally.[44]

In order for God and the Soul to exist, there must be duality or separation between them, and this duality, says Eckhart, obscures the primordial Godhead:

When I still stood in my first cause, there I had no God and was cause of myself. There I willed nothing, I desired nothing, for I was a pure Being in delight of the truth. There I stood, free of God and of all things. But when I took leave from this state and received my created being, then I had a God.[45]

This Godhead (or what Eckhart also calls "God beyond God") is radically free of any finite or created thing, whether of matter or nature or mind or visions or Soul or God. Eckhart refers to this completely transcendental, free, or unmanifest state by words such as "Abyss," "unborn," "formless," "primordial origin," "emptiness," "nothingness."

Empty yourself of *everything*. That is to say, empty yourself of your ego [or any sort of separate-self sense, soul, or oversoul] and empty yourself of all things and of all that you are in yourself and consider yourself as what you are in God. God is a being beyond being and a nothingness beyond being. Therefore, be still and do not flinch from this emptiness.[46]

This "emptiness" is not a theory. Even less is it "poetry" (which I have often heard). Nor is it a philosophical suggestion. It is a direct apprehension (direct "experience" is not quite right, since it is free of the duality of subject and object, and since it never enters the stream of time and thus is never "experiential" in any typical sense)—free of thoughts, free of dualities, free of time and temporal succession:

> I speak therefore of a Godhead from which as yet nothing emanates and nothing moves or is thought about. Even if the soul were to see God insofar as he is God or insofar as God can be imagined or insofar as he is a thought, this same insufficiency would be there. But when all images of the soul are taken away and the soul [is] only the single One, then the pure being of the soul finds resting in itself the pure, formless Being of the divine. . . .
>
> Neither space nor time touch this place. Nothing so much hinders the soul's understanding of God as time and space. Time and space are parts of the whole, but God is one. So if the soul is to recognize God, it must do so beyond space and time. For God is neither this nor that ["*neti, neti*"] in the way of the manifold things of earth, since God is one. If the soul wants to know God, it cannot do so in time. For so long as the soul is conscious of time or space or any other [object], it cannot know God.
>
> Know then that all our perfection and all our bliss depend on the fact that the individual goes *through* and *beyond* all creation and all temporality and all being, and enters the foundation that is without foundation. They must arrive at forgetfulness [of objects] and no-self consciousness—and there must be absolute silence there and stillness.[47]

In this state of formless and silent awareness, one does not *see* the Godhead, for one *is* the Godhead, and knows it from within, self-felt, and not from without, as an object. The pure Witness (which Eckhart calls "the essence of the Subject") cannot be seen, for the simple reason that it is the Seer (and the Seer itself is pure Emptiness, the pure opening or clearing in which all objects, experiences, things and events arise, but which itself merely abides). Anything *seen* is just more objects, more finite things, more creatures, more images or concepts or visions, which is exactly *what it is not*.

> It is free of all names and barren of all forms, totally free and void, just as God is void and free in himself. It is totally one and simple, just

as God is one and simple, so that we can in no manner gaze upon it [see it as an object; it is the Seer, not anything seen; and the Seer is pure Emptiness, out of which seen objects emerge].

There the "means" is silent, for neither a creature nor an image can enter there. The soul knows in that place neither action nor knowledge. It is not aware in that place of any kind of image, either from itself or from any other creature.

You should love him as he is, a not-God, not-mind, not-person, not-image—even more, as he is a pure, clear One, separate from all twoness. You should love God mindlessly, that is, so that your soul is without mind and free from all mental activities, for as long as your soul is operating like a mind, so long does it have images and representations. But as long as it has images, it has intermediaries, and as long as it has intermediaries, it has neither oneness nor simplicity. And therefore your soul should be bare of all mind and should stay there without mind.[48]

Following Saint Dionysius, Eckhart refers to this "mindless" or "unknowing presence," or pure formless awareness without mental intermediaries, as "Divine Ignorance."

Whoever does not leave all external aspects of creatures can neither be received into this divine birth nor be born. The more you are able to bring all your powers to a unity and a forgetfulness of all the objects and images you have absorbed, and the more you depart from creatures and their images, the nearer and more receptive you are. If you were able to become completely unaware of all things, attain a forgetfulness of things and of self, the more [there is] the silent darkness where you will come to a recognition of the unknown, transbegotten God. For this ignorance draws you away from all knowledge about things, and beyond this it draws you away from yourself.[49]

Like Eckhart, Sri Ramana Maharshi, India's greatest modern sage, *begins* by merely giving us some verbal pointers and information about the Self and its relation to God (and Godhead). But he will soon, we will see, go beyond mere chatter and point directly to the unknown and unknowing Source. So here he speaks in "positive" terms, before drawing us into Divine Ignorance.

The Self is known to everyone but not clearly. The Being is the Self. "I am" is the name of God. Of all the definitions of God, none is indeed so well put as the Biblical statement I AM THAT I AM. The Absolute Being is *what* is—It is the Self. It is God. Knowing the Self, God is known. In fact, God is none other than the Self.[50]

And here Ramana clearly means "Godhead," as he himself often pointed out: "Creation is by the entire Godhead breaking into God and Nature."[51]

As for the pure Self/Spirit or Godhead, Ramana constantly repeats, in words virtually identical to Eckhart's (and to the sages of this level the world over) that the Self is not body, not mind, not thought; it is not feelings, not sensations, not perceptions; it is radically free of all objects, all subjects, all dualities; it cannot be seen, cannot be known, cannot be thought. "In that state there is Being alone. There is no you, nor I, nor he; no present, nor past, nor future. It is beyond time and space, beyond expression. It is ever there."[52]

The Self is "not this, not that," which in Sanskrit is the "*neti, neti*" I bracketed in Eckhart's quotation. The Self is not this, not that, precisely because it is the pure Witness of this or that, and thus in all cases transcends any this and any that. The Self cannot even be said to be "One," for that is just another quality, another object that is perceived or witnessed. The Self is not "Spirit"; rather, it is that which, right now, is witnessing that concept. The Self is not the "Witness"—that is just another word or concept, and the Self is that which is witnessing that concept. The Self is not Emptiness, the Self is not a pure Self—and so on.

> There are neither good nor bad qualities in the Self. The Self is free from all qualities. Qualities pertain to the mind only. It is beyond quality. If there is unity, there will also be duality. The numerical one gives rise to other numbers. The truth is neither one nor two. It is as it is.
>
> People want to see the Self as something. They desire to see it as a blazing light, etc. But how could that be? The Self is not light, not darkness, not any observed thing. The Self is ever the Witness. It is eternal and remains the same all along.[53]

Ramana often refers to the Self by the name "I-I," since the Self is the simple Witness of even the ordinary "I." We are all, says Ramana, perfectly aware of the I-I, for we are all aware of our capacity to witness in the present moment. But we *mistake* the pure I-I or pure Seer with some sort of object *that can be seen* and is thus precisely *not* the Seer or the true Self, but is merely some sort of memory or image or identity or self-concept, all of which are objects, none of which is the Witness of objects. We identify the I-I with this or that I, and thus identified with a mere finite and temporal object, we suffer the slings and arrows of all finite objects, whereas the Self remains ever as it is, timeless, eternal, unborn, unwavering, undying, ever and always present.

> The I-I is always there. There is no knowing it [as an object]. It is not a new knowledge acquired. The I-I is always there.
>
> There is no one who even for a trice fails to experience the Self. In deep sleep you exist; awake, you remain. The same Self is in both states. The difference is only in the awareness and the non-awareness of the world. The world rises with the mind and sets with the mind. That which rises and sets is not the Self.
>
> The individual is miserable because he confounds the mind and body with the Self. It is the nature of the mind to wander. But you are not the mind. The mind springs up and sinks down. It is impermanent, transitory, whereas you are eternal. There is nothing but the Self. To abide as the Self is the thing. Never mind the mind. If the mind's source is sought, the mind will vanish leaving the Self unaffected.[54]

The "mind vanished" is, of course, Eckhart's "mindless awareness" (and Zen's "no-mind," etc.). Ramana therefore counsels us to seek the *source* of the mind, to look for that which is aware of the mental or personal "I," for *that* is the transpersonal "I-I," unchanged by the fluctuations of any particular states, particular objects, particular circumstances, particular births, particular deaths.

> Tracing the source of "I," the primal I-I alone remains over, and it is inexpressible. The seat of Realization is within and the seeker cannot find it as an object outside him. That seat is bliss and is the core [the ultimate depth] of all beings. Hence it is called the Heart. The mind now sees itself diversified as the universe. If the diversity is not manifest it remains in its own essence, its original state, and that is the Heart. Entering the Heart means remaining without distractions [ob-

jects]. The Heart is the only Reality. The mind is only a transient phase. To remain as one's Self is to enter the Heart.

The Self is not born nor does it die. The sages see everything in the Self. There is no diversity in it. If a man thinks that he is born and cannot avoid the fear of death, let him find out if the Self has any birth. He will discover that the Self always exists, that the body which is born resolves itself into thought and that the emergence of thought is the root of all mischief. Find the source of thoughts. Then you will abide in the ever-present inmost Self and be free from the idea of birth and the fear of death.[55]

As one pursues this "self-inquiry" into the source of thoughts, into the source of "I" and the "world," one enters a state of pure empty awareness, free of all objects whatsoever—precisely Eckhart's "completely unaware of all things"—which in Vedanta is known as *nirvikalpa samadhi* (*nirvikalpa* means "without any qualities or objects"). In awareness, there is perfect clarity, perfect consciousness, but the entire manifest world (up to and including the subtle) simply *ceases to arise*, and one is directly introduced to what Eckhart called "the naked existence of Godhead." Sri Ramana:

> If you hold to the Self [remain as Witness in all circumstances], there is no second. When you see the world you have lost hold of the Self. On the contrary, hold the Self and the world will not appear.
>
> By unswerving vigilant constancy in the Self, ceaseless like the unbroken flow of water, is generated the natural or changeless state of nirvikalpa samadhi, which readily and spontaneously yields that direct, immediate, unobstructed and universal perception of Brahman, which transcends all time and space.[56]

For Ramana and Eckhart (and not them alone), the causal is a type of ultimate omega point (but it is not, as we will see, the end of the story). As the *Source* of all manifestation, it the *Goal* of all development. Ramana: "This is Self-Realization; and thereby is cut asunder the Knot of the Heart [the separate-self sense]; this is the limitless bliss of liberation, beyond doubt and duality. To realize this state of freedom from duality is the *summum bonum* of life: and he alone that has won it is a *jivanmukta* (the liberated one while yet alive), and not he who has merely a theoretical understanding of the Self or the desired end and aim of all human behavior. The disciple is then enjoined to remain in the beatitude of Aham-Brahman—'I-I' is the Absolute."[57]

THE NONDUAL

Such is the formless causal. But the causal is not, as Eckhart put it, the "final Word." When one breaks through the causal absorption in pure unmanifest and unborn Spirit, the entire manifest world arises once again, but this time as a perfect expression of Spirit and as Spirit. The Formless and the entire world of manifest Form—pure Emptiness and the whole Kosmos—are seen to be not-two (or nondual). The Witness is seen to be *everything* that is witnessed, so that, as Ramana puts it, "The object to be witnessed and the Witness finally merge together [and disappear as separate entities] and Absolute consciousness alone reigns supreme." But *this* nondual consciousness is not other to the world: "Brahman is the World."

This move from causal unmanifest to nondual embrace, Ramana refers to as the development from nirvikalpa to *sahaj samadhi,* which means "unbroken and spontaneously so," a "state" in which "the whole cosmos [Kosmos] is contained in the Heart, with perfect equality of all, for grace is all-pervading and there is nothing that is not the Self. All this world is Brahman."[58]

Meister Eckhart explains both of these movements (utterly transcending the world, utterly embracing it):

First, "Be asleep to all things": that means ignore time, creatures, images [causal]. And then you could perceive what God works in you. That is why the soul says in the Song of Songs, "I sleep but my Heart watches." Therefore, if all creatures are asleep in you, you can perceive what God works in you [as Godhead].

Second: "Concern yourself with all things." This has three meanings. That means, first, seize God in all things, for God is in all things.

The second meaning is: "Love your neighbor as yourself." If you love one human being more than another, that is wrong. If you love your father and mother and yourself more than another human being, that is wrong. And if you love your own happiness more than another's, that is also wrong.

The third meaning is this: Love God in all things equally. For God is equally near to all creatures. And among all these creatures God does not love any one more than any other. God is all and is one. All things become nothing but God [Nondual].[59]

When all things are nothing but God, there are then no things, and no God, but only *this*.

No objects, no subjects, only this. No entering this state, no leaving it; it is absolutely and eternally and always already the case: the simple feeling of being, the basic and simple immediacy of any and all states, prior to the four quadrants, prior to the split between inside and outside, prior to seer and seen, prior to the rise of worlds, ever-present as pure Presence, the simple feeling of being: empty awareness as the opening or clearing in which all worlds arise, ceaselessly: I-I is the box the universe comes in.

Abiding as I-I, the world arises as before, but now there is no one to witness it. I-I is not "in here" looking "out there": there is no in here, no out there, only this. It is the radical end to all egocentrism, all geocentrism, all biocentrism, all sociocentrism, all theocentrism, because it is the radical end of all centrisms, period. It is the final *decentering* of all manifest realms, in all domains, at all times, in all places. As Dzogchen Buddhism would put it, because all phenomena are primordially empty, all phenomena, just as they are, are self-liberated as they arise.

In that pure empty awareness, I-I am the rise and fall of all worlds, ceaselessly, endlessly. I-I swallow the Kosmos and span the centuries, untouched by time or turmoil, embracing each with primordial purity, fierce compassion. It has never started, this nightmare of evolution, and therefore it will never end.

It is as it is, self-liberated at the moment of its very arising. And it is only *this*.

The All is I-I. I-I is Emptiness. Emptiness is freely manifesting. Freely manifesting is self-liberating.

Zen, of course, would put it all much more simply, and point directly to just *this*.

> Still pond
> A frog jumps in
> Plop!

THE END OF HISTORY

Does history, then, have a final omega point, the Omega of all previous and lesser omegas? Is there an actual End to History as we know it? Where all beings unite in their conscious realization of Godhead? Are we all being drawn to that "one, far-off Divine Event" that dissolves its own trail?

Many mystically inclined writers have made this assumption; it does make a certain amount of first-blush sense. From the "Aquarian Conspiracy" to Teilhard's "final Omega-point," from the dawn of a "New Age" to "Timewave Zero"—the millenarian End of History has been exuberantly announced. Such theorists as Terence McKenna and José Arguelles have even been good enough to calculate the actual date of this final omega point, and it is December 2012—"Timewave Zero."[60]

Nor, of course, is this the first time we have seen such "End of History" notions, and in chapter 2 we saw why such notions actually make a certain amount of sense and have some degree of truth to them. To summarize that discussion:

Every senior dimension acts as a transformative omega point for its junior dimension, exerting a palpable pull of the deeper and wider on the shallower and narrower. A holon's regime is the transformative omega-point for its *own* growth and development, facilitated perhaps by morphic resonance from the sum total of similar forms acting as omega. In *self-transcendence*, however, the emergent and senior level exerts omega pull on junior dimensions, something that neither they themselves, nor their morphically resonating partners, could do alone.[61] And short of reaching its immediately senior omega, that lesser dimension suffers the slings and arrows of an outrageous fortune of partialness, division, alienation. Each deeper and wider context condemns the lesser to suffering (or rather, the narrower suffers from the boundaries of its own lacerating limitations). And evolution, in the broadest sense, is a sensitive flight from the pain of partiality.

Each deeper and wider context in the Kosmos thus exerts an omega pull on the shallower and narrower contexts, and when that particular wider depth is reached, that particular omega pull subsides, with the new depth finding that it now exists in a yet-wider and yet-deeper context of its own, which now exerts an unrelenting omega force to once again transcend, to once again embrace more of the Kosmos with care and consciousness.

In short, no holon rests happy short of finding its own immediately deeper context, its own omega point, which means that each holon rushes to the End of its own History.

In the West, since the time of the Enlightenment, the great omega point or End of History has generally been pictured as some form of *rationality* (either formop, as in the classical Enlightenment, or vision-logic, as with Hegel)—and modernity believed this to be the case because that was, indeed, its present state of development, to which all lesser occasions had pointed, and from which rationality had finally and triumphantly emerged. "Remember the cruelties!"—Voltaire's battle cry of the tortures that magic and mythic had inflicted on history, the tortures of lesser omega points brutally cutting into each other in search of . . . reason.

This omega point of rationality can therefore be seen permeating the theories of virtually all developmentalists in the wake of modernity. We see it in Freud: magical and mythic primary-process cognition gives way, after much reluctance and turmoil, to the secondary (mature) process of rationality. We see it in Marx: rationality, as a worldcentric mode of cognition, will, with its economic developments, overcome egocentric and ethnocentric class divisions and usher in a true communion of equally free subjects. We see it in Piaget: preop to conop to formop, with each previous stage suffering the limitations of its own incapacities. Kohlberg and Gilligan: egocentric to sociocentric to worldcentric reason. Hegel: Self-positing Spirit returns to itself in the form of global Reason, the culmination of History itself. And Habermas: mutual understanding in unrestrained communicative action unfolded by rationality is the omega point of individual and social evolution itself.

The list is virtually endless. And, as we discussed earlier, they are all, *as far as they go*, essentially correct in many important ways, and they each can teach us much about expanding the circle and the context of care. (Previously, the mythic structure had said the *same thing* about *itself* in relation to archaic and magic: it had claimed that the coming of the mythic God was an end to all tribal history—and that was also true enough.)

Francis Fukuyama recently caused an international sensation with the publication of *The End of History*, in which he asks "whether, at the end of the twentieth century, it makes sense for us to speak of a coherent and directional History of humankind? The answer I arrive at is yes, for two separate reasons. One has to do with economics, and the other has to do with what is termed the 'struggle for recognition.' "[62]

The "struggle for recognition" is simply the theme, developed from Hegel to Habermas to Taylor, that *mutual recognition*—what we have also been calling the *free exchange* of *mutual self-esteem* among all peoples (the emergence of the rational-egoic self-esteem needs)—is an omega point that pulls history and communication forward toward the free emergence of that mutual recognition. *Short of that emergence*, history is a brutalization of one self or group of selves trying to triumph over, dominate, or subjugate others.

When, on the other hand, human beings universally recognize each other "as beings with a certain worth or dignity," then *history* in that sense "*comes to an end* because the longing that had driven the historical process—the struggle for recognition—has been satisfied in a society characterized by universal and reciprocal recognition. No other arrangement of human social institutions is better able to satisfy this longing, and hence no further progressive historical change is possible."[63] The End of History.

Echoing Hegel, Fukuyama notes that "this does not mean that the natural cycle of birth, life, and death would end, or that important events would no longer happen, or that newspapers reporting them would cease to be published. It means, rather, that there would be no further progress in the development of underlying principles and institutions, because all of the really big questions had been settled."[64]

"The really big questions" would be settled in this sense: once we have arrived at the worldcentric rational structures that both allow and demand (1) free and equal subjects of civil law, (2) morally free subjects, and (3) politically free subjects as world citizens (worldcentric agents in worldcentric communions)—once we have arrived at that, what more, specifically, *could* there be to do in *that* domain? In that domain, "the really big questions would have been settled." And I believe that is indeed true.

We would, of course, continue to fine-tune the ways to implement these freedoms, and help ensure their global equity. And we might, indeed, find *new* freedoms, and new ways to extend the old freedoms. But these three factors would surely form an important platform for any new developments. And to the extent that History up to that point has been the clash of factions that refused those three factors, then this would indeed mark the End of that History.

All of which, as I said, can be true (and is true, I think) and still leave open—and still demand—that further *historical* changes are indeed possible, however much they will build upon the platform of mutual egoic self-esteem and self-recognition—and for the simple reason that there are indeed structures of consciousness *beyond the egoic*, structures that, in their own turn, exert new and subtler omega pulls on the already actualized self-esteem needs. And these new omega pulls will most assuredly *destabilize* the apparently "secure" structure of universal egoic recognition (which will be *preserved*, for sure . . . but also painfully *negated*, in the future paradigm wars set to rock the globe).

In other words, there are indeed some more "really big questions" that would still need to be settled: History would not have ended, only Egoic History.

Which brings us to the topic of the yet deeper or higher structures (beyond the ego), and whether the whole developmental process will ever actually culminate in a genuine and utter End of all possible History. For that is what is implied in the millenarian themes of Timewave Zero: the absolute Omega of all omegas, the End of all ends, the evaporation of disunity, the disappearance of appearance into the utter Abyss, the restaurant at the end of the universe, where the truly Last Supper will finally be served. . . .

Recall that, in figure 5–1, I included the developments in the four quadrants up to vision-logic/centauric/planetary, which is, as it were, the leading edge of the World Soul's evolution at this point in time (or so I would maintain). And thus, at this point, any higher developments, in all four quadrants, have to occur through an individual's own efforts (UL), evidenced in individual bodily transformations (UR), practiced in a microcommunity or sangha of the similarly depthed (LL), with its own microsocial structure (LR).

As these higher potentials begin to emerge collectively—in the decades, centuries, millennia ahead[65]—we can only guess at what the actual surface structures will be like, for none of that is predetermined. These higher structures (psychic, subtle, causal, nondual) are simply potential worldspaces, pre-ontological worldspaces, that are given only as potentials, not as fate.

But we do know that they are *structural potentials* of the human bodymind, because since its emergence in its present form (around fifty thousand years ago), that bodymind has indeed supported realizations across the spectrum (has supported the realizations of a Buddha, a Gaudapada, a Dame Julian, a Lady Tsogyal). In other words, that bodymind has *already* supported psychic, subtle, causal, and nondual realizations; and thus, by a reconstructive science, we know those potentials are already available. This is not an *a priori* Hegelian deduction; it is an *a posteriori* conclusion.

The structural potentials are therefore available; but how they unfold will depend upon the mutual interaction and interplay between *all four quadrants*—intentional, behavioral, cultural, and social—as all four continue to evolve in history, and *none* of that is predetermined in any strong sense. Just as, for example, when the human bodymind with its complex triune brain emerged in its present form (again, around fifty thousand years ago), that brain already possessed the potential (or the hard-wiring) for symbolic logic, but that potential would have to await cultural, social, and intentional developments before it could display its form and function, just so with the higher potentials: how they will unfold remains to be seen. But the fact that they are there is demonstrated by the fact that they have already unfolded in some individuals (Buddha to Krishna), and they are thus already available to any individual, at any time, who chooses to continue his or her own evolution within and beyond.

And so the question remains: given all of that, is there still any sense in which a collective humanity would eventually evolve into an Absolute Omega Point, a pure Christ Consciousness (or some such) for all beings? Are we heading for the Ultimate End of History, the Omega of all omegas? Does It even exist?

And the answer is that It does exist, and we are not heading toward it. Or away from It. Or around It. Uncreate Spirit, the causal unmanifest, is the nature and condition, the source and support, of this and every moment of evolution. It does not enter the stream of time at a beginning or exit at the end. It upholds all times and supports all places, with no partiality at all, and thus exerts neither push nor pull on history.

As the utterly Formless, it does not *enter* the stream of form at *any* point. And yet, as Ramana said, there is a sense in which it is indeed the *summum bonum*, the ultimate Omega Point, in the sense that no finite thing will rest short of release into this Infinity. The Formless, in other words, is indeed an ultimate Omega, an ultimate End, but an End that is never reached *in* the world of form. Forms continue endlessly, ceaselessly, holarchically forever (unless the universe collapses in on itself, retreating back along the path it came—to start anew, one presumes).

Forms continue endlessly, holarchically—holons all the way up, all the way down—the universe as a self-reflexively infinite hall of mirrors. This is why the subtle level, which does indeed act as a manifest omega pull to its lesser and junior dimensions, is said nonetheless to contain literally an infinite number of subtle levels within subtle levels within subtle levels—in billions and billions of other universes!

Thus, in the world of Form, the ultimate Omega appears as an ever-receding horizon of fulfillment (the ever-receding horizon of the totality of manifestation),[66] forever pulling us forward, forever retreating itself, thus always conferring wholeness and partialness in the same breath: the wholeness of this moment is part of the whole of the next moment: the world is always complete and incomplete in any given moment, and thus condemned to a fulfillment that is never fulfilled: the forms rush and run forward to a reward that retreats with the run itself.

But at any sufficiently developed point in an individual's development, a radical leap (Eckhart's "breakthrough") into the Formless can occur. The higher the development, the easier and more likely the jump will occur. Yet the Formless itself is not the result of that jump, nor does it then come to be. It is there, from the start, as one's own Original Face, the Face one had before the Big Bang, the Face that looks out from each and every sentient being in each and every universe, calling out to each and all for mutual Self-, and not just self-, recognition.

Abide as Emptiness, embrace all Form: the liberation is in the Emptiness, never finally in the Form (though never apart from it). And thus even if I realize the *summum bonum*, even if I cut abruptly off the path of endless form and find myself in the Formless, still, still, and still the world of form goes on—into the psychic, into the subtle, into the billions and billions of universes of form available and available and available, endlessly, ceaselessly, dramatically.

Evolution seeks only this Formless *summum bonum*—it wants *only* this ultimate Omega—it rushes forward always and solely in search of *this*—and it will *never* find it, because evolution unfolds in the world of form. The Kosmos is driven forward endlessly, searching in the world of time for that which is altogether timeless. And since it will *never* find it, it will *never cease* the search. Samsara circles endlessly, and that is always the brutal nightmare hidden in its heart.

And the twenty tenets are the form and function, the structure and the pattern, of this endless dream.

BOOK TWO

Among the cultural inventions of mankind there is a treasury of devices, techniques, ideas, procedures, and so on, that cannot exactly be reactivated, but at least constitute, or help constitute, a certain point of view which can be very useful as a tool for analyzing what's going on now—and to change it.

—MICHEL FOUCAULT

Some have declared that it lies within our choice to gaze upon a world of even greater wonder and beauty. It is said by these that the experiments of the mystics are related to the transmutation of the entire Universe. This method, or art, or science, or whatever we choose to call it (supposing it to exist, or to have existed), is simply concerned to restore the delights of Paradise; to enable men and women, if they will, to inhabit a world of joy and splendour. It is perhaps possible that there is such an experiment, and that there are some who have made it. —HAMPOLE

9

The Way Up
Is the Way Down

I T HAS ALWAYS struck me as odd that so much of our Western tradition is supposed to be a series of footnotes to Plato, and yet the
crucial book to which we are all footnotes . . . was never written. "No
treatise by me concerning it exists or ever will exist."

Scholars generally agree that Plato was here referring to the mystical
knowledge of the One, the Good "beyond Being" (as Plato also calls it).
There, apparently, was the heart of the Platonic message, and yet *that* he
never committed to print (whereas he had no trouble writing volumes

on ethics, on the archetypal Forms, epistemology, politics, love, and so forth).

But on that central point, Plato was silent, as silent he could only be. That "knowledge" or "divine ignorance" is not verbal but transverbal; it is not of the mind but of "no-mind"; it is not part of a "discursive philosophy" or merely "talking religion" but a "contemplative flash of truth in the soul." And while this truth or sudden illumination can be directly and injunctively *shown* (by long contemplative practice in community, as Plato put it), it cannot be fully *said* or verbally passed on (without the corresponding developmental signified in any individual). The true footnotes on *that* knowledge would be merely a series of empty circles and fingers pointing to the moon.

I would like to pursue this legacy of Plato for several reasons. One is that, if Western civilization is a series of footnotes to Plato, the footnotes are fractured.

Another is that we see in Plato one of the first clear descriptions of two movements related to the unspoken One, or two "movements" related to Spirit itself (to the extent it can be verbalized at all). The first movement is a *descent* of the One into the world of the Many, a movement that actually creates the world of the Many, blesses the Many and confers Goodness on *all* of it: Spirit *immanent* in the world. The other is the movement of return or *ascent* from the Many to the One, a process of remembering or recollecting the Good: Spirit *transcendent* to the world.

For, as we will see, while Plato emphasized *both* movements, Western civilization has been a battle royale *between* these two movements, between those who wanted only to live in "this world" of Manyness and those who wanted to live only in the "other world" of transcendent Oneness—*both* of them equally and catastrophically forgetting the unifying Heart, the unspoken Word, that integrates both Ascent and Descent and finds Spirit both transcending the Many and embracing the Many.

In Plato, as we will see in a moment, the two movements are given equal emphasis and equal importance, because both were grounded in the unspoken One of sudden illumination. But when that unifying One is forgotten, then the two movements fall apart into warring opposites, into ascetic and repressive Ascenders, on the one hand, who seem willing to destroy "this world" (of nature, body, senses) in favor of anything they imagine as an "other world"; and, on the other hand, the shadow-hugging Descenders, who fuss about in the world of time looking for the

Timeless, and who, in trying to turn the finite realm into an infinite value, end up distorting "this world" as horribly as do the Ascenders, precisely because they want from "this world" something that it could never deliver: salvation.

These two strategies—denying creation, seeing only creation, the Ascenders and the Descenders—have been the two main forms of fractured footnotes to Plato that have plagued Western civilization for two thousand years, and it is with these fractured footnotes that the West (and not it alone) has deeply and cruelly carved its initials on the innocent face of Heaven and of Earth.

And finally, I want to examine Plato's legacy because it is in these two movements—the One descending into the Many, the Many ascending back to the One—that we will find some of the last clues we need to understand the masculine and feminine faces of God, if such indeed exist.

THE TWO LEGACIES OF PLATO

Arthur Lovejoy, in his highly acclaimed book *The Great Chain of Being*, has brilliantly traced out these two opposite and conflicting legacies of Plato—the Descending and the Ascending—which we may also call the "manyness strategy" and the "oneness strategy." The former emphasizes the created world of manyness, the latter the uncreated source or origin—and both, taken in and by themselves, are dualistic through and through, no matter how much they might call themselves monistic, nondual, all-encompassing, holistic, and whatnot.

Drawing on Lovejoy,[1] but not confined to his interpretations, I would like to trace out the history of this dualism, the dualism of which all other Western dualisms are merely an incidental subset, and the dualism that, in just recent times, is showing signs of being overcome.

Most people tend to remember "Plato the Ascender," striving for the "other world" of eternal oneness, and "ignoring" the temporal world of the created many, the mere shadows in the Cave. And, indeed, Plato clearly maps out the Ascending path from the Many to the One. In treatises such as *The Republic* and *The Symposium*, Plato traces the journey of the soul from its infatuation with the material realm of the immediate senses, through the mental realm of higher Forms, to a spiritual immersion in the eternal and unspoken One.

In this Ascent, it certainly appears that Plato is describing—and it is quite certain that his Neoplatonic successors were describing—the general movement that we have called the evolution or development from matter to body to mind to soul to spirit, culminating in the Good "beyond Being" disclosed in "sudden illumination," which is both the *summit* and the *goal* of the soul's journey in time. As Lovejoy puts it, "The Good is the universal object of desire, that which draws all souls toward itself [ultimate Omega]; and the chief good for man even in this life is nothing but the contemplation of this absolute or essential Good."[2]

This Good transcends all possible manifestation—Plato even says "it actually transcends existence. . . . The whole soul must be wheeled round from that which is subject to becoming [the entire world of time and manifestation] until it is able to endure the contemplation of *that which is* . . ." (*The Republic*).

Plato refers to this "contemplation of that which is" as being "the Spectator of all time and existence"—a pure Spectator (Witness) which is itself "not this, not that." Whatever metaphoric or mythic or positive attributes Plato occasionally assigned to the Absolute, there can be little doubt that what he finally had in mind was "*neti, neti,*" or pure transcendental (formless) awareness, the summit and source of all being, and about which, strictly speaking, not even that could be said, as Lovejoy makes very clear:

> The interpreters of Plato in both ancient and modern times have endlessly disputed over the question whether the absolute Good was for him identical with the conception of God. If [God] be taken for the *ens perfectissimum*, the summit of the hierarchy of being, the ultimate and only completely satisfying object of contemplation, there can be little doubt that the Idea of the Good *was* the God of Plato; and there can be none that it became the God of Aristotle, and of most of the philosophic theologies of the Middle Ages, and of nearly all the modern Platonizing poets and philosophers.
>
> The attributes of God were, in strictness, expressible only in *negations of the attributes of this world.* You could take, one after another, any quality or relation or kind of object presented in natural experience, and say, with the Sage in the Upanishad: "The true reality is *not this*, it is *not that*. . . ."[3]

Lovejoy, to his great credit, states clearly that Plato's unspoken One is not primarily a philosophical theory or argument; it is not poetry or myth; and it is not, in the last analysis, a rational argument for Forms—it is a *direct mystical experience* (Lovejoy calls it a "natural religious experience"), cultivated in contemplation among the like-spirited and *transmitted directly* from teacher to student "like a flame kindled by a leaping spark," as Plato said.

Taken in and by itself, this Ascending side of Plato deals, I have suggested, with a mystical or transcendent awareness "beyond Being," beyond manifestation, beyond all qualities (causal/formless), and next to which the entire manifest world appears a shadow, a copy, an illusion. And, indeed, Plato saw the entire manifest world as a pale image of a Reality and Light beyond the Cave of Shadows, the Cave in which the troglodytes are chained, the Cave of fleeting sensory impressions and fluctuating mental opinions.

All Truth, all Goodness, all Beauty was to be found, finally and fully, only in the contemplative absorption in the eternal and unspoken One. No other release is possible; no other release can be desired; and those without this sudden illumination, Plato tells us, are forever lost in the suffering and turmoil of the fleeting and finite and temporal world. No matter how much "beauty" or "joy" or "wonderment" we might find in the manifest realm, it remains only a shadow (Emerson's "symbol") of the Beauty beyond, which is revealed in direct contemplative experience and confirmed by all who have the eyes to see.

This Ascending aspect of Plato, in other words, can be fairly well indicated in the standard realization of causal-level awareness wherever it is found: "The world is illusory (shadowy); Brahman alone is real." Put otherwise: flee the Many, find the One.

And thus was set the standard Ascending Goal of Western civilization. It would become the God of Aristotle and of Augustine, and therefore of virtually all of Christianity, both in popular and esoteric forms (Eastern, Roman, and Protestant). It would become the Goal of the Gnostics and Manichaeists, so much so that any trace of the finite and shadowy world was equated flat-out with *evil*. And when religion itself became passé, the same Goal simply jumped to the Age of Enlightenment, and continued in unabated and undented fashion its ever-ascending yearnings for the light. And for those who found Reason obnoxious, it would nevertheless equally and just as intensely fire every poetic and artistic imagination that found life as given unbearable and eternity a release.

We will return to this history in a moment, but such have been the Ascenders, in just a few of their many guises.

But the absolutely crucial point is that that was only *half* of what Plato set in motion. "Now if Plato had stopped here," says Lovejoy, "the subsequent history of Western thought would, it can hardly be doubted, have been profoundly different from what it has been." Indeed, most of Plato's critics—or rather, those who see him merely as an Ascender, and those who decry his "otherworldliness"—seem to think that he actually did stop here. "But," Lovejoy points out, "the most notable—and the less noted—fact about his historic influence is that he did *not* merely give to European otherworldliness its characteristic form and phraseology and dialectic, but that he also gave the characteristic form and phraseology and dialectic to precisely the contrary tendency—to a peculiarly *exuberant kind of this-worldliness*."[4]

This "exuberant this-worldliness" is what so many of Plato's ecological critics have missed. They set up a straw Plato and then triumphantly, boisterously knock it down, and all congratulate themselves on their this-worldly victory. But Plato is not so easily manhandled.

> For his own philosophy no sooner reaches its climax in what we may call the otherworldly direction [Ascending] than it reverses course. Having arrived at the conception [the "natural religious experience"] of a pure perfection alien to all the categories of ordinary thought and in need of nothing external to itself [causal awareness], he forthwith finds in just this transcendent and absolute Being the necessitating ground of *this world*; and he does not stop short of the assertion of the *necessity* and *worth* of the existence of *all conceivable kinds* of finite, temporal, imperfect, and corporeal beings.[5]

The manifest realm, far from being a world of shadows in the Cave, is now seen as the realm and very embodiment of the Radiance of Spirit itself, suffused with Goodness and with Love. "The world of sense could no longer be adequately described as an idle flickering of insubstantial shadow-shapes, at two removes from both the Good and the Real. Not only did the Sun itself [the causal One] produce cave, and fire, and moving shapes, and the shadows, and their beholders, but in doing so it manifested a property of its own nature not less essential—and, as might well appear, *even more excellent*—than that pure radiance upon which no earthly eye could steadfastly gaze. The shadows were as needful to the Sun as the Sun to the shadows; *their existence was the very consummation of its perfection.*"[6]

And thus their existence was cause for an exuberant this-worldly celebration and embrace!

And so, after establishing an ascending "return to the One," Plato then sets forth a genuinely *creation-centered* spirituality, an effulgence and embrace of the radiant splendor of the Many. The purely Ascended One, far from being a complete Perfection in itself, is viewed as decidedly *inferior* to a One that also flows out of itself and into all manifestation, into all creation. Indeed, for Plato, an Absolute that cannot create is no Absolute at all, and thus true Perfection means a Perfection whose superabundance flows out and into all beings without exception.

In the *Timaeus*—the book that would have such a singular impact on all subsequent Western cosmologies—Plato traces out the creative superabundance and overflowing of the One through a creator God and world of archetypal Forms into humans (mind) and then other living creatures (body) and the world of physical existence (matter).

Thus, as the Neoplatonists would spell it out: from spirit to soul to mind to body to matter—each level is said to be an outflowing of its senior dimension, so that the One Source and Ground of all is reflected in each, to its own degree. The entire universe Plato therefore likens to a giant "superorganism," with all parts interwoven with one another and with their eternal Ground (Plato being an articulate spokesman for Gaia). The entire manifest world—*this world*—Plato calls a "visible, sensible God."

Thus, Plato's Self-Sufficing Perfection is also, and at the same time, a Self-Emptying Fecundity. It is not only the *summit* and *goal* of ascent, it is the *source* and *origin* and fundamental reality of all descent, of all manifestation, of all creation. As Lovejoy puts it, "A timeless and incorporeal One became the ground as well as the dynamic source of the existence of a temporal and material and extremely multiple and variegated universe."[7]

And, Plato leaves no doubt whatsoever, this Spirit-in-Creation (or One-in-Many) is much more complete and full and perfect, as it were, than unmanifest Spirit alone. He implies, and on occasion actually states, that the merely unmanifest Spirit (causal) is in a type of tension or "envy" with manifestation, and that this tension is overcome only as it is seen that all of manifestation, of every possible type and degree in every possible world—all of it is the Radiance of Spirit.

So there is Plato's final stance, and he means it in a very strong way: Spirit is *more* perfect *in* the world than *out* of it. Thus the completion, the all-necessary third line of Nondual realization: "Brahman is the world" (or again, "Nirvana and samsara are not two").

And so it was that Plato at this point united or integrated the path of Ascent with the path of Descent, giving an equal emphasis to the One and to the Many, to nirvana and to samsara. Or, as Lovejoy puts it, *"the two originally distinct strains in Plato's thought are here fused."*[8]

We can therefore summarize Plato's overall position in words that would apply to any Nondual stance wherever it appears (as we have already seen it apply to Eckhart and Ramana): flee the Many, find the One; having found the One, embrace the Many *as* the One.

Or, in short: Return to One, embrace Many. The exuberant and loving and unconditional embrace of the Many is the *fruition* and *consummation* of the Perfection of the One, and without which the One remains dualistic, fractured, "envious."[9]

This integration may be thought of (very crudely and, as always, somewhat misleadingly) as a Great Circle. The descending or manifesting or creative path moves from the top of the circle to the bottom, and the ascending or returning path from the bottom to the top—both arcs traversing the *same dimensions*—which is why, as we will see, "The way up is the way down."

Thus, Descent is not bad, unless taken in and by itself; on the contrary, it is the Creative Source and Matrix of all that is and the fruition of Perfection itself. Likewise, Ascent is not bad, unless taken in and by itself; on the contrary, it is the realization of the Summit and Goal of all that is. The point, we might say, is that the circle of Ascending and Descending energies must always be unbroken: "this world" and the "other world" united in one ongoing, everlasting, exuberant embrace.

WISDOM AND COMPASSION

The Platonic and Neoplatonic traditions (and similar Nondual traditions in the East) therefore maintained that the "Good" or Perfect One is expressed in and as the "Goodness" of all creation. Those two terms—the "Good" and "Goodness"—are extremely important, because we will see these actual terms repeated again and again in subsequent history: The path of Ascent is the path of the *Good*; the path of Descent is the path of *Goodness*.

Indeed, wherever the Nondual traditions would appear—traditions uniting and integrating the Ascending and Descending paths, in the East and in the West—we find a similar set of themes expressed so constantly as to border on mathematical precision. From Tantra to Zen, from the Neoplatonists to Sufism, from Shaivism to Kegon, stated in a thousands different ways and in a hundred different contexts, nonetheless the same essential word would ring out from the Nondual Heart: the Many returning to and embracing the One is Good, and is known as *wisdom*; the One returning to and embracing the Many is Goodness, and is known as *compassion*.

Wisdom knows that behind the Many is the One. Wisdom sees through the confusion of shifting shapes and passing forms to the groundless Ground of all being. Wisdom sees beyond the shadows to the timeless and formless Light (in Tantra, the self-luminosity of Being). Wisdom, in short, sees that the Many is One. Or, as in Zen, wisdom or *prajna* sees that Form is Emptiness (the "solid" and "substantial" world of phenomena is really fleeting, impermanent, insubstantial—"like a bubble, a dream, a shadow," as *The Diamond Sutra* puts it). Wisdom sees that "this world is illusory; Brahman alone is real."

But if wisdom sees that the Many is One, *compassion* knows that the One is the Many; that the One is expressed *equally* in each and every being, and so each is to be treated with compassion and care, not in any condescending fashion, but rather because each being, exactly as it is, is a perfect manifestation of Spirit. Thus, compassion sees that the One is the Many. Or, as in Zen, compassion or *karuna* sees that Emptiness is Form (the ultimate empty Dharmakaya is not other to the entire world of Form, so that prajna or wisdom is the birth of the Bodhisattva and karuna or compassion is the motivation of the Bodhisattva). Compassion sees that "Brahman is the world," and that, as Plato put it, the entire world is a "visible, sensible God."

And it was further maintained, in East and West alike, that the *integration* of Ascent and Descent is the *union* of wisdom (which sees that Many is One) and compassion (which sees that One is Many). The love we have for the One is extended equally to the Many, since they are ultimately not-two, thus uniting wisdom and compassion in every moment of perception. (This is the secret of all tantric spirituality, whether of the East or of the West, as we will see in volume 2.)

We will be going over these details as we proceed; for now, the essential point is simply that either the Ascending or Descending path, taken in and by itself, is catastrophic. Start with Descent or Outflow or Effulgence: creation is not a sin, creation is not evil (as the Ascenders, the Gnostics, the Manichaeists, the Theravadins would all maintain)—creation is not a sin, getting lost in creation is. For at that point, the creatures that *express* the final perfection of Spirit *now* become merely the shadows in the Cave *obscuring* Spirit.

In Plato's version: the soul, once having been grounded in the radiant glories of the unspoken One, "falls away" from the One, through the mental Forms or archetypes and into the material and bodily realm of the senses, and finds itself there, so to speak, lost and confused. In the *Phaedrus*, Socrates speaks of the soul ("uncreate and immortal") as a "winged charioteer" that "taken as a whole . . . traverses the entire universe [Kosmos]; when it is winged and perfect it moves on high and governs all creation." "The region of which I speak," he says, "is the abode of the reality with which true knowledge is concerned, a reality *without color or shape*, intangible but utterly real, apprehensible only by the pilot of the soul."[10] Here souls can make the "circular revolution" through all of manifestation and "back to the starting point" without being caught in it, lost in it, bound by it. And the "two horses" of the chariot—vitality and appetite (representing the body)—are given "ambrosia" and "nectar" to drink—these horses (the body) are no impediment to salvation at all.

But some souls get "carried away," not by the body or the two horses, but by an imbalance or unruliness between the horses and the charioteer (the soul), and the fault lies squarely with the charioteer. They begin to "see parts, but not the whole," Socrates says. "Great is the confusion and struggle and sweat, and many souls are lamed and many have their wings all broken through the feebleness of their charioteers." If the soul can retain its vision of eternal Truth, then, Socrates says, "it can remain unscathed, and if it can continue thus for ever shall be for ever free from hurt"—free, we might say, from the suffering of the round of birth and death, which Socrates speaks of at length ("the great circuit of past lives").

"But when a soul fails to follow and misses the vision, [it] sinks beneath the *burden of forgetfulness. . . .*" And it is this *forgetfulness* of the Origin, and not the existence of manifestation, that constitutes the Fall. "Every human soul by its very nature has beheld true Being. . . . Beauty was once ours to see in all its brightness. Whole were we who celebrated that festival, and whole and unspotted and serene were the objects [creation itself] revealed to us in the light of that mystic vision. Pure was the light and pure were we. . . ."[11]

The Fall is reversed, then, not by erasing creation, but by reversing our forgetfulness, of both the Good of the pure One and the Goodness of the unstained Many. *Recollection* (or remembrance of Source) is thus the path of Return, the path of Ascent. "But," says Socrates, "it is not every soul that finds it easy to use its *present experience* [which is a direct radiance of Spirit] as a means of recollecting the world of reality. It is only by the right use of such aids to recollection, which form a continual initiation into the perfect mystic vision, that a man can become perfect in the true sense of the word." Not only is creation *not* the cause of the Fall, it is the continual *initiation* into the perfect Redemption.

In other words, this world is not a sin; *forgetting* that "this world" is the radiance and Goodness of Spirit—there is the sin.

It is not my intention in these sections to give extensive cross-cultural references to the same themes, but perhaps we should at least notice that, in the East "recollection" or *smriti*, "mindfulness," is the beginning of virtually all paths of contemplation, the aim of which is the *remembering* that one's true nature is Buddha-nature, that Atman is Brahman. And this is most definitely said to be a remembering or a recognition or a recollection of that which one already is, but has simply forgotten or ignored (*avidya* = "ignorance," "forgetting"). Enlightenment or awakening (*bodhi, moksha*) is not a bringing into being of that which was not, but a realizing of that which always already is. ("Do not pretend that by meditation you are going to become Buddha," says Huang Po. "You have always been Buddha but have forgotten that simple fact. Hard is the meaning of this saying!") Likewise, Christ's injunction, "Do this in *remembrance* of me," is a remembrance that "not I but Christ liveth in me" (yet another version of Atman = Brahman).

As Philosophia said to Boethius in his distress, "You have forgotten who you are."

EROS AND THANATOS

That which was dis-membered must be re-membered. This re-membering or re-collecting or re-uniting is the Path of Ascent, which, Socrates says, is driven by Eros, by Love, by the finding of greater and greater union—a higher and wider identity, as we have been putting it. By means of Eros, says Socrates, the lovers are taken out of themselves and into a larger union with the beloved, and this Eros continues from the objects of the body to the mind to the soul, until the final Union is re-collected and re-membered.

Eros, as Socrates (Plato) uses the term, is essentially what we have been calling self-transcendence (tenet 2c), the very motor of Ascent or development or evolution: the finding of ever-higher self-identity with ever-wider embrace of others. And the *opposite* of that was regression or dissolution, a move downward to less unity, more fragmentation (what we called the self-dissolution factor, tenet 2d).

And here I will give one last comparison: Freud, it is well known, finally came to see all psychic life governed by two opposing "forces"—Eros and Thanatos, which are usually referred to as sex and aggression, although that is not the final way that Freud defined them. In *An Outline of Psychoanalysis* Freud gives his final statement: "After long hesitancies and vacillations, we have decided to assume the existence of only two basic instincts, *Eros* and *the destructive instinct.* The aim of the first of these basic instincts is to establish ever greater unities—in short, to bind together." There is no mistaking the meaning of that: it is pure Eros. "The aim of the second is, on the contrary, to undo connections and so to destroy things. In the case of the destructive instinct we may suppose that its final aim is to lead what is living into an inorganic state [matter]. For this reason we also call it the *death instinct.*"

Freud, of course, was severely criticized by virtually everybody, including his own followers, for proposing the death instinct (thanatos), but clearly it is exactly what we have in mind with the self-dissolution factor. It is simply the impulse to move to a lower level in the holarchy, and its final aim is therefore insentient matter, exactly as Freud said.

But the only point that I wish to emphasize is that even Freud—one of the West's greatest psychologists, and certainly its arch-antimystic—found that human misery could be reduced to a *battle* and *disharmony* between the Path of Ascent and the Path of Descent, as it appears in each of us. As is well known, Freud found no solution to the discontent, and remained profoundly pessimistic about the human condition.

Precisely because Freud did not, like Plato, carry Ascent to its conclusion in the One, he had no way whatsoever to unite it with a radiant Descent into the Many. No way, that is, to unite Eros and Thanatos—to unite the way up and the way down—to overcome their eternal strife.[12] Freud clearly and accurately saw Eros; he clearly and accurately saw Thanatos; and perhaps more clearly than anybody in history, he saw that so much human misery is and always will be a battle between the two, and that the only solution to our suffering is a union of Eros and Thanatos—and yet there is precisely nothing Freud could do about it. There he was stranded, and there he left us stranded.

Lacking the unifying Heart, the unspoken One, that joins Ascent and Descent in the everlasting Circle of Redemption and Embrace, Freud simply remained as one of the many, and certainly one of the greatest, of the fractured footnotes to Plato.

PLOTINUS

From Plato the torch of nonduality, the integrative vision, still intact, passed most notably to Plotinus (205–270 CE), who gave it one of the most complete, most compelling, most powerful statements to be found anywhere, at any time, in any form, ancient or modern, East or West. "No other mystical thinker," says William Inge, "even approaches Plotinus in power and insight and profound spiritual penetration." Even Saint Augustine stood back in awe: "The utterance of Plato, the most pure and bright in all philosophy, scattering the clouds of error, has shone forth most of all in Plotinus, who has been deemed so like his master that one might think them contemporaries, if the length of time between them did not compel us to say that in Plotinus Plato lived again."

Whereas other Platonic teachers were often referred to as "divine," the superlative "most divine" was always reserved for Plotinus. And even if we are one of the many moderns to whom "divine" is meaningless, then we still have to come to terms with the fact that, as Whittaker put it, Plotinus is "the greatest individual thinker between Aristotle and Descartes," and that, according to Benn, "no other thinker has ever accomplished a revolution so immediate, so comprehensive, and of such prolonged duration."[13]

Besides being a profoundly original philosopher and contemplative sage, Plotinus was a synthesizing genius of unparalleled proportion, and therein lies much of his importance. Born in Egypt at the beginning of the third century (nobody knows where, exactly, and he never said), he spent much of his formative years in Alexandria, studying with his main teacher, the remarkable Ammonius Saccas ("This is the man I was looking for," he exclaimed after hearing his first lecture; he stayed with him for ten years). From there he went to Rome, where the emperor Gallienus and his wife, Salonina, showed him great favor, and there he taught for the rest of life, attracting such disciples as Amelius and Porphyry (the great Proclus would eventually carry on the lineage). "Simple in his habits," says Porphyry, "though without any harsh asceticism, he won all hearts by his gentle and affectionate nature [this apparently is no exaggeration, for during his entire time in Rome, he made no known enemies, an almost impossible feat], and his sympathy with all that is good and beautiful in the world. His countenance, naturally handsome, seemed to radiate light and love when he discoursed with his friends."[14]

Plotinus, by his own account, had numerous profound experiences of the "all-transcending, all-pervading Godhead,"[15] and this, of course, was the central pillar of his teaching, along with specific contemplative *practices* (injunctions) to actualize this realization in one's own case. In addition to recollection or contemplation (going within and beyond), he describes what amounts to a type of karma yoga like that of the *Bhagavad Gita*: *grounded in Unity, do your duty*, and do so in a thoroughly "this-worldly" fashion. Karl Jaspers writes that "he was called upon as an arbiter in quarrels, but never had an enemy. Distinguished men and women on the point of death brought him their children to educate and entrusted him with their fortunes to administer. Consequently, his house was full of boys and girls."

A house full of orphaned children? And this is the man about whom Jaspers says, very accurately and with much admiration, "No philosopher has lived more in the One than Plotinus"! Clearly, when Plotinus said that the true One embraces the Many, he meant it.

While Plotinus was visiting a country house in Campania, fatal symptoms appeared, and his friend and physician Eustochius was sent for, which apparently took a bit of time, for when Eustochius finally arrived,

Plotinus said only, "I was waiting for you, before that which is divine in me departs to unite itself with the Divine in the universe," whereupon he closed his eyes and ceased breathing.

Alexandria of the third century was extraordinary for the cross-currents of intellectual, philosophical, and spiritual teachings that poured in, literally, from all over the world—there has probably been nothing else quite like it in the history of the West. Clement and Origen (arguably the two most important of the early Church Fathers) were fellow townsmen of Plotinus. In Alexandria, one had direct access to at least the following teachers or their schools: the Goddess cult of Isis, Mithra worship, Plutarch (eclectic Platonism), the Neo-Pythagoreans, the Orphic-Dionysian mysteries, Apollonius of Tyana, the extraordinary Jewish mystic Philo, Manichaeanism, the all-important Stoics, Numenius, the great African novelist Apuleius, much of the Hermetic writings, the Magi, Brahmanic Hinduism, early Buddhism, and virtually every variety of Gnosticism (which had originated in Syria, and eventually found its greatest proponent in Valentinus)—not to mention two of the most important founders of Christianity.

It is only a slight exaggeration to say that Plotinus took the best elements from each school and jettisoned the rest (in fact, at age thirty-nine, he deliberately joined Emperor Gordianus's Eastern campaign in order to familiarize himself with whatever wisdom traditions he might find); and, based on his own profound contemplative experiences, fashioned the whole thing into what can only be called an awesome vision, as coherent as it is beautifully compelling.

(Even Bertrand Russell, fairly jaded in all things spiritual, maintained that we could judge a philosopher according to the True, or the Good, or the Beautiful. The True belonged, presumably, to Bertrand; the Good belonged, he said, to Spinoza. But the Beautiful . . . ah, the Beautiful, that belonged to Plotinus.)

It is with Plotinus that the Great Holarchy of Being receives its first comprehensive presentation, although the notion itself, of course, goes back directly to Plato and Aristotle (and earlier). In this chapter I have been using the simple matter-body-mind-soul-spirit, but figure 9-1 is a more complete version of the holarchy as given by Plotinus. All of the words used in figure 9-1, including those in parentheses (but not in brackets), are taken directly from William Inge. And let me repeat that although we have to write these "levels" down in a "linear" fashion, a better repre-

Absolute One (Godhead)	Satchitananda/Supermind (Godhead)
Nous (Intuitive Mind) [subtle]	Intuitive Mind/Overmind
Soul/World-Soul [psychic]	Illumined World-Mind
Creative Reason [vision-logic]	Higher-mind/Network-mind
Logical Faculty [formop]	Logical mind
Concepts and Opinions	Concrete mind [conop]
Images	Lower mind [preop]
Pleasure/pain (emotions)	Vital-emotional; impulse
Perception	Perception
Sensation	Sensation
Vegetative life function	Vegetative
Matter	Matter (physical)
PLOTINUS	AUROBINDO

Figure 9-1. The Great Holarchy according to Plotinus and Aurobindo[16]

sentation might be a series of concentric spheres, expanding and envelop-ing, and that development through the "expanding spheres" is not a lin-ear or unidirectional affair, but may best be represented as, say, a spiral staircase, with all sorts of ups and downs, but an overall and unmistak-able direction: transcend and embrace. All of these ideas, we will see, are strongly emphasized in Plotinus.

For comparison, I have included the developmental holarchy as given by Sri Aurobindo, generally regarded as the greatest synthesizer of the philosophies and psychologies of India (all of the words are Aurobindo's, including those in parentheses but not brackets). That the greatest synthe-sizers of the West and the East are in such fundamental agreement is, I think, no real surprise.

The higher stages (psychic, subtle, and causal/ultimate) in both Plotinus and Aurobindo are essentially as I described them in the previous chapter. We might simply note, for Plotinus, that in the subtle level (Nous or God), "In knowing God, spirit knows itself" and we find "ourselves and the whole one with God." But that union only finds *its* ground in the causal Godhead "beyond existence and beyond knowledge." To reach this form-less state, says Plotinus, "Strip thyself of *everything*. We must not be sur-prised that that which excites the keenest of longings is *without any form*, even spiritual form."[17]

This emptying of self and objects (and even of spiritual form) results, once again, in formless (nirvikalpa) samadhi, as Inge words it, "the mystical experience of formless intuition." "The Soul," as Plotinus puts it, "will not allow itself to be distracted by anything external [any object], but will ignore them all, as at first by not [getting lost in] them, and then at last by *not even seeing them* [unmanifest absorption]; it will not even know itself; and so it will abide as the One."[18]

But this "divine darkness" or "unknowingness" is not a blank state or trance, not a loss of consciousness but an intensification of it. "The One in Plotinus," says Inge, "is not unconscious, but superconscious; not infra-rational but supra-rational." Plotinus himself seems to have clearly in mind (or in no-mind) that ever-wakeful awareness or Witness that is present even in dream and deep sleep: he says that all objects, even spiritual Beauty, "parade by in front of it," but that it itself remains a *wakefulness beyond Being*," a "sleepless light" which "is inborn and present to us even when we sleep," and "directed not on the future but on the present, or rather, on the eternal Now and the always present." We cannot see it as an object because there is no duality here—"it is not an object but an atmosphere."[19]

The One is not a numerical one, as so many of Plotinus's interpreters imagine. As he patiently but dryly puts it, "the One is not one of the units that make up the number two." To say that "the whole world is one and undivided" would be to miss the point entirely, for that itself is merely a concept, and a dualistic concept at that. The "real" One is the ever-present Wakefulness that is aware of any concept, including "One," but is *not itself* that or any other image, thought, or object, but embraces all, equally and fully, with ever-present Wakefulness.

And thus (to lapse back into language) this "One," being Nothing in itself, can be, and is, the Ground of all that arises. Plotinus speaks of it (metaphorically) as "the fountain of life and the fountain of God, the source of Being, the cause of the Good, the root of Soul. These flow out of the One, *but not in such a way as to diminish it*."[20] Wakefulness is wakefulness, no matter what "parades by," and so Wakefulness is never diminished by any object that it witnesses, and yet is never apart from what it witnesses either, just as the reflections on an empty mirror are never apart from it.

And thus Plotinus can easily make the nondual leap: "Spirit not only engenders all things; *it is all things*"[21] (the all-important third line of the Nondual realization: the Many are illusory; the One alone is Real; the

One is the Many—Brahman is the World). Which brings him right back to the unconditional embrace of this and any world, and a houseful of wayward orphans.

Scholars usually take Plotinus's system (and Aurobindo's and all such similar transpersonal holarchies) to be primarily a form of philosophy or "metaphysics": the various levels, particularly the higher ones, are imagined to be some sort of theoretical constructs that are deduced, logically, or postulated, speculatively, to account for existence and manifestation.

But in fact these systems are, through and through, from top to bottom, the results of actual contemplative apprehensions and direct developmental phenomenology. The higher levels of these systems cannot be experienced or deduced *rationally*, and nobody from Plotinus to Aurobindo thinks they can. However, *after the fact* of direct and repeated experiential disclosures, they can be rationally reconstructed and presented as a "system." But the "system," so called, has been discovered, not deduced, and checked against direct experience in a community of the like-minded and like-spirited (it is no accident that Inge refers to Plotinus's spirituality as being based on "experimental verification"—"faith begins as an experiment and ends as an experience").[22]

Not a single component of these systems is hidden to experience or nestled safely away in a "metaphysical" domain that cannot be checked cognitively with the appropriate tools. There is absolutely nothing "metaphysical" about these systems: they are empirical-phenomenological developmental psychology at its most rigorous and most comprehensive, carried straightforwardly and openly into the transpersonal domains via the experimental instrument of contemplation.[23]

In short, they follow all three strands of valid knowledge accumulation—injunction, apprehension, confirmation/refutation. And one can "dismiss" these higher levels of development only on the same grounds that the Churchmen refused to look through Galileo's telescope: dogmatic stubbornness tells me that there's nothing to see.

THE WAY UP IS THE WAY DOWN

For Plotinus (and Aurobindo), we find that *on the Path of Ascent*—or what Plotinus calls Reflux (return)—each successive level goes beyond and yet subsumes or "envelops," as Plotinus says, its predecessors—the familiar concept of development as successive holons.[24] All of the lower is in the higher, he says, but not all of the higher is in the lower (for in essence it transcends the lower). But all of the higher "pervades" or "permeates" the lower ("there is nothing transcendent that is not also immanent"). For Plotinus, all development is envelopment.

Thus, a typical Plotinian statement is (to use Inge's phrasing), "The World-Soul is not in the world; rather, the World is in it, embraced by it and moulded by it." Or again, "The Soul is not in the Body, but the Body is [in the Soul,] enveloped and penetrated by the Soul which created it." Plotinus means this for each level, and this is, of course, the standard asymmetry of emergence that we have seen at every stage of development. But the net result is that, as Inge puts it:

> Nature presents us with a living chain of being [holarchy], an unbroken series of ascending or descending values. The whole constitutes a harmony, in which each grade is "in" the next above. Each existence is thus vitally connected with all the others, a conception that asserts the right of [all] existences to be what and where they are.[25]

The Path of Ascent or Reflux thus traces, in reverse order, the Path of Creation or Descent or Efflux, for, as Heraclitus had pointed out, "The way up is the way down, the way down is the way up." As we said earlier, both paths traverse the same dimensions. But if that is so, then when did the Path of Descent occur?

Plotinus would answer, it is occurring right now. In this timeless moment, all things issue forth from the Absolute One, ceaselessly. Each grade or dimension of being is a stepped-down version of its senior dimension; each has its ground, its reality, and its explanation in the level above it (thus, in Efflux, destroy any higher dimension and all lower dimensions would lose their ground, but not vice versa; the higher is "in" the lower as its *ground*, *not* as its *component*).[26]

And yet men and women, according to Plotinus, are *unconscious* of any of the levels or *potentials* that are right now available above their own level of *development*. All of those italicized words are given prominence in Plotinus.[27] Indeed, historians have often traced the first clear conception of the unconscious to Plotinus. Although we have a lower unconscious that contains mainly images and fantasies, says Plotinus, all of the *higher levels* of being are for most people merely *potentials* waiting to be *actualized* in their own case and manifested in their own being.

Thus "sin," for Plotinus, is not a "no" but a "not yet"—we have "not yet" realized our true potentials, and so we are given to "sin." Sin is thus overcome not by a new belief but a new growth. An acorn is not a sin; it is simply not yet an oak. And this growth or actualization occurs precisely through a *process* of development, namely, the Path of Ascent, because "the self is not *given* to start with."[28]

A KOSMOS OF EROS AND AGAPE

Most significantly, at each stage of Ascent, according to Plotinus, the lower has to be "embraced" and "permeated," so that Descent and embrace should, if all goes well, *occur with each stage* of Ascent and development (up to one's present level). In Christian terms, *Eros* or transcendental wisdom (the lower reaching up to the higher) has to be balanced with compassion or *Agape* (the higher reaching down and embracing the lower)—at each and every stage.

This general notion—of a multidimensional Kosmos interwoven by Ascending and Descending patterns of Love (Eros and Agape)—would become a dominant theme of all Neoplatonic schools, and exert a profound influence on virtually all currents of subsequent thought, up to (and beyond) the Enlightenment. Through Augustine and Dionysius, it would permeate all of Christianity, in one form or another, from Boethius to Jakob Boehme, from the great Victorine mystics (Hugh and Richard) to Saint Catherine and Dame Julian, from Saint Teresa and Saint John of the Cross to Tauler and Eckhart. Through Nicholas Cusanus and Giordano Bruno it would help jolt the Middle Ages into the Renaissance. Through Novalis and Schelling, it would be the roots of the Romantic and Idealist Rebellion against the flatland aspects of the Enlightenment. It would find its way into Leibniz and Spinoza and Schopenhauer, and make its way to Emerson and James and Jung. Even Locke would operate within its broad framework, though he would collapse the frame considerably. Indeed, when Lovejoy traces the influence of the Great Chain, and refers to it as "the dominant official philosophy of the greater part of civilized mankind through most of its history," the hand of Plotinus lurks, virtually without exception, there in the background.

Thus, to give only a quick example now, the Cambridge Platonists, who would have such an interesting hand in the molding of modernity, had "their roots in the Platonism of the Renaissance, as developed in the fourteenth century by Ficino and Pico. This was a Platonism very influenced by Plotinus. It was a doctrine in which love played a central part; not only the ascending love of the lower for the higher, Plato's *Eros*, but also a love of the higher which expressed itself in care for the lower, which could easily be identified with Christian *Agape*. The two together make a vast circle of love through the universe."[29] The Great Circle, that is, of refluxing Eros (the Many returning to the One: wisdom and the Good) and effluxing Agape (the One becoming Many: Goodness and compassion).

In this general conception (which is how I will use the terms from now on), Eros is the love of the lower reaching up to the higher (Ascent); Agape is the love of the higher reaching down to the lower (Descent). In individual development, one *ascends* via Eros (or expanding to a higher and wider identity), and then *integrates* via Agape (or reaching down to embrace with care all lower holons), so that balanced development *transcends* but *includes*—it is negation and preservation, ascent and descent, Eros and Agape.

Likewise, the love of the Kosmos reaching down *to us* from a higher level than our present stage of development is also Agape (compassion), helping us to respond with Eros until the *source* of that Agape is *our own* developmental level, our own self. The Agape of a higher dimension is the omega pull for our own Eros, inviting us to ascend, via wisdom, and thus expand the circle of our own compassion for more and more beings.

(And not just in the West is Agape stressed. Many tantric and yogic schools—Aurobindo's for example—put prime emphasis on "the descent of the supermind," the agape of the supermind that "comes down" in order to pull us up to an identity with it, so that we then express that agape or compassion for all beings now "in" us; as usual, Agape and Eros are united only in the nondual Heart.)[30]

PHOBOS AND THANATOS

As we will eventually see in much detail, when Eros and Agape are *not* *integrated* in the individual, then Eros appears as *Phobos* and Agape appears as *Thanatos*.

That is, unintegrated Eros does not just reach up to the higher levels and transcend the lower; it alienates the lower, represses the lower—and does so out of *fear* (Phobos), fear that the lower will "drag it down"—always it is the fear that the lower will "contaminate it," "dirty it," "pull it down." Phobos is Eros in flight from the lower instead of embracing the lower. Phobos is Ascent divorced from Descent. And Phobos, we can see, is the ultimate force of all *repression* (a rancid transcendence).

Or, to say the same thing, Phobos is Eros without Agape (transcendence without embrace, negation without preservation).

And Phobos drives the mere Ascenders.

In their frantic wish for an "other world," their ascending Eros strivings, otherwise so appropriate, are shot through with Phobos, with ascetic repression, with a denial and a fear and a hatred of anything "this-worldly," a denial of vital life, of sexuality, of sensuality, of nature, of body (and always of woman).

They are dangerous people, these Ascenders, for the violent hand of Phobos lurks always behind the "love" of the higher that they profess to all and sundry. With tears streaming down the face and upward-turned eyes, these Ascenders are ready to destroy this world—or at the very least, neglect it to death—in order to get to the Promised Land, a land that, however vaguely it might actually be conceived, is clearly enough understood to be *anything but* this land, definitely not *this* world, which is shadows to the core, deceptions in depth, illusions at best, demonic at worse. The Ascenders are destroying this world, because it is the one world they are all certain that they thoroughly despise.

Thanatos, on the other hand, is Descent divorced from Ascent. It is the lower in flight from the higher, compassion gone mad: not just embracing the lower but *regressing* to the lower, not just caressing but remaining stuck in it (fixation, arrest)—cosmic reductionism run amok. And the end game of that reductionistic drive is death and matter, with no connection to Source. Thanatos is Agape in flight from the higher instead of expressing the higher. It preserves the lower but refuses to negate it (and thus remains stuck in it). And as Phobos is the source of repression and dissociation, Thanatos is the source of regression and reduction, fixation and arrest. It attempts to save the lower by killing the higher.

In other words, Thanatos is Agape without Eros.

And Thanatos drives the mere Descenders.

"Away with all other worlds," they joyously proclaim, as their downward-turned eyes rivet on the wonders of multiplicity, and their infinite joy begins the gruesome task of fitting itself into a finite receptacle. "Away with all other worlds," as they match their joy to the shadows, kiss and hug the spokes in samsara's grinding wheel, marry the source of their misery. And their failure to find final release in the Cave, their rage at the finite cage, is turned merely on any of their poor fellows who happen to disagree with them and their love of shadows.

The higher doesn't embrace the lower; the higher is killed in the name of the lower: not Agape, but Thanatos, and the hand of death touches every love that the Descenders profess for all and sundry, tears also streaming down the face with "compassion" written all over it.

They are dangerous people, these Descenders, for in the name of Agape and compassion, otherwise so appropriate, they mistakenly destroy all higher in a frantic attempt to embrace the lower. And more dangerous still: in their attempt to make *this* poor finite world into a world of infinite value—and every Descender does exactly that in a thousand different ways—they slowly, painfully, inevitably, destroy this very world, by placing on it a burden the poor beast could never carry. The Descenders are destroying this world, because this world is the only world they have.

PLOTINUS'S ATTACK ON THE GNOSTICS

The need to balance and unite Ascent and Descent, Eros and Agape, wisdom and compassion, transcendence and immanence—this Nondual integration is the great and enduring contribution of Plotinus, and it will stand always, I believe, as a luminous beacon to all those who tire of the violence and brutality of the merely Ascending or merely Descending trails.

Plotinus was uncompromising with those who wanted to glorify either this world or the other world—they were both missing the point entirely. Each expansion of the self, Plotinus says, brings more and more of the "external world" *into* ourselves; it doesn't shut more and more of it out. World denial—the denial of any existent, for that matter—is for Plotinus the perfect sign of sickness.[31]

Nowhere is this more forcefully seen than in Plotinus's extraordinary attack on the Gnostics, who were archetypal Ascenders, viewing all of manifestation as nothing but shadows, and evil shadows at that. The Gnostics had indeed achieved a causal-level intuition ("The world is illusory, Brahman alone is real"), but they had not broken through to the Nondual ("Brahman is the world"). They thus taught that the world is evil, the body is a tomb, the senses are to be despised. This apparently infuriated the usually even-tempered Plotinus, and he eloquently responded in a passage that justly became world-famous:

> Do not suppose that a man becomes good by despising the world and all the beauties that are in it. They [the Gnostics] have no right to profess respect for the gods of the world above. When we love a person, we love all that belongs to him; we extend to the children the affection we feel for the parent. Now every Soul is a daughter of the [Godhead]. How can *this world* be separated from the *spiritual* world? Those who despise what is so nearly akin to the spiritual world, prove that they know nothing of the spiritual world, except in name. . . .
>
> Let it [any individual soul] make itself worthy to contemplate the Great Soul by ridding itself, through quiet recollection, of deceit and of all that bewitches vulgar souls. For it let all be quiet; let all its environment be at peace. Let the earth be quiet and the sea and air, and the heaven itself waiting. Let it observe how the Soul flows in from all sides into the resting world, pours itself into it, penetrates it and illumines it. Even as the bright beams of the sun enlighten a dark cloud and give it a golden border, so the Soul when it enters into the body of the heaven gives it life and timeless beauty and awakens it from sleep. So the world, grounded in a timeless movement by the Soul which suffuses it with intelligence, becomes a living and blessed being. . . .
>
> It [Spirit/Soul] gives itself to every point in this vast body, and vouchsafes its being to every part, great and small, though these parts are divided in space and manner of disposition, and though some are opposed to each other, others dependent on each other. But the Soul is not divided, nor does it split up in order to give life to each individual. All things live by the Soul *in its entirety* [i.e., ultimately there are no degrees, no levels, but simply pure Presence]; it is all present everywhere. The heaven, vast and various as it is, is one by the power of the Soul, and by it is this universe of ours Divine. The sun too is Divine, and so are the stars; and we ourselves, if we are worth any-

thing, are so on account of the Soul. Be persuaded that by it thou can attain to God. And know that thou wilt not have to go far afield. . . .[32]

Plotinus goes on to point out that those who would find an "other world" apart from "this world" have missed the whole point. There is no "this world" or "other world"—it is all a matter of one's perception. There is not even any "going up" or "coming down." No movement in space takes place. "Spirit and Soul are everywhere and nowhere." We are in "Heaven" whenever "we in heart and mind remember God"; we are "immersed in Matter" whenever we forget God. Same place, different perception. Plotinus explicitly and often states that we *will arrive at the All without change of place.* Those who talk of "this world" or an "other world" have *both* missed the point—they are the Descenders or the Ascenders, not the Whole-Hearted.

William Inge gives an altogether extraordinary summary of the world-view of Plotinus, and it is worth quoting at length (all italics are mine):

> Plotinus conceives the universe as a living chain of being, an unbroken series of ascending and descending values and existences. The whole constitutes a harmony; each grade is "in" the next above; each existence is vitally connected with all others. But those grades which are inferior in value are also imperfectly real, *so long as we look at them in disconnexion* [the "imperfection" and the "inferior value" appear only when *we* disconnect them from Spirit and thus fail to see that each is a perfect expression of the Divine—again, in a nonanthropocentric or "decentered" fashion]. They are characterized by impermanence and inner discord, until we see them in their true relations. Then we perceive them to be integral parts of the eternal systole and diastole [efflux and reflux] in which the life of the universe consists, a life in which there is nothing arbitrary or irregular, seeing that all . . . act in accord with their own nature.
>
> The perfect and timeless life of the Divine Spirit overflows in an incessant stream of creative activity, which spends itself only when it has reached the lowest confines of being [matter], so that every possible manifestation of Divine energy, every hue of the Divine Radiance, every variety in degree as well as in kind, is realized somewhere and somehow.
>
> And by the side of this outward flow of creative energy [Agape and

Efflux] there is another current which carries all the creatures back toward the source of their being. It is this centripetal movement [Eros and Reflux] that directs the active life of all creatures. This aspiration, which slumbers even in unconscious beings [is in all holons as the self-transcending drive], is the mainspring [also] of the moral, intellectual, and esthetic life of humankind.[33]

As we will see in chapter 11, this is the Kosmos of Eros and Agape that would meet the Enlightenment, where the empirical/representational version of rationality (monological or Right-Hand reason) would collapse the Kosmos into a flatland interlocking order of holistic elements, with the embarrassed subject dangling over the flatland holistic world with absolutely no idea how it got there.

Because all of the Kosmos does indeed have some sort of correlates in the Right-Hand dimension, it thus understandably appeared that scientific positivism and naturalism could and would cover all possible bases, because science does indeed register so many significant alterations in the Right-Hand world. In this collapse of the Kosmos (subtle reductionism), the rich vertical and horizontal Holarchy, driven by Eros and Agape, was reduced to a merely horizontal holarchy, driven by surfaces and "its" alone.[34]

The necessary collective evolution from mythic to rational was the Enlightenment's great achievement; the unnecessary collapse of the Kosmos to the holistic flatland was its great and enduring crime.

For it was precisely this nondual Kosmos that broke in two, crippled and fallen, in the coming nightmare of Western spirituality and philosophy and science. The fractured footnotes to Plato began to litter the landscape with their partialities and favored dualisms, and it is now, just now, only now, that we have begun to pick up the pieces.

10

This-Worldly, Otherworldly

There is thus an incessant multiplication of the inexhaustible One and unification of the indefinitely Many. Such are the beginnings and endings of worlds and of individual beings: expanded from a point without position or dimensions and a now without date or duration.

—A. K. COOMARASWAMY, summarizing the essence of both Hinduism and Buddhism

THE GREAT DUALISM of all dualisms, I have suggested, is between "this world" and an "other world." It has infected our spirituality, our philosophy, our science; it runs as equally through the repressive Ascenders who wish only the "other world" of eternal release, as through the shadow-hugging Descenders who want salvation solely in the passing glories of "this world." It slices through every Age of Enlightenment with its upward-yearning Reason and every Romantic reaction that seeks instead to explore every downward-turning darkness and depth. It governs where we seek our salvation, and which "world" we will ignore or destroy in order to get it.

It is the cause of bitter, bitter acrimony between the two camps, with each formally accusing the other of being the epitome and essence of *evil* (literally): The Ascenders accuse the Descenders of being lost in the Cave of Shadows, of being materialists, hedonists, pantheists, reductionists, and "nothing-morists" (they believe in "nothing more" than can be grasped with the senses). To the Ascenders, "this world" is, in form and function, illusory at best, evil at worst—and the Descenders are the primary representatives of that evil.

The Descenders accuse the Ascenders of being repressive, puritanical, life-denying, sex-denying, earth-destroying, and body-ignoring. In their attempts to rise above "this world," the Ascenders have in fact done more to destroy this world and introduce evil into this world than any other force, so it is precisely the "otherworldly" stance, according to the Descenders, that spews evil into *this* world, and the Ascenders are the primary representatives of that evil.

And they are both right. Or, we might say, they are both half right and half wrong—the fractured footnotes to Plato. I can make this clearer by restating the conclusions of the last chapter and introducing our new topics:

According to the Nondual schools (Plato/Plotinus, Eckhart, Vedanta, Mahayana Buddhism, Tantra, etc.), reality—the "real world"—is neither this world nor the other world; although it cannot easily or accurately be *described*, it can directly be *shown*, or directly apprehended in immediate awareness through contemplative practice in community. And if we must speak about it, then we must include at the least the following three points, not because these points actually describe Reality, but because they act as a curb and restraint to our always-inadequate theorizing.

1. *The One is the Good to which all things aspire.* The Absolute is the Summit and Goal of all evolution, all ascent, all manifestation. It provides the motive, the action, the "pull" of all things to actualize their own highest potential, whatever that potential might be. According to Whitehead, this is the God that acts "through gentle persuasion" for all creative emergence. As Aristotle would put it, the Good is the *final cause* toward which all beings, however imperfect, are pulled. The One as the *Good* is the final *omega point* of all ascent and all wisdom. It is the return of the Many to the One. (Taken in and by itself, this, of course, is the God of the Ascenders.)

2. *The One is the Goodness from which all things flow.* It is the Origin and the Source of all manifestation, at all times, in all places. There is a timeless creativity or overflow of the One into the Many. All things, high or low, sacred or profane, yesterday or tomorrow, issue forth from the divine Fountainhead, the Source of all, the Origin of all. The One as *Goodness* is the *first cause* of all causes; it is the *alpha point* of all worlds;

and as such, all worlds express the Goodness, the compassion, the love and superabundance of the Divine. This creative superabundance of the One (as alpha point) is an uncontainable outflowing that results in the Plenitude and variety and multiplicity of this world, so that this world is itself a "visible God," expressing compassion and Goodness through and through, and this world is to be fully embraced as such. It is God as the Many. (Taken in and by itself, this is the God of the Descenders.)

3. *The Absolute is the Nondual Ground of both the One and the Many.* It is equally and both Good and Goodness, One and Many, Ascent and Descent, Alpha and Omega, Wisdom and Compassion. Both Ascending and Descending Paths express profound truths—neither of their truths is denied (in fact, both are strongly asserted). But neither path alone expresses the whole Truth, and neither path alone imbibes fully of Reality. Reality is not just Summit (omega) and not just Source (alpha), but is Suchness—the timeless and ever-present Ground which is equally and fully present in and as every single being, high or low, ascending or descending, effluxing or refluxing.[1]

In the Great Circle of Descent and Ascent, the Nondual can be represented as the paper on which the entire circle is drawn; or again, it can be represented as the center of the circle itself, which is *equidistant to all points* on the circumference ("the center which is everywhere, the circumference, nowhere"). Beings *can* be said to be closer to or farther from the Summit or the Source (that is the meaning of the Great Chain of Being); but no being is closer to or farther from Suchness; there is no "up" or "down." Each individual being is, fully and completely, just as it is, precisely just as it is, the One and the All.

This is why, in nondual Suchness, it is absolutely *not* that each being is a *part* of the One, or participates in the One, or is an aspect of the One. In other words, it is not, as in pantheism, that each is merely a *piece* of the "One," a slice of the pie, or a strand in the Web. "As Plotinus tells us repeatedly, individual spirits are not part of the One Spirit."[2] An individual holon is not part of the One Spirit because each individual holon *is* the One Spirit *in its entirety*—the Infinite, being radically dimensionless, is fully present at each and every point of spacetime (or, as Zen would have it, Emptiness "in its entirety" is completely present in each and every Form: Emptiness is not the sum total of Form, it is the essence of Form; it is therefore present in the sum total as well, but nothing has been added by the totality that was not present in each individuality).

We remember Plotinus saying, "It [infinite Soul/Spirit] gives itself to every point in this vast body, and vouchsafes its being to every part, great and small, though these parts are divided in space and manner of disposition, and though some are opposed to each other, others dependent on each other. But the Soul is not divided, nor does it split up in order to give life to each individual. All things live by the Soul *in its entirety*; it is all present everywhere." The One is fully present in Each. We don't have to add up each finite thing to get a "systems spirit"—each individual thing, just as it is, is fully expressive of Spirit.

This is why the Nondual traditions are so radically *non*instrumental in their value orientations: individual holons do not have value merely because they are *parts* of a great web; they have value because, just as they are, they are perfect manifestations of the primordial Purity of Spirit. In the systems web-of-life spirituality, all things derive their value secondarily, derive their value extrinsically and instrumentally as strands in the extended web, which alone is supposed to be finally "real."

On the contrary, as Plotinus puts it, "Each part of the All is Infinite." Notice three very different things here: the One, the Each, and the All. Each finite thing exists as part of All finite things, but the "spirituality" is not found in the All, but in the One that is fully present in both Each and All. (The web-of-life systems theorists get the All but miss the One.) But because the One is indeed fully present in both Each and All, then additionally, as Plotinus says, "All is each, and each is all, and infinite the glory"[3]—the web of life is there, it's just very partial, and very secondary to the infinite empty Ground that is the Source and Suchness of every part and every whole, regardless of how much we add up the never-ending series of whole/parts (a series that in itself is just Hegel's "bad infinity").

To finish our summary: Reality is Summit (omega), and Source (alpha), and their common ground as Suchness (Nondual).[4] It is in the integration of the Ascending and Descending paths that the Nondual awareness flashes: "The world is illusory; Brahman alone is real; Brahman is the world."

Anything short of that realization degenerates into just the Ascenders (Brahman alone is real) or just the Descenders (Brahman is the world). And I am suggesting that that is not just a theoretical nicety or an esoteric contemplative concern. I am suggesting that that is precisely what happened in the ensuing series of fractured footnotes to Plato that is called Western civilization.

EARTH IN THE BALANCE

It is often said that Plato was the otherworldly philosopher and Aristotle was the this-worldly philosopher. And yet once again, as Lovejoy demonstrates, this is exactly backward.

It is true that Aristotle spent much of his time in this-worldly analysis and thought, but the God of Aristotle, unlike Plato's, is a purely Ascending God. Aristotle's God was the *summit* and highest *goal* of all creation (omega), but was *not* the source or origin or fountainhead of creation (alpha). In fact, Aristotle's God was not in creation at all, except as the final cause of all beings, which means as the goal which they would always strive toward and never reach. Things did not *come from* Aristotle's God, they are only *going toward* it. In other words, Aristotle understood very well and very accurately the Perfect One as the Good, but not at all the manifestation of that One as Goodness or creative Plenitude. Lovejoy:

> The reverse process [that of creative Efflux, described by Plato and extended by Plotinus] finds no place in the system of Aristotle. His God generates nothing. Except for a few lapses into the common fashions of speech, Aristotle adheres consistently to the notion of self-sufficiency as the essential attribute of Deity [causal]; and he sees that it precludes any [relationship to finite beings that would be implied by its creating them]. It is true that this Unmoved Perfection is for Aristotle the cause of all motion and of all the activity of imperfect beings; but it is their final cause only. The bliss which God unchangingly enjoys in his never-ending self-contemplation is the Good after which all other things yearn and, in their various measures and manners, strive.
>
> But the Unmoved Mover is no world-ground; his nature and existence do not explain why the other things exist, why there are so many of them, why the modes and degrees of their declension from the divine perfection are so various. He therefore cannot provide a basis for the principle of Plenitude.[5]

Aristotle, then, quite apart from his extraordinary contributions to "this-worldly" understanding, was at root the West's archetypal Ascender. And thus the first great fractured footnote to Plato. Plato, we have seen, championed both Ascent and Descent, Summit and Source. These two currents—Plato's God both in the world and beyond it, Aristotle's

God only beyond it—would enter the currents of Western civilization with two diametrically opposed agendas: befriend the world, begrudge the world.

A fractured Plato could be called on to support *either view*; Aristotle could be called on *only* for the latter. The weight of opinion, then, was already precariously tipped in favor of the Ascenders. *If* the whole Plato was not evoked, there was precious little left to hold anybody on earth.

And such precisely was the platform—wobbling now between this world and the other world—upon which Western culture was about to be built.

THE UNEARTHLY TRINITY

Plotinus had held the Ascending and Descending currents together admirably, expressing and polishing the original Platonic nonduality.[6] And the nondual school of Plato/Plotinus would have almost certainly carried the day—as the essentially similar Nondual systems of Shankara and Nagarjuna would do for Hinduism and Buddhism, respectively[7]—were it not for one overwhelming and utterly decisive factor: the entrance on the scene of mythic-literal Christianity.

With its mythic-dissociated God ontologically divorced from nature and human nature—what Tillich called "the strict dualism of a divine sphere in heaven and a human sphere on earth"—there was no way to finally *ground* God in *this* world. And thus *no way that a human being's final destiny could be realized in this life, in this body, on this earth.*[8] Thus the other-worldly Ascenders, fractured through and through—and driven in part, as all exclusive Ascenders are, by Phobos (fear of earth, body, nature, woman, sex, and sense)—could and would dominate (although never exclusively rule) the Western scene for a millennium.

But the ironic point is that this mythically ascended God wasn't even a truly ascendent or transcendent God—it wasn't even a Gnostic or causal-level One (except for a very few realizers). It was, through and through, a mythic-level production: a geocentric, egocentric, anthropocentric local volcano god—Yahweh by name—whose true structural colors were shown precisely in the fact that He touched human history only by interfering with it, either to "reward" or "punish" His "chosen" peoples or, more often, to miraculously smite the enemies of His chosen people, or to otherwise spend His time turning spinach into potatoes.

There is simply no mistaking the structure of consciousness that authors such productions, and it has led to some very harsh judgments from the more profound thinkers of the Christian (and Judaic) tradition itself, starting with Clement and Origen (and Philo), all of whom would concur with Paul Tillich: "Things like miraculous interventions of God, special inspirations and revelations are beneath the level of real religious experience. Religion itself is *immediacy* [by which he means precisely the immediacy of basic Wakefulness or pure Presence, which is Spirit *in* us, as Tillich himself makes very clear]. The supernaturalistic heritage about the suspension of the laws of nature for the sake of miracles collapses completely."9

But, as I have tried to make clear, it was not the simple existence of the Roman Christian mythic empire that was problematic. That mythic-rational structure acted as the basic principle of cultural meaning and social integration for virtually a thousand years; it was phase-specifically *appropriate enough*, bringing with it all the advances (and all the disasters) that we discussed in the mythic-rational section of chapter 5, particularly new forms of sociocentric morality and the beginning belief (postconventional) in the *equal* status of *all* citizens under the (Roman) law and in the eyes of God—replete, of course, with all the empire-building and imperialism and military "globalizing" attempts inherent in the mythic-rational structure.

When we discuss the systems of Plato or Plotinus (or Nagarjuna or Shankara), we are speaking of the most advanced or leading edge of consciousness development for that period; I have no expectation that that would or could have become the average-mode consciousness at that point in time, nor do I judge the average-mode consciousness of that time with the standards of today's average mode. It was not the arrival of the Roman Christian mythic-military empire that was the particular problem; it was what the worldview of that average-mode development did to the leading-edge worldview: it pronounced it anathema, condemned it, and condemned it thoroughly, rigorously, even viciously at times (particularly in the Counter-Reformation), and in ways that (with very few exceptions) happened nowhere else in world history on quite that scale.

Every structure of consciousness is suspicious of all higher structures, structures lying within and beyond it, structures that are in fact its own inherent potential, but structures that require a frightening death and re-birth to unfold in each case. And societies, it seems, can be arranged along a continuum of tolerance for those structures that exist higher than the structure of its own principles of social organization and cohesion. The very success, and the constant threats, to the mythic-military Christian empire put tolerance—never a strong point in mythic structures—virtually out of the question. Any outspoken person who evidenced a structure of consciousness higher than the mythic-rational was, correctly enough, viewed as a *political* threat and condemned, in effect, for treason.[10]

The condemnation was often pandemic: the structures of Reason (and science) were condemned because they demanded *evidence* (reason was therefore allowed only in service of Dogma). The psychic level of nature-nation mysticism was condemned because it brought God "too much into" this world, it "dragged God down" from His celestial throne and the Heavenly City above. Subtle-level mysticism was condemned, or at best barely tolerated, because it brought the soul *up too close* to God. And the Church became absolutely apoplectic if anybody expressed a causal-level intuition of supreme identity with Godhead—the Inquisition would burn Giordano Bruno at the stake and condemn the theses of Meister Eckhart on such grounds.

But that was an old story for causal-level Realizers at the hands of mythic believers, starting with Jesus of Nazareth, whose own causal-level realization ("I and the Father are One") would not be treated kindly. "Why do you stone me?" Jesus asks. "Is it for good works?" The pious reply: "No, it is not for good works; it is because you, being a man, make yourself out God." His reply that "we are *all* sons (and daughters) of God" was lost on the crowd, and that realization led him, as it would al-Hallaj and Bruno and Origen and a long line of subsequent Realizers, to a grisly death for both political and religious reasons—it was simultaneously a threat to the state and to the old religion.

Church dogma handled the case of the extraordinary Realizer from Nazareth in a very ingenious way, using all the powers of rationality to prop up the myth. It was true, they granted, that Jesus was one with God (or, as they would later put it, God is one substance with three Persons—Tertullian's *trinitas*—and the Person of Jesus has two Natures: Divine and Human). *But let the causal-level Ascension stop there.* No other person shall be allowed this Realization, even though, as everybody plainly knew at the time, Jesus never made a single remark suggesting that he alone had or could have this Realization, and he explicitly forbade his followers to use the term "Messiah" in reference to him.

But, as many commentators have pointed out, if the Nazarene had in fact realized a Godhead that belongs to all, equally and fully, then there was no way he could be made the sole property of an exclusive mythology. Put bluntly, there was no way to market him. So Jesus was made, not the suffering servant of all humankind, which is all he ever claimed, but the Sole Son of Jehovah, *literally*. In other words, he was tucked downward and seamlessly into the prevailing mythology, and seen as yet another (but much greater) instance of a miraculous and supernatural *intervention in history* to save a new group of *chosen* peoples: those who embraced the Church, the one true way and only salvation for all souls (which meant: the only way for imperial-political cohesion of the mythic empire).

The realization of the Nazarene was thus placed on a pedestal and made an utterly *unique property* of the Church (and not directly a property of the Soul).[11] It should be remembered that at this stage in development, the moral and political spheres (church and state) had not yet been fully differentiated (which is true for all mythological structures—the head of state gains legitimacy, we saw, by claiming mythogenic status, by claiming to be specially connected to, descended from, or one with the gods/goddesses: Cleopatra *is* Isis). As Tillich explains, "This meant that the person who breaks the canonic law of doctrines is not only a *heretic*, one who disagrees with the fundamental doctrines of the church, but he is also a *criminal* against the state. Since the heretic undermines not only the church but also the state, he must be not only excommunicated but also delivered into the hands of the civil authorities to be punished as a criminal."[12]

The Church would produce *many* great philosophers (reason), and *many* great psychic and subtle mystics, but no matter how much these realizers tried to downplay the myths, no matter how much they allegorized them or as-iffed them or interpreted them *away*, there was always the one fundamental dogma that hung like a weight around their attempts to transcend, that crashed down on their shoulders and pinned them to the ground and never but never budged an inch: *the utterly unique and nonreproducible realization of Jesus*.[13]

The Ascension itself was immediately mythologized, following the very old mythic motif of the three-day-dead-and-resurrected lunar consort of the Earth goddess (in the pagan rituals, as well as in the Christianized version, one would "eat the flesh" and "drink the blood" of the consort, thus to participate in its resurrected powers).[14] Individual Christians who shared the proper mythic belief (or faith) would therefore also be resurrected, after death, on Judgment Day, in another world, where their bodies would be reassembled ("Um, excuse me, isn't that my fourth metacarpal you've got there?") to sit forever with Jehovah, His Son, and Company. There was no way for individuals to find enlightenment or ascension *in this life, on this earth.* Any and all who claimed otherwise were thus immediately both heretics *and* criminals.

Again, I have no quarrel with that phase-specific mythic-rational structure and the interpretation that it (necessarily) gave to the Realization of the Adept from Nazareth. It was a crucial component of social integration and cultural meaning at that point in development, and it apparently served its purposes quite well. The problem, rather, was the degree to which and the fury with which this Realization was so thoroughly *reduced* to mythic levels. Rarely has a causal-level realization been translated so dramatically *downward.* Rarely has such a powerful realization been allowed to produce *so few same-level realizations in its followers.*

Not Buddha, not Shankara, not Lao Tzu; not Valentinus, not Numenius, not Apollonius; not Dogen, not Fa-tsang, not Chih-i; not Garab Dorje, not Tsongkapa, not Padmasambhava—none would be so thoroughly *reduced.* It is simply astonishing. Myths would, of course, grow up around all of those realizers, precisely for those who relate to reality in that degree; but their final teaching, causal/nondual, was available to all who embraced the practice, engaged the injunctions, went beyond myth and reason and psychic and subtle, and discovered the Empty Ground in their own case. And to any student who awakened to discover that he or she was actually One with the infinite Ground, in formless *identity,* the reply came back, in all cases: "Congratulations! You finally discovered who you are!"

The reply that came back from the Church was: you shall now be toast.

The peculiar and net effect of all this was that, although the God of the Church was primarily an Ascending God—in this aspect, otherworldly to the core—*there was no way to consummate the Ascension,* not even for the leading-edge few; only Jesus had done that.[15] And while we might to varying degrees "participate" in Christ's nature, there could be no true and whole-bodied Realization and Ascension until after death, at some other time, and certainly in some other world. *This* world is merely a runway for the real takeoff.

Precisely because there was no way to consummate the Ascension in this body, in this life, on this earth, there was *no way whatsoever* for the Way Up to spill over into the Way Down. No way, that is, for causal-level Oneness with the Good to *issue forth* in all-pervading, all-embracing Goodness, grounded thoroughly in this world, resplendent *as* the entire world.[16] That extraordinary reversal that Plato (and all nondual realizers) had made—transcend this world, awaken as this world—could not be called upon as compass.

Precisely because the Ascension could not be consummated, the West was locked into a perpetually frustrated Ascendent yearning—a yearning for a Goal that would *never* be officially allowed, and therefore a perpetual yearning that could *never* be satisfied and let go of—the perpetual itch in the Western psyche that was never allowed to be scratched and then forgotten—the ascendent carrot on a very long stick held above and in front of the collective donkey that assured that the poor beast would always lurch forward and never be allowed to eat.

It was these *frustrated* Ascenders, these frozen Ascenders, who would call on (1) the other-worldly Aristotle, (2) the other-worldly *half* of Plato, and (3) the mythic other world of the Only Begotten and Only Ascended Son of God. And there was the true Holy Trinity, the unearthly Trinity, that carved out the next thousand years of Western culture, a Trinity that, if it never totally ruled the scene, always officially dominated it.

THE TWO GODS

Without direct access to the unifying and unspoken One, the two paths, Ascending and Descending, became logically incompatible and utterly irreconcilable. Since true and full Ascent was blocked (even theoretically), then true and full Descent could be neither directly experienced nor correctly formulated (even theoretically). The One World of Plato/Plotinus fractured into "this world" versus the "other world," and this dualism produced two diametrically opposed and absolutely incompatible World Stances, with two very different (and irreconcilable) views of the "good life," of men and women's ultimate place in the universe, of the object and nature of human destiny, of where self-fulfillment was to be found, of the very ideals that one ought to have in this life, and, most important, of the types of practical goals that one ought to strive to fulfill in this life.

It produced, in fact, two utterly irreconcilable Gods.

It was agreed, by both sides, that the final good for a human being was the *imitatio Dei*—the imitation or contemplation or absorption in God. *But which God?* The ascending God that takes all things back to the One, or the descending God that delights in the diversity of the Many? Without the unifying Heart, they seem utterly incompatible, and therefore do I shun the world and seek only the One, or do I follow the creative God and embrace the Goodness of all creation and find my God fully expressed there?

The Nondual answer has always been: do one, then the other, embracing finally both. As only one example, Aurobindo: "The spiritual transformation culminates in a permanent ascension from the lower consciousness to the higher consciousness, followed by an effective permanent descent of the higher nature into the lower." Likewise, in the Zen oxherding pictures, which express the ten "stages" to enlightenment, the first seven pictures depict an exclusive search for the One and a withdrawal from the world—the ascent to the Good. The eighth picture is an empty circle—pure formless Emptiness (causal). The final picture shows the person entering the marketplace "with open hands"—the embrace of the Many, the effulgent Path of Compassion and "creation-centered" spirituality.

The important point is that what perfectly unites the two paths—the way up and the way down—is the eighth picture, the empty circle, which is not merely a theory of nonduality but the direct realization of formless Emptiness (causal Spirit). Lacking access to that, the two paths have *no direct experiential point of contact*, and they therefore fall merely into theoretically disjointed affairs, one *versus* the other, and there is no way whatsoever to unite them in theory or in fact.

And that, we will see, is precisely what happened.

WESTERN VEDANTA

From Plotinus the One World vision, embracing fully both Ascent and Descent, had passed, virtually unscathed and intact, to the remarkable

Saint Dionysius (who, because he or she was mistakenly thought to be Paul's first Athenian convert, had a profound impact on all subsequent Christian mysticism; the actual author remains unknown; hence the name "pseudo-Dionysius"). The mysticism of the pseudo-Dionysius is virtually identical to that of Plotinus, except now it is fitted quite seamlessly into Christian theology and Christian terminology: the same wine, slightly new casks. There is, of course, the perilous Ascent straight to the formless Godhead, *and* a perfect Descent to a loving embrace of the entire world of the Many—the standard message of all Nondual schools: *transcend* absolutely every single thing in the Kosmos, *embrace* absolutely every single thing in the Kosmos—with choiceless compassion or love. Dionysius:

> Love which works good to all things, pre-existing overflowingly in the Good, moved itself to creation, as befits the superabundance by which all things are created. The Good by being extends its Goodness to all things [without exception!]. For as our sun, *not by choosing or taking thought* but by merely *being*, enlightens all things, so the Good by its mere existence sends forth upon all things the beams of its Goodness.[17]

Saint Augustine (354–430 CE) brilliantly carried on much of the Plotinian tradition, with a few subtle twists and additions. There is much in Augustine's teaching that is clearly nondual and profound in its implications, particularly as concerns the union of Eros and Agape, the utter primacy of introspection for a knowledge of God, the creative power of Amor (love), and the central importance of a transformation of consciousness (a transformation of will) in order to awaken to the Divine.

Most important of all, I believe, is that with Augustine we see the first flowering of a Western form of Vedanta that would, perhaps, have a more profound impact on subsequent Western philosophy (and civilization) than any other single idea, bar none—and which I will try to summarize in a paragraph:

Plato's "Spectator" and Plotinus's "ever-present Wakefulness" are developed by Augustine into a full-fledged conception of the interior Witness (Augustine would say the Soul) as that which cannot, under any circumstances, be doubted. Start with general doubt, says Augustine, and doubt absolutely everything you can. You will find that you can doubt the reliability of logic (it might be wrong), you can even doubt the reality of sense impressions (they might be a hallucination). But even in the most intense doubt, you are aware of the doubt itself; in your *immediate awareness* there is *certainty*, even if it is only a certainty that you are doubting— and you can never shake that certainty. Any truth in the *exterior* world can be doubted, but always there is the *certainty* of interior *immediateness* or basic Wakefulness; and God, said Augustine, lies in and through that basic Wakefulness, whose certainty is never, and can never be, actually doubted.

Thus the existence of God, as the immediacy of self-presence (or basic Wakefulness), is an unshakable certainty. God is even the immediateness of the very presupposition, or actual ground, behind the doubt about God's existence. If you are aware of the thought "I don't believe in God," well, says Augustine, that is already God. Belief in God, and doubt in God, both presuppose God.

Further, in the intimacy and immediateness of presence, there is no subject-object duality. I do not stand back from this immediate presence and look at it; if I did, that would be merely another seen object, not the immediacy in which this seeing occurs. God is the ultimate *prius* of self and of world. "God is seen in the soul. God is in the center of humans, before the split into subjectivity and objectivity. God is not a strange being whose existence or nonexistence one might discuss. Rather, God is our own *a priori*. In God the split between the subject and object, and the desire of the subject to know the object, are overcome. There is no such gap. God is given to the subject as nearer to itself than it is to itself."[18] God, for Augustine, is what you know *before* you know *anything* else, and upon which *everything* else depends, and something that can *never* actually be doubted.

God as the ground, not just of all being, but of our own immediate and primordial awareness—this is the call of Augustine. How similar to the Eastern traditions! "The ordinary mind, just that is the Tao." In Zen, the ordinary mind means the mind in its simple immediacy, before any effort or duality. In Tibetan Buddhism, the ultimate Dharmakaya is *rigpa*, the clarity of the immediacy of present awareness, prior to subject and object. And we already saw Ramana discourse on the impossibility of ever being without the Self: if you think you have not found or seen the primordial Self, the awareness of that lack *is itself* the supposedly lacking Self. These traditions are not saying that you have Buddha-nature but don't know it: you know it but won't admit it.

This Western Vedanta, as I have called it,[19] passed, in its various forms, from Augustine to Descartes, Spinoza, Berkeley, Kant, Fichte, Schelling, Hegel, Husserl, Heidegger, and Sartre, to mention a few. Basic Wakefulness, immediate consciousness, is what we have and all we have; everything else is phenomenal (Kant), or deductive (Descartes), or exterior (Spinoza), or twice-removed (Husserl, Sartre). Spirit, as basic Wakefulness, is not something that needs to be proven, but something that even the existence of doubt *always presupposes* as its own ground. And thus, Spirit (or consciousness or pure Ego or transcendental Self or basic openness) is not something hard to find but rather impossible to avoid.

We have seen that each stage of development is a going within-and-beyond. This is clearly spotted by Augustine, and he states it in such a profound and persuasive form that, according to Charles Taylor, this notion of "inwardness" would literally color an entire civilization. Careful and sober scholars (such as Tillich and Taylor) are simply astonished at what this man wrought. Tillich says Augustine is "the man who is more than anyone else the representative of the West; he is the foundation of everything the West had to say."[20]

In regard to this going within-and-beyond, Augustine picks up the argument from Plotinus that the interior transcends and envelops the exterior, and the interior itself then dissolves in the One, so that, as Gilson summarizes it, Augustine's path is one "leading from the exterior to the interior and from the interior to the superior," which Taylor phrases as "Going within . . . takes me beyond."

This had a profound historical impact because it took the notion of *interiority* present in Plotinus (and hinted at in Plato) and put it fully center stage. The path to ultimate Reality is not outside; it is inside. *Starting* with reason, one goes *within* reason, to the basic immediacy at its base, and that immediacy takes me *beyond* reason to the Ground of the Kosmos itself. So that finally, and ultimately, the Truth is not in me or inside me or egoically locked up in me, but is rather beyond me altogether: the ground of intimate presence opens up beyond me to the timeless and eternal Being of all beings. Going *within* me, I am finally *free* of me: and that is a timeless liberation from the fetters of being *only* me.

Taylor calls this new *emphasis on the within* "radical reflexivity." And, he says, "it is hardly an exaggeration to say that it was Augustine who introduced the inwardness of radical reflexivity and bequeathed it to the West."[21] Indeed, Taylor traces three main aspects of the "inwardness" or "radical reflexivity" characteristic of Western civilization up through modernity and postmodernity (namely, self-control, self-expression, and self-commitment), and finds all three of those roads lead to . . . Augustine (with roots in Plotinus).

This radical reflexivity would play a profound role in the West (as it did in the East) because even in its distorted forms it spoke a certain truth—the superior lies through the interior, the beyond lies through the within. The Enlightenment would continue to take the "turn within" . . . and then forget the beyond. The disengaged ego of the Enlightenment turned within . . . and got stuck there, dangling haplessly over a holistic flatland that had no room for it: not within and beyond, but within and withdrawn. "Knowledge closed in on itself," as Foucault put it, and there was no way to actually reconnect the self with a "holistic cosmos," since *that* cosmos of interlocking exteriors excluded the interior on its own terms (collapse of the Kosmos to the cosmos, or subtle reductionism).

But this enclosing of knowledge, within and withdrawn—going within, getting stuck there, and then describing the external world in interlocking system terms that left out the subject altogether—this is not *truth* for Augustine; this is precisely *evil*. "Evil is when this reflexivity is enclosed on itself"[22]—within and withdrawn. There was no way to get to the *superior* because the *interior* was excluded from the interlocking order of *exteriors*, and thus all that remained was a system of harmonious shadows and a wonderful web that ground up all depth in interwoven surfaces.

But Augustine's immediate impact was in precisely the opposite direction. If the Enlightenment got lost in the interior and was thus stuck with nothing but exteriors, the influence of Augustine up to the Renaissance was on reaching the superior and downplaying, forgetting, ignoring the exteriors altogether. "Since truth is something which we can find only in the interior of the human soul, physics is useless for ultimate truth."[23]

Indeed so, in a sense, but here creeps in Augustine's dualistic legacy. For Plotinus the Descended world of physics and the Ascended world of the Soul were both integral components of the One World. True, we could never reach the Soul or Spirit through a study of physics (in fact, that would blind us to the Soul), but nonetheless physics was part of the visible, sensible God.

Augustine is no ontological dualist, but he does start to offer us a brutal choice: "Everyone becomes like what he loves. Dost thou love the earth? Thou shalt be earth. Dost thou love God? Then I say, thou shalt be God."

And is there no way to be *both*?

For the *Bhagavad Gita*, as for Plotinus, "Even on earth the world is transformed by those whose minds are established in the vision of Oneness," as the *Gita* has it. For Plotinus (and most of the nondual Eastern traditions), one *starts* immersed in the Earth, immersed in Gaia, interlocked in exteriors (egocentric, biocentric, geocentric); one then finds the interior, transcends Gaia, and is immersed in Heaven (Soul, Spirit); which then dissolves the separate self altogether and embraces equally both Heaven and Earth as the glorious radiance of the Divine (this would be the tantric and Mahayana revolution that would overthrow the exclusively Ascending ideal of the yogic and Theravadin gnostics).

But this road Augustine cannot take. The dogmatic belief in the future resurrection of the body put an effective end to that vision. The future resurrection of the body, and the radically unique (and nonrepeatable, nonreproducible) realization of the Nazarene *blocks* the final liberation of the soul in this body, in this life, on this earth. No longer can the Ascending and Descending Paths be held together in the immediacy of this moment. There is a fracture in the face of Love, and a perplexed God now stares blankly in two diametrically opposed directions.

The net effect was: pick one—the way up or the way down. The two could no longer be united, and historically the Ascenders and the Descenders went virtually their separate ways.

In short, with Augustine the tension between the "two Gods" is becoming unbearable. Partly because of his ten years with the dualistic Manichaeans, partly because of his own Gnosticism, and overwhelmingly because of his unflinching embrace of the Church dogma of the "one and only begotten," Augustine feels that no true Ascent can occur in this life, in this body, on this earth—the mystical awareness of God, yes; final liberation in this body, on this earth, no.

For Plotinus, "Ascent" does not mean a change of place or a change in location or a change from "this world" to an "other world." It means a change in perception so that more and more of this world is perceived *as* the other world—more and more of this world is perceived as Perfectly Divine, until there is *only* the Perfectly Divine in *all* perception, "this world" and "that world" being utterly irrelevant, and the way up and the way down meeting *in every single act* of loving and choiceless and nondual awareness.

For all of Augustine's undoubted brilliance—for all of that, he cannot shake his dualistic dogma that this world is merely a preparation for the next world; he is locked into the myth of the *future* resurrection of the body, not its present divinization by a change in perception; and thus even such an ardent devotee of Augustine as Paul Tillich can summarize his view as: "On the one hand, there is the city of God, on the other the city of earth or the devil."[24] This is no spiritual ontology; this is mythic dissociation.

I do not mean to pin this dualism solely on Augustine; it is a mythic dissociation inherent in the worldview which, for other reasons, he felt compelled to embrace. Augustine, I am tempted to say, is actually the very best one can possibly do with the Plotinian system if it must be strained into a mythic worldview. But both Augustine's fans and detractors have pointed to this unstable dualism in this system, and to his "otherworldly" bent, the stamp of *frustrated* Ascent (and therefore crippled Descent).

And if "Augustine is the foundation of everything the West had to say," well then, everything the West had to say was: frustrated Ascent, spiritus interruptus. And thus no true Descent, no true divinization of this Earth, of this body, or of this life.

THE SCHIZOID GOD

Lovejoy, who does not understand the contemplative and experiential nature of this split (between Descended this-worldly and Ascended other-worldly), nonetheless is again the perfect reporter of what did in fact happen, which is that the West was saddled with two absolutely incompatible Gods:

> The one God was the goal of the "way up," of that ascending process by which the finite soul, turning from all created things, took its way back to the immutable Perfection in which alone it could find rest. The other God was the source and the informing energy of that descending process by which Being flows through all levels of possibility down to the very lowest. . . .
>
> There was no way in which the flight from the Many to the One, the quest of a perfection defined wholly in terms of contrast with the created world, could be effectively harmonized with the imitation of a

Goodness that delights in diversity and manifests itself in the emanation of the Many out of the One.[25]

Once true Ascent was denied, true Descent was likewise lost, and their
secret marriage remained deeply a secret. The pessimistic Ascenders
dourly pursued an otherworldly Goal they were assured of never reaching, and the optimistic Descenders giddily embraced a this-worldly creation whose Source they celebrated but never experienced.

And this produced, as I said, two Gods and two World Stances that
were different in every conceivable way. It was not just a difference in
"theoretical" orientations (although that was constantly debated); it was
through and through a difference in paradigmatic injunctions and exemplars, in practical lifestyles, in the very goal of what it meant to be human
and to lead "the good life."

The Ascending program demanded a withdrawal from all "attachment
to creatures." It recommended ascetic, sometimes harsh discipline, always
oriented toward a withdrawal of attention from the senses, from the
body, from the earth, and above all from sexuality (and therefore
woman)—all of which were looked upon as temptations or much worse
(even though all were "theoretically" creations of the Good). That is, they
all became objects of Phobos, not Eros.

This "oneness strategy" was introspective and highly *introverted*; and
wherever it appeared it brought a thoroughgoing worldly *pessimism* (one
of its key features, whether in religion or philosophy) and a *contemptus
mundi*. "My Kingdom is not of this world" and "Lay not your fortune
upon this earth" were its earmarks. Worldly pursuits were ultimately for
the deluded, who could "participate" in the Good through the sacra-
ments, but were otherwise lost in the Cave of Shadows. The "cloistered
life" was the only good life, and the only life that could lead to a true
imitatio Dei.

The Descending program led in precisely the opposite direction, following precisely the opposite God. It summoned men and women to participate, in some finite measure, in the creative passion of God, and to collab-
orate consciously in the processes by which the *diversity* of things, the
Plenitude and Fullness of the universe, is achieved. It found *its* beatific
vision in the joy of beholding the splendor of creation; or in excitedly
tracing out the detail of its infinite variety. It placed the active life above
contemplation. It was altogether *extroverted*, and brought always in its

wake an *optimism* that would over and over again be codified in the phrase "This is the best of all possible worlds" (meaning that it all comes straight from God and expresses Goodness).[26]

And as for its *imitatio Dei*, it often viewed the activity of the creative artist (in its many forms) as the mode of human life *most like the Divine*, because the artist participates in the creative process, a process that is similar to, or even one with, the Superabundance and Effulgence of Spirit that brings forth the glories of the Many from the One. And yet, so driven was this path to go outward and outward that it fell constantly into the clutches of downward and downward: not Agape, but Thanatos; not an embrace of the lower by the higher, but a simple embrace of the lower, period—with all height and all depth reduced, flattened, leveled, extinguished.

These two Gods wrestled for the soul of Western humanity for the next thousand years (and, in fact, have done so right up to today, as we will see). Since these Gods were, at this point, utterly incompatible, the Church officially had to choose one. Under sway of the unearthly Trinity, it chose, of course, the former, the path of (frustrated) Ascent.

> It was the Idea of the Good [Ascending], not the conception of a generative Goodness [Descending], that determined the ethical teaching of the Church (at least in her counsels of perfection) and shaped the assumptions concerning man's chief end which dominated European thought down to the Renaissance, and in orthodox theology, Protestant as well as Catholic, beyond it. The "way up" alone was the direction in which man was to look for the good. . . .[27]

Such was the thousand-year reign of the mythically Ascended God, about which, in this short survey, we need say not much more. I have wished only to suggest its origin and its internal structure, what it accomplished (which was much to the good), and what it tore asunder (which was equally far-reaching).

Because then, most interesting of all: starting with the Renaissance and running through the Enlightenment, there occurred what we might call "the great reversal." Suddenly, very suddenly, the Ascenders were out, the Descenders were in—and the transition was bloody, arguably the bloodiest cognitive transformation in European history.

THE GREAT PLENITUDE

In order to understand the coming dominance of the Descenders, it is necessary to understand how the nondual system of Plato/Plotinus viewed the Descending Path of Efflux or Plenitude, which was only *half* of the equation, but which may be summarized in the modern phrase "Biodiversity is good."

And not just biodiversity. Diversity per se was the direct evidence of the extraordinary Superabundance and Goodness of the One, and therefore *the more diversity, the greater the Goodness.*

I emphasize that phrase because it is, we will abundantly see, the key to the God of the Descenders (which itself was "half" of the God of Plotinus, very crucial but very partial).

In the *Timaeus* (the intellectual West's Bible of Effulgence), Plato had already maintained that we must *not* suppose that "the world was made in the likeness of any Idea that is merely partial; for nothing incomplete is beautiful. We must suppose rather that it is the perfect image of the whole of which all animals—both individuals and species—are parts. For the pattern of the universe contains within itself the intelligible forms of all beings just as this world comprehends us and all other visible creatures." Thus, creative Effulgence resulted in *"this world* like the most fairest and most perfect of intelligible beings—one visible living being containing within itself all other living beings of like nature"—which is exactly why Plato referred to the entire manifest universe as "a visible, sensible God" and a "superorganism." In order, he says, "that the Whole may be really All, the universe was filled completely" and thereby became "a visible God—the greatest, the best, the fairest, the most perfect."[28]

Thus, in this view, the existence of *all possible kinds* of beings, no matter whether *we* judge them to be good or bad, higher or lower, better or worse—all are an integral and necessary part of the Kosmos, and thus each has *intrinsic* worth just as it is. Put bluntly, the world is better, the more things it contains.

Plotinus, carrying on this nondual vision, says that we are like people who do not understand art and would thus like to eliminate certain colors from a painting simply because we judge them "the less beautiful," not realizing that that would destroy the whole painting at the same time. "To eliminate any characters," says Plotinus, "would be to spoil the beauty of the whole; and it is by means of them that it becomes complete." Thus, he says, those who would "eliminate from the universe" what they think of as "inferior beings" would simply "eliminate Providence itself," whose nature it is to "produce all things and to diversify all in the manner of their existence."[29]

Lovejoy demonstrates that wherever the Descending current was acknowledged—under the general name of the principle of Plenitude—it was always of the form: the Goodness of the One is reflected in the "variety of its parts"; the *greater the diversity, the greater the Goodness.* Maximization of diversity is, so to speak, what Plenitude seeks.

We will see, in the next chapter, what an extraordinary influence this doctrine of Plenitude had in the rise of modernity, from its sciences to its politics to its ethics (and, of course, it is still loudly echoed by theorists from multiculturalists to ecophilosophers, most of whom evidence no idea of its original source).

Again, the easiest way to summarize Plenitude (or Efflux) is to say, not just that diversity is good for *us* humans—a very geocentric, anthropocentric, egocentric stance—but that diversity is the very Goodness of the One itself, quite apart from what it may or may not have to do with humans. This was part of Plotinus's devastating attack on the Gnostics, who, viewing humans as the top of the scale of being, in effect ignored and devalued the rest of the scale itself, and thus could not see "that the sun is Divine, and the earth Divine, and all manner of beings Divine."

This realization came, as I said, directly from Plotinus's radically *decentered* and nonanthropocentric stance. He does not develop biodiversity because it is good for humans, or because human existence depends upon it (although he recognizes that), or because it's nice to go out and have a "wilderness experience"—all of those, for Plotinus, are anthropocentric to the core, and serve only the ego in men and women, not the Divine in each and all.

On the Ascending side, according to Plotinus: as long as we are still Ascending and have not yet actually *discovered* the formless One, which is *equally* present in all beings—*then* the world indeed *appears* as a great hierarchy of beings, from the lowest to the highest, a scale kept in place *by our own ignorance,* but a scale or ladder that must *first* be completely climbed before we can throw it away; those who throw it away without climbing it are merely the Descenders, the troglodytes who worship the Shadows without seeing the Light.

And the central point of this Great Holarchy or Great Chain of (Apparent) Being was summarized in the phrase: *there are no gaps in nature.* Although there were indeed (apparent) grades of existence, from the lowest to the highest, each shaded into the other with "the least degree of possible difference" (Aristotle). No "links" in the Great Chain could be left empty, or the universe would not be "Full," would not express the *plenum formarum* that delights in diversity.

And thus, precisely because the universe was a great holarchy, completely "Full" as an expression of the Goodness of Spirit, each "level" or "grade" was perfectly continuous with both its senior and junior dimensions (Aristotle: "Things are said to be continuous whenever there is one and the same limit of both wherein they overlap and which they possess in common")—there were no gaps, no holes, no missing links in the universe: the Kosmos was one tightly woven fabric of Goodness.

"If there is between two given natural species a theoretically possible intermediate type, that type must be realized—and so on *ad indefinitum*; otherwise, there would be gaps in the universe, the creation would not be as 'full' as it might be, and this would imply that its Source was not 'good,' in the sense of the *Timaeus*."[30] This was not primarily a logical deduction, as Lovejoy supposes, but a contemplative experience of a superabundance that, like water, would overflow into any and every nook and cranny, leaving nothing dry—hence, "there are no gaps in nature."

At the same time, this did not prevent Plotinus (and his nondual descendents) from also seeing that "nature makes leaps everywhere." A higher level, recall, was defined (by Aristotle) as one that had the essentials of the lower *plus* an extra something (an emergent) that defined the new level and was not present in the junior dimension. But the idea was that, at some point, they shared a common boundary, so that nothing was radically disjointed in the universe, even though it was everywhere shaded or graded. Either point taken alone—"no gaps" versus "leaps everywhere"—would lead to an unbalanced view.[31]

The upshot of all this was that the manifest universe was viewed as a huge organic Totality, graded in terms of levels of diversity, but with each and all an integral link in the Kosmos, and "the perfection of the universe is attained essentially in proportion to the diversity of natures in it" (Saint Thomas). Precisely because of the Great Holarchy, the manifest world was one and unbroken. Nicholas Cusanus (one of the great representatives of the nondual during the late Middle Ages):

All things, however different, are linked together. There is in the genera of things such a connection between the higher and the lower that they meet in a common point; such an order obtains among species that the highest species of one genus coincides with the lowest of the next higher genus, in order that the universe may be one, perfect, continuous.[32]

Plenitude expresses itself in apparent gradation—everything can't burst on the scene all at once!—but a gradation that is "full" and continuous and unbroken, with humans occupying one slot among an infinite number of slots. Far from placing men and women at the center of the Kosmos—as is obnoxiously supposed by many ecological critics—this conception made them acutely aware of how insignificant their actual position was. Already Maimonides in the twelfth century had written (in *The Guide for the Perplexed*, the same title that Schumacher deliberately used in his book extolling the Great Holarchy):[33]

And if the earth is thus no bigger than a point relatively to the sphere of the fixed stars, what must be the ratio of the human species to the created universe as a whole? And how then can any of us think that these things exist for his sake, and that they are meant to serve his uses?

All of this *radical* nonanthropocentrism came, as we saw, precisely because of the radically *decentered* nature of Nondual mysticism. Thus Giordano Bruno would state flat out, "Thou canst not more nearly approach to a likeness of the Infinite by being a man than by being an ant; not more nearly by being a star than by being a man."[34] This would likewise allow Bruno to realize that "because of the countless grades of perfection in which the incorporeal divine Excellence must needs manifest itself in a corporeal manner, there must be countless individuals such as are those great living beings of which our divine mother, the Earth, is one."[35]

This did not sit well with the Church, which did not take kindly to a God who was equidistant from an ant and a man, and who had already situated Motherhood quite elsewhere. But the crucial point, as Lovejoy expertly summarizes it, was that:

No creature's existence was merely instrumental to the well-being of those above it in the scale [the great holarchy]. Each had its own *independent* reason for being; in the final account, none was more important than any other [ultimately, as Plotinus said, each was fully the One; gradation is apparent]; and each, therefore, had its own claim to respect and consideration, its own right to live its own life and to possess all that might be needful to enable it to fulfill the functions and enjoy the "privileges and perquisites" of its station. Each link in the Chain of Being exists, not merely and not primarily for the benefit of any other link, but *for its own sake*; and therefore the true *raison d'etre* of one species of being was never to be sought in its utility to any other.[36]

And thus, *all of the following were present in the Descending Side* of the Plato/Plotinus system (balanced and integrated, of course, with its counterpart on the Ascending Side): a radically decentered vision, nonanthropocentric to the core; an apparent gradation of being with no missing links, no gaps, but emergents at a common boundary; a necessity for biodiversity and a delight in the manifold nature of the creative process; a desire to participate fully in the passion of Spirit's creative Goodness; an exuberant embrace of diversity as "the Goodness of the One"; and a celebration of the earth and all its creatures as "a visible, sensible God" and our "divine Mother."

That is the half of the equation that had been downplayed, even repressed, under the rule of the unearthly Trinity—from the time, roughly, of Augustine to Copernicus. All of those ideas were thus present, simmering, hidden under the weight of the rather exclusively Ascending Ideal of the age.

And whereas the Ascenders had dominated the scene up to the Renaissance, all it took was a decisive shift in consciousness to unleash the Descending Path, a path which, bursting forth from its thousand-year confinement, exploded on the scene with a creative fury that would, in the span of a mere few centuries, remake the entire Western world—and in the process substitute, more or less permanently, one broken God for the other.

11

Brave New World

All fixed, fast-frozen relations, with their train of ancient and venerable prejudices and opinions, are swept away, all new-formed ones become antiquated before they can ossify. All that is solid melts into air, all that is holy is profaned, and man is at last compelled to face with sober senses his real conditions of life, and his relations with his kind. —KARL MARX, on modernity

"MODERNITY": ALL THAT is solid melts into air. We are still living in its shadow, still under the sway of powerful currents unleashed three centuries ago, still trying to situate ourselves in a Kosmos profoundly shaken by the events of the Enlightenment, still wondering exactly what it all meant.

But this much seems certain: from the Renaissance to the Enlightenment (to today): the Ascenders were out, the Descenders were in.

And the catalyst was the widespread emergence of Reason (formal operational), not just in a few individuals (which had often happened in the past), but as a basic organizing principle of society itself (which had never happened in the past)—a Reason that was in fact an actual ascending or transcending of myth, but a Reason that, fed up with a millennium of (frustrated) upward-looking, turned its eyes instead to the glories of this manifest world, and followed that descending God who finds its passion and delight—and its perfect consummation—in the marvels and the wonders of diversity.

MODERNITY: GOOD NEWS,
BAD NEWS

The movement of modernity (from the Enlightenment to today) contained, and contains, two very different trends. And in trying to situate ourselves in relation to modernity—and in trying to decide whether modernity was "good news" or "bad news"—both of these trends, I believe, need to be kept firmly in mind.

The first trend that defined modernity was "No more myths!" (and the Enlightenment philosophers would use exactly that phrase in describing their own endeavors). The Enlightenment mentality, with its rational demand for *evidence*, burst asunder the closed circle of the mythological world and deconstructed its cultural worldspace in no uncertain terms—and did so by asking, in each and every case, "How do you know Moses parted the Red Sea?" "Because it says so in the Bible"—where for a thousand years that would have been an irrefutable and unchallengeable answer, it now impressed no one (no one "in the know," that is).

And so some two thousand years after a mythic-membership society had forced the first great proponent of Reason to drink hemlock, the first Reason-oriented societies now turned on their mythic predecessors with a vengeance.

The mythic dominator hierarchies (God and Pope and King on top, with shades and grades of servitude stretching out beneath), and the politico-religious forms of social organization that went with them, were dismantled with frightening and often bloody speed. "Remember the cruelties!" cried Voltaire—"Liberty or death!" the American revolutionaries screamed—as the dominator hierarchies, often vicious and cruel, everywhere began to topple, precisely because they were top-heavy with a depth they did not earn or deserve or in fact possess (however much they did indeed serve their phase-specific task of social integration).

As we put it earlier, the relatively widespread emergence of an *ego* identity from a *role* identity (and the switch from a conventional/sociocentric to a postconventional/worldcentric morality) was a major evolutionary transformation and transcendence (driven by Eros, or the self-transcending drive of the Kosmos). This brought with it an eventual demand for (1) free and equal subjects of civil law, (2) morally free subjects, and (3) politically free subjects as citizens of the democratic state. And to say "free subjects" means only that the rational-ego possessed more *relative* autonomy (tenet 12d) *compared with* the tightly bound mythic-membership roles in dominator hierarchies (even if the rational-ego imagined its "autonomy" to be much more absolute). But all of that, with its own ups and downs, was part of the overall "good news" of modernity: a genuine transformation in the average mode of consciousness.

But "No more myths!" came *also* to mean—and here is the second major trend that defines modernity—"No more Ascent!" Since the myths were all upward-yearning, "No more myths!" seemed to demand no more upward-yearning of any sort (frustrated or genuine or anything in between). Understandably fed up with a millennium or two of (frustrated) upward-yearning and "pie in the sky" aspirations, Reason threw out the transcendental baby with the mythic bathwater.

This jettisoning of higher Ascent was allowed, even demanded, by the *positivism* of the evidence now recognized by rationality. Since rational science could represent every event in the Kosmos as a Right-Hand empirical, observable, natural phenomenon, no higher or "metaphysical" states were needed (and thus, Ockhamly, none were allowed). Since every holon in the Kosmos does indeed possess a Right-Hand component, empirical science could honestly, even decently, but nonetheless mistakenly, imagine that in registering the empirical component it had covered all the bases.

Thus the Kosmos was collapsed into the Right-Hand path, and a flatland empirical holism (with a few flatland atomistic cranks) replaced a multidimensional holarchy as the guiding paradigm (the fundamental Enlightenment paradigm, as we put it earlier). The same Reason that allowed the deconstruction of the dominator hierarchies inherent in the mythic structure could also, with the same tools, collapse the Kosmos in favor of the cosmos. This "flatland," I will argue, was the essence of the "bad news."

And these are the two major trends ("No more myths!" and "No more Ascent!") that gave modernity its "good news, bad news" character (or so I will try to show).

From the very beginning of modernity—from the beginning of the changes wrought from Descartes to Kant—there have been its many vocal critics (none more perceptive and brutal than Hegel, for whom the Enlightenment in general, and Kant in particular, was a "vanity of the understanding" and a "monster of arrested development"). From Nietzsche to Bataille to Foucault, from Heidegger to Derrida to Lyotard, the critics have continued their assault, with the postmodern poststructuralists being merely the most recent, although possibly the loudest, of the long line of antimodernists.

But it's the "good news, bad news" nature of modernity that most of these critics all too often seem to miss, and they especially miss (or ignore) anything resembling "good news." But after all, the students in Tiananmen Square were screaming the names of Locke and Jefferson; they were not screaming the names of Foucault and Derrida.

Charles Taylor surely captures the essence of the situation when he says, "I find myself dissatisfied with the views on this subject [the nature of modernity] which are now current. Some are upbeat, and see us as having climbed to a higher plateau; others show a picture of decline, of loss, of forgetfulness. Neither sort seems to me right; both ignore massively important features of our situation. We have yet to capture, I think, the unique combination of greatness and danger, of *grandeur et misère*, which characterizes the modern age. To see the full complexity and richness of the modern identity is to see, first, how much we are all caught up in it, for all our attempts to repudiate it; and second, how shallow and partial are the one-sided judgements we bandy around about it."[1]

This is, as Taylor indicates, an enormously rich and complex topic; so much so that the entire third volume of this series (*The Spirit of Post/Modernity*) is devoted entirely to it. What I am doing in this chapter and the next is simply summarizing and condensing an enormous number of factors into these two broad trends, "No more myths!" (the good news), and "No more Ascent!" (the bad news).[2]

The former was, so to speak, a step up, a shift in the center of gravity of society from mythic-membership to rational-egoic; it was a major step in Ascent, driven by Eros. Most important of all, it brought a differentiation of the Big Three (science, art, and morality), with the now-disclosed possibility of actually advancing each of their truth claims in a worldcentric and universal pluralism of tolerance and mutual recognition. And even though that was collectively an evolutionary advance, a moving forward of Eros, like every stage in development it nonetheless brought its own problems that could not be solved on its own level (the dialectic of progress).

But quite apart from all that, as a separate line of analysis, there was "No more Ascent!"—that is, "Ascend to Reason, but no further!"—and this brought *nothing but* bad news. It not only flattened the Kosmos to a one-dimensional, monological affair, it sealed out the possibility of deeper and wider developments that alone could defuse its own insoluble dilemmas. If, for example, rationality had finally *differentiated* the Big Three, so that art and science and morality could fortify and enrich their own pursuits without dogmatic interference, nonetheless, without a further ascent to vision-logic, there was no way that the Big Three could be *integrated* (as Schelling and Hegel would both point out). The *differentiation* of the Big Three thus degenerated, by the end of the eighteenth century, into the *dissociation* of the Big Three (a point made by Habermas)—which *in turn* allowed the Big Three to be rudely reduced to the Big One of it-language: the flatland, one-dimensional, monological systems theory of the representational and production paradigms, nightmares through and through, with all interior depth, whether in humans or animals or any sentient being, reduced to being merely a strand in the endlessly flat and faded system.

Thus, reason *itself* was an increase in interior depth compared to myth (formop transcends and includes conop). But reason *confined to flatland* (reason recognizing only Right-Hand, empirical, sensorimotor objects) is a consciousness confined to a lesser reality than most myth acknowledged (e.g., a Divine domain). Thus, under the "No more Ascent!" agenda, reason abandoned higher, other worlds altogether, and scrunched itself down to what it could grasp with its senses—scrunched itself down, that is to say, to some of the most fundamental, least significant holons in existence: exterior, sensorimotor, empirical objects, and its.

To put it all in a sentence, modernity brought a *deeper* subject in a *shallower* world: and thus, I have some good news, and some bad news.

THE AGE OF REASON

In this chapter, the good news.

From the Renaissance to the Enlightenment, the names say it all: Copernicus, Kepler, Galileo, Leonardo da Vinci, Botticelli, Michelangelo, Raphael, Machiavelli, Alexander Pope, Erasmus, Francis Bacon, Jan van Eyck, Giotto, Shakespeare, Isaac Newton, John Locke, G. E. Lessing, Descartes, Spinoza, Voltaire, Leibniz, Rousseau, Diderot, Hume, Kant, Thomas Paine, Thomas Jefferson.

What they all had in common (other than the fact that they were all men—that is part of the next, "bad news" chapter) was that they shared, in various degrees and in different ways, the space of possibilities disclosed and created by rationality. ("Reason," with a capital *R*, technically refers to vision-logic or dialectical-network reason, as opposed to simple formal-operational reasoning, or "the understanding." But when historians refer to the "Age of Reason," they mean formal operational as we have defined it. The context will make clear which usage is intended.)

The "emergence of Reason" and the "Age of Reason" do not mean that individuals prior to this time had no access to rationality. We saw that the space of rationality was accessible, to the few, even when the average or typical mode was magical; and some individuals—the Eckharts and Teresas and Platos—had access not only to rationality but beyond it to subtle and causal occasions.[3]

Rather, the "Age of Reason" simply means that access to the space of reason was now common enough that the "center of social gravity"—the basic organizing principles of society—shifted from mythic-rational to rational structures (as evidenced legally, politically, institutionally), with, again, some people falling below the norm, others above it. Reason became both a basic principle of social organization and the highest expectable potential for the average individual.

And while we eventually want to move beyond rationality (and its own inherent limitations and grave problems), it is crucial to acknowledge what it *did* manage to accomplish, and what its evolutionarily phase-specific task was.

I believe it is a profound truth of human development that one can fully transcend any level only if one fully honors it first (thus allowing embrace/Agape). Otherwise one's "development" is simply a reaction to, a reaction against, the preceding level, and thus one remains stuck to it with the energy of disapproval—Phobos, not Eros.

Let us come to bury Caesar; let us come to praise him first. If we look at the above-mentioned list of names, virtually all of their works, and all of their accomplishments, can be read directly off the list of the characteristics of formal operational awareness that we have already discussed:

1. *Rationality is hypothetico-deductive or experimental*—Tycho Brahe, Kepler, Galileo, Newton, Darwin, Charcot, Kelvin. Rationality demands evidence and reasons, not dogmas and myths, which is precisely why (as Whitehead pointed out for the West and Needham for the East) the mystics and the newly rising scientists often joined hands against the

Church (or exoteric religions), since both mysticism and science are experimental and evidential. It was the demand for evidence that, more than any other component of the Age of Reason, would shatter the closed world of mythology in an irreversible and irrevocable way: that toothpaste could never be put back in the tube.

But let us note the intricacy of what was actually involved: it wasn't simply that the mythic worldview had no recourse to evidence of any sort; its evidence, rather, was grounded in the obviousness (and "unchallenge-ableness") of its own worldspace, and was *true enough* in that worldspace.[4] *Within* that mythic worldspace, there was evidence aplenty: the gods and goddesses exist ("How do you know they exist?" "Are you crazy? Where do you think the rain comes from? Look at the creation all around you! Can't you see? Are you blind?"). And the gods and goddesses have given us their words; the pharaoh/king/emperor is the descendant of the gods, and thus we owe our allegiance to this temporal representative of the gods/goddesses; the evidence for this is everywhere, you have simply to look!

Thus, to say that the mythic worldspace was not fallibilist misses the point. Any given interpretation *within* the mythic worldspace could indeed be *rejected* by a community of interpreters, and this happened *all the time*. It is true that the mythic worldspace was not fallibilist from the depth of rationality (i.e., was not fallibilist to the *types* of evidence that rationality would *disclose*): but *that* evidence simply could not be seen in the first place, and so didn't (and couldn't) enter the picture to dislodge any particular interpretations. But the *bad* interpretations *within* the mythic worldspace were indeed rebuffed by a community of competent interpreters in tune with the nuances of the mythic worldspace (and they were right more often than not, in that space). These structures weren't *stupid* (no structure is; and they all produced their own bona fide geniuses). Rather, each structure weighs carefully the evidence that it *can* see.

In other words, the hermeneutics of any worldspace is closed and perfectly evidential for that worldspace. No new interpretation will step outside the worldspace. Rather, a developmental supersession will suddenly, via emergence, disclose new depths and wider perceptions that themselves pass judgment on yesterday's relative blindness. It is transformation that negates the old translations, and not anything the translations themselves could see from within their own horizons (and bad interpretations *within* those relatively true horizons can be, and are, soundly rejected by the evidence that the eyes of that structure *can* see, and see quite clearly, as far as it goes).

Likewise, Reason also imagines that *its* evidence is simply *the* evidence, obvious and incontrovertible per se (i.e., its *type* of evidence is likewise stamped with "unchallengeableness"). Bad interpretations *within* its rational horizon (which also means within its structural worldspace) will be soundly rebuffed from within that horizon by the community of those whose eyes can see the depths disclosed by rationality. Interpretations from the transrational structures will not be rebuffed: they will simply not be *seen* in the first place, and thus will be met with the standard "What, are you crazy?" (which is not a rebuff but a retreat).

But the evidence disclosed by rationality will indeed be able to outcontextualize, and thus outtrump, the evidence disclosed within the horizons and the structures of mythic awareness (just as mythic masterfully outtrumped and outcontextualized magic). It is not that a greater and absolute truth overthrows a falsehood, so much as that a great truth supersedes a lesser truth (no epoch lives, or can live, simply on falsehoods). And this is what the Age of Enlightenment brought to bear upon the Age of Myth: a new horizon of evidence that outcontextualized the old.

And outcontextualize, it did indeed. We tend to forget the evidence that the mythic worldspace took to be perfectly obvious. Here is one of the widely accepted "refutations" of Galileo's discovery of the moons of Jupiter: "There are seven windows given to animals in the domicile of the head, through which the air is admitted to the tabernacle of the body, to enlighten, to warm and to nourish it. What are these parts of the *microcosm*? Two nostrils, two eyes, two ears, and a mouth. So in the heavens, as in a *macrocosmos*, there are two favorable stars, two unpropitious, two luminaries, and Mercury undecided and indifferent. From this and many other similarities in nature, such as the seven metals, etc., which it were tedious to enumerate, we gather that the number of planets is necessarily seven."[5]

That is an excellent and very *accurate* description of the insides of the mythological worldspace, where syncretic wholes define the nature of interconnectivity in the Kosmos *as disclosed at that depth*. Rationality can explain the interconnectivity from a deeper dimension, and thus marshal more *types* of evidence, but rationality is not itself operating from the final Archimedian point of all possible worlds.

A thousand years from now, when atoms are understood to be pinpoint

388 | BOOK TWO

entries of a massive intelligence from the eighth dimension (or whatever the discoveries are that will shock and shake our present views, as shock they most certainly shall), we will look not so bright and gifted in our firm assurances of what the Kosmos is "actually" like. Atoms will still be atoms (no truths will be lost); but they will be set in deeper contexts that will render our narrower perspectives and interpretations as silly-sounding as the poor soul who lit into Galileo all self-assured and confident.

But my present point is simple enough: the types of evidence rationality could disclose had it all over the mythic structure; a shallower context gave way to a deeper disclosure; and the death rattle of an old and honorable worldspace released its last breath into the rising wind blowing now across a strange new landscape.

2. *Rationality is highly reflexive and introspective*—as witness the constant introspective analyses of Descartes, Locke, Berkeley, Hume, Kant (descendants each and all of Augustine's rational and radical reflexivity).

Take, as an example, the study of psychology, the study of an *individual* psyche. The word "psychology" itself wasn't even used until Freigius in 1575; but by the closing decade of the sixteenth century, it was common to treat *anthropologia*—the study of human nature—as composed of two parts: *psychologia* and *somatologia*, reflecting the ever-clearer differentiation of mind and body (typical of formop). The first widespread use of the term *psychology* was by Christian von Wolff, successor of Leibniz and predecessor of Kant, in such works as *Psychologia Empirica* (1731) and *Psychologia Rationalis* (1734). The term was picked up by Hartley in England (1748) and Bonnet in France (1755), and by the end of the eighteenth century it was in widespread use, reflecting, it appears, a newly emergent entity to be studied: the individual ego.

The first modern "school" of psychology was the associationist school of David Hartley, John Stuart Mill, and Alexander Bain (in England); and the first experimental approach was the great laboratory work of Weber, Fechner, Wundt (in Germany) and Titchener (in England)—the point being that all of this was made possible, in large measure, by the structural worldspace of rationality and its reflexive/introspective modes.

3. *Rationality grasps multiple perspectives* (or is universal as pluralism). This perspectivism/pluralism showed up in everything from political theory to art. John Locke wrote three separate essays on tolerance (or the acceptance of different perspectives and points of view, except those that

condemn others), concluding, among other things, that no one had a right to harm another in "life, health, liberty, or possessions"—the influence of which on subsequent history was simply incalculable.

In art, to give a very different but very related example, perspectivism began to be actually depicted, for the first time ever, in painting, beginning in fifteenth-century Florence and eventually associated with the names of Brunelleschi, Alberti, Donatello, Leonardo, Giotto. As Jean Gebser put it, "An entirely new spatial consciousness was beginning to break out of the soul into the world. By virtue of the *perspective*, not only is space made visible and brought out into the daylight of waking consciousness; man himself thereby attains visibility. The gradual possession of the perspective, which became the principal concern of Renaissance man, had the effect of expanding the world image. . . ."[6] And this "visibility of man" is directly related to:

4. *Rationality brings forth an ego identity from the previous role identity.* Morality is, or can now be, postconventional and *worldcentric*, not merely conventional and ethnocentric and sociocentric (or mythic-membership). Moral decisions thus rest with the *individual*, who must assume responsibility for his or her own relatively autonomous choices (and who may or may not choose to live up to rationality and act from a worldcentric space of universal pluralism—but the relatively autonomous choice is nonetheless present).

And, indeed, *autonomy* became one of the great and enduring themes of the Enlightenment. Kant would define it as the courage to think for oneself and not rely on socially given rules and dogmas. (That this would play into the disengaged subject of flatland holism is a point we will pursue in the next chapter.)

In political theory (and practice), this resulted, as we have seen, in the conception of men (and soon women) as autonomous agents, which meant free and equal subjects of civil law, morally free subjects, and politically free subjects or democratic citizens of the state.

In art, this meant, for the first time, that there was an actual person (ego-bearer) to paint: the beginning of portraiture (such as by Jan van Eyck). And if there was now an ego to be painted, there was likewise a story to be told, a narrative of, by, and about egos: the novel was born.

5. *Rationality is ecological or relational.* Along with autonomy, the great theme of the Enlightenment was *systems harmony*, which did not mean "peaceful" or "without conflict," but rather that the "whole sys-

tem" of reality would balance itself out *in spite of* the particular conflicts and discords that individual components might undergo. As we have seen (and will see again in the next chapter), it was a thoroughgoing systems view of reality, where the totality "self-regulated itself" despite all the ups and downs of individuals. And this systems view, despite what misuses it might suffer, was nonetheless a great advance over mythology in understanding the interconnectivity of the Kosmos, for it replaced belief in syncretic wholes (seven orifices mean seven planets) with an understanding of holons, or whole/parts that are contextually situated in a way that does not do violence to individual wholes or parts.

(That this systems view would also involve the collapse of the Kosmos into the empirical holistic cosmos of monological flatland—this is a separate issue, part of the "bad news" of the next chapter; and that this holistic worldview would paradoxically be one of the greatest contributors to an ecological crisis: that, too, the topic of the next chapter.)

In the meantime, the relational mode of systems thinking was all the rage. It was literally everywhere, and was greeted with cries of Eureka across the Continent and the Isles. We see it in, for example, Adam Smith's "invisible hand," where, in spite of the fact that each individual might act in a greedy fashion, the overall result was good and beneficial for the whole (it was believed). We see it in Leibniz's "preestablished" harmony of all monads. We see it powerfully in John Locke's notion of "interlocking orders." In politics, it shows up most influentially in Rousseau's idea of the "general will" of the people, which supposed that if each individual follows his or her own reasons, a majority or consensus will emerge that will be to the benefit of all. This *volonte generale* was a direct precursor to the American and French revolutions. But the point is that "all the philosophers of the Enlightenment use this concept of harmony directly or indirectly, explicitly or implicitly,"[7] and all were given to exaltations of the "perfection of the great Universal System."

6. *Rationality is nonanthropocentric.* The shocking decentering that rationality/science brought is now notorious: Copernicus, Darwin, Freud—men and women are merely links in that vast, vast chain of existence. (This had always been obvious to the likes of Cusanus or Bruno or Plotinus; we are now discussing what rationality brought to the average-mode and what thus came to be very commonly accepted views outside of the mythic holdovers.)

As we will see in the next chapter, this anti-anthropocentric stance became so intense at times that ethics virtually stalled altogether, so paralyzed were people with the thought of their utter insignificance. To step outside of one's imperfection was viewed as pride, the great sin of the eighteenth century, and there emerged a moral atmosphere that "may be described as a counsel of imperfection—an ethics of prudent mediocrity."[8] But all of this was a direct result of the fact that reason could set aside its own viewpoint, its own egocentricities, and see the world through the eyes of the other, and thus begin to honor the other in unprecedented ways, and, for better or worse, to profoundly question its own right to existence.

7. *Rationality brings a new space of deeper feeling and greater passion*—the explosive idealism of the true dreamer who can cognitively imagine all the possibilities. This new passion—coupled with reason's capacity for postconventional morality and its capacity to criticize the given state of affairs in light of a *possible future*—meant that this was not just the Age of Reason, it was the Age of Revolution.

> We must understand what this reason was. It was not [just] a calculating reason which decides whether to do this or that, depending on which is more advantageous. Rather, it was a full, passionate, revolutionary emphasis on man's essential goodness in the name of the principle of justice [postconventional morality]. The revolutionary fought against feudalism and the authoritarian churches. He had a passionate belief in the [goodness or harmony] of reality and was convinced that the human mind is able to re-establish this structure by transforming society. We could therefore call it revolutionary reason as well as critical reason. Because of its *depths* this critical revolutionary reason overcame the prejudices of the feudal order, the heteronomous [sociocentric] subjection of people both by the state and the church. It could do so because it spoke in the name of [pluralistic or postconventional] truth and justice. The philosophers of the Enlightenment were extremely passionate. They were not positivists [merely one of the possibilities of reason]; they were not interested in merely collecting facts. They became martyrs for the passion. . . .[9]

This was particularly a passion, we might say, for the equality found only in the noospheric space of reason. The result, of course, is the existence of today's pluralistic democracies, forged by a passion for tolerance—a tolerance that, since not all of its citizens can always live up to it, is therefore firmly embedded in legal institutions, and a tolerance that, from its inception, has always been driving toward the emergence of planetary vision-logic with its genuinely world-class citizens. And one can only stand in utter awe and admiration for the many men and women who fought and died, from America to France, under the slogan "I disagree with what you say but will defend to the death your right to say it."

THE LIBERATION MOVEMENTS

The widespread emergence of reason meant, as always, a new differentiation demanding a new integration. We have already seen that some of the new and important differentiations included a separation of church and state; the differentiation of the Big Three (an important point we will return to in a moment); and the increasingly clearer differentiation of the noosphere and the biosphere.

Most significantly, this differentiation of noosphere and biosphere meant, of course, that the differentiation could go too far into dissociation, and this dissociation would show up in an extraordinary number of guises. First and foremost, the stage was set for a possible ecological crisis, a possibility that of late has become an increasingly gruesome actuality. The ecological crisis, as we earlier put it, is a worldwide, collective psychoneurosis—a denial, alienation, dissociation of the biosphere by the noosphere.

The ecophilosophers maintain that the ecological crisis is the result of human beings not realizing that they are merely a part of a larger biosphere. But that, as we have seen, gets the picture exactly backward, and cannot even explain how a part can dominate the whole. No, we have seen quite clearly that the human compound individual is not a part of the biosphere. Rather, a *part* of the human compound individual is a *part* of the biosphere, and the biosphere itself is a *part* of the noosphere (because the noosphere transcends and includes the biosphere). And for just that reason, repression can set in; for just that reason, the noosphere can dissociate the biosphere; and for just that reason, the poisoning of our roots means the death of our branches as well.

But that is more of the "bad news" (next chapter). At this point, we should note instead that in the immediate wake of the differentiation of the noosphere and the biosphere, biological determinants were no longer the strong (and often overwhelming) factor they had been up to that point in history. In the biosphere, might makes right. In the biosphere, big fishes eat little fishes. In the biosphere, muscle rules. And in the biosphere, men dominate other men, and men dominate women.

Feminist and orthodox researchers both agree that, in virtually all known societies to date, power between the sexes has been asymmetrical in a very specific and very widespread (virtually universal) fashion. That is, all known societies can be arranged along a continuum from relatively "equalitarian" (males and females share power in the public sphere roughly equally) to male-dominant (males alone govern the public sphere), but *never* the other way around, never tending toward societies of female dominance of the public/productive domain.

In other words, the *most* feminine societies (such as early hoe/horticultural societies) were not ones dominated by females, but ones where females shared public/productive power roughly 50-50 with men; whereas the "most masculine" societies (such as the plow/agrarian) were ones where men possessed virtually 100 percent of public/productive power (with women relegated to the private/reproductive sphere).

As researchers from Janet Chafetz to Riane Eisler to Rae Blumberg to Joyce Nielsen have pointed out, there are virtually no known exceptions to this scheme—it goes from 50-50 male/female to 100 percent male, but never toward 100 percent female. There is obviously a set of very strong, universal factors that are tilting the scheme in one direction only (from balance to male dominance, never from balance to female dominance). What are these virtually universal factors?

"To answer the question 'Why are women nowhere superior to men in their access to scarce and valued societal resources?' one must look to the key sets of intervening variables," says Janet Chafetz.[10] "As a category, men nowhere [in no known societies] specialize solely in the reproductive/private sphere activities, nor do women as a group specialize solely in the productive/public sphere. Stated otherwise, women's activities are either specialized in nonproductive roles, which leads to inferior status, or divided between the two realms, which may afford them relatively equal status"—in other words, at best 50-50, and that against the odds.

"This pushes the question back one step: why do women as a category never specialize solely in productive/public sector roles? Since this is a constant [no known exceptions], logically it must be explained in terms of one or more constants. In this case, *a set of biological constants* helps to answer the question raised: the women carry babies in their bodies and lactate, which circumscribes their physical mobility. While this can be *minimized* [through low birth rates, etc.], it cannot be *eliminated* and, in fact, for most human societies throughout most of human history, such restriction due to pregnancy and nursing has been far from minimal."

This has far-reaching, even massive, consequences. "Given these biological facts, most societies find it more *efficient* if women also do the bulk of the caretaking of young children who are no longer nursing. That is, on the basis of *expediency*, the nurturance role is typically extended beyond the biologically based phenomenon of breast-feeding. And to the extent that these factors keep women in the proximity of the domicile more than men, it becomes more efficient to extend their domestic role to encompass other household tasks, such as food preparation and maintenance of family possessions. Since many societies have had a large share of their productive activities moved away from the homesite, these fetters on female mobility restrict women's opportunity to specialize in productive roles."

According to Chafetz, these biological givens, universal in nature, are part of the reason that societies universally do *not* drift toward 100 percent female dominance of the public/productive sphere, but, at best, allow a 50-50 sharing, and, at worst, allow a drift toward 100 percent male advantage.

And this drift toward male dominance of the public/productive sphere is aggravated by a universal or constant biological factor on the part of the males as well. "An additional biological constant is probably in force: male superiority in physical strength. In many societies, especially pre-industrial, some or many productive activities require—or at least are handled more efficiently by—greater strength. Again, females as a category find themselves somewhat disadvantaged in competing for such roles."

The conclusion, then, is rather straightforward: "In short, females are not found to have superior status to males because their relative disadvantage in mobility and/or strength makes it more efficient for men to monopolize some, many, or all productive roles and for women to assume at least some of the household labor. Thus, women as a category are never free to specialize in or monopolize productive/public sector roles. Where men do specialize in or monopolize such roles, they are vastly superior to women in status [100 percent male advantage]. Where the sexes more equally divide the two realms, their statuses approach equality [roughly 50-50]. Females as a group do not have vastly superior status to males [it never drifts in the other direction] because, in all the history of our species, women have never been able to divest themselves of some involvement in the reproductive/private sphere of activity."

Now, Chafetz's overall theory (which is very comprehensive, and which is examined in detail in volume 2) does not simply rest with these biological constants (or "strong universals in the sexual scheme"). These biological givens are simply the background constants against which numerous other factors play themselves out, variables such as modes of technology, environmental threats, family structures, types of productive activity, population density, nature of work organization, degree of separation of work- and homesites. In all, Chafetz isolates a dozen factors that are involved in the overall "gender status" of men and women in various societies.

But given the biological constants, all of these other factors affect whether the given society will be relatively equalitarian (50-50), or whether, for reasons of expediency and efficiency, it will drift toward a male dominance of the public/productive sphere (but never drift in the other direction).

And the conclusion, backed by extensive empirical data, is that whenever these other variables do indeed shift—whenever there are environmental threats or disasters, food scarcities, warfare, social threats, and intense stresses—whenever these factors occur, then the 50-50 balance is inevitably disrupted, a premium is placed on male physical strength/mobility, and *the sexes polarize dramatically* (with males dominating the public/productive sphere and females dominating the private/reproductive, but never the reverse).

And, most important, when the sexes polarize dramatically, a horrible stress is placed on *both* men and women. (In fact, according to Chafetz, men have it worse in "male-dominant" societies than do women, starting with the fact that they alone are conscripted for defense). *Oppression* as a causal explanation is deficient and inadequate in almost every respect, since, among other things, it simply does not fit the data curve. "These oppression theories," says Chafetz, "are based on vaguely defined concepts often ill suited to operationalization, such as 'patriarchy,' 'female subordination,' and 'sexism.' The use of such emotion-laden but unclear terms, combined typically with a heavily normative approach to the topic of sex inequality, results in a maximum of rhetoric but a minimum of clear insight."

No, this polarization of the sexes—with males dominating the public/productive sphere and females dominating the private/reproductive, to the detriment of both—has virtually nothing to do with male oppression and female sheepdom/subjugation. It has everything to do with life in the biosphere.

Thus, with the *differentiation* of the noosphere and the biosphere, these biological constants are not eliminated, but they are *subordinated* (preserved and negated). They are no longer necessarily the major determinants of the form of social organization. Biology is no longer necessarily destiny.

Even in the relatively "equalitarian" societies of the past (such as early horticultural and some foraging), the "equality" was kept in place, not by legal and moral (or noospheric) determinants, but by luck. The hoe placed public/productive power in women's hands, not by conscious and intentional design, but by sheer happenstance. It turns out that a simple hoe or digging stick can be physically handled as easily by a woman as by a man; and further, in hoe societies, the worksite is often right next to the homesite, so that pregnant women are not disadvantaged when it comes to physical mobility (they, like the men, can walk to work). And, indeed, in such societies, women produced about 80 percent of the foodstuff (the men still often hunted), and consequently the societies remained relatively close to that ideal 50-50 balance—but a balance entirely at the mercy of circumstances.

And thus, just as the hoe had placed power in women's hands, not by conscious design but by happenstance, so the plow would yank that power away, not by oppression or oppressive design, but by the same biological givens subjected now to entirely different (and equally nonintentional) circumstances. (Women who operated plows suffered increased rates of miscarriage, as pointed out by Chafetz, so it is to their Darwinian advantage *not* to plow; further, the plowed fields were always at a considerable distance from the homesite, discouraging a pregnant woman from getting to work, as pointed out by Nielsen and others).[11] The same biological constants that under one set of circumstances simply happened to advantage women, under different circumstances simply happened to disadvantage them.

There is neither oppression on the one hand, nor virtue on the other, operating here.

With the differentiation of the noosphere and the biosphere, these biological constants, although not eliminated, are no longer necessarily the major determinants in the *public* status of men and women. In the biosphere, might makes right, and status follows physical function. In the noosphere, right makes might, and status follows the rights of free agents in the worldcentric space of universal perspectivism. Whether or not a woman is producing foodstuffs in the public/productive sphere is *irrelevant* to her *rights* in that domain—and *that* idea was revolutionary and absolutely unprecedented on any sort of large scale.

With the differentiation of the noosphere and the biosphere, women could secure rights as cultural-noospheric *agents*, *regardless* of biospheric factors or functions or constants, which were judged, for the first time in history, to be *irrelevant* to one's public status as a citizen and a bearer, not just of children, but of rights. The noosphere, not the biosphere, would determine rights and responsibilities. And these rights were not previously *repressed*, they were previously *meaningless* (they were not oppressed, they had simply *not yet* emerged).

To repeat what was said in chapter 5 (for now I think it will make more sense): in this way, and this way only, can we maintain the otherwise paradoxical and confusing truth—a paradoxical truth that has crippled feminism vis-à-vis "female liberation"—that women now stand open to and in need of liberation, but previously were *not* acting in an unliberated (or duped) fashion. The widespread emergence of the women's movement in the eighteenth and nineteenth centuries occurred precisely because the noosphere and the biosphere were finally differentiated.[12] And this inescapably means that the widespread emergence of the women's movement was not primarily the *undoing* of a nasty state of affairs that easily could have been different, but rather it marked the *emergence* of an altogether *new* state of affairs that was in significant ways *unprecedented*.

What had grounded *both* men and women in narrowly based biological roles was an evolutionary process that was itself, until fairly recently, grounded in the biosphere, and that only recently is now in the process of liberating *both* men and women from confinement to those particular roles—necessary at the time but now outmoded.

Historically, with the emergence of pluralistic rationality—with the differentiation of the noosphere and the biosphere—right slowly began to replace might. Wherever pluralistic rationality emerged, social relations based on physical power—particularly slavery and the polarization or dissociation of the sexes—were found intolerable to reason, to pluralism and perspectivism—put yourself in the slave's shoes (or lack of them) and see how it feels! This was all radically new. . . .

And thus, in the West and in the East, wherever rationality emerged, feminist movements and antislavery movements first emerged as well. As only one example from the East, the real revolution that Gautama Buddha brought to India lay not so much in his actual spiritual practices—his yoga was already afoot in India—as in the fact that he denied the caste system. Much more than a religious pioneer, he was a social revolutionary, and this rocked India profoundly—yet another example of the "revolutionary passion" brought by reason. Even though Gautama was transrational as well, his religion—one of the first "rational religions" in the world—was *not* based on mythic gods and goddesses. He denied *all* of that, and thus he simultaneously denied the dominator hierarchies, the caste system, inherent in the mythological worldview: this is what was so revolutionary about his system.

And my point is that it wasn't so much that these various *liberation movements* (of slaves, of women, of the untouchables) had tried to surface in the past but were rudely oppressed; it was mostly that they previously made no sense whatsoever, to men or women or slaves or masters, and so were for all intents and purposes nonexistent. Power was earned or taken; to *share* it was insanity, according to the evidence that these structures *could* see. Only where the noosphere began to emerge and differentiate itself from the biosphere—only where right began to bear down on might—did these "slave situations" come to be perceived as *problems*. Otherwise, the problem was how to *exercise* power strategically, not how to *share* it. For the warrior ethic, for the duty ethic, and for the ethnocentric ethic, universal compassion was a sign of *weakness*. The *value* of compassion was *not* suppressed or oppressed; it simply was *not seen* in the first place.

But with the differentiation of the noosphere and the biosphere, there *could*—and therefore *would*—arise a Mary Wollstonecraft, a John Stuart Mill, a Harriet Taylor (just as similar voices, although isolated, had arisen in Greece and Rome and India and China, wherever pluralistic reason had lifted up its universal head).

After the death of her first husband, Harriet Taylor married John Stuart

Mill (he reports he had been in love with her for twenty years), and to-gether they collaborated on the famous *On Liberty* (1859). The year it was published was the hundredth anniversary of the birth date of Mary Wollstonecraft, whose *Vindication of the Rights of Women* (1792) is gen-erally regarded as the first major feminist treatise anywhere in history.

"Feminism as a modern ideology did not emerge until the middle of the nineteenth century," points out Riane Eisler. "Although many of the philosophical foundations for feminism had been articulated earlier by women like Mary Wollstonecraft, Frances Wright, Ernestine Rose, George Sand, Sarah and Angelina Grimke, and Margaret Fuller, its formal birthday is July 19, 1848, at Seneca Falls, New York. Here, at the first convention in recorded history held for the express purpose of launching women's collective struggle, Elizabeth Cady Stanton made a pivotal state-ment. 'Among the many important questions which have been brought before the public,' Stanton said, 'there is none that more vitally affects the human family than that which is technically called *Women's Rights.*' "[13]

In the noosphere, right makes might. William Wilberforce, in a cam-paign forged with his lifelong friend William Pitt, spearheaded a move-ment that resulted, in 1807, in the abolition of the slave trade in the British empire. In the States, a war fought in part for antislavery motives would grind up as many men in single battles as were lost in all of Viet-nam: 51,000 were killed in three days at Gettysburg alone. The president at the time would remind the world, in an address at that site, and in a mere 269 words, that this battle was fought because the nation was "dedi-cated to the proposition that all men are created equal," a proposition scorned by nature and by all societies embedded in it.

That *proposition*—that rational idea—very soon detached itself from remnants of biospheric power and extended to women as well. In less than a century from Harriet Taylor's death, women in democratic socie-ties had secured the legal protection, *institutionalized*, of (1) free and equal subjects of civil law, (2) morally free subjects, and (3) politically free subjects as citizens of the democratic state. That these freedoms and protections still need to be extended and fortified does not denigrate the fact that they are structurally *in place*.[14]

With the differentiation of the noosphere and the biosphere, women could still claim a rich heritage in the biospheric qualities (mostly, embod-ied communion) that differentiate them from men (mostly, hypera-gency)—this is the claim of radical feminists. But likewise, they could also claim *noospheric equality* and demand that it be protected by law—this is the claim of liberal feminists.

I believe both are true, and *can* be true, now that the noosphere and biosphere have been differentiated and enfolded, embraced, in the same human compound individual. And this is part of the rich heritage of the various "liberation movements" (of women, of slaves, of untouchables) that emerged around the globe with the emergence of rational perspectivism and universal pluralism.

In the same way, the female value sphere, freed by evolution from its confinement via the biosphere, could now begin more readily to infuse the public domain with its own care and concerns. "For it was through the impact of the 'female ethos' embodied by women like Florence Nightingale, Jane Addams, Sojourner Truth, and Dorothea Dix, who were now beginning to enter the 'public world' en masse, that new professions like organized nursing and social work emerged, that the abolitionist movement to free slaves gained massive grass roots support, that the treatment of the insane and mentally deficient became more humane."[15]

Not to mention that it was in the "salons"—in the *atmosphere*—created by women such as Ninon de Lenclos, Madame Geoffrin, Madame Rambouillet, and Madame du Chatelet, that so many of the ideas that we would recognize as "humanist" and "Enlightenment" arose and were nurtured, an atmosphere that can only be called "liberational."[16]

All of which points unmistakably, as Riane Eisler puts it (and as Habermas has often argued): "What becomes evident is that the great transformation of Western society that began with the eighteenth-century Enlightenment did not fail but is merely incomplete."[17]

(And the next "liberation movement"? Understanding that the real urgency is still to implement more fully the liberations brought by rationality and the ego/centaur, then next up is: the Eco-Noetic Self, the Over-Soul that is the World Soul, which demands a liberation not just of peoples but of life, of all sentient beings, not as having equal rights, but as worthy of care and respect and honor, cherished as manifestations of Spirit. We will return to this topic of "environmental ethics" in the last chapter. My point for now is simply that every Eros or Ascent brings a liberating force which can then, barring fixation or repression, be embodied in a wider Agape or compassion.)

THE BIG THREE

I would like to point in particular to Immanuel Kant's three critiques (*Critique of Pure Reason, Critique of Practical Reason,* and *Critique of Judgment*), because they demonstrate more clearly than anything else that the three spheres of science, morality, and art were now irrevocably differentiated. In mythic and mythic-rational syncretism, science and morality and art are still in significant ways globally fused; and far from being "holistic," this indicates, as in all syncretisms, that the individual whole/ parts are distorted, or not taken at face value, in order that they may be bound and forced into the syncretic "whole." A scientific "truth" is true only if it fits the religious dogma, and all true art depicts some aspect of that mythic organization.

With Kant, each of these spheres is differentiated and set free to develop its own potentials without violence from a "forced fit" in a global indissociation. These three spheres, we have seen, refer in general to the dimensions of "it," of "we," and of "I." (These are the four quadrants—the exterior and the interior of the individual and the collective—with the two Right-Hand quadrants treated as one since both of them are described in similar it-language; both deal with seen exteriors.)

The sphere of empirical science deals with those aspects of reality that can be investigated in a relatively "objective" fashion and described in an "it" language (descriptive, monological, propositional truth, a state of affairs, the Right half); this involves the exteriors or the surfaces of holons, since these can be seen with the senses or their instrumental extensions. Practical reason or moral reason refers to the sphere of "we," of how you and I can pragmatically interact and interrelate in terms of something common that we can find between us, not as shared surfaces but as a common depth of mutual understanding ("we," justice, the good, the Lower Left). And the sphere of art or aesthetic judgment refers, in the broadest sense, to how I express myself and what I express of myself, the depth in the individual ("I," sincerity, expressivity, the Upper Left).

That Kant differentiates these realms means that "I" and "we" are differentiated (I need no longer automatically go along with society's rules and norms; I can norm the norms; what the church and state say are not necessarily the true or the good). It means that "I" and "it" are clearly differentiated (my subjective wishes do not constitute reality or its scientific study). It means that "we" and "it" are clearly differentiated (the state or church does not have the right to dictate what reality is).

Each of these three realms, then, can be *judged separately* to see if it is telling "its kind of truth." In the realm of "itness" or empiric-scientific truths, we want to know if propositions more or less accurately match the facts as disclosed ("Is there life on Jupiter or not?"). In the realm of "I-ness," the criterion is *sincerity*: you want to know if I am telling the truth, if I am being sincere about how I feel, if I am being truthful about expressing my inner state (if I am reporting my depth sincerely). And in the realm of "we-ness," the criterion is goodness, or justness, or relational care and concern: we want to know if our actions with each other show kindness and non-egocentric caring, or at the very least, mutual understanding; we want to act with a worldcentric fairness (act as if the maxim of one's actions might be universal—the categorical imperative).

These differentiations of modernity are irreversible, irrevocable. Even those who intensely criticize them do so from within their space; seven human orifices will simply never again automatically mean there must be only seven planets: that doesn't even make any sense to someone in the modern worldspace. The differentiation of the Big Three can be criticized, but the criticism itself presupposes the differentiation.

What is required, of course, is not a retreat to a predifferentiated state (that's not even *possible*, although retro-Romantics often make the recommendation); what is required is the *integration* of the Big Three. And that, indeed, is what might be called the *central problem of postmodernity*: now that science, art, and morality have been irreversibly *differentiated*, how does one *integrate* them?

In the Third Critique, Kant attempts to integrate propositional and pragmatic truth via telos/organism and via the aesthetic dimension (by which he meant, in this case, the sensory-aesthetic apprehension). This attempt was immensely rich and suggestive, and theorists from Schiller to Schelling would, directly under Kant's influence, look to the aesthetic dimension for their attempted integration; and, indeed, modernity and postmodernity would henceforth find something immensely "healing" about the aesthetic/artistic dimension.

And yet, in the end, this attempt would fail. The sensory-aesthetic dimension is indeed a type of "connecting link" between the empirical phenomena of sensorimotor cognition ("science") and practical ethics ("morality"); in the developmental spectrum, as a matter of fact, this aesthetic or "endoceptual" cognition lies between sensorimotor cognition and conventional moral structures.[18] It is indeed their "missing link" or "connecting link"; but a connecting link is *not* an integration. It's like saying that the fourth grade lies between the third grade and the fifth grade and connects them, which it does; but the *integration* of the knowledge in those three grades is to be found in the sixth grade, not in the third. The merely *intermediate* or connecting link cannot be an *integrating* link, since that connecting link is a link *between* the lower and the higher, and is not itself the super-higher which would integrate them all.

In other words, the integration of the Big Three would await (and is still awaiting) the emergence of vision-logic. Kant would point out one of the connecting links between moral reason ("we") and reason tied to objective sensorimotor phenomena ("it"), namely, aesthetic apprehension (a lower portion of the "I" dimension). That would indeed show some of the connections between the Big Three, but that could not finally integrate them, an integration that would demand not an intermediate but a yet-higher level.

What rationality had put asunder, vision-logic would unite. That, at any rate, is the potential and the promise and the struggle of postmodernity. *What modernity differentiated, postmodernity must integrate.* And if rationality did the differentiation, then postrationality must do the healing (and postrationality, I have been maintaining, is preeminently vision-logic). This is the postmodern *integrative vision* we have mentioned often before, in connection with Gebser ("integral-aperspectival"), Habermas (whose theory of communicative action is specifically designed to integrate the Big Three), and Heidegger's centauric being-in-the-world (also an attempt to reweave the fragments); we see it in Foucault's systematic use of vision-logic to map the exteriors of epistemes (and then his including the interiors as well in dispositifs, looking at the dimensions of truth, of power, of ethics, and of self; even Foucault, a "Nietzschean," returned to Kant again and again, and, in a surprise move, finally identified himself broadly with Kant's lineage). We have mentioned this integration in terms of "mind and body are both experiences of an integrated self"; and we mentioned it in terms of the integration of the noosphere, the biosphere, and the physiosphere. We have seen it in connection with integrating male and female in the noosphere. And now we see it as an integration of science, morality, and art.

Those are all simply different aspects of the overall attempts to integrate the now irreversible differentiations of the Big Three. And whatever we might think of any of those theorists and their proposed solutions (or my proposed solutions), the point is that *all* of them came in the wake of the differentiation of the Big Three, and all were (and are) responding in various ways to that differentiation—it was an epochal development, a world-shaking, world-making emergence. Kant was well aware of this (particularly in his essay "What is Enlightenment?"), and his immediate successors, and critics, are all *inescapably* oriented now to a worldspace in which art, science, and morality are not only differentiated, but are in danger of flying apart completely.

It was Schelling, we will see in the next chapter, who first responded most vigorously to this new world that was not only differentiating but rapidly dissociating. And Schelling was simply the first of many "doctors of modernity"—from Hegel to Marx to Schiller to Freud to Weber to Heidegger—who would try desperately, in their various ways, to heal the fragments that began to cascade around them. For not only does modernity (and postmodernity) have to worry about how to integrate what are now irrevocable differentiations in the growth process, it has to deal "therapeutically" with those differentiations that *go too far* into actual dissociation. We already saw, as only one example, that the differentiation of mind and body can go too far into a repression of the body by the mind, a repression that spews neurotic symptoms everywhere in its wake. Thus, at this particular point in history, if Freud did not exist, evolution would have had to invent him in order to doctor some of the pathological dissociations in its own processes of new and emergent growth.

But I am again drifting into the "bad news." All I would like to emphasize, at this point, is that I personally believe that the new integrations sought by modernity (and postmodernity)—the integrations particularly of I, we, and it; of self, culture, and nature; of art, morals, and science; of consciousness, values, and facts—are already being slowly effected by vision-logic. Science, values, and subjectivity are being approached as so many aspects of the centauric being-in-the-world.

Volume 3 is devoted to just this topic, but at this point we might simply note that evolution has never produced a differentiation that it could not eventually integrate (unless the differentiation goes utterly too far into dissociation, at which point simple extinction ensues). Extinction seems to be a genuine possibility for humans, particularly with reference to the threatening dissociation of the biosphere and the noosphere, the body and the mind. But assuming, as I do, that extinction is not altogether a foregone conclusion, it follows that *integration* is *possible*, and that possibility lies, I believe, in centauric vision-logic, the step beyond the rational-ego, the postmodern counterthrust to modernity (and, as I said, the burden of volume 3 is to render this thesis plausible).[19]

I mention all of this now only to point out that, with reference to the differentiations that have defined modernity, the solution, if there is one, lies in the marriages of tomorrow, not in the syncretisms of yesterday.

THE DEATH OF GOD

> *ZEUS:* Orestes, I created you, and I created all things. Now see! Impudent spawn! So I am not your king? Who, then, made you?
>
> *ORESTES:* You. But you blundered; you should not have made me free. You are God and I am free; each of us is alone, and anguish is akin. Human life begins on the far side of despair.
>
> *ZEUS:* Well, Orestes, all this was foreknown. In the fullness of time a man was to come, to announce my decline. And you're that man, it seems.
>
> —JEAN-PAUL SARTRE, *The Flies*

Before we look more closely at exactly why a Descending God replaced an Ascending God (and then "God" dropped altogether or, finally, reduced to *monological nature*), it is necessary, I think, to understand as clearly as possible the historical relation of the Age of Reason to the preceding Age of Mythology, for that is exactly how the philosophers of the Enlightenment viewed the situation. "No more myths!" was the philosophical battle cry of an entire age, and it soon came to be summarized in Nietzsche's famous dictum, "God is dead."

And my conclusion will be that in throwing out a prerational, anthropomorphic, mythic God figure, the "modern West" also tossed out any transrational, nonanthropomorphic, superconscient Godhead. Gone was a mass of bathwater; gone too a precious baby.

The problem, for the Church, from its inception, was how to deal with a complex of literal-concrete myths that it had inherited and to which it owed a certain allegiance. And in the life and figure of Origen (185–254 CE)—the Church's greatest theologian[20]—we see encapsulated the entire "problem of mythology" and its subsequent influence on an entire civilization.

Origen, recall, was a fellow townsman of Plotinus in Alexandria, where he had attended lectures by Plotinus's teacher, Ammonius Saccas. He was steeped in a causal/nondual worldview and clearly sought to bring this worldview into line with the causal realization of the Nazarene, which would not be hard in the least—were it not for a set of concrete-literal mythologies in which the Nazarene's realization was quickly being clothed.

Recall also that this was during the early flowering of the rational structure (widespread glimmerings of which had begun, according to theorists from Gebser to Campbell, around the sixth century BCE in Greece, although, as Habermas points out, formop was occasionally available to rare individuals much earlier). And even though the contemplative sages of the time had, in their own cases, explored even further into the transrational domain, there was a concerted effort to state the discoveries in *rational* terms. (This can be seen in a particularly clear case with Gautama Buddha, c. sixth-century-BCE India. Where most previous systems of Indian religion would *begin* with the gods and goddesses, and then prescribe *rituals* and *incantations* to appease and persuade them, Gautama radically dispenses with *all* of that, and recommends instead a direct and intensive introspection into the mind stream, into the *subject*, and he expresses this in eminently rational terms, even if the result of his injunctions and paradigms was the deconstruction of the rational stream and its supersession by transpersonal awareness.)

And wherever the contemplative philosopher-sages met *mythology*, either they dismissed it directly (which was an extremely delicate procedure; as Socrates discovered, the gods are not so easily tossed aside, especially when they serve the function of cultural "gluons"); or they began a long and slow process of *rationally reinterpreting* the myths. This allowed them to *preserve* the myth while *negating* its original meaning. This is surely one of the most fascinating features of this entire era, whether it was the Gnostics ("When you read the systems of the Gnostics, you have the feeling that they have rationalized mythology. And this feeling is accurate");[21] Philo ("He is an upholder of the verbal inspiration of the Old Testament, which nevertheless he turns into a moral and metaphysical romance by his theory of . . . rational allegorism");[22] or Plato and Aristotle ("Plato had maintained strongly that religion must be mythological *in its earlier stages . . .* a 'medicinal lie'; he warns us never to take them literally").[23]

Origen was of the same mind. He openly derides the literal myths in the harshest of terms. "Origen speaks with contempt of those Christians who take literally the temporal promises and threats of the Old Testament."[24] Origen does not believe in the survival of the individual soul or ego, or that salvation consists of anything like egoic perpetuity at all (that would in fact be hell; salvation, for Origen, consists in discovering that "God is all in all": Origen is, through and through, of a Neoplatonic mind).

Boldest of all, he does not believe in the future resurrection of the body. "He cannot restrain his impatience at the crude beliefs of traditionalists about the last day and the resurrection of the dead. The Gospels cannot [be taken] literally. How can material bodies be recompounded, every particle of which has passed into many other bodies? To which body do these molecules belong? So, he says scornfully, men fall into the lowest depths of absurdity, and take refuge in the pious assurance that 'everything is possible with God.' "[25]

And yet, Origen was a Christian (by the age of eighteen he was already head of the Catechetical School in Alexandria), and thus constrained to some degree by Scripture. And so he hit upon a brilliant solution, a solution for which he is justly famous (with Philo) and justly regarded a genius, and a solution that would henceforth be used whenever mythology needed to be both negated and preserved: the allegorical method.

Any myth, he says (referring to the Old Testament) can be interpreted on three levels: literal, ethical, and allegorical, each successive reading being "higher." The literal is just what the myth says; the ethical is taking the myth and rationally reworking it to apply to a present ethical situation; the allegorical is using the myth to mean (virtually any) mystical or spiritual or transrational thing we want it to.

The brilliance of this scheme is that it takes a prerational myth (literal) and reworks it at both a rational (ethical) and a transrational (mystical) level, so that "the myth" can be made to say whatever it is necessary to make it say, quite regardless of how its originators actually meant it. In other words, the interpretation takes the myth quite beyond itself—first into the space of reason, and then into the space of spirit. The myth is thoroughly preserved—and utterly negated.

This allows Origen to put into the myths whatever meanings from a higher level he wishes to put into them, so that he can both claim scriptural authority and basically ignore it at the same time. And every subsequent "mythologizer" in history, from the Catholic Church to Joseph Campbell, would use exactly this method whenever it was necessary to make the myths mean something other than what the myths themselves plainly said they meant.

Even though the Church would adopt this method without fail—what possible choice did it have?—it nevertheless found Origen's emphasis on the "higher" meanings to be very dangerous (Origen was busy pouring into the myths the entire corpus of the Neoplatonic outlook; Saint Jerome says he wrote eight hundred works). It was one thing to say that the myths *also* had higher meanings; quite another to basically *deny* any substantial reality to the lower or literal meanings. The Church suspected, rightly, that Origen had just blown a hole in their myths below the waterline. During the persecution of Decius (c. 250 CE), Origen was imprisoned, tortured, and pilloried; he was eventually condemned as a heretic by Justinian.

Origen referred to those who "took the myths literally" as possessing "mere faith," the "simple ones." And as harsh as he could be on the philosophers who believed in the "silliness," nonetheless "Origen extends to the popular, half mythological beliefs of the uneducated Christian [a great deal of] tolerance. The Logos teaches men in various manners, according to their capacities; some must be fed with milk, others with strong meat."[26] His idea was to start people at one level of development and, using the transcendent or allegorical method, work them up to gnosis. At (causal) gnosis, both the myths (literal) and the rationality (ethical) would be transcended, having thus served their purpose.

The Church, in generally denying a pure Ascent to causal gnosis, denied that the myths could be transcended (for all the reasons that we discussed). Other meanings could be *added to* the literal myths, but the myths themselves could never be negated. It was a victory for the "simple ones." Transrational gnosis was, once again, caught and frozen in prerational forms—the literal myths.

And it is the literal myths that the philosophers of the Enlightenment railed against, as had Origen. Virtually every single "scientific claim" of the Bible (and thus the Church) could be demonstrated, by disclosed evidence at a deeper depth, to be woefully inadequate (which was taken to mean: simply and plainly wrong). "The chief affront of Copernicanism to theological orthodoxy lay, not in any fundamental discrepancy between it and the more philosophical parts of the traditional scheme of the universe, but in its apparent irreconcilability with certain details of that body of purely historical propositions [myths taken literally] which Christianity had, to an extent matched in no other religion, incorporated in its creed."[27]

Church religion, across the board, became the laughingstock of the philosophers of the Age of Reason; it was met everywhere with scorn and derision. Are we, quipped Thomas Paine, "to suppose that every world in the boundless creation had an Eve, an apple, a serpent, and a Redeemer?"

The upshot of all this is that virtually the entire corpus of mythological belief was thrown overboard. And since, additionally, the great philosopher-sages (from Plotinus to Eckhart), who had themselves significantly transcended both myth and reason, were nevertheless *speaking out from mythically situated cultures*, virtually *everything* the great contemplatives had to say was jettisoned as well.

The baby with the bathwater. Reason could (and can) just as easily look up as it can look down. Reason—for Plato, for the Neoplatonists, and for the all-important Stoics—meant *seeing one's place in the Kosmos*; to be "rational" meant to see the great holarchy of existence and joyfully take one's proper place in it (quite apart from any further or transrational or mystical developments, where one directly identified with the Kosmos in toto). To be rational meant *attunement with the Kosmos* (as a preparation for identity with the Kosmos, with the All itself).

But with the rejection by modernity of the Ascendent God in virtually *any* form, there was only the Descended God left, the God of a marvelous and creative nature, a nature that was everywhere not to be denied, but a nature in which *alone* could salvation now be found. The doctrine of Plenitude and the Many completely eclipsed that of the Good and the One: God in any form was pronounced dead, and nature alone alive.

Reason, in reaction to myth, thus chose to look almost exclusively downward, and in that withering glance, the modern Western world was born.

PLENITUDE AS A RESEARCH PROGRAM

It is common knowledge that the early Renaissance scientists believed strongly that they were looking for the "laws of God"—that is, for the

Logos (or archetypes) of a Creative and Effluxing and Descending God, archetypes that expressed themselves in the *regular patterns* of a harmonious nature. The "laws of nature" were in "the mind of God" (or the Ideas of God), and since that same Mind created us as well, we ought to be able to see and understand and formulate those laws.

That they believed that is true enough; but it is much more specific than that. Lovejoy demonstrates, quite convincingly I think, that many of the "scientific discoveries" from the Renaissance to the Enlightenment were not originally empirical and "surprising" discoveries, but were in fact *a priori* assumptions derived directly (and consciously) from the two main tenets of the Platonic/Plotinian doctrine of Plenitude: nature is graded in a holarchy of being, and there are no missing links *anywhere*.

In other words, the Descending or Plenitude side of the originally Nondual system—the side that had been previously suppressed in the Middle Ages—was now isolated and emphasized and pressed into service as an actual *research agenda*, and it was this agenda that then *subsequently* began to produce the scientific discoveries for which the age is now famous, and in whose "modern shadow" we all now stand.

> In bringing about the change from the medieval to the modern conception, it was not the Copernican hypothesis, nor even the splendid achievements of scientific astronomy during the following two centuries, that played the most significant and decisive part.
>
> In the cosmography that by the beginning of the eighteenth century had come to be commonly held among educated men, the features which differentiated the new from the old world-picture most widely—these features owed their introduction and, for the most part, their eventual general acceptance, not to the actual discoveries or the technical reasonings of astronomers, but to the influence of those originally Platonistic [conceptions] which had, though potent and persistent, been always repressed and abortive in medieval thought.
>
> The more important features of the new conception of the world, then, owed little to any new hypotheses based upon the sort of observational grounds which we should nowadays call scientific. They were chiefly derivative from Platonistic premises [from the Descending side]. They were, in short, manifest corollaries of the principle of Plenitude.[28]

If we look first at astronomy, the doctrine of Plenitude—the infinite creativity of an infinite Source—seemed to demand not just the existence of this Earth and this inhabited world, but a virtual infinity or plurality of other worlds (in the concrete sense of other solar systems, for example), and there was no reason those other worlds shouldn't be inhabited by conscious beings, some perhaps as or more intelligent than we.

This notion of an infinity or plurality of other worlds profoundly shook the medieval mind, and shook it much more catastrophically than the simpler Copernican revolution. In fact, the heliocentric theory of Copernicus/Kepler had been taken rather easily in stride. After all, even if that theory displaced the Earth from the center of the universe, the Earth was still held to be unique, and still held to be the *sole* arena of rational beings watched over by a loving God, and the *sole* theater of the descent and ascension of the One and Only Begotten. The heliocentric theory, to the late medieval mind, might have displaced Gaia from the center of the solar system, but it did not displace Gaia from the center of attention; and, as we would say today, it barely made the headlines.

But a plurality of other inhabited worlds, with the Earth but one of many? This moved Gaia (and the Nazarene) out of the picture almost totally, and it was this *acentric* notion, derived not from any scientific evidence at all (we still don't have any scientific evidence for life on other planets) but derived rather from the doctrine of infinite Plenitude, that more than any other conception jolted the medieval mind out of the Middle Ages and into the Renaissance. "The change from a geocentric to a heliocentric system was far less momentous than the change from a heliocentric to an acentric one."[29]

And its chief proponent? It is no surprise: Giordano Bruno. The notion of a plurality of worlds was implicit (and occasionally explicit) in the Plotinian system, and it had received its first "modern" expression by—again, no surprise—Nicolas Cusanus. But "it is Giordano Bruno who must be regarded as the principal representative of the doctrine of the decentralized, infinite, and infinitely populous universe."[30] The Gaia-aggrandizing systems, whether geo- or heliocentric, were shattered (and with them, the always correlative egocentric, anthropocentric stance).

Cusanus and Bruno were thus the first to bring the Plenitude or Descending side (previously suppressed) into prominence in astronomy,[31] and the astronomers of Bruno's own and succeeding generation—Tycho Brahe, Kepler, Galileo—then dutifully began looking for that which was already assumed to be the case. And it was the Brunonian revolution, not the Copernican, that jolted Europe into the modern age (and that, not incidentally, brought Bruno to the attention of the Inquisition, and he was eventually burned at the stake). From Bruno the conception passed most notably to Descartes, and by the end of the seventeenth century, the belief in a plurality of inhabited worlds was accepted by most educated men and women, and the Gaia-centric mythology of the Church never recovered (though it would often be reactivated, right down to today).

The point is that these ideas long antedated the discovery of the scientific evidence for any of them; they were "becoming familiar in very widely read writings before the middle of the eighteenth century; and their development seems to have been chiefly due to the influence of the principles of Plenitude and continuity."[32]

If there could be no "gaps" or "missing links" in the heavens or macroworld, the same held for the micro-world. Quite before van Leeuwenhoek actually began peering down his microscopes, representatives of Plenitude were writing, as Leibniz did, that "I am convinced that there must be such creatures, and that natural history will perhaps some day become acquainted with them, when it has further studied that infinity of living things whose small size conceals them from ordinary observation and which are hidden in the bowels of the earth and the depths of the sea." Leeuwenhoek's discoveries came after the fact, so to speak, and were of little surprise to most, least of all to Leibniz.

But by far the most momentous, and certainly the most comical, search for "missing links" occurred in the "middle realm" of humans themselves: the search for the missing links between species, and particularly between humans and apes. This search was of crucial importance in the beginnings of the science of anthropology. Locke and Leibniz had already made it quite clear that what appeared to be gaps or missing links in nature were due solely to our ignorance, and thus yet another research agenda was set for the naturalist by the doctrine of Plenitude. As the great Bonnet would put it, "It is here above all that it is impossible to fail to recognize the graduated progression of beings; it is here above all that is verified the famous axiom of the German Plato [Leibniz], Nature makes no leaps. . . ."

Not only specialists but the general public was enthralled by the search, which looked for ever-earlier forms of human life and their linkages to higher forms of apes. Rousseau asserted (in 1753) that humans and higher

apes were in fact members of the same species, since language was probably not originally natural to humans—a notion that closed the gap genetically. But who other than the notorious P. T. Barnum could put his finger
on the pulse of an entire generation so accurately? In 1842 he advertised
that among the exhibits in his museum were

> the Ornithorhincus, or the connecting link between the seal and the
> duck; two distinct species of flying fish, which undoubtedly connect
> the bird and the fish; the Mud Iguana, a connecting link between rep-
> tiles and fish—with other animals forming connecting links in the
> great chain of Animated Nature.

And this was nearly two decades before the publication of the *Origin of
Species*!

The doctrine of Plenitude as a research agenda: there was the Descending God of Infinite Efflux released from its confinement in the Middle
Ages and set loose with all its delight in glorious diversity. "Every discovery of a new form could be regarded, not as the disclosure of an additional
unrelated fact in nature, but as a step toward the completion of a systematic structure of which the general plan was known in advance."[33]

Although this systems-theory research agenda had started, in the hands
of Cusanus and Bruno, as but one half of the total picture of the Many to
the One and the One to the Many, it soon became the entire picture (as
we will see), and attention centered solely on the Many, the All, the manifest Whole as a harmonious system quite apart from its Source and
Ground. The guiding agenda for *this world* was still the same—fill in the
gaps in the Whole System—but with no mention of the Source, the Good,
the One. It was to the manifest Whole, the Great System of Nature, that
research was confined, and the research was indeed, through and through,
a systems theory approach ("overall harmony of the whole"), whether in
economics, political theory, or science.

"Everything in nature is linked together," began the *Encyclopédie* entry
under the advancement of knowledge, and "the art of the philosopher
consists in adding new links to the separated parts."[34] Scientists and philosophers alike "were wont, when they reached the height of their argument, to discourse with eloquence on the perfection of the Universal System as a whole."[35] As Pascal would put it, "Its parts all are so related and
interlinked with one another that it is impossible to know the parts with-

out knowing the whole or the whole without knowing all the parts." And Pope would memorialize the temperament of an entire age in search of "strong connections, nice dependencies":

> He who through vast immensity can pierce,
> See worlds on worlds compose one universe,
> Observe how system into system runs,
> What other planets circle other suns,
> What vary'd being peoples every star,
> May tell why Heaven has made us as we are.
> But of this frame, the bearing and the ties,
> The strong connections, nice dependencies,
> Gradations just, has thy pervading soul
> Look'd thro?

But in all of this, something very, very strange was occurring, and had in many cases already occurred. In the multidimensional Plotinian Kosmos, the One gives birth to the Many—to the All—and the All return to the One, with Each individual a perfect embodiment of the infinite One itself. The One underlies both Each and All.

However, in the new and truncated conception, since the infinite One was no longer admitted ("No more Ascent!"), then the Descending or finite or manifest All or Whole was the *sole* residence of Providence and Harmony. It was no longer the relation of Each and All to the One, but merely the relation of Each to the All.

Systems theory was born.

Put bluntly, the *sum total* of the shadows in the Cave was now confused with the Light beyond the Cave. And it was dearly felt and firmly believed that *salvation*—both individual and social—could be secured if only we could demonstrate with ever greater clarity that the shadows are not atomistic but rather holistic—that they hang together as a great harmonious system, that the universal invisible hand of holism draws Each to the All in functional perfection.

Salvation was no longer Each discovering that it *is* the One (and thus capable of embracing the All). Rather, salvation now demanded that Each somehow identify merely with the All (bypassing the One)—that Each would *functionally fit* with the All, that Each would take its instrumental place in the great and glorious web, whether that web be nature, the state, or the cosmos, the idea being that if shadows snuggle together tightly enough, Light will somehow be produced.

The research agenda was thus to *demonstrate the interlocking order and holistic nature of the shadows* (as if that would do anything to help one out of the Cave). It emphasized, as I said, the descending systems theory of Plenitude, which is correct as far as it goes, but is only half the story. But it was that half-story that would soon claim to be the alpha and omega of the universe (and that is the "bad news" of the next chapter).

In the meantime, the Great Holarchy, in its Effluxing side, was the *research agenda* for the new and rising culture of modernity. And the Plenitude of Efflux would come to define, as it does to this day, the wondrous new world that had just unfolded, and whose God would be defined solely by the maxim: The greater the diversity, the greater the Goodness.

A NEW PLACE IN NATURE

Nothing shows more clearly that the Great Holarchy of Being was not a product of the medieval mind than the fact that, by the middle of the eighteenth century, although virtually every single aspect of mythic and mythic-bound Scholastic religion had been derisively rejected by the philosophers of the Enlightenment, virtually every single one of them embraced the Great Holarchy. It quite obviously, to them, had nothing to do with mythic religion or the medieval mind at all, or it would have been as thoroughly and as quickly abandoned.

> There has been no period in which writers of all sorts—men of science and philosophers, poets and popular essayists, deists and orthodox divines—talked so much about the Chain of Being, or accepted more implicitly the general scheme of ideas connected with it, or more boldly drew from these their latent implications. Addison, King, Bolingbroke, Pope, Haller, Thomson, Akenside, Buffon, Bonnet, Goldsmith, Diderot, Kant, Lambert, Herder, Schiller—all these and a host of lesser writers not only expatiated upon the theme but drew from it new consequences. Next to the word "Nature," "the Great Chain of Being" was the sacred phrase of the eighteenth century.[36]

The Great Chain of Being was palatable to the Age of Enlightenment, not just because it apparently expressed certain profound truths, but precisely because it was not a mythic-level production (even if it also, in various forms, was present there as well). The Church had appropriated the Great Chain straight from Plotinus via Origen and Dionysius, but had, as we saw, precisely by mythologizing it, put a limit on the upper reaches of Ascent and generally repressed Descent altogether. And the theme of this (and the next) chapter is that the Enlightenment would also appropriate the Great Chain, in an equally lopsided but reverse fashion. It is fascinating, though, that both the mythic and the rational mind, and the coming vision-logic, would all embrace the Great Holarchy; it clearly was one of the very few conceptions that could withstand the complaints of each.

Of course, the Enlightenment would not accept the complete Circle of Ascent and Descent, as we will further see in just a moment. But the net effect of the "virtually unanimous" acceptance of the Great Chain itself was a radically altered view of the place of men and women in the universe, a view almost diametrically opposed to that of the Middle Ages (where its mythologized version was in sociocentric service). In this particular regard, the "new" and "modern" view was precisely that held by Plotinus or Bruno or Cusanus or any other upholder of the doctrine of a radically decentered universe, and a view we have mentioned before: a radical *nonanthropocentrism*.

It is necessary to mention this point one last time, simply because of the misinformation displayed by most ecological critics of the Great Holarchy. Let Lovejoy say it for us one last time: "It was implied by the principle of Plenitude that every link in the Chain of Being exists, not merely and not primarily for the benefit of any other link, but for its own sake; they all had an equal claim to existence; and therefore the true *raison d'être* of one species of being was never to be sought in its utility to any other."[37]

It is true that all beings were held to be different degrees of manifestation, and could therefore be ranked according to degree, with the higher embracing the lower but not vice versa. But the reason that all beings were held to have ultimately equal claim to existence was that the ranking was apparent only; it was the product of a not-fully-awakened awareness. For, in the purely Nondual Ground, *all points* on the Great Circle, whether high or low, ascending or descending, were equidistant from the Center—equidistant from an ant, a star, or a human, as Bruno put it.

While this view still accepted a *relative* or apparent-only ranking, and thus allowed for pragmatic value judgments in the relative world (better to kill an apple than an ape), this was profoundly different from a *mere* ranking or an absolute ranking, which would hesitate not at all in making all "lower grades" *only* and always instrumental means. And it was exactly the medieval view—anthropocentric, geocentric—and its stubbornly persistent remnants that the philosophers of the Great Chain were so intent on fighting. A principal text of Scholastic religion/philosophy had stated:

> As man is made for the sake of God, namely, that he may serve Him, so is the world made for the sake of man, that it may serve him.

From a Protestant theological text much admired at the time:

> If we consider closely what constitutes the excellence of the fairest parts of the Universe, we shall find that they have value only in their relation to us, only in so far as our soul attaches value to them; that the esteem of men is what constitutes the chief dignity of rocks and metals, that man's use and pleasure gives their value to plants, trees and fruits.

It was precisely *against* this worldview "that the logic of the conception of the Chain of Being worked potently."[38] And the chief opponent of anthropocentrism in all its forms was . . . Descartes. Leibniz and Spinoza (a dedicated Cartesian) picked up the fight, and Goethe represented them all in his poem *Athroismos*: "Every animal is an end in itself." Or, as Schiller would so clearly put it, every being "has an unchallengeable right of citizenship in this larger understanding of creation." But perhaps, once again, Pope said it best:

> Man walk'd with beast, joint tenant of the shade;
> The same his table, and the same his bed;
> No murder cloath'd him and no murder fed.
> In the same temple, the resounding wood,
> All vocal beings hymn'd their equal God!

That, indeed, was good news.

And so it was that Reason rose up to greet the new dawn, unaware that its very strengths would soon turn on it. The sheer power of its capacity to represent the Kosmos empirically—a power unleashed by the differentiation of the Big Three—would allow it inadvertently to collapse the Kosmos merely to its monological, empirical, Right-Hand aspects, a collapse that would lock men and women into a purely Descended world, flatland to the core.

And strangest of all, the new Reason would appropriate even its own worst critics, take them up and hide them in itself, and thus ensure that even the newly rising proponents of Gaia would harbor the flatland ontology that is still destroying Gaia to this day.

For it would soon become obvious that the Descenders were destroying this world, because this world was the only world they had.

12

The Collapse
of the Kosmos

*Everything that from eternity has happened in heaven and earth,
the life of God and all the deeds of time simply are the struggles
for Spirit to know itself, to find itself, be for itself, and finally unite
itself to itself; it is alienated and divided, but only so as to be able
thus to find itself and return to itself. . . .*

*As existing in an individual form, this liberation is called "I"; as
developed to its totality, it is free Spirit; as feeling, it is Love; and
as enjoyment, it is Blessedness.* —HEGEL

IN THE WEST, from the time roughly of Augustine to Copernicus,
the mythic-rational structure, for reasons I tried to suggest in the last
two chapters, emphasized quite heavily the Ascending current, and
theoretically held out the Goal of perfect Ascent in Christ (while at the
same time officially prohibiting it).[1] This combination locked the West
into an almost exclusively Ascending Ideal for over a thousand years.
With rare individual exceptions, the great omega point was promised always, delivered never: *Ascendus interruptus* at the heart of it all, leaving
the West with a truly *peculiar* spiritual hunger, a hunger found nowhere
else, really, with quite the same sort of desperate face. . . .

With the widespread emergence of Reason as the center of gravity of
social organizing forces, the mythic-rational structure was itself transcended (in all the ways we discussed in chapter 5), which brought a step
up in Ascent (Eros), with a correlative increase in the capacity for Em-

brace (Agape; evidenced, for example, in universal benevolence, world-centric perspectivism, and the many liberation movements).

The rational structure itself could then, depending on contingent and cultural factors, emphasize, as an ideal, the Ascending current, the Descending current, or both (preferably). And I have suggested that—largely as a *reaction* to a millennium of (frustrated) Ascending Ideal—the Age of Reason, quite apart from the fact that it was itself a new step of Ascent, almost wholeheartedly threw itself over to the Path of Descent and the glories of Creation and Efflux, turned its back on its own Source and Origin, and pledged allegiance to a visible, sensible God. Unlike the Ascending God, *that* God could be bedded. That God could be had. That itch, scratched . . . indefinitely.

Instead of an infinite *above*, the West pitched its attention to an infinite *ahead*. The vertical dimension of depth/height was ditched in favor of a horizontal expansion, an emphasis not on *depth* but on *span*—and the standard God of the modern Western world was set. It would become the God of the bourgeois as well as of the dedicated scientist; the God of the materialist as well as of the social reformer; the God of the Greens and the "back to nature" movement wherever it appeared; the God of democracy as well as the God of the Marxists and Maoists—what they all had in common is the God of all that is visible, and all that can be seen, and all that can be grasped with the hands (grasped, that is, with the Right Hand).

An "other world" of *any* sort was thrown over; and the eyes of men and women settled steely on the horizons not above but in front of them, settled coldly on this world, and this world, and this world again. If salvation could not be found on this small Earth, it could not be found at all.

THE COLLAPSE OF THE KOSMOS

Even though virtually all of the Renaissance and Enlightenment philosophers (and the Romantic philosophers, too, as we will see) subscribed to the Great Chain, something profound had happened to it, and especially to its "upper reaches." Most educated men and women assumed that humans were somewhere in the middle of the Great Chain, assumed that there were as many (or more) dimensions of "higher development" above them as there were below them. "We have reason to be persuaded," said

John Locke, "that there are far more species of creatures above us, than there are beneath; we being in degrees of perfection much more remote from the infinite Being of God, than we are from the lowest state of being, and that which approaches nearest to nothing."

This nonanthropocentric arrangement contributed to a not altogether unhealthy diminution of human pride. Formey's lament was typical of the time:

> How little cause have I to exalt myself above others, and whence can I derive motives for pride? Heretofore I used to conceit myself one of the most excellent of God's creatures, but I now perceive how great my delusion was. I find myself toward the lowest part of the Scale [of Being], and all I can boast of is, that I have a small pre-eminence above irrational creatures; and this is not always so, there being many things in which they possess advantages which I have not. On the contrary I see above me a multitude of superior intelligences.[2]

As Lovejoy notes, virtually every Renaissance and Enlightenment philosopher made that assumption. But quite apart from whatever salutary humility the whole notion might have imbued in men and women, it is quite apparent that, for them, the "upper reaches" of the Great Chain were simply being *postulated* to exist ("reason to believe"). The higher or deeper dimensions were *assumed* to exist, they were a rational hypothesis put forward to "fill in the gaps" between humans and Deity. Since, according to the principle of Plenitude, there could be no gaps in existence, and since there was definitely a gap between humans and the infinite Source, then that gap simply had to be filled by an almost infinite number of superior grades of intelligence, grades that, if they weren't exactly "creatures," were at least a series of infinitely receding higher Ideas and forever unattainable Perfections utterly beyond human reach.

In the Middle Ages, the average mode had *mythologized* these "higher grades" as angels, archangels, and various heavenly agents stretching from humans to the ultimate Divine. And now these "superior creatures" were instead being *rationalized* as superior grades of intelligence and virtue and wisdom, but none of which could be reached by the finite human mind. And however true that might be in a certain sense (they cannot be reached by the conventional mind), this is not at all the way these higher grades were viewed by the originators of the system itself.

For Plotinus, Dionysius, Eckhart, and company, these higher dimensions of existence are, above all else, *potentials* inherent or present in each and every being (since each and every being issues forth from them), and thus these higher potentials can be *realized* or consciously *actualized* for any being who can find the requisite height/depth in his or her own soul. These higher levels are *not* angels out there, or metaphysical notions merely postulated to exist somewhere, or philosophical and logical assumptions used to fill in any "gaps" in the world. They are potentials that can be *directly experienced* and *directly realized*. There is absolutely nothing *other* about them, except the otherness created by our own lack of inner awareness.

For a child who has not yet learned language, the world of symbols and concepts seems to float around all about it and above it; language is "over its head"; language is an "other world," full of powerful magic and populated with beings who seem to understand things that the child does not, beings who seem to possess a great deal of power incomprehensible to the child—there are gods and goddesses out there, angels out there, "creatures" out there, against which the child can feel only its own miserable impotence. But once language emerges as a developmental component of the child's own being: *voilà!* the linguistic powers, the conceptual angels, are all within.

For Plotinus (and similar nondual sages everywhere), the higher grades of consciousness are not otherworldly or metaphysical in the least; they are *not* philosophical postulates; there is nothing about them that cannot be directly experienced and directly realized in one's own case, given the proper growth and development (following the proper paradigms, injunctions, and exemplars in a community of shared social practice).

And while the mythic mind and the rational mind will construe these higher dimensions—their *own* higher potentials—in terms that reflect their own stage of development (I have no quarrel with that), nonetheless to even theoretically turn these higher stages into merely metaphysical postulates is a disaster. It denies the very essence of these higher stages: they are all experimental, contemplative, experiential realities that are directly disclosed to immediate awareness under the proper laboratory conditions.

The Enlightenment mind was, I have suggested, following two very different but fused agendas. One was "No more myths!," which was phase-specifically appropriate enough. But since the myths were almost exclusively upward-looking (although frustrated), the second agenda, quite unnecessary, was "No more Ascent!" The upper reaches of the Great Chain were thus converted into mere *metaphysical postulates*—and subsequently dropped altogether.

The result was that Ascent—transcendence—of *any* higher sort was looked upon with gravest suspicion ("No more Ascent!" meant "Ascend to Reason, but no further!"). Incredibly, the higher stages of the Great Chain became exactly those virtues and powers that one *could not reach* and *should not even try*. The very *attempt* to reach them, in fact, was equated with *pride*, which, we will see, was the dreaded sin of the century.

The Great Chain, which was *always* supposed to be a map of one's own highest potentials—a map of the course that Eros would follow in the blossoming of a love that would embrace the entire Kosmos with Agape and caritas—that extraordinary map of liberation became instead a cage into which humans were locked with no possibility of higher/ deeper development. As Rousseau (in *Emile*) would exhort an entire generation: "O Man! confine thine existence within thyself, and thou wilt no longer be miserable. *Remain in the place which Nature has assigned to thee in the chain of beings*, and nothing can compel thee to depart from it . . ." (my italics).

Pride thus became any attempt to "rise above one's station," to "counterwork the Universal Cause," to "disturb the very system of the universe." The way up, in any form, in any variety, in any guise, was the arch-sin, the sin of horrid pride. Any hint of "otherworldliness" was openly scorned, severely derided—not to mention completely distorted and misunderstood. Pope again:

> Go soar with Plato to the empyreal sphere,
> To the first good, first perfect, and first fair,
> Or to tread the mazy round his follow'rs trod,
> And quitting sense call imitating God;
> As Eastern priests in giddy circles run,
> And turn their heads to imitate the sun.

So the way up—any form of Ascent beyond Reason—was viewed as not just a bad idea, but as a literal crime against nature, a crime *against* the Great Chain and its allotted spot for men and women.

> In pride, in reas'ning pride, our error lies;
> All quit their sphere and rush into the skies.
> Pride still is aiming at the blest abodes. . . .

The theme of the modern age thus became, as Pope again would have it: "Trace Science then with modesty thy guide." Do not soar, do not look up, do not transform, but follow downward-looking Science with modesty thy guide.

The Deists—Toland, Tindal, Voltaire, Rousseau, Jefferson, Franklin—would try to salvage what they could of some sort of a God—a God that originally created the world, "wound it up," and then retired from the scene entirely. Unlike Aristotle's Pure Omega God, from which nothing issues but to which all aspires, the Deists' God was the pure Alpha Source, from which all issued but toward which nothing moves. The two paths of Ascent and Descent at this point had utterly and totally separated, dissociated, divorced, anemic and fractured, with no point of contact, no point even of discourse, let alone integration. And the Deists' God, it soon became obvious, served precisely no function whatsoever, either morally, practically, or even theoretically—and was understandably dropped altogether.

The wonderful world of Plenitude was no longer from God; Plenitude itself was God, the only God. In all directions there was only Descent. In all directions, only Shadows. In all directions, only Flatland. And the only real sin? To try to leave, to try to escape from the Cave, to try to see some reality more than the eye of flesh might manage to take in, to try to transform instead of merely translate. All of that was . . . pride.

The realm of the senses, guided by Reason: there was the foundational reality. The wonders of creation—who needs a source for them? The bigness of creation!—who could ask for more? Let me bask in the riches of the sensory world, and reason upon it where necessary, but why ask behind the scenes for more than this display? Let cosmic emotion and piety carry the day, and let me weep with joy at any passing sight in nature that strikes the slightest chord of egoic sentiment, and let me spend my days and nights suckling shadows that are dear enough to me. And let me never, never, never sin by asking beyond the shaded nooks and crannies. Let me always be known for saving and defending the wonderful appearances, and with eyes so modestly reverted from that all-embracing Source, let me never look up and never step out of my allotted spot on that great, great Chain.

And so, after more than two millennia, *it had finally come to this*: the path of liberation ended up the sin of pride. The Great Chain, the map of what we could become, became a map of what we should not even try. The Great Chain was used to deny the Great Chain. The way out of the Cave was used to imprison men and women in it. The Great Chain was no longer the map of escape, but the plan of the prison cell with a sentence of life.

THE GREAT INTERLOCKING ORDER

With the collapse of the Kosmos into the merely and purely Descended world, all that was left of the Great Holarchy showed up as an empirical, flatland *interlocking order* of surfaces, or exteriors, or Right-Hand components. In place of the Kosmos, flatland systems theory.

I would like, in this section, to explore carefully this "great interlocking order" (as John Locke would have it)—how and why it arose, what it accomplished, and what it aggressively violated. Because, one way or another, in one guise or another, the only God the modern world would henceforth acknowledge, if it chose to acknowledge a God at all, would be the God of the great interlocking order.

And so let me say it one last time: for Plotinus or Aurobindo or Eckhart or Shankara, the great interlocking order was true but partial; it was only half the story (the "no gaps in Plenitude" half, the Descending or Effluxing aspect). It represented the mutual interpenetration of the Each and the All. But the ground of both the Each and the All was the nondual One, found not merely in the Descent of Goodness but in Ascent to the Good. But when this half-story became the "whole" story, there precisely was the birth of the nightmare known as modernity.

And modernity's collapse of the Kosmos occurred right alongside its genuine accomplishments, which, I believe, are not to be denied (the good news / bad news nature of modernity). We have seen that one of the great defining marks of modernity—what has been called its genuine *dignity*—was the clear differentiation of the Big Three, the differentiation of science (it), morals (we), and art (I), so that each could pursue its own course and establish its own truths without domination by the others. "Cultural modernity's specific dignity is constituted by what Max Weber called the differentiation of the value spheres in accord with their own logics."[3] And "by the end of the eighteenth century, science, morality, and art were even institutionally *differentiated* as realms of activity in which questions of truth [science], of justice [morals], and of taste [art] were autonomously elaborated, that is, each of these *spheres of knowing* [was pursued] under its own specific aspect of validity."[4]

But if the *dignity* of modernity was the differentiation of the Big Three, the *disaster* of modernity would be that it had *not yet* found a way to *integrate* them. Indeed, with no unifying center, the Big Three had already begun to move into *dissociation*—to exist as radically separate, "hyper-autonomous" realms, with nothing to say to each other, bound only by a thoroughgoing and often mutual disdain. "This separation and self-sufficiency, which, considered from the standpoint of philosophy of history, paved the way for emancipation from age-old dependencies [mythic-membership syncretisms], were experienced at the same time as abstraction, as *alienation* [dissociation] from the totality of an ethical context of life. Once religion had been the unbreakable seal upon this totality; it is not by chance that this seal has been broken."[5]

In the absence of the as-yet-undisclosed integration, the three spheres went virtually their separate and lonely ways. In particular, the subjective and moral spheres—the Left-Hand path, the interior path—pursued its own logics rigorously, and the Right-Hand path, that of the study of the objective and empirical exteriors, pursued its own course in isolation as well. Both paths were making absolutely world-shaking discoveries, but discoveries that spoke to each other virtually not at all.

It was this *dissociation* of the Big Three that, more than anything else, allowed the sweeping advances of the natural sciences (the it-domain) to overwhelm the interior, subjective, moral, and cultural domains (I and we), and this is what I mean when I say the Big Three were reduced to the Big One, or reduced to the Right-Hand path alone, the path of objective it-language (the representation/reflection paradigm, and its converse, the production paradigm, and the overall validity claims only of propositional truth and/or functional fit). The very *power* of monological reason (ever so important and ever so partial), which was *released* from mythic syncretism precisely by the *differentiation* of the Big Three, was allowed an overwhelming dominance because of the *dissociation* of the Big Three, and thus their subsequent reduction to the Univocal One, the flatland world of monological, empirical-analytic, positivistic "its," where only the entities studied by empirical-natural sciences are "real."

In subsequent volumes, I trace a large part of this dissociation and resultant emphasis on flatland to the strong influence of the techno-economic base of *industrialization* and the machine mentality (which is similar to, but not quite the same as, the more common analysis that traces it to capitalism): the techno-economic base supported instrumental-purposive activities, and in a way all out of proportion to the instrumental-purposive rationality that did in fact build it: a positive feedback loop that sent calculative rationality spinning out of control, precisely in the avowed purpose of gaining control.

In the meantime, Donald Rothberg gives an excellent summary of these broad trends, which is quite consistent with my presentation: "For a variety of reasons, these emergent potentials of modernity, these three modes of knowing and acting [art, science, morality: the Big Three], were not evenly developed. Rather, as many of the main critical analysts of modernity have suggested, modernity as it actually unfolded (and still unfolds) was (and is) heavily weighted toward the knowing and manipulation of the first world [the it-world, Right Hand, objectivistic], of a 'disenchanted' and objectified world dominated by an 'instrumental' or 'technical' rationality (Berman, Habermas, Heidegger, Horkheimer, Marcuse, Merchant, MacIntyre)."

Thus, Rothberg points out, "The content of the other two worlds (the intersubjective [we] and subjective [I] worlds) was increasingly organized according to the structures of the empirical sciences and instrumental or calculative rationality; Habermas links this unbalanced development especially with the powerful influences of the forces of capitalism. Under the 'scientistic' influence of positivism and empiricism, there were claims of a unified empirical science that encompasses all 'real' knowledge, excluding the various attempts to claim an autonomous status for the emerging human sciences of cultural and subjective reality"[6]—the Big Three reduced to the Big One.

The flattening, the leveling, the collapse of the Kosmos. The universe was pushed through a strainer of objectification, and the result was thin soup indeed. All that was left of a richly multidimensional Kosmos was simply the sensory/empirical exteriors and outlines and flatland forms, much as if a sphere had been projected onto a plane surface, producing only a series of flat circles—all span, no depth—at which point we say, "What sphere?"

The vertical *and* horizontal holarchies—the interior *and* exterior dimensions—which, we might say, "interlock" in a rich network of values, degrees of consciousness, and qualitative distinctions, none of which can be measured in terms of *physical size* or empirical inches or material span: these were reduced to their sensory/empirical and physical correlates, to the physical forms and positivistic exteriors that *could* be seen (the Left reduced to its correlates in the Right), and these indeed could be measured primarily by physical size or other extensional attributes (mass, velocity, momentum, number, span).

And where qualities and values are measured in terms of better and worse (compassion is usually better than murder), empirical sizes are *not* better or worse, only bigger or smaller (a star is not better than a planet, only bigger). And thus the great interlocking order of sensory surfaces and exterior forms—the final result of pushing the universe through the strainer of objectification—was inexorably the *disqualified universe*.[7]

The Kosmos, literally, was a shadow of its former self. The vertical and interior holarchies (of I and we) were ditched in favor of the merely horizontal and exterior holarchies (of the great interlocking order of valueless its).[8] *Qualitative* distinctions were replaced merely with *quantitative* distinctions and technical measurements. "What does it mean?" was fundamentally replaced with "What does it do?" Intrinsic value disappeared into empirical action terms. "What worth?" was replaced with "How much?" *Greater* was replaced with *bigger*. Cultural meaning drifted into functional fit and holistic interaction of surfaces. Morals melted into systems theories. Eros was converted to instrumental-functional *efficacy*, and Agape melted down into an affirmation of *ordinary* Descended life.

In short, *depths* that required *interpretation* were largely ignored in favor of the interlocking *surfaces* that can simply be *seen* (empiric-analytic)—valueless surfaces that could be patiently, persistently, accurately mapped: on the other side of the objective strainer, the world appeared *only* as a great interlocking order of sensory surfaces, empirical forms, process its.

The massive empirical mapping game—the epistemological game of modernity—was afoot.

We heard Taylor summarize this whole shift as: "The universe as a meaningful order of qualitatively differentiated levels [vertical holarchy] gave way first to a vision of mathematical order, and then finally to the 'modern' view of a world of ultimately contingent correlations, to be patiently mapped by empirical observation."[9]

And, as we saw in chapter 1, it wasn't just that this empirical mapping game was partial: taken in and by itself, it was perfectly self-contradictory. On the other side of the objective strainer, there was simply no evidence whatsoever that this thin objective soup was the *only* evidence worth knowing. No, *that* assertion (Right-Hand data alone are real) was a Left-Hand judgment that denied all Left-Hand judgments. The objectivistic worldview could not account for its *own status* (the "view from nowhere"): it could prove objects but could not prove the exclusiveness of objects—could not prove itself—but rather was taken, literally, on blind faith, a faith *blind* to the entire Left half of the Kosmos, even as it surreptitiously dipped into that dimension for its own hidden judgments, judgments which it forcefully and vehemently made and then flat-out denied making. "Empirical knowledge alone is true knowledge"—and where is the *empirical* proof for *that*?

The net effect was that the Great Chain was tipped on its side, so to speak—an infinite within and beyond was ditched in favor of an infinite in front of and ahead, and the West began to scratch that itch in earnest.

I want to briefly interrupt the narrative at this point to mention a simple but crucial item. This will take us ahead of our story, but it's important, so let us at least briefly note the following before returning to the main narrative.

With the postmodern overthrow of the Univocal One—that is, with the return of an investigation of the Left-Hand dimensions, including multicultural interpretation and deep hermeneutics, introspection and interior disclosures, the existence of intersubjective discursive formations and cognitive paradigms, chains of signification and depths of communication, the demand for qualitative distinctions and the search for worth and meaning—with, in short, the beginning return of the Big Three as opposed to monolithic flatland, interest can once again turn (and has turned) to the depths of the subjective I and the intersubjective we: depths that range from Heidegger's clearing or opening (the pure *transcendens*), to hermeneutics' restless search for the depths, to the mystical openings found even in Nietzsche and Bataille and Derrida and, yes, in Foucault's intense search for limit experiences and the "mad mystical" poets. Depths that have sparked an explosion of interest in everything from humanistic and transpersonal psychology to Eastern mysticism and yoga. Depths that call out even to the postmodern poststructuralists, who, if nothing else, are quite certain that truth is not merely a clean game of positivistic and empirical mapping.

For all postmodern currents have at least one thing in common: mere empiricism is dead. The surfaces are shot. The flat and faded system is anemic to the core. The thin soup is causing malnutrition: epistemological, ontological, motivational.

The whole point is that once the weight of the Univocal One is lifted from the shoulders of awareness, the Big Three jump instantly back into focus, and interior depths once *forbidden to serious discourse*—depths forbidden with the shuddering collapse of the Kosmos—now unfurl before the mind's inward eye: the surfaces are not all, the shadows hide something else. The appearances don't just reveal, they conceal: something other is going on. There are nutritional supplements that this thin soup of shiny surfaces does not contain, secrets not apparent to the senses, items that are more . . . interesting.

And among those *waiting for us* in this renewed investigation of interior subjective and intersubjective depths are precisely the original pioneers in that endeavor: the great mystic philosopher/sages from Plotinus to Dionysius to Augustine to Teresa to Eckhart to Emerson, pioneers that struck out to investigate a vast and unchartered continent of consciousness, a voyage of interior discovery that continued right up to the historical point—*this* point in our narrative—where that particular continent was proclaimed nonexistent and all explorers were banned (by "cultural marginalization") from even discussing their findings. The interior dimensions were no longer considered *serious discourse*: all of that was part of the "merely mythic" past (no more myths, no more Ascent). As Foucault would put it of this entire period (of collapse), if something couldn't be laid out on a *flat table of representation*, then it *didn't exist* (and, said Foucault, it seemed not to bother the classical Enlightenment that the drawer of the table—subjective consciousness itself—could not itself be laid out on the table). So much for the extraordinary pioneers of the interior (from Plotinus to Ficino to Eckhart . . .).

But they left us their early maps, maps of the interiors of what now amounted to a New World awaiting (re)discovery. And most of these maps were of the general form: turn left at mind, cut straight into the formless, rest in Spirit, embrace the world; and all of the steps required for that extraordinary journey (maps such as figure 9-1).

But, indeed, many of the specific details of these maps are less than accurately drawn. These early maps are in many ways like the early maps of the Americas: Cuba is the size of Texas, Greenland extends forever, Florida is a huge continent. But the fact that some of these maps are crude or initial or outdated does not mean that Cuba and Florida and Greenland don't exist.

Thus, it is not necessary that we embrace every detail of their maps. Not only are some of them hastily drawn, the whole point of consciousness development is that it also *creates territories* as it goes, and doesn't simply discover *totally* pregiven landscapes. The fact that certain broad domains of interior consciousness are available to us (as deep structures) does not mean that the surface structures are given once and for all (which is why even excellent interior maps, such as Plotinus's, have to be refined and updated). Even if Columbus had drawn a perfectly accurate geographical map of Florida, it would still be woefully inadequate for most of today's uses (it wouldn't show the new cities, roads, towns, developments, etc. Where are all the subsequently *created* surface structures?). Each level of consciousness is actually a *four-quadrant affair*, molded equally by intentional, behavioral, cultural, and social factors, all of which have continued to evolve and develop since the time of the great pioneers.

Rather, these early maps are much along the lines of what Foucault suggested: "Among the cultural inventions of mankind there is a treasury of devices, techniques, ideas, procedures, and so on, that cannot exactly be reactivated, but at least constitute, or help constitute, a certain point of view which can be very useful as a tool for analyzing what's going on now—and to change it."[10]

Thus, as we will see shortly, when the Idealists began the overthrow of flatland, the overthrow of the collapsed and disqualified universe, they had recourse to precisely these early maps of the interior Worlds. And, as we will also see, if there was one person who could be said to be seminal to this entire movement, that person was . . . Plotinus.

But that, indeed, is quite ahead of our story. All I wish to emphasize here is that, with the postmodern *possibility* of the return of the Big Three, we do indeed have the early maps of the interior New World left by these extraordinary pioneers. And relying on those maps—at least initially—is not a regressive yearning for yesteryear, nor a mythic devolution, but rather the only sensible course of action for a new breed of pioneers trying once again to plumb the depths that were utterly disqualified with the modern collapse. Refine the maps, yes; redraw many of their outlines, surely; but thank the stars for the guts and glory of those who went before, and left a trail, clearly marked, for all those souls sensitive enough to follow.

A CALCULUS OF PLEASURE

To return to our narrative. With the collapse of the Big Three to the Big One: the collapsed and disqualified world was *still a hierarchy*, still a *holarchy*, but now played out strictly and only in the *horizontal* plane (any vertical move was the sin of pride). So of course the universe was still viewed as a massive interlocking order, a great net of systems theory, but a net that was now merely empirical and physical, that could be seen with senses or their extensions, a net of the Right-Hand path.

Thus, not also depths to be interpreted and interiors to be accessed, but simply exterior surfaces to be seen and "patiently mapped by empirical observation"—the so-called "reflection" or "representation paradigm," where a pregiven, unproblematic subject simply *reflects on* or *represents* the objects in the pregiven, sensory, empirical world.[11]

The Upper-Right quadrant, for example, was still a holarchy—atoms to molecules to cells to organisms, and so on. *This wasn't denied.* What *was* denied was that the *Left-Hand dimensions* needed to be explained in any terms *other* than those found in the *Right Hand*. This is subtle reductionism. (A few ingenious cranks then went further—into gross reductionism—and claimed everything in the Right Hand could actually be reduced to atoms, the old Epicurean move.) But in both gross and subtle reductionism, reality is merely monological and empirical and representational (reflection on the empirically available), and thus *no inner transformation* (no interior Ascent or vertical development) is required in order to have access to "all the truth that is fit to know."

And this fit perfectly but *perfectly* with the arch-sin of the age. To attempt to rise above one's station—that was, of course, the dreaded pride. Well, the Right-Hand path doesn't require anybody to rise above anything: just look more carefully at the empirical world that is more or less available. Not arduous *transformation*, but simply more precise *translation* (mapping of the empirically accessible).[12] The Great Chain tottered off into empirical observables, degrees of depth collapsed into degrees of span, interpretive dimensions disappeared into empirical action terms, and qualitative distinctions melted down into functional interrelations: the great interlocking order of empirical its.

And after all, every event in the real Kosmos does indeed have its Right-Hand correlate, its empirical or sensorimaterial footprint (its exterior form), and so some sort of *evidence* for the empirical and positivistic course could *always* be supplied. Thoughts have their correlates in brain physiology. Intentions have their behavioral aspects. Culture has material social components. The Right-Hand path is never just simply wrong.

Just very partial. Plenitude without Spirit is systems theory. And systems theory, in its various guises, was now all the rage (from the Enlightenment down to today; in its various forms, the subtle reductionism of systems theory is the arch-paradigm of the Right-Hand path, where it does bitter battle with the gross reductionists—atomists, mechanists—neither of them formally acknowledging the Left half of the Kosmos).

In the great systems paradigm, which swept through the Enlightenment like wildfire (based, as both were, on monological rationality): instead of the relation of Each to the One (and thus to the All), one's salvation depended upon how the Each fit functionally and instrumentally into the All, into the whole functional system. The *holistic* world was an *instrumental* world, to the core. Since all things were merely strands in the great empirical web, all things had merely extrinsic and instrumental value in the overall system.

Accordingly, *quality* (interior depth) was measured in terms of *quantitative fit* with the great interlocking order (exterior span), and this was, incredibly, true for both the natural *sciences* and the *religions* of the time. The "great instrumentalization" was in motion.[13] "It is interesting to note," says Taylor of this collapse, "that this conception itself had undergone the transformation which made the *instrumental* central. Our grasp of this [holistic] order was referred to by the term 'technologia,' and the unity of God's order was seen not as a structure to be contemplated but as an interlocking set of things calling for actions which formed a harmonious whole."[14]

Taylor is discussing the central strands in religion from the Reformation to the Puritans to the Deists. How one's *actions* fit into the harmonious whole: this was crucial to salvation. "The harmony between the parts [of the harmonious whole] was captured in the term '*eupraxia*'; it was more a matter of the coherence of the occasions for action than of the mutual reflection of things in an order of signs." In other words, a harmonious whole of surfaces and exteriors involved in action ("What does it *do?*") and not *also* interior contemplation and an order of signs ("What does it *mean?*").

This conversion of qualitative value to instrumental action, Taylor shows, marked the entire era. "This means that the instrumental stance toward the world has been given a new and important spiritual meaning. . . . *Instrumentalizing* things is the spiritually essential step. . . ."[15] The *instrumentalizing* consisted precisely in converting all things into *strands* in the *web* of a now purely descended Plenitude, therein to take their instrumental place in the great universal system of sensory surfaces ("reflected on" by instrumental rationality).

No less than John Locke—"the teacher of the Enlightenment"—would stamp this holistic instrumentalizing with his seal of approval. "Instrumental rationality, properly conducted, is of the essence of our service to God. This is how we participate in God's purposes. Not through instinct, like the animals, but through conscious calculation, we *take our place in the whole*."[16] As Taylor summarizes it, "Locke helped to define and give currency to the growing Deist picture, which will emerge fully in the eighteenth century, of the universe as a *vast interlocking order of beings, mutually subserving each other's flourishing*. . . ." And *instrumental* awareness was thus "our avenue of participation in God's will."

(It should be obvious by now that if we substitute "Gaia" for "God's will," we have the "new paradigm" of many of the ecophilosophers of today, who extend in an unbroken lineage to the originators of the collapse of the Kosmos. Far from being a new paradigm, it is still covertly following the fundamental Enlightenment paradigm, still attempting to instrumentalize the world subservient to a web-of-life ideology, still caught in Foucault's biopower, still doing bitter battle with the *gross* reductionists, which is fine, but still missing the larger violence they have perpetrated on the Kosmos in the process of their own *subtle* reductionism—all topics we will investigate further below.)

Thus, the Great Holarchy (with "No more Ascent!" allowed) was converted, by a flatland monological and instrumental reasoning (technologia), into mutually interlocking and interpenetrating surfaces and exteriors, into a coherent and harmonious whole of mutually subserving strands in the great universal and objective system, a structural-functional system defined in procedural and action terms ("What does it *do*?")—in short, into a flatland web of life that lacked all interiority and degrees of depth—lacked, as Taylor puts it, all qualitative distinctions: I and we reduced to interwoven its.

It was thus a short step to "the greatest good for the greatest numbers"—a *calculus* of interlocking exteriors described in action terms (the great hallmark of the Right-Hand path). With no depths to be accessed, there are only surfaces to calculate: the more, the merrier. In a disqualified universe, "quality" is found by addition. And who can doubt, in such an atmosphere, that bigger must be better?

On the one hand, we can appreciate, I think, that "the greatest good for the greatest numbers" was a genuinely *worldcentric* notion (it often even included animals). And all of that was due in large measure to the positive emergence of postconventional and worldcentric rationality. So far, so good.

But with the collapse of the Kosmos from both Left and Right dimensions to merely Right-Hand empirical and sensory markers, there was no room left for, there was no way to register, differences in the *types* or *qualities* of the "good" that is supposed to be extended to all beings. There is *only* the functional fit of surfaces, and more or less of *only* that.

In other words, the *flatland* world means a *homogenization of motives* for all beings, and the ethical imperative then becomes to extend this homogenized mess to as many beings as possible. And this means as well that the homogenized motive would have to be the *lowest common denominator* for all beings: namely, sensory pleasure/pain, hedonic happiness.

Already, by 1742, Hutcheson would declare that "that action is best which accomplishes the greatest Happiness for the greatest Numbers." And never mind that there might be different types or levels of happiness. In a flatland cosmos, all happiness is the same, all happiness is monohappiness, sensory happiness, hedonic happiness, and thus a flatland calculus simply says to *extend the span* of monohappiness in all directions, in accord with the great interlocking order: not also degrees of depth but simply endless addition of sensory span, a truly bad infinity in an expanding nightmare of the *great interlocking order*.

"The most important notion Hutcheson and Locke share is that of the great interlocking universe, in which the parts are so designed as to conduce to their mutual preservation and flourishing."[17] Hutcheson himself would clearly explain that "our Affections were contrived for the whole; by them each Agent is made, in great measure, subservient to the good of the whole. [All] are thus linked together, and make one great System, by an invisible Union. He who voluntarily continues in this Union, makes himself happy: He who does not continue this Union freely, but affects to break it, makes himself wretched; nor yet can he break the Bonds of Nature."[18]

Become anything other than an instrumental strand in the great interlocking order and System of Nature—look beneath the surfaces, look into the depths or heights—and you are not only guilty of pride but doomed to wretched unhappiness. You have broken the hedonic calculus, have *subtracted* from the Sum Total of what should be uniformly happy Egos fitting snug as a bug into sensory flatland: you have disrupted the cavalcade of smiley-faced Shadows.

For the Cambridge Platonists, who at the time were fighting a losing battle, the key to salvation was *theiosis*—one becomes the Divine, and thus participates in the depths of the All, embracing Each as a perfect manifestation of the One. Not just fitting into the System as a part, but discovering the Source of the System as the One. On the other hand, for the purely Descended worldview that was now developing, God's Goodness was found *solely* in the plenitude of benevolence shown in the interlocking order of beings—in the great System—each part of which was assumed to be pursuing essentially the *same* monohappiness of sensory pleasure.

This is why, as Taylor points out, the instrumental and interlocking-order theorists (from Locke to Bentham) always had recourse to a *hedonistic theory of motivation*. In an empirical-sensory world, there is only empirical-sensory motivation (what else *could* there be?). Since we are all strands in the great empirical web, we must all share something in common, and what all living beings have in common—the lowest common denominator, in fact—is the pursuit of physical pleasure and the avoidance of physical pain.[19] The homogenization of motives, the reduction of all motives to the lowest common denominator.

(Closely related to physical pleasure/pain, in terms of motivation, is simple survival or *self-preservation,* and subsequent systems theorists would often make self-preservation of the system—autopoiesis—the mono-motivation that all holons possess. This is partially true in an abstract fashion—it's tenet 2a—but it is incapable of differentiating Left-Hand qualities, since any that exist are equally autopoietic. Autopoiesis would thus become another common mono-motivation in flatland.)

In other words, the disqualified universe still needed some sort of quality to move it into action, it needed some sort of push for its action terms—it still needed some type of motivation to turn the gears, jiggle the web—and the lowest common denominator was all that remained. Hedonic pleasure, so to speak, was the furthest into the interior that would be ventured, and this "sole" quality—because it was essentially the "same" for all—could then be easily "quantitized" and thus "calculated." (Different qualities cannot easily be *added* precisely because they are different types of entities—we say, "It's apples and oranges." But if you can get rid of the differences in quality, if you can disqualify the universe by converting it into more or less of only one quality, convert it into nothing but apples—well, apples can be added.)

Of course, it is one thing to say that we are all linked at that *fundamental* level of pleasure/pain (for indeed we are, at the sensorimotor level, and for all holons that possess that level). It is quite another thing to say that we all have *only* that level of motivation. But already Helvetius would announce that "Physical pain and pleasure are the unknown principles of all human actions."[20] Significance is here completely collapsed to fundamentalness, and the interlocking-utilitarians thus arrive at the common weakest-noodle motivation, and this weakest-noodle motivation then *defines* the single, sole, mono-happiness that we are supposed to *extend* throughout the flatland order as an *ethical* imperative: the *more* hedonic happiness for each, the *better* for all.

The idea seemed to be that if I can just extend my mediocrity to more and more beings, then a collective good and a great happiness would somehow result. But it was exactly the expansion of this weakest-noodle motivation, *calculated* by a monological and instrumental reason, that would allow me to take my rightful place in the great instrumental whole. Since it is only in reason that I can comprehend the great interlocking order, and since we are all strands in that order, then reason demands that I extend the same monohappiness throughout the entire monochrome system.

Even Aristotle had ranked motivations in a holarchy of deepening significance. Not everything we do is *equally* valuable, or equally good, or equally just. In Aristotle's version, there are several "goods" in life, but some of them are "more good" than others. Least significant was sensual pleasure, or hedonistic involvement. This was still a "good," and not to be condemned per se, but there were also other and more significant pursuits.

Deeper and wider than sensual pleasure was the life of ordinary production and consumption of material goods. Deeper and wider than production and consumption was the citizen life, the life of intersubjective relationship and interpersonal friendship. And deepest/highest of all, was the contemplation of the timeless (we approach the Divine most closely by contemplating the unchanging).[21]

This is, of course, just another version of the Great Holarchy. For Aristotle and the Stoics, this hierarchy was an order of goals, and an order of motivation. For Plato and Plotinus, it was also the structure of the Kosmos (which is why Plato's "science" and his "ethics" are of a piece).

But in all cases, it meant that the ethical life involved significant *transformations* in consciousness (or in one's being). The deeper or higher motivations are not simply *given* from the start; one has to grow, develop, and unfold to effect these changes (just as in Maslow's needs hierarchy, which is precisely the same Great Holarchy in modern form). Above all, one had to *change one's perceptions*, because the deeper and wider and more encompassing motivations are not just lying around waiting to be seen by the senses or their extensions. *Transformation* was mandatory for a truly *ethical* life.

But with the collapse of the Kosmos to the empirical Right-Hand path and to the weakest-noodle motivation of pleasure/pain (the very shallowest motivation on Aristotle's and Plato's holarchy), a profound and altogether far-reaching shift has occurred: since the weakest-noodle motivation is *already* present in most adults (it is present, in fact, in most sentient beings, a true fundamental)—since it is *already* present in awareness, then one *does not have to transform* in order to live *this* "ethical" life: one simply extends one's present state to as many others as possible. One *translates* more widely; one does not have to *transform* more deeply. (It is interesting that, when Foucault switched his studies to ethics, the major criticism he had of modern ethical theories compared with the premodern ones was that, as he put it, the modern theories by and large had no *ascesis*, no transformative practice: one merely thought differently, one didn't have to transform.)

In fact, aiming for any deeper and transformative motivation (which is also actually much wider and more encompassing)—this is our old friend "pride." Seeking a deeper meaning and motivation must now be viewed as stepping outside of the great interlocking order of weakest-noodle motivation. It was literally labeled, as we saw, a "crime against nature" and a crime against "the great Universal System."[22]

And thus, to come full circle, it was through weakest-noodle motivation, calculated by an instrumental-holistic rationality, that *we could embrace the flatland whole.* "The importance of instrumental reason then comes from its being the way that we are intended to play our part in the design . . . of the great interlocking order of the universe."[23] It is a concord of interlocking exteriors described in action terms, not an embrace of *both* interiors and exteriors demanding an inner *transformation.*

In this withering monological gaze, Agape is fast becoming Thanatos— not just a higher embracing a lower, but a higher regressing, being reduced to, a lower: the drive to dismantle the higher into its lowest common denominator parts (thus killing the higher in the process: Thanatos). There is no longer a deeper or higher happiness, only a wider happiness. No longer a vertical happiness that would embrace all beings as the One, but merely a horizontal happiness that would *reduce* Each to a sensory component in the All, and then attempt to ameliorate the devastating damage by an endless addition of broken parts, this bad infinity being the sole "happiness" of Each in the flattened All.

An old Catskills joke: Two elderly women, vacationing in the Catskills, are having dinner in the resort dining room. The food arrives, and one woman says, "This food is just awful." The other says, "Yes, and such small portions!"

The utilitarian ethic of this age was devoted to the proposition that bigger portions of bad food would somehow constitute the social good.

The instrumental-holistic and utilitarian atmosphere that dominated the age would simply level moral distinctions, and make it impossible to see that there are altogether different degrees of moral virtue, make it impossible to see why it is "better to be an unhappy Socrates than a contented pig." Rousseau and Kant would tear into this homogenization of motives with a vengeance, and Hegel would tear into them for not being nearly nasty enough. But in the meantime, the mono-motives of flatland would make it impossible to see that deeper human happiness might reside in a profound transformation of consciousness, and not just in an expanding translation of one's present state.[24]

The great advantage of the Descended God: I can already see *that* God clunking around in my visual field. Pleasure and pain already govern that world. They are given to me. That God is obvious, present, demands no work, no ascesis; the pleasure is there with each sunrise. And I shall extend the pleasure—and *only* that—indefinitely, thus never stepping out of the great interlocking system.

Accordingly, "sensuality was given a new value. Sensual fulfillment . . . seems to be one of the irreversible changes brought about by the radical Enlightenment. The promotion of ordinary life [nontranscendental], already transposed by Deists into an affirmation of the pursuit of happiness, now begins to turn into an exaltation of the sensual. *Sensualism was what made Enlightenment naturalism radical.*"[25]

Aristotle's lowest good is here the only good!—because it is the only good that fits the weakest-noodle net. *For a merely empirical-sensory world, a merely empirical-sensory motivation.* No deeper stages of growth, transformation, or development are necessary, and certainly none were allowed.

Heaps of sensual mono-happiness, not wholes of higher transformation: this would be a crucial ingredient in the newly forming Descended worldview—a world where there are no depths to be interpreted, only surfaces to be seen. Thus no inner transformation required, only exterior and empirical mapping. No qualitative distinctions to be forged, only quantative extensions to be measured. No better or worse inherent in the Kosmos, only bigger and smaller found in the cosmos. Not also an interlocking order of interpretive depth, but only a great interlocking order of empirical span. Not also degrees of value and motivation, but merely degrees of functional fit, driven by pleasure and survival. Not also "What does it mean?" but basically "What does it do?"

A Descended worldview where, to summarize bluntly, an expansion of span replaced a deepening of depth.

The Ego and the Eco

We earlier noted the altogether extraordinary paradox of the Enlightenment paradigm: the holism of nature produced the atomism of the self.

That is, so holistic was this great empirical interlocking order and inter-woven net, so perfect was the "great Universal System," that the *subject* perceiving this net, the rational-ego, was left *disengaged* and *dangling* as its own "autonomous" and "self-defining" agency, with no way, in participatory terms, that the *subject* could actually fit into the interlocking world of *objects* it had so seamlessly described. The I and we could find no room in the inn of interwoven its.[26]

This massive rift in the holistic world, in the *Système de la Nature*, was rapidly becoming all too apparent. "The Enlightenment developed a conception of nature, including human nature, . . . as a harmonious whole whose parts meshed perfectly. But the rift was still there between nature, whether as plan or instrument, and the will which acted on this plan"[27]— the will or agency which could not be described in the same terms as the net of its in which it found itself, a subject that could not fit into the world of empirical objects.

Thus, the newly emergent rational-ego might simply *say* it was merely a strand in the great empirical web of nature, but that was to reduce the subjective sphere to an empirical dimension (reduce the Upper Left to a merely Right-Hand dimension: this is precisely what systems theorists continue to do to this day). But that did not explain the fit of the subject into the empirically holistic net; rather, that explained away the subject as being merely another object. It erased the very terms it was supposed to explain.

The holistic flatland world left no point of insertion for the *subject with depth* (no room for interiors, for I's or we's, for genuine depth in any holons anywhere, animal, human, divine, or otherwise).

And thus arose what has been called the *central problem of modernity*: human subjectivity and its relation to the world. Not only was it the central problem of the Enlightenment and the Romantic-Idealist rebel-lion, it has continued to be a central, often *the* central, topic in theorists from Habermas to Foucault to Taylor. (Foucault, for example, said, "The goal of my work during the last twenty years . . . has not been to analyze the phenomena of power, nor to elaborate the foundations of such an analysis. It is not power, but the subject, which is the general theme of my research.")[28]

What is of most interest to us now is that the two poles of the funda-mental Enlightenment paradigm—namely, the autonomous subject (or the rational-ego) and the holistic world—were a mutual discord in essentially the *same* flatland ontology.[29] Both poles accepted the collapse of the Kosmos into fundamentally empirical and naturalistic terms. It was no longer a battle between the Ascenders and the Descenders. The Descenders had fundamentally triumphed—in rejecting the bathwater of myth they had tossed out the baby of Ascent: no deeper or higher stages than the rational-ego and its empirical world were acknowledged. On this, all sides were in virtual agreement.

No, it was now a battle between what was left over: it was a battle, at the egoic-rational level, between the *agency* and the *communion* of that particular level: a battle of the ego-subject that was supposed to be self-defining, self-generating, and self-autonomous versus the objective/empirical world that was supposed to encompass and govern the ego-subject as well (as *part* of the great universal web)—and both of those assertions obviously cannot be unadulteratedly true.

Thus, this was *not* a battle of higher *transformation* (both sides had ruled that out), but a battle of *translations*, in the form: *given* the rational ego-subject, does the good life consist in (1) following the autonomous *agency* of the rational-ego in order to generate its own self-assured morals and aspirations, or in (2) connecting the ego with the wider ground of its shared *communions* in the natural world and thus finding something "larger" than the isolated ego?

This battle—it is still with us today—I will call the battle between the Ego and the Eco. Both were flatland to the core, which prevented their ever being synthesized in a higher union (they cannot be vertically integrated if both sides deny vertical holarchy). Rather, in a flatland world, the *more* there is of one, the *less* there must be of the other, and so the Ego camps and the Eco camps lined up as mortal enemies, each accusing the other, once again, of being the essence of Evil.

The battle: Did the good life reside in the Ego's *agency* or in its *communions*? In the Ego or the Eco?

Plotinus, of course, had a coherent way to say *both* (namely, the subjective agency of Each and the objective communions of the All are transcended and included in the nondual One). But since both sides had now thoroughly rejected holarchic Ascent to the Nondual One, there was only the thorny question of how to relate the finite and supposedly autonomous subject (the Each) with finite and supposedly all-inclusive net of objects (the All), without reducing one to the other and thus violently and syncretically sabotaging the hard-won integrity of each.

The two camps, in other words, were completely and mutually incompatible (this they both happily—and aggressively—acknowledged). The *more* the Ego was emphasized as the true ground of a rational morality and an ethical will not found in nature, then the more the Eco was devalued, the more it came to be perceived as a source of *heteronomy* ("other-direction") opposed to the rational autonomy of the ego-subject. Nothing in the system of nature seeks universal compassion or mutual understanding or ethical restraint, the argument ran, and thus the task of the Ego is to free itself from the nasty, short, and brutish net of nature. The task of the Ego was to *disengage from the Eco*. To establish its agency over the communions of the great and grinding system. In short, and in all ways: the less Eco, and the more Ego, the better.

The Eco camp watched all of this with absolute alarm. The Ego wasn't just establishing its agency, this camp maintained; it was severing and repressing its connections and communions with nature—both external nature and, more important, internal nature (sensual, desirous, sexual, vital). And the more the Ego succeeded in its goal of devaluing the Eco, then the more abstract, arid, dry, and desiccated it became. The very emergence of the Ego, they maintained, was now destroying the rich fabric of communions on which the Ego itself depended. And the solution seemed equally obvious: the more Eco, and the less Ego, the better for all.

Without a genuine holarchy, the two partial views could not be superseded: they both *claimed* to encompass the other camp, but they both did so only through aggressive subtraction. They were in fact locked into a battle royale that, in many ways, was the archbattle, and remains the archbattle, of modernity and postmodernity.[30]

In *this* world, this Descended world, the Ego and the Eco battle for the good life, doomed as they are to mutual repugnance.

It would be useful, then, to look more closely at the flatland Ego and the flatland Eco—both what they accomplished (which was much) and what they mutilated (which was much more), since, as I indicated, this fruitless battle, which cannot be solved in a flatland ontology, is still with us today, and still contributing equally to the despoliation of the real world, the world they both claim to be saving.

A DIAGRAMMATIC OVERVIEW

Because this topic is so crucial, and also somewhat intricate, I would like in this section to give a very simplified overview using a handful of diagrams. As is always the case with diagrams, they can cause more problems than they solve; but as long as we keep in mind the very schematic nature of these figures, they might highlight several of the central issues.

Figure 12-1 is a summary of the traditional Great Chain of Being, in a simplified form, matter to body to mind to soul to spirit. Since each level transcends and includes, I have drawn it more accurately as the Great Nest (or Great Holarchy) of Being. Moving upward from the center (matter, the most fundamental) is the process of evolution (Reflux or Ascent, driven by Eros), and moving downward from spirit (the most significant) is involution (Efflux or Descent, driven by Agape). Each senior level is an emergent, marked by properties not found in its juniors. Spirit is both the highest level (which transcends all, includes all), and the equally present Ground of each level (equidistant from a star, an ant, a man).

But what the traditional Great Nest failed to do, on any sort of significant scale, was to clearly differentiate the four quadrants. The material brain, for example, was simply placed, with all matter, on the bottom

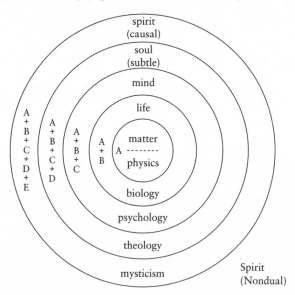

Figure 12-1. The Great Nest of Being. Spirit is both the highest level (causal) and the nondual Ground of all levels.

rung of existence, instead of seeing that the material brain is the exterior correlate of interior states of consciousness (so that the brain is not simply part of the lowest of all levels, but rather is the exterior correlate of some very high levels). Instead of seeing that consciousness is intimately associated with the material brain, consciousness seemed to hover above all matter, transcendental and metaphysical and completely other-worldly. (The discovery that states of consciousness have correlates in brain states—that all Left-Hand events have Right-Hand correlates—was a devastating blow to the Great Chain and all metaphysics as traditionally conceived, and rightly so, even though it went too far and contributed to the collapse of the Kosmos into nothing but scientific materialism.)

Likewise, the traditional Greaet Nest embodied little understanding of the profound ways that cultural contexts (LL) mold all perception; of the ways that the techno-economic base (LR) strongly influences individual consciousness; of the evolution of worldviews, individual consciousness, modes of production, and so on. In all these ways and more, the Great Chain was severely limited. (This did not stop individuals from using the Great Chain as a perfectly adequate map for individual spiritual development in the Upper-Left quadrant, which is what many did. But it did severely limit the sensitivity of the Great Nest to those aspects of reality that we are calling the four quadrants, because they were not differentiated on a large scale by the average mode.)

All of this would change with modernity and the widespread differentiation of the Big Three. This can be represented as in figure 12-2 (and simplified as in fig. 12-3), which is the Great Nest differentiated into the four quadrants (or the Big Three). This *integral* vision is what *could* have emerged, and what might yet still emerge, but what in fact did *not* emerge in any enduring fashion. Instead, the quadrants were differentiated just fine, but then eventually collapsed into their material (or empirical or objective) correlates, to result, soon enough, in flatland, in scientific materialism, in sensory mononature, in the great interlocking order. Tracing this collapse of the Kosmos is the crux of our story in this chapter.

This collapse could occur, I have suggested, because every Left-Hand event does indeed have a Right-Hand or empirical correlate in the material (or objective) world—and thus it is always tempting to reduce the former (depths that require arduous interpretation) to the latter (surfaces that can be easily seen). This is shown more explicitly in figure 12-4. Each interior state of consciousness (from bodily consciousness to mental

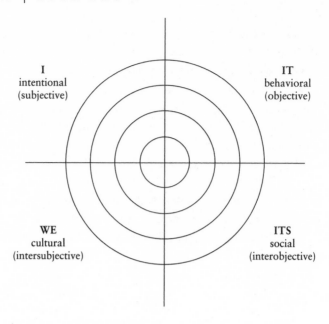

Figure 12-2. The Great Nest with the Four Quadrants

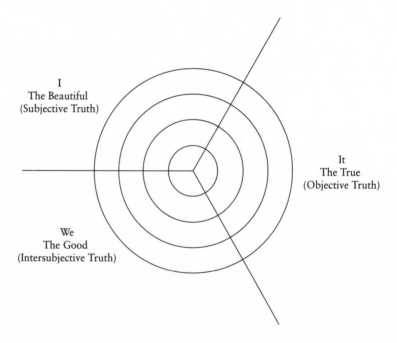

Figure 12-3. The Big Three

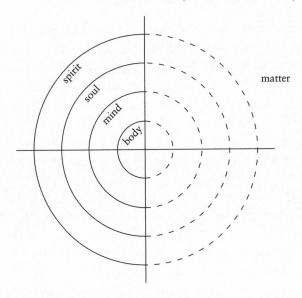

*Figure 12-4. Correlations of Interior (Consciousness) States
with Exterior (Material) States*

consciousness to spiritual consciousness) has some sort of correlate in the material/objective brain and organism (Upper Right). Thus, for example, even somebody who is having a spiritual experience (UL) will show changes in objective brainwaves that can be measured on an EEG machine (UR).

But conscious states, values, depth, and intentions cannot be *reduced* to material brainwaves because, although one value is better than another, one brainwave is not. You experience compassion and you know it is better than murder; while you are thinking that, your brain will be producing brainwaves that can be registered on an EEG. But although you know that compassion is better than murder, there is nothing on the EEG machine that says, "This brainwave is more valuable; this brainwave is more moral; this brainwave is more beautiful." The EEG machine can *only* show that one brain state is *different* than another; it cannot say that one state is *better* than another. To make that judgment, you have to rely on interior consciousness, depth, and value recognition—holarchies of *quality*—while the EEG machine can only register holarchies of *quantity*. To reduce the Left to the Right is to reduce all quality to quantity and thus, as we have seen, land directly in the disqualified universe, also known as flatland.

Which is precisely what happened. And within flatland, as we were saying, there arose two camps: the Ego and the Eco. Both were fundamentally oriented to flatland, to the Right-Hand world of empirical, sensory, objective Nature, which was the "really real" world. Both, that is, were involved in subtle reductionism to some degree or another. But with a difference: the Ego camps held on to the Left-Hand rational Ego and its agency, and the Eco camps wanted to go all the way and insert the Ego into the great web of Nature and its communions altogether. This state of affairs can be represented as in figures 12-5 and 12-6.

In figure 12-5, we see the basic story as we have told it thus far, up to and *before* the collapse into flatland. Cultural evolution (LL) has moved from archaic to magic to mythic to rational (in the individual or UL: body/sensorimotor to preop to conop to formop), with a correlative moral evolution from egocentric to ethnocentric to worldcentric, and a social evolution (LR) from foraging to horticultural to agrarian to industrial, accompanied by a social organization from tribes to villages to empires to nations (i.e., social systems). The four quadrants (or simply the Big Three of I, we, and it) have been differentiated, and although they are not yet genuinely integrated, nonetheless they are not yet dissociated; at this point in time each is pursuing its own truth unhampered by domination from the others. (Fig. 12-5 is also a map of what the Big Three, when integrated, might look like, with each quadrant given its own respected and honored place: a task of vision-logic to which we will return.)

But then begins the one-century or so slide into flatland, into the great interlocking order of empirical realities, into subtle reductionism, into the holistic web of interwoven its, the great *Système de la Nature*. This is represented in figure 12-6, which is a general map of flatland itself.

The most conspicuous item about that map is that virtually *everything* is missing in the Left-Hand domains except the rational Ego. That is a stark and graphic representation of the dangling, disengaged, alienated, hyper-autonomous subject that was the Enlightenment ego.

In the Upper-Left quadrant, the rational Ego was cut off from its own lower levels (interior nature), and it was cut off from the higher levels of its own Spirit, its own higher Self ("No more Ascent!"). In the Lower-Left quadrant, it was ignorant of the vast role played by cultural contexts in disclosing truths and molding values, and thus it tended to mistake its own cultural givens as universal truths (when they were often merely the preferences of propertied white bourgeois males). And although the ratio-

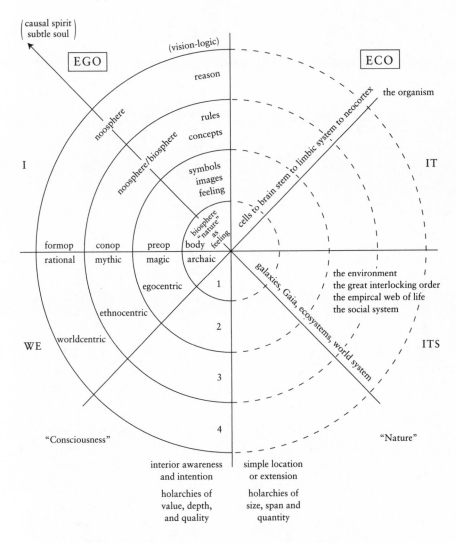

Figure 12-5. The Ego and the Eco before the Collapse

nal Ego would, for the most part, consider the Right-Hand world of empirical Nature and the great interlocking order to be the fundamental reality, the Ego had cut itself off from communion or union with that Nature because it imagined that its job was impartially and in detached manner to *reflect on* Nature (the representation paradigm, the "mirror of Nature"). All of this is quite accurately represented by the Ego dangling in midair in the Upper-Left quadrant.

This plight of the rational Ego was partly innocent. It was not con-

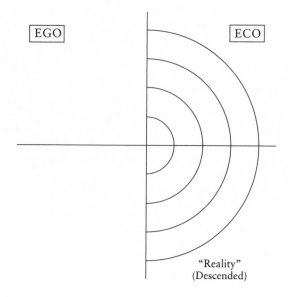

EGO

ECO

"Reality"
(Descended)

Figure 12-6. Flatland

nected to its own interior roots (the developmental stages leading up to the rational Ego) because, in many ways, these simply had not yet been discovered. Most people of today use reason without really knowing the ontogenetic stages that produced it—namely, the stages of sensorimotor, preop, conop, and then formop. It is simply not immediately obvious to reason that reason itself developed or evolved. And yet reason is the first structure that can impartially reflect on the world. Thus, the natural stance of reason is to simply assume that it is apart from the world and can innocently reflect on it. This part of the Cartesian dualism is completely understandable, if mistaken. And most of the Enlightenment philosophers, from Locke to Kant, made this assumption by failing to grasp the actual developmental stages that lead up to reason. In fact, one of Hegel's main criticisms of Kant was that consciousness is not simply given, but rather "can only be conceived as one *that has developed*"—and tracing that development was one of the extraordinary gifts of Idealism. Still, it was only in the last half of this century that the stages of consciousness development were outlined in any sort of rigorous fashion backed by research—Baldwin, Werner, Graves, Maslow, Piaget, Loevinger, and so on. Thus, what I have drawn in figure 12-5 is how we of today can understand the quadrants, with all of the research that has gone into them.

Although the Enlightenment philosophers had full access to the quadrants, research in each was just beginning, and this simple and innocent ignorance is partly responsible for the world looking like figure 12-6, which is how the Enlightenment generally viewed the world.

The stages beyond the rational Ego were also denied (or ignored)—the higher transrational stages of subtle soul and causal spirit—precisely because, as we have seen, the Enlightenment screamed "No more Ascent!" This was part of the noble attempt to move beyond ethnocentric religious mythology into the worldcentric rights of man (and soon woman)—"No more myths!" But the transrational baby was tossed with the prerational bathwater, and thus the Ego was cut off from both its roots in earth and its branches in heaven.

But not all of the Ego's isolation was innocent or noble. After all, Plotinus fully possessed reason and he did not gut the interiors. Something else was afoot in this game of collapsing the Kosmos, and, as we will continue to see, that game was the reduction of all Left-Hand events to their Right-Hand correlates, the reduction of all realities to the great interlocking order of the system of Nature. And once that reduction was complete, the Ego as subject had no room to fit into the world of holistic its—and thus once again, as shown in figure 12-6, the rational Ego was left dangling.

The Eco camps agreed with the Ego camps that the great interlocking order and the system of Nature was the "really real" world (both camps subscribed to Descended flatland); but the Eco camps were alarmed at the Ego's detachment and its rational distancing ("the disenchantment of the world"), and desired instead a sympathetic union with the great Web of Life: it desired union with flatland, not a reflection on flatland. And in this desire for "union with Nature," we see the archetypal philosophical problem of modernity (which is to say, of flatland), namely, the mind/body problem, which is another name for what we called "the central problem of modernity," or how the subject ("mind") fits into the world ("body").

THE MIND/BODY PROBLEM

The mind/body problem is a product of flatland. It has proven so intractable because, in the wake of the modern collapse, both "mind" and "body" have at least two completely different meanings, giving us four problems hidden in one. But it's not as bad as it sounds, and you can easily follow all of this using figure 12-5.

To begin with, "body" can mean the biological organism as a whole, including the brain (the neocortex, the limbic system, the brain stem, etc.)—in other words, "body" can mean the entire Upper-Right quadrant, which I will call "the organism."

But "body" can also mean, and for the average person does mean, the subjective feelings, emotions, and sensations of the felt body. When the typical person says, "My mind is fighting my body," he means his will is fighting some bodily desire or inclination (such as for sex or food). In other words, in this common usage, "body" means the lower levels of one's own interior. On figure 12-5, I have numbered the levels 1 through 4, so in this usage "body" is level 1 in the Upper-Left quadrant, and the other meaning of "body" is all of the Upper-Right quadrant (the total organism).

Moving from body to mind, many scientific researchers simply identify "mind" with "brain," and they prefer to speak only of brain states, neurotransmitters, cognitive science, and so on. I will use the term "brain" to cover that meaning, which refers to the upper levels of the Upper-Right quadrant (e.g., the neocortex).

On the other hand, when the average person says, "My mind is fighting my body," he does not mean that his neocortex is fighting his limbic system. By "mind" he means the upper levels of his own interior, the upper levels of the Upper-Left quadrant (although he might not use exactly those terms)—in other words, his rational will is fighting his feelings or desires (level 4 is fighting level 1). The mind is described in first-person phenomenal accounts and I-language, whereas the brain is described in third-person objective accounts and it-language.

(There is another general meaning for mind/body: "mind" can mean the interior dimension in general—or the Left Hand—and "body" the exterior dimension in general—or the Right Hand; I will specifically indicate that usage when it comes up.)

Those various definitions of mind and body have been used in juxtaposition to both create and solve the mind/body problem. The reductionist simply reduces the mind to the brain, and since the brain is indeed part of the organism, there is no dualism: the mind/body problem is solved! And that is correct—the brain is part of the organism, part of the great web of life and the empirical interlocking order, so there is no dualism; nor are there any values, consciousness, depth, or divinity anywhere in the resultant universe. And that reductionism is exactly the "solution" that the fundamental Enlightenment paradigm would impose on reality, a "solution" still rampant in most forms of cognitive science, neuroscience, systems theory, and so on: reduce the Left to the Right and then claim you have solved the problem.

But the reason most people, even most scientists, are uneasy with that "solution"—and the reason the problem remains a problem—is that, even though materialistic monism announces that there is no dualism, most people know otherwise, because they *feel* the difference between their mind and their body—they feel it every time they consciously decide to move their arm, they feel it in every exercise of will. And the average person is right, in a general sense: there is a differentiation of mind and body, noosphere and biosphere, and this can be felt in the interior or Left-Hand domains. It is not a rigid dualism, as some philosophers have postulated, but rather a case of "transcend and include," and almost every rational adult has a sense of the transcend part, in that the mind can, on a good day, control the body and its desires. And all of that is phenomenologically true for the Left-Hand domains (in more precise language, formop is indeed different from sensorimotor: formop transcends and includes sensorimotor, and thus can operate on it). But *none* of those *interior* stages of development are captured when "body" means Right-Hand organism and "mind" means Right-Hand brain—all of those important distinctions are completely lost in material monism, which does not solve the problem but obliterates it.

On the other hand, an "all-quadrant, all-level" approach allows a place for each of those definitions of mind and body and brain and organism, because each of them represent genuine and nonreducible realities. This integration of the four quadrants (and the levels in each)—a task of vision-logic—helps to defuse the many forms of the mind/body problem. It allows (and demands) an equal inclusion of *first-person* phenomenal accounts ("I"), *second-person* structures ("we"), and *third-person* organismic systems ("it"). [See *Integral Psychology*, "The 1-2-3 of Consciousness Studies."] Moreover, the ultimate mind/body problem—the relation of interior-subjective to exterior-objective—is solved only in nondual awakening, which transcends and includes the quadrants. My claim, then, is that an "all-level, all-quadrant" view substantially handles the mind/body problem in its major forms.

The details of that claim would obviously require a large book in itself; I mention it now only because it directly relates to the topic at hand, namely, the relation of the Ego and the Eco. For the Eco camps, in claiming that the great Web of Life covered all of reality, slipped into the standard flatland notion of "reality" (namely, the Right-Hand world), and this was reflected in their use of the term "Nature" and consequently in their recommendations for "reenchanting" the world, recommendations that are still with us.

NATURE

The word "nature" has at least three meanings (paralleling those of "mind/body"). It can mean the entire Kosmos, or the Great Holarchy of Being, including both Right- and Left-Hand dimensions. Call that "NATURE." It can mean the entire sensory and empirical world, the world disclosed to the senses—in other words, the entire Right-Hand world. Call that "Nature," which is so listed on the Right Hand in figure 12-5. And it can mean nature as opposed to culture, or nature as opposed to history, or the biosphere as opposed to the noosphere (a meaning which is similar to sensory body versus rational mind). Call that "nature," which is listed as level 1 in figure 12-5.

All three of those definitions *represent actual realities,* and thus all three of them are acceptable. But the problems start when one does not specify which is meant. (Henceforth I will generally indicate which usage is meant, but the context should also be the guide.)

What we will see happening is this: since the interior dimensions—and especially the higher, transcendental dimensions—were denied fundamental reality in flatland, NATURE was ruled out of the picture altogether (by both the Ego and the Eco camps). Instead, reality came to be "Nature," or the *entire sensory and empirical world*—this was the great interlocking order and system of Nature. And since all interiors have correlates in exteriors, and since the *sum total* of exteriors was called Nature, then if you reduce all interiors to exteriors, it appears that *Nature includes everything* (whereas it simply includes the Right-Hand world).

Once we focus on Nature (or the entire sensory and empirical world), it becomes very hard to grasp how the noosphere transcends the biosphere, which transcends the physiosphere. For those distinctions rest in part on *interior* domains (see fig. 12-4). In the world of material monism or scientific materialism, there is no reason to make any of those distinctions. They are all just variations on complex matter/energy systems. Noosphere, biosphere, and physiosphere are all viewed as simply complex forms of the physiosphere itself. What we call the noosphere, for example, is simply reduced by scientific materialism to the functions of the neocortex, itself a complex system of matter/energy and information fluctuations. And since the brain is part of the organism, which is part of Nature, then obviously the noosphere is just a part of Nature (or just a part of the "biosphere" in the broad sense of "Nature"). And so obviously the noosphere does not transcend and include the biosphere—the noosphere is part of the biosphere, or so it seems to flatland. *When all interiors are reduced to exteriors,* one can no longer recognize degrees of interior depth, and thus everything becomes *equally* a strand in the great interlocking web of valueless its. *Everything* is part of Nature. . . .

On the interior dimensions, on the other hand, the distinctions between biosphere, noosphere, and theosphere show up, for example, as the distinctions between bodily feelings and desires (prerational), the reasoning mind (rational), and soul and spirit (transrational)—all of which can be known directly and immediately in consciousness, and all of which have objective correlates in the world of Nature (as shown in fig. 12-4), but *none of which can be reduced to Nature* without ending up precisely with the flatland shown in figure 12-6. Which is, of course, what happened.

There, in the world of flatland, "transcend and include" is still the general rule (the twenty tenets are true in all four quadrants), and thus the neocortex still transcends and includes the limbic system, which transcends and includes the reptilian stem, just as its cells transcend and include molecules, which transcend and include atoms, and so on. Destroy the lower, and the higher is also destroyed (but not vice versa; tenet 9). That is already enough to show us that the higher transcends and includes the lower. But as we have repeatedly seen, in that Right-Hand scheme there are no qualitative rankings, only quantitative: there is no better or worse, only bigger or smaller, and thus to reduce NATURE to Nature is to land squarely in flatland. Which is exactly what both camps did.

Thus, both the Ego camps and the Eco camps ended up embracing the general reality depicted in figure 12-6. And with that, we can pick up the story.

THE EGO

In this and the next section, we will look more carefully at the strengths and the weaknesses of both the Ego and the Eco camps.

As I indicated, this was a battle that was fought fundamentally *within* the Descended and flatland world, with the internal disengaged Ego attempting to figure out its relation to the external holistic Eco, both monological to the core. As we will see, however, the Ego camps were constantly breaking in the direction of a pure Ascending Path, attempting to disengage the Ego altogether in the direction of a Pure Self, and this would culminate in Fichte's *infinite Subject* as the transcendental One. And the Eco camps were constantly breaking toward a pure and radical Descent, embracing fiercely the purely Descended world of the diverse and immanent Many, epitomized in Spinoza's *pure Substance* (which was often taken to mean pure *immanence* or Descent, as Fichte's infinite Subject was taken to mean pure *transcendence* or Ascent).

And by the end of the eighteenth century, this tension—between the Ego and the Eco, and in its extreme forms, between pure Ascent and pure Descent—this tension had become extremely and even unbearably acute, and the watchwords of the age became "unite Fichte and Spinoza!"—unite Ego and Eco, unite Ascent and Descent, and thus heal the two-thousand-year-old schizoid God that had dominated Western culture virtually from its inception. And a very special individual would rise to that occasion and attempt the long-sought integration. . . .

But in the meantime, the Ego and the Eco squared off as mortal enemies, and the battle, as I said, was: Did the good life reside in the Ego's *agency* or in its *communions*? In the Ego or the Eco?

I will divide each section into "positive" and "negative." In the positive, I will sympathetically state the case for each camp, and include what seem to be the genuinely enduring truths of each. In the negative, I will give the criticisms that each camp failed to answer satisfactorily, criticisms that exposed the real weaknesses in each camp, weaknesses that would historically drive Idealism to attempt the long-awaited integration.

Ego-positive: Freedom

We can correctly situate the Ego camp(s) only if we remember the genuinely beneficial gains that the rational-ego had indeed accomplished in its emergence from the role identity of mythic dominator hierarchies.

(And recall the confusing terminology: the rational-ego was not, per se, "egotistical," but something of the *opposite*: the self had developed from egocentric to ethnocentric to worldcentric capacities—to the postconventional worldview of universal pluralism, tolerance, and perspectivism, as shown in fig. 12-5. To the extent the ego lived up to its rational potential, that mature ego was worldcentric—a decentered view of universal pluralism. It is this "mature ego" that I mean by "the Ego" camp, and it was this mature ego of worldcentric reason that its proponents had in mind.)

This was indeed part of the *dignity* of modernity, part of the liberating and enlightening move that strove, ideally, to free itself from ethnocentric prejudice and mythic imperialism, and the violence inherent in those shallower engagements. No more myths!

"In the tradition of the Enlightenment," Habermas reminds us, "enlightened thinking has been understood as an opposition and counterforce to myth. As *opposition*, because it opposes the unforced force of the better argument to the authoritarian normativity [dominator hierarchy] of a tradition interlinked with the chain of generations; as *counterforce*, because it . . . breaks the spell of collective powers. Enlightenment contradicts myth and thereby escapes its violence."[31]

The power—and violence—of mythology is due in large measure to the fact that it is making claims that cannot be exposed to deeper evidence . . . without destroying the authority of the claim itself. That is the definition of ideology: hidden interests, hidden power claims, parading as truth, a truth that cannot be *exposed to evidence* without robbing it of its *power*. "That behind the back of the theory there lies hidden an inadmissible mixture of power and validity, and that it still owes its reputation to this—they are confused because validity claims are determined by relationships of power."[32]

And at its best, the Enlightenment committed itself to exposing these power relations and to dismantling the dominator hierarchies that had formed the core of the social, cultural, and religious institutions of the entire mythic and mythic-rational epochs. Liberty, equality, fraternity: "No more myths!" meant "No more dominator hierarchies!," and the Age of Reason and Revolution set out to prove just that.

"Despite all the efforts of tyranny, despite the violence and trickery of the priesthood, despite the vigilant efforts of all the enemies of mankind," cried an impassioned Holbach, "the human race will attain enlightenment; nations will know their true interests; a multitude of rays, assembled, will form one day a boundless mass of light that will warm all hearts, that will illuminate all minds" (*Essai sur les prejuges*). *Ecrasez l'infame!*

The utilitarians may have been operating within a purely flatland ontology (which now goes without saying), but even they were driven by a goal of universal benevolence extended to all humans (and often animals), regardless of race, creed, color (and soon gender). Bentham: "Is there one of these my pages in which the love of humankind has for a moment been forgotten? Show it me, and this hand shall be the first to tear it out."

All of this, indeed, was radically new, on any sort of collective scale. That reason could rise above the narrow interests of an egocentric or ethnocentric stance, and attempt to discover and affirm what was good and fair for all beings, and not merely for me or my tribe or my nation— this was one of the enduring drives of the rational Ego.

And in order to rise to this worldcentric pluralism, I, as Ego, must fight, in myself, any tendencies to act in egocentric or ethnocentric ways. I must fight my own inclinations, my own desires, my own natural impulses, to the extent that they deter me from a worldcentric fairness.

The rational Ego, then, has to *battle* those currents both in nature and in culture that pull it away from its worldcentric moral stance. Reason, and reason alone, delivers this moral vision to me, and thus I, as Ego, must distance myself from those powers and pulls that do not live up to this rational ethics of universal care.

And, the Ego camp would argue, if the rational-ego were really just another strand in the great natural web, then all of its actions would already be determined by the web itself. We would then be driven merely by pleasure and pain and self-preservation, and thus we would have no resources in ourselves that could deny our own egocentric drives in the name of a deeper and wider compassion for others. It is certain the fox never goes out of its way to help a stranded chicken (unless it's on the menu), and why should it? But is *that* the ideal model of my ethical response?[33]

No, the great web of Nature, the Eco-world, can offer me nothing for a worldcentric moral response. Rather it is reason, and reason alone, that allows me to step outside of my own natural inclinations and act for the benefit of others, and to treat others as I would myself be treated.

The central figure in all of this is, of course, Immanuel Kant. The moral subject, the rational-ego, is *free* precisely to the degree that it can *disen-*

gage itself from the *lesser* pulls of egocentric desire, merely natural self-preservation, and dispersed inclinations, and reside as moral freedom.

One of Kant's central aims was to vindicate this moral freedom and independence of the ego-subject—to establish its *autonomy* as against the pull of *heteronomy*, or "other-directedness," by which he meant especially the other-directedness of ethnocentric culture, of mythic (metaphysical) religion, and of lower natural inclinations and desires. The truly moral will draws its worldcentric perspective from *within*, from its own reason, and not from any external source whatsoever, whether that source be church or state or nature.

Thus, for Kant, a person is truly *free*—that is to say, truly *self-determined*, and not pushed by some external other—when, and only when, the person acts from a determination imposed by his or her own rational will. When I act from a worldcentric perspective, then I am truly *free*, free of the lesser and meaner stances that wound me in their shallowness. When I act from a worldcentric stance, I am truly free—and that worldcentric stance is itself generated and understood only by rationality (by universal perspectivism).

And thus, I am truly free *only* when I am acting in accord with my own moral will in its worldcentric capacity. When I act from that rational stance, *then* I am acting in the pure freedom of the rational will, since that is *my own* deepest and truest self (so to speak). I am not being forced by some other to act in this way (heteronomy), because the worldcentric stance is generated only by my own rationality: I am fully self-determined and thus free in the truest sense.

Thus, I am not really free when I am acting egocentrically—that is simply being a slave to my desires. Nor am I free when I am a slave to the herd mentality and mass opinion (of the ethnocentric and conformist mode). Further, Kant would say, neither am I free in amoral nature—there, I am merely a cog in the interlocking web. Rather, I am free only when I act from the worldcentric stance given by my own rationality, free because I transcend these lesser and shallower engagements, engagements that proceed without benefit of universal compassion and categorical care, and thus engagements that hurt me as well.

In moral freedom, I declare my independence from nature and its amoral ways. "Such independence," says Kant, "is called *freedom* in a strict, i.e. transcendental sense." And conversely, to be bound by external authorities, traditional customs, Church dogma, or nature and nature's net: all of that drags me down from a worldcentric stance and into one or another partiality, meanness, narrowness, theft.

"This is the central, exhilarating notion of Kant's ethics. Moral life is equivalent to freedom, in this radical sense of self-determination by [one's own] moral will. This is called 'autonomy.' Any deviation from it, any determination of the will by some external consideration, some inclination, some authority, even as high as God himself, is condemned as heteronomy. The numinous which inspired awe was not God as much as the moral law itself, the self-given [but worldcentric] command of Reason . . . the moral law which he gives to himself as rational will. So that men were thought to come closest to the divine, to what commands unconditional respect, not when they worship but when they act in moral freedom."[34]

There is much truth in Kant's position, I believe, and certainly in the general terms as I sketched it. Not only has Kant carefully differentiated the Big Three, he is fighting mightily the reductionistic flatland trend of the Age.[35] Kant would point out (to use our terms) that the noosphere is not *in* the biosphere; therefore the noospheric moral imperative *cannot be found empirically* (but only interpersonally or practically). And he is clearly fighting the homogenization of motives so prevalent with the utilitarians; for Kant, expanding sensual inclinations and desires is just expanding heteronomy, expanding slavery, expanding addictions.

But Kant's view is also hobbled by his inability to finally integrate the Big Three (despite his efforts in the *Third Critique* via art and organism; and indeed, this integration or "reconciliation" is what Schelling and Hegel took upon themselves to provide, as we will see shortly). For among other things, this deficiency left the intractable problem: how are the Ego and the Eco related?

But in the meantime, some of the obvious truths in this type of position were the driving force behind the Ego camps in their many forms. The greater the agency of the Ego—that is, the more the Ego can disengage from the Eco (and heteronomy in general)—the greater the knowledge and freedom of the self.

And thus arose the subjective pole of the fundamental Enlightenment paradigm, variously referred to as the "self-defining subject," the "autonomous ego," the "disengaged self," the "punctual self," the "philosophy of the subject," "self-sufficient subjectivity."[36] Taylor refers to this overall emergence as "a revolution in moral consciousness." Foucault calls it a "sudden, radical restructuring" and a "profound upheaval, an archaeological mutation." It is, in essence, the emergence of modernity. It is the emergence of the Ego.

We could say, on the positive side, that it is an ethic of self-responsibility and self-esteem needs (Maslow) arising out of conformist and belongingness needs. But its general trend is clear: from Descartes to Locke to Kant to Fichte: the more Ego, and the less Eco, the better.

And here the Ego camps began to veer into the negative.

Ego-negative: Repression

For each of those theorists (and for a host of others in the rational-egoic camp), the disengaged subject does indeed *disenchant* the world, and so much the better! It is precisely by disenchanting the world that the knowing subject liberates itself!

(Apart from a definite truth in that stance—namely, reason does indeed *disenchant* mere myth: grownups no longer need to believe in Santa Claus—nonetheless, this is the point where the Eco camps would generally begin their critiques. Even if a large dose of disenchantment was called for, the Eco camps would maintain, it had gotten quite out of hand, even wreckless and vicious, and disenchanted was fast becoming disemboweled.)

In the meantime, the disengaged Ego-subject, in its cruder forms, simply set about to unambiguously reflect on the world in purely objective terms (since the two were radically divorced, this did not seem to present any problem: one simply looks). The disengaged subject stares at the world, objectifies it, and describes the results—the so-called "reflection paradigm"—a representational epistemology that has been one of the central tenets of modernity from Descartes and Hobbes and Locke down to Quine—what Habermas calls "the Enlightenment's proud culture of reflection" and Taylor summarizes as "To know reality is to have a correct representation of things." And if it wasn't merely the Ego *reflecting* on the Eco, then it was the Ego *working* on the Eco—the so-called "production paradigm" (Marx). As shown in figure 12-6, all reality is Right-Hand, and the Ego simply *reflects on* those empirical objects: the mirror of Nature.

In this scheme, language is thought to be very simple: it is a set of conventional signs that straightforwardly represent concrete things (and we see this from Hobbes and Condillac to today's empiricist linguistics, all of which overlook the fact that signs themselves possess meaning only by virtue of the intersubjective system of other signs, and, further, only the least interesting signs "point" to things; most signs point to other signs: language creates worlds, it doesn't just reflect pregiven worlds).

In all of these accounts, and in the reflection paradigm in general, what we see is the world reduced unambiguously to monological terms, to it-language, to objective surfaces with no depth requiring *interpretation* and no dialogue requiring *mutual understanding*. (As shown in fig. 12-6, there is no Lower Left as an irreducible reality.) The world is simply a collection of objectively interlocking events, and knowledge consists in correctly representing those events. Knowledge is of those objects, and not also the mutual understanding between subjects (and certainly not the transcendence of both subjects and objects in the nondual One). This knowledge is not translogical; it is not dialogical; it is monological.

Again, it is not that representational and objective (or empirical) knowledge is simply wrong, but rather is extremely partial. It reduces all dimensions to their Right-Hand correlates (which is subtle reductionism; in gross reductionism, the Right-Hand dimension is itself further reduced to its atomistic elements; both are monological). And since all dimensions do have some sort of Right-Hand correlate, the project can always seem to be making headway—there is always some sort of evidence one can point to (even though it involves a reduction). It genuinely appears, within this prejudice, that the entire world can be captured in it-language.

Thus is the Kosmos reduced to the neutral flatland cosmos. And whereas the Kosmos possesses interior depth—and thus *intrinsic value* inheres in the *depth* of each and every holon (animal, human, divine)—the flatland cosmos itself (or so it appears to it-language) is completely devoid of meaning, value, awareness, prehension, consciousness—there are no Left-Hand components at all! As Whitehead's famous summary has it, the cosmos is now "a dull affair, soundless, scentless, colourless; merely the hurrying of material, endlessly, meaninglessly." (To which he added, "Thereby, modern philosophy has been ruined.")

But that is how all surfaces *correctly* appear anyway; and thus if the *only* knowledge one allows is objectifying knowledge—knowledge of surfaces that can be seen with the senses or their extensions—then the results of that type of investigation are pre-determined: look at any value in an objectifying fashion, and you find no value, just neutral surfaces, at which point you say, "What value?"

As we have seen, this empirical, objective, mutually interlocking and holistic world of "its" left no room for the subject (I's and we's). And so, in the strangest twist in this already strange tale, beginning around the eighteenth century (with a few earlier efforts), the subject attempted to get at itself, to know itself, by *turning this monological gaze inward upon itself.* It *retroflected* the monological gaze. That is, the subject attempted to categorize, describe, and explain itself in objective terms, in it-language. Not in intersubjective terms of mutual understanding and mutual recognition, but in the neutral, naturalistic, "scientific" terms of it-language.

And thus was born what Foucault called "the Age of Man"—the merely "objective" study of humans. And in a beautiful phrase that would summarize the net effect of the entire Enlightenment paradigm (right down to today's systems theory), Foucault points out that with the "Age of Man," humans became "the object of information, never the subject in communication."[37]

Much of Foucault's work is a cataloguing of the nightmares that descended on citizens as they became merely objects (and thus "subjugated subjects") in the holistic net of monological knowledge (particularly in the biologized form of "biopower").

This is likewise what Habermas refers to as "the totalitarian characteristics of an instrumental reason that *objectifies* everything around it, *itself included.*"[38] In this respect, Habermas is in complete agreement with Foucault. "It is an idea that attains dominance at the same time as subject-centered reason: that killing off dialogical relationships transforms subjects, who are monologically turned in upon themselves, into objects for one another, and *only objects.*"[39]

Further, says Habermas, "The same structure [retroflected monological gaze] is to be found at the cradle of the human sciences"—the so-called "dehumanizing humanism" and the "sciences of man" (toward which so much of the postmodern attack has rightly been focused). Habermas, noting his point of agreement with Foucault, is specific: "A *gaze* that objectifies and examines, that takes things apart analytically, that monitors and penetrates everything, gains a power that is structurally formative for these institutions. It is the gaze of the rational subject who has lost all merely intuitive bonds with his environment and torn down all the bridges built up of intersubjective [and dialogical] agreement, and for whom in his monological isolation, other subjects are only accessible as the *objects of nonparticipant observation.*"⁴⁰ Habermas makes clear that this is yet another version of . . . monological systems theory, where the great monological systems net descended on citizens for their own "benefit and welfare."

The bad news continues. These objectifying and systematic "sciences of man" are in fact pseudo-sciences, according to both Foucault and Habermas, and pseudo-sciences driven by self-aggrandizing *power.*⁴¹ "Because the human sciences with their borrowed models and alien ideals of [exclusive] objectivity, became involved with a human being that was for the first time turned into an object of scientific investigation by the modern form of knowledge, an impulse could prevail in them unawares, which they could not admit without risking their claim to truth: just that restless pressure for knowledge, self-mastery, and self-aggrandizement. . . ."⁴²

In other words, the "sciences of man" and the new "dehumanizing humanism" did not just study the objective (and monological) aspects of human beings (which would be fine), they *reduced* human beings to their merely objective and empirical components (which was the crime). Humans were not "subjects in communication" but merely "objects of information." And because that *reduction* is not supported by the Kosmos, because that reduction is a violation of the richness of the universe, then it must be driven by something other than truth: it must be driven in large measure by self-aggrandizing *power*, according to both Foucault and Habermas.

And with this, the whole dark side of the Enlightenment comes lurching to the fore. The catastrophe was not the emergence of reason, but reason *confined* to and initially *captured* by empiric-analytic modes—objectifying, monological, positivistic—modes which see *only* the Right-Hand dimensions and never the Left-Hand (even though their own operation depends intimately on them). It was a reason that differentiated the Big Three only to let them fall into dissociation, with the resultant emphasis solely on the Right-Hand (the Big Three reduced to the Big One of it-language).

Or again, it was a reason that went from an altogether commendable worldcentric understanding, to a virtual *repression* of anything not automatically of its ilk. It was a higher emergence used for altogether rude purposes, turning everything into objects of the monological gaze, and severing, alienating, repressing the rich communions that allowed its agency to function in the first place.[43] If *only* the Right-Hand dimensions are acknowledged, then the entire Left half of the Kosmos has to be re-pressed, denied, reduced, distorted.

It was the *repressive* side of the Enlightenment that its critics have espe-cially focused on (in various forms, their names would soon be legion: Schelling, Hegel, Nietzsche, Heidegger, Bataille, Foucault, Derrida). But few of the Enlightenment's critics have been more acute than Hegel. We have already heard him brand the Enlightenment as a "vanity of the un-derstanding"—"the understanding" being empirical-categorical, reflec-tive, and monological reason, as opposed to (or rather subsumed in) ma-ture Reason (vision-logic), which is dialectical, dialogical, and process-oriented. Hegel felt that "in the modern world emancipation became *transformed into unfreedom* because the unshackling power of [monolog-ical] reflection had become autonomous and now achieved unification only through the violence of a subjugating subjectivity . . . it posited as absolute that which is conditioned."[44]

In short, reason that has become monological holds all nonreason at arm's length, marginalizes it, excludes it, dominates it, objectifies it, disin-fects it with disgust.

Put a final way: it was reason shot through with Phobos. The great Enlightenment rationality did not just transcend and include, and meet the world with an Agape to balance its Eros. Instead of embrace, there was all too often a distancing gone mad, disengagement gone extreme, alienation and repression and dissociation—all out of an understandable but ultimately insane aversion to anything that looked like "heter-onomy."

Not just commendable Eros but a prickly Phobos was at work in the Ego's cherished autonomy, an autonomy and an agency that therefore all too often *severed its communions* with nature, body, sensuality, commu-nity—and sent scattered fragments running to the winds where union and integration were supposed to be gloriously forthcoming.

(The Ego's isolation—from its own inner depths, its culture, and Na-ture—is well illustrated in fig. 12-6. Notice that this is a precise example of a holarchy gone pathological, which we discussed in chapter 1—an arrogant holon usurps its place in the normal holarchy, thus dominating/repressing those holons upon which its own existence intimately depends. Not holarchy, but *pathological* holarchy, with hyper-agency severing and dominating—precisely because fearing—its communions: there was rationality run amok, and there was Phobos driving the repression.)

Thus, even Habermas, proponent of reconstructed modernity, could recognize that objectifying reason "denounces and undermines all unconcealed forms of suppression and exploitation, of degradation and alienation, only to set up in their place the unassailable *domination of rationality*. Because this regime of a subjectivity puffed up into a false absolute transforms the means of consciousness-raising and emancipation into just so many instruments of objectification and control, it fashions for itself an uncanny immunity in the form of a thoroughly concealed *domination*."[45]

And thus, "The permanent sign of enlightenment is domination over an *objectified* external nature and a *repressed* internal nature. Reason itself destroys the humanity it first made possible."[46]

The *repressive* side of the Enlightenment, the tendency of monological reason to marginalize everything that is not of its ilk, Eros gone catastrophically into rampant Phobos—this cancer was eating its way into the Age of Reason even while its nobler ideals were remaking Europe. Holbach could indeed sing—and who would not join him?—"How this portrait of mankind, free of all these chains, no longer under the enemies of progress, and walking with a sure and certain step on the path of truth, of virtue and happiness, presents to the philosopher a sight which consoles him for the errors, the crimes, the injustices which still sully the earth. . . . He then dares to bind these efforts to the chain of human destiny: there he finds virtue's true reward, the pleasure of having created an enduring good, which fate will no longer destroy with a deadly compensation by bringing back prejudice and slavery. Living in thought with a humanity re-established in the rights and dignity of its nature, he forgets the one which is corrupted and tormented by greed, fear, or envy; it is there that he lives . . . , which his love for humanity has embellished with the purest enjoyments."[47]

There, truly, the noble aspirations of the Age, and never so nobly expressed.

What Holbach could not know, as he wrote those words, was that he had but a few months to live. He was hiding in Paris, 1793, where a warrant for his arrest had been issued by the Jacobins, whose main leader, Robespierre—by all accounts a man of great integrity, courage, and devout republicanism—would soon oversee the Reign of Terror, in which many thousands would be drowned or guillotined for the revolutionary cause of a reason rising up to build the radiant future—only to succumb to its own limitations mistaken for the absolute, and set Phobos loose with terror and retaliation in its hardened heart.

THE ECO

The Eco camps would seize most immediately upon the repressions and exclusionary practices that reason could, and often did, bring in its wake. The representational/reflection paradigm left the subject staring out at an alien world, with an unbridgeable gap, hiatus, rift running right down the middle of the Kosmos. The Eco camps found this rift intolerable and set out to heal the split between the ego and the eco, between the self and Nature, which was then—and is now—as noble a goal as one can imagine.

Eco-positive: Wholeness

The intentions, as always, were honorable (as were those of the Ego camps), and the approach contained much that was (and is) undoubtedly true. First and foremost was the attack on the absolute status of the reflection paradigm, where a disengaged and dangling subject stared out mutely, dumbly, at a monochrome world. The Enlightenment, the Eco camps felt, had introduced "a false world of representation which cuts man off from the real living sources."[48] And it was only in a return to nature, to the rich fabric of the great organic life stream, that the "living sources" could be resurrected, recontacted, celebrated, renewed. (In this section, "nature" means both nature and Nature, or the empirical-sensory world within and without. In the next section we will use our more precise definitions for an overall summary.)

As Taylor summarizes the situation, "If the Enlightenment anthropology recommended itself through the sense of freedom, even exhilaration, of self-definition, the reaction to it [the Eco camp] experienced this picture of man as dry, dead, as destroying life. For the sense of freedom as a self-defining, reasoning subject was won by objectifying nature, and even our own nature in so far as we are objects for ourselves [retroflected monological gaze]. It was won at the expense of a rift between the subject who knows and wills, and the given: things as they are in nature."[49]

The originators of the broad Eco movement—Rousseau (1712–1778), Herder (1744–1803), the various Romantics (the Schlegels, Schiller, Novalis, Coleridge, Wordsworth, Whitman)—took it upon themselves to heal this rift. Herder: "See the whole of nature, behold the great analogy of creation. Everything feels itself and its like, life reverberates with life."

The mode of knowing most appropriate to this type of vision was not disinterested thought or representation, but rather a type of ecstatic feeling, what Wordsworth called "the spontaneous overflow of powerful feelings." As Herder put it, "Impulse is the driving force of our existence, and it must remain this even in our noblest knowings. Love is the noblest form of knowing, as it is the noblest feeling." And as for those who believe detached representation is the truest form of knowledge, they must be, says Herder, either "liars or enervated beings."

Thus, the truest and purest use of language, the Eco camp maintained, was not merely to reflect or represent an objectified and stultified reality, but rather to *express* the wondrous currents of life—and for this poetry and art were most ideally suited. (This was an important move, with far-reaching repercussions, to reintroduce the Upper-Left dimension of sincerity and *truthfulness* in expression, and not just the Right-half dimension of the *truth* of propositions; this was a truly profound contribution, even if it was often overblown to silly proportions.)

The idea wasn't to stand back and impartially and disinterestedly *reflect* on nature; the idea was to *participate* with nature and *express* nature, to join in the great currents of *communion* that nourish each and all, and not merely glorify one's own self-generated, self-independent *agency*. "This feeling cannot stop at the boundary of my self; it has to be open to the great current of life that flows across it. It is this greater current, and not just the current of my own body, which has to be united with higher aspiration . . . if there is to be unity in the self. Thus our self-feeling must be continuous with our feeling for this larger current of life which flows through us and of which we are a part; this current must nourish us not only physically but spiritually as well. Hence it must be more than a useful interchange of matter. It must be experienced as a communion."[50]

Thus, what the Ego camp had taken as proof of self-freedom and self-determination—namely, the objectification and disenchantment of nature—the Eco camp took as a sign of a deep inner cleavage, a profound alienation *within* the subject that expressed itself as a cleavage or rift *between* the subject and nature as well. Where the Ego camps had developed a passion for what they saw as self-determination, independence, and self-responsibility, the Eco camps displayed an equal passion for what they saw as unity, wholeness, harmony. For the one, self-responsible *agency*; for the other, self-abnegating *communion*.

"There was a passionate demand for unity and wholeness. The [Eco] view bitterly reproached the Enlightenment thinkers for having dissected man and hence distorted the true image of human life in objectifying human nature. All these dichotomies distorted the true nature of man which had rather to be seen as a single stream of life, or on the model of a work of art, in which no part could be defined in abstraction from the others. These distinctions thus were seen as abstractions from reality. But they were more than that, they were mutilations of man. . . . It was a denial of the life of the subject, his communion with nature and his self-expression in his own natural being."[51]

And thus, as Taylor summarizes the aspirations of an entire age of Eco-thinkers, "What they themselves yearned for was unity with self and communion with nature—that man be *united in communion with nature*." And this was to be accomplished by a "sympathetic insertion into the great stream of life of which they are a part."[52]

Eco-negative: Regression

And just there was the problem. *Granted* that this insertion was desirable, just how does the ego insert itself into something larger without losing the positive side of the beneficial gains that it had just made, and for which it had so valiantly fought? By giving up ego, the Ego camps asked, do you mean giving up autonomy and going back to heteronomy? Back to conformist determination by the church or state, since they are definitely "larger"? Back to personal impulse and inclination and natural feeling, no matter how self-serving or self-absorbed they might be? Back to the amoral currents of nature that, however lively and vital, still do not have to construct intersubjective cultures and do not therefore provide a model for human interaction? Back to fox eats chicken—which is supposed to be good simply because it is *natural*?

The problem was that we cannot simply *say* the subject and the object are one in the great stream of life. We have to show how this is so, and how it can be so, without resorting to facile verbal flourishes. The Ego camps were quite persuasive in this particular regard. If we are supposed to insert ourselves into the great stream of nature, they asked, then you—the Eco camps—must tell us how we can do this without losing the moral stance of worldcentric benevolence and postconventional compassion. Since the worldcentric stance is achieved only in the space of rationality—it is found nowhere in nature—then you must tell us how to dissolve reason into life-feeling without destroying the very stance that *allows that sympathy to exist in the first place.*

The Ego camp, from Kant to Fichte to Hegel, rather brutally tore into the Eco-minded "mushiness." Standing in nature and egocentrically emoting, they maintained, and seizing any sentiment that makes me weep at my wonder—that hardly constitutes the "noblest knowing, noblest feeling."

To be more precise, using our three definitions (NATURE, or the entire Kosmos, the Great Nest of Being; Nature, or the Right-Hand world and great Web of Life; nature, or bodily feelings and prerational awareness). At their best, the Eco-Romantics desired a union with NATURE, a union with the Kosmos (an Eco-Noetic Self or higher, the Over-Soul that is the World Soul). But for all the reasons we have discussed, the Eco camps drifted into an equation of NATURE with the great Web of Life, with the vast sensory-empirical world in all its organic richness and vitality. Under the intense gravitational pull of flatland, they reduced NATURE to Nature. And to Nature they pledged their undying allegiance.

The best of the Eco-Romantics understood, quite correctly, that NATURE (or Spirit) can only be grasped by an awareness that moves *beyond the rational-ego.* But in their understandable passion to go *trans*rational, they often ended up recommending anything *non*rational, including much that was frankly *pre*rational; not postconventional, but preconventional; not transpersonal, but prepersonal; not ascending above the ego, but descending—regressing—below it. Not beyond reason, but beneath it.

In other words, in order to embrace Nature, the great Web of Life, they often recommended a simple return to nature (or preconventional sensory awareness). Not a transcendence from biosphere to noosphere to theosphere (or prerational to rational to transrational, or nature to mind to spirit), but a movement in exactly the opposite direction. Under a pre/trans fallacy and the pull of flatland, they confused postconventional spirit with preconventional nature.

They thus ended up with the same general flatland as shown in figure

12-6, but instead of standing back from Nature with rationality, they would insert themselves into Nature with feeling. The great Web of Life, which in any event is the world as known to the senses, is to be grasped not by a rationality that distances itself from Nature, but by an intense feeling or communion that inserts itself into Nature, that is close to Nature, that gets back to Nature, back to the immediate richness of the sensory-empirical world and the preconventional awareness that more immediately discloses it.

In short, the Eco-Romantics sought to become one with Nature by trying to get back to nature. ("Back to Nature" and "Back to nature" thus came to mean essentially the same thing, because "Nature" is the entire Right-Hand, empirical-sensory world, and "nature" is the empirical-sensory awareness that knows that world. To recontact my own interior nature is thus to become closer to Nature, since they are both essentially the same empirical-sensory world.) And thus, the Eco camps maintained, the job of the rational-ego is to insert itself into the seamless web of Nature/nature, to become one with the great, holistic, empirical Web of Life. The job of the rational-ego is to get back to Nature/nature; the job of the noosphere is to get back to the biosphere. In any event, "to get back to" was the operative phrase.

The great Romantic regressive slide was about to begin. Confusing postconventional spirit with preconventional nature, the Romantics set out on their search for a visible, sensible God. They would not stand back and reflect on flatland; they would become one with it.

In this move there was an almost complete confusion of *differentiation* and *dissociation*. Since the Eco-Romantics were seeking Wholeness, any sort of differentiation was looked upon with suspicion. Instead of seeing that differentiation is the necessary prerequisite for a new and higher integration (and thus a deeper and wider Wholeness), the Romantics took any differentiation as a sign of dissociation and fracture, a sign that a prior "union" had been torn asunder, a sign of a *lost paradise*.

And thus, instead of going *forward* to a higher but not-yet-emerged integration, we are supposed to go back to nature *prior* to the alleged crime—back to a time before the differentiation of the noosphere from the biosphere, back to those cultures that do not differentiate the Big Three, back to those idyllic times of the noble savage and the pristine communion with nature, back in any case to some sort of Eden that *must* have existed because differentiation is now thoroughly mistaken for alienation.

In other words, instead of seeing that differentiation is the necessary prelude to a deeper or higher and *emergent* integration, it was seen, in all cases, as a disruption, a division and destruction, of a prior harmonious state. The oak was somehow a violation of the acorn.

And in this confusion—this pre/trans confusion—all true critical edge was lost, because the cure for the actual *dissociations* that had indeed beset modernity was mistakenly thought to be a *regression* to a state prior to all *differentiation* whatsoever.

In other words, where the Ego camps were marked by *repression*, the Eco camps were shot through with *regression*.

Both camps would tear aggressively, often brutally, into the fabric of the Kosmos with their preferred ideology, precisely because, as we will continue to see, they were (and are) both locked into the same flatland paradigm, where the more of Ego means the less of Eco, and vice versa—with no way to consolidate their equally important claims in a new and emergent and integrative growth.

In the meantime, the Eco-Romantic atmosphere was all aglow with yesterday. Central to most of the Eco-theorists was the notion that something had gone horribly wrong, that somebody or something had made a terrible mistake, that history was primarily the shadow of a great and heinous Crime, and that salvation, above all, was the resurrection of the dead and dismembered Paradise Lost.

Various theorists would set their Way Back machine to different dates, depending primarily, it seemed, on what they personally disliked most about modernity. Because differentiation and dissociation are now confused, *culture* itself becomes, not also a mode of emergent transformation that discloses deeper and wider worlds, but primarily a *distorting* force that conceals and contorts both a *pure* nature and my truly *pristine* self. Culture is not a transformative emergence of greater differentiation/integration that *can* go too far into dissociation/alienation. No, culture is primarily an alienating force that necessarily and nastily separates humans from nature and me from myself.

And so above all, I must get away from culture and get back to nature, back to a wilderness that expresses a pure nature and my own most authentic impulses. Culture is fundamentally a crime of distortion, and I must seek behind this distortion for the truth of pure nature and the truth of my pure self.

(That "back to nature" might be headed in a preconventional and not postconventional direction did not seem to dawn on the Eco-Romantics, or at any rate did not deter them in the least. When differentiation is confused with dissociation, regression is confused with salvation.)[53]

And so the search for Paradise Lost commenced. First stop on the Way Back was the medieval age, which was eulogized by many Romantics (then and now) as an "organically seamless society" whose members were tightly woven together in a net of meaning that did not yet sever culture and nature. This society certainly did not differentiate the Big Three—seven body orifices demanding seven planets is indeed a tight organization. On the other hand, there was that nastiness of the dominator hierarchies of King and Pope and hypermasculine God: perhaps, after all, 'tis not the Paradise sought.

The next, and by far the most popular stop on the Regress Express, was classical Greece, the admiration for, even worship of which, reached fevered proportions. "For the ancient Greeks represented to this Age a mode of life in which the highest in man, his aspiration to form and clarity was at one with his nature and with all of nature. It was an era of unity and harmony within man, in which thought and feeling, morality and sensibility were one, in which the form which man stamped on his life whether moral, political or spiritual flowed from his own natural being, and was not imposed on it by the force of raw will."[54]

Never mind that the force of raw will had in fact made approximately one out of every three people into slaves, and assured a virtually identical fate for all women. This, of course, did not stop Schiller from singing:

> When poetry's magic cloak
> Still with delight enfolded truth
> Life's fulness flowed through creation
> And there felt what never more will feel.
> Man acknowledged a higher nobility in Nature
> To press her to love's breast. . . .

The Eco camp continued unabated (and virtually unchanged) into many of today's ecophilosophies and "new paradigm" movements, which have taken up the Regress Express but in even more enthusiastic forms.

The ecofeminists do not at all approve of ancient Greece (patriarchal), but rather prefer the immediately preceding period of horticultural society, ruled by the Great Mother, where women, who did most of the productive work (women produced about 80 percent of the food in such societies), at least shared in many of the public policy decisions. This often produced what has been called "equalitarian" societies, or what Riane Eisler calls "partnership" culture.

But that partnership, we have seen, was not established by conscious choice in the noosphere, but rather by luck and happenstance in the biosphere: it was the product of a hoe, not a mutual recognition. Such predifferentiated societies are no model for an integrative partnership culture; they are models for biological slavery. And thus the actual statistics tell the real story: as Lenski's massive data makes obvious, an astonishing 44 percent of these societies engaged in frequent warfare and over 50 percent in intermittent warfare (which effectively challenges the notion that the Great Mother societies were "peace-loving"); 61 percent had private property rights; 14 percent had slavery; and 45 percent had bride price. And let us delicately ignore the fact that many horticultural societies practiced ritual human sacrifice, which was required, among other things, to insure crop fertility (one such site revealed eighty one-year-old girls sacrificed to the Great Mother). These horticultural societies were anything but "pure and pristine," as the ecomasculinists themselves have aggressively pointed out.

On the other hand, today's ecomasculinists have pushed the Way Back machine to its human limits. They do not at all approve of horticultural societies (the heaven of the ecofeminists)—in fact, the beginning of *farming* is for them the beginning of the great Crime of Humanity, of which industrialization is merely the inevitable endgame of planting seeds and thus "interfering" with "pure" nature. (This is one of the reasons that ecomasculinists and ecofeminists often do not get along well—one's "pure and pristine nature" is the other's "primal crime against pure and pristine nature"—the reason is that "pure and pristine nature" is pure ideology to which both camps equally succumb, and thus each is irritated that the other is infringing on its territory.)

Ralph Metzner, representing a long line of ecomasculinists, explains that "in tracing back the historical origins of the split between human consciousness and nature"—and notice already that in speaking only of a "split," he is not distinguishing between differentiation and dissociation, and so he is going to have to push back to the *very beginning* in order to get beneath the Crime, because every differentiation is going to be interpreted as a distortion—"it is possible to discern further layers of distortion, further choice-points and cultural transformations that lead to deeper dissociation and alienation. The first movement toward a controlling, altering attitude toward nature came with the neolithic [horticultural] domestication, ten thousand to twelve thousand years ago, when the hunting-gathering way of life that we had practiced for hundreds of thousands of years, attuned to the changing seasons and rhythms of life, gave way to herding and farming economies, settlement in small villages and towns, the accumulation and hoarding of surpluses, raiding and warring upon neighboring tribes, and so forth."[55]

And so there we have it. The Regress Express crashes into ground zero, beyond which it cannot go and still call itself human. There is the ultimate Lost Paradise: prior even to the invention of the wheel, the pristine and pure state is to be found in an updated version of the noble savage in small foraging tribes. All of this is told in the most idyllic terms. The typical day was a nice nature hike, then a little male bonding in the great hunt, then home to the women, barefoot, pregnant, around the campfire; smoke some herb, a little singing ("Nature loves me, this I know, for the raindrops tell me so . . .").

And let us conveniently forget that, as Lenski reports, the average life expectancy was 22.5 years; one out of three infants were "thrown away" for population control; 10 percent of these societies had slavery, 37 percent had bride price, and 58 percent engaged in frequent or intermittent warfare. . . .

And so the search would go, scraping more and more layers of depth off the Kosmos in search of a pristine and *foundational* paradise, beyond which all evolution was then thought to be a grotesque mistake. The representatives of this Paradise were then in possession of a *critique of culture*, they believed, that gave them immense power in condemnatory rhetoric for all things modern, or rather, all things past the chosen paradisaical epoch. Not only did this condemnatory rhetoric plot the course of humanity's horrific Fall, it gave these critics the key to Salvation, because the Paradise Lost, from which we had egregiously deviated, would also be the Promised Land, to which we must all return (following the directions of those who possessed the chosen critique).

In all of these particular Eco camps (from the Romantic era to today), the result of the regressive slide—of the failure to understand the difference between differentiation and dissociation—results usually in a rather dualistic, Manichaean, almost Zoroastrian, worldview. There is the *good* spirit (pristine nature) and the *bad* spirit (human intervention interpreted as any deviation from "harmony with nature").

But "harmony with nature" is impossible to define or even indicate. We already saw the ecofeminists eulogize farming societies as being *most* in harmony with the seasonal currents of nature and the Great Mother, while ecomasculinists damn farming altogether as *the* primordial Crime against nature. The ecomasculinists, of course, then claim that the even earlier tribal and foraging life represented pure and pristine harmony with nature. But these tribes cooked, built shelters, created clothing, entered into rituals to influence the outcome of the hunt, and so forth. These are all "interferences with nature" of one sort or another.

It soon becomes obvious that, since "harmony with nature" is impossible to define in any way that allows humans to be anything other than primates, then each Eco-theorist simply and arbitrarily decides what he or she doesn't like about the current of humanity's necessary differentiations.

Like its Ego cousin, this Eco-stance has its own sad pathos and absolute irony. The Ego-Enlightenment set out to free Eros from its heteronomy, its immersion in conformist and herd mentalities; and it sought this autonomy with such force that it went too far into alienation, and ended up with Eros degenerating into Phobos: seeking freedom, it found only fear and alienation, which bound it even more tightly to that which it wished to transcend.

And the Eco camps, so intensely desiring insertion into a Larger Life, an Agape that embraced the depth of the Kosmos with joyful Love and Care, ended up scrapping layers and layers of depth off the universe in search of the primal ground where this insertion could occur: it reduced the deeper and higher to the lower and shallower, a reduction and regression and leveling that, by any other name, is Thanatos. In search of a larger life, they found only a morbid death (i.e., a lesser depth), a rancid leveling of just those differentiations that allowed their search in the first place.

"This leveling," says Habermas, "can also be seen in the diachronic comparison of modern form of life with pre-modern ones. The high price earlier exacted from the mass of the population (in the dimensions of bodily labor, material conditions, possibilities of individual choice, security of law and punishment, political participation, and schooling) is barely even noticed."[56]

What is noticed, in these Eco-schools, is that today's world is God-forsaken. At some point in the historical (or prehistorical) *past*—at some point in yesterday's *time*—we have actually *deviated* from the Great Spirit, we have dissociated from the Great Goddess. And it is this dualistic notion—the notion that we *can* profoundly deviate from the actual Current of the Kosmos—that often sets up a regressive Zoroastrian worldview, a worldview that overlooks the fact that the Great Spirit, if it really is Great, must be behind even those moves that look to us like deviations. All of which overlooks, as Zen would have it, that "that which one can deviate from is not the true Tao."[57]

The Ego and the Eco: the flatland twins locked in the dance of ironic self-destruction, both contributing equally to the *failure* of integration. The one absolutizes the noosphere, the other absolutizes the biosphere, neither of which alone can integrate the other. And that which does not contribute to the integration of both contributes directly to the destruction of each.

It was in this climate of mutual repugnance—Ego versus Eco—that two events would emerge that altered the course of history irrevocably.

One was the discovery of evolution.

13

The Dominance
of the Descenders

It is the time of the gods that have fled and *of the god that is coming. It is the time of* need, *because it lies under a double lack and a double Not: the No-more of the gods that have fled and the Not-yet of the god that is coming.* —MARTIN HEIDEGGER

BOTH THE EGO camps and the Eco camps, starting out with the best and even noblest of intentions, had paradoxically brought about precisely what they had set out to eradicate. The reason was fairly simple: since both camps were (and still are) playing on the same flatland field, then the more one pole succeeds, the more the excluded pole returns in paradoxical forms—the more the "return of the repressed (or regressed)" comes to haunt the house of its originator. Lacking a holarchical Ascent to (and Descent from) the nondual One—which negates and preserves both Ego and Eco in a Nondual transcendence and embrace—there was only a mutual *negation*, and mutual *repugnance*, of one camp for the other.

The *integration* of the Big Three (person, culture, nature), now that they had been finally *differentiated*: this was (and is) the single greatest task facing modernity (and postmodernity). In simplified form, what was demanded was the integration of the interior or subjective worlds (I and we) with the exterior or objective (Nature); or again, the integration of Ego and Eco.

The great differentiation had already, irreversibly, irrevocably occurred. What the world awaited, anxiously, was a voice signaling integration.

THE PARADOX OF DAMAGE

Both the Ego camps and the Eco camps had responded with their own healing agenda, only to find that both agendas aggravated what they were meant to soothe and inflamed what they were meant to cure. Both camps were their own worst enemy, and both camps contributed to a violent fracturing of the world they were supposed to heal.

We already saw that the rational Ego took as its goal the transformation from egocentric inclinations and ethnocentric dominator hierarchies to a worldcentric stance of universal pluralism, altruism, benevolence, and freedom. It was a "declaration of independence" in more ways than one: independence from religious and mythic domination, from state-imposed interference in personal life, from conformist modes of the herd mentality, and from nature conceived as a source of not-yet-moral drives and inclinations. The autonomy of the rational Ego had to be fought for, had to be actively secured against all those forces of heteronomy that constantly were at work to pull it down from its worldcentric stance of universal tolerance and benevolence.

These values—in particular, universal pluralism, altruism, and freedom—are what Taylor calls the three "hypergoods" (goods that are felt to be better than other goods) that have most deeply characterized modernity (and postmodernity), and they are values that no doubt constitute much of the true dignity of modernity. But no sooner had the rational Ego secured some notion of these hypergoods than it began not just to differentiate from other goods, but to alienate them, dissociate them, repress them, seal them off.

And in all directions:

1. The rational Ego totally refused transrational and *transpersonal* occasions—"No more Ascent!" In understandably wanting to preserve the autonomy it had fought so hard to secure, it latched onto it with a death grip, and pressed its own autonomous agency in an extreme and exorbi-

tant form—what we will call "hyperagency"—and this meant, above all else, not handing its newly won freedom over to anything that looked even remotely like a God or Spirit.

2. It completely ignored, almost totally forgot, its own *interpersonal* dimension, the dimension of dialogical and intersubjective communication, in favor of the merely monological and objectifying mode, which is also a very *hyperagentic* mode, in that the *communions* of intersubjectivity are ditched in favor of the monologues of individual power and agency. The Age of Man: the disengaged (hyperagentic) subject staring at other humans as "objects of information" and not "subjects in communication" or communion.

3. And where it could have transcended and included the *prepersonal* dimension of élan vital, biotic roots, organic richness, it all too often simply dissociated and repressed them (in a hyperagentic fashion: nothing would touch its "freedom"). The disenchantment of the natural world, both without and within. (See fig. 12-6.)

Each of those dissociations and alienations—of the transpersonal, the interpersonal, and the prepersonal—would severely curtail its cherished freedom. And thus the paradox of damage: in aggressively pursuing intense freedom, it manufactured massive unfreedom.

In seeking freedom—autonomy not bound by any sort of heteronomy—it ended up cut off from spirit, cut off from nature, cut off from its own body, cut off from its fellows. And the worse it felt, the more it intensified its hyperagentic withdrawal and disenchantment: the unfreer it became. Not within and beyond, but within and withdrawn.

In short, in all those domains (spirit, nature, body, civics), corpses lay strewn everywhere on the road to a reckless independence. The rational-ego, hyperagentic and hyperindependent, took its own relative autonomy (which had indeed increased significantly), and blew it up to absolute proportions. In understandably wishing to increase freedom and liberty, it paradoxically left massive road kill everywhere on the highway to rational heaven.

But the same paradox of damage beset the Eco camps. In starting out with the express intention of *decentering* the Ego, of inserting it back into the larger currents of Life and Love, the Eco camps ended up—inadvertently, paradoxically—championing modes of knowing and feeling that were often egocentric and flagrantly narcissistic. In wishing to overthrow the Ego—and still being stuck, with their opponents, in monological flatland—the Eco camps introduced the modern world to a glorification of *divine egoism*: the outrageous return and exaltation of that which it expressly set out to overcome.

It started innocently enough. It *started*, as all modern and postmodern currents do, with a worldcentric stance secured in rational space—a stance of universal pluralism.

THE ECO SLIDE INTO DIVINE EGOISM

Rationality, or worldcentric and universal perspectivism, operates in a worldspace that allows, as all structures do, several different and apparently contradictory stances, which can be confusing if we don't look carefully at what is actually occurring.

Thus, on the one hand, rationality can tend toward a rather rigid version of "universal" truth, operating in a very abstract and mechanical and formulaic fashion, ignoring all individuality and all particulars and all differences. "Universal" comes to mean rigidly "uniform," and an otherwise admirably worldcentric perspective comes to mean "must be the same for everybody everywhere." And, we might say, rationality tends first to emerge in this fashion, in a monological and objectifying universalism that is, in fact, a *uniformitarianism* (and that is certainly the way it first emerged in the Age of Reason). It understands nicely the worldcentric perspective, but in its excitement it goes much too far and imagines that the worldcentric perspective is, in all its particulars, necessarily the same for everybody everywhere, with all colorful differences ironed out of the abstraction.

Thus, in the Age of Reason and the Enlightenment—with the exhilarating excitement of newly emergent reason, but reason captured almost exclusively by the monological and objectifying mode—*uniformitarianism* was all the rage. The central and by far the dominant reason that mythic-religions were rejected by the Age of Enlightenment was that these religions were everywhere *different* from each other—this god here, that goddess there, one creation story here, another there. Since they all disagreed with each other, how could they possibly all be true? What conceivable use could they have except to divide and fragment humanity? Precisely because they all lacked a true universality, a true worldcentric perspective, then they must be rejected: "No more myths!" meant "No more isolated and contradictory truths!"

There thus arose a positive mania for the universal and "common truths" of humankind, truths that could speak to everybody, and thus truths that must be "truly true," deeply true, for all peoples. All merely individual preferences and tastes, all peculiarities, all local differences, were dismissed as not being part of a common and universal humanity. "La réligion naturelle," proclaimed Voltaire, can only include "les principes de morale communs au genre humain." And Dr. Samuel Clarke spoke for an entire age when he said, "What is not universally made known to all men is not needful for any."

This search for a universal and unvarying set of codes and "natural law" (meaning universal law) swept through every branch of learning, science, political and social theory—and reached right down into the arts, into poetry, into painting, into architecture.

Dr. Johnson's pronouncements on uniformitarianism in aesthetic theory are well known; it is, he says, "a general rule of poetry that all appropriated terms of art should be sunk in general expressions," which meant, as he put it, "poetry is to speak a universal language." A universal language, a common humanity.

Thomas Wharton, the English poet and literary historian, came to view his youthful embrace of Gothic architecture as a "deviation" into a merely particular style, and recanted his grievous error with this exclamation, offered in 1782:

> Thy powerful hand has broke the Gothic chain,
> And brought my bosom back to truth again.
> To truth, by no peculiar taste confined,
> Whose universal pattern strikes mankind.

And thus, *for nearly two centuries*, "the efforts made for improvement and correction in beliefs, in institutions, and in art had been, in the main, controlled by the assumption that, in each phase of his activity, man should conform as nearly as possible to a standard conceived as universal, uncomplicated, immutable, *uniform for every rational being*."[1]

Spinoza, no doubt, summed it up best: "The purpose of Nature is to make men uniform, as children of a common mother."

A universal language, a common humanity, a uniform pattern. It was a noble enough cause, and there are some important truths in it. For we all, all of us, share the capacity for images and symbols, for concepts and desires, for aspirations and hopes; we are all born from a womb and find our way to a tomb; we all eat and labor and love and learn.

And thus, during its first large-scale emergence, the admirably world-centric stance of rationality took all the various perspectives of different peoples, places, races, and creeds, and attempted to abstract those items that they all agreed upon, that were not merely idiosyncratic or egocentric or ethnocentric.

But in its frantic rush to find and speak a "universal language"—whether in science, politics, art, or religion (Deism)—it simply steamrollered over every single trace of individuality wherever it appeared. Humans were monologically objectified, and only objectified, to find their commonalities (and this, we have seen, was a central project of the Ego-Enlightenment camps).

But rationality can rise to its fullest potential and find its fairest expression when it *also* pursues more even-handedly the actual space of universal perspectivism. The whole point of rationality and its capacity for multiple perspectives is not simply to abstract the commonalities (important as that might be, for example, in medicine), but to put oneself in the shoes of others and thus find a mutual enrichment and appreciation of differences. This is the *same* worldcentric rationality, but now celebrating all the multiple perspectives and not merely steamrollering them into monotonous uniformity.

And it was precisely this *celebration of diversity* that was behind so much of the Eco-Romantic rebellion against the Ego-Enlightenment. Eco-Romanticism often called itself "antirational," but in this particular regard it was actually "anti-uniformitarian." It started from the *same rational stance* of universal perspectivism, but it emphasized the "perspectives" instead of the "universal."

This trend is altogether unmistakable in the early Romantics. Notice how Schlegel moves from the rational-universal perspective seamlessly into the cherishing of cultural differences: "One cannot become a connoisseur without universality of mind, that is, without the flexibility which enables us, through the renunciation of personal likings and blind preference for what we are accustomed to, to transpose ourselves into that which is peculiar to other peoples and times, and, so to say, to *feel* this from its centre outwards."

It was precisely the decentering or worldcentering capacity of rationality that allowed cultural differences to be cherished and even glorified, something that had *never* happened on a large scale anywhere before in history (it is simply inconceivable, for example, that one horticultural society would celebrate the specific values of another). People, wrote Wackenroder at the time, "always think of the point at which they stand as the centre of gravity of the universe; and similarly they regard their own feeling as the centre of all that is beautiful in art, pronouncing, as from the judge's seat, the final verdict upon all things, without remembering that no one has appointed them to be judges. If you are unable to enter directly into the feelings of so many beings different from yourself and, by penetrating to their hearts, *feel* their works, [then] strive at least, by using the intellect as a connecting bond, to attain such an understanding of them indirectly."

This was not simply a tolerance for different points of view, although that was certainly part of expansive rationality's worldcentric project (even the Ego camps would champion tolerance). The Romantic move was that and much more: it was rather a direct attempt to cultivate an understanding and a sympathetic valuation of different points of view, *both* as a means of enriching one's own inner life *and* as a way to give objective validity to different cultures and valuations. Thus, where Kant would search for the universal in all humans, the Romantics, standing in the *same* open space of worldcentric rationality, would search out all the differences.

Accordingly, an ethic of individuality, of diversity, of idiosyncrasies, of the exotic, of the deeply personal, even of the strange and weird, of the anything-out-of-the-ordinary, rushed in to displace the previous uniformitarianism. "It is precisely individuality," as Friedrich Schlegel would put it, "that is the original and eternal thing in men. The cultivation and development of this individuality [is] one's highest vocation." Or Novalis on art: "The more personal, local, peculiar, of its own time, a poem is, the nearer it stands to the center of poetry." Schlegel on literature: "From the romantic point of view, the abnormal species of literature also have their value—even the eccentric and monstrous—as materials and preparatory exercises for universality"—universality in the sense that *everything* is to be honored. As he puts it, "Does admiration for one really require of us depreciation of the other? Can we not grant that each is in its own way great and admirable, even though the one is utterly unlike the other? The world is wide, and many things can coexist in it side by side."

The idealization of diversity. This was, in very large measure, the direct influence of the Great Chain theorists, who, we have amply seen, saw diversity and Plenitude as marks of the superabundant Efflux and Goodness of the Divine—the visible, sensible God: the greater the variety, the greater the Goodness. The early Romantics, just like the Enlightenment philosophers, were themselves dedicated theorists of the Great Chain, and traced much of their heritage to Plotinus (as we will see), so it is not surprising in the least that they would eulogize and celebrate, as Akenside put it, "the fair variety of things."

Most of the "politically correct multiculturalists" of today would, no doubt, be shocked and horrified to learn that their ideas come directly in an unbroken lineage from Plato and Plotinus (the dreaded Eurocentric, logocentric Plato, not to mention dead, white, and male). But Lovejoy is surely accurate in his appraisal: "The discovery of the intrinsic worth of diversity, with all of the perils latent in it, was one of the great discoveries of the human mind; and the fact that it, like so many other discoveries, has been turned by man to ruinous uses, is no evidence that it is in itself without value. In so far as it was historically due to the age-long influence, culminating in the eighteenth century, of the [Great Chain] principle of Plenitude, we may set it down among the most important and potentially the most benign of the manifold consequences of that influence."[2]

At its very best, then, the Romantic idealization of diversity tended "to nothing less than an enlargement of human nature itself—to an increase of men's, and nations', understanding and appreciation of one another, not as multiple samples of an identical model, but as representatives of a legitimate and welcome diversity of cultures and of individual reactions to the world which we have in common."[3]

"The fair variety of things." And thus, it is only a slight exaggeration to say that where the Ego-Enlightenment project of modernity would emphasize *universal uniformitarianism*, the Eco-Romantic project, searching for a larger Life and Love beyond the self-defining Ego, would emphasize *universal diversitarianism*.

Both were *universal* stances, because both stood in the same postconventional, worldcentric space of reason. But their aims were diametrically opposed, and their methods as well. The Ego camp was still intent upon securing its self-defining freedom, autonomy, and *agency* in an absolute and monologically aggressive enterprise; the Eco camp, on the other hand, was desperately reaching out for a larger and more variegated Life and Love, enriched with the *communions* of all imaginable variety—in nature and in other cultures—and promoting diversity and egalitarianism with a passion.

This diversitarianism would continue unabated right down to today's "postmodern" and "multicultural" movements. And as much as "postmodern poststructuralists" like to claim wild originality for their musings, it is all right there in the original Eco-Romantic agenda (including, of course, the emphasis on individual uniqueness and "wild originality").

Read carefully Lovejoy's summary of the Eco-Romantic project (and remember, Lovejoy wrote this in 1932; he could not possibly have had the "poststructuralists" and "multiculturalists" in mind, even though his summary reads with frightening accuracy their standard agenda, and not it alone; moreover, Lovejoy is describing the agenda as it first developed *a century and a half* ago):

> The immense multiplication of genres and of verse-forms; the admission of the aesthetic legitimacy of the *genre mixte*; the *goût de la nuance* [relishing of differences]; the naturalization in art of the "grotesque"; the quest for local color; the endeavor to reconstruct in imagination the distinctive inner life of peoples remote in time or space or in cultural condition; the *étalage du moi* [display of self]; the demand for particularized fidelity in landscape-description; the revulsion against simplicity; the distrust of universal formulas in politics [or social theory]; the feeling of "the glory of the imperfect"; the cultivation of individual, national, and racial peculiarities; the depreciation of the obvious and the general high valuation (wholly foreign to most earlier periods) of originality, and the usually futile and absurd self-conscious pursuit of that attribute.[4]

Where the Ego-Enlightenment counseled each and all to be uniformly universal, the Eco-Romantics' deepest desire was to be absolutely and even radically different, that is, utterly and radically unique. "It is precisely individuality," we heard Friedrich Schlegel announce, "that is the original and eternal thing in men." To which he adds, "The cultivation and development of this individuality, as one's highest vocation, would be a divine egoism."

Imagine one's *highest vocation* being, not the alleviation of human suffering, not the promotion of universal tolerance, not selfless service, not duty or devotion to family or kin or nation or globe, not compassion and not care: but the cultivation of divine egoism!

There is a worldcentric cherishing gone catastrophically egocentric. There is the Eco-Romantic slide into forms of ego-absorption that no Ego camp would even dream of! In fact, this divine egoism was precisely the *heteronomy* that the Ego camps were fighting!—and fighting because this self-absorbed and self-glorifying stance actually pulls me down from the higher and wider stance of worldcentric embrace and compassion—pulls me from worldcentric to sociocentric to egocentric—pulls me away from the *freedom* of a deeper and wider self to the confines of a shallower and narrower engagement, an engagement that hurts me as well as others, because it is so much less than what is available, and the lesser always tears at the seams, and leaves an agony where joy ought to sing.

And that indeed is *heteronomy*—other-directedness—because the egocentric self is not my true self, not my higher and deeper worldcentric self, but rather an addiction to a lower impulse, rendering me a slave of my own me-ness. Divine egoism is a sickly other within my own deeper possibilities. And thus, in divine egoism, I am other-directed, off-centered, out of my true self and out of my right mind.

The Eco-slide into an absolute anarchy of the particular, a riot of individual differences, a heap of the necessarily isolated and fragmented, with no commonalities at all, save a mutual divine egoism, which, at bottom, despises the existence of all other egoists as well—this fanatical diversitarianism "promoted a great deal of sickly and sterile introversion in literature—a tiresome exhibition of the eccentricities of the individual Ego, these eccentricities being often, as is now notorious, merely conventions painfully turned inside out. Thus the wheel came full circle; what may be called a *particularistic uniformitarianism*, a tendency to seek to universalize things originally valued because they were not universal, found expression in poetry, in a sort of philosophy, in the policies of great states and the enthusiasms of their populations. It has lent itself all too easily to the service of egotism, and especially—in the political and social sphere—of the kind of collective vanity which is nationalism or racialism. The belief in the *sanctity of one's idiosyncrasy*—especially if it be a group idiosyncrasy, and therefore sustained and intensified by mutual flattery—is rapidly converted into a belief in its superiority. More than one great people, in the course of the past century and a half, having first made a god of its own peculiarities, good or bad or both, presently began to suspect that there was no other god."[5]

The Eco-diversitarian slide into egocentric glorification. Precisely what it had set out to avoid, it now preeminently embodied.

And nowhere could this be more clearly seen than in the Eco-Romantic approach to what it now called "nature."

A DENATURED NATURE

In the eighteenth century, the whole approach to nature itself underwent a monumental shift (and by "nature" in this section, I mean both Nature and nature, or the empirical-sensory world without and within. We will, in the next section, use our more precise definitions for an overall summary). Nature was no longer a syncretic indissociation, as in magic and also, to a lesser but still important degree, in mythic-membership (and this newly emergent differentiation was itself a step up, so to speak: it was part of the differentiation of the Big Three, part of the dignity of modernity). *But neither* was nature an embodiment of Spirit or an expression of the One (as in Plato or Plotinus or Eckhart—or Lao Tzu or Huang Po or Padmasambhava).

Rather, nature became, first and foremost, the source of egoic sentiment. Nature became "nature as it makes me feel."

Although this was of course called "spiritual"—the "pure and pristine" encounter with a spiritual nature away from the confines of culture—it was in many cases, as we will see, the simple biosphere as reflected in divine egoism—nature as narcissistic reflecting pond—a monological nature that was, in all respects, the photographic negative of the monological ego. The Ego and the Eco were still battling it out on the flat and faded landscape of the great interlocking order—both of them monological to the core.

We have already traced this general development—the collapse of NATURE (or Spirit) to Nature (or the empirical Web of Life). But Charles Taylor comes to a similar conclusion based on a meticulous reading of texts of the time, and thus it is important to touch bases with his presentation. Indeed, part of Taylor's task, in *Sources of the Self*, is to trace this extraordinary shift in our encounter with nature, and he is not alone in his conclusions. "I have been tracing a movement, indeed, a massive shift in the notion of nature."[6] One aspect of this shift, he points out, is what we have been calling "the collapse of the Kosmos"—the Great Holarchy of Being was collapsed into a monological and flatland holism of observable exteriors, namely, the great interlocking order.

In flatland holism, as we have seen, nature and reality are still conceived as hierarchical or holarchical (or composed of interlocking wholes), but these wholes are now all *empirical* (the great interlocking order, or the universal system, is simply how the Great Holarchy looks when empirically and monologically collapsed; it is the result of *subtle reductionism*).

Since the wholes and the holarchies are now all empirical, then *inner transformation* is no longer needed to see a truer reality. One simply pokes around, with more sophisticated instruments, in one's already given empirical field. No ascesis, no yoga, no shiktan-taza, no personal injunctions or transformations are required to get at the truth—one simply opens one's eyes and sharpens one's monological gaze.

The radical Enlightenment had offered instrumental or calculative reason as a way to orient the Ego to the great interlocking order; the Eco-Romantic rebellion chaffed at the too-often exclusionary reason, and offered instead *sentiments* as a way to "feel" ourselves into the great interlocking whole. It offered, we might say, not instrumental reason but instrumental feelings, feelings that allowed us to fit into the Web of Life as a part and not try to dominate on the whole. Nature—the "true" or "real" nature—was thus disclosed to me primarily in feelings; nature was the sum total of what it evoked or awakened in me, as my own inner nature, and as my own pure and undistorted feelings.

This is what Taylor means when he says that "this new orientation to nature was . . . concerned with the sentiments which nature awakens in us. We return to nature, because it brings out strong and noble feelings in us: feelings of awe before the greatness of creation, of peace before a pastoral scene, of sublimity before storms and deserted vastnesses, of melancholy in some lonely woodland spot. Nature draws us because it is in some way attuned to our feelings, so that it can reflect and intensify those we already feel or else awaken those which are dormant. We turn to nature, as we might turn to music, to evoke and strengthen the best in us. But always the point lay in the *feelings*" (his italics).[7]

Taylor's acute insight is that this "nature as sentiment" presupposes a profound denial of the Platonic/Plotinian notion of nature as the *embodiment of Spirit* (or what Taylor calls an "ontic logos"—another way of saying that this "new" nature was not *part* of the Kosmos but was *itself* the *entire* cosmos of which *we* are now simply a *part*: we are merely a strand in the great Web, instead of seeing that the Web is merely a strand in the Kosmos, whose Self I-I can become: the Web itself is not Spirit, but is simply a manifestation of Spirit).

According to this new understanding, he says, "the meaning that the natural phenomena bear is no longer defined by the . . . Ideas [or Spirit] which they embody. It is defined through the effect of the phenomena on us, in the reactions they awaken. The affinity between nature and ourselves is now mediated not by an [ontic or Kosmic] order but by the way that nature resonates in us. Our attunement with nature no longer consists in a recognition of ontic hierarchy [the Great Holarchy], but in being able to release the echo within ourselves."[8] *The egoic echo . . .*

It is the *same* flatland Web of Life, the great interlocking whole, but now approached in egoic sentiment and not egoic logic. Taylor's well-argued point is that they (the Ego and the Eco) are, in essence, the photographic negatives of each other. In the Platonic/Plotinian view, says Taylor, "our being is identified with our being attuned to the order of things"—that is, attuned to the multidimensional Kosmos, to the Great Holarchy that "interlocks" not only horizontally in empirical wholes, but also vertically in inner transformations aspiring to the nondual One. "But in the feeling for nature which we see emerging in the eighteenth century and since, this [order] is fundamentally broken and then forgotten"[9]—the collapse of the Kosmos into the flatland interlocking order, into a monological reason and an *equally monological nature*—a mononature, itself conceived as absolute (just as the Ego considered its "reason" to be absolute).

"In other words," summarizes Taylor, "*this modern feeling for nature* which starts in the eighteenth century *presupposes* the triumph of the new identity of disengaged reason over the premodern one embedded in an ontic logos." The Ego and the Eco as flip sides of the same flat and faded coin, both oriented toward the same great interlocking order, however much their responses to it indeed varied (the Ego trying to get out of it, the Eco trying to get into it).

This is why Taylor can point out that, for the Eco camps, "our nature is no longer defined by a substantive order of purposes [ontic Kosmos], but by our own inner impulses and our place in the interlocking whole"[10]—the same interlocking whole of the Ego camps! The Ego and the Eco: the Tweedledum and Tweedledee of interlocking flatland.

We might summarize the overall shift in this way:

The differentiation of the Big Three, which released the rational-ego from its confinement to syncretic mythic-membership, fell into *dissociation* with the capture of reason by its objectifying, monological, and instrumental forms (a capture concomitant with industrialization). The Big Three were reduced to the Big One of mononature: the empirical flatland world, the great interlocking order of the système de la nature.

The Ego camps approached the great interlocking order of nature with an objectifying and monological analysis (empiric-analytic)—a "dry and abstract" approach against which the Eco-Romantic camps aggressively rebelled, preferring instead to find a unity with self and communion with nature via a more feeling-oriented awareness.

But Taylor's point is that in both cases, the nature that is approached is essentially the same nature, and this "nature" is a denatured nature in comparison with the Platonic/Plotinian view, where nature is a direct expression and embodiment of Spirit. In the Ego camps, nature is the final and ultimate reflected reality, and in the Eco camps nature itself gets elevated to Spirit, simply because real Spirit can no longer be found.

In both cases the same flatland nature is the object of awareness; in both cases, an ontic Order has been "broken and forgotten"; in both cases, mononature is serving a purpose it cannot ultimately sustain; and both cases absolutely presuppose the same collapse of the Kosmos.

THREE NATURES

In short, and using our more precise definitions: with the collapse of the Kosmos, NATURE (or the great Nest of Being) was reduced to Nature, or the great Web of Life. The Ego approached Nature with reason, the Eco approached Nature with nature (or preconventional feelings). In both cases, Nature reigned.

We have already seen that the Ego camps would approach this mononature in an objectifying and analytical fashion: Nature was the fundamental reality to be known by reflection (the "mirror of Nature"; Nature was the ultimate "reflectee"). The Eco camps would approach the same mononature in an attempt to unite with it, commune with it, return to it, spiritualize it. Communing with or even uniting with flatland became the ultimate spiritual occasion. The way you embrace Nature is by returning to nature, returning to wilderness feelings.

We have seen that in the modern world, nature has two related but slightly different meanings, which we have called "Nature" and "nature." Both refer in general to the empirical-sensory world. But, due to the collapse of the Kosmos, the empirical-sensory world appears in two paradoxical guises (both of which can be seen in fig. 12-5). In the great evolution from physiosphere to biosphere to noosphere, there are the *interiors* of those domains, and the *exteriors*. Nature, in common usage, refers *both* to the sum total of those exteriors ("Nature")—or the entire Right-Hand empirical world—and to those aspects, whether interior or exterior, that are just the *biosphere* ("nature") and *not* the noosphere (mind or culture). In the latter usage, "nature" as biosphere means the body as opposed to the mind, or nature as opposed to culture, or empirical as opposed to rational, or sensory as opposed to conceptual.

Both Nature and nature are different views of the same empirical-sensory domain: they are both the world as disclosed to the senses (both without and within). But the massively confusing item is that aspects of the noosphere exist in Nature, but not in nature. (This is identical to the mind/body problem, which is the same problem created by flatland.)

We can follow this easily on figure 12-5. (Instead of matter, body, and mind, I will treat the physiosphere as a part of the biosphere, which it is; so we will simply be dealing with the general relation of biosphere and noosphere, or body and mind.) In figure 12-5, the biosphere is level 1 (in both interiors and exteriors); the noosphere is level 4 (in both interiors and exteriors); and levels 2 and 3 represent stages where the noosphere and biosphere are differentiating (and possibly dissociating). Level 1 is the biosphere or nature, and the entire Right Hand is Nature.

In the interior or Left-Hand domains, the biosphere exists as prerational feelings, emotional-sexual impulses, life vitality, sensory richness, and all those items that people generally mean by the "body" as opposed to the "mind." Likewise, in the interior, the noosphere refers to all of those "mental" items as contrasted with bodily ones, such as rationality, symbolic culture, language and linguistic communication, and so forth. The interior noosphere also houses a moral development that moves from egocentric (preconventional) to ethnocentric (conventional) to worldcentric (postconventional), none of which are found in the sensory body or the biosphere or nature. And, of course, the noosphere transcends and includes the biosphere (formop transcends but includes sensorimotor). That is, *the biosphere is a part of the noosphere*, because if you destroy the biosphere, you destroy the noosphere, but not vice versa. In other words, nature (not Nature, but nature) is a part of mind (in the sense shown in fig. 12-5).

The difference between noosphere and biosphere—or mind and body, or culture and nature—shows up in common speech when we say that humans are destroying nature or ruining the biosphere. And when we look at the empirical world out there, we tend to think of trees as being part of nature, but we do not think of automobiles as being part of nature. In other words, we make a distinction between the products of the biosphere and the products of the noosphere.

So far, so good. The confusing part is that the *exterior* of the noosphere exists in Nature, but *not* in nature. And thus, the relation of the mind to the natural world depends upon whether you mean nature or Nature. For all of nature is in the noosphere, but the exterior of the noosphere is in Nature (as shown in fig. 12-5).

Thus, if you reduce all interiors to exteriors, if you reduce the Kosmos to Nature, it then appears that *Nature includes everything*, so of course the mind is just part of Nature! (for that reduces mind to brain—the mind/body problem is isomorphic to the Kosmos/Nature problem).

The *interior* of the noosphere is the mind, or your own consciousness right now, but the *exterior* of the noosphere is the neocortex, and the neocortex exists in Nature. The neocortex is an objective system of structures and functions that can be seen with the senses or their extensions; it possesses simple location and is built of holarchies of spatial extension. The brain exists in Nature.

But the mind does not exist in Nature. The mind is not a sensory object but an interior awareness. As we have seen, concepts and ideas and meanings and values do not have simple location; they are not running around out there in the sensory world; they cannot be seen anywhere in the empirical Web of Life. They are interior values known by the eye of mind, not exterior objects seen with the eye of flesh.

Thus, in an individual, nature (the biosphere or the body) is *in* the noosphere or the mind (reason transcends but includes feelings): *nature is in the mind, but the brain is in Nature*—and the latter makes the former almost impossible to see. There is the paradox of flatland and another version of the world-knot of the mind/body problem (whose unraveling we discussed in the previous chapter).

In Nature, or the Right-Hand world, the noosphere still transcends the biosphere, in the sense that more complex forms of evolution (e.g., neocortex) transcend and include the less complex (e.g., reptilian stem). And *all* of those exteriors exist in the world of Nature. The brain, the organism, the limbic system, cells, molecules—all of those exist in the vastness of Nature. They are all exteriors appearing in the world of the senses.

Thus, even though mind and feelings and interiors cannot be found in nature, all of their exteriors can. And thus, as I said, if you reduce the Kosmos to empirical flatland, it appeard that *Nature includes everything*, whereas it is just the Right-Hand world, or the great empirical Web. It then appears that, if we want to achieve Wholeness with the Kosmos, we must become one with Nature! (Because Nature is now all that is left of the Kosmos.)

But—and here is where the confusing part about Nature/nature reenters the Romantic agenda: not all of Nature is pure and pristine. If we look around in the empirical-sensory world (Nature), we can see not only the exteriors of the biosphere (trees and lakes and nature), but also the exteriors of the noosphere (such as airplanes and automobiles), and those exteriors are often problematic. The exteriors of the noosphere include things like industrialization, pollution, ozone depletion, and so on, which we do not count as part of nature. Even though those are part of Nature, they are *not* part of nature.

Already you can see that the Romantic recommendation to *become one with the natural world* has an inherent dilemma: do we become one with Nature, or one with nature? And, of course, the Eco-Romantics generally recommended that we must get back to nature.

Thus, with the modern collapse of the Kosmos to Nature, the Eco-Romantics did not embrace all of Nature, because those aspects of Nature that were produced by the noosphere (and especially by the noospheric items as industrialization) were shunned. The Romantics embraced Nature in general, but nature in particular. They aimed for NATURE, got Nature, and ended up glorifying nature.

"Back to nature" was thus the calling card of every good Eco-Romantic. Both Nature and nature are the sensory-empirical world, but nature (or the biosphere) is that part of Nature untouched by culture (or the noosphere). And thus, when the modern era had ensconced Nature (the descended, monological, Right-Hand, empirical Web of Life and great interlocking order) as the only God, the Eco-Romantics attempted to embrace the purer aspects of Nature by returning to nature.

Both Nature and nature I will often refer to as *mononature*, because both were equally monological. Taylor's conclusion can thus be stated very simply: there was no longer NATURE (or Spirit), but only Nature as disclosed in nature. No longer Spirit, but simply an empirical Web of Life disclosed in feelings and sentiment: Nature as it makes me feel. The theosphere and the noosphere reduced to the biosphere; postconventional spirit confused with preconventional nature.

We already heard Emerson strenuously arguing against this flatland Eco-view (which equated mononature with "spirit"): "Beauty in nature is not ultimate. It is the herald of inward and eternal beauty, and is not alone a solid and satisfactory good." Rather, he says, "nature is a symbol [or expression] of Spirit. Before the revelations of the Soul, time, space and nature shrink away. Vast spaces of nature, the Atlantic Ocean, the South Sea; long intervals of time, years, centuries, are of no account. Let us stun and astonish the intruding rabble of men and books and institutions by a simple declaration of the divine fact. Bid the intruders take the shoes from off their feet, for God is here within. Let our simplicity judge them, and our docility to our own law demonstrate the poverty of nature beside our native riches."[11]

But now, with the collapse of the Kosmos, nature *was itself* spirit, was the ultimate *source* of spiritual sentiment and the ultimate *foundation* of a salvation to be found away from the distortions and crimes of culture.[12] The Ego was to immerse itself in nature, in the simple biosphere, and therein was to be found its liberation from the stultifying grip of rationality.

Thus, in this new and flattened orientation, during a pure encounter with nature, spirit would flow *from* nature and *into* me, instead of flowing *through* me and *into* nature. No longer did Spirit or the Over-Soul shine through me and illuminate nature with a spiritual radiance, disclosing nature as a perfect manifestation of Spirit. Rather, now I am flooded by feelings released in my self by and from a mononature.

And the more immersed in this sensory mononature I can become, the closer to "spirit" I will be. I am no longer an opening or clearing *through which* the Divine can flow, baptizing nature and all manifestation with the Goodness of the Divine; I am now the receptacle of feelings coming *from* nature and *into* me, carrying me away in floods of sentiment and communions of sensory drenching.

Which is why Emerson would continue and say, "To the senses and the unrenewed understanding, belongs a sort of instinctive belief in the absolute existence of nature. In their view [the Eco camp's view] man and nature are indissolubly joined. Things are ultimates, and they never look beyond their sphere. His mind is imbruted, and he is a selfish savage."

But true intuition breaks this divine egoism, says Emerson, because "the best moments of life are these delicious awakenings of the higher powers, and the reverential withdrawing of nature before its God."

But nature would not now withdraw before its God, because *nature was now the only God.* Monological thought (by collapsing the Kosmos) had converted nature into God (for the Eco camps), and thus "God" was to be known by the impact that *nature* released *in me.* Nature was valued because of the thrill it sent swishing through the ego in "pristine" wilderness encounters. Not a transparency to the Divine, but the divine ego reflected to itself in monological feeling. This new nature was not *translogical*, and certainly it was not the dreaded *dialogical* intersubjectivity of culture—instead it was away from all of that and back to pristine and unadulterated nature: the same *monological* gaze, settled now on the simple biosphere.

In place of NATURE (or Spirit as the nondual One), there was now only sensory nature, and sin is therefore anything that separates me not from NATURE but from nature. Thus return to *biocentric* impulse and feeling is and must be my salvation, with rationality and culture being fundamentally the forces that stifle and contort this nature, both within me and without. And in immersing myself in this nature, I will find my true self.

(Culture is thus not a development *on the way* to conscious realization of Spirit; rather, culture is the crime that hides and distorts and negates nature, that negates "spirit." Of course, what culture is actually negating is biocentric and egocentric perception, and this the divine ego will not countenance.)

In mononature, away from dialogical culture, I can do as I please. In mononature, my pristine self reemerges. In mononature, I expand my glory and freedom without the stultifying presence of cultural demands, a culture that anyway embodies the great Crime of Humanity against the système de la nature. In this nature, I find my God or Goddess in the mute purity of the monological gaze.

And I can find my release and my salvation by an immersion in and union with the world so disclosed. To *unite* with the Shadows: that would be the Light. To become one with a denatured nature: there would be salvation. To unite with flatland: there would be the glory. To return in feelings so as to be *closer* to the Shadows: there would be my true self.

It wasn't that true spiritual insights weren't available in this new encounter with nature. But even Wordsworth knew the true source of the spiritual intuitions and illuminations that might be evoked in nature: "An auxiliar light, Came from my mind, which on the setting sun, Bestowed new splendour."

But in the new flat and faded cosmos, that splendor would immediately be *interpreted* as flowing *from* the sunset and *into* me, and I would then imagine that the sunset, that nature, is the *source* of the spiritual illumination. But that "splendor" is not lying around "out there" waiting to be perceived by the senses. Rabbits and weasels and foxes and company do not stare for hours at the splendor and beauty of the sunset, even though in many cases their senses are sharper.

No, nature is not the *source* of this spiritual splendor but rather its *destination*. When I can relax my egoic contraction (and many people appropriately find nature a fit and inviting space for this more easily to happen), then I can relax into that great Over-Soul, and then *through me* comes rushing the spiritual splendor of the One which outshines even the setting sun and bestows on it new glory.

But as long as I am locked into the flatland world of empirical mononature, as long as I *interpret* nature as the *source* of the Divine, then to just that extent I am locked out of any deeper or truer spiritual illuminations and intuitions: in glorifying merely the golden eggs, I am ignoring to death the goose that laid them.

Moreover, I am thus completely locked out of any sort of integrative vision. If the biosphere is the Divine, then the noosphere must be the

Crime. I cannot integrate nature and culture in Spirit (or true NATURE), I can only recommend *back to nature*. I must recommend, that is, *regression*.

Away from the transpersonal, away from the interpersonal—and back to the prepersonal, back to biocentric immersion—perfectly geocentric and perfectly egocentric—back to the prerational divine egoism that mistakes unconfined sensory immersion for transpersonal release. Back to a sensory nature that, as Taylor puts it, all too often confuses self-indulgence with self-transcendence.

And thus the horrible realization dawns: back to nature is away from NATURE. And the Eco camps, like the Ego camps, end up bringing about precisely that which they wished most to avoid: the glorification of the ego and the denial of true NATURE.

The Ego and the Eco will never be integrated in this scheme, with the one absolutizing the noosphere and the other absolutizing the biosphere. Neither of them will allow a true integration because both are rabid about privileging their chosen fragments. Both insist on converting the other by sharing their disease.

Under these circumstances, we will never be able to take monological nature up and into dialogical culture, and then take both nature and culture up and into translogical Spirit. The Ego and the Eco, the noosphere and the biosphere, will never be integrated in this scheme. Each will remain forever the great Devil in the other's eyes.

And so the paradox of damage. Playing on the same flatland field, locked in mutual incompatibility and mutual repugnance, both camps would bring about precisely that which they most wanted to avoid. The Ego camps, seeking first and foremost autonomy and independence, sought this as a freedom from the heteronomy of nature and conformist modes, only to find that its own intense demand for a higher freedom choked off its roots in nature and in cultural communions, and these repressions would return to sabotage and dramatically curtail its own freedom, a rancid transcendence (alienation and repression) driven by a Phobos bound ever more tightly to the objects of its disgust.

And the Eco camps, seeking first and foremost a freedom from the Ego, embraced the world of Eco-nature with such exuberance, and moved so intensely to do away with the rational Ego, that they ended up with prerational impulses in nature's display, and were alarmed to find that the prerational modes are precisely the most egocentric. *The closer you get to*

nature, the more egocentric you become. In setting out to do away with
Ego, they ended up locked into their own sentiments, the echo within
of a mononature without, both glorified as Divine. From worldcentric
pluralism to divine egoism and biocentric sensory immersion—at one
with sentimental nature in my own self-reverberating feelings—this was
the other endgame of flatland holism, a morbid embrace driven by a
Thanatos that, in the way of all deception, whispered always of the wonders of ever-shallower engagements.

SEXUALITY AND MODERNITY

Small wonder that both the Ego's view of nature as the fundamental reflected reality and the Eco's attempt to glorify nature as spirit would both
meet in an absolute obsession with: sexuality. (Henceforth, unless necessary, I will not distinguish nature and Nature, since both refer essentially
to the empirical-sensory domain. In any event, context will indicate.)

As Foucault ably demonstrates,[13] the rise of modernity was marked,
not so much by a repression of sexuality (although he does not deny that
happened) as by an *obsession* with sexuality. The entire era (beginning a
century before Freud) saw an explosion of interest in all things sexual,
from medicine to psychiatry to literature to education. Never, says Foucault, had an epoch's discourse been so dominated by sexuality in so
many domains. "Never, it seems, had so much attention been focused on
every aspect of the body and every dimension of its sexuality. Sex became
the object of a major investment of signification, of power, and of knowledge."[14]

We needn't go into Foucault's explanation (or rather, description) of
this obsessive sexuality, or how I would reconstruct it; this is dealt with
in volume 3. In the meantime, I have a simple take on the whole situation:

If nature is the fundamental and in many respects the final reality (according to the reflective Ego) and nature is the source of deep, ultimate,
and often hidden truths (according to the Eco), then how better to unite
them than in the belief that *sexuality holds the secret to human personality*? When nature alone is the final reality, what other secrets are there?

There was something (and still is something) in the obsession with sexuality that satisfied nearly all parties in the flatland world: hedonic happiness for the utilitarians; a source of hidden interpretations for the psychol-

ogists, interpretations that nevertheless safely remained quite shallow and flatland, very empirical-sensory (a hermeneutics that hugs the surface); a finding *in nature* (in the biosphere) the *ultimate* and *most profound* impulses of human life, which delighted the Romantics.

In general, the differentiation of the Big Three (and the differentiation of the noosphere and biosphere) meant that now there was indeed a differentiated body/nature/libido *to be obsessed with* (and likewise repressed in novel ways). And since, with the collapse of the Big Three to the Big One—to the objective and empirical interlocking order of mononature—sexuality came very close to being *all things to all people*: the ultimate hidden secret of human personality; the ultimate motivation that pushed all human beings; the great Life Force into which we can immerse ourselves for liberation and freedom ("sexual freedom," "liberation from repression," "sexuality as cultural anarchy," "free love"), a great Life Force that is, in and by itself, the ultimate spirit of the universe: sex as God, sex as Goddess, nature alone can rule supreme: the push of the bios is the only push around.

Georg Groddeck's *The Book of the It* (which was based in part on Eastern mysticism, the great "It" being essentially the great Tao that moves us all) was explicitly acknowledged by Freud to be the source of what he also started calling "the it" (translated as "the id"). The id is what moves all psychic life, and disrupting the id is the source of most pathology (and unhappiness). And the id is biospheric sex and aggression, the great organic and vital-somatic life of nature, the *pranamayakosha*. Not Spirit but sex is the *great secret*, and digging down into the depths of the human soul we will ultimately find the hidden, deepest, darkest and profoundest secret of all: not Ur-Geist, but orgasm.

The Tao reduced to the id: there is only the most notorious example of the collapse of the Kosmos into a flatland mononature, with sexuality the new and virtually sole (or certainly "most important") drive of the flatland world.

It's not just that Plotinus or Shankara or Garab Dorje would have found that notion preposterous; it was a perfect example of the great collapse, where the most superficial (the shallowest, the least-depthed) holons were proclaimed *the most profound*, precisely because the collapse of the Kosmos left no place else to situate profundity. Like so many other features of this collapse, most fundamental was confused with most significant.

Approaches such as the Freudian were called "depth psychology" precisely as a *misnomer*: it was actually "shallow psychology." (Although not, of course, as shallow as behaviorism, in contrast to which psychoanalysis did indeed seem "deep." In the nine-layered onion of the self—the nine Fulcrums—where matter is the first skin or sheath, libido or prana is the second layer or sheath, and this appears "deep" only *in comparison* with behaviorism-positivism-empiricism, which refused to peel beneath even the first skin. Psychoanalysis was thus a bold and daring plunge into the "depths" for the flatland world! And never mind the seven deeper/higher sheaths beyond psychoanalysis!)

And thus, in "digging down" into the libido, one was not actually digging toward any great depths, but more or less the opposite: one was *suspending the deeper or higher cognitive faculties* in order to temporarily *regress* to a lower and *shallower* and more ontologically surface level. One intentionally surrendered the deeper/higher cognitive faculties of intersubjective and worldcentric reason ("secondary processes") and moved to shallower and more narcissistic feeling-toned and imaginary processes ("primary process"), thus recontacting these shallower occasions, which, precisely because they were more fundamental (and less significant), could cause so much *subsequent developmental damage*. The point being (true enough) that when these lower/shallower levels are traumatized, then deeper/higher developments are often hobbled.

Thus, getting in touch with these earlier, shallower occasions was (and is) quite *difficult*, not only because they were laid down in earlier development, but mostly because they are indeed so altogether *shallow*. Staying with the libido is like having to stay afloat on the surface of the ocean, instead of relaxing and sinking into the real depths (of higher consciousness). One has to thrash around incessantly to *remain on the surface*: as everybody knows, hedonism is absolutely exhausting.

(One of the often misunderstood aims of Tantra is to take sexual orgasmic thrill and release it from its exhaustion in surface forms by opening it to deeper and higher spiritual occasions, where thrill is converted to bliss and exhaustion to rejuvenation. Sex is not apart from or against Spirit, but is simply one of the lowest or most fundamental of Spirit's *expressions*, and so a person's sexual nature is one of the easiest of the many threads that can be used to return to Spirit. There is indeed Life Force or prana or "cosmic libido," as it were, but it is only one of the lower of several sheaths of Spirit: the sheath found in nature, in the biosphere.)

But for a reality that is *only nature*, and is thus only moved by sex, pleasure, pain—and wonderfully strange Marquis-de-Sadean combinations thereof—then a thrashing exhaustion that demanded constant obsession was the only basic course (and curse) available to the collective mind: mononature is the ultimate reality, the surface is supreme, and the "deepest deep" of those surfaces is sexuality—and that exhausts the secret motivations of nature and captures the key to human personality.

And thus captures the key to human liberation. When monological thought turns nature into God, then spiritual liberation consists not in the developmental emergence of a transcendental Spirit, but in digging up a repressed sexuality. What other hidden god is there? Where else could liberation be found?

In its Romantic version: when nature is the ultimate reality, what other kind of liberation could there be, except a liberation from anything (such as culture) that "obstructs" nature and sexuality from its free play? Not progression to superconscious Spirit but regression to infraconscious nature; not recollecting a Spirit prior to manifestation but digging up an impulse prior to culture; not postconventional transcendence but preconventional immersion: sexuality is the great *Deus abscondus*, the great repressed and absent God, a God buried and suppressed by the Crime of Culture, by the wretched "smothering" of the life force that is known as civilization, a smothering that brings only the winter of our discontent.

And thus the libidinal God must be resurrected, a resurrection that will confer salvation, in the form of liberation from repression, on all who participate in the Great Libidinal Body as it rises from the dead night of repression and confers free orgasmic release on all who eat of that body and drink of that blood. Not Spirit, but sexuality, is the gateway to the Kingdom; and the liberation offered is not a freedom from finitude and mortality and temporal suffering, but a freedom from stultifying culture by finding orgasmic potency.

What's wrong with this picture? It's not that sexuality wasn't being repressed and couldn't use a little "loosening up." Nobody (including Foucault) is denying that culture can "go too far" and repress sexuality, and that a certain amount of "de-repression" is called for. Precisely because the noosphere and biosphere had been differentiated, they could indeed be dissociated (repression), and this no doubt accounted for a good deal of the explosion of interest in sexuality that marked modernity.

But something else was going on. Sexuality was invested with a force, a power, a mystique, an aura, an authority, all out of proportion to anything that could actually be dug up from the libido itself. The actual Deus abscondus—authentic spiritual realization of the truly deeper or higher disclosures—was being dramatically sought in the hidden and oppressed thrill of orgasmic release, because, again, what other kind of release can be found in a flatland world where mononature alone is real?

Thus the basic and otherwise quite different interests of the Ego and the Eco came together in a proclaiming of the fundamental and even ultimate reality of mononature (the flatland interlocking order), and consequently both their interests intersected in an obsession with sexuality, which is, after all, the great mover of the natural world.

Put differently, when monological thought became the God of the Ego—an Ego which confidently and supremely reflected on reality—then empirical nature (the reflectee) became equally absolute (it was the final and ultimate reality "reflected on" by the Ego): the monological thought of the Ego "absolutized" empirical nature (the Big One). And in similar fashion, the *same monological thought*, absolutizing nature, would, in the hands of the Eco-Romantics, simply proclaim this new and absolute mononature to be the ultimate God/Goddess/Spirit (the Ego and the Eco as the twin poles of the monoworld, both created by the same collapse, both oriented to the same denatured nature: this was Taylor's point). Thus both camps were united by their worship, in very different ways, of monological nature. And the great secret of nature is sex.

And so began the fabulous search of modernity for sexual release, upon which so many hopes were pinned. The Marquis de Sade became an enduring (and quite influential) point of reference, and Freud became Everyperson's doctor. As modernity unfolded, "free sexuality" would spread into Reichian forms (where culture was converted into muscular armor and liberation identified with orgasmic potency); into political forms (where sexual "de-repression" joined forces with Marxist "de-oppression": "free love" would not only undo repression, it would bring down capitalism: fascism supports genital repression, and orgasmic potency would end dictatorships, whether dictatorships of capital or of authority); it would drive student rebellions (Paris, May 1968: frenetic shouts of "Marx, Mao, Marcuse"); it would revitalize the retro-Romantic movements and invade literary theory (Norman O. Brown's salvation through libidinal "polymorphous perversity," all the way to Roland Barthes's *Le Plaisir du texte* and Bataille's shock-value orgasmic release and even Foucault's sadomasochistic pursuits, a "new economy of bodies and pleasure"). When an authentic spiritual realization is no longer part of *serious discourse*, then endless digging into the personal libido is one of the only and lonely substitutes.

The orgasmic well would be dipped into time and time again; strange that the release obtained never seemed to measure up to the liberation promised. And yet these "liberators" would still keep looking in the same direction, the hidden libidinal and biospheric god, the lunar goddess of sexuality, which had, it was fervently believed, been so brutally suppressed by modernity (completely overlooking the other side of the coin: the obsession with sexuality was not an escape from modernity but a mark of modernity. This *obsession* was not simply and solely due to the fact that real sexuality had been repressed, but was in part a substitute for a release no longer recognized; and thus in pursuing their intense sexual liberation as an escape from modernity, the "liberators" were pursuing something quintessentially modern—another version of the paradox of damage: they were once again presupposing and pursuing exactly that which they claimed to be overthrowing).

And when the liberation was not in fact forthcoming, when orgasm after orgasm left the participants not enlightened but exhausted, what was the recommendation? Try harder, fuck longer, come wilder, scream louder: the liberation has to be in there *somewhere*.

All these movements, and so many more, bound by the notion: the secret to human liberation is the liberation of sexuality.

Scholars have puzzled for decades over the fact that Freud was both a supreme rationalist (Ego camp) and a profound Romantic (Eco camp). He was said to possess "the mind of a moralist" and yet also called the "last great Romantic." He has been seen as primarily Victorian, repressive, and harshly puritanical; and yet he is also the father of virtually every Reichian, free love, antirepression movement of modernity. He has been seen as the reactionary upholder of culture and the status quo, and yet he has inspired virtually every progressive "sexual anarchy" movement of the last century. He has been proclaimed as the great prophet of the pleasure principle, and yet also as the firm and rigid upholder of the reality principle.

My point is that this is not surprising in the least. Freud was a prime example of the perfect intersection of the Ego and the Eco in an obsession with sexuality, the only real "juice" left in the flatland world. Since development now moved only from prerational libido to rational culture (and no genuinely transrational Spirit was officially acknowledged), then a profound and agonizing ambivalence developed around libido itself: if nature is the absolute (is even "God"), and if libido is the primary and "paradisaical" state free of repression, nonetheless an exclusive allegiance to that God has to be surrendered for culture to develop at all. Libido might be God (the Romantic position), but that God has to be broken (the Ego-rationalist position), and this leaves everybody, Freud included, in an altogether depressing dilemma: we want our God, but we really can't have it.

The Eco-Romantics would forever lament the loss and wail at modernity for killing their God, and take from Freud the urgent "necessity" to de-repress sexuality and free nature from the suffocation of culture (they would thus remain forever fascinated with sexuality as the great liberating force that would *undo modernity*, a way for the de-repressed biosphere to throw the noosphere off its back once and for all).

The Ego-rationalists, on the other hand, would remain just as fascinated and obsessed with sexuality, but take from Freud the "necessity" to build against the libido the bulwark of civilization (while meanwhile producing an endless scientific and medical literature on the subject of sexuality, which assured the scientific community of its own right to investigate every nook and cranny of individuals' sexual lives, rather like the antipornographer who nonetheless carries with him always the X-rated photographs to demonstrate the dangerous force that he is nobly trying to control. Foucault's work on "biopower"—the modern rational Ego's obsession with sexual and bodily information as a way to control it—is merely one of numerous indicators).

In short, when monological thought became God, sex became God's obsession.

And when monological thought likewise converted nature into God, sex became the key to human liberation and salvation.

In both cases, monological thought collapsed the Kosmos into flatland nature, which was the one, final, ultimate reality "reflected on" by the Ego and exalted to Spirit by the Eco, which left them both obsessed with a nature whose prime mover is sex. And in this obsession Ego and Eco were secretly united—mononature was all, sex was the secret.

The Ego would control it, the Eco would liberate it. But the "it," in both cases, was sex.

And so it came about that "modernity" and "sexual obsession" would be another two names for the same headache. And modernity rarely answered, Not tonight, dear. . . .

EVOLUTION

We could, indeed, embrace the whole in the single principle of development; if this were clear, all else would result and follow of its own accord.
                                                        —HEGEL

The Ego and the Eco—apart from their shared obsession with sexuality and their worship, in very different ways, of mononature—remained nevertheless in a theoretical standoff, locked otherwise in mutual contempt: approach nature by disengaging from it versus uniting with it.

And here the stalemate might have remained, were it not for the discovery of evolution.

There were many forerunners to the doctrine of evolution—Thales, Empedocles, Anaximander, Aristotle, Saint Gregory. Yet for fifteen centuries any sustained inquiry into evolution was effectively halted by the literal Genesis myth. But by the sixteenth and seventeenth century, upholders of the Great Chain began to realize that the Holarchy itself, exactly as presented by Plotinus, could in fact be not just a map of individual growth and development, but a map of the growth and development of the Kosmos at large. If the Great Holarchy was a general map of the dimensions of Reality, what difference would it make *when* they unfolded? And, indeed, philosopher-sages such as Origen had long ago proposed that Plotinus's hierarchy was actually played out over cosmic epochs, one epoch after another, stage by unfolding stage.

Thus, a century before Darwin, theories of cosmic and human evolution were springing up everywhere, theories that, as we saw, were looking for "missing links," but missing links that were now believed to have unfolded in time and history. Leibniz maintained that "the species of animals have many times been transformed," and that, for example, at some previous time "many species which have in them something of the cat, such as the lion, the tiger, the lynx, may have been of the same race, and may now be regarded as new sub-varieties of the original cat-species." The entire universe, he suggested, may have been one that *developed*, because we see everywhere the creative advance of nature—what he called "transcreation" (transformation).

Again, long before any scientific evidence had been marshaled for any of these ideas, the conclusion had been quite firmly reached that, as Lovejoy puts it, "the world is as yet incomplete, and the Chain of Being must be construed as a process in which all forms are gradually realized in the order of time."[15] Already by 1693 Leibniz was deducing from this the necessity for the past extinction of certain species and the "transcreation" of others. By 1745, the president of the Berlin Academy of Sciences, Maupertuis, had advanced the thesis that all species alive today had derived from a very small number, perhaps a single pair, of original ancestors, and this thesis was echoed by Diderot, the major editor of the *Encyclopédie*, in 1749. (This was one hundred and ten years before the publication of *Origin of Species*.)

Kant, as is well known, soon produced a rather profound theory of cosmic evolution. The universe began relatively simple and undifferentiated, he says, and became progressively more various, complex, organized. Even *matter*, he says, "has even in its simplest state a tendency to organize itself by a natural evolution into a more perfect constitution"— surely one of the clearest descriptions of the self-organizing property of evolution (reaching into even the physiosphere). Thus evolution, he says, "is forever busy achieving new ascents of nature, bringing into existence new things and new worlds." And the Fecundity (Plenitude) of Nature has one law, he maintains: *progressive diversification* (what we would call progressive complexification or differentiation).

Nature's unfolding, in short, has a *direction*—greater and greater creative diversification. From Robinet to Bonnet to Schelling to Bergson, "the pageant of evolution is the progressive manifestation of the expansive, self-differentiating energy, the creative urge. . . ."[16]

But not just differentiation; each ascent, it was maintained, also brings a new and higher union or integration. As Kant put it, "the creation, or rather the development, of Nature spreads by degrees with a continuous advance to an ever greater breadth. . . ." Schelling, writing in 1800, was specific:

Succession itself is gradual, i.e., it cannot in any single moment be given in its entirety. But the farther succession proceeds, the more fully the universe is unfolded. Consequently, the organic world also, in proportion as succession advances, will attain to a fuller extension and represent a greater part of the universe. And on the other hand the farther we go back in the world of organisms, the smaller becomes the part of the universe which the organism embraces within itself [notice the distinction between significant and fundamental].[17]

The greater the advance and differentiation, the greater the embrace and integration. As with Plotinus, development *included* diversity, took diversity with it, into higher unions, higher embraces—it is not the One versus the Many, it is the One-in-the-Many, played out in successive differentiations and integrations, each unfolding more of the universe and therefore enfolding more of the universe.

And here we have our old friend Eros, ever striving for greater unions, greater integrations, the binding love that unfolds more, enfolds more (Eros/Agape), the omega force hidden even in matter, driving it into higher levels of self-organization and self-transcendence, the omega force that will not let even matter sit around in unorganized heaps. We will return, in a moment, to this part of the story, and to Darwin's drudging and dutiful accumulation of evidence for a conception that was in the main already known in advance.

But how the new evolutionary understanding would fit into the prevailing flatland worldview was not at all obvious. Ego and Eco were still staring at each other across an unbridgeable gulf, and the two absolutisms were altogether incompatible.

THE AGONY OF MODERNITY:
FICHTE VERSUS SPINOZA

The Ego camp had developed into the end limit of Fichte's Pure Ego or infinite Subject, and however much the Pure Ego was indeed partly a genuine intuition of the Over-Soul (and even the Witness),[18] and however much the Pure Ego or infinite Subject was said to have posited nature out of itself, nonetheless the pragmatic and ethical point was clear enough: the more of the Ego, and the less of the Eco, the better for all.

This was indeed a genuine attempt to escape flatland and introduce Spirit, but a Spirit found by radicalizing the Subject or Ego (and thus a path breaking toward pure Ascent, free of any descended nature, free of any objects).[19] And indeed, to an entire age Fichte stood for Spirit as pure freedom in the transcedental and Absolute Subject. It was the height of the aspirations of everything that the Ego camps had stood for: a radical autonomy that found pure freedom in the Pure Subject.

The Eco camps, on the other hand, all had recourse to Spinoza, suitably interpreted for their purposes (this recourse to Spinoza is still quite common in ecomasculinist circles).[20] Spirit was here viewed as the absolute and total *Objective System*, into which the ego was to insert itself seamlessly, and so the ethical imperative was just as clear: the more of the Eco, and the less of the Ego, the better for all.

This, too, was in essence an altogether noble attempt to introduce Spirit, but a Spirit found by radicalizing the Eco (and thus a path embodying pure Descent, pure immanence).[21] To the same age, Spinoza stood for Spirit as eternal and total objective *Substance* (or infinite and Absolute Object, as Fichte stood for infinite and Absolute Subject). It was the essence of everything the Eco camps had stood for: a dissolution of the ego into the Greater Order of infinite Life and Love, Spirit present *as* the total diversity of the manifest world.

The great interlocking order of nature: the Ego camps wanted out of it (tending toward the Pure Ascending Subject, driven, at its best, by Eros), and the Eco camps wanted into it (as pure Descended Union, driven, at its best, by Agape). For the Ego, the more it could disenchant and disengage from nature and communions, the greater the knowledge and freedom of the self. The more agency, and the less communion, the freer and more liberated it would be. And the purest freedom would be the Pure Ego, the pure Subject, *transcending* nature and objects altogether, and finding ultimate ethical clarity. And apart from the paradox of damage (the Ego's reckless repressions), there was much truth in this stance of mature self-responsibility and self-determination.

For the Eco camp, self-isolated agency, however important, had sealed off the rich networks of unions and communions with nature and other cultures, and thus the more reenchanted the world, the better. The more communion, and the less agency, the freer and the more whole it would be. And the purest wholeness would be the Pure Eco, free of separate egos altogether. And this was Spinoza's great system as an Objective and Eternal Substance, purely immanent (Descended). And apart from the paradox of damage (the Eco's reckless regressions), there was much profound truth in the Eco's insistence on reengaging with the world of cultural and natural communions.

As Frederick Copleston, Taylor, and numerous other scholars have pointed out, an entire intellectual age was devoted to, or at least obsessed with, the integration of Absolute Subject and Absolute Object, of Fichte and Spinoza (or Kant and Spinoza, or Fichte and Goethe)—it came to the same thing: how can you possibly unite Subjective Freedom with Objective Union?[22]

That is: how can you unite a path of radical freedom and detachment heading in the direction of the ascending One, versus a path dedicated to descending union and communion with the diverse Many? How to heal this deepest of dualisms that had cruelly carved the landscape of Western culture for two thousand years? How to end the schizoid fracture in the massive footnotes to Plato? How to join the two broken Gods that had dominated Western history from its inception? And that had finally found their endpoint champions in Fichte and Spinoza? And that showed up in individuals as Subjective Freedom versus Objective Union?

We can't simply add them together with glib platitudes about subject and object being one, because subjective autonomy and objective heteronomy are absolutely incompatible, which is why, if consistently pursued, these positions lead inevitably to Fichte *or* Spinoza. Since the gain of one is the loss of the other, it's a win-lose situation, never a win-win.

By the end of the eighteenth century, the split between the ascending Ego and the descending Eco, although fundamentally the same basic split, showed up in an extraordinary number of dualisms, of which Taylor summarizes some of the more pertinent: "The opposition between thought, reason, morality, on one side [Ego], and desire and sensibility on the other [Eco]; the opposition between the fullest self-conscious freedom [autonomy/agency], on one side, and life in the community [civic communions], on the other; the opposition between self-consciousness [Ego] and communion with nature [Eco]; and beyond this the separation [and] the barrier between the Kantian subject and the Spinozist substance. . . ."[23]

Notice that the dualisms Taylor mentions include the split between mind and body, human and human, human and nature, and nature and spirit. Both camps claimed to have handled these massive dualisms; both camps lied. The Big Three had indeed been differentiated, but there was no apparent way to fashion their integration without privileging one or another domain. The "I" split from the "we," and the "I" and "we" split from the "it" . . .

. . . and in particular, there was that Sovereign I (Kant/Fichte) and that holistic It (Spinoza), the total I and the total It, the Ego and the Eco. "And the ambition becomes general in the 1790's: autonomy must be reconciled with unity with nature, Kant and Spinoza must be united; these were the watchwords."[24]

In order to honor the positive and true contributions of both the Ego and the Eco camps, we would have to *simultaneously* separate from the Kosmos (to preserve autonomy) and unite with the Kosmos (for wholeness), and neither camp had the slightest clue as to how to actually proceed with that necessary but seemingly contradictory project. How to reconcile Ego-agency with Eco-communion, reconcile a path tending toward pure Ascent cut off from all, and a path tending toward pure Descent immersed in all: It was an agony for the entire era, and it was not at all obvious how to proceed.

And indeed, it might have remained a mystery to this day, were it not for an altogether extraordinary individual.

SPIRIT-IN-ACTION

> *Schelling is lecturing to an amazing audience, but amidst so much noise and bustle, whistling, and knocking on the windows by those who cannot get in the door, in such an overcrowded lecture hall, that one is almost tempted to give up listening to him if this is to continue. During the first lectures it was almost a matter of risking one's life to hear him. However, I have put my trust in Schelling and at the risk of my life I have the courage to hear him once more. It may very well blossom during the lectures, and if so one might gladly risk one's life—what would one not do to be able to hear Schelling?*
>
> *I am so happy to have heard Schelling's second lecture— indescribably. The embryonic child of thought leapt for joy within me when he mentioned the world "actuality" in connection with the relation of philosophy to actuality. I remember almost every word he said after that. Here, perhaps, clarity can be achieved. This one word recalled all my philosophical pains and suffer-*

ings.—And so that she, too, might share my joy, how willingly I
would return to her, how eagerly I would coax myself to believe
that this is the right course—Oh, if only I could!—now I have put
all my hope in Schelling. . . .

—SØREN KIERKEGAARD, letters from
Schelling's Berlin lectures (1841)[25]

Friedrich Wilhelm Joseph von Schelling (1775–1854) was born at Leon-
berg in Wurttemberg, and, always something of a boy wonder, was at age
fifteen admitted to the University at Tübingen, where he became friends
with Hölderlin and Hegel (both were five years older). He began publishing in his late teens and, though younger than Hegel, was the senior part-
ner in their period of collaboration at Jena (where, in 1798, he was appointed to a chair at the age of twenty-three; his writings had already
won commendation from Fichte and Goethe; he had Hegel called there in
1800).

Unfortunately, Schelling is usually treated summarily as the transition
from Fichte to Hegel (that is, from Kant's critical philosophy to absolute
idealism). This is a great mistake, I believe, because in many ways he was
eventually more influential, in my opinion, than even Hegel (Heidegger,
for one, felt Schelling was always the most *profound* of the Idealists).
Hegel's system was so massive and so coherent that when it began to
unravel, which it did, it rather totally collapsed, whereas Schelling ended
up planting numerous small time bombs that are still going off.

Schelling began, in a sense, by reacting to the Enlightenment notion
that rationality alone is the highest Ascent to which we can aspire. For if
it is true that the Enlightenment had succeeded in differentiating mind
and nature, the noosphere and the biosphere, it had also tended to forget
the transcendental and unifying Ground of both, and thus it tended to
dissociate mind and nature (the great potential pathology of reason). This
dissociation of mind and nature, Ego and Eco, with mind "mirroring"
nature in scientific inquiry (the reflection paradigm) was, of course, well
under way; but reflection had, Schelling pointed out, introduced a rift or
cleavage between nature as external object and the reflecting self as subject (which also, he said, made humans objects to themselves—the retroflected monological gaze, as we earlier put it). And when reflection is
made an end it itself, it becomes "a spiritual malady."

At the same time, Schelling realized that the dissociation could "not be overcome by a return to the immediacy of feeling, to the childhood, as it were, of the human race."[26] There was no going back to Eco-nature, and Schelling knew it.

Rather, he maintained, we have to go forward *beyond* reason in order to discover that mind and nature are both simply different movements of one absolute Spirit, a Spirit that manifests itself in its own successive stages of unfoldment and enfoldment. As Hegel would soon put it, Spirit is not One apart from Many, but the *very process* of the One expressing itself in successive unfoldings in and through the Many—it is infinite activity expressing itself in and as the finite process of development itself (or evolution).

The Absolute is thus both the Alpha and the Omega of development. Schelling: "I posit God as both the first and the last, as the Alpha and the Omega, as the unevolved, *Deus implicitus*, and the fully evolved, *Deus explicitus*"—in other words, both Source and Summit—and *present* throughout the entire ascending or developmental process as telos (Eros), as the self-organizing and self-transcending drive of the *whole process*. And, as Schelling put it, it is this "*organizing principle* which makes the world a system . . . a self-organizing totality."

Whatever else we might say, the world does hang together, and evolution does have a direction: Eros as Spirit-in-action. Thus, both Schelling and Hegel would maintain that evolution is not simply the drive *toward* Spirit; it is the drive *of* Spirit toward Spirit, manifested in a series of increasing wholes and integrations (holons) that express increasing degrees of Spirit's own *self-realization* or *self-actualization*. Eros is fully present at each and every stage of the process, *as the very process itself*. As Fichte had put it, Being is one "endlessly self-developing life which always advances toward a higher self-realization in a never-ending stream of time." Or Hegel: the Absolute is "the process of its own becoming; it becomes concrete or actual only by its development."

As Hegel explained in his *Lectures on the Philosophy of History*: "That the history of the world, with all the changing scenes which its annals present, is this process of development and the realization of Spirit—only *this* insight can reconcile Spirit with the history of the world—that what has happened, and is happening every day, is not only not 'without God,' but is essentially God's work" (which is precisely why "that which one can deviate from is not the true Tao").

Each stage of development (or evolution) is thus Spirit's *knowledge of itself* through the structures (and limitations) *of that stage.* Each stage is therefore a thesis (Fichte, Hegel) that eventually runs into its own *limitations* (Fichte: antithesis; Hegel: contradictions; Schelling: checking forces), which triggers a *self-transcendence* to a new *synthesis* (Fichte, Hegel; Schelling: organic unity), which both *negates* and *preserves* its predecessor (Schelling, Hegel). This dialectic, of course, is Eros, or Spirit-in-action, the drive of Spirit to unfold itself more fully and thus unify itself more fully.

Thus, Spirit tries to know itself first through sensation, then perception, then impulse.[27] At this point Spirit is still un-self-conscious. Both Fichte and Schelling make much of the notion of the unconscious, quite along the lines used by Plotinus: the higher stages are implicit in the lower, they are unconscious potentials of the lower, and as they develop they envelop.

Thus the whole of nature Schelling refers to as "slumbering Spirit." Moreover, nature is not a mere inert and instrumental backdrop for mind. Rather, nature is a "self-organizing dynamic system" that is "the objective manifestation of Spirit"—"nature as a unified, self-developing super-organism"[28]—precisely Plato's "visible God," but now set developmentally or evolutionarily afloat.

Thus nature is most definitely not a static or deterministic machine; it is "God-in-the-making." All life processes are manifestations of the Divine Life, unfolding in space and time. But it is, says Schelling, Spirit *slumbering* because Spirit has not yet become self-conscious, the Kosmos has not yet begun to consciously reflect on itself.

With the emergence of mind, Spirit becomes self-conscious. Spirit seeks to know itself through symbols and concepts and reason, and the result is that the universe begins to think about the universe—which produces the world of reason and, in particular, the world of conscious morals. Thus, says Schelling, where nature was objective Spirit, mind is subjective Spirit.[29]

And it is here, he says, that mind and nature can seem to drift apart, to be totally unrelated, to stare blankly and uncomprehendingly across the subject/object dualism at the alien beings on each side of the divide. The Ego and the Eco.

These two "apparent absolutes," as he calls them, are *synthesized* in the third great movement of Spirit, which is the transcendence of *both* nature and mind and thus their radical synthesis or union "in which these two absolutenesses (absolute objectivity and absolute subjectivity) are again one absoluteness," as he puts it (with Fichte and Spinoza in mind). This is also the identity of subject and object in one timeless act of self-knowledge, of Spirit *directly knowing itself as* Spirit, a direct mystical (or contemplative) intuition, says Schelling, that is *not mediated* through *any forms*, whether those forms be the *feelings* of objective nature or the *thoughts* of subjective mind.

There is an unmistakable and profound glimpse of the formless and nondual groundless Ground; Schelling would often have recourse to the "indifference" and the "Abyss," precisely in the lineage of Eckhart and Boehme (and Dionysius). "In the ultimate dark Abyss of the divine Being, the primal ground or Urgrund, there is no differentiation but only pure identity."[30] Precisely what the Sufis would call the Supreme Identity. And, as Schelling says, "A division, a difference must be posited, if we wish to pass from essence to existence" (and thus start the whole process of what we have called involution or Efflux).

Thus, for both Schelling and Hegel, Spirit goes out of itself to produce objective nature, awakens to itself in subjective mind, and then recovers itself in pure Nondual perception, where subject and object are one pure act of nondual awareness that unifies both nature and mind in realized Spirit. Spirit *knows itself objectively* as nature; *knows itself subjectively* as mind; and *knows itself absolutely* as Spirit—the Source, the Summit, and the Eros of the whole sequence.

(We would also recognize these three large movements as subconscious, self-conscious, and superconscious; or biosphere, noosphere, and theosphere; or prepersonal, personal, and transpersonal—the three large domains that we have met many times in the previous chapters.)

The whole evolutionary sequence is thus holonic through and through. Each stage (of Ascent) builds upon and incorporates its predecessors— "the last is the result of all earlier: nothing is lost, all principles are preserved" (Hegel). What *is* lost is the narrowness of the predecessor, or its claim to be the Whole (Hegel calls this "the negation of the negation"). Each stage incorporates its predecessor but negates its partiality. Says Hegel: "To supersede is at once to negate and to preserve."

Schelling's view is likewise a precise and perfect summary of development or evolution as emergent holons within emergent holons: "The lower is the necessary foundation for the higher, and the latter subsumes the former in itself. But there is also the emergence of something new, and this new level explains the level which it presupposes."[31]

This is a truly stunning vision, a profound integration of Ego-mind and Eco-nature, of Spirit descending into even the lowest state and ascending back to itself, with Spirit nonetheless *fully present at each and every stage* as the *process* of its own self-realization and self-actualization, its own self-unfolding and self-enfolding development, a divine play of Spirit present in every single movement of the Kosmos, yet finding more and more of itself as its own Play proceeds, dancing fully and divine in every gesture of the universe, never really lost and never really found, but present from the start and all along, a wink and a nod from the radiant Abyss.

We will return to Schelling's vision in the next chapter, and point out more clearly just how it managed to unite Ego and Eco, Ascending and Descending, without sacrificing the gains of either.[32] But for the moment, we should note that if all of this sounds like *Plotinus temporalized,* that would not be far wrong. As one historian of the period put it:

> If we are to speak of a "key" to early Romanticism, it is to be found in one of the thinkers of antiquity, Plotinus. For this Neoplatonic philosopher not only inspired the entire system of Novalis, and many of the ideas of Schelling in his middle period; his arm reached further: through Novalis and Schelling he exercised an influence upon both Schlegels, and without a knowledge of this fact many a passage in the "Dialogue Concerning Poetry" and in the Berlin lectures [of W. Schlegel] remains an enigma.[33]

The "genius Novalis," as he is usually referred to, wrote in a letter to Friedrich Schlegel, "I do not know if I have already spoken to you of my beloved Plotinus. Through Tiedemann I have been initiated into this philosopher, born expressly for me, and I have almost been frightened by his resemblance to Kant and Fichte. He pleases me more than either of them."

He certainly pleased Schelling, and Schelling's influence was enormous, and not only as the starting point for Hegel's system. Schelling was friends with Hölderlin, Goethe, Schiller, the Schlegels. He directly influenced Friedrich Schleiermacher, who revolutionized the concept of religion by suggesting that all genuine spirituality is a result of direct experience, not beliefs. Schelling's notion of organic unity, and his belief that perception is not "mirroring" but "actioning," anticipated much of Gestalt psychology (and autopoietic cognitive theory). In his idea of Eros as unlimited activity, he set off a whole line of "will" theorists, from Schopenhauer to Bergson to Nietzsche. Paul Tillich traces this line of influence: "The most important pupil of Schopenhauer was Nietzsche. The line then runs from Nietzsche to Bergson, the French voluntarist, Heidegger and Sartre, and to Whitehead, the great metaphysician of our century—all of this came from Schelling."[34]

With his theory of art and aesthetic perception—namely, that in great art we have a chance to transcend the subject/object duality—he became, with Schiller, the father of virtually all Romantic aesthetic movements. And his theory of symbolic forms was a forerunner, as Cassirer pointed out, of his own work and much of Jung's analytic psychology.

Schelling's (Plotinian) notion of the unconscious passed to Carus, whose book *Psyche* (1846) maintained that the key to conscious mental life is to be found in the unconscious. From there it reappears in Schopenhauer, one of whose students, Eduard von Hartmann, wrote the enormously successful and influential *Philosophy of the Unconscious* (1869)—it went into ten editions, and appeared two decades before Freud and Breuer's *Studies in Hysteria* (1895).

Connected with the notion of the unconscious was Schelling's introduction of the extremely influential concept of *alienation*, which specifically meant that Spirit "loses" itself in manifestation, and evolution or development is the overcoming of this self-estrangement, an overcoming of alienation by the return of Spirit to Spirit, or a direct rediscovery of the Nondual (what we called dis-memberment re-membered). Men and women, caught in the subject/object duality, are thus *alienated* from Source and Summit—and thus alienated from the All—and it is only in overcoming duality and existential alienation that men and women can find genuine happiness. Among those attending these Berlin lectures were Søren Kierkegaard, Jakob Burkhardt, Michael Bakunin, and Friedrich Engels, collaborator with Karl Marx.

But the immediate effect of this "developmental philosophy" (or the philosophy of Eros) was that the notion of evolution was everywhere "in the air"—and this was still six decades before Darwin.[35] "A theory of emergent evolution [was] demanded by Schelling's view of the world as a self-developing organic unity. Indeed, he explicitly refers to the possibility of evolution. He observes, for instance, that even if man's experience does not reveal any case of the transformation of one species into another, lack of empirical evidence does not prove that such a transformation is impossible. For it may well be that such changes can take place only in a much longer period of time than that covered by man's experience."[36]

Thus, by 1810, Schelling's friend and student, the naturalist Lorenz Oken, maintained that "the philosophy of Nature is the science of the eternal transformation of God into the world," and outlined the research agenda:

> It has the task of showing the phases of the world's evolution from the primal nothingness [doesn't sound quite so silly anymore, does it?]: how the heavenly bodies and the elements arose, how these advanced to higher forms, how organisms finally appeared. . . . These phases constitute the history of the generation of the universe.[37]

Plotinus temporalized = evolution. Lovejoy, ever on the scene with a perceptive report, points out that "the appearance of the general notion of an evolutionary advance long antedated the discovery of most of the scientific evidence for that hypothesis; it was becoming familiar in very widely read writings before the middle of the eighteenth century. Thus one of the effects often attributed to the influence of biological evolutionism had in fact come about long before the establishment and general diffusion of that doctrine, and quite independently of it."[38]

When Darwin came along and dutifully supplied some of the empirical evidence for biological evolution, it shocked nobody except the remnants of the mythic believers in the literal Genesis myth; but they were already shocked by what Schelling and other developmentalists were doing anyway.

For the past century or so, it has been scientific dogma that Darwin's great and enduring contribution was the discovery, not that evolution occurred (this, as we have seen, was an almost century-old hypothesis by then), but *how* it occurred: namely, through natural selection (codiscovered by Wallace). In an updated version: chance mutation leads to genetic alterations, the vast majority of which are maladaptive, but a few of which contribute to a greater (or at least different) capacity for adaptation and survival, and thus these mutations are selected and genetically carried forward in the gene pool. Whatever else this theory did or did not accomplish, it preeminently allowed science to ignore any sort of Eros, any sort of "gentle persuasion" or transcendent/emergent drive, and thus men and women could "trace science then with modesty thy guide."

But it is now almost universally recognized by scientists that, although natural selection can account quite well for "microevolution" (or variation within a given range of possibilities), it can account not at all for macroevolution (or the emergence of new ranges of possibility).[39] Add to that the fact that the Big Bang has made Idealists out of virtually anybody who thinks about it, and the result is that most philosophers of science now openly admit—and even champion—the fact that evolution has some sort of self-transcending drive tucked within its own processes— what we called tenet 2c, the self-transcending capacities of all holons. In other words, Eros.

The lasting contribution of Darwin's theory, then, was not that it discovered a mechanism for macroevolution, for it did not; rather, it obscured for over a century the fact that a genuine theory of evolution demands something resembling Eros. Darwin's *lasting* contribution was primarily a massive obscuratanism. Scientists all cheery and self-content began to scrub the universe clean of anything resembling love and its all-encompassing embrace, and all congratulated themselves on yet another victory for truth (Wallace, as is well known, did *not* think natural selection could replace Eros; evolution was itself, he thought, "the mode and manner of Spirit's creation," and Darwin himself notoriously wavered).

But the world did not waver. As we will see in the next chapter, with the collapse of virtually any form of Idealism, the Western world nestled comfortably into the Descended domain of the naturalistic flatland, with its altogether low center of ontological gravity, like a beanbag chair already settled. And within this cozily low and snug—some would say cowardly—stance, the mono-Ego and mono-Eco battled it out in mortal combat, fighting for the royal right to dominate a cardboard cosmos.

And there the Western world would remain until only the most recent of times.

> It is the time of the gods that have fled *and* of the god that is coming. It is the time of *need*, because it lies under a double lack and a double Not: the No-more of the gods that have fled and the Not-yet of the god that is coming.

14

The Unpacking of God

*This awareness, free from an inside or an outside, is open like
  the sky.
It is penetrating Wakefulness free from limitations and partiality.
Within the vast and open space of this all-embracing mind,
All phenomena of samsara and nirvana manifest like rainbows
  in the sky.
Within this state of unwavering awareness,
All that appears and exists, like a reflection,
Appears but is empty, resounds but is empty,
Its nature is Emptiness from the very beginning.*
  —TSOGDRUK RANGDROL, *The Flight of the Garuda*

THE GOD THAT was to come. The Descent of the all-pervading
World Soul. The coming of the Over-human, the Over-Soul, the
transpersonal dawn, *Homo universalis*. And not merely in rare,
individual, isolated cases, but as a *center of social organizing forces*—just
as magic and mythic and mental had previously emerged at large and
organized cultures around the world according to their basic patterns.
The coming of the World Soul, blessing each and all with intuitions of the
Over-Soul, joined each and all in the council and communion of all sentient beings, the community of all souls, likewise institutionalized in structures that guard its preciousness the way the worldcentric rationality is
now institutionalized and guarded in law and education and government
and community. The integration of the physiosphere and biosphere and
noosphere in each and every compound individual, not as a theory but as

a *central identity in consciousness* (just as the ego or person is the central identity institutionalized in rational cultures of today).

The coming of the Over-Soul that is the World Soul, touching each and all with its Goodness and its Glory, baptizing each with its Brilliance and its Blessing. The coming of the World Soul, trailing clouds of wonderment, singing songs of liberation, dancing madly and divine in splendor and salvation. The long-sought coming of the World Soul, changing every "it" and every "we" and every "I" it touches: in a moment, in the twinkling of an eye, we will be changed, we all will be changed.

The loveless, beaten, battered self will let go the torment and the torture of its self-embracing ways, tire of that marriage to a special misery that it had chosen over loneliness, to nurse it through the long brutality of a life that doesn't care, surrender the murderous love affair with its own perplexed reflection, which had itself pretended to the throne of the Divine, and find instead its soul in Grace and drenched throughout with a luminous God that is its own true Being—its always and only Original Face, smiling now from the radiant Abyss, unreasonably happy in the face of every sight, set helplessly afloat on the Sea of Intimacy, adrift in currents of Compassion and caressed in unrelenting Care, one with each and one with all in mutual Self-recognition, dancing in the dawn that heralds now the Self of all that truly is, and the Community of all that well might be, and the State of all that is to come.

And every I will sing of the Self, and every We will resonate with worship of the Divine, and every It will radiate the light of a Spirit happy to be seen, with dialogue the abode of the Gods and perception the home of Grace, and gone the lonely loveless self, the god of its own perception, and gone the Godless destiny of time and separation.

The blessed, blessed Descent of the World Soul: in a moment, in the twinkling of an eye, we will be changed, we all will be changed.

Perhaps it will happen after all.

And perhaps it will not. I believe that the intuition of the Over-Soul, the World Soul, is indeed increasing in frequency and in intensity in more and more individuals around the globe; and in the next two volumes in this series we will look more closely at the evidence, quite impressive, I believe, that this is indeed happening.

But as always, we have to *make* the future that is *given* to us. And it is my strongest conviction that the Descent of the all-pervading World Soul is facilitated, or hindered, to precisely the degree that we unpack its intuition adequately.

As we have seen, all *depth* must be *interpreted*. And how we interpret depth is crucially important for the birth of that depth itself. New depth allows us new interpretations; the new interpretations cocreate and give birth to that depth, help unpack that depth. Unpacking the depth is the emergence of that depth.

And thus, let us oh-so-carefully unpack this precious Gift of spiritual intuition.

Many individuals intuit the Over-Soul (or higher) and yet unpack that intuition, interpret that intuition, in terms merely or solely of the Higher Self, the Inner Voice, the care of the Soul, interior Witnessing, the Universal Mind, pure Awareness, transcendental Consciousness, or similar such Upper-Left quadrant terms. And however true that aspect of the intuition is, this unpacking leaves out, or seriously diminishes, the "we" and the "it" dimensions. It leaves out the social and cultural and objective manifestations: it fails to give a seamless account of the types of community and social service and cultural activity that are *inherently* demanded by a higher Self; it ignores or neglects the changes in the techno-economic infrastructures that support each and every type of embodied self (whether higher or lower or anything in between); it ignores the overall objective state of affairs or objective reality that does not *detract* from the Self but is an unavoidable *aspect* of that Self's very manifestation.

The idea seems to be that if I can just contact my higher Self, then everything else will take care of itself. But this fails miserably to see that Spirit manifests always and simultaneously *as* the four quadrants of the Kosmos. Spirit (at any level) manifests as a self in a community with social and cultural foundations and objective correlates, and thus any *higher* Self will inextricably involve a *wider* community existing in a *deeper* objective state of affairs. Contacting the higher Self is not the end of all problems but the beginning of the immense and difficult new work to be done.

Thus, intuiting Spirit but interpreting it merely in "I" terms as a Higher Self or whatnot (Upper-Left quadrant)—this ends up being a merely Ascending Path, with no sincere and well-conceived interpretations of exactly how to go about embracing the manifest world in all its dimensions. It is the old Fichtean move. It is still caught in one pole of the fundamental Enlightenment paradigm: the pure Ego, the pure Self, will solve all problems.

And thus the lingering suspicion that soaks these approaches: if you really care about the world, or worry about the world, then you can't actually be contacting the Higher Self (because the Higher Self would take care of that: your "concern" or "worry" for the world shows you haven't found the true Self). Whereas in fact, it is just the opposite: the more you contact the Higher Self, the *more* you worry about the world, as a component of your Very Self, the Self of each and all. Emptiness is Form. Brahman is the World. To *finally* contact Brahman is to *ultimately* engage the World.

Thus, a true and profound intuition is corrupted, or distorted, or tilted, by a well-meaning but not very adequate interpretation. The intuition of Spirit is unpacked poorly, hastily, and the Gift is damaged in the process. Conversely, a more graceful unpacking facilitates further and deeper intuitions, intuitions touching the I and the We and the It domains: not just how to contact the higher Self, but how to see it embraced in culture, embodied in nature, and embedded in social institutions.

Realized, embraced, embodied, embedded: a more graceful interpretation covering all four quadrants (because Spirit itself manifests as all four quadrants) facilitates the birth of that Spirit which is demanding the interpretation. Graceful interpretation midwifes Spirit's birth, Spirit's descent. The more adequately I can interpret the intuition of Spirit, the more that Spirit can speak to me, the more the channels of communication are open, leading from communication to communion to union to identity.

And Spirit's interpretation merely as a Higher Self is not very graceful, I don't think.[1]

Likewise, there are many good souls who have a profound intuition of Spirit but unpack that intuition in merely "It" terms, describing Spirit as the sum total of all phenomena or processes interwoven together in a great unified system or net or web or implicate order or unified field (Lower-Right quadrant). All of which is true enough, but all of which leaves out entirely the interior dimensions of "I" and "we." This less-than-adequate interpretation is *monological* to the core, flatland through and through.

It is the old Spinozist move, the other pole—the Eco pole—of the fundamental Enlightenment paradigm (in the form of the Romantic rebellion). It thinks that the enemy is atomism, and that the central problem is simply to be able to *prove* or demonstrate once and for all that the universe is a great and unified holistic System or Order or Web. It marshals a vast amount of scientific evidence, from physics to biology, and offers extensive arguments, all geared to objectively proving the holistic nature of the universe. It fails to see that if we take a bunch of egos with atomistic concepts and teach them that the universe is holistic, all we will actually get is a bunch of egos with holistic concepts.

Precisely because this monological approach, with its unskillful interpretation of an otherwise genuine intuition, ignores or neglects the "I" and the "we" dimensions, it doesn't understand very well the exact nature of the *inner transformations* that are necessary in the first place in order to be able to find an identity that embraces the manifest All. *Talk* about the All as much as we want, nothing fundamentally changes.

And nothing changes because the "proof" or the "new paradigm" or "great system" is still being put in monological it-language. This doesn't engage the process of inner transformation. The profoundly transformative question is *not*: is the world holistic or atomistic? The transformative question is: who or what is aware of both holistic and atomistic concepts? (The move from the exterior to the interior.) And *then*: having rested in the Witness of those concepts, a Witness that itself is neither holistic nor atomistic, see here the Witness dissolve in an Emptiness that embraces the entire Kosmos. (The move from the interior to the superior.)

To remain arguing about the exteriors is to effectively seal out the interior and thus the superior. Focusing on the exterior world being holistic thus misses the central transformative occasion.

Long ago Nagarjuna devastated such approaches, pointing out that holistic or atomistic, or both or neither, are all beside the point (and, he added, they're all false anyway). It is the radical deconstruction of all conceptualizations whatsoever that paves the way for pure intuition (prajna) of Shunyata (Emptiness or Openness). If we meet even the Buddha and are supposed to kill him, guess how holism will fare.

Thus, however true the original intuition of Spirit is (and I do not doubt that it is true), it is not facilitated by an interpretation that reduces all of Spirit's dimensions merely to the Right-Hand path, merely to descriptions of the great interlocking order (even if those descriptions are true enough).[2] Those interpretations, taken in and by themselves, block the transformative event; those interpretations, driven originally by a true intuition of the very Divine, do not facilitate the further descent of that Divine; those interpretations are unskillful to midwife the birth of Spirit (which is nonetheless the original source of what turns out to be less-than-adequate interpretations).

And worse: when the intuition of Spirit is unpacked merely or solely in it-terms, in monological terms, the approach ends up being a purely Descended worldview. Spirit is simply identified with the Sum Total of exteriors, the Sum Total of the shadows in the Cave. So intent are we on proving that the shadows are one great interlocking order that we never move from these exteriors to the real interior, and thus we never find the genuine superior.

This less-than-adequate interpretation makes it appear that the most urgent problem in the modern world is to teach everybody systems theory (or some version of Gaia's web-of-life notions, or some version of the "new physics"), instead of seeing that what is required is an understanding of the *interior stages* of consciousness development. Gaia's main problem is not toxic waste dumps, ozone depletion, or biospheric pollution. These global problems can *only* be recognized and responded to from a global, worldcentric awareness, and thus Gaia's main problem is that not enough human beings have developed and evolved from egocentric to sociocentric to worldcentric, there to realize—and act on—the ecological crisis. Gaia's main problem is not exterior pollution but interior development, which alone *can* end exterior pollution.

Preconventional/egocentric and conventional/ethnocentric could care less about the global commons; only postconventional/worldcentric can fully see, and effectively respond to, the *universal* dimensions of the problem (only formop and vision-logic can grasp universal perspectives). Thus, the more we emphasize teaching a merely Right-Hand map of systems theory or a Gaia Web of Life, instead of equally emphasizing the importance of *interior development* from egocentric to sociocentric to worldcentric, then the more we are contributing to Gaia's demise. Global problems demand global awareness, and only interior and Left-Hand stages of development—precisely the domain ignored by flatland Eco approaches—can even begin to handle the problem.

The pure Eco approach, like its cousin the pure Ego (or pure Self) approach, simply reduces all of Spirit's dimensions to one privileged domain, crippling the full emergence and descent of Spirit itself. Just as the idea in the pure Self camp is that if we simply contact the Higher Self, all our problems will be solved, so the idea in the pure Eco approach is that if we can just demonstrate, once and for all, the unified and holistic nature of the pure Eco, then that will solve all our problems. That forceful demonstration (that Spinozist *proof* in objective terms) will itself *compel* all individuals to transform to ultimate unity, or so it is maintained.

Thus, the only "interior" change this camp recognizes is the *single* change from holding an atomistic conception to holding a holistic conception. Because this camp centers on exteriors, it has an incredibly naive and anemic conception of all the inner and interior developments necessary in order to be able to embrace the All. Further, in recognizing only this "single" change from atomistic to holistic concepts, it misses the crucial fact that change of beliefs does not mean change of consciousness. It means simply a new translation, not necessarily a new transformation.

The standard response from the Eco camps is that if people truly learned and really understood the holistic oneness of reality and the great web, that would force them to give up their egos and they would indeed truly transform. But, as we have seen, it is actually quite the contrary: embracing a monological worldview in flatland terms results precisely in divine egoism.

(Because the Right-Hand-only view of systems theory lacks an understanding of the interior stages of consciousness development—from preconventional/egocentric to conventional/sociocentric to postconventional/worldcentric—or more precisely, because it lacks an understanding of all nine fulcrums of interior unfolding, it has no way to carefully gauge consciousness evolution and development. For this reason, flatland cannot spot the difference between infrarational consciousness and suprarational consciousness—because it cannot spot interior consciousness at all—and thus it constantly falls prey to massive pre/trans fallacies. This allows its proponents to embrace preconventional interiors as if they were postconventional realities: allows them to fall into preconventional/egocentric enthusiasms, even as they champion exterior holism and the great Web of Life, precisely as we saw with the Romantics.)

This approach, then, focuses almost exclusively on Dharma, or the objective Truth, and not enough on Buddha (subjective) and Sangha (intersubjective), or how that Truth refracts as well through the psychological and cultural domains (how that Spirit manifests as all four quadrants of the Kosmos). And when objective truth is made the "total truth"—when it is really thought to be "utterly holistic" and "all-encompassing"—then Buddha and Sangha are violently reduced to flatland objectivist terms, and an approach that originally wishes us to transcend and include, ends up being a merely and purely Descended worldview that effectively blocks transcendence altogether.

My point is that less-than-skillful interpretations of otherwise genuine Spiritual intuitions do not facilitate the birthing of further intuitions, and thus a special care and thoroughness is required in the unpacking of this Gift of God, this precious spiritual intuition.

I would like to spend the rest of this chapter exploring this idea, and end with a brief review and summary of the difficulties in both the Ego and the Eco views. And we will finish our account of the Idealists and their enduring legacy. But I would like to start with one last reason that an approach to Spirit based on monological exteriors is a losing proposition.

INCOMPLETE OR UNCERTAIN

Addition 2: Every holon issues an IOU to the Kosmos.

A general theme running throughout the Idealist writers—and indeed, a theme found in virtually all of the mystically or contemplatively oriented philosopher-sages the world over—is that finite things, finite holons, are somehow profoundly *lacking*, or even profoundly *contradictory*, in and of themselves. "All finite things are contradictory," as Hegel put it. Nagarjuna would maintain the same for all finite phenomena (both thought and things are self-contradictory). From Eckhart to Bradley, from Shankara to Ramana, from Abinavagupta to Gaudapada—the general notion is that, as Hegel put it, "All things in themselves are contradictory."

These types of statements have often stirred much controversy in philosophical circles, and many philosophers are either annoyed or puzzled by what they mean (or even can mean). But the reason these types of statements ("All holons are self-contradictory") come from mystically oriented philosopher-sages is that they have glimpsed the eternal, tasted infinity, and thus all finite things *by comparison* are pale, incomplete, uncertain, shifting, shadowy. And thus to be *merely* finite is not only a constriction, it is ultimately self-defeating: to be merely finite is to deny infinity, and this is self-contradictory in the deepest sense because it denies one's deepest reality.

This is why, I think, Hegel says, "For anything to be finite, is just to *suppress itself* and *put itself aside*." And thus, "all that is determinate and finite is *unstable*." And it is this incompleteness, this instability, that drives the agitated movement of the entire finite and manifest universe: "Only insofar," he says, "as something has contradiction in itself does it move, have impulse, or activity." (And Eros, as we saw, is this agitated movement to find higher unions and thus overcome or escape the instability, and Eros therefore finally rests only in the infinite, which itself has no contradiction and thus is the "timeless," the "changeless," the indestructible Abyss, the Vajra or Diamond nature—although all of those terms are not descriptive per se, but only developmental signifiers.)

I mention all of this because, as I said, virtually every mystically oriented philosopher ends up making some sort of these types of statements—none, perhaps, with more exuberance than Francis Bradley: "Relation, cause, space, time, thing, and self are self-contradictory." And Nagarjuna's powerful dialectic is an intense and unrelenting bearing-down on every single category of thought imaginable, all with the same result: they are totally self-contradictory and, if consistently pressed, they totally self-destruct (leaving Emptiness, leaving the formless infinite; the deconstruction of the phenomenal leaves prajna).

The deconstructionists have picked up certain of these lines of thought (mostly from Hegel, whom Derrida uncharacteristically treats with much respect), but after Bradley and Nagarjuna, the deconstructionists are very thin soup indeed, and, depressingly, they almost always miss the punch line: if you don't want to be a complete self-contradiction, then you must rest in infinity.

But admittedly, the philosopher-sages' explanations of why all holons are "self-contradictory" are not very clear, and have caused a great deal of confusion. I think the situation can be explained more easily and a bit more clearly.

The point, as I would put it, is that every holon is a whole/part. There are no wholes and no parts anywhere in the manifest universe; there are only whole/parts. If actual wholes or actual parts really existed somewhere, then they could rest; they would simply be what they were; there would be no massive instability, no internal "self-contradiction."

But every holon is *simultaneously* a whole/part. It has a dual tension inherent in its very constitution. As a *wholeness*, it must achieve a degree of *coherence* and *consistency* in order to endure at all as the same entity across time (this is its regime, code, agency, relative autonomy, and so on). But as a *partness*, as a part of some other holon, it must embrace its partialness, embrace its incompleteness, or else it will simply not fit in, will not be a part but will always drift off into its own isolated wholeness. In order to be complete, or to complete itself, it must join with forces larger than itself. As a whole/part, there is thus a constant tension between coherency or *consistency*, on the one hand, and *completeness*, on the other.

And the more of the one, the less of the other—neither "force" can win without destroying the holon, and so the holon remains in constant instability. The more consistent (self-contained), then the less complete (the less in communion).

And thus the second Addition: *All holons issue an IOU to the Kosmos*, where *IOU* means "Incomplete Or Uncertain," and which specifically means, the more complete or encompassing a holon, the less consistent or certain, and vice versa. To say a holon can be complete or consistent, but not both, is also to say that every holon is therefore incomplete or inconsistent (uncertain), and thus: every holon issues an IOU to the Kosmos.

Put more simply, since every holon is incomplete or inconsistent, every holon issues a promissory note to the universe, which says, in effect: I can't pay you now, I can't achieve certainty and stability and completeness and consistency today, but I will gladly pay you tomorrow. And no holon *ever* delivers, or *can* deliver, on that promise.

This IOU principle has, of course, started to become very obvious (and very famous) in certain branches of knowledge, particularly mathematics, physics, and sociology (to name a few). In mathematics, it shows up as Tarski's Theorem and Gödel's Incompleteness Theorem, both of which are taken to mean that in any sufficiently developed mathematical system (mathematical holon), the holon can be *either* complete *or* consistent, *but not both*. That is, if the mathematical system is made to be *consistent* (or self-certain), there remain fundamental truths that *cannot* be derived from the system itself (it is *incomplete*); but if the system is made to *include* these truths and thus attempts to become *complete*, then it inevitably (and inherently) contradicts itself at crucial points—it becomes *inconsistent*.

Perhaps a simple example from sociology will illustrate what is involved. The United States and Japan are often taken as examples of two very different types of social organizations. Japan is an extremely coherent or very tightly woven society (it is consistent); but it achieves this consistency only by excluding foreign races (Japan's xenophobia being rather notorious). In other words, it is very *consistent* but very *incomplete* (very partial or very exclusionary).

The United States, on the other hand, attempts to be as *complete* as possible, attempts to open its doors to any and all (the "melting pot"), but it does so at the cost of being rather incoherent and unstable: at times, the U.S. seems so willing to embrace various cultures that it is in danger of flying apart at the seams. It achieves a great deal of *completeness* at the cost of being *inconsistent* or incoherent or uncertain, of having no tightly knit unifying regime or common principle.[3] In other words: complete *or* coherent, and the more of one, the less of the other—IOU.

In mathematics, the introduction of the IOU principle (in various ways, by Russell, Tarski, Gödel), the introduction of "the paradoxes" in set theory, initially caused an uproar, almost panic, because it meant that set theory and arithmetic (and by implication, the whole of mathematics) were on very shaky ground—that is to say, on self-contradictory ground. And in a sense that is true, but mathematics "escaped" the paradoxes by postulating unendingly expanding sets ("transfinite").

We saw, in chapter 2, that this was just another example of "holons all the way up, all the way down." And we can see now that this also means that mathematics simply issued a transfinite IOU to the Kosmos. Put rather simplistically, the only way for mathematics to avoid profound self-contradiction is to postulate a yet higher level of inclusion, which avoids the paradoxes of one level—but then faces the same paradoxes on its own level. Another yet-higher level is thus postulated, and this continues *endlessly* ("transfinitely").

Thus, the mathematical paradoxes and IOUs of one level can be *superseded* at the next higher level (the next more-encompassing set), but that set then faces its own IOU (it is either incomplete or inconsistent), and that continues . . . *forever*. The sets must be postulated to expand forever, because the moment they stop, mathematics becomes self-contradictory. Nobody ever actually sees all of these transfinite sets: they are just a promissory note that allows mathematics both to keep going and to get going in the first place.

Thus the IOU: it says, I cannot pay you now, but I will gladly pay you tomorrow. It will even gladly pay with lots of interest, because the point is: it can never *actually* pay. The debt is *never* settled. Mathematics, like all holons, lurches forward forever in an attempt to get over its inherent limitations, its "self-contradictions." (Recall Hegel: "Only insofar as something has contradiction in itself does it move, have impulse, or activity.")

The point: *all* holons issue an IOU to the Kosmos, and the debt is *never* redeemed.

IOUs, then, represent the tension or instability (incomplete or inconsistent) *inherent* in *all* manifest holons by virtue of the fact that they are all whole/parts. Another way to say *exactly* the same thing is to say that all holons are agency-in-communion (wholes that are parts of other wholes). As *agency* (as wholeness), a holon must seek consistency in order to endure. But as *communion* (as partness), it must seek union with other holons in order to be more complete. And the more of one, the less of the other (on the same level; a supersession embraces both, but then the new level faces its own IOU).

In other words, the more agency, the less communion, and vice versa (another version of the IOU), and neither side can finally win without destroying the other. They are mutual antagonists that depend on each other for existence (which, we might say, makes samsara an unpleasant place to be caught).

In volume 3, I attempt to explain this IOU principle in much greater detail, and this is meant only as a very brief introduction. I attempt to show that agency across time and communion across space, as the two fundamental "drives" of all holons (on a given level), show up in various fields as, respectively: time and space, coherence and correspondence, rights and responsibilities, metaphor and metonym, intrinsic value and extrinsic value, determined and probabilistic, necessity and chance, consistency and completeness, consciousness and communication—the list is virtually endless. But the central point is that these typical dualisms (such as coherence versus correspondence in epistemology) are dual partners forever fated to battle it out with each other . . . and never, never win.

The final example I will mention of an IOU principle that has become quite famous is the Heisenberg Uncertainty Principle, whose many forms all state the same thing: the more we can know of the time-component of a wavicle (agency), the less we can know of the space-component (communion)—momentum and position, for example—with the crucial point being: the wavicle doesn't know either. The wavicle issues an IOU to the Kosmos. The more one aspect is pinned down, the more the other fades into obscurity.

EMPTINESS

Now the real reason I mention all these examples of the IOU principle, and the real reason the mystically oriented philosopher-sages are always bringing it up (Hegel to Bradley to Nagarjuna to Shankara), is very simple:

Addition 3: All IOUs are redeemed in Emptiness.

Emptiness is neither a Whole nor a Part nor a Whole/Part. Emptiness is the reality of which all wholes and all parts are simply manifestations. In Emptiness I do not become Whole, nor do I realize that I am merely a Part of some Great Big Whole. Rather, in Emptiness I become the opening or clearing in which all wholes and all parts arise eternally. I-I am the groundless Ground, the empty Abyss, that never enters the stream of endless IOUs; that never lives the transfinite nightmare of ceaseless self-contradiction, the perpetual agony of samsara's cascading wholes and parts; the empty Ground that never bows to the terrors of time and the perpetual motion sickness known as space; that steps off the relentless torment of unending IOUs—only to embrace them all, to embrace the All, as a mirror one with its many reflections, and nothing comes to rest except the insane thought that I ever could have found liberation in endless IOUs, or that salvation had anything to do with the unending debts and promissory notes that samsara calls "reality," the sickness-inducing self-contradiction in the very heart of every finite thing.

And there is the message of the mystics, ever so simply put: Emptiness, and Emptiness alone, redeems all IOUs. In Emptiness alone, my debt is paid to the Kosmos, because in Emptiness, I-I am the Kosmos. Redemption of debt, erasure of guilt, a balancing of the Kosmic books, a release from transfinite insanity.[4]

Not in Emptiness, but as Emptiness, I am released from the fate of a never-ending addition of parts, and I stand free as the Source and Suchness of the glorious display. I taste the sky and swallow whole the Kosmos, and nothing is added to me; I disappear in a million forms and nothing is subtracted; I rise as the sun to greet my own day, and nothing moves at all.

THE LEGACY OF THE IDEALISTS

The Idealist movement was, in the West, the last great attempt to introduce true Ascent and, most important, to integrate it believably with true

Descent—the Ego and the Eco both taken up, preserved and negated, honored and released, in all-encompassing Spirit. The true heirs of Plotinus.

And although the great Idealist movement finally failed, its enduring contributions will, I believe, be part of any graceful unpacking of spiritual intuition for the modern and postmodern world. Its enduring contributions will, I think, be a necessary component in any sort of truly comprehensive worldview, a worldview still begging to emerge, a World Soul still struggling to break into collective consciousness and heal the fragments of a modernity gone slightly mad.

The details of the Idealist system are explored in volume 3. For the moment, I think we could fairly summarize the thrust of the Idealist movement as: *an intuition of the transpersonal domain expressed in vision-logic.* And there was both its great strength and its fatal weakness.

As *vision-logic*, it was a developmental evolution beyond simple formal operational rationality, a move beyond instrumental and ego-centered rationality (*Verstand*) into dialogical, dialectical, intersubjective reason (*Vernunft*), carrying with it a unifying of opposites and a reconciliation of fragments.

As *holarchic* vision-logic, it saw neither isolated wholes nor abandoned parts: each stage of development was a whole/part that preserved and negated its predecessors. Idealism pointed out, in effect, that since every holon is simultaneously both a subholon and a superholon, then all agency is *always* agency-in-communion (*the* insight utterly lacking in the Enlightenment's "proud culture of reflection," that "monster of arrested development"). Thus no dualism has a final place to rest; no whole, and no part, can ever repose triumphant; Spirit speaks throughout the evolutionary process, as the process itself—self-unfolding, self-enfolding, self-realizing: "nothing is lost, all is preserved."

As vision-logic, the integration of the Big Three could be seriously contemplated and in many ways completed: nature is not just a collection of phenomenal interlocking processes or "its," but of potential I's; and I's exist only in an intersubjective space of we. Accordingly, science, morality, and art are all integral expressions and moments of Spirit; none can claim isolated authority; each is to be honored in its respective task. And I can *integrate* them because I am vehicle of Spirit, and not any *particular* manifestation.

Most important, these integrations were not just verbal formulas; up to a certain point, the Idealists genuinely managed to "square the circle," as Taylor summarizes the seemingly impossible task of reconciling Ego and Eco, Fichte and Spinoza, Mind and Nature, Ascent and Descent, Subjective Freedom and Objective Union.

The requirements were inordinately intense: "If the aspirations to radical freedom [Ego] and to integral expressive unity with nature [Eco] are to be *totally fulfilled together*, if man is to be at one with nature in himself and in the cosmos while also being most fully a self-determining subject, then it is necessary first, that my basic natural inclination spontaneously be to morality and freedom; and more than this, since I am a dependent part of a larger order of nature, it is necessary that this whole order within me and without tend of itself *toward spiritual goals*, tend to realize a form in which it can *unite with subjective freedom*. If I am to remain a spiritual being and yet not be opposed to nature in my interchange with it, then this interchange must be a *communion* in which I enter into relation with some spiritual being or force. But this is to say that spirituality, tending to realize spiritual goals, is of the essence of nature. Underlying natural reality is a spiritual principle striving to realize itself."[5]

And those were precisely the requirements met by Schelling and hammered out by Hegel. Nature and Mind are both taken up and integrated in Spirit, an integration that can occur precisely because the Spirit that is *awakened* in the *integration* is the *same* Spirit that was *present throughout* the entire process of unfolding and enfolding itself.

"So that while nature tends to realize spirit, that is, self-consciousness, men and women as conscious beings tend toward a grasp of nature in which they will see it as spirit and one with their own spirit. In this process men and women come to a new understanding of self: they see themselves not just as individual fragments of the universe, but rather as vehicles of spirit. And hence men and women can achieve at once *the greatest unity with nature*, i.e., with the spirit which unfolds itself in nature, *and* the fullest *autonomous self-expression*. The two come together since man's basic identity is as vehicle of spirit. [This] provides the basis of a union between finite and cosmic spirit which meets the requirement that men and women be united to the whole and yet not sacrifice their own self-consciousness and autonomous will."[6]

Recall that the dilemma—and the agony—was how to "unite Fichte and Spinoza"—that is, how to be both *autonomous* (or self-defining) and *whole* (or one with nature and therefore defined and determined by nature, which means surrendering autonomy): how can you possibly do *both,* since they are mutually exclusive? On the one hand, there was the desire to preserve the worldcentric moral autonomy and capacity for compassion for which the Ego had fought so long: to merely become "one with nature" would be to lose that autonomy and be governed by natural laws such as fox eats chicken—and there goes moral compassion. On the other hand, there was an intense desire for wholeness and completion, a desire to be part of something bigger than my ego, a desire to fit into the great Eco. How to unite the Ego and the Eco? the *self* and *nature?* It certainly appears that the more of the one, the less of the other—the IOU principle haunting all manifest holons!

The great Idealist realization was that both the Ego and the Eco are manifestations of Spirit, and Spirit redeems all IOUs. On the one hand, the Spirit (or highest Self or pure Ego or I-I) of the Kosmos is *perfectly autonomous,* because it is the pure Self (which is self-positing and self-defining and thus *completely free* or autonomous); and, on the other, it is *perfect wholeness* at the same time, because there is *nothing outside of it:* both Mind and Nature are expressions of its own Being and Becoming. And thus, to the extent that I rise to a intuition and identity with Spirit, then *both* the free will of Mind and the union with Nature are given to me *simultaneously.* I-I am *everything* that is arising, and thus I am autonomous *and* whole, free *and* determined, ascended and descended, one and many, wisdom and compassion, eros and agape.

Thus, Spirit redeems all IOUs, because in Spirit I am not a whole/part, but the *infinity* in which all whole/parts arise, remain a bit, and pass. And thus, by *developing* from nature to mind to Spirit, I can embrace the entire Kosmos in a free and complete fashion, for I-I am that Kosmos for all eternity: which is to say: *right now,* when seen in the eye of Spirit, for which both mind and nature are simply integral chapters in my own continuing story.

United in Spirit without erasing differences—there is the One-in-the-Many as my truest Self (the ultimate I or Buddha), and as the highest Truth (the ultimate It or Dharma), and as the all-encompassing Community of all sentient beings (the ultimate We or Sangha).

Further, according to the Idealists (and nondual sages everywhere), the extraordinary and altogether paradoxical secret is that the Final Release is always already accomplished. The "last step" is to step off the cycle of time altogether and find the Timeless there *from the start,* ever-present from the very beginning and at *every point* along the way. The great far-off spectacular climax . . . is right now. "The Good," says Hegel, "the absolutely Good, is eternally accomplishing itself in the world; and the result is that it need not wait upon us, but is *already in full actuality accomplished.*"

As J. N. Findlay, one of Hegel's greatest interpreters, puts it: "It is by the capacity to understand this that the true Hegelian is marked off from his often diligent and scholarly, but still profoundly misguided misinterpreter, who still yearns after the showy spectacular climax, the Absolute coming down . . . accompanied by a flock of doves, when a simple return to utter ordinariness is in place [the *language of ordinariness*, as we saw, always marks the Nondual]. Finite existence in the here and now, with every limitation, is, Hegel teaches, when rightly regarded and accepted, identical with the infinite existence which is everywhere and always. To live in Main Street is, if one lives in the right spirit, to inhabit the Holy City."[7]

As Plotinus knew and Nagarjuna taught: always and always, the other world is this world rightly seen.

This is a truly staggering synthesis and integration. (Credit for this type of developmental synthesis often goes to Aurobindo; without in any way detracting from Aurobindo's magnificent contributions, the pioneering credit belongs to Schelling, and by a century.) And yet, in the end, the Idealist movement failed to live up to its bright promise and huge potential. Although, as I mentioned, the details will be taken up in volume 3, I think we can quickly summarize the failure in two parts.

The first was a failure to develop any truly *injunctive practices*—that is, any true paradigms, any reproducible exemplars. Put differently: no yoga, no contemplative practices, no meditative paradigms, no experimental methodology to reproduce in consciousness the transpersonal insights of its founders. The great Idealist systems were mistaken for metaphysics—which in this sense, they were, alas—and suffered rightly the fate of all mere metaphysics.

The second failure: although profound intuitions and insights into the genuinely transpersonal domains were clearly some of the major, I would say *the* major, driving forces behind the Idealist movement, these intuitions and insights were expressed almost totally in and through *vision-logic*, and this burdened Reason with a task it could never carry.[8] Particularly with Hegel, the transpersonal and transrational Spirit becomes wholly *identified* with vision-logic or mature Reason, which condemns Reason to collapsing under a weight it could never carry.

In 1796, Hegel wrote a poem for Hölderlin, which says in part: "For thought cannot grasp the soul which forgetting itself plunges out of space and time into a presentiment of infinity, and now reawakens. Whoever wanted to speak of this to others, though he spoke with the tongues of angels, would feel the poverty of words."

Would that Hegel had remained in poverty (with Plato: "No treatise by me concerning it exists or ever will exist"). But Hegel decided—in part in reaction to the Eco camp's calamitous slide into regressive feeling and divine egoism—that Reason could and should develop the tongues of angels. This would have been fine, *if* Hegel also had more dependable paradigms, more reproducible injunctions, for the developmental unfolding of the higher and transpersonal stages. As we said, Zen masters talk about Emptiness all the time! But they have a practice and a methodology (*zazen*) which allows them to discover the *transcendental referent* via their own *developmental signified*, and thus their words (the signifiers) remain grounded in experiential, reproducible, fallibilist criteria.

The Idealists had none of this. Their insights, not easily reproducible, and thus not fallibilistic, were dismissed as "mere metaphysics," and gone was a priceless opportunity that the West, no doubt, will have to attempt yet again if it is ever to be hospitable to the future descent of the all-embracing World Soul.

In the meantime, the collapse of Idealism left the Descenders virtually unchallenged as the molders of modernity.

DARK SHADOWS

Darwin's "naturalism"—this denatured nature—seemed, to the dominant flatland worldview, to cover all the necessary bases. Mononature alone was all the "spirit" men and women needed, and all the "spirit" there was.

But it wasn't the Darwinian "revolution" that would play most decisively into the hands of the Descenders. After all, the Darwinists could always be seen, whether they wished to or not, as simply supplying empirical evidence for a scheme already known and accepted, namely, evolution as God-in-the-making, Eros not simply seeking Spirit but expressing Spirit all along via a series of ever-higher ascents, which Darwin had merely chronicled in a not-very-surprising fashion.

Darwin was *never* the problem. It was Carnot, Clausius, and Kelvin. At about the time that Darwin was on HMS *Beagle*, William Kelvin and Rudolf Clausius were formulating two versions of the second law of thermodynamics, versions later shown to be equivalent, and both of which were taken to mean: the universe is running down. In all real physical processes, disorder increases.

This split the world of science into two utterly incompatible halves: a biology describing the world winding up, and a physics describing a world winding down. The "two arrows" of time . . .

With this, of course, we have come full circle, come to the point where we began our account in chapter 1. This wasn't just subtle reductionism—the collapse of the Kosmos into the empirical interlocking order—this was *gross* reductionism—the further collapse of the interlocking order into its atomistic components, a situation that can fairly be described as complete psychotic insanity. (If the Kosmos is the wondrous multidimensional reality anything similar to that described from Plotinus to Schelling, from Mahayana Buddhism to Vedanta Hinduism, or even from Kant to Rousseau, imagine the blindness and the violence of the mentality that acknowledges only atoms.)

The laws of thermodynamics seemed to totally undermine any sort of "organic unity" or "self-organizing nature" or "the total and unitary process of self-manifestation" (Schelling, Hegel)—let alone any sort of Eros operative "even in the lowest state of matter" (Kant). And if the Idealists couldn't even get the first floor of evolution right, why should we believe them about any "higher stages"? There is no Eros in the physical world, and if the physical world is part and parcel of Spirit's manifestation, then there is and can be no Eros in Spirit. Which is to say, there is no Spirit, period.

This completely undercut any sort of idealist or spiritual view of evolution or manifestation in general. The first floor of the magnificent Idealist edifice crumbled, and the higher floors almost immediately followed suit (Idealism would survive in any viable form no more than a few decades after Hegel's death).

The net effect of the utter collapse of this noble attempt to escape the Cave was that, once again, Ascent was looked upon very suspiciously. Everywhere the cry "Back to Kant!" arose, which eventually meant: back to rationality and its grounding in the senses, back to "tracing science then with modesty thy guide."

Back to the phenomena, back to the shadows. Back to "this world." The collapse of Idealism left the Descenders virtually unchallenged as the holders and molders of modernity. Not only was any sort of transcendental Spirit abandoned, *vision-logic itself* was scaled back to various forms of instrumental, subject-centered rationality. "As a consequence," Habermas reminds us, "in the period after Hegel, only those who grasp reason in a more modest fashion have any options." Modernity then, he says, "pruned reason back into understanding and rationality back into purposive rationality"—that is, vision-logic back into the reflection paradigm and back into subject-centered instrumental-rationality.[9]

Ludwig Feuerbach, a student of Hegel, would soon announce that *any* sort of spirituality, *any* sort of Ascent, was simply a *projection* of men and women's human potentials onto an "other world" of wholly imaginative origin. And, according to Feuerbach, it is exactly this projection of human potential onto a "divine" sphere that cripples men and women, and is the true cause of self-alienation. Any real social progress, then, is not spirit's return to spirit, but man's return to himself. Feuerbach: "Religion is the separation of man from himself: he sets God over against himself as an opposed being. God is not what man is, and man is not what God is. God is the infinite Being, man the finite; God is perfect, man is imperfect; God is eternal, man is temporal; God is almighty, man is powerless; God is holy, man is sinful. God and man are extremes."

However well that does indeed describe aspects of *mythic dissociation*, it has precisely nothing to do with one's own transpersonal potentials, *vertical* potentials that if *they* aren't actualized are merely *projected horizontally* into utterly futile schemes to turn this finite world into a utopian world of infinite wonderment: an infinite above collapses into an infinite ahead, and whether that infinite ahead be endless scientific progress or boundless material possessions or political utopianisms as ultimate salvation, they are all fundamentally ways of fussing about in the finite looking for the infinite, and doing irreparable harm to the finite world in the process by placing demands upon it which it could never fulfill.

But that is exactly what Feuerbach would recommend: "Politics," he says, "must become our religion." Karl Marx and Friedrich Engels were paying very close attention. "Apart from Nature and human beings," Engels would write, "nothing exists; and the higher beings which our religious fantasy created are only the fantastic reflection of our own essence. The enthusiasm was general; we were all for the moment followers of Feuerbach."

And the entire modern (and postmodern) world is, in effect, the followers of Feuerbach.

THE CONTOURS OF THE CAVE

And so it was, I believe, that the Descenders became the inheritors of frustrated Ascent—the ever-frustrated desire to move up became the ever-frustrating desire to move forward. One crippled God replaced the other, and Ascent pitched headlong into the endless world of scurrying shadows.

With "No more Ascent!" the infinite vertical yearning toppled forward, and that correct intuition of infinity was now misplaced and displaced onto the horizons of a temporal and agitated and insatiable hunger for all things material and "natural," a hunger driven by the same power of infinite yearning, but now under the gaze of a different God.

And there remains the modern—and the postmodern—mind. People had, and have, two basic choices in these modern times: myth or naturalism. Remain at a mythic level of development, and embrace this or that imperialistic fundamentalism with its (promised and frustrated) Ascent; or evolve to rationality, shed the myths, and embrace a world of pure Descent, basking in the shadows, never getting metaphysical sunburn of any imaginable variety, "tracing science then with modesty thy guide," but seeking always to reform the phenomena, tinker with the shades, eat the shadows, endlessly.

And within naturalism, there are still the same two-centuries-old "paradigms" engaged in the same shopworn arguments: the Ego versus the Eco, both still dragging behind them the same pitiable paradox of damage. The Ego, for all its otherwise admirable worldcentric stance, still has no idea how to go about integrating the biosphere in a higher synthesis of its own interests; still leaves everywhere the corpses of its reckless repressions; still totally confuses transrational Spirit with prerational myth; and thus still runs breathlessly away from the Source of its own salvation.

If the Ego camps still absolutize the noosphere, the Eco camps are still absolutizing the biosphere, utterly unaware that this contributes every bit as much as the Ego camps to the destruction of the biosphere itself. "Saving the biosphere" depends first and foremost on human beings reaching mutual understanding and unforced agreement as to common ends. And that intersubjective accord occurs only in the noosphere. Anything short of that noospheric accord *will continue to destroy the biosphere.*

Thus, if the Ego is destroying the biosphere by a sin of commission, the Eco is destroying it by a sin of omission.

As we saw, Gaia's primary problems and threats are *not* pollution, industrialization, overcultivation, soil despoliation, overpopulation, ozone depletion, or whatnot. Gaia's major problem is lack of mutual understanding and mutual agreement *in the noosphere.*

The problem is *not* how to demonstrate, in monological terms and with scientific proofs, that Gaia is in desperate trouble. The general evidence of this serious trouble is already and simply and absolutely overwhelming. Anybody can grasp the data. But most just don't *care.*

In other words, the real problem is *not* exterior. The real problem is *interior.* The real problem is how to get people to *internally transform* from egocentric to sociocentric to worldcentric consciousness, which is the *only* stance that *can* grasp the global dimensions of the problem in the first place, and thus the *only* stance that can freely, even eagerly, embrace global solutions.

And about these interior transformations, and all the intricate psychological changes necessary in order to effect them, the Eco camps have virtually nothing to say. They are so focused on exterior, monological, reflective "paradigms" that their understanding of *interior* dynamics and development is incredibly anemic; thus they are contributing little to the real changes that have to occur in order to "save Gaia."

Hence, the only "transformation" most ecophilosophers talk about is having everybody change their objective and monological views of reality and accept a "web-of-life" conception, as if that would effect a genuine interior transformation. But not only is the web-of-life ontology regressive (its end limit is always biocentric feeling in divine egoism), but, more tellingly—and this is the only point I would like to emphasize—even if the web-of-life ontology were absolutely true, nonetheless change in *objective belief* is *not* the primary driving force of *interior development.*

(For example: all ecological forecasts are forms of as-if and what-if scientific projections, often computerized. These can only be grasped by formop. Teach these what-if scenarios to preop or conop individuals, and no matter how much they repeat the words, they do not possess the developmental signified and thus they have no real idea of the actual referent. Genuinely global or worldcentric consciousness is not possible short of formop. In other words, global consciousness is not an *objective belief* that can be *taught* to anybody and everybody, but a *subjective transformation* in the interior structures that *can* hold the belief in the first place, which itself is the product of a long line of inner consciousness development.)

We have an enormous amount of information about how and why those interior psychological transformations occur (egocentric to sociocentric to worldcentric), but the Eco camps by and large display no awareness of, and no interest in, those inner dynamics, fixated as they are on describing exterior mononature in "holistic" terms.

This is most naive, and belies the inadequacy of attempting to change people by altering the object instead of growing the subject. Focusing merely on monological and objective and exterior and scientific terms—no matter how *utterly true*—beyond a certain point simply detracts away from the fundamental problem, hides the fundamental problem, which is not exterior pollution but interior development.[10]

Everywhere the paradox of damage: absolutizing the biosphere contributes inexorably to its destruction.

Everywhere the same old story: the Ego and the Eco as the congenital twins in the dance of Gaia's demise.

THE EGO AND THE ECO: PHOBOS AND THANATOS

And thus, to return to my point at the beginning of the chapter: stuck with these two variants of flatland ideology, any intuitions of deeper and higher occasions become immediately misinterpreted in terms of one or the other of these monological paradigms. People continue to have strong and powerful intuitions of the Over-Soul and World Soul, but they immediately *misinterpret* these otherwise correct intuitions in terms of either a disengaged "Higher Self" (à la Fichte) or a "reengaged biospheric union" (à la Spinoza).

The "Higher Self" camp is notoriously immune to social concerns. Everything that happens to one is said to be "one's own choice"—the hyperagentic Higher Self is responsible for *everything* that happens—this is the monological and totally disengaged Ego gone horribly amok in omnipotent self-only fantasies. This simply *represses* the networks of communions that are just as important as agency in constituting the manifestation of Spirit.

This is not Eros; this is Phobos—a withdrawal from social engagement and intersubjective action. All of this totally overlooks the fact that Spirit manifests not only as Self (I) but as intersubjective Community (We) and as an objective State of Affairs (It)—as Buddha, Sangha, Dharma—each inseparably interwoven with the others and interwoven in the Good and the Goodness of the All.

The Eco camps likewise too often misinterpret the intuition of the World Soul, but in the other direction, as some sort of Gaia-self, but still and equally framed in monological and flatland terms. Not NATURE, but nature, is their beloved God/dess. Actual hierarchies of any sort are denied in the name of a diversitarian stance that explicitly denies that which its own stance implicitly presupposes. In their understandable zeal to go transrational, they often embrace any prerational occasion simply because it is nonrational—any occasion that looks biocentrically oriented, from horticultural planting mythology to rampant tribalism to indissociated magic and sensual glorification of a sentimental nature, all in the name of saving Gaia.

The Eco-Noetic Self is thus often misinterpreted as a merely ecological self (the absolutizing of the biosphere)—completely overlooking the fact that no ecological self can take the role of other. The resultant regressive slide is just as disheartening as the Ego's aggressive repression. This is not Agape; this is Thanatos. An attempt to save the lower by killing the higher.

Under these conditions—Ego versus Eco—we will never take monological nature up and into dialogical culture, and then take both nature and culture up and into translogical Spirit. And yet that, I will argue in the next two volumes of this series, is exactly the task whose initial steps will define postmodernity.

ENVIRONMENTAL ETHICS: HOLONIC ECOLOGY

In my opinion, one of the first steps toward an integral postmodernity is the development and establishment of a genuine environmental ethics, or a moral and ethical stance to nonhuman holons.

Nowhere are the difficulties of the Eco camp more obvious than in its attempts to develop an environmental ethics. With some notable exceptions (such as Birch and Cobb's wonderful *The Liberation of Life*, Michael Zimmerman's work, and a handful of others),[11] most approaches center predominantly on the principle of "bioequality," a reworking of the tenets of the Descending path of Plenitude (divorced and dissociated from any true Ascent).

The point, it is argued, is that all holons, or certainly all life forms, have equal value and equal worth (another qualitative distinction that denies all qualitative distinctions). This environmental ethics—noble enough in itself, and often driven by a profound intuition of the World Soul, the Eco-Noetic Self—is nonetheless crippled by operating within the flatland paradigm.

In volume 2, I will examine each of the major ecophilosophies in great detail, and then present an alternative—holonic ecology—which may be very briefly summarized as follows:

1. All things and events, of whatever nature, are perfect manifestations of Spirit. No holon, whether conventionally considered high or low, sacred or profane, simple or complex, primitive or advanced, is closer or farther from Ground, and thus all holons have equal ultimate value or equal *Ground-value*. All Forms are *equally* pure Emptiness, primordial Purity.

2. But in addition to Ground-value, all holons, we have seen, are also both particular wholes and particular parts. *As a wholeness*, any holon has "whole-value." It has a value *in itself*, and not merely for something else. It is an *end* in and for itself, and not merely a *means* for something other. It has autonomous value, and not merely instrumental value. This is usually referred to as "intrinsic value," which I accept; I will also call in-itself value "autonomous value" or "wholeness value." All holons, as wholes, have *intrinsic value*.

From which it follows, the greater the wholeness, the greater the intrinsic value. Wholeness-value, in other words, is the same as depth-value. The greater the depth, the greater the value (and the greater the potential depth, the greater the potential value).

This also means, as we have seen repeatedly, that there are *levels of significance*: the greater the depth, or the greater the wholeness, then the more *significant* that wholeness is for the Kosmos, because the more of the Kosmos is embraced in that wholeness, embraced in that depth.

(Thus, cells are more significant than molecules, because cells contain molecules, thus embracing and signifying more of the Kosmos. An ape is more significant than a cell, and so on.)

Likewise, as a wholeness (as agency), all holons have *rights* which express the conditions that are *necessary* for that holon to *remain whole*. The greater the wholeness, the more rights necessary to maintain it (i.e., the more significant the wholeness, the greater the network of rights required to sustain the significance. An ape has more rights than an ant). These rights are not something *added* to the holon; these rights are a simple statement of the *conditions* (objective, interobjective, subjective, and intersubjective) that are necessary to sustain the wholeness of the particular holon. If those rights are not met, the wholeness dissolves or disintegrates.

3. All holons are also *parts*, and as a part, all holons have *instrumental value* (also called *extrinsic* value). That is, all holons have value *for others*. All holons have part-value, or partness-value (as part of a larger whole, and that whole and its members *depend upon* each part: each part is thus *instrumental* to the existence of the whole, each part has extrinsic value, value not just in and for itself but for others). And the more partness-value a holon has—that is, the greater the number of wholes of which that holon is a part—the more *fundamental* that holon is for the Kosmos, because the more of the Kosmos there is that contains that part as a necessary constituent. An atom is more fundamental than an ape. (From which it follows, as we put it earlier, the more fundamental, the less significant, and vice versa.)

Likewise, as a part (in *communion*), each holon exists in a network of care and responsibility. Like rights, *responsibilities* are not something *added* to holons. Networks of responsibility simply define the *conditions* necessary to support the whole of which the part is indeed a part. Further, the greater the depth of a holon, then the *more networks* of communion it is involved in, and thus the *greater its responsibilities* in communion (correlative with its greater rights of agency).[12]

Thus, each and every holon in the Kosmos has equal *Ground-value* as a pure manifestation of Spirit or Emptiness. Further, as a particular *wholeness*, each holon possesses *intrinsic value*, depth value, which is valuable precisely because it internally embraces aspects of the Kosmos as part of its own being (and the more aspects it embraces, then the greater its depth and the greater its intrinsic value, the greater its significance).

And finally, as a *part*, each holon possesses extrinsic or *instrumental value*, because other holons external to it nonetheless depend upon it in various ways for their own existence and survival (and the more networks and wholes of which the holon is a part, then the greater its extrinsic value, the greater its fundamentalness: its existence is instrumental to the existence of so many other holons).[13]

My point in mentioning this example now is simply that this type of multidimensional environmental ethics is subverted by the flatland holism so prevalent in the Gaia-centric approaches. The fact that all holons have *equal Ground-value* is often confused with the notion that they must therefore all have *equal intrinsic value* ("bioequality"), and this paralyzes any sort of pragmatic action at all.

It is *much better* to kill a carrot than a cow, *even though* they are *both* perfect manifestations of Spirit. They both have equal Ground-value, but one has more intrinsic value because one has more depth (and therefore more consciousness). And the Eco camp's *general* attempt to "save the biosphere" by privileging it and leveling any distinctions in it (bioequality) paralyzes any actual pragmatic steps that might be taken to reform our anthropocentric stance.

Worse, in a flatland world, *intrinsic value* is often given *only* to the web-of-life or the system as a whole (the great interlocking order)—and thus we are all fundamentally nothing but strands in the great web. This attempt to introduce "wholeness" actually *instrumentalizes* all of us, instrumentalizes each and every individual living being, because now living beings only have part value, extrinsic value, instrumental value. Holism instrumentalizes everything! (Another example of the paradox of damage inherent in flatland: trying to holisticize everything converts everything into parts, into fragments, with no individual wholeness value, no intrinsic value apart from the great web.)[14]

In short: this approach confuses equal Ground-value with equal instrinsic value ("bioequality"), which leaves *only* the biosphere as a whole with instrinsic value, and this converts every individual living being into merely a part, a strand, an instrumental means in the glorious web (leaves, that is, what many critics have called ecofascism).

As I said, I believe in many cases the original intuition is true and good, namely, all holons have equal Ground-value (I believe that is the primary intuition behind the noble attempts to deanthropocentrize ethics). But in unpacking this noble intuition merely in terms of a flatland holism (the

Right-Hand path, which *always* embodies a qualitative distinction that *denies* all qualitative distinctions: in this case, "bioequality"), we end up instrumentalizing everything: precisely the fundamental Enlightenment paradigm most responsible for despoiling Gaia is still claiming, in its new guise, to save that which it is inexorably destroying.[15]

Thus, a correct and often noble intuition of the World Soul and Eco-Noetic Self is misinterpreted in terms of a flatland holism that, in leveling qualitative distinctions, paralyzes actions that would further the descent of that World Soul. And in this case it is still true: that which is not part of the solution is most definitely part of the problem.

EGO AND ECO: THE ASCENDERS AND THE DESCENDERS

> *There is nothing that can cure the senses but the soul;*
> *And nothing that can cure the soul but the senses.*
> —OSCAR WILDE

My point, then, is that the World Soul is very often being correctly and profoundly intuited but not very gracefully interpreted. It tends to be interpreted as either a Higher Self or a Gaia-self, both dualistic to the core, and both trying to convert the other by proselytizing their wounds. They represent respectively not Eros and Agape, but Phobos and Thanatos.

It comes to the same thing to say that, to the extent that individuals in modernity even attempt a genuine spirituality, they still tend to fall into either a purely Ascending Higher Self, or a purely Descending biospheric self.

The Ascenders have recourse to various forms of Gnosticism, Theravadin Buddhism, a type of Advaita Vedanta, a "higher Inner Voice," the inner Holy Spirit, archetypal psychology, the care of the Soul, contacting the Higher Self—all of which are true enough and *altogether important*; but in their *partiality*, in their exclusively Ascending bent, they attempt to get out of flatland by denying it altogether: Phobos, the fear-laden hand of earth-denying, community-denying, body-denying, sensory-denying escape.

The Descenders, on the other hand, have recourse to the visible, sensible God/dess—but that *profound truth*, cut off from its complementary

Ascending current, degenerates into geocentric, egocentric, highly individualistic and anti-authoritarian stances, desiring to preserve themselves in the free play of uncoerced shadows.[16] By becoming one with the Shadows and seamlessly inserting ourselves into a denatured nature, salvation will finally be found—and if it doesn't seem to be working, just insert harder.

The idea still seems to be that if I can just kiss the Shadows, I will see the Light. All transcendental occasions are reduced to this-worldly embrace: Thanatos, the dead and heavy hand of reduction and regression.

And these purely Descended approaches absolutely despise the Ascending paths, and blame them for virtually all of humanity's and Gaia's problems. But not to worry, the loathing is mutual: the Ascenders maintain the Descenders are simply caught in self-dispersal and outward-bound ignorance, which is the real source of all humanity's turmoils.

The Ascenders and the Descenders, *after two thousand years*, still at each other's throat—each still claiming to be the Whole, each still accusing the other of Evil, each perpetrating the same fractured insanity it despises in the other.

The Ascenders and the Descenders. Still crazy after all these years.

AT THE EDGE OF HISTORY

The great quest of postmodernity, I will argue, is for the bodymind integration of the centaur anchored in worldcentric vision-logic, and there are signs everywhere that this is in fact occurring. But most important, from within and beyond the centaur are now emerging glimpses of the Over-Soul as the World Soul (and I believe the evidence for this, too, is quite compelling).[17]

But Spirit's misinterpretation does not further its descent; its flattening and leveling in monological terms sabotages its very birth and blessing. We cannot integrate the Ascending and Descending currents in the emerging World Soul, but merely play them off against each other in a brutal torment of chosen ideology.

But the crucial components have already been won through for us. The breakthrough to Nonduality has already been accomplished by Plotinus in the West and Nagarjuna in the East (and not them alone). And the evolutionary understanding of Nonduality has already been won by

Schelling in the West and Aurobindo in the East (and not them alone). What they (and their many descendants) have accomplished is already the foundation for a World Federation and a council of all beings. And whereas they were, in their times, the rare and isolated, the average mode is now catching up with them: their paths can be ours, collectively, now.

Much work, of course, remains to be done. But the foundation is there, it has been won through for us; the basics are in existence. These roads are open to us: roads present but untraveled; paths cut clear but not chosen.

Can we not embrace these roots? Can we not see Spirit as the Life of Evolution and the Love of Kosmos itself? Does not the Good of Spirit, its Eros, release both Nature and Mind from the torments we have inflicted on them in vain attempts to make them each the source of infinite value? Does not the Goodness of Spirit, its Agape, embrace both Mind and Nature in a loving caress that heals the self-inflicted wounds? Does not the refluxing movement of God and the effluxing movement of the Goddess embrace the entire Circle of Ascent and Descent? Can we not round out the original insights and see that Spirit always manifests in all four quadrants equally? Is not Spirit here and now in all its radiant glory, eternally present as every I and every We and every It? Will not our more adequate interpretations of Spirit facilitate Spirit's rescue of us?

As Plotinus knew: Let the world be quiet. Let the heavens and the earth and the seas be still. Let the world be waiting. Let the self-contraction relax into the empty ground of its own awareness, and let it there quietly die. See how Spirit pours through each and every opening in the turmoil, and bestows new splendor on the setting Sun and its glorious Earth and all its radiant inhabitants. See the Kosmos dance in Emptiness; see the play of light in all creatures great and small; see finite worlds sing and rejoice in the play of the very Divine, floating on a Glory that renders each transparent, flooded by a Joy that refuses time or terror, that undoes the madness of the loveless self and buries it in splendor.

Indeed, indeed: let the self-contraction relax into the empty ground of its own awareness, and let it there quietly die. See the Kosmos arise in its place, dancing madly and divine, self-luminous and self-liberating, intoxicated by a Light that never dawns nor ceases. See the worlds arise and fall, never caught in time or turmoil, transparent images shimmering in the radiant Abyss. Watch the mountain walk on water, drink the Pacific in a single gulp, blink and a billion universes rise and fall, breathe out and create a Kosmos, breathe in and watch it dissolve.

Let the ecstasy overflow and outshine the loveless self, driven mad with the torments of its self-embracing ways, hugging mightily samsara's spokes of endless agony, and sing instead triumphantly with Saint Catherine, "My being is God, not by simple participation, but by a true transformation of my Being. My *me* is God!" And let the joy sing with Dame Julian, "See! I am God! See! I am in all things! See! I do all things!" And let the joy shout with Hakuin, "This very body is the Body of Buddha! and this very land the Pure Land!"

And this Earth becomes a blessed being, and every I becomes a God, and every We becomes God's sincerest worship, and every It becomes God's most gracious temple.

And comes to rest that Godless search, tormented and tormenting. The knot in the Heart of the Kosmos relaxes to allow its only God, and overflows the Spirit ravished and enraptured by the lost and found Beloved. And gone the Godless destiny of death and desperation, and gone the madness of a life committed to uncare, and gone the tears and terror of the brutal days and endless nights where time alone would rule.

And I-I rise to taste the dawn, and find that love alone will shine today. And the Shining says: to love it all, and love it madly, and always endlessly, and ever fiercely, to love without choice and thus enter the All, to love it mindlessly and thus be the All, embracing the only and radiant Divine: now as Emptiness, now as Form, together and forever, the Godless search undone, and love alone will shine today.

THE ONE THAT WAS TO COME

Today's world of modernity gone slightly mad: Myth for the peasants, flatland naturalism for the intelligentsia. It is more than a little ironic, then, that it was science, Descended science, that in the closing decades of the twentieth century would rediscover the self-organizing and self-transcending nature of evolution itself. More than a little ironic that in uniting the "two arrows" of time it would make Eros the single and all-pervading principle of manifestation. More than a little ironic that it would pave the way for an *evolution beyond rationality*, since it has clearly demonstrated that evolution stops for nobody, that each stage passes into a larger tomorrow. And if today is rationality, tomorrow is transrationality, and there is not a single scientific argument in the world that can disagree with that, and every argument in favor of it.

And so there we stand now, at rationality, poised on the edge of trans-rational perception, a *scientia visionis* that is bringing here and there, but ever more clearly, to all sorts of people in all sorts of places, powerful glimmers of a true Descent of the all-pervading World Soul.

Notes

CHAPTER 1. THE WEB OF LIFE

1. See Roger Walsh's perceptive discussion of this in *Staying alive*.

2. Warwick Fox has given a very useful summary of these historical developments; see *Toward a transpersonal ecology*.

3. Capra, *Turning point*, p. 16.

4. In Diamond & Orenstein, *Reweaving the world*, pp. 156, 161.

5. In Devall & Sessions, *Deep ecology*, pp. 8, 20.

6. Laszlo, *Evolution*, pp. 9, 4.

7. Ibid., p. 5.

8. Exactly how these domains are related, whether they are in fact three spheres or instead just aspects of one encompassing sphere, and how they might relate individually or collectively to the "theosphere"—the Divine Domain: these are questions we will want eventually to address. In the meantime perhaps we can accept these three domains as provisionally given.

To present only one example now: In *Ecology, community, and lifestyle*, Arne Naess points out that "humankind is the first species on earth with the intellectual capacity to limit its numbers *consciously* and live in an enduring, dynamic equilibrium with other forms of life," but that "a global *culture* . . . is now encroaching upon all the world's milieux, desecrating *living conditions* for future generations" (p. 23; my italics). This is surely true, and the fact that human culture can consciously either *accord* with the biosphere (living conditions) or *deviate* from it shows precisely that culture is *not* in all ways *the same thing* as the biosphere—it is differentiated from it in some significant ways, even though it depends upon it for its own existence; and that differentiation we call the noosphere. We will be returning to this topic time and time again throughout this book, each time refining it; so I ask my ecological friends, many of whom are profoundly suspicious of anything resembling a "noosphere," to bear with me until the arguments unfold more fully.

9. Bertalanffy, *General system theory*, p. 87.

10. Lovejoy, *Great chain of being*, whose themes we will explore in detail in chapter 9.

11. Ibid., p. 26.

12. Laszlo, *Evolution*, p. 14.

13. Ibid., p. 13.

14. Thus, for example, the following from Ilya Prigogine: After quoting Ivor Leclerc, who says, "In our century we are suffering the consequences of the separation of science and philosophy which followed upon the triumph of Western physics in the 18th century," Prigogine goes on to say: "However, I believe that the situation today is much more favorable in the sense that the recent rediscovery of time [i.e., irreversibility] leads to a new perspective. Now the dialogue between hard sciences on one side, human sciences and philosophy on the other, may become again fruitful as it was during the classic period of Greece or during the 17th century of Newton and Leibniz"—this being exactly the point after which what we might call the "great fracture" (between life and matter, or more accurately, interiority and exteriority) occurred. *Nobel Prize conversations*, Saybrook, 1985, p. 121.

15. Gardner, *Quest for mind*, p. 172.

16. Jakobson, *On language*, p. 11. This is the editor's summary of Jakobson's view.

17. Gardner, *Quest for mind*, p. 172.

18. *Encyclopedia of philosophy*, ed. Paul Edwards, vol. 2, p. 474.

19. Koestler, *Ghost in the machine*.

20. Eisler, *Chalice and the blade*, p. 205.

21. Taylor, *Sources of the self*, pp. 19, 20.

22. Ibid., p. 31, my italics.

23. Ibid., pp. 27, 78.

24. Ibid., pp. 22, 98.

25. Ibid., p. 88.

26. In "The evolution of consciousness? Transpersonal theories in light of cultural relativism," Michael Winkelman marshals the cultural relativist arguments against ranking any sort of consciousness achievement as higher or lower than another across cultures. No cultural position is or can be superior to another, he says; it has to be decided on its own merits, and on those alone: no position is intrinsically superior. He then proceeds to explain why his position is intrinsically superior.

Sensing a contradiction, he then *exempts* his own global-theorizing stance from having any *adaptive* value. He must do this, because if he admits that a worldcentric, global perspectivism has *adaptive* advantage over narrower perspectives, then he must admit that *his* cultural stance of universal-global perspectivism is *superior* to those cultures that he studies that do *not* share his universal pluralism.

Thus, Winkelman's performative contradiction shows up most obviously in the fact that he is convinced that his overall stance is superior to alternative and rival approaches, but his own theory cannot state why this is so. Instead, he simply lashes out at those who profess any sort of ranking system (and he does so using *his* implicit ranking system). This performative contradiction is then extended to all of religion, with the same results: nobody can say anything is higher or deeper; whatever works for one culture is absolutely as good as any other, as long as it is *functioning* in an *adaptive* fashion.

Besides being an incorrect assumption (are the Nazis as good as anybody else? We're not allowed to judge them with our own standards), the position is also secretly biased. The collapse of *cultural values* into adaptive or *functional fit* is, as we will see in great detail, simply part of the fundamental Enlightenment paradigm, and as such, is in its own way very ethnocentric (we will examine this in chapters 4, 11, 12, and 13).

What little theoretical evidence Winkelman presents for his position comes mostly from synchronic structuralism (e.g., Lévi-Strauss), which is now widely acknowledged to be unable to account for development (or diachronic structures) at all. Lévi-Strauss (e.g., *The savage mind*) believed that early cultures displayed hidden structural patterns that, when *objectively represented*, were every bit as complex as patterns displayed by advanced cultures, and thus the cognitive richness of early cultures was not fundamentally lacking with respect to modern cultures. This led him likewise to claim that there was no fundamental difference in cognition between a five-year-old and a scientist.

But this is like saying that since the Schroedinger wave equation, which represents the probability of an electron's behavior, is every bit as complex as the formulas of Aristotle's formal logic, then there is no fundamental difference between an electron's cognitive capacity and Aristotle's. The claim that the scientist and the five-year-old both possess incredibly complex structures is not enough to show that there aren't also profound differences in capacity as well. And, indeed, Lévi-Strauss subsequently withdrew his claim about five-year-olds and scientists being essentially equivalent in capacity, and this recanting of merely *synchronic* (unchanging) structuralism marked the end of a lack of *developmental* (diachronic) sensitivity. Culture is not as static, synchronic, and ahistorical as the early structuralists imagined, but rather involves deeply historical and developmental currents, and cultural/historical *development* shows broad *learning* trends, in cognitive, moral, legal, and technological aspects (as we will see).

Thus, if we acknowledge (with Winkelman) that universal perspectivism is better (in any sense of the word) than narrow ethnocentrism (and this is the *true* part of all cultural relativists' stance), and since this worldcentric perspectivism is not simply given to cultures at the start but develops and evolves slowly over the millennia, then we are justified in examining the developmental stages that lead to a capacity to take a worldcentric stance. This leads inexorably to theories of communication and the evolution of societies, and this is precisely the path that Habermas and others have taken.

In so doing, Habermas arrives at a series of universal validity claims that are cross-cultural and extra-linguistic, and open to fallibilist criteria. There is, so to speak, no other way to proceed if we want to actually acknowledge the moments of truth in cultural relativism, moments of truth that, if pursued sincerely (and not merely exempted from their own claims), lead inexorably to universalist validity claims (as we will see). These claims do not rank so much between cultures, but within cultures: the cultures themselves recognize higher and lower, or deeper and shallower, or better and worse values. And, *not* the values per se but the *types* of values can indeed be ranked—and are ranked by the societies themselves (criteria we will be investigating as the book proceeds).

We will, in the succeeding chapters, be explaining Habermas's views in great detail, and their significance will, I believe, become quite apparent. (For a critique of Winkelman's position in light of these more recent developments, see note 26 to chapter 5, page 600.)

27. Thus, universal and worldcentric pluralism is a very difficult, very rare, very special, very elite stance. In many ways I happen to agree with that elite stance. But I am not impressed when these elitists call it anti-elitist.

28. We will be going into this in detail in chapters 9 and 10, so for the moment a few very brief examples will suffice. In his book *Forgotten Truth*, Huston Smith summarizes the world's major religions in one phrase: "a hierarchy of being and knowing." Chögyam Trungpa, Rinpoche pointed out in *Shambhala* that *the* essential and background idea pervading all of the philosophies of the East, from India to Tibet to China, lying behind everything from Shintoism to Taoism to shamanism, is "a hierarchy of earth, human, heaven," which he also pointed out is equivalent to "body, mind, spirit." And Coomaraswamy in *Hinduism and Buddhism* (New York: Philosophical Library, 1943) noted that the world's great religions, bar none, "in their different degrees represent a hierarchy of types or levels of consciousness extending from animal to deity, and according to which one and the same individual may function on different occasions."

Chapter 2. The Pattern That Connects

1. We will be returning to this topic throughout the book; I can't state my final position until we discuss the IOU principle, in chapter 14. Suffice it to say, for now, that trying to decide *absolutely* in favor of "unvarying laws" or "learned habits" involves us in various contradictions. To say that there are no fixed laws, only learned habits, is itself a fixed law. But to say that there are only fixed laws is to ignore those laws that do in fact develop.

2. Hofstadter, *Gödel, Escher, Bach,* pp. 142, 146. Likewise, as Jakobson pointed out, "An important structural particularity of language is that at no stage of resolving higher units into their component parts does one encounter informationally pointless fragments." *On language,* p. 11.

3. Edwards, *Encyclopedia of philosophy,* vol. 5, pp. 202–3.

4. Bataille, *Visions of excess,* p. 174.

5. Ibid.

6. Quoted ibid., p. xi.

7. Culler, *On deconstruction,* p. 215. My italics.

8. Ibid., p. 123.

9. *Philosophical discourse of modernity,* p. 197. As I point out in the next para-

graph of the text, sliding contexts do not invalidate the existence of truth, but simply situate it. And there are many nonarbitrary ways that contexts are held still in order for the situating to occur. Habermas points out that in communication, for example, the presumption of identical meaning must be made or else the communication does not, and cannot, get started in the first place: and that stops the slide.

10. Varela, in Thompson, *Gaia,* p. 50.

11. Koestler, *Ghost in the machine,* p. 63, my italics.

12. Whitehead's "Category of the Ultimate" includes three concepts: creativity, many, and one. In a sense, he could have reduced that to creativity and holon, since holon is one/many (and since many and one never exist separately). Whitehead's prehensive unification is the present, subjective holon passing as object into the succeeding present, subjective holon, so that every holon prehends its entire actual universe, and lives on in the prehensive unification of all its descendants ("causality")—with, of course, the whole series showing gradation (hierarchy), depending upon the degree of creativity injected into the stream at any given moment.

This is why "emergence" as used in science doesn't really *explain* anything, it only *describes* what in fact happens. The *explanation* has to lie in something like Whitehead's ultimate category of creativity, a feature of reality itself that accounts for emergence and cannot itself be accounted for. We will be returning to this topic throughout this book, each time refining it. But it is clear that some sort of Eros is involved in the process, or it could never get going in the first place. Eros/creativity shows up especially in the self-transcending capacity of every holon.

13. In self-transformation or self-transcendence, the agency of the subholons are subsumed in the new agency of the superholon, and that new agency then exists in its own networks of new types of communion.

14. Jantsch, *Self-organizing universe,* p. 11.

15. Ibid., p. 183, my italics.

16. Laszlo, *Evolution,* pp. 76, 78, 35, 36.

17. Murphy's summary of Simpson's view. Murphy, *The future of the body,* p. 28.

18. Ibid., pp. 28–29.

19. Jantsch, *Self-organizing universe,* p. 49.

20. Quoted in Gardner, *Quest for mind,* p. 199–200.

21. Foss and Rothenberg, *Second medical revolution,* p. 151.

22. Hofstadter, *Gödel, Escher, Bach,* p. 308.

23. Laszlo, *Evolution,* p. 36.

24. Varela et al., *The embodied mind,* pp. 88, 90. We will be returning to this important work at several places in the following chapters, drawing on its considerable strengths (and suggesting a few possible weaknesses). I can't reference their work as fully as I would like in this chapter because I need to introduce several new topics before the significant contours of their valuable contribution can be fully appreciated. But I believe the general outlines presented in this chapter are consonant with the main themes of their work on enactive cognition and structural coupling, and when

we later return to their work, these parallels will, I believe, be more apparent. See notes 49 and 52, below; chap. 4, notes 13 and 43; chap. 14, note 1.

25. Mayr, *Growth of biological thought*, p. 63.

26. We can see *why* a holon acted in a particular way, but not that it would act in only that way.

27. Natural, monological sciences study holons that, through a relatively free emergence, have settled into habits so fixed, so stable, with such a minimum of creativity, that their behavior approaches (and thus mistakenly appears to follow) immutable laws. This, I take it, is the essence of Sheldrake's position.

28. Laszlo, *Evolution*, p. 196.

29. Bertalanffy, *General system theory*, pp. 74, 87.

30. Goldsmith, *The Way*, p. 28.

31. Sheldrake, *New science of life*, p. 74.

32. Varela, *Principles of biological autonomy*, p. 86.

33. Naess, *Ecology, community, and lifestyle*, p. 58.

34. Ibid.

35. Jantsch, *Self-organizing universe*, pp. 33, 16.

36. This does not mean that every transcendence necessarily includes every predecessor in every detail, but simply that each transcendence builds upon some of the fundamental features of its predecessor(s). We will see numerous examples of this as we proceed.

37. Varela, *Principles of biological autonomy*, p. 86.

38. Mitchell (ed.), *Nobel Prize conversations*, p. 59.

39. This turns out to be very important in development, where the failure to surrender an exclusivity structure results in fixation, or pathologically narrowed identity. In human development this is the difference between, for example, an oral drive and an oral fixation: the need for food is a basic structure that remains in existence in subsequent development, but if the infant exclusively clings to it, an oral fixation results; it refuses to be negated and preserved, and wants only preservation, hence fixation—Hawaii wants to be part of the Union but won't surrender its right to print its own money.

All development is fundamentally the conversion of exclusivity structures to basic structures, which is another way of saying that one still possesses the function but is not exclusively identified with it.

40. Sheldrake, *New science of life*, pp. 83, 87, 88; my italics.

41. Laszlo, *Evolution*, p. 54.

42. Sheldrake, *The presence of the past*, pp. 120–21.

43. Koestler, *Ghost in the machine*, p. 50.

44. In the same vein, some critics object to the use of the terms "vertical" and "horizontal" (not to mention "higher" and "lower"), because they feel that those notions are, among other things, either anthropocentric or patriarchal (androcentric) or "old paradigm." And this goes hand in hand with these critics' entire rejection of

the notion of holarchy itself, which seems confused, in that most of these critics also claim to be holistic.

But that confusion stems, as I will try to show later, from mistaking great span with great depth, mistaking "wider" with "higher" (which then leads them to reject the concept of "higher" or "vertical" altogether, since they believe that it is not necessary, being merely subsumed under the concept of "wider"). This is, I will suggest, a form of reductionism, and we will have much to say about the inherent problems of that approach in our discussion of ecofeminism and deep ecology.

For the moment, then, let me repeat that I believe that what these critics are really (and rightly) objecting to is not holarchy per se (as Laszlo said, "the empirical evidence for this is indisputable"), but rather pathological holarchy wherever it appears, and particularly as it appears in anthropocentrism and androcentrism, where agency is drastically out of balance with communion, and where certain arrogant holons have usurped their role in the Kosmos. This is one of the major topics in volume 2 of this series.

45. Transcription is at work already in the models of subatomic particles: how does what we recognize as, say, an electron (its deep structure) actually manifest as a given particle? The collapse of the wave function is an example of transcription (after all, the electron manifests as an electron and not a frog). In the biosphere we see transcription in an explicit form with, to give only one example, nucleic acid reproduction, which involves both translation and transcription (which are so named); transformation in this domain is called "mutation." In the noosphere, cognitive behaviorists and linguists, from Piaget to Chomsky, have dealt with various transcription rules, whereby the potentials of the deep structure are unfolded in actual surface structures. And much of the work on "worldviews"—from Gebser to Heidegger to Habermas—involves elaborating the pre-articulate deep structures of the worldview (or pre-understanding) that tends to govern the actual surface perceptions of its inhabitants. We will see many more examples as we proceed.

46. Of course, both translation and transformation actually deal with whole/parts (there are only holons); but transformation deals with emergent holons that subsume those of its predecessors, and thus is "more holistic." The wholes of the previous level are now parts of the senior level, so the "whole units" of the previous translation are now "parts" of the new: the senior has more depth and is thus more encompassing.

Differentiation divides and produces more regimes on a given level, but integration produces a new level, a new wholeness, a new regime, a new deep structure—it is a transformation, not a mere translation (we will return to this topic of differentiation and integration in tenet 12).

We heard Jantsch explain that evolution was not simply expanding communion (or self-adaptation), for that produces merely more of the same, not something novel. Likewise, Laszlo reminds us that expanded agency, even in its living, self-organizing form (as autopoiesis), cannot account for evolution either: "Autopoiesis is not evolution, however, even if autopoietic cellular automata models can simulate certain evolutionary phenomena such as the convergence of cells into multicellular systems. In order to understand evolution one needs also to understand how discontinuous, nonlinear change takes place in real-world dynamic systems" (*Evolution*, p. 39). In other words, one needs not merely translation models but transformation models.

As for translation itself, the fact that all holons are involved in translation means, in the broadest sense, that all holons are signs. Borrowing from Peirce, I define a sign as: any aspect of reality that stands for another, to another. Thus, even atoms translate the physical forces around them into terms that they respond to: an empty orbital shell is a sign, to an electron, that it may enter the shell. It is an aspect of reality (the empty shell) that stands for another aspect (it may be entered), to another aspect (the electron). All holons are signs because all agents are always agency-in-communion (or agents-in-context), and all signs are sliding because contexts are boundless.

And finally, this is why no signs, in any domain, are *merely* representational, or merely one holon representing another holon; signs are always a holon representing a holon to a holon. See Jakobson's interesting discussion on this in relation to Peirce: "all signification is but the 'translation of a sign into another system of signs' "[*Framework of language*, p. 10.]—which, among other things, is why linguistic theory can never define a sign without reference to the system in which the sign occurs (de Saussure), and why poststructuralism has been able further to point to endlessly sliding signification (although that is not so totally "deconstructive" as they imagine, but simply normal life in any world). And this also prevents knowledge from being *merely* representational; an interpretive moment is built into cognition (which we will examine at length in chapter 4). [For an "integral semiotics," see *The Eye of Spirit*, chap. 5, note 12.]

47. All types of holons at the same level would be destroyed; the different types that are also destroyed are higher; the different types not destroyed are lower.

48. This principle also holds for all types of human developmental sequences. For example, if we destroyed all conventional moral stages, we would also destroy all postconventional stages, but not preconventional stages. Destroy all concepts, and all rules are gone, but images still remain—and so on.

This principle also points up the bankruptcy of pure heterarchy. If there were *only* a horizontal dimension of *pure* heterarchy, where everything exists *solely* by virtue of its internal connections or relations to *everything else*, then if I destroyed all of one type of anything, I would simultaneously destroy all types of everything, since there is no ontological gradation of any sort in a *pure* heterarchy. This is where flatland ontologies fall apart; they ignore evolution and the meaning of emergence. There are aspects of evolution that are perfectly continuous, and aspects that are quantized or discontinuous, and emphasizing *exclusively* one or the other leads to pathological heterarchy and pathological hierarchy, respectively.

49. That agency is always agency-in-communion is a central feature, as I construe it, of Varela's *enactive* paradigm. As I would phrase it, a holon's *agency* ("intrinsic

self-organizing properties") enacts a worldspace (brings forth a domain of distinctions) and does so relatively autonomously, with the crucial addition that a holon's agency is partly a result of "structural couplings" with the appropriate worldspace (all of which we will be explaining in more detail as we proceed). Agency as agency-in-communion thus explains both relative *autonomy* and micro/macro *codetermination*. As Varela, Thompson, and Rosch put it:

"The crucial point here is that we do not retain the notion of an independent, pregiven environment but let it fade into the background in favor of so-called intrinsic factors [agency]. Instead we emphasize that the very notion of what an environment is cannot be separated from what organisms are and what they do. This point has been made quite eloquently by Richard Lewontin: 'The organism and the environment are not actually separately determined. The environment is not a structure imposed on living beings from the outside but is in fact a creation of those beings. . . .' "

They continue: "The key point, then, is that the species brings forth and specifies its own domain of problems to be solved by satisficing; this domain does not exist 'out there' in an environment that acts as a landing pad for organisms that somehow drop or parachute into the world. Instead, living beings and their environments stand in relation to each other through *mutual specification* or *codetermination*. Thus what we describe as environmental regularities are not external features that have been internalized, as representationism and adaptationism both assume. Environmental regularities are the result of a conjoint history, a congruence that unfolds from a long history of codetermination" (*Embodied mind*, p. 198).

This is also why the "unit" of evolution is, basically, a micro/macro unit, which is to say: almost any holon in existence. "A full list of units looks rather formidable: DNA short sequences, genes, whole gene families, the cell itself, the species genome, the individual, 'inclusive' groups of genes that are carried by different individuals, the social group, the actually interbreeding population, the entire species, the ecosystem of actually interacting species, and the global biosphere. Each unit [holon] harbors modes of coupling and selection constraints [communion], has unique self-organizing qualities [agency], and so has its own emergent status with respect to other levels . . ." (pp. 192–93).

For more on the enactive paradigm, see note 52 below; chapter 4, notes 13 and 43; and chapter 14, note 1.

50. Jantsch, *Self-organizing universe,* p. 85.

51. A thousand years from now, quarks will have been found to contain as many new levels of particles as its investigators have discovered new levels of consciousness in themselves; but that belief, needless to say, is not necessary for this argument.

52. In *Up from Eden,* I tried to demonstrate that this was the only way we could really understand developmental pathologies; namely, as disturbances in relational exchanges with a social environment *at the same depth*, with disturbances on one level (physical, emotional, linguistic) reverberating throughout the entire system, inclining other levels to reproduce the distortion developmentally.

And, as we will see throughout this book, same-level relational exchange does not mean that the micro exchanges with a pregiven macro: they co-create each other in emergent worldspaces. What this means will become more obvious as we proceed. For the moment, note Varela et al.: "The key point is that such systems do not operate by representation [of a pregiven world]. Instead of *representing* an independent world, they *enact* a world as a domain of distinctions that is inseparable from the structure embodied by the cognitive system" (or, more generally, by the deep structure or agency of the holon; *Embodied mind*, p. 140).

53. Jantsch, *Self-organizing universe*, p. 75.

54. Laszlo, *Evolution*, p. 25.

55. Lowe, *Understanding Whitehead*, p. 36. Lowe's wording.

56. Ibid.; the last phrase is Lowe's.

57. Derrida, *Positions*, p. 101.

58. Coward, *Derrida and Indian philosophy*, p. 40.

59. Ibid., pp. 40, 135, 137, 148.

60. Derrida, *Positions*, p. 26.

61. Culler's translation, in Sturrock, *Structuralism and since*, p. 164.

62. Or, as Saussure put it, "Language is a system of interdependent terms in which the value of each term results solely from the simultaneous presence of the others"— that is, it is never simply present on its own. Quoted in Hawkes, *Structuralism and semiotics*, p. 26.

As for the differentiating process of language, Saussure's famous quote puts it in a strong form: "In language there are only differences without positive terms. Their most precise characteristic is in being what the others are not. Whether we take the signified or the signifier, language has neither ideas nor sounds that existed before the linguistic system, but only conceptual and phonic differences that have issued from the system" (p. 28). And that is where integration is crucial, as Hawkes summarizes: "Language stands as the supreme example of a self-contained [relatively autonomous] relational structure whose constituent parts have no significance unless and until they are integrated within its bounds."

Saussure's great breakthrough was to treat language as a relatively autonomous holon, such that its structure or regime constituted the differentiating and integrating patterns that governed every element (subholon) in the system, so that a heap of otherwise entirely meaningless elements come together to form meaningful signs—by virtue of the system, the higher holon that confers meaning on the subholons by holding them in a common relationship. Saussure did not quite win through to a totally holonic view, but it was a major step in the right direction, and had an enormous historical impact (virtually all structuralists, and therefore all poststructuralists, and many semioticians, trace their lineage to Saussure). We will return to his thoughts on linguistic signs in chapter 7.

Finally, that "nothing is ever simply present" has almost nothing to do with mystical Presence, as we will see later. Nor is it to be confused with Buddhist Emptiness.

63. Habermas, *Philosophical discourse of modernity,* p. 359.

64. Ibid., pp. 345-46. Habermas makes clear that he is, in part, picking up certain Hegelian themes and pursuing them along a road not taken, that of reason as communicative action. Hegel's entire philosophy is, in a sense, the philosophy of holons, and the differentiating-and-integrating dialectic that drives all development.

65. Dreyfus and Rabinow, *Michel Foucault,* p. 55.

66. Laszlo, *Evolution,* p. 35.

67. Futuyma, *Evolutionary biology,* p. 289, my italics.

68. This is so for the same reason, as Varela pointed out, that a level in a holarchy is external to its juniors and internal to its seniors.

69. The whole of literary theory can be seen as a rather spirited attempt to decide just what we will identify as a *literary holon,* and *therefore* where we can find or locate the *meaning* of a text. It used to be that "meaning" was something the author created and simply put into a text, and the reader simply pulled it out. This view is now regarded, by all parties, as hopelessly naive.

With the arrival of psychoanalysis, which undermined the autonomous ego with organic drives (i.e., there are unconscious drives motivating the supposedly "autonomous" ego), it was recognized that some meaning could therefore be unconscious, or unconsciously generated, and this unconscious meaning would find its way into the text even though the author was unconscious of it. It was therefore the job of the psychoanalyst, and not the naive reader, to pull this hidden meaning out.

The "hermeneutics of suspicion" (Ricoeur), in its many forms, thus came to view texts as repositories of unconscious meaning that could be pulled out only by the knowing critic. Any repressed, oppressed, or otherwise marginalized context would show up, disguised, in the text, and the text was a testament to the repression, oppression, marginalization. Marginalized context was hidden subtext.

The Marxist variation was that the critics themselves existed in the context of capitalist-industrial social practices of covert domination, and these hidden contexts (and therefore meanings) could be found in (and therefore pulled out of) any text written by any person in *that* context. Similarly, texts would be read in the context of racism, sexism, elitism, speciesism, jingoism, imperialism, logocentrism, phallocentrism.

Various forms of structuralism and hermeneutics fought vigorously to find the "real" context which would, *therefore,* provide the real and final *meaning,* which would undercut (or supersede) all other interpretations. Foucault, in his archaeological period, outdid them both, situating both structuralism and hermeneutics in an episteme (later, dispositif) that itself was the cause and context of the type of people who would even want to do hermeneutics and structuralism in the first place.

In part in reaction to some of this, the New Criticism (actually many decades old now) had said, basically, to hell with all that. The text, in and by itself, is the autonomous literary holon. Ignore the personality (conscious or unconscious) of the author, the time, the place, and look solely at the structural integrity of the text (its regime, its code). "Affect-response" or "reader-criticism" theory (Fish) reacted strongly to all that and maintained that since meaning is only generated in reading (or in viewing the artwork), then the *meaning* of the text is actually found in the *response* of the reader. The phenomenologists (e.g., Iser, Ingarden) had tried a combination of the two: the text has gaps ("spots of indeterminancy"), and the meaning of the gaps can be found in the reader.

And deconstruction came along and said, basically, you're all wrong (it's very hard to trump that)—meaning is context dependent, and contexts are boundless.

While I agree with that latter statement of deconstruction, I do so for almost opposite reasons, because what interests me is not why all of those theories of meaning are wrong, but why all of them are right. They are all (relatively) true snapshots of a particular context, and they can all be relatively accurate because contexts are always contexts within contexts, making room for each and all (*except* where they deny other contexts).

The study of holarchy, in short, is the study of nested truths. Deconstruction in its American form simply takes the photographic negative of that and declares any approach (other than its own) to be the study of nested lies, which is precisely why that attitude lands it squarely in nihilism, whereas the study of nested truths leads not to nihilism but to Emptiness, the creative plenum of the Kosmos. [See *The Eye of Spirit*, chaps. 4 and 5.]

70. Laszlo, *Evolution*, p. 41.

71. Ibid. Compare Varela et al. "What all these diverse phenomena have in common is that in each case a network gives rise to new properties, which researchers try to understand in all their generality. One of the most useful ways of capturing the emergent properties that these various systems have in common is through the notion of an 'attractor' in dynamical systems theory." Varela et al., *Embodied mind*, p. 88.

72. Laszlo, *Evolution*, p. 43.

73. Ibid.

74. Ibid., p. 46.

75. Ibid.

76. Laszlo, *The choice*, p. 93.

77. Or Jonas Salk: "It is in the nature of the organism to be oriented for the change that occurs. The intrinsic nature of the organism [regime, deep structure] influences the range and direction of change that can occur; the change that occurs becomes added to others, all of which together seem to be 'causes' toward which the developing organism is drawn, and the word 'cause' in this context obtains the philosophical meaning of 'end or purpose'." All quotes from Jakobson, *On language*, pp. 481, 482, 483.

And, of course, Sheldrake's immensely thoughtful application of morphogenetic fields (which he now calls simply morphic fields) is an attempt to explain many of the incontrovertible facts of teleological development that simply cannot be explained by materialistic reductionism.

78. Gardner, *Quest for mind*, pp. 187–88. I have substituted "development" for "genesis."

This also points up a fact that all psychological growth theorists (especially the humanistic and transpersonal psychologists) have known for a long time: most neurosis is due not primarily to past conditioning per se, but to a future omega prevented from emerging. Even Freud's fixations, *when* they occurred, were aborted omega points: libido cannot distribute in its normal and preferred pattern, but instead is traumatized in its growth, much like stepping on an acorn. These aborted omega drives *then* become conditioned and thus appear as *past* fixations, but that is not how they *started*. Recovery therapies in general ("regression in service of the ego") thus attempt to return to the point where the past conditioning was an aborted future growth, then take the boot off the acorn, and thus allow the growth to proceed forward in a more normal pattern, following its more natural omega. (Of course, it's a little more complicated than that—the acorn has continued to grow in slightly twisted ways, and this can't simply be cleanly undone; but the point is clear enough.) If the past conditionings didn't have a preferred pattern (omega), they couldn't have been aborted in the first place.

79. Quoted in Jakobson, *On language,* p. 483.

80. Ibid., pp. 481, 20.

81. Habermas, *Philosophical discourse of modernity,* p. 347.

82. Harold Coward's wording, *Derrida and Indian philosophy,* p. 92, my italics.

83. Because holarchy is the study of nested truths, no matter how much we expand our contexts, this does not invalidate the relative truths of smaller contexts. It *negates* their exclusiveness (or their ultimateness), but *preserves* their moment of truth, their context-dependent truth. Even if we go beyond Freud (which I trust we will), even if we expand our contexts beyond the isolated ego to the communitarian society, or to the whole biosphere, or even to God Thunderous and Almighty, this will not change the fact that if I have a really vicious oral fixation, I am not going to have an altogether fun time in life.

CHAPTER 3. INDIVIDUAL AND SOCIAL

1. Capra, *Belonging to the universe,* p. xii.

2. In Diamond & Orenstein, *Reweaving the world,* p. 227.

3. From *The self and its brain,* p. 17.

4. Jantsch, *Self-organizing universe,* p. 117, 118.

5. Evolution is indicated not necessarily by increasing size but by increasing depth, or degree of structural organization. In *Gödel, Escher, Bach,* Hofstadter makes the point that, given that evolution is transcendence and inclusion ("embrace"), what is included in all cases is the code or canon or regime, and not always the physical material; or sometimes, the physical material is "shrunk" while the code is included. Thus, for example, the paleomammalian brain includes and envelops the reptilian brain, but dinosaurs are much bigger than mammals. What the mammals include is not the size but the capacities and functions of the reptilian brain and its codons.

Thus, while size is sometimes an indicator, that fact obscures what is really going on: increasing depth of embrace, not increase in spatial extension.

Further—and this is really the most important point, but a point I can't make until the concepts of the next chapter are introduced—the size factor, or physical extension, applies only to the exterior or surface forms of holons, and does not apply to their interiors. In cognition, for example, concepts include symbols but are not physically bigger than symbols. Interiors deal with degrees of intentions, not extensions, and trying to convert all evolutionary changes into physical size is simply part of the flattening of the Kosmos, part of the brutalization of qualitative distinctions, that has marked the instrumentalism of all flatland ontologies.

6. In other words, communities get smaller between levels, larger within levels (up to eco-equilibrium).

7. We already heard Varela point out that a given dimension in a holarchy is internal to its seniors and external to its juniors—that's the same as saying the cosmos is internal to the bios and the bios is external to the cosmos: literally, out of the physical world: transcends and includes.

8. Jantsch, *Self-organizing universe*, p. 157.

9. Quoted in Koestler, *Ghost in the machine*, pp. 277–78.

10. Ibid., p. 279.

11. Ibid., p. 283.

12. Jantsch, *Self-organizing universe*, p. 167.

13. Ibid., p. 168.

14. Ibid., p. 169.

15. In Jantsch and Waddington, *Evolution and consciousness*, p. 173.

16. And notice that these social holons become larger, not smaller, in physical size, because once on a new level, that level expands up to its sustainable limits.

Chapter 4. A View from Within

1. Teilhard, *Phenomenon of man*, p. 57.

2. My own claim, however, is that the distinction interior/exterior is *not* an emergent quality, but rather exists from the *first moment* a *boundary* is drawn; exists, that is, from the moment of creation. What most panpsychists mean by consciousness or mind is *not* what I mean by consciousness, which is depth. Because consciousness is depth, it is itself literally *unqualifiable*. It is depth, not any particular, qualifiable level of depth (such as sensation or impulse or perception or intention)—those are all forms of consciousness, not consciousness as such.

In other words, depth isn't a quality, like sensation or impulse or idea, but a relationship (or opening) among holons. Thus, I have never been satisfied with any of the panpsychical theories, because they *qualify* depth with particular manifestations of depth (such as sensations or feelings or intentions), and these do *not* exist throughout the holarchy of being, but emerge only at *particular* levels of depth, whereas depth itself is present from the start (or wherever there is a boundary). I am a pan-depthist, not a pan-psychist, since the psyche itself emerges only at a *particular* level of depth.

This is why I keep saying that I don't really care how far down the reader wishes to push "consciousness," and why I'm not really concerned with whether plants have sensations, etc., because by "consciousness" most people mean some favorite form of consciousness (sensations or intentions or desire, etc.), and none of those go all the way down. But depth does go all the way down, and depth is unqualifiable.

When I say that consciousness or depth is unqualifiable, I mean, in a strong sense, to evoke the Mahayana Buddhist notion of *shunyata*, or pure Emptiness, and for the moment I am further following the Yogachara Buddhist notion that pure Emptiness and pure Consciousness are synonymous. Consciousness is not a thing or a process—we can just as well, with William James, deny that it even exists, because it is ultimately Emptiness, the opening or clearing in which the *form* of beings *manifest* themselves, and not any *particular* manifestation itself, a type of Zen reconstruction of Heidegger that dispenses with Dasein per se and de-anthropocentrizes depth. The Being of beings is depth, which, being unqualifiable as such, is finally Emptiness as such (consciousness as such); but since Dasein does participate self-reflexively in depth, Dasein can realize Emptiness (I will be returning to this topic as the story unfolds into the transpersonal domains).

3. Teilhard, *Phenomenon of man*, p. 60, note 1.

4. The interiority of one stage is taken into the interiority of the next and thus becomes an *external* form *within* that interiority, a three-dimensional-chess detail that we will get to later.

5. Most systems theorists, that is, absolutize "B"—absolutize living systems (or the biosphere). As Habermas tellingly points out, they are involved, not in metaphysics (absolutizing A), but metabiology (absolutizing B). It is with this metabiology that they then attempt to explain cultural evolution (Luhmann's approach being by far the most sophisticated), but they leave out precisely those things that make culture culture and not merely life.

Many ecotheorists further take this metabiological absolutizing and not only attempt to explain culture with its terms, but also necessarily see culture as a lamentable deviation from those terms: all conclusions guaranteed by the prior absolutizing. We will return to this topic in later chapters, when we trace the rise of this metabiological absolutizing back to its source in the Enlightenment paradigm.

6. Jantsch, *Self-organizing universe*, p. 56.

7. Sheldrake, *Presence of the past*, p. 105.

8. Murphy, *Future of the body*, p. 27.

9. See especially *The Philosophical discourse of modernity, Communication and the evolution of society*, and *The theory of communicative action*. After giving Habermas's views, I will sometimes translate them into the more familiar terms that we have been using, without, at that point, implying that Habermas would necessarily agree with all subsequent discussion; in any event, Habermas's own views will be clearly indicated.

10. Wuthnow et al., *Cultural analysis*, pp. 1, 3, 4, 2.

11. Ibid., pp. 7, 6.

12. As we will see, Foucault's archaeological method simply focused on the *behavior* of serious speech acts (discourse-objects) in a discursive formation. Not only did he bracket the *truth* of these statements (the standard phenomenological bracketing), he bracketed, or attempted to bracket, the *meaning* of the statements as well; and then he simply described, in a distantiated stance, the behavior of these discursive formations, much as one would describe the behavior of gas particles or an ant colony. He then sought the rules of rarefication (transformation) that governed the systematic discursive formation itself, a sort of existential structuralism.

It wasn't that this approach was simply wrong, but that it depended upon factors that Foucault failed to take into account, not the least of which was that serious speech acts could not themselves be identified for study without at least a rudimentary grasp of their meaning (their interiors). Some sort of hermeneutics or interpretive measure had to be brought into the study alongside of the archaeological "quasi-structuralism": the detached study of exterior linguistic formations had to be supplemented with an interior understanding of their meaning, and thus Foucault's archaeology gave way to his more balanced "interpretive analytics." The reasons for this will become clearer as the narrative proceeds.

13. As I already briefly suggested, to the extent that we recognize that holons have an interior, and to the extent that all holons exist by virtue of relational exchange with same-depth holons, then to say that a group of holons share a common physical space (which everybody accepts) is also to say that they share a common interior space.

"Worldview" is too pan-psychic and suggests that, for example, cells share a developed cognitive map of the external world, which is a bit much (although Varela's biological cognition comes close, and I am comfortable with the way he presents it). But for most purposes, I will usually adopt the more general term "worldspace," which means the sum total of stimuli that *can* be responded to (i.e., that have actual meaning or impact or registration). This cannot be determined by a mere empirical analysis of action systems, because we all *already* exist in the *same* physical universe, so physical parameters alone cannot explain the *differences*. Worldspaces are *a priori* to physical parameters (but not ultimately *a priori* to awareness, as we will see).

In other words, worldspace, as I use the term, does not have the ordinary pan-psychic connotations or implications. Typical panpsychism confuses consciousness with a *particular level* of consciousness (perception or intention or feeling) and then is *forced* to push *that* "consciousness" all the way down. I am completely uninterested in that approach. Do atoms possess an actual prehension (Whitehead) or perception (Leibniz)? I don't know; that seems a bit much. But they do possess depth, and therefore they do share a common depth. And a common depth is a worldspace, a worldspace created/disclosed by a particular degree of shared depth. (I will continue to use Whitehead's "prehension," but only in this considerably changed context.)

I earlier noted that a sign is any aspect of reality that represents another, to another. So that all holons, existing by virtue of a network of relationships with other holons— all holons are signs. That is, the regime of any holon *translates* only a particular *range* of *signs* (it *registers* only a circumscribed band of stimuli); the band of common translatable signs is a holon's worldspace. (I will in a moment add the crucial qualification that the meaning of "registerable signs" therefore implies that the signs are in fact codetermined by the registration, and that codetermination is what I mean by a worldspace proper.)

Likewise, if a holon is to enter a deeper worldspace, it will have to *transform*, not merely *translate*, and this transformation opens it to an entirely deeper and wider range of signs: a new world, a new worldspace, within which it will then translate according to its own regime, or basic self-organizing principles, in structural coupling (agency-in-communion).

With reference to Varela's enactive problematic, a holon's agency *enacts* a worldspace (brings forth a domain of distinctions), and does so relatively autonomously, with the added understanding that a holon's agency is partly a result of its historical "structural couplings" with the appropriate milieu. Agency as agency-in-communion thus enacts a worldspace codetermined by subject and object.

Varela, Thompson, and Rosch: "Emergent states [are] constrained by a history of coupling with an appropriate world. By enriching our account to include this dimension of *structural coupling* [codetermining communion], we can begin to appreciate the capacity of a complex system to enact a world. . . . The result is that over time this coupling selects or enacts from a world of randomness a domain of distinctions. In other words, on the basis of its autonomy the system selects or enacts a domain of significance" (*Embodied mind*, pp. 151, 155–56).

Thus, as I earlier mentioned, a worldspace is not simply pregiven and then merely *represented* via a *correspondence* of agency with its allegedly separable communions (other agencies). Rather, the coherency of its agency (autonomy), structurally coupled with other communing agencies, enacts a worldspace mutually codetermined. Using a cellular automata (named Bittorio) as an example: "We can say that a minimal kind of interpretation is involved, where *interpretation* is understood widely to mean the enactment of a domain of distinctions out of a background. Thus Bittorio, on the basis of its autonomy (closure), performs an interpretation in the sense that it selects or brings forth a domain of significance out of the background of its random milieu. . . . These regularities constitute what we could call Bittorio's world [worldspace]. It should be apparent that this world is not pregiven and then recovered through a representation. . . . Bittorio provides, then, a paradigm for how closure [agency] and coupling [communion] suffice to bring forth a world of relevance [a worldspace] for a system [a holon]" (p. 155). (For more on the enactive paradigm, see notes 49 and 52 for chapter 2; note 43 below; and note 1 for chapter 14.)

I will have much to say about the representational paradigm later in chapter 13 (and throughout the book), and its relation to interpretation. Indeed, much of the historical portion of this book is an examination of the rise (and eventual dominance) of the representation (or reflection) paradigm (e.g., chap. 12: The Collapse of the Kosmos); we will begin this examination later in this chapter under the heading "The Fundamental Enlightenment Paradigm."

Worldspace, as I use the term, also has a rich philosophical background, beginning most notably with Leibniz and Kant, and running through the hermeneuticists to Nietzsche, Heidegger, Gebser, Foucault, Gadamer, Piaget, Habermas, the structuralists, etc. (as will become apparent later in this chapter). My own position can be succinctly summarized: the agency of each holon establishes an opening or clearing in which similar-depthed holons can manifest to each other, for each other: agency-in-communion (all the way down). These points will become clearer as we proceed.

One last note: in previous works I used the term *pleromatic* for the physical worldspace, or the sum total of stimuli the physical world can enact; *protoplasmic* for cellular worldspace; *uroboric* for reptilian; *typhonic* for mammalian (limbic) . . . which then runs into magic, mythic, and mental . . . and then the higher or transpersonal worldspaces. I will continue to use that terminology here.

14. See Wilber, *Eye to eye*, chapter 2.

15. For Foucault, see notes 12, 17, 23, and 28, this chapter.

16. Donald Rothberg gives an excellent short summary of what amounts to the Right- and Left-Hand paths: "Contemporary epistemology generally recognizes two main interrelated forms of knowledge: (1) naturalistic and (2) interpretive. (Some, of course, maintain the priority or even the exclusivity of one of the two forms.) Naturalistic inquiry aims at empirical *explanation*, conceived of as the development of theories that identify lawful or lawlike regularities and causal connections between variables. . . . Interpretive inquiry aims at the *understanding* of meanings, whether subjective or intersubjective. The goal is to understand the meaning of an individual's action (e.g., 'What did that mean to her?' or 'What did that mean in that specific context?'); the often only implicit rules of a group or society (e.g., 'What implicit and explicit rules do we follow in social interaction as students, as co-workers, or as husband and wife?'); and the implicit or explicit meanings of texts and other expressions of human creative activities. . . . Such meanings cannot be reduced to the identification of causal connections and require accounts making use of intentional language rather than simply descriptive language." Rothberg, "Contemporary epistemology," in Forman, *Problem of pure consciousness*, pp. 175–77.

17. As for its being almost that simple (phenomenologically, that is), take for example Heidegger's hermeneutics (Left-Hand, or interior) and Foucault's archaeology (Right-Hand, or exterior), the latter being a variant of the structural/functional paradigm applied to linguistic structures, or an attempt to describe discursive structures "purely" from the *outside*. Dreyfus and Rabinow: "This devotion to the description of concrete structures understood as conditions of existence bears a striking similarity to what Heidegger, in *Being and Time*, calls an existential analytic. But there is an important difference. For although both Heidegger and Foucault attempt to . . . relate

the 'factical' principles which structure the space [worldspace] governing the emergence of objects and subjects, Heidegger's method is hermeneutic or *internal*, whereas Foucault's is archaeological or *external*. Foucault is explicitly rejecting both Husserlian phenomenology and Heideggerian hermeneutics when he opposes to the exegetical account the *exteriority* of the archaeological attitude" (*Michel Foucault*, p. 57; my italics).

Foucault himself stated that the archaeologist isolates statements "in order to analyze them in an *exteriority*. . . . Perhaps we should speak of 'neutrality' rather than exteriority; but even this word implies rather too easily a suspension of belief, whereas it is a question of rediscovering that *outside* in which, in their deployed space, enunciative events are distributed" (pp. 57, 51; my italics).

In more detail: "Foucault and the hermeneuticists agree that practices 'free' objects and subjects by setting up what Heidegger calls a 'clearing' [worldspace], in which only certain objects, subjects, or possibilities for actions can be identified and individuated. They also agree that neither the primary relations of physical and social causality, nor the secondary relations of intentional mental causality can account for the way practices free entities. But they differ fundamentally in their account of how this freeing works. According to the hermeneuticists, who describe the phenomenon from the *inside* [Left-Hand], nondiscursive practices 'govern' human action by setting up a horizon of intelligibility in which only certain discursive practices and their objects and subjects make sense. Foucault, the archaeologist looking from *outside* [Right-Hand], rejects this appeal to meaning. He contends that, viewed with external neutrality, the discursive practices themselves provide a meaningless space of rule-governed transformations in which statements, subjects, objects, concepts and so forth are taken by those involved to be meaningful. . . . The archaeologist studies mute statements and thus avoids becoming involved in the serious search for truth and meaning he describes" (pp. 79, 85, my italics).

These are all familiar ploys from the Left-Hand and Right-Hand paths. Foucault's exterior approach, his bracketing of truth and meaning, his confinement to "mute" statements (monological), his "happy positivism"—these are all maneuvers of the Right-Hand path, applied not to the bone-crunching concreteness of physical-social realities but to the exterior, material, archaeological remnants of discursive practices: language looked at from the outside as a rule-governed system. (Even into his genealogy phase, "Genealogy avoids the search for depth. Instead, it seeks the surfaces of events . . . ," p. 106).

Foucault's archaeology and genealogy are legitimate endeavors, but, as he soon came to realize, they cannot stand alone. In fact, that approach, by itself, is perfectly contradictory: since it brackets truth (truth is merely something so labeled in a discursive system, or so employed in service of power), then this approach cannot itself claim that what it is saying is true. It hovers above the ground with no reason to be taken seriously. Rather, Foucault came to see that it has to be supplemented with Left-Hand approaches and a more balanced overview, including not only nondiscursive social practices but also hermeneutic interiors (or, at the least, a better interpretation of interpretation). Dreyfus and Rabinow: "What Foucault offers in *The History of Sexuality* is an incisive example of what a better interpretation looks like." As Gilles Deleuze would remark, Foucault came to "thinking of the past as it is condensed on the *inside*." Dreyfus and Rabinow conclude that Foucault's approach might be called "interpretive analytics," which seems fine, though they do note, with many critics, that while Foucault's work was immensely suggestive in this area, he nonetheless "owes us an interpretive description of his own right way to do interpretation. He has not provided us one yet" (p. 183). Alas, his death removed that possibility.

18. Subtle reductionism reduces Left-Hand to Right-Hand dimensions; it often further reduces the Upper Right to the Lower Right, and emphasizes purely the holistic system over individuals; systems theory does indeed often do this, and most New Age paradigms and ecotheorists follow this version of subtle reductionism.

19. Habermas, *Philosophical discourse*, pp. 363, 367.

20. Taylor, *Hegel*, pp. 10, 22, and *Sources of the self*, p. 233.

21. *Essay on man*.

22. Lovejoy, *The great chain*, p. 211.

23. As we noted, in Foucault's earlier work, especially the archaeology, he bracketed *both* truth and meaning ("double phenomenology"), and he consequently was himself disdainful of anything resembling "depth" languages (I actually agree with many of his conclusions, but often for very different reasons, as we will see). His double bracketing ("a phenomenology to end phenomenology") therefore was excluded from depth and interpretation from the start: just the exteriors. Nonetheless, the sciences that he saw as beginning to escape the "Age of Man" and "humanism" were precisely those sciences that began to reintroduce the notion of depth (psychoanalysis, ethnology, linguistics). And when Foucault later moved from archaeology and genealogy to ethics, he himself began a more judicious use of "understanding from the inside," or a reconstructed hermeneutics ("interpretive analytics").

But even when Foucault was rejecting depth, as a methodological Right-Hand ploy, he nonetheless had his own versions of it (or else he couldn't have formed any sort of judgments in the first place). He himself describes his approach thus: "Whereas the interpreter is obliged to go to the depths of things, like an excavator, the moment of interpretation [his genealogy] is like an overview, from higher and higher up, which allows the depth to be laid out in front of him in more and more profound visibility; depth is resituated as an absolutely superficial secret" ("Nietzsche, Freud, Marx").

What Foucault is here calling "height" is a version of what I am calling depth, and what he derisively calls depth is indeed rather shallow and superficial, just as he claims. Freud's "depth psychology," for example, digs "down" to the libido: a very fundamental, but not very significant, aspect of the human being—it is very superficial, very shallow (chapter 13). That human beings would spend their lives looking into their ids and feel something incredibly significant was afoot, this was for Foucault laughable, and who cannot agree? There were higher (deeper) views available.

As we will see later, the languages of the centaur level are preeminently the *languages of depth* (and dialectics). The languages of the rational-ego are the languages of representation, monological explanation, and the understanding (cf. *Vernunft* vs. *Verstand*). My thesis will be that one legitimate way to read Foucault's early work is that the Enlightenment and the subsequent Age of Man ("dehumanizing humanism") was the widespread emergence of the egoic-rational mode out of the mythic-membership mode, which brought many genuine benefits, but, captured by what Hegel called a "vanity of the understanding" (or empiric-analytic and monological thought), it was committed to a flattening and leveling of qualitative distinctions (and the retroflection of that flatland mode upon the subject itself produced the more brutalizing aspects of the "sciences of man"). The subsequent emergence of the centaur, with its vision-logic (dialectics of depth), was the more mature reason (*Vernunft*) in intersubjective communicative exchange (dialogical, not just monological), and this opening of dialogical and depth-understanding marked the beginning of the "end of man."

This also brings Foucault's accounts more into line with those of Habermas and Taylor, as we will see in chapters 12 and 13.

Incidentally, beyond the centaur, the languages of depth have less and less applicability, giving way to languages of vision and vibration (psychic), then languages of archetype and illumination (subtle), then languages of emptiness and dream (causal), then languages of extraordinary ordinariness (nondual), all of which we will examine later (see note 58 for chap. 8 for an extended discussion). This present volume is intentionally written in the centaur-level language of depth.

24. It might be obvious that the Left-Hand/Right-Hand is another version of the mind/body, or mind/brain, or consciousness/form distinction. It is not a rigid duality, in my system, because they mutually interact; nor is it a *final* duality, because in the transpersonal domains, it becomes obvious that "form is not other than emptiness [consciousness], emptiness is not other than form." We will return to this important point later and look at various transpersonal solutions, from Plotinus to Aurobindo to Mahayana Buddhism.

Likewise, the linguistic *signifiers* (or the *material* components of a sign, the written symbol or the physical air vibrations of the spoken word) are all Right-Hand components, whereas the *signifieds* (the interior meanings that a person associates with a word) are all Left-Hand occasions, and definitely, à la deconstruction, signifiers and signifieds slide all over each other—yet another version of the mind/body "interaction" (and another version of exteriors being enveloped in development).

At the same time, postmodernism's anxious attempt to supplant modernity's program does not succeed in precisely the area that it imagines it does. There is, in the postmodern poststructuralists, a barely disguised attempt to shift emphasis from the interior signified to the material signifier, coupled with the notion that signifiers are free to alight pretty much wherever they want (subject only to shifting paradigms of power and discursive practices). We saw that the significant insight of the postmodern poststructuralists was that meaning is context-dependent and contexts are boundless.

But that has quickly degenerated into the assertion of the primacy of the material signifier as viewed exteriorly. This merely exterior, Right-Hand reductionism runs from Foucault's archaeology to Derrida's grammatology, and as such, as a *subtle reductionism*, is precisely heir to the Enlightenment paradigm. (Habermas reaches the same conclusion with a similar analysis, seeing both of them as infected with remnants of the Enlightenment philosophy of the subject).

In Habermas, the Left-Hand and the Right-Hand appear, among other places, as the distinction between linguistically generated intersubjectivity (Left) and self-referentially closed systems (Right), which, he says, "are now the catchwords for a controversy that will take the place of the discredited [form of] the mind-body problematic" (*Philosophical discourse*, p. 385).

As for the mind/body problem itself, I will later argue that its precise nature cannot be solved, only (dis)solved in contemplative awareness, where emptiness and form disclose their ultimate relationship. In the meantime, I refer to my position as "interactionist," not so much because I believe that position really solves anything, but as a way to disavow the other "solutions" (identity, parallel, dualist, preestablished) that, in my opinion, have even worse difficulties.

As noted, my use of "interactionism" or "codetermination" is in some ways quite sympathetic with Varela's "enaction," but also differs on a few points, as we will see (note 43 below; note 1 for chap. 14).

25. The collapsing of Left to Right means that interior-and-exterior is collapsed into and confused with physical inside-and-outside (not to mention better-and-worse collapsing into bigger-and-smaller). This collapse fails to grasp the fact that something can be simultaneously "inside" the skin boundary and still an "exterior" dimension. The *within* or actual *interior* of the brain is *not inside* the brain. *Inside* the brain is just more brain physiology. Empirically examine the brain all you want, with microscopes, EEG, etc., and you will still only have more of the insides of the brain, more inside surfaces, and not the interior or the within of the brain, which is depth or consciousness (and which you can find only by *talking* to me and *interpreting* what I say). And the brain itself is *exterior* to consciousness, it is another (potential) form in consciousness. Not inside but interior to the brain is consciousness, or depth, or the within, and the brain is exterior to that within, to that interior.

In other words, consciousness is not *inside* the brain, and not *outside* it either. It cuts at right angles to all that, and moves in the dimension of interiority, which is not found or measured in terms of physical form, and therefore moves wherever it likes without ever leaving the brain because it was never *in* the brain to begin with (and never apart from it either). How else explain, for example, the phenomenon of individual and group identity? If I (consciousness) am merely my brain, how is it that psychological development of identity moves from a bodyego to an egocentric to a sociocentric to a worldcentric stance? How could that happen if consciousness were simply in the brain? Forms of identity transcend the skin boundary in dozens of profoundly important ways, which an account that gives primacy to exteriors cannot even begin to adequately frame. Out-of-the-body experiences are some of the least surprising (and least interesting) things that consciousness can do, and can do precisely because it is not inside or outside the brain (the insides and outsides of the brain are *in* consciousness).

In note 4 above I mentioned that "the interiority of one stage is taken into the interiority of the next, and thus becomes an *external* form *within* that interiority, a three-dimensional-chess detail that we will get to later." The three-dimensional part involves differentiating between several different axes of both translation and transformation, so that both the insides and outsides of a lower level are, to a senior level, enveloped in its own interiority, and thus, conversely, the senior level appears "external" to its components and "internal" to its successors.

Language is obviously a real problem here, because we are actually discussing three axes, all of which involve being "in" and being "out." Call them inside/outside, internal/external, and interior/exterior. On a flatland ontology, you simply have inside and outside a boundary. But if several holons transformatively join together—say, many cells into an organism—then the inside *and* the outside of each of the cells are now all *internal* to the organism (all are inside the organism, except not on the *same level* of insides as the original cells, which is why I call this "internal," and why the organism, to the cells, appears *external*—Varela's point).

But all of that is *still* a series of *exterior* forms—not outside, and not external, but *exterior* to the interiority that *alongside* all that, at each point, is consciousness. And consciousness does not enfold previous interiors in terms of size (unlike the cell which actually and physically contains the molecules, or the organism which actually contains the cells); rather, consciousness enfolds previous interior intentions or prehensions, which exist *alongside* all those exteriors. As such, the various consciousness structures themselves also have insides and outsides and internals and externals—for example, something is "inside" a mental structure, not in terms of spatial enclosure, but in terms of whether it follows the deep structure or the regime or the codon of the mental structure (much as you are in a checkers game, not if you are physically inside anything, but if you are following the rules of the game). When two interior structures are brought together via transformation, then they are both internal to the new superstructure, and the superstructure appears external to them, and so on.

This is why I think it is such a terrible mistake to view consciousness as an *emergent quality*, instead of seeing it as *depth*, itself *unqualifiable*. If we think that consciousness is a *specific quality*, then we have only two major choices in how to situate it: (1) define the quality that consciousness is supposed to be (feeling, intention, perception, or whatever) and then push that quality all the way down to atoms. This is the standard panpsychic approach, which I find unacceptable (even though, for convenience, I sometimes resort to that terminology). But even less acceptable is: (2) define consciousness as an *emergent quality*, and then pick the point in evolution that we think it is supposed to have jumped out and emerged (and haggle endlessly over terms: Do apes have self-consciousness or just rudimentary consciousness? Do the shrimp we boil for dinner really have feelings?).

But consciousness is not an emergent quality, because the Left-Hand dimension does not *emerge from* the Right-Hand dimension. That is, the interior does not emerge from the exterior; obviously, they *both emerge together*, hand in hand, with the *first boundary* creating a universe, from the very first differentiation of the Kosmos into an inside and an outside (which is, at the first boundary, an interior and an exterior). The Left-Hand dimension does not emerge from the Right-Hand dimension, but rather *goes with it*, as the within, the interior of the exterior, *at every stage*. It is the interior depth of every stage, and not something that surprisingly pops out at some bizarre stage down the line. *Forms of consciousness* do indeed emerge (as forms of matter do), but consciousness itself is simply *alongside* all along, as the interior of whatever form is there (from the moment of creation). Because surfaces are surfaces *of* depth, forms are forms *of* consciousness, and the interior and exterior, as the Buddhists have it, are coemergent from the start, and not something leaping into being somewhere down the road.

Put differently, consciousness is the openness (or Emptiness) of the holon's regime—an openness *itself* that is not an emergent quality, but what allows qualities *to emerge* (*as* the various regimes and worldspaces).

And finally, the exteriors can enter consciousness: the interior embraces the exterior (and never vice versa). Leibniz's view was quite similar to this "3-D chess" (which is what makes him such demanding reading), except he opted out of the actual relation between interior and exterior with a preestablished harmony (but, I might add, that is no worse than most other alternative solutions). I'll return to all this when we suggest the contemplative (dis)solution of the mind-body problem.

26. I am not saying behaviorism is wrong; I am saying it is an accurate study of only the Upper-Right quadrant; it faithfully reports everything it sees there, and what it sees is real enough. It's just one-fourth of the story, more or less.

Likewise, medical psychiatry is the study of pharmacological intervention in the Upper-Right quadrant. What the psychoanalyst (UL) sees as rage at the abandoning father converted into depression and treatable with talk, medical psychiatry sees as a problem with serotonin uptake at the neural synapses and treatable with Prozac. Both are accurate enough in their quadrant (we will return to this topic throughout this volume).

27. Also known as the reflection paradigm or the representation paradigm, whose historical rise we will trace in chapters 12 and 13. Propositional truth is generally a Right-Hand path, as indicated, but often it is confined to the Upper Right, with functional fit the criteria for the Lower Right (as I will explain in the text), and there it vies with the production paradigm (which we will also trace in chapters 12 and 13).

Also, since the objective stance can be taken with regard to any of the quadrants, propositional truth has its important applications in the Left-Hand dimensions; the point is that it does not exhaust those domains nor capture what is essential and defining for those Left-Hand dimensions, as we will see.

28. And if we look at the research in each of the four quadrants, we find that each

of them has a "surface" and a "deep" dimension. Or, using "truth" in the broadest sense, each quadrant has a surface truth and a deep truth (although, as we will see, the accounts vary as to precisely what constitutes the truth in each).

We might start with the general Left-Hand and Right-Hand dimensions (and subdivide them into upper and lower, or individual and social, as we proceed):

Left-Hand hermeneutics and Right-Hand structural/functionalism *both* recognize a *surface* and a *deep* dimension (or manifest and latent), the former being mostly *conscious* and the latter being mostly *unconscious* (although the unconscious or hidden or latent dimension can be brought into awareness through special means, as we will see).

In hermeneutics this "surface" and "deep" interpretation was most carefully presented by Heidegger (in Divisions I and II of *Being and time*). A surface interpretation is an interpretation that cultural natives might not be fully conscious of in day-to-day living (or in their personal life), but if you point it out to them, they would usually recognize it and respond with "That's right, that's just what that means." This "hermeneutics of the everyday" is somehow known (either consciously or preconsciously) but not necessarily articulated; and the surface layer of hermeneutics deals with articulating these common interpretations (or making manifest the various aspects of the cultural *background* that support the meaning in the everyday practices; the background cannot be made fully manifest or intellectually and discursively articulated in an *exhaustive* fashion, however, because much of the background is nondiscursive and therefore cannot, à la Husserl, be conceptually exposed). This "hermeneutics of the everyday" has been fruitfully taken up by such theorists as Garfinkel, Geertz, Kuhn, and Taylor. It simply seeks to articulate and explain the common *meanings* of cultural practices via a sympathetic communion or participant-observer inquiry.

Deep hermeneutics, on the other hand, attempts to dig beneath surface interpretations (beneath the hermeneutics of the everyday), because deep hermeneutics suspects that our everyday understandings and interpretations might in fact be quite distorted, partial, or deluded; and might (as Heidegger suggested) actually be motivated to hide the truth, to obscure or cover up deeper (and more frightening) truths. "The hermeneutics of suspicion" (as Ricoeur called these various approaches) therefore seeks to dig beneath the surface and expose deeper interpretive truths and meanings.

And various deep truths have in fact been offered: underneath the surface and everyday consciousness lies Freudian libido, Marxist class struggle, Heidegger's groundlessness, Gadamer's tradition-carried Being, Nietzsche's power, and so on.

In each of these cases, our common everyday meanings and truths and interpretations (which are all more or less *conscious*) are nonetheless said to be often designed precisely to hide or obscure or distort the deeper and more painful truths lurking beneath the surface (truths which thus tend to remain *unconscious* or *deeply* hidden: they do not just fall into the unconscious, they are pushed there. What we might call a dynamic psychological pain barrier—a repression barrier of some sort or another— seals these disturbing truths from easy awareness, and, according to these theorists, lifting this barrier is difficult, painful, shocking, and deeply disturbing).

In each of these cases of deep hermeneutics, the cultural native must confirm the deeper truth, once it is dug up, by *acknowledging* it, and this journey is *not* likely to happen without the help of a professional guide, so to speak, who has made it his or her business to travel that deep path. The cultural native, in fact, has a great deal of *resistance* to the deeper truth (because of the pain barrier), and much of the everyday interpretations and practices were in fact generated precisely to hide this deep and often painful truth. However, deep hermeneutics maintains that once the person loosens the repression barrier, exposes this deeper truth, and acknowledges it, then a certain *liberation* is gained, a liberation from the distortions, lies, and delusions that were constructed to hide the truth.

Deep hermeneutics is obviously a tricky venture: many of the "deep truths" offered to humanity have actually been deep ideologies and profound prejudices designed primarily to contort others' awareness to conform to one's own power drives. We have seen what the "deep truths" of Marxism have done to the human spirit, and what Freud's deeply reductionistic "truths" managed to destroy. Both Marx and Freud, we may grant, were on to some sort of deeper and *very important truths*: but are these truths the whole truth, so to speak? And how does one know when a deeper truth is indeed *true* but *partial*? And when does shoving a partial truth down someone's throat—under the guise of helping them to "de-repress"—when does that step over the line and become itself a new form of repression? When does the *partialness* of the *cure* begin to *repress* the *rest* of reality?

Deep hermeneutics, in other words, faces two extremely delicate problems: First, is the proposed deeper truth really true? And second: is this deeper truth the only truth? Or is it itself a partial truth hiding yet deeper disclosures? How do we know when to stop?

Using Freud as an acknowledged example, the first problem is something like: does the libido exist? Is Freud's deep truth really true? In very general terms, do cultural and conscious ideas often hide, distort, or repress unconscious emotional-sexual drives? Most psychotherapeutic theorists would acknowledge that that is often the case; in various ways, people can and do become out of touch with their deeper feelings and drives and therefore express these drives in less than straightforward ways. In other words, Freud's deeper truth is more or less true: there is something like the libido, and it can be repressed/distorted in ordinary life, causing often painful emotional disturbances. And lifting the repression barrier (however it is conceived) can indeed help people get in touch with these deeper feelings and express them more adequately, thus liberating them from the constrictions of distorted emotions.

But the second problem is: granted there is some sort of unconscious, is the libido all we will ever find down there? Is Freud's deep truth—which, we have already acknowledged, is indeed a truth—is it the whole truth? Is my psychological anguish due solely or even primarily to repressing the libido? Might there not be other truths that, when denied or repressed or obscured, would cause even greater psychic anguish? What if the deeper truth I am distorting involves Jung's collective unconscious and not just my personal id? What if it involves Marx's social unconscious, and the society itself is rather sick and maladjusted? Lifting the repression in that case would involve not psychoanalysis but political revolution. What if the deeper truth is Heidegger's utter groundlessness, where all of these other truths are desperate attempts to fix and solidify that which is shifting and shapeless? Further yet, what if the deeper truth is Hegel's Spirit, and I cannot de-repress until I find God?

As I suggested in chapter 2, each of these theorists, at their best, is trying to point to a larger context that, when discovered and acknowledged, helps liberate us from a narrower and shallower imprisonment. We are all holons within holons within holons forever, and each of these theorists helps orient us to deeper and larger engagements, therein to find deeper and wider meanings and liberations. But each of those theorists tended to go from solving the first problem to failing the second. That is, after demonstrating that their newly discovered deep truth was in fact true (and I personally believe each of them sufficiently demonstrated that), they went on to claim that *other* deep truths were in fact *distortions* of *their* deep truth: that other "deep" truths were really just *surface* truths that their analysis should have ended forever.

Thus their partial truths, promoted to the whole truth, stepped over the line and into a new form of repression of the rest of the Kosmos. It is one thing to say that material-economic conditions are formative for many cultural and spiritual endeavors; quite another to say that God is nothing but an excuse for material oppression. It is one thing to say that repressed libido can power a fanatic belief in Spirit; quite another to say that Spirit is simply repressed libido.

But if you believe that Spirit per se is indeed repressed libido, then a client's belief in Spirit *must* be interpreted as part of the defense mechanisms and repression barriers that are hiding the deeper truth of sexual lust. Hegel's *deeper* truth is instantly interpreted as a *surface* truth that is hiding the *real* deep truth of libido.

And so the battle of partial truths would go, solving the first problem but failing miserably at the second. That many of these theorists favored their own deeper context, usually to the exclusion of others', condemns their partiality, but not the deep hermeneutic project itself, nor the *truth* of the partial truths.

But indeed, the misuses of deep hermeneutics are legion ("I know what you really need"), and the many successes of these theorists—and their many failures—stand as sharp lessons on this delicate journey. Surfaces can be *seen* but depths must be *interpreted*, and in that interpretation, you and I can be *mistaken*, and thus some sort of the hermeneutics of suspicion will always accompany us into the depths.

Turning now to the Right-Hand dimensions (moving from "What does it *mean?*" to "What does it *do?*"): in structural-functionalism (e.g., Parsons), the distinction between surface and deep usually shows up as "manifest" function (what the cultural natives *say* they are doing) and "latent" function (what the researcher/authority discovers the "real" function to be in terms of social systems and functional fit, a function unknown and unconscious to the natives). This, too, is a useful and legitimate (if partial) inquiry (a Right-Hand approach that depends upon an objective, distancing stance and a relatively detached observer).

And, since we are all holons within holons within holons forever, the structural-functionalist approach runs into the same types of "limit" problems or "boundary" problems as the hermeneutic or Left-Hand approach, namely, how do you know when to stop? Granted individual behavior is to some important extent determined by the wider social holon. But by the *same logic*, what about the social holon itself? It is set in numerous contexts within numerous other contexts . . . forever.

Thus, structural-functional accounts are always drifting into wider contexts: the individual is set in a family which sets certain broad parameters; the family is set in a subculture; the subculture is set in a larger culture; the culture is set in a community of cultures, which are set in larger geopolitical blocks, which . . .

Consequently (and just as in the hermeneutic approach), the *boundaries* that are set are often simply *pragmatic*. I will arbitrarily set the boundaries of my investigation at the border of the family, or the border of the town, or the village, or the nation . . . , and simply pick up the story at that point (just as the hermeneutic therapist might say, "Granted society is sick. But you, my client, have come to me for help, and since we can't fix society right now, let's just work with you").

Although I will not, in this summary endnote, go into details, we might simply note that structural-functionalism (in any of its forms) therefore faces the same two delicate problems in dealing with its own versions of surface and deep: first, is the deep functional truth (the latent truth of the particular functional fit) really true? And second, is it the whole truth? (Or is it the *most important* context for the particular problem addressed? I'll return to this in a moment.)

Here, then, is a brief summary of the types of "unconscious truths" found in each of the four quadrants and dealt with by these various approaches:

In the Left-Hand dimensions ("What does it mean?"), we have surface, everyday meanings and interpretations, which can be understood and articulated by the "hermeneutics of the everyday." The meanings of my everyday reality depend upon a whole series of background contexts: both unarticulated foreknowledge and nondiscursive cultural practices, some of which are conscious and some of which are not directly conscious (but could easily become conscious if pointed out to me, although some of the overall background simply shades away into inaccessibility). This is true both individually (UL) and collectively (LL).

But beneath the everyday interpretations there are various types of deeper truths, deeper realities, deeper meanings, that the everyday interpretations often attempt to cover up, obscure, or distort. These deeper meanings are dynamically *resisted*, because their acknowledgment would involve some sort of deep pain or shocking recognition. Both cultural practices at large (LL) and my own personal consciousness (UL) can contain repressions or obscurations of these deeper meanings (e.g., there is mass hysteria and collective collusion just as often as individual repression, and indeed, the two are often conjoined: cultural repression is internalized as part of the individual superego, which is often the bridge for various Freudian/Marxist integrative theories).

These deeper but painful meanings can be brought to light by the "hermeneutics of suspicion," in which an authority who has traveled the deep path can help the individual (and possibly the society) relax the particular repression barrier and acknowledge the deeper truth hidden and obscured by the hermeneutics of the everyday. Contacting the deeper meaning is a liberation from the false and distorted surface meaning, and this liberation is experienced as type of freedom from a prior suffering or constraint.

In the Right-Hand dimensions ("What does it do?," or general structural-functionalism), we have surface or *manifest* meaning, which is simply whatever it is that the cultural natives say they are doing. But the social scientist finds that the everyday practices and meanings are actually serving a *deeper function* (the *latent* function), a function unknown to the cultural natives, but a function that the scientist can discover.

But notice that, according to the structural-functionalist, the cultural natives are *not repressing this deeper truth* (the truth of functional fit); they are simply *unaware* of it. This "functional unconscious" is simply not yet known; it is not a prior knowledge that was once in awareness and then repressed and resisted and rendered unconscious. Any cultural native can pick up this knowledge by a simple study of social systems theory.

All of which is fine. But the reason that there are no theories of repression in the exclusive Right-Hand paths is that there is no consciousness in the Right-Hand dimensions (no irreducible interiors), and thus there is nothing to institute the repression in the first place (there is no interior repression because there are no interiors, period). In the Right-Hand path, there is simply the knowledge of how objects work or *lack* of that knowledge (and that is what "unconscious" means here: you simply lack the knowledge but you could easily gain it), but there are no real *subjects* that could *repress* anything. The Right-Hand path is just an objective description of how objective exteriors function, and thus no "black box" problems arise. The only "problems" are caused, not by a battle in the interior depths, but by a simple lack of the objective knowledge that the natural social scientist will be glad to give you if you simply ask.

On this Right-Hand path, we have the division into Upper and Lower (individual and collective). I have just described the Lower Right, the social systems theory with its *manifest* meaning (anything interior: consciousness, values, repressions, etc.) and its latent function (the foundational or real reality of functional fit).

In the Upper Right, we have the exteriorly observed functions of the individual: brain physiology, cognitive science, neuroscience, etc. Here, too, there is no real consciousness: only the computational mind, or mindless neurotransmitters, or fireworks at the synapses. All of this is totally *unconscious* to typical cultural natives, in two senses: we can never really directly experience neurotransmitters per se (nor the computational processes), but we *can* become aware of them theoretically if we just study neuroscience (we can remove the unconsciousness in that sense: we can gain the objective theoretical awareness of the nonconscious processes, just as we can gain scientific knowledge of subcellular biochemical processes, and so on).

And thus here, too, in the Upper Right, there are no theories of repression, or resistance to the truth, because there is no consciousness to resist anything. There are only completely nonconscious processes scurrying about, intent on carrying out their primary task, which is to create the illusion of consciousness being aware of any of this.

Once you have swallowed that foundational nonsense (consciousness is the illusion that pronounces itself an illusion, which is an awful lot of friskiness for a ghost)— once you have swallowed that (once you have ignored the entire Left Hand of the Kosmos), then once again, as with any Right-Hand path, there are no more problems to solve, except to continue to discover more objective facts, more mindless objects, more exterior functions.

The only "problems" here are thus problems of technical know-how. If life is meaningless, we have a serotonin functional deficiency, and so forth (numerous examples of this are given in the text).

So the "unconscious" in the Upper Right means, to begin with, any of the physiological/computational processes of the physical brain, processes that we are normally not aware of (and that we can never directly experience per se), but processes and functions we can know about through a study, not of our introspected mind, but of objective brain theory. And no doubt, those processes are indeed there, and indeed can be disclosed in empiric-analytic research and formal theorizing.

But "unconscious" also takes on the same general and sweeping sense as in any Right-Hand path, namely, all of our *interior* problems are caused by a lack of knowledge of *exteriors*. We are "unconscious" in the sense that we don't yet know enough technical and objective facts to settle the issue and solve all the problems. And as soon as we know all the objective facts, that will solve all our problems. While that is certainly true for the exterior functions of holons, it is suicidal for the interior dimensions. (Not to mention self-deconstructing: all of our problems come down to the fact that we are not conscious of the fact that there is no consciousness, which is all the Right Hand *can* tell us.)

Finally, there are the approaches that look at the Left-Hand or interior dimensions using some of the tools and the objective-like stance of the Right-Hand or exterior paths. And they do so because they are based on a careful analysis of individual and cultural interiors as they become manifest in exterior behavior or linguistic form: e.g., they will look at the structure of *material signifiers*, or *written forms*, or *spoken behavior*, not with a view to reducing the interior to these exteriors (although this is sometimes done, unfortunately), but in an attempt to reconstruct the interior pattern or structure that gives rise to, or supports, the exterior behavior. (Thus, for example, when I use *formop*, which is an exterior description of the form of a particular mental structure, I mean not only that structure described in an exterior fashion, but also, and especially, the *interior lived experience* and *actual awareness* that occurs *within* that structure, which is why it is listed in the Upper-Left quadrant.)

For the general structural approach (including its much modified forms in Piaget, Kohlberg, Foucault, Habermas, and aspects of my own work), there are *surface struc-*

tures known to the cultural natives, and *deep structures* discovered by systematic analysis among various sets of surface structures. (I am using these terms—*surface* and *deep*—in a general sense, and not, for example, in the very narrow and specific sense given them by Chomsky, a use he now finds too narrow.)

The deep structures themselves, like any external-descriptive structure, are *not repressed*, but are simply *not known* to most individuals; they could easily be recognized if pointed out. The various surface structures, however, can indeed be dynamically repressed or otherwise submerged: for example, one does not repress the overall capacity for images, but one can repress particular images—and that is the repressed or submerged unconscious (hermeneutically recovered). But the *structural unconscious* refers instead to the deep pattern displayed by all the various surface structures within that pattern, and it is unconscious not because it is repressed but simply not yet known.

Thus, a surface structure is what the cultural natives consciously recognize, and a deep structure is a (usually universal, but at any rate nonindividual) pattern that is common to different surface structures (but not consciously known to the individual surface structures themselves, although this can be pointed out).

Researchers readily acknowledge that each of these deep structures is itself set in contexts of other deep structures, against which they themselves might be surface. Thus, for example, Jane Loevinger will state that cognitive structures are the deep structures against which interpersonal structures are surface; interpersonal structures in turn are the deep structures against which moral structures are surface, and so on. I tend to avoid this particular usage because it can imply that moral structures are only local; but the general point is simply that to say a structure is deep is not to deny its own contexts.

This *structural unconscious* (the deep pattern followed by the individual surface structures) is indeed normally unconscious because it does not typically or directly enter into the immediate awareness of the cultural natives, even though they are in fact following it (just as most natives can speak the particular language without being able to articulate the actual rules of grammar governing the usage). This is why various structural approaches move beyond a simple phenomenology of immediate experience and also attempt to analyze, in a Right-Hand fashion, the patterns or forms or rules that the various immediate experiences display (rules that themselves *do not* appear to ordinary experience).

This, for example, was Piaget's charge against Husserl and traditional phenomenology—various phenomena follow patterns that do not themselves show up as phenomena (much as the rules of a card game do not appear on any of the cards). The *subjective space* in which experienced phenomena arise is itself constructed and follows patterns that do not ordinarily show up in naive (or even bracketed) experience; and further, the subjective space itself develops only via *intersubjective* patterns of dialogical and interpretive cognitions, *none of which show up in a simple phenomenology of immediate lived experience.* (See note 17 for chapter 14.)

The existence of these ordinarily hidden patterns (in various forms) is one of the enduring contributions of structural linguistics (de Saussure), structuralism (Lévi-Strauss), neostructuralism or archaeology (Foucault), genetic structuralism (Piaget), and hermeneutic and developmental structuralism (Habermas). And this structural unconscious can become conscious in the sense that any native can study and learn the patterns "behind" or "beneath" conscious experience.

To summarize this overall discussion in a paragraph: the various types of "unconscious" refer to the various validity claims in each of the four quadrants, and to the presence or lack of *valid awareness* that can infect each of the quadrants in its own quite different ways: lack of awareness of deeper propositional truths (UR), deeper functional fits (LR), deeper cultural meanings (LL), deeper truthfulness or sincerity (UR). All of which ultimately rest on the fact that we are contexts within contexts, holons within holons, forever. . . .

And, of course, I believe all of those approaches have something important to tell us; they become ridiculous only when they try to corner the market on the hidden dimension.

29. And I will, of course, have a great deal of empirical "evidence" to "support" this contention that consciousness simply *is* fundamental brain physiology. If you challenge me, I will simply say, "Well then, where is *your* empirical evidence?," thus covertly slipping in an empirical or monological test for that which is, by definition, nonempirical or dialogical, a sneaky move that converts "nonempirical" to "nonexistent" (and never mind that the position itself is perfectly self-contradictory, as we saw in chapter 1—it is a qualitative distinction that denies the existence of qualitative distinctions. How can a brain that doesn't involve values produce a researcher that does? To say that values *emerge* from brain states is to say they cannot be reduced to, or accounted for, or explained by them—which is then promptly denied by a demand for empirical evidence).

30. Technically, the interior and exterior holarchies of depth and span were collapsed into only the exterior holarchies of depth and span.

In other words, holarchies were still recognized: the Upper-Right and Lower-Right domains are both holarchical, and this was not denied; and these Right-Hand holarchies, like all holarchies, technically possessed both a depth and a span—but *no interiors* (no interior depth or span). Rather, the only holarchies acknowledged were holarchies of empirical/positivistic *exteriors*, of surfaces that could be seen empirically with the senses or their instrumental extensions, and thus any values (any Left-Hand correlates) were *automatically excluded*. Since the interior dimensions were denied, there were no holarchies of qualitative distinctions, of value, of consciousness, of goods or hypergoods (Left Hand or interior); no transformations of consciousness required to get at the truth, and no truth other than what could be analytically formulated around empirical, value-free data.

Thus, even though the Right-Hand path acknowledges holarchies with depth and span, the *depths* of these Right-Hand holarchies all have the *same value*, namely, *nothing* (that is, they are "value-free" surfaces). And thus, from the view of the interior and Left-Hand path, all of the "empirical depth" of exterior holarchies are no more than variations on the same span (namely, empirical or functional).

Thus, when interior depths (of "I" and "we") were reduced to their correlates in exterior depths of "its" (e.g., values reduced to limbic system signals), then depth itself (of any sort) was severed from its intrinsic connection to consciousness. Greater depth no longer meant greater consciousness, only greater physical complexity. No longer was there a Kosmos where conceptual values transcended and included emotional values which transcended and included drives, but simply a cosmos where the neocortex physically enwrapped a limbic system which physically enwrapped a brain stem. Depth no longer means deepening awareness, it means more complex matter. Holistic matter, to be sure, but "its" none the same. And matter does become more complex—but only as a *correlate* of increasing depth and consciousness, all of which was conveniently ignored.

When addition #1 says that depth is synonymous with consciousness, it doesn't specify whether in an individual that is Left-Hand depth or Right-Hand depth, because the point is that they are isomorphic: more of one is more of the other in homologous ways. But with the collapse of the Kosmos into the Right-Hand cosmos, exterior depth no longer had a correlate in interior depth (or consciousness), and thus the *intrinsic* connection between increasing consciousness, increasing depth, and increasing physical complexification became severed at a fundamental level. The flatland world of holistic "its" had no room for holistic "I's" and "we's," and the worldview of scientific materialism began its long and imperial career.

(Even though all four quadrants are holarchies with depth and span, I often use "depth" to mean "interior depth" or consciousness precisely because of the intrinsic correlation between consciousness and depth in any natural holarchy: more physical depth in the Right-Hand means more consciousness depth in the Left-Hand. With the collapse of the Kosmos, "depth" often came to mean something merely "inside" the person, associated derogatorily only with interiors and subjectivity, and all three were denied irreducible reality.)

All of these distinctions are examined in detail in chapter 12; this is meant as a short introduction to this general topic of "the fundamental Enlightenment paradigm." Its more precise dimensions are outlined in chapters 12 and 13.

31. Charles Taylor: "Final causes and the related vision of the universe as a meaningful order of qualitatively differentiated levels [holarchical Kosmos] gave way first to a . . . vision of mathematical order, and then finally to the 'modern' view of a world of [quantitative] relations to be mapped by empirical observation" (*Hegel*, p. 4).

As for the reduction of the Left Hand to the Right Hand, which was the fundamental Enlightenment paradigm, generating *both* a self-defining disengaged subject *and* a holistic natural worldview (or systems theory), Habermas sees this in terms of the "philosophy of the subject" that easily slid into a systems approach, because the "self-organizing system," with a few adjustments, simply replaced the "self-defining subject" (the two poles of the Enlightenment paradigm). Systems theory is, he says, "an ingenious continuation of a tradition that has left a strong imprint upon the self-understanding of early modernity in Europe. The cognitive-instrumental one-sided-ness of cultural and societal rationalization was expressed in philosophical attempts to establish an objectivistic [it-language] self-understanding of human beings and their world" (*Philosophical discourse*, p. 384).

This flatland approach was initially, he points out, painted in *physicalistic* and mechanistic terms, but these were quickly replaced with *biologizing* systems theory (i.e., metabiology replaced metaphysics). "To the extent that systems theory pene-trates into the lifeworld, introducing into it a metabiological perspective from which it then learns to understand itself as a system in an environment-with-other-systems-in-an-environment, to that extent there is an objectifying effect. In this way, subject-centered reason is replaced by systems rationality . . ." (p. 385).

Taylor's version is similar, in the sense that the multileveled Kosmos of qualitative distinctions was collapsed, by the "disengaged subject," into "interlocking orders of being" *quantitatively* and *empirically* representable, a grid through which value and constitutive goods necessarily slip. And he then proceeds, as saw in chapter 1, to show that the very belief in this systems view is self-contradictory.

32. In short, the fundamental Enlightenment paradigm, which was the reduction of the Kosmos to the Right-Hand path, had two poles: the disengaged subject and the flatland network-world. The flatland world itself had two warring camps: atomism (or gross reductionism) and holism (subtle reductionism). For all of its positive contri-butions, which I will heavily emphasize later, Enlightenment thought was nonetheless a series of bizarre oscillations between these two poles and two camps (we see a version of these oscillations in Foucault's aporias). This is the main topic of chapters 12 and 13.

As for the reduction of subjective and intersubjective to exterior functional fit, Ha-bermas is a particularly vocal critic. It is a massive reductionism, he says, that col-lapses truth and meaning into functional capacity, and reduces intersubjectivity to rather crass egocentrism. "Systems theory lets cognitive acts, even its own, meld into the system's achievement of mastering complexity. Reason as specified in relation to being, thought, or proposition is replaced by the self-enhancing self-maintenance of the system. . . . Functionalist reason expresses itself in the ironic self-denial of a reason shrunk down to the reduction of complexity—'shrunk down' because the metabiolog-ical frame of reference does not go beyond the logocentric limitations of metaphysics. . . . With a concept of meaning conceived in functionalist terms, the internal connec-tion between meaning and validity dissolves. The same thing happens with Foucault [in his archaeology, which, as we saw, was basically a structural-functionalism]: The interest in truth (and validity in general) is restricted to the effects of holding-some-thing-as-true" (*Philosophical discourse*, pp. 371–73). This is what I meant when I said that structural-functionalism collapses propositional truth and intersubjective un-derstanding to interobjective fit or functional fit, and ironically dissolves its own truth status into monological mesh.

The conversion of interior depth to functional fit also collapses any chance of genuine intersubjective understanding and communion, because without degrees of depth, you cannot unite isolated elements. "Thus," says Habermas, "no common denominator can be built up among different psychic systems, unless it be an autocatalytically emergent social system, which is immediately locked again within its own systemic perspectives and draws back into its own egocentric observational standpoints" (p. 381). Again and again Habermas hits the *egocentrism* of the systems perspective, which also makes sense when we understand that the systems view was the other pole of the self-defining and self-glorifying underside of the fundamental Enlightenment paradigm, a remnant of the philosophy of the subject.

The egocentrism of the systems approach can be seen in some of the "ecological/holistic" theories that today call themselves "new paradigm." These theories maintain that (1) each of us is an integral part of the Whole System, strands in the overall Web of Life; and (2) we are to find our liberation by an identification with the Whole System that would (3) simultaneously be a Self-realization.

But according to systems theory, no part is ever, or can ever be, aware of the whole. The only thing that is aware of the whole is the *actual regime* of a holon, its "regnant nexus" (Whitehead) or "dominant monad" or nucleus of awareness. The only way for all three of the above-listed tenets to be true is if an individual human being were in fact the dominant monad of the entire Kosmos—and out comes the egocentrism (and anthropocentrism) of the systems theory view.

I believe there is indeed such a thing as Kosmic consciousness, or mystical awareness (as we will see in chapter 8); my point is that it cannot be gracefully explained by systems theory, which covers only the LR quadrant.

33. This is Charlene Spretnak's summary of Berman's position. See Spretnak, *States of grace*, p. 298.

34. Habermas points out that the three basic modes of rationality (cognitive, moral, aesthetic) result from the intersection of the three basic attitudes (objectivating, normative, expressive) applied to the three domains (objective, social, subjective). In very simplified terms, this means that each of the Big Three domains can be studied with the "attitudes" of the others (e.g., the subjective domain can be studied objectively, although not captured objectively without remainder).

I agree with this general position. And, indeed, this volume is vision-logic used in its objectifying mode, presenting a propositional map of the various quadrants and domains. This is not merely propositional or positivistic, however, because parts of the map explain the parts that cannot be merely represented on the map itself (e.g., it is propositionally true that intersubjective understanding is not merely propositional).

35. Taylor, *Sources of the self*, p. 105. Taylor refers to the cultural worldspace as the "common space" of the moral good we collectively inhabit, a "conception of the human good as something realized between people rather than simply within them."

36. Habermas, *Philosophical discourse*, pp. 313–14.

37. Ibid., p. 313.

38. And its converse, the production paradigm (e.g., Marx). In the reflection paradigm, disengaged thought reflects the cosmos in an attempt to construct, out of its own disengaged devices, a correct "map." In the production paradigm, the material cosmos constructs thought as a superstructure to its own production.

39. Cf. Rorty's *Philosophy and the mirror of nature*. However, I agree with Habermas's sharp criticism of Rorty's overall stance: the fact that cultural values and "truths" are not reducible to reflections of "unchanging" nature (truth is not reducible to a merely representational correspondence with an unvarying and pregiven objectivistic world) does not mean that cultural productions are shifting conversations with no validity claims of their own. There are better ways to steer a path between mindless objectivism and objectless subjectivism.

40. This reductionism is particularly (and I think altogether apropriately) excoriated by Taylor in *Sources of the self*.

41. Typically, the "purest" reflection paradigm is positivism, and the "purest" form of functional fit is structural-functionalism. For pure positivism, the only things that are *real* are physical objects and numbers (Quine), and a pure propositional truth is thus a mathematical assertion referring finally to individual physical objects (all propositions refer ultimately to UR alone; numbers are allowed to be real in order to have something *with which* to assert, or point to, or signify physical objects). This is pure Upper-Right reductionism, as are the more typical (but less rigorous) forms of empiricism.

In functional fit, all reality is finally reduced to LR terms (the social system), and so all other validity claims (from propositional truth to cultural meaning to personal integrity) are judged ultimately in terms of their capacity to serve the autopoietic functioning of the social system. In other words, all other truth claims are, at best, given a "manifest value," while the "latent value," the *real* value, is functional fit. This is pure Lower-Right reductionism. All qualitative distinctions are reduced to terms of expediency and efficiency; nothing is "true," because all that enters the equation is usefulness.

In order to escape the performative contradiction that results when functional fit is itself claimed to be true, many functional-fit theorists have taken the only route left open to them: they claim their theories are themselves the way the system is preserving itself: the theorists themselves are merely manifest functions of the real latent drive of the system to autopoietically self-maintain: the theory of functional fit thus functionally fits (and here they rest their argument).

This finally leads to the implicitly reactionary stance of these theorists: no holon is better or worse, only efficient or not. In these theories, there is no reason to get rid of Nazis: they are very efficient, very functionally fitting, very expedient. If Stalin's regime is *working adequately*, then it is valid, and who are we to question the murder of twenty million Ukrainians? It works!

But this is just the endgame of describing exteriors, where no values and no qualitative distinctions can ever be found.

42. That medical psychiatry, attempting to be "scientific" (Right Hand), often falls into "adaptation" as *the* criterion of mental health can be seen all the way back to Heinz Hartmann, generally regarded as the grandfather of both object relations theory and ego/self psychology, as was indicated in the title of his masterpiece, *Ego psychology and the problem of adaptation.* "We call a man well-adapted if his productivity, his ability to enjoy life, and his mental equilibrium are undisturbed . . . and we ascribe failure to lack of adaptation." In other words, behavioral and functional fit (the Right-Hand path), with no inquiry as to whether the social situation is worth fitting into or not. If housewives' lives are dreary and meaningless, the prescription is thus not more cultural rights, but more Valium.

Harry Guntrip, in his much-admired *Psychoanalytic theory, therapy, and the self,* fought a valiant (but losing) battle against just such reductionism in psychiatry. Adaptation, says Guntrip, is altogether inadequate as a measure of "mental health," largely because it is, he points out, based solely on biological factors, and not on specifically human values (I would amend that more broadly to say, the one is Right Hand, the other is Left Hand—naturalistic versus hermeneutic—but the general point is quite similar). A frog doesn't question whether the environment that it must adapt to is a worthy environment: it adapts or dies. And thus, when this biological adaptation is made into the criterion of human mental health, then likewise the worthiness of the culture is not profoundtaly questioned: you are happy and healthy if you fit in, and sick and disturbed if you do not. Adaptation covertly buys the values it is supposed to adapt to, and thus reduces the Left-Hand to Right-Hand functional fit.

But, protests Guntrip, "The ego must be more than just an organ of adaptation. This would aim not [just] at physical survival but at preserving the *integrity* of the person and the defense of his *values*"—both Left-Hand qualities. Then, in a beautiful phrase that captures the essence of the difference between exterior and interior approaches, Guntrip says that "in studying human living, 'adaptation' is replaced by a higher concept, that of a *meaningful relationship* in terms of *values*" (his italics).

Thus, according to Guntrip, adaptation has an important but limited use (e.g., Right-Hand), but more significant are both personal integrity/values (Upper Left) and mutual understanding in meaningful relationships (Lower Left). "Adaptation, strictly speaking, can only express one-sided *fitting in* [functional fit]. But personal relations involve mutual self-fulfillment in communication and shared experience." And the disaster, he says, is that when psychoanalysis (or psychiatry in general) is reduced to mere adaptation, its true essence is lost: "Psychoanalysis is then treated as being about the human organism adapting to the structure of the external world [UR adapting to LR], instead of being about the human psyche realizing its inherent potential for unique individuality as a person relating to other persons [UL relating to LL]."

Guntrip even spots the fact that the difference between the Left- and Right-Hand paths involves values versus action terms, or being versus doing: "Psychoanalysis has to understand the person, the unique individual as he or she lives and grows in complex meaningful relationships with other persons who are at the same time growing in their relationships with them. This involves that being is more fundamental than doing, quality more fundamental than activity, that the reality of what a man does is determined by what he is, as when a middle-aged woman on a British television program said, 'I plunged early into marriage and motherhood, trying to substitute doing for being.' In this lies the difference between *adapting* and *relating* . . ." (!).

He finishes with a wonderful reminder: "Adaptation is one-sided [monological] and is certainly a matter of action. But personal relations are essentially two-sided [dialogical understanding versus monological action]; they are mutual by reason of being personal, and not a matter of mutual adaptation merely, but of mutual appreciation, communication, sharing, and of each being for the other" (all quotes from pp. 105–12).

Let me repeat that I myself am not arguing for the primacy of any quadrant over the others; I am arguing against the reduction of any quadrant to the others. Medical psychiatry and brain physiology, and behaviorism and adaptation, are crucially important; it is just that certain violent forms of aggression result when one quadrant is reduced to another.

43. That this differentiation went too far into dissociation is a topic we will discuss in chapters 12 and 13. That is, the story of the fundamental Enlightenment paradigm (the reduction of the Kosmos to the Right-Hand path: the entire Left-Hand dimensions were simply taken for granted, and consciousness was assumed to be a punctual, disengaged subject staring out unproblematically and transparently at a pregiven world, and the job of this disengaged subject was to reflect, represent, describe, categorize, and analyze the objective, pregiven world) could occur because the noosphere had finally differentiated from the biosphere, a differentiation that went too far into dissociation (Hegel's "vanity of the understanding," Taylor's "monster of arrested development"), so that the dissociated noosphere was merely the disengaged and self-defining subject staring at, hovering over, divorced from, the holistic flatland of the natural world. The fundamental Hegelian project was the healing of this "diremption" or divorce.

And, I will argue, the healing of that diremption is still the good, and the true, and the beautiful aspect of the post-Enlightenment ("postmodern") endeavor.

In the meantime, the dominant epistemological paradigms (especially in Anglo-Saxon countries) remain firmly committed to some form of exclusive embeddedness in the Right-Hand dimensions of the Kosmos, still following the fundamental Enlightenment paradigm—either in the form of a simple representation or reflection paradigm, or, at a slightly more sophisticated level, some form of connectionism, wholism, emergentism, dynamical systems theory—both thoroughly grounded in Right-Hand reductionism.

We can take, as a prime example, the rising field of *cognitive science*. In the history of this increasingly influential field, we can see the move from a simple representational paradigm to an emergence/connectionist paradigm (both nonetheless thoroughly Right-Handed), with a recent struggle to begin integrating the Right and Left dimensions in the fledgling enactive paradigm.

Indeed, just that historical progression (representationism to connectionism to enactivism) is traced in *The embodied mind: Cognitive science and human experience*, by Varela, Thompson, and Rosch. "We begin," they say, "with the center or core of cognitive science, known generally as *cognitivism*. The central tool and guiding metaphor of cognitivism is the digital computer. A computer is a physical device built in such a way that a particular set of its physical changes can be interpreted as computations. A computation is an operation performed or carried out on symbols, that is, on elements that *represent* what they stand for. . . . Simplifying for the moment, we can say that cognitivism consists in the hypothesis that cognition is *mental representation*: the representation of a pregiven world by a pregiven mind" (pp. 8–9).

Thus, in cognitivism "we then have a full-fledged theory that says (1) the world is pregiven; (2) our cognition is of this world . . . , and (3) the way in which we cognize this pregiven world is to represent its features and then act on the basis of these representations" (p. 135).

That, of course, *is* the fundamental Enlightenment paradigm. Varela et al. thus point out that, contrary to Rorty's conclusion, cognitivism does indeed continue the "mirror of nature" notion. "Contrary [to Rorty], a crucial feature of the [representational] image remains alive in contemporary cognitive science—the idea of a world or environment with extrinsic, pregiven features that are recovered through a process of representation. In some ways *cognitivism is the strongest statement yet of the representational view of the mind* inaugurated by Descartes and Locke [the history of which we will examine in detail in chapters 12 and 13; my italics]. Jerry Fodor, one of cognitivism's leading and most eloquent exponents, goes so far as to say that the *only* respect in which cognitivism is a major advance over eighteenth- and nineteenth-century representationism is in its use of the computer as a model of mind" (p. 138).

And, indeed, the same representationism is the primary paradigm in neurophysiology, as I previously indicated, and as Varela et al. point out: "The basic idea that the brain is an information-processing device that responds selectively to features of the environment remains as the dominant core of modern neuroscience and in the public's understanding" (p. 44).

Although the representational/computational paradigm can handle many facets of cognition (much of our knowledge, after all, is indeed representational after the fact, even if that is not the whole story—or even the most basic story), nonetheless, recalcitrant problems with its overall adequacy has lead to two major alternative accounts of cognition. "The first alternative, which we call *emergence*, is typically referred to as connectionism. This name is derived from the idea that many cognitive tasks (such as vision and memory) seem to be handled best by systems made up of many simple components, which, when connected by the appropriate rules, give rise to global behavior corresponding to the desired task. Connectionist models generally trade localized, symbolic processing for distributed operations (ones that extend over an entire network of components) and so result in the emergence of global properties resilient to local malfunction" (p. 8).

Related to this approach is the more sophisticated version of "the society of mind" (Minsky)—mind as a society of agents and subagents. "The agent motifs are fairly complex processes. Each of these may be thought of as composed of subagents, or more accurately as themselves agencies composed of agents. The task, then, is to organize the agents who operate in these specific domains into effective larger systems or 'agencies,' and these agencies in turn into higher-level systems. In doing so, mind emerges as a kind of society" (a holonic view, although "self" and "consciousness" remain problematic, as we will see). "The model of the mind as a society of numerous agents is intended to encompass a multiplicity of approaches to the study of cognition, ranging from distributed, self-organizing networks to the classical, cognitivist conception of localized, serial symbolic processing" (pp. 116, 106).

But the crucial point is that these approaches (cognitivist, connectionist, society of agents) all remain firmly rooted in Right-Hand reductionism (as Varela et al. point out in their own terms). The connectionist and society models, in fact, are prime examples of subtle reductionism: in championing emergence, networks, global properties, holarchies, and so forth, they loudly battle the gross reductionism of simple atomistic representationism—only to ensconce their own subtle reductionism in a way that makes it all the harder to spot (and thus doubly reinforces the primary problem).

But Varela and company are clearly on to them: the connectionist model simply replaces symbol with global state, which then itself represents the pregiven world: "For connectionists a representation consists in the correspondence between such an emergent global state and properties of the world" (p. 8). Thus, nothing fundamental has changed in this "blindness deeply entrenched in our Western tradition and recently reinforced by cognitivism. Thus even when the very ideas of representation and information processing change considerably, as they do in the study of connectionist networks, self-organization, and emergent properties, some form of the realist assumption remains. In cognitivism, the realism is at least explicit and defended; in the emergence approach, however, it often becomes simply tacit and unquestioned [always the great problem with subtle reductionism]. This unreflective stance is one of the greatest dangers facing the field of cognitive science; it limits the range of theories and ideas and so prevents a broader vision and future for the field" (p. 133). Not to mention the society at large.

Nowhere can the unpleasant effects of this subtle reductionism be better seen than in the cognitive science approach to consciousness itself (or simply lived experience in general). From the field that claims to be the final authority on such matters, what we learn about consciousness and lived experience is this: basically, it doesn't exist. Or, as Jackendoff words it, "consciousness is not good for anything" (Varela et al., p. 108).

Jackendoff's particular cognitive model postulates (as most forms of cognitive science do, although in rather different ways) a "computational mind" embedded in some largely unspecified fashion in the brain's physical structure, and this computational mind, which operates completely *nonconsciously*, performs all the required functions of cognition. Consciousness or awareness or personal lived experience is thus viewed as a puzzling (ultimately unnecessary) epiphenomenon, projection, or causal result of the nonconscious computational mind. "The actual [cognitive] subsystems," as Dennett puts it, "are deemed to be unproblematic *non*conscious bits of organic machinery, as utterly lacking in point of view or inner life as a kidney or kneecap" (Varela et al., p. 50).

The result is that, although consciousness is viewed as the effect/projection of nonconscious processes, nonetheless cognitivism, says Jackendoff, "offers no explication of what a conscious experience is. . . . leaving totally opaque the means by which these effects come about" (Varela et al., pp. 52, 231).

But then, what else is new? There is no consciousness, awareness, or lived experience anywhere in the Right-Hand, objectivistic dimensions—which register only the exterior forms of interior consciousness—and thus if you look at the Kosmos only in an objectivating, exterior, monological/computational fashion, you will *never* find a subjective lived experience anyway. Far from being "an empirical discovery of no small importance," it is a discovery of no big deal: it is absolutely an a priori *fait accompli* of the representational/computational paradigm.

I am not saying there isn't a computational mind (just as I am not denying the existence of neurotransmitters); but the fact that consciousness can't be found in either of them (or in kidneys or kneecaps) is a forgone conclusion, not a dramatic new discovery. Such approaches purport to be "consciousness explained," whereas they are "consciousness explained away": they take a sphere, cut it into circles, and say, "See! No sphere."

"Consciousness is not good for anything." Such a strange stance. Jackendoff's *theory* of the computational mind leads him to the mental conclusion, which he *directly experiences* in his own consciousness, that his consciousness is a puzzling and ultimately unnecessary byproduct of his nonconscious computational mind. The theory may or may not be true. But either way, does the theory actually change anything in Jackendoff's real life? Does he love his friends and mates any the less? Treat his kids any differently? Wish or will or hope any less? And what if the theory is eventually shown to be completely wrong? Will anything about his own consciousness be experienced any differently at all? Does the theory change or alter anything in his lived, actual experience? Why, based on Jackendoff's real life, and if we are *forced to choose*, then we would have to come to the real conclusion: "The computational theory is not good for anything." (Since, about consciousness, it changes nothing, and it is only in consciousness that changes count, i.e., have value.)

To his credit, Jackendoff is quite sensitive to these issues, and his approach is one attempt to wrestle with the "interaction" of the phenomenological mind (lived experienced) and the computational mind (nonconscious mental mechanism) and the physical body-brain structure—but still locked within the representational paradigm, which ensures the failure of that particular synthesis from the outset. And it is exactly these types of difficulties that led Varela et al. to propose the *enactive* paradigm of cognition.

As I would word it, the enactive paradigm is a direct and explicitly stated attempt to integrate Left- and Right-Hand approaches to cognition, uniting lived experience and theoretical formulations. "We wish to emphasize the deep tension in our present world between science and experience. In our present world, science is so dominant that we give it the authority to explain even when it denies what is most immediate and direct—our everyday, immediate experience" (p. 12).

And yet, "when it is cognition or mind that is being examined, the dismissal of experience becomes untenable, even paradoxical"—becomes, actually, the performative contradiction inherent in all Right-Hand reductionisms, a sawing-off of the branch on which one is happily perched.

How to heal this rift between experience and science (Left and Right) is thus paramount. "The tension comes to the surface especially in cognitive science because cognitive science stands at the crossroads where the natural sciences and the human sciences meet. Cognitive science is therefore Janus-faced, for it looks down both roads at once: One of its faces is turned toward nature and sees cognitive processes as behavior [standard Right-Hand approach]. The other is turned toward the human world, or what phenomenologists call the lifeworld, and sees cognition as experience [Left Hand]" (p. 13).

It is the *mutual codetermination* of Left and Right, or their "fundamental circularity," that is the core of the enactive paradigm. "In such an approach, then, perception is not simply embedded within and constrained by the surrounding world; it also contributes to the enactment of this surrounding world. Thus as Merleau-Ponty notes, the organism both initiates and is shaped by the environment. Merleau-Ponty clearly recognized, then, that we must see the organism and environment as bound together in reciprocal specification and selection" (p. 174).

Although many phenomenological, hermeneutical, and existential approaches have emphasized the importance of the immediate lifeworld (UL)—and the authors specifically recognize the value of their contributions—nonetheless there is a tendency in many of these writers to drift merely into theoretical and objectivist *discourse about* the lived lifeworld: they "attempted to grasp the immediacy of our unreflective experience and tried to give voice to it in conscious reflection. But precisely by being a theoretical activity after the fact, it could not recapture the richness of experience; it could be only a discourse about that experience" (p. 19).

And thus, although the enactive paradigm is sympathetic to many of the lifeworld theorists, it seeks to remain much closer to the lived experience itself, and let that experience unfold in an open-ended fashion, especially through such experimental/ investigative techniques as mindfulness/awareness training, which, we might say, is not an attempt to introspect and categorize awareness (theoretical analysis), but rather a direct approach to simply experiencing experience (a type of experiential analysis). Thus, "if science is to continue to maintain its position of de facto authority in a responsible and enlightened manner, then it must enlarge its horizon to include mindful, open-ended analyses of experience, such as the one evoked here" (p. 81).

I am in substantial agreement with this important work. However, since it broaches the topics of meditation, contemplation, and mindfulness/awareness training, I will reserve my final comments (and my few disagreements) until after these topics have been introduced in the text. See note 1 for chap. 14.

In the meantime, for a summary of structural coupling, the proposed process of linkage in circularity, see note 13 above.

44. We should note that Habermas is discussing the interrelations between the Upper Left (I), Lower Left (we), and Right Half (it), so he uses terminology from all three. But he is particularly, at this point, focusing on the correlations between the individual structures of consciousness (UL)—such as individual cognition, individual moral development, and individual identity—and the cultural structures (LL), such as collective morality, worldviews, and group identities. But both of these also become embedded in, and interact with, the concrete, material, social institutions and physical forms (LR).

In this section, and in the next chapter, we will particularly be looking at the correlations ("homologies") between the individual structures of consciousness (UL) and the cultural structures, worldviews, and collective identities (LL). (In volume 2 of this series, we will look more closely at their interrelation with the material-institutional forms, forces of production, modes of technology, and so on—the LR).

45. Habermas, *Communication and the evolution of society*, pp. 98–99.

46. Ibid., p. 99. My italics.

47. Ibid.

48. Isaac Asimov, *New guide to science* (New York: Basic Books, 1984), p. 771.

49. Arieti, *Intrapsychic self*, p. 6.

50. Jantsch, *Self-organizing universe*, p. 150.

51. *Up from Eden* and *The Atman project* are two books that cover the micro and macro branch of human evolution in, respectively, phylogeny and ontogeny. The center of both books is (1) a sequence of developmental stages or structures (holons) in the evolution of consciousness; (2) a developmental logic internally explaining and partially driving the unfolding of those stages (a version of the twenty tenets); and (3) a dynamic of Spirit-telos and Spirit-substitutes. Quite secondarily those stages were assigned actual dates of emergence based on available evidence. This is "secondary" because, in all developmental sequences, the nature and sequence of the abstract stages are not altered by the actual times that they appear, because these times may vary from person to person and from culture to culture. Thus, for example, in the sequence oral-anal-phallic, the sequence itself is universal but the precise time schedule can vary. The fourth central claim of those books is that (4) there are homologous structures in onto and phylo development (except, of course, for the earliest stages; since they are prehuman, there were never any *human* societies evidencing those).

I have continued to refine and revise the actual dates of emergence in both onto and phylo development (as well as their general correlations). In *Eden* I had underestimated the cognitive complexity required for magic and mythic productions, and I corrected this in *Eye to eye*, where the correlations are as I am presenting them here (e.g., preoperational/magic, concrete operational/mythic, and so on); these readjustments were reflected in *Transformations of consciousness*. But the *stages* themselves, and their *sequence*, and their *dynamic* remain essentially the same, which is why I still stand strongly behind the main conclusions of those early books; indeed, subsequent research has made them even more, not less, plausible. [See *Integral Psychology*.]

In Winkelman's (1990) review of *Up from Eden*, he misses the first three essential points and instead argues about dates, an argument that, where it is not simply wrong, is irrelevant, because I had already readjusted them in *Eye to eye* and *Transformations*. He further takes an essentially synchronic structuralist view, which is now almost universally acknowledged to be unable to account for development at all (see note 26 for chap. 1). He states that the use of onto/phylo parallels is completely outmoded and nobody is using that anymore, overlooking, for example, the extensive evidence collected and presented by Habermas, Sheldrake, etc. He collapses all validity claims to functional fit. And he conducts his overall attack from within a performative contradiction that dissolves his own position; he claims a substantive stance that itself is merely rhetorical.

For a critique of Winkelman's position as being essentially a reductionistic collapse of all validity claims to functional fit, see note 26 for chapter 5.

52. Neither Habermas nor I *started* with the *assumption* of onto/phylo parallels. It is rather our *conclusion* after an analysis of the structures of self, morality, notions of causality, types of cognition, and so on, that emerge at various stages of development. As Rothberg (1990) pointed out, both Habermas and I are following a reconstructive science, which means a reconstructing of the structures of consciousness *after* they have emerged in a population large enough to study empirically/phenomenologically/dialogically. These are our conclusions, not our assumptions. And the conclusion is that "certain homologies can be found."

CHAPTER 5. THE EMERGENCE OF HUMAN NATURE

1. "In the present context we are naturally interested in the question, whether the concept of social labor adequately characterizes the form of reproduction of human life. Thus we must specify more exactly what we wish to understand by 'human mode of life.' In the last generation anthropology has gained new knowledge about the long (more than four million years) phase during which the development from primates to humans, that is, the process of hominization, took place; beginning with a postulated common ancestor of chimpanzees and humans, the evolution proceeded through homo erectus to homo sapiens. This hominization was determined by the *cooperation of organic and cultural mechanisms of development* [biosphere and noosphere, or biogenesis and noogenesis; his italics]. On the one hand, during this period of anthropogenesis, there were changes—based on a long series of mutations [transformations]—in the size of the brain and in important morphological features [exterior forms].

"On the other hand, the environments from which the pressure for selection proceeded were *no longer determined solely by natural ecology* [by the biosphere], but through the active, adaptive accomplishments of hunting bands of hominids [my italics]. Only at the threshold to homo sapiens did this mixed organic-cultural form of evolution give way to an exclusively *social* [cultural] *evolution* [the all-important differentiation of the noosphere and the biosphere]. The natural mechanism of evolution came to a standstill. No new species [of humans] arose. Instead, the exogamy that was the basis for the societization of homo sapiens resulted in a broad, intraspecific dispersion and mixture of the genetic inheritance. This internal differentiation [tenet 12b in interiors] was the natural basis for a cultural diversification evidenced in a multiplicity of social learning processes. It is therefore advisable to demarcate the sociocultural stage of development—at which alone social [human sociocultural] evolution takes place (i.e., society is caught up in evolution [that is, in its own noospheric evolution])—from not only the primate stage—at which there is still exclusively natural selection (i.e., the species are caught up in evolution [that is, biospheric evolution]—but also from the hominid stage—at which the two evolutionary mechanisms are working together, the evolution of the brain being the most important single variable" (Habermas, *Communication and the evolution of society*, pp. 133–34).

2. Ibid., p. 134.

3. Ibid., pp. 134–35.

4. Ibid., p. 135.

5. McCarthy's wording, McCarthy, *Critical theory of Jürgen Habermas*, p. 238; my italics.

6. Habermas, *Communication*, p. 135.

7. Lenski, *Human societies*, chap. 7.

8. Habermas, *Communication*, pp. 135–36.

9. Volume 2 deals with the various schools of feminism (liberal, radical, social, Marxist), the masculine and feminine faces of Spirit, ecofeminism and ecomasculinism (deep ecology), transpersonal ecology, Marxist and techno-economic factors in the generation of worldviews, and an introduction to currents in postmodern thought, as set in the context of the spectrum overview.

Also: The term *sex* refers to biological givens and *gender* to sociocultural roles. *Male* and *female* refer to sex, *masculine* and *feminine* to gender. I will not usually distinguish between *sex* and *gender* until volume 2. Also, *sexual asymmetry*, which simply means sexual differentiation, is sometimes taken to mean "male dominator hierarchy," and I won't differentiate that until volume 2 either.

10. In those horticultural societies that were matrifocal, the Big Three had not yet differentiated; that is, the noosphere had not yet clearly differentiated from the biosphere, and thus "matrifocality" did not *integrate* those spheres but simply left them in undifferentiation or indissociation, which is why they provide no model for today's world. Women's power was still biologically determined, which is no real power at all but the fate of circumstances (all of which I will explore in volume 2).

11. As we will see, the historical forerunners of the women's movement occurred in precisely those places (such as Greece) where the noosphere and biosphere had first begun to differentiate clearly (where reason had first begun to emerge).

This means that feminism *as an issue* would not arise, and could not arise, as a widespread historical movement *until the noosphere finally differentiated from the biosphere*—and this is, of course, what in fact happened.

12. Charlene Spretnak's *States of grace*, as only one of many examples, gives an unrelenting history of Woman as Eternal Victim, which is presumably meant to empower women, but actually and profoundly dissolves the power of women in the very first step by defining them as primarily molded by an Other. Instead of giving women power by seeing how both men and women co-created all previous stages of development, these types of approaches scan history for any response, on the part of women, that does not happen to fit the present-day ideology of the particular feminist, whereupon that response, since it can't have been what women "really" wanted, is promptly ascribed to male oppression, thus *subtracting* it from the women.

Thus, in all these approaches, the very attempt to liberate women disempowers them by definition. These approaches are not insulting to men; they are profoundly insulting to women.

Women do not have to take back their power because they never gave it away; they co-selected, with men, the best possible societal arrangements under the extremely difficult environmental and social conditions of the time (the average life span of individuals in the eulogized horticultural societies, for example, was twenty-five years [Lenski]). As we will see in volume 2, there is now an entirely new wave of feminist research that has relocated and rediscovered female power in the patriarchy, research that demonstrates that female power was indeed a strong, conscious, intelligent co-creator of the patriarchy, a co-creator of that (and every) stage of human development.

With the differentiation of the noosphere and the biosphere, female power could now be deployed in completely novel and emergent ways; this was not a recovery or exhumation of lost power but the deployment of the same power in new and unprecedented ways. Victim feminists, on the other hand, are indeed caught in unending rounds of disempowerment, trying to "take back" their power by first giving it away.

This trading of real power for the pale substitute of extorted guilt is beginning to boomerang badly on the victim feminists, who are still having a difficult time assuming co-responsibility for their own history and their own choices. But the lesson of evolution is that most of today's problems are the result, not of something good yesterday that has been lost, but of something good tomorrow still struggling to emerge. Many victim feminists don't want to assume co-responsibility for their state because that state is still "mean" in many ways; but the cure for the meanness lies not in a *recovery* of a yesterday fashioned by an ideology of blame, but in an *uncovery* of tomorrow fought for in the face of recalcitrant emergence. The "enemy" is not something men did to women yesterday, but something that not-yet-enough evolution has done to them both.

13. *Patriarchy* has two rather different meanings in feminist literature. (1) For many radical feminists, patriarchy (as male domination) has been present from day one, from literally the beginning of the human race (Alison Jaggar takes this view); women who don't feel dominated have been brainwashed to accept their condition. (2) For the ecofeminists, who feel that horticultural societies dominated by the Great Mother were in many ways ideal (Gimbutas, Eisler), the patriarchy refers to the shift from matrifocal (or equifocal) to patrifocal modes, and thus has been with us for the last five thousand years or so.

Patriarchy 1 is, in effect, the familization of the male via the role of the father, and that indeed has been with us from day one; that was not avoidable. Patriarchy 2 occurred across the globe as horticulture gave way to plow and horse societies, which selected for physical strength/mobility; nor was that avoidable either, except by developmental arrest (which is precisely the societies eulogized by ecofeminists). In none of these societies were the noosphere and biosphere integrated; they were simply undifferentiated, which is why they are no role models at all, despite the rhetorical use to which they have been put.

14. Habermas, *Communication*, pp. 104, 121.

15. This type of cognition as "simple representation" is, of course, the cognition still favored by empiricist theories of knowing. Being "close to the body," or "close to sensations," it is a cognition that registers mostly surfaces and exteriors, and so it is naturally the cognition favored by the reflection paradigm (monological, representational). But empiricist *theories* themselves are actually making massive use of formal operational space, whose mechanisms, being not empirically obvious, are simply ignored (or rather, implicitly assumed even as the theory denies their existence).

16. McCarthy, *Critical theory*, p. 250.

17. This doesn't mean magic doesn't serve crucially important functions. In its *cultural* capacity (LL) it is the center of cultural *meaning* and *value*, and in its *social* capacity (LR), the locus of *social integration*. Radical egalitarians who object to the term "magical" imagine that the term somehow denies all these important functions.

18. Habermas, *Communication*, pp. 111–12.

19. Eisler, The Gaia tradition, in Diamond & Orenstein, *Reweaving the world*, p. 32.

20. Dubos, *A god within*, quoted in Novak, Tao how?. Novak's article is highly recommended.

21. Roszak, *Voice of the earth*, pp. 226, 69. This book is discussed at length in note 32 for chapter 13.

22. Habermas, *Communication*, p. 162. Habermas and his associates are referring to tribal societies the way they first existed, not necessarily how they exist now. As Roszak points out, "As anthropologists warn us, it is always risky to infer from contemporary to prehistoric tribal groups. Existing tribal societies have experienced as much history as the rest of the human race . . ." (*Voice of the earth*, p. 77).

Thus, when members of tribal or indigenous societies existing today are given Piaget's test of cognitive development, they show a capacity for formal operational thinking, just like anybody else. But the point, as Habermas demonstrates, is that the first tribal societies *as a whole* did not evidence formal operational cognition in any of their actual, social structures: not in legal codes, not in conflict resolution, not in arbitration, not in modes of group or collective identity, not in worldviews, and so on. Whether some individuals did (or did not) reach formal operational in those societies is beside the central point, which is that the basic organizing principles of the primal societies themselves did not evidence formal operational, but rather followed the primary structures of preoperational cognition (and preconventional dispute arbitration, etc.). *Some individuals developed above, some below, that average level,* but the basic average level was clearly preoperational and preconventional in many fundamental ways—such is the conclusion of this line of research put forward by Habermas and associates (similar types of arguments are put forward by Gebser, Campbell, Piaget, Kohlberg, Bellah, Lenski, etc.). It is the tribal *structure*, not particular tribal members, that is being addressed.

At the same time, both Habermas and I believe that at least some individuals in foraging societies developed the capacity for rationality; and I believe some went even further into the transrational (namely, the shamans), if not as a permanent adaptation, then as a peak experience or altered state. We will discuss these transrational realms in chapter 8. [See *Integral Psychology*.]

23. Habermas, *Communication*, p. 161.

24. Ibid., p. 112.

25. Ibid., p. 104.

When Gary Snyder made the remark, much admired in certain ecological circles, that when humans invented gods and goddesses of an "otherworldly" nature, all of humanity's troubles began, he betrayed the ecomasculinist preference for a reinterpreted tribalism. There was no way for tribal social organization to move beyond isolationism without finding a source of commonalities beyond kinship, and this the mythic motifs provided. Lenski reports that most foraging tribes had a carrying capacity of about forty people, and these tribal units could not be integrated into larger conventional sociocentric collectives without the transcendental power of a common mythic ancestry and group identity that extended beyond the limited bonds of kinship unions. Far from being "global," tribal consciousness is considerably fragmented when it comes to modes of intersubjective bonding and unification (see also note 26 below).

26. Detailed evidence for these correlations is presented in *Up from Eden*.

As for the nature of the shamanic vision, the only book I can unreservedly recommend is Walsh's *The spirit of shamanism*, an excellent overview of this tricky and difficult topic. Walsh is one of the leading lights of the transpersonal movement, and his balanced and sane overview, decisive in many ways, is a major contribution to the field. Walsh's presentation is valuable because it manages both to acknowledge the pioneering spiritual insights of shamanism and simultaneously set it in a larger perspective, an accomplishment which rightly earned it the praise of most scholars.

In the course of Walsh's succinct but comprehensive presentation, he points out that the shamanic journey (a general phenomenon that is perhaps the central and most defining characteristic of the complex of shamanic practices) has usually been assumed to be simply the same general state as yogic, Buddhist, mystical, or even schizophrenic states. Walsh first presents a "horizontal" analysis of various factors (such as degree of conscious control, awareness of environment, concentration, ability to communicate, identity, and arousal), all of which, taken together, quite clearly differentiate the shamanic journey from yogic, Buddhist, mystical, and schizophrenic states (which are also quite clearly differentiated from each other). "This approach," he points out, "allows us to move from unidimensional comparisons to multidimensional comparisons and to distinguish between states with greater sensitivity."

Walsh doesn't stop there. Having clearly differentiated the shamanic state in his horizontal grid, he then applies a "vertical" or developmental grid, using standard and well-accepted criteria for developmental stages in general. His conclusion is that of the three major spiritual stages of development (subtle, causal, and nondual, each of which we will investigate in chapter 8), "shamans were perhaps the earliest masters of the subtle realm." He acknowledges that a few shamans very likely transcended their own path and disclosed causal and nondual occasions, but the central and most defining characteristics of shamanism seem to be quite clearly subtle-level phenomena.

Since many forms of yogic and Buddhist meditation—by the same criteria—reach beyond the subtle and into causal and nondual states, Walsh's clear implication is that these contemplative endeavors (and others like them) are in some ways more advanced (or more disclosing of greater depth) than the typical shamanic journey, and this "ranking" riled a few cultural relativists, on the one hand, and some shamanic advocates, on the other.

Representing the cultural relativists, Winkelman (1993) advanced the reasons that no stance can be intrinsically superior to another (except, of course, his stance, which is claimed to be intrinsically superior to the alternatives). Winkelman takes the stance that it is universally true that there are no universally true stances.

Having thus exempted his own position from the criticism he levels at others, Winkelman then proceeds to give the standard *functionalist* analysis of social systems, and concludes—as all functionalist accounts must—that if a sociocultural practice is *adaptive*—that is, if it possesses *functional fit*—then it cannot be judged inferior to any other production, because it fits just as it is (and thus, he says, the yogic systems would not have worked in shamanic cultures, and therefore the yogic systems cannot be said to be "deeper" or "more advanced" or any such qualitative distinction).

But nobody is claiming that the yogic (or Buddhist or contemplative) systems would fit into shamanic cultures (in fact, they would not; among numerous reasons, the base of the shamanic cultures was preconventional, with a scarce resource of *power*, whereas the base of the cultures of the contemplative traditions was conventional with beginning postconventional, whose scarce resources were legal *membership* and beginning self-recognition).

Thus, Winkelman's conclusion, when it comes solely to *functional fit*, is quite right (and both Walsh and I have long agreed with that conclusion, as far as it goes); the point is, it is quite irrelevant as well for the central issue. Winkelman's purely Right-Hand approach (Lower-Right, to be specific) reduces all validity claims (of truth, truthfulness, cultural meaning, and functional fit)—reduces all of those to *functional fit* and monological systems markers (reduces all Left-Hand dimensions to Lower-Right terms), and so of course his approach cannot spot qualitative distinctions in the first place (which is why he must *implicitly* bury *his own* value ranking in a stance that *explicitly* denies its own possibility—the standard performative contradiction in all merely Right-Hand paths).

Functional fit, as we have seen, recognizes only one validity claim: does the system work or not? Is it functioning adequately in its environment? Is it adaptive? Is it pragmatically functional? *Does it work?* And to this single, monological validity claim, functional fit has only one answer: yes, it is working, or no, it is not. *If it is working* (in its own way, in its own place), then that *ends the discussion* about values, about which system is higher or lower, deeper or shallower, better or worse, more encompassing or less encompassing. If it is working well, then it is as good as any other system, and who are we to go judging further?

Never mind that that judgment of universal pluralism is itself a *judgment* that *ranks* its own stance as better than narrow ethnocentrism—and rightly so!, except that, since this functional relativistic stance collapses all Left-Hand *qualitative distinctions* (including those that *allow its own stance*) into Right-Hand functionalist markers, then it cannot even account for the truth or goodness of its own stance, and thus it thoroughly *exempts itself* from any open discussion and evaluation of qualitative distinctions, of the Left-Hand dimensions in general (it makes a poorly-thought-out Left-Hand judgment that denies the significance of the Left-Hand in toto).

As Taylor put it, this stance therefore buries its own tracks, and then spends its time denouncing all those who do not likewise conceal their own position. It is a *value ranking* (partly correct in my opinion) that nonetheless obstinately and officially *denies* all value ranking, and the only way it can do so is by refusing to acknowledge its own now *hidden* value ranking (it claims all values are *equivalent* if they functionally fit: but if all values were really equivalent, then tolerant pluralism could in no way be any better than narrow-minded ethnocentrism and racism: this position misses *its own* deeper and prior value judgment—correct in my opinion—upon which its pluralistic tolerance necessarily rests), and this deep obfuscation then allows it to *collapse all values* into functional fit, whereupon it runs smack into its own absurdities: if it *works* for the Nazis, who are we to judge? Auschwitz, all agree, worked quite efficiently, and by Winkleman's stated criteria, we can make no further judgments.

Thus, the bulk of Winkelman's critique reduces all values and all validities to functional fit. His functional analysis is accurate enough, as far as it goes; it is just, as I indicated, extremely partial. His analysis tells us that, in the Lower-Right quadrant, shamanism was perfectly adapted to (and functionally fit) the societies that grounded it, and thus it cannot, in those terms, be judged inferior—and that is certainly true. By the criteria of *functional fit* (legitimacy), the yogic/contemplative system was indeed neither better nor worse than the shamanic (actually, we all agree it would have been worse: it would not have fit).

Nobody (not Walsh, not I) has ever claimed otherwise.

But by virtually any criteria from the Left-Hand dimensions (developmental unfolding of structural potentials that display deeper and wider qualitative shifts), the typical shamanic journey was a pioneering engagement that would nevertheless be deepened and widened in the contemplative endeavors that would build upon it in subsequent spiritual evolution, an evolution that would lead from subtle to causal to nondual disclosures (all of which will be discussed in chapter 8). [See also *Integral Psychology*.]

Winkelman himself gives an example of what he considers to be his major point. A cigarette lighter, he says, may be more advanced than a simple flint, but there are times when the flint is more practically useful, so that neither lighter nor flint can be intrinsically better, since if they fit their practical/adaptive function, they are equally satisfactory. That, of course, is the simple functional fit argument, and we have already seen that Walsh (and I) have accepted that truism from the start. Let us therefore complete the analogy and note the other (and neglected) fact of Winkelman's own analogy: as he says, the lighter is technologically *more advanced*. The lighter includes a flint, *plus* other more sophisticated cognitive/technical discoveries (it includes, for example, a wheel, which early foraging societies had not discovered).

Thus Winkelman's main analogy, with which he ends his critique, points to these two facts: there are circumstances where both the flint and the lighter were equally appropriate, or possessed equal *functional fit*; *and* the lighter is nonetheless more cognitively and technically advanced. Winkelman's analysis covers only the former; Walsh's analysis covers both; and thus Walsh's presentation supersedes the rather narrow functionalist endeavor.

Winkelman smuggles his own values into his analysis by claiming (quite rightly, I believe) that worldcentric perspectivism "has advantages in providing multiple perspectives which are *more encompassing and complete*"—an altogether correct stance that he must nonetheless officially deny, since most of the ethnocentric cultures he studies do not possess a worldcentric pluralism, and thus he would have to admit that his (modern, Western) stance is superior to ethnocentric narrowness (and this he is not allowed to do by his official stance that no cross-cultural universal judgments of intrinsic merit are possible). See note 26 to chap. 1, p. 554.

Many approaches to shamanism assume that there are, basically, ordinary reality on the one hand and "mysticism" (or nonordinary reality) on the other. This simplistic distinction (ordinary vs. nonordinary) allowed Gary Doore, for example, to state that "shamans, yogis and Buddhists alike are accessing the same state of consciousness," and Holger Kalweit to state that the shaman "experiences existential unity—

the samadhi of the Hindus or what Western spiritualists and mystics call enlightenment, illumination, unio mystica." But, as Walsh points out, "These claims seem to be based on relatively superficial similarities."

Walsh's multidimensional grid analysis (both horizontal and vertical) has irrevocably shattered that simplistic approach. It is simply impossible to look at all of his variables (which differ considerably among the various states) and not be impressed with the gross and subtle differences among these states. Further, armed with this finer analysis, it is impossible not to spot developmental factors in the unfolding of these states themselves: like everything else in the Kosmos, they seem to grow and evolve: the acorn does not show up as a fully formed oak on the spot. And Walsh's horizontal and vertical analyses, while preliminary by his own accounts, have nonetheless advanced the field enormously.

The second criticism of Walsh's analysis came from those theorists who view shamanism as accessing the same qualitative levels of depth disclosure as the subsequent contemplative systems. But by the finer analysis that Walsh gives, it now seems likely that the typical shamanic journey was a pioneering breakthrough into the transpersonal/spiritual domain, a breakthrough that disclosed levels of depth undreamt of in the conventional world, and a depth that subsequent contemplative endeavors would deepen and extend: building upon the spiritual disclosures of the great shamanic voyagers, and standing thus on the shoulders of giants, they would subsequently probe even further into the extraordinary depths of the great Beyond.

27. The "mythic-rational" structure is a specific term for late concrete operational, but I also use it, in a general sense, to refer to the rationalization of any of the mythic structures.

28. Habermas, *Communication*, pp. 112–13.

29. Ibid., p. 105.

30. Ibid.

31. Ibid., p. 114.

32. The few women's movements that briefly emerged in earlier societies, such as in Greece, emerged in precisely those places where rationality had also precociously emerged, and where ego identities were also tentatively emerging from role identities. But since the differentiation of the biosphere and the noosphere was not fully complete, the biospheric identities eventually sucked these movements out of the noosphere and back into the bodily or biological determinants.

And, as I earlier mentioned, in the horticultural societies where women were a large portion of the productive work force, a type of egalitarian arrangement was indeed at work, but this was secured not by stable legal and noospheric determinants, but simply by biospheric contingencies. The hoe placed power in women's hands, just as the plow would remove it. The biosphere and noosphere had not been differentiated in these societies, and thus the social determinants always reverted to biospheric selection in times of stress, defense, or turmoil. In no case were they models of what an egalitarian and *differentiated/integrated* society would look like. (All of this is covered in great detail in volume 2).

Also, as we will see in volume 2, the role of industrialization (the historical LR correlate of UL egoic-rationality) was instrumental in liberating human labor (and the female) from the determinants of the biosphere (and could go too far into dissociation or eco-crisis). I am in no way ignoring industrialization or the LR quadrant; they are the main focus of volume 2.

As I will point out in the text, I think many ecofeminists feel they have to be able to point to a past society where women were "equal" in order to show that such equality is a possibility of the human condition. But my point is that the possibility is an emergence, not a resurrection. And if we take the resurrection model—that a past glory was lost and needs to be regained—then we have to get into the blame game: the nasty men had to have taken it away, and the women have to have been stupider and/or weaker to have let it happen. The resurrection (or Romantic) model necessitates the pigification of males and the sheepification of females, whereas the emergence (or growth) model portrays both males and females as essentially co-creative.

33. There were, of course, many forerunners to feminism throughout history (in Europe, Greece, India, northern Africa, etc.); one might particularly mention Christine de Pisan's *Book of the cities of ladies*. Nonetheless, as Riane Eisler points out, "Feminism as a modern ideology did not emerge until the middle of the nineteenth century" (*Chalice*, p. 165). We will return to the emergence of the women's movement in chapter 11.

Cf. Elise Boulding, *The underside of history*; Sheila Rowbotham, *Women, resistance and revolution*; Marilyn French, *Beyond power: On women, men, and morals*; Susan M. Stuard (ed.), *Women in medieval society*; Nawal El Sadawii, *The hidden face of Eve: Women in the Arab world*; Nancy Cott and Elizabeth Pleck (eds.), *A heritage of her own*; Gerda Lerner, *The majority finds its past: Placing women in history*; La Frances Rodgers-Rose (ed.), *The black woman*; Martha Vicinus (ed.), *Suffer and be still: Women in the Victorian age*; Jeanne Achterberg, *Woman as healer*.

34. This claim of a special connection between woman, nature (earth), and body is examined in detail in volume 2. The claim itself has been put forward in a strong and positive light by many radical feminists and ecofeminists, and equally condemned by many liberal feminists. The difficulty, as I see it, is that *if* we are going to make that claim, then how can it be done without simultaneously "pigeonholing" women or simply relegating them to the emotional/feeling domain? In other words, a way has to be found to preserve and negate the claim. I believe this can be done, and volume 2 explores this possibility.

For the moment, we have the following (quite different) stances toward the "special connection" between woman, nature, and body: the special connection (1) actually exists, (2) was claimed to exist by some women for positive value, (3) was claimed to exist by some men as a prelude to oppressing that entire sphere. Any of those versions of the "special connection" will suffice for my comments in the text; my own specific opinion, and the actual details, are covered in volume 2.

35. Habermas, *Communication*, p. 114.

36. Ibid., p. 197.

37. Gebser's *Ever-Present Origin* is now available in English (see bibliography). I have taken all quotes in this section from Feuerstein, *Structures of consciousness*, which is by far the best presentation of Gebser's work. Feuerstein also correlates Gebser's stages with my own.

38. All quotes from Feuerstein, *Structures of consciousness*. [For the precise details of vision-logic and its postformal substages, see *Integral Psychology*.]

39. Several terms in the Lower-Left are carried over from *Up from Eden* and *Eye to eye*: pleromatic (physical/material), uroboric (reptilian/brain stem), and typhonic (paleomammalian/limbic), although these are all used in a very general sense (both here and in *Eden*), to indicate worldviews where these elements, although by no means the sole determinant, loom large in its primary concerns, often defining its most general atmosphere (and sometimes its most scarce resource). And the "archaic," as indicated in the text, is a catchall phrase for any or all structures prior the emergence of concepts (and sometimes, as with Jung, to refer to any of the "inherited" or "archetypal" early concepts, images, mythic forms, etc.).

As I have previously indicated, if the reader does not feel comfortable extending "consciousness" (or prehension of any sort) to any "lower" holons—and consequently does not believe that they share any sort of common worldspaces—then one can interpret the entire Lower-Left quadrant *merely* as the ontogenetic unfolding of today's human being: it starts out as a single-celled zygote (protoplasmic), which itself embraces and enfolds molecules, atoms, and subatomic particles (pleromatic); and it then develops through differentiated life functions (vegetative) and into a reptilian brain stem (uroboric, complete with gill slits), and a paleomammalian limbic system (typhonic), and a slowly unfolding neocortex/neurosystem, with its worldviews then developing from archaic to magic to mythic to rational.

My point is that these two interpretations (human ontogeny and worldspaces of previous holons) are fundamentally saying the same thing. Even bifurcating dissipative structures show "systems memory" of the path of their buildup (because they retreat along that path when they dissolve, as we saw in chapter 2). That human beings would possess such systems memory seems not even a little surprising (memory stored in brain cells is not the only memory the Kosmos has generated!). The previous worldspaces are enfolded in us, precisely as our ancestral systems pathways.

I am by no means basing my argument on the following example, but it is illustrative that human beings given such conventional-reality-blowing drugs as LSD often report memories not only of their own birth, but "memories" of plant and animal identification, even atomic resonance: the worldspaces are there, enfolded in us, subholons in our own makeup, speaking to us still from the depths of our own Kosmic family photo album.

40. The Upper-Right quadrant is also meant to include and imply the possibility of the existence of subtler bio-energies that numerous researchers (from William Tiller

to Hiroshi Motoyama) feel are *holarchically* enveloping the bodymind. But these energies are still *exterior* and can be *monologically* perceived with appropriate equipment (or subtler senses), and thus belong to this quadrant (although they would have various correlates in the other quadrants as well, as all holons do).

Likewise, I would like to point to the work of Da Avabhasa, who has outlined several stages of the growth and development of human consciousness (which we will explore in chapter 8); without reducing these stages to bodily components, he nonetheless points out that *all* of them have bodily correlates (even the supra-individual causal Witness has its bodily correlate in the heart region on the right). All of these bodily correlates are Upper-Right quadrant, and reinforce my central point: every holon (no matter how transcendental) has these four quadrants, these four correlates.

Finally, I would like especially to single out the work of Michael Murphy, whose book *The future of the body* is a magnificent study of the bodily correlates of a transforming and evolving consciousness—yet more evidence that all manifest holons *anywhere* possess the four quadrants. Murphy almost single-handedly has been representing the great importance of the Upper-Right quadrant in human transformation (without merely reducing human evolution to the Upper Right).

In volume 2 we will return to this topic of the interaction between the UR and UL quadrants in both conventional and higher development. We see this interaction not only in such fields as medical psychiatry (where the role of neurotransmitters in normal functioning and mental illness is becoming more obvious) but also in such "mystical" areas as psychedelics and the newly developing "brain/mind machines" capable of inducing a broad range of "meditative" states. Exactly what all this means will be examined in detail.

41. Nonetheless, Marx's own writings are not quite as reductionistic as they are often made out to be, nor as reductionistic as they often became in practice. In volume 2, the more subtle aspects of historical materialism are examined and set in the context of evolving and developing worldviews. The great strength of the Marxist system on the whole, was that it kept alive a *developmental* scheme of human evolution, and this is why so many serious (and nonreductionistic) theorists, from Lenski to Habermas, have returned to it time and again for useful insights (all of which, as I said, is examined at length in volume 2).

42. An example of which is Habermas's work on unrestrained communicative exchange, which alone can secure *unforced agreement* on *any* of these issues in the first place. We will return, in chapters 12, 13, and 14, to the failure of most of the eco-philosophies to deal nonreductionistically with the noosphere and its distributions and distortions.

43. That is, the overall *movement of cultural transformation* is from the Upper-Left of individual cognitive potential to the Lower-Left of collective worldview, at first marginalized, but finally embedded in Lower-Right social institutions, at which point these basic institutions automatically help reproduce the worldview (LL) and

socialize the individual (UL and UR) in succeeding generations, acting as "pacers of transformation"—a transformation first started or begun in a moment of individual creative emergence and transcendence. This process is discussed in much detail in volume 2.

44. The great "world religions" (Hinduism, Buddhism, Christianity, Islam, etc.) that still exist today (and which we will examine beginning in chapter 7) all arose in the general era of *mythic-imperialism*, and all of them remain clothed to some degree in *surface structures* (and ethics) that are two major technological epochs behind the times. None of those religions (nor the previous tribal religions) arose from within a global culture, and thus none of them can (or will) speak to the rising world culture, however important their specific spiritual practices are (and will remain).

Rather, as a global culture is created, and as its surface structures begin to be part of a common language, then from within that global culture will arise the new religion(s), speaking in part this global language, working within this common discourse, and pointing as well beyond it. The era when one specific religion (say, Christianity or Buddhism) could move into a new territory and simply convert it (by force of arms or force of truthful persuasion) has long gone. That the new religion(s) will have roots in the previous religions I do not doubt; that they will simply come from any of them I do not believe.

The tribal Earth religions arose within a worldview of magical indissociation and upon a techno-economic base of wooden club and spear (with scarce resource of power), and bound together only kinship lineages. The great world religions arose within a worldview of mythic and mythic-rational, and upon a techno-economic base of horse and plow (herding/agrarian)—with scarce resource of membership and thus with corresponding (and very rigid) hierarchies of power, membership, wealth, and access to the Divine. The specific moral injunctions in these religions are thus often (and still) bound to surface structures that simply make little (if any) sense in today's world: don't eat pork, don't eat with your right hand, don't sleep on wool, don't allow women in the temple. . . .

However much the esoteric or mystical components of those religions did indeed penetrate beyond such outward exoteric forms, nonetheless there are limits to just how far out of one's skin one can jump. Moving from the agrarian to the modern, and from the modern to the postmodern: there is too much ground to make up, and too much distance to travel, for the traditional contemplative endeavors to cover on their own (and too much dross that needs to jettisoned without being condemned as a sinner).

No, the new global religion(s) will come from within the new global culture and will not be anything simply grafted on from the past. The new global religion(s) will be at home in contemplative awareness, but awareness that also speaks naturally and natively a digital language of the silicon chip, and sees itself as clearly in virtual reality as in the play of the wind and rain; its global perspective and universal pluralism will be taken for granted, and Spirit will move through circuits of fiber optics as well as through flesh and blood; and all of that will be natural, and normal, and alive: and from within that global network the new voices of transcendence will begin to attract those sensitive to the Divine. From within the informational neural network of a global commons will come the voices indicating liberation.

The various New Age movements claim to herald such a worldwide consciousness revolution. But, as this book will make obvious, I think, these movements are insufficient: they lack any sustained vision-logic of *both* exterior and interior dimensions, they lack a consistent technology of access to higher interior dimensions, they lack a means (and even a theory) of social institutionalization (in other words, they lack an all-four-quadrant analysis and engagement). Moreover, most New Age and new paradigm approaches, despite their claims to be post-Cartesian, are simply extending the types of phenomena described as possible in the Cartesian world; they do not decisively challenge the fundamental Enlightenment paradigm. (We will be returning to this topic frequently throughout the book.)

As such, most of these New Age movements do not engage the rational worldview in a way that can transcend and include it; rather, many of them end up regressing to various forms of mythic-imperialism (and even tribal magic). These movements put a premium on a self-actualization that all too often reverts to magical egoism; and this magical narcissism is worked into a mythology of world transformation that barely conceals its imperialistic thrust.

45. Habermas, *Communication*, pp. 164–65.

46. Ibid., pp. 165–6.

47. For an interesting discussion of these themes, see Laszlo's *The choice: Evolution or extinction?* As usual, Laszlo's presentation is admirably straightforward and to the point, and we can all applaud when Laszlo says (p. 109) that "the culture of interexistence [which he advocates] has an inclusive logic: It is you *and* I, they *and* we. It replaces the logic of egotism and exclusion, which says me *or* you, we *or* they. The new logic can enable people and societies to play postive sum win-win games. . . . The great advantage of the culture of interexistence is that, with its inclusive logic, it could harmonize the current forms and facets of diversity."

It could indeed; the "inclusive logic" is vision-logic. But Laszlo is silent on the interior stages leading up to this vision-logic, and thus the crucial steps necessary for the "great solution" remain obscure. Laszlo is quite right, I believe, when he says, "Though these factors are subjective and cultural [Left Hand] rather than objective and physical [Right Hand], we must not ignore them nor underestimate their import. The more a person is developed (not, of course, merely in the economic sense, but in the sense of being a mature and responsible citizen), the more his or her society has a chance to develop" (pp. 140–41).

But about this "more developed," we hear nothing more. I am very sympathetic with Laszlo's general presentation in *The choice* (in particular, his emphasis on the Left-Hand dimension is in sharp contrast with his previous and almost exclusive emphasis on Right-Hand systems theory, even if he continues to frame his thoughts in objectivistic and systems holism; and I entirely support his notion of "modernization, not Westernization"). But without a clearer understanding of the actual stages of personal (UL) and cultural (LL) transformation that result in a pluralistic worldcentric awareness capable of "interexistence," Laszlo's call for "education," "communication," and "more information" remains unfulfilled.

48. For a particularly chilling account of this retribalization and its growing influence in the immediate future, see Robert Kaplan's "The coming anarchy" in the February 1994 *Atlantic*. Kaplan also sees the world heading toward globalization, but with an extended transition period of retribalization: "Whereas the distant future will probably see the emergence of a racially hybrid, globalized man, the coming decades will see us more aware of our differences than of our similarities."

Kaplan ties his thesis to the work of Van Creveld's *Transformation of war*, Homer-Dixon's environmental studies, and Huntington's thoughts on culture clash: Under various intense environmental and demographic stresses, numerous state mechanisms of governance will fragment into ethnic tribal bands. And, Kaplan points out (quoting Van Creveld), future "armed conflict will have more in common with the struggles of primitive tribes than with large-scale conventional war" (i.e., regression to tribal warfare prior to the state warfare about which von Clausewitz theorized).

But *war* because, unless restrained, a large number of humans love the thrill of battle (men to fight, women to offer up sons and husbands to the glory of the battle). Van Creveld's book begins "by demolishing the notion that men don't like to fight." As Van Creveld points out, "Throughout history, for every person who has expressed his horror of war there is another who found in it the most marvelous of all the experiences that are vouchsafed to man, even to the point that he later spent a lifetime boring his descendents by recounting his exploits." And with extremely acute insight, Van Creveld knows exactly why: "By compelling the senses to focus themselves on the here and now, battle can cause a man to take his leave of them."

And thus, Kaplan points out, "As anybody who has had experience with Chetniks in Serbia, technicals in Somalia, Tontons Macoutes in Haiti, or soldiers in Sierra Leone can tell you, in places where the Western Enlightenment has not penetrated and where there has always been mass poverty, people find liberation in violence." And "only when people attain a certain economic, educational, and cultural standard is this trait tranquilized."

As tribalized warfare increases: "Because the radius of trust within tribal societies is narrowed to one's immediate family and guerilla comrades [largely preconventional and egocentric], then truces arranged with one commander may be broken immediately by another." Likewise, "when cultures [ethnicities], rather than states, fight, then cultural and religious monuments are weapons of war, making them fair game."

I deny the possibility of none of that; it is altogether consistent with regression downward along the same broad lines of previous transformation upward. "Tribalization" in today's world is "bad," where previously (and prehistorically) it was the norm (was "good"), precisely because it is *now* regressive: it now moves below the horizon of universal perspectivism, whereas previously it was simply struggling to get up to it.

And that *regression* infects not just social movements, it is becoming rampant in many types of "new paradigms," from tribal "eco-wisdom" to magical and mythical New Age imperialisms, to biocentric and ecocentric immersion in precisely the sphere that cannot itself take universal perspectivism. All of these, of course, are sincerely offered as "global" salvation, but most of them are simply splinters of a widespread, regressive retribalization.

CHAPTER 6. MAGIC, MYTH, AND BEYOND

1. As we will see in much greater detail in chapter 8, Spirit is not merely or even especially the summit of the scale of evolution, or some sort of Divine omega point (although that is part of the story). Spirit is preeminently the empty Ground, or groundless Emptiness, fully present at each and every stage of evolution, as the openness in which the particular stage unfolds, as well as the substance of that which is unfolded. Spirit *transcends* and *includes* the world: *transcends*, in the sense that it is prior to the world, prior to the Big Bang, prior to any manifestation; *includes*, in the sense that the world is not other to Spirit, form is not other to Emptiness. Manifestation is not "apart from" Spirit but an activity of Spirit: the evolving Kosmos is Spirit-in-action.

Many ecophilosophers, with a preference for a God that can be seen with their empirical eyes, attempt to deny any transcendental aspect to Spirit and paint it instead in totally immanent terms, and thus completely confuse finite and infinite, or confuse the Formless with any merely passing and finite form (confuse the Nirmanakaya and Dharmakaya). They often point to Spinoza as supporting this position, which is incredible. Spinoza maintained that Spirit has infinite dimensions, only two of which we can know (extension and thought), so that Spirit infinitely *transcends* yet utterly *includes* the manifest world—a version of the position I am presenting here.

On the other hand, those who emphasize *only* transcendence or *only* immanence are extremely dualistic and divisive in their ontology, which therefore often degenerates into ideology. The only-immanence stance is now quite popular; it fits well with the common ecomasculinist preference for tribalisms and the common ecofeminist preference for horticulture: empirical indissociation mistaken for transcendence-and-inclusion.

2. Technically, preop and conop are one broad stage, but it is customary to treat them separately since many of their characteristics are quite different. The ages of emergence are simply averages.

3. Blanck and Blanck introduced the term *fulcrum of development* to refer to the separation-individuation of the infant's self from the emotional (m)other, based especially on the pioneering work of Margaret Mahler. In Wilber et al., *Transformations of consciousness*, I suggested that this fulcrum was but one of numerous quite

distinct fulcrums, each representing a qualitatively new and distinct differentiation/ integration (or transcendence-and-inclusion). Each of these fulcrums (and I generally outline nine or ten of the most important fulcrums) establishes a new, important, and very different *type* of self *boundary*. A failure to grasp these different boundaries (and their very different functions) leads to what I have called the single-boundary fallacy, a fallacy that has hobbled many psychological theories and most mystical/psychological theories (as we will see in subsequent chapters; see, for example, note 32 for chap. 13; and note 17 for chapter 14). For students of this approach the technical correlations are as follows:

The first fulcrum (F-1) covers the early sensorimotor period (0–1 yr), beginning with the initial primary indissociation or "protoplasmic consciousness," whose worldview is "archaic," and is resolved with the first differentiation of the physical self from the physical environment ("hatching"), and is finally consolidated with the attainment of physical object constancy (around 18 months, at which point the second fulcrum is under way). F-1 is particularly concerned with the establishment of physical boundaries (the physical self differentiated from physical objects).

The second fulcrum (F-2), the phantasmic-emotional, covers the late sensorimotor period and the beginning of preop (1–3 yrs), and begins with an initial indissociation of emotional self and emotional objects ("archaic-magic"), and is resolved with the general attainment of emotional or libidinal object constancy (24–36 months; differentiation of emotional self and other, the establishment of emotional boundaries).

The third fulcrum (F-3) begins with the initial indissociation of mental signs and referents (and mind-body indissociation), covers the preop period (2–7 yrs), and is resolved with the emergence of conceptual object constancy (5–7 yrs, the establishment of conceptual self boundaries). This period is divided, with Piaget, into early preop (2–4 yrs), governed by images and symbols (with a worldview of "magic"), and late preop (4–7 yrs), governed by concepts (with a worldview of "magic-mythic").

The next three fulcrums (F-4, F-5, and F-6) refer to the differentiation/integrations that occur during conop, formop, and vision-logic (which we will explore later in this chapter). The worldviews of conop are divided into mythic and mythic-rational; formop into rational and rational-existential; and vision-logic into existential and existential-psychic, all of which we will investigate later. Higher or transpersonal fulcrums (F-7 through F-9) will be dealt with in chapters 7 and 8.

Fulcrums, in other words, refers to the *self-stages* and their differentiation/integration as they negotiate the *basic waves* or levels of consciousness development (which themselves are unfolding developmentally, and are necessary but not sufficient for self-stages). *Worldviews* (archaic, magic, magic-mythic, mythic, etc.) refers to the cognitive map of the world created at each basic level or wave (both individually and collectively).

Neither Piaget nor the psychoanalytic developmentalists (such as Mahler) explicitly distinguish the first and second fulcrums, but rather they treat them together as one general process of differentiation (separation-individuation). Nonetheless, their own

data clearly suggest that these are two very different stages or fulcrums of self development, the first referring to the differentiation of the physical self and physical environment (a physical boundary), the second to the differentiation of the emotional self
and emotional environment (an emotional boundary). Before infants have negotiated
the first fulcrum, they are immersed in protoplasmic indissociation; after the first fulcrum, they can clearly differentiate physical self and other, and clearly ground themselves in physical constancies, but they are then in the initial stages of the second
fulcrum, marked by *emotional indissociation* (a lack of differentiation between emotions of self and emotions of other). It is the preponderance of such indissociations
and "adherences" in both fulcrums that lead these researchers to treat the first two
fulcrums as being quite similar, but the statements of both of these researchers make
it very clear that they recognize two different movements here, with, among other
things, different possible pathologies (psychotic and borderline). Other researchers,
such as Kernberg, explicitly recognize these different stages. But all orthodox researchers stop their research at either the fifth or sixth fulcrum, and ignore the transpersonal fulcrums (7–9).

Likewise, at the other end of the spectrum, there is a large body of (very controversial) theory and evidence for the intrauterine state and the birth process (and birth
trauma). Without entering into the intricacies of this debate, I have simply allowed
that some of this prenatal research might be genuine, and I have referred to the whole
period from conception to birth as fulcrum-0.

This fulcrum follows the *same general features* as any other fulcrum, namely: (1)
an initial state of undifferentiation or indissociation, (2) a period of intense and often
difficult differentiation (in this case, the birth process/trauma itself), and (3) a period
of postdifferentiation consolidation and integration (in this case, postuterine), in preparation for the next fulcrum or round of differentiation/integration (in this case,
F-1).

And, as always, a developmental malformation *at any of the specific subphases* of
a fulcrum (subphase 1, 2, or 3; i.e., a malformation or disruption at the indissociation,
differentiation, or integration subphase) can result in very specific pathologies that
correspond with the characteristics of that subphase formation (and malformation).
(All of these general features of a fulcrum are examined in detail in *Transformations
of consciousness*, chaps. 3, 4, and 5).

In the case of F-0, a fixation at the fusion/indissociation subphase might predispose
an individual to "somatic mystical" fusion with the world; a disruption at the differentiation subphase might dispose the individual to "hellish no-exit" vital shock, intense sadomasochistic activity, involutional depression; and fixation to the integration
stage might lead to delusional messianic complexes (to give only a brief, random
sampling of possible pathologies at these subphases of F-0; I will return to this research in a moment).

Likewise, the formations (and malformations) at this F-0 would, *as is the case with
all structuration*, incline (but not cause) *subsequent* development to tilt in the same

direction: the birth trauma, for example, could infect (and malform) subsequent development with its own pathological scars. Thus, a person with profound "no-exit" differentiation subphase malformation might hit subsequent developmental fulcrums with a strong predisposition to tilt *those* fulcrums in the direction of depression, withdrawal, inhibitions, etc. So that, in general, a fulcrum formation (and malformation) would, rather like a grain of sand in the formation of a pearl, dint or "crinkle" subsequent developmental layers, forming a "complex" of similar, layered malformations, the core of which might indeed reach back to a particular F-0 subphase.

In this regard, the research of Stanislav Grof has led him to postulate three broad "realms of the human unconscious," which he refers to as Freudian or biographical (or the individual unconscious), Rankian or existential (the psychological matrices laid down by the actual process of birth and the birth trauma), and the transpersonal domains (which involve collective, supra-individual archetypes, experiences, insights, etc.). Grof maintains that in several instances of intense stress (especially physically or drug-induced), the individual tends to regress through the Freudian or biographical stages (experiencing individual psychodynamic material) to the Rankian birth stages (reexperiencing the actual birth trauma in its several major substages), then occasionally entering the transpersonal domains (experiencing a broad range of mystical, collective, archetypal, supra-individual, transegoic phenomena).

Grof places particular emphasis on the second broad realm, the Rankian/birth matrices. These "basic perinatal matrices" (BPMs) follow essentially the fulcrum-0 subphases as I briefly outlined them: BPM I is subphase 1, or the oceanic indissociation state, both in its undisturbed and disturbed states. BPM II is the beginning of subphase 2, or the differentiation process, with an intensification of "cosmic engulfment" and "no-exit" hellish pressure, etc. BPM III is the latter stage of subphase 2, with the beginning of the expulsion from the womb, an intensification of "volcanic" pleasure/pain, dismemberment distress, ecstatic/masochistic, etc. And BPM IV is subphase 3, the postbirth, postpartum neonatal state, which must integrate its newly separate bodily being apart from the uterus (although this self-sense still cannot stably distinguish its own self-boundaries from the those of the physical world around it: it is now beginning F-1).

Grof has also presented putative evidence (still quite controversial) that these basic perinatal matrices form a latticework grid upon which subsequent psychological development will unfold. All I would like to point out is that his BPMs (and the subsequent COEX systems—or "systems of condensed experience") are not incompatible with the model presented here (the subphases of F-0 and the standard pattern of subsequent developmental influence). Should research support Grof's model, its specific stages can be accommodated.

Theorists have often noted that Grof's map and my own share some common features but do not match up, as it were, on the chronological aspects (or the "sequence" of the unconscious disclosures). But this is simply because Grof's map is recovered from a *regressive* series of patterns: under drug or stress induction, individuals *regress* from ordinary ego to Freudian (and childhood) traumas (or recovery of biographical material), and from there continue *regression* into birth trauma and intrauterine states; at which point they may cease identifying with the physical bodymind altogether and thus fall into transpersonal, supra-individual states. Grof's typical "sequence" is thus: everyday ego, Freudian, birth, transpersonal.

My map, on the other hand, is based primarily on broad-scale *growth* and development patterns, and thus it runs in the other direction, so to speak, but covers the same general territory: my "sequence" is birth trauma to Freudian to ordinary ego to transpersonal (since these cover the order in which these domains actually enter awareness as a stable adaptation and not as a temporary experience). However, using my map, if typical individuals at the ordinary ego indeed begin to *regress*, this map would predict the *same general sequence* as the Grof map: it would predict ego, Freudian, birth, transpersonal.

There are, however, a few important differences. Between the ego and the transpersonal (in the direction of growth, not regression) I place the general existential (centaur) level (F-6), and this is not at all to be confused with Grof's "existential" perinatal matrices (subphase 2 of F-0), although they often "intermix," so to speak. In Grof's standard "sequence," the BPM separate the individual and the transpersonal, whereas in the growth and development sequence, the "dividing line," as it were, is the centaur (as we will see in subsequent chapters). This has led some theorists to attempt to simply equate the BPM and the centaur, and this will not do: the centaur is the entrace to the transpersonal through the front door, the BPM, through the back door, so to speak.

Some "existential crises" might indeed reactivate a perinatal matrix (or vice versa), as Grof maintains, but the general existential crisis of F-6 is caused, not by the separation of mother and infant, but by the separation of subject and object, which will occur after even the most idyllic of birth circumstances, and *this* existential crisis cannot be alleviated by regression to birth experiences, but only by the undoing or transcending of the gross subject/object duality, and although this transcendence might happen as a byproduct of the experienced birth regression, the transcendence itself is not *caused* by that regression nor is that regression *necessary* for that transcendence. [See *The eye of spirit*, chap. 7, for an extensive critique of Grof's model.]

I will return to these issues (and to Grof's work) after the transpersonal dimensions have been introduced in the text, and pinpoint a few more of our agreements as well as our differences. See note 17 for chap. 14.

4. I am giving, in this chapter, the development of the Upper-Left quadrant; each development in this quadrant will have correlates in the other three quadrants. The other individual quadrant, the Upper Right, consists particularly of the correlative changes in brain physiology as each of the interior (Piagetian, Freudian, Jungian) stages is negotiated, and in some cases, the physiological component can be dominant. Strong (sometimes conclusive) evidence has been found for genetic predispositions in psychoses, manic-depression, obsessive-compulsive disorder, and various phobias

(with other dysfunctions, no doubt, to follow). In theses cases, genetics loads the gun, development pulls the trigger. A genetic predisposition tilts the outcome of a particular fulcrum toward pathology, and that genetic factor is often the major factor in the syndrome; nonetheless, since not everybody with the gene(s) for, say, manic-depression develops the syndrome, developmental factors still come into play.

Likewise, these syndromes (and many others) can be treated with medication. That is exactly part of the Right-Hand path. But what that medical approach (in and by itself) does not do is further a person's self-understanding: it does not help a person *interpret* their own depth and thus come to some form of self-clarity and self-insight and self-responsibility (all of those are the Left-Hand path). I am a strong supporter of both paths, judiciously balanced.

Most mental health workers intuitively understand that both paths (medication and therapy) are useful and important, but there is a constant tension between the two as to which is "really real" and most important. My point is that neither one *can* be reduced to the other, and a thus judicious balance is not just a pragmatic concession but a theoretical necessity.

Likewise, these individual developments unfold in correlation with (and in the context of) the social and cultural holons of the individual. Cultural meanings (LL) and social institutions (LR) powerfully interact with the unfoldment of individual patterns (mediated principally through the family structure). This shows why "adaptation to conventional reality" is not a very good yardstick of mental health, because the society itself might be "sick." Cultural meanings that flagrantly violate the fabric of the Kosmos (from Nazism to Serbian ethnic cleansing) are nothing that one would particularly want to adapt to in the first place; likewise, alienated and alienating techno-economic infrastructures (from actual slavery to slave wages) place a severe strain on the individual holon and its development. To say that in the modern world slavery disrupts brain chemistry, devastates self-esteem, embodies sick cultural meanings, and is a dehumanizing institution—is to say the same thing; they are four quadrants of the same holon (once the world soul has reached a collective level of rational pluralism; in mythic structures, slavery is the normative rule).

Thus, all four quadrants are crucially important in the development of the "individual." That I am centering, in this chapter, on the Upper-Left quadrant does not mean that I am in any way privileging this quadrant.

Finally, "mental health" can be defined as "adaptation to reality" if, and only if, we have an adequate definition of reality in the first place. *Conventional reality* will never do as a yardstick, as we said, because adaptation to being a happy Nazi is no mental health at all. The very definition of mental health is thus inescapably philosophical to the core (which is something medical practitioners and Right-Hand path specialists simply do not want to get involved with).

But pursuing this inescapable line of thought brings us to a very Platonic conclusion: mental health is attunement with the Kosmos—a Kosmos that includes matter, body, mind, soul, and spirit. Anything *less* than that *attunement* is pathological. A culture that has less than that attunement is a sick culture. If Plato, Plotinus, and the vast majority of the world's philosopher-sages are right, and all these dimensions (matter, body, mind, soul, spirit) are available to men and women, then not honoring them would be tantamount to malnutrition—which, by any other name, is an illness.

5. P. 132. All quotations are from *The essential Piaget* unless indicated otherwise.

6. Quoted in Flavell, p. 285.

7. Pp. 132–33.

8. Pp. 134–35.

9. P. 151.

10. Pp. 151–52.

11. Pp. 139–40.

12. These two figures of speech, based on similar agency (metaphor) and similar communion (metonym)—or simply similarity (agency) and contiguity (communion)—are, as linguists have pointed out, the most basic holons of linguistic communication (along with a hybrid, synecdoche, which substitutes parts for wholes).

Interestingly, these two basic figures of linguistic cognition (agency and communion) are so *fundamental* that Jakobson found, in the two major forms of childhood aphasia (loss of power to understand speech), that in one the child looses the capacity to understand metaphor but grasps metonym quite well, whereas in the other the child cannot understand metonym but perfectly grasps metaphor. Jakobson actually called them "similarity disorder" and "contiguity disorder"—in other words, agency disorder and communion disorder. "As a result it becomes possible to propose that human language in fact does exist in terms of the two fundamental dimensions [of agency and communion], and that these dimensions crystallize into the rhetorical devices on which poetry characteristically and preeminently draws. Both metaphor and metonymy can be subdivided into other figures (simile is a type of metaphor; synecdoche is a type of metonymy) but the distinction between the two modes remains fundamental, because it is a product of the fundamental modes of language itself: it is how language works" (Hawkes, *Structuralism and semiotics*, pp. 78–79).

Given that, for Freud, the primary-process cognition operates primarily with metaphor and metonym, and given that neurosis is basically a failure to outgrow the primary process in certain important ways, it is a small step to Lacan's formulations, under the banner of structuralism, in which symptom is a metaphor and desire a metonym.

Likewise, the extremely *fundamental* nature of metaphor and metonym have allowed certain postmodern poststructuralists, particularly the deconstructionists, to play fast and loose with all higher and more *significant* cognitions. To say that all higher thought is based on metaphor and metonym is one thing; to say it is nothing more than metaphor and metonym is profoundly reductionistic, and serves mainly the purpose, in these antitheoreticians, of leveling the distinctions between science and poetry, on the one hand, and philosophy and literature, on the other. They simply claim there are no significant differences between those endeavors (and particularly no difference between science and poetry), which nonetheless does not prevent them, when they become ill, from running to a doctor and not the local poetry reading.

More developed and more significant cognitive holons will place restraints on contexts of metaphor and metonym in order to decrease egocentrism and expand the network of mutual understanding and reciprocity via intersubjective exchange. Metaphor and metonym remain the roots, but are not the branches.

And at this point in our narrative of ontogenetic development, we are at the point where only the roots have emerged.

Incidentally, the actual meanings of the various terms that different authors use—*condensation, displacement, juxtaposition, syncretism*, etc.—sometimes vary, depending on the theoretical perspective they adopt. *Condensation* thus refers sometimes to metaphor and sometimes to metonym; likewise *syncretism*. But the two main figures, agency and communion, are the same.

Thus, as Lacan uses the words, displacement is a metonymy that marks the nature of the subject's desire, which is primarily a lack; condensation is a metaphor that traces the repressed meaning of desire through sliding chains of signification, found in symptoms. Same fundamental figures of speech (metaphor and metonym), which I accept, but tied to a "single boundary" theory of "lack" that I reject entirely (see note 32 for chap. 13, for a critique of the single-boundary theory).

13. Pp. 140, 152.

14. P. 146.

15. Piaget describes this shift as follows: "During the first stage [up through magic], all the explanations given are psychological, phenomenistic, finalistic, and magical. During the second stage [magic-mythic] the explanations are artificialist, animistic, and dynamic, and the magical forms tends to diminish" (p. 143).

16. Cowan, *Piaget with feeling*, p. 168.

17. Freud, *An outline of psychoanalysis*.

18. Ibid.; Frey-Rohn, *From Freud to Jung*.

19. Campbell, *Primitive mythology*, pp. 86–87.

20. Since images are the first forms or first holons in the noosphere, and since mammals, starting perhaps with horses, also form images, these animals are also starting to exist in the noosphere. Jantsch believes this begins, indeed, with horses, and from that point on, as he notes, the merely *ecological symmetry* is *broken forever*, because images may, or may not, accurately reflect the biosphere. With the emergence of images there is a "symmetry break," a break that, in humans, is dangerously out of control. But again, the cure is not regression but integration.

21. Since this entire third fulcrum is still close to magical cognition, every neurotic symptom, Freud would point out, contains displacement and condensation, or metaphor and metonym. As Ferenczi put it, every neurotic symptom is an unconscious belief in magic.

22. Op. cit., p. 222.

23. The client comes in fully aware of the cognitive maps ("I'm rotten to the core,

I never succeed at anything," etc.); the maps just happen to be wrong and distorted. Thus most cognitive therapists do not do much digging into actual past history, but rather attack the false maps head on in the present.

This head-on attack does not work as well with the more primitive libidinal holons aimed at by *uncovering therapies* (fulcrum-3 therapies, such as psychoanalysis), because in those cases the client is *not* already aware of the material that needs to be addressed; directly confronting the client with shadow material tends to increase the resistance to it. The "deeper" one probes, the longer and more delicate the process.

Most people can benefit so immediately and so greatly from simple cognitive and interpersonal therapy (fulcrum-4 therapies), that these and similar-type therapies are rapidly replacing the more arduous uncovering techniques. This does not lessen the importance of fulcrum-3 therapies in the least; but pragmatically, they are losing ground; the time and expense make them prohibitive for most people. (Not to mention the fact that in America, many of the things classical psychoanalysis considered pathologies are now considered virtues: narcissistic self-promotion, immediate gratification, impulsive acting out, etc., with cognitive therapy brought in to help one achieve these "spontaneities" without feeling any guilt, a rather interesting alliance.)

Finally, fulcrum-2 therapies are referred to as "structure building" therapies (to differentiate them from F-3 uncovering therapies). The pathology with F-2 is not repression, but the fact that the self is not strong enough to repress in the first place. There is no "uncovering" the shadow or "digging it up": because there isn't much of a self-structure, there isn't much of a shadow. Therapy involves instead a strengthening of the ego so that it can "get up" to repression!—involves the differentiation of self and object representations and the strengthening of boundaries.

And, as I have argued in *Transformations*, meditation is not primarily a structure-building technique, nor an uncovering technique, nor a script-rewriting technique, although all of those things might come into play with meditation. But in itself, meditation is fundamentally a further growth and unfolding of structural potentials. If there is any "uncovering," it is an uncovering of tomorrow, not yesterday; an uncovering of eternity, not infancy. See *The Atman Project* in Volume Two of the *Collected Works*.

24. Accordingly, we could accurately speak of a spectrum or continuum of "ego" states or "self" states or "I" states, each of which is a transcendence of its predecessor, and thus *each* of which can correctly be defined as being either "pre-egoic" or "trans-egoic" depending on whether it is being compared to its successor or its predecessor (that is, the ego or self of one stage is "trans" the ego of the previous stage and "pre" the ego of the next stage), and this continues until all egos are transcended in pure Ego, or until Atman dissolves in Brahman. Although that is a technically correct usage, it confuses almost everybody, because it counters both the orthodox and the New Age usages. So I will follow the usage I outline in the next paragraph in the text.

25. The ego as the actual organizing process of the psyche I refer to as "the self-system." It is present at all stages of growth, except the extreme limits. It is, for exam-

ple, the self-system (or simply the "self") that navigates, at each stage of growth, between the "four forces" of self-preservation, self-adaptation, self-transcendence, and self-regression.

Thus, the self-system in the early stages is pre-egoic, in the middle stages egoic, and in the transpersonal stages trans-egoic, where it converges on the Self, which opens into pure Emptiness. The self-system, in other words, is the regime or codon of the interior human holon, and like all regimes, it is the opening or clearing in which correlative holons can manifest: it is Emptiness looking out through a separate self until that self simply reverts to Emptiness per se.

26. By "pure Self" I do not mean disembodied or nonsituated. It is not a hyperagentic autonomy disconnected from manifestation, but the Source and Suchness of manifestation, appearing as Buddha-Dharma-Sangha (I, it, we)—all of which we will discuss later in detail.

27. Op cit., p. 270. The notion that rationality is "linear" and therefore "anti-ecological" is a cornerstone of many "new paradigm" notions, and nothing could be further from the truth. What these theories actually mean by "rationality" is any theory that disagrees with their rationality.

28. Ibid.

29. [See *Integral psychology*, in Volume Four of the *Collected Works*.] The way Kohlberg defined postconventional was *relative* to the conventional stages, which were grounded in late conop and early formop. Anything beyond those stages were thus "postconventional." When Kohlberg added a seventh stage, the universal-spiritual, that stage was thus "post-postconventional," and even higher stages would be post-post-post. All of these "posts" simply come from taking conventional/conop as the reference point. If we take formop as the reference point, then the higher stages are simply postrational, or transrational, or transpersonal. Purely a matter of semantics.

30. Campbell, *Primitive mythology*, pp. 84–85.

31. Ibid., p. 27.

32. Ibid., p. 28.

33. Piaget demonstrated, and Campbell acknowledges in another context, that the capacity to understand "what if" and "as if" statements emerges only with formop. Concrete operational, as its name implies, is too concrete-literal to frame possible worlds and hypothetical worlds; which is why, indeed, myths—generated by *concrete* operations—are taken so concretely, so literally, as *empirical* truths, and not as as-if or hypothetical possibilities used metaphorically or symbolically. This is why science (or the extensive framing of hypothetical possibilities checked against real world evidence) does not emerge in the mythological worldview.

34. Campbell, *Creative mythology*, p. 630.

35. Ibid., p. 4.

36. Campbell, *Primitive mythology*, p. 18.

37. All quotes from *Creative mythology*, pp. 4–6, 609–23.

38. The magical worldview can be instantly reconstituted whenever conop fails

and regression to preop occurs (since holons regress downward along the same lines evolution progressed upwardly—tenet 2d). It is not, in my view, that the magical structure *itself* remains present but unconscious; this is true only if there is fixation to parts of that exclusivity structure: those repressed or alienated pockets of consciousness indeed remain fully magical, and this constitutes part of their pathology (which is why these types of neurotic symptoms are "an unconscious belief in magic"). The preop structures remain fully present and active, but now as sub-holons in conop; but the exclusivity structure itself (magic) is, barring fixation, *replaced* by the new exclusivity structure (namely, mythic).

This formulation, I believe, handles some very recalcitrant problems about what one can "find" in the unconscious. See *Transformations of consciousness*.

39. The same thing happens with moral development. In this briefest terms: preop underpins a preconventional (egocentric) morality, conop underpins a conventional (sociocentric) morality, and formop underpins a postconventional (worldcentric) morality. A person at formop has complete and simultaneous access to the structures of conop (roles, rules) and the structures of preop (images, symbols, and concepts), and indeed, formop includes all of those as necessary components in its own makeup. But a person at postconventional morality *does not* have simultaneous access to all the previous stages of moral development, because these were generated by the exclusive identification with a lower stage, and when a higher identity emerges, the lower one must be released and replaced. You cannot genuinely be acting with a postconventional/global perspective and a preconventional/egocentric perspective at the same time.

40. *From Freud to Jung;* Frey-Rohn's excellent book contains several lengthy and perceptive discussions of this theme (loaded, on occasion, with elevationist interpretations).

41. Jung also came to include any typical, common structures—such as the shadow, the persona, and the ego—as being archetypal. These structures, these basic imprints, are also collectively inherited. This, too, is fine; and this, too, has nothing to do with transpersonal (see notes 44 and 45 below).

42. Jung, *Archetypes of the collective unconscious*, p. 173.

43. Spirituality can be expressed through those—or any other—forms, but those forms are not its *source*. [See *The eye of spirit* for a full discussion of Jungian archetypes.]

44. In Sheldrake's morphic resonance theory, an archetype is viewed as a past form that constant repetition has reinforced, and which reaches out, therefore, to guide the formative process of subsequent similar holons. The more one holon is repeated, then the more "archetypal" and powerful it is.

Thus, the number of "mother encounters" is enormous since every human being ever in existence had that encounter, and thus the sum total resonance from that "archetype" is indeed profound (and so on with the birth archetype, the father arachetype, the shadow archetype, etc.).

I agree with all that. My point is that, compared with the number of mother or father encounters, the number of genuine mystical encounters is pitifully small. These past mystics have indeed set up a certain holonic resonance for all of us who follow. But as probably the *rarest* and *least* common experience of humanity, it would be the *weakest* and *least powerful* archetype of all—if explained *only* by that theory (or any version of Jungian inheritance theory). But when real Spirit descends, it blasts to smithereens the mother archetype, the father archetype, and every other itty bitty finite archetype—it is coming from the other direction with the force of infinity, and not some merely past and finite evolutionary habit.

The more anybody "plugs into" the higher, transcendental, and actual archetypes, lying now as structural potentials, the easier it is for subsequent individuals to likewise "plug in"; and that is what the great past heroes of the transpersonal have done. But what *they* were plugging into was *not* more past common and typical and repetitive patterns, but future and higher possibilities. They inherited the future, not the past.

Sheldrake acknowledges this by pointing out that morphic resonance does not account, and is not meant to account, for creative emergence—but in the creative emergence often lies Spirit. The creative emergence is from the structural potential of the higher and future; morphic resonance is from the past and is established only *after* it creatively emerges in the first place. After a morphic field has emerged, it then acts as an omega point (up to its own level of depth) for subsequent and similar-depth holons: it is a past actual acting as a future perfect for similar development.

45. There are collective prepersonal, collective personal, and collective transpersonal structures, just as we all collectively inherit ten toes, two lungs, etc. Collective is not necessarily transpersonal.

Further, as I pointed out in *Eye to eye*, the *basic structures* of human consciousness are collectively inherited *potentials*, but that does not mean the potentials were manifest in the past. The human neocortex, for example, possesses the potential for rationality or hypothetico-deductive thought, but the neocortex, which emerged in today's form about fifty thousand years ago, did not emerge with rational thought in full blossom. When the neocortex first emerged, it carried the *structural potential* for logic, but the logic itself was not an inheritance from the past; it was a future potential acting as an omega point of mutual understanding drawing culture forward and upward. Just so with the spiritual potentials of the human structure; we inherit the potential, which in any epoch has, so far, been manifested in the very few, and which *therefore* continues to exert, as a higher and wider context, an omega pull on each of us.

CHAPTER 7. THE FARTHER REACHES OF HUMAN NATURE

1. Blanck and Blanck, *Ego psychology.*
2. Piaget, *The essential Piaget*, p. 133.
3. Gardner, *Quest for mind*, p. 63.

4. Ibid., p. 64.

5. Cowan, Piaget, p. 275.

6. Fowler's empirical and phenomenological research, executed as a reconstructive science, found that individuals moved through six or seven major stages of the development of spiritual faith (or spiritual orientation); and his findings match very closely aspects of the map I am here presenting.

Briefly: stage 0 is "preverbal undifferentiated" (our archaic), and stage 1 Fowler calls "projective, magical" (our magical), which he also correlates with preop.

Stage 2 he calls "mythic-literal," correlated with early conop, where faith extends to "those like us." Stage 3 is "conventional," which involves "mutual role taking" and "conformity to class norms and interests," late conop and early formop (stages 2 and 3 being our mythic and mythic-rational, respectively, with both stages being our overall mythic-membership).

Stage 4 is "individual-reflexive," as "dichotomizing formop" (and the ego) emerges, and involves "reflexive relativism" and "self-ratified ideological perspective". Stage 5 is "conjunctive faith," as "dialectical formop" emerges and begins to "include groups, classes, and traditions other than one's own"; this involves "dialectical joining of judgement-experience processes with reflective claims of others and of various expressions of cumulative human wisdom." (Postconventional, universal rationality and universal pluralism, mature ego, beginning of worldcentric orientation.)

Stage 6 is "universalizing," which is "informed by the experiences and truths of previous stages" (centauric integration of previous stages, which Fowler also calls "unitive actuality" and "unification of reality mediated by symbols and the self"— that is, the integrated centauric self). This self is "purified of egoic striving, and linked by disciplined intuition to the principle of being," involving a "commonwealth of being" and a "trans-class awareness and identification," correlated with "synthetic formop" (our vision-logic, the fruition of the worldcentric orientation, and the beginning of transpersonal intuition; the higher contemplative stages themselves were not investigated by Fowler, given their rarity, but the fit up to that point is quite close, often exact). Fowler, *Stages of faith.*

7. Broughton's original research, a doctoral dissertation at Harvard, is summarized in Loevinger's *Ego development.* All quotes, unless otherwise indicated, are from Loevinger's summary. All italics are mine.

8. The self could theoretically be reduced to its components, it was believed, precisely because the world was mistaken as a great interlocking empirical order (flatland holism), so that the subjective self, as a strand in the wonderful web and the relativistic system, had to be spliced into (and thus reduced to) an empirically observable occasion.

Most Anglo-Saxon psychology, starting with John Locke, took this cynical misstep as an actual paradigm of the human sciences. When the Kosmos was reduced to the cosmos, then the self, as self, was cut out of the picture. The self was merely a cybernetic cog in the display, reducible to its present experiences (behaviorism), and thus ultimately *isolated* from any sort of transcendental bond with other selves. No strand in the web is ever aware of the whole web, which is why empirical holism ends up divisive, dualistic, and isolationist.

9. The differentiation from (and integration of) (1) the physical environment, (2) the emotional-libidinal biosphere, (3) the (early) mind from the body, (4) sociocentrism from egocentrism, (5) the rational-ego from sociocentrism, and (6) the centaur from the ego. Each is a deeper identity with a wider embrace.

10. In *A sociable God*, I distinguished between the legitimacy and the authenticity of a spiritual engagement. Legitimacy is a measure of the degree that a spiritual engagement facilitates *translation*, and authenticity is a measure of the degree that it facilitates *transformation*. Mythic-religions often demonstrated a very high degree of legitimacy, for reasons explained in the last chapter, but were not very profound on the authenticity scale. Reason is more authentic than myth, but often faces its own intense legitimation crises, especially when confined within the borders of nations. The transpersonal domain is, in turn, more authentic than reason, and usually faces even more difficult legitimation, at least in the modern Western world.

11. Both the Left- and Right-Hand dimensions start from immediate apprehension or immediate experience, and that immediate experience, as William James explained, is properly called a datum, whether it occurs "internally" or "externally." Even if the data themselves are mediated by other factors (culture, mental sets, instruments, etc.), nonetheless at the moment of awareness, the immediacy of the experience is the pure datum in that occasion. Whether I am perceiving a tree "out there" or a desire for food "in here," they are both, at the moment of awareness, conveyed to me in pure immediacy (even though that immediacy may have mediating factors).

The Right-Hand path simply sticks to sensory immediacy, and ties its theorizing to the aspects of holons that can be detected with the external senses or their extensions: its data are sensory, which always means external senses (it doesn't even trust interior senses, since those are "private" and supposedly "not shareable"). Monological and empiric-analytic modes use an enormous amount of interior, conceptual, *a priori* cognitions, but eventually this approach ties all of them to the immediacy of externally perceived exteriors. And, as I said, that can be done, and is an important (if limited) mode of knowledge.

The Left-Hand path starts with the same immediacy of given experience, or immediate apprehension (it is the same immediacy as the external or Right-Hand dimension, because in the moment of direct apprehension, there is neither subject nor object, neither Left nor Right, as James explained). But unlike the Right-Hand path, which sticks to the objective/exteriors of holons as the basic immediacy gives way to subject and object, the Left-Hand path investigates the interior/subjective dimension of that basic immediacy as it gives way to subject and object.

And that is where and why interpretation enters the Left-Hand scene (and not so much the Right, where the blank stare at surfaces is most valued). Although in my own interior, I know depth directly by acquaintance (the immediate experience on any level is known directly, whether that be sensory impulses or archetypal light), nonetheless if you and I are going to share these experiences, we must communicate our depth to each other. And whereas the external surfaces are "out there" for all to see, the interior apprehensions can only be shared by communicative exchange which requires that we each interpret what the other is communicating. And that is why interpretation (hermeneutics) is inescapable for the Left-Hand paths.

The Right-Hand path of course uses interpretation in its own theorizing—it interprets the data—but the data are all ultimately data that do not respond communicatively or interiorly (the data are just the exterior surfaces, whether that be rocks, brain chemistry, or suicide rates: they don't complicate the scene by talking back: no nasty dialogues here).

Thus, what really irks the Right-Hand path about the Left Hand is not that its data aren't immediate (because they are; it is the same immediacy that is also the starting point for the Right-Hand path), but rather the necessity for introspection, interiority, and interpretation, all of which are rejected as not publicly shareable, which is of course nonsense: what the Right-Hand path itself accepts as data is determined only communicatively.

In short, both the Right- and Left-Hand paths start with the same immediacy of given experience, but the Right Hand sticks to "external surfaces" that, as objects of investigation, do not have to respond communicatively (it is monological), whereas the Left Hand goes a step further and investigates those holons that do respond communicatively (dialogically, hermeneutically, interpretively). The one sticks to surfaces, the other investigates the depths. And since surfaces can be seen, the one asks, "What does it do?"; but since depth does not sit on the surface, the other must ask "What does it mean?" ("What is under the surface?").

In the following discussion (in the text), I am staying primarily with the first immediacy of experience, whether that experience be sensory experience, mental experience, or spiritual experience. The "three strands of valid knowledge accumulation" that are discussed refer primarily to the acquisition of immediate data in any domain (sensory, mental, spiritual). The further interpretive steps taken by the Left-Hand path, and the actual theorizing of the Right-Hand path, also follow these three strands, but as an additional application, as it were (see Wilber, *Eye to eye*, for a fuller discussion of this notion).

12. Foucault came to see that both Right-Hand and Left-Hand approaches are necessary. His (early) archaeology of *actual* existence was a neostructuralist reworking of the traditional structuralist's analysis of *possible* types of experience, but it still placed emphasis on the exterior surfaces and structures of discursive formations and the transformation rules (of rarefication and exclusion) that individuated serious speech acts. This neostructuralism avoided—indeed, scorned—any attempt to get at the interior meaning of the discursive formations. (Indeed, Foucault bracketed not only the truth of linguistic utterances—the standard phenomenological move—but

their meaning as well, which is the ultimate exterior or monological move: you absolutely never have to talk to the bearers of the linguistic formation because you don't even care what their utterances mean; this is simply the endgame of structuralism: just the exteriors of the structures, with no hermeneutic nastiness.)

In his later and more balanced view, the discursive episteme was replaced by the dispositif, or overall context of social practices (encompassing, as it were, the episteme), whose meaning could still only be seen in the coherence, but whose "insides" also had to be hermeneutically entered. "This new method," comment Dreyfus and Rabinow, "combines a type of archaeological analysis which preserves the distancing effect of structuralism [the exterior, objectifying approach, or the Right-Hand path], and an interpretive dimension which develops the hermeneutic insight that the investigator is always situated and must understand the meaning of his cultural practices from within them [the Left-Hand path]." *Michel Foucault*, p. xii.

13. Even if words are necessary to structure the space that does capture it. That is, I am not saying that the lifeworld and worldspace are not linguistically structured in large measure; I am saying that *specific* experiences, themselves linguistically structured in many ways, are not captured in signifiers without a corresponding lifeworld signified. I am certainly not drawing a distinction between "the map and the territory," which is a crude representational-empiricist theory of language. Only the least significant signifiers represent physical things. In the mind, maps are the territory, and they do not simply *represent* the exterior world, they *present* or create new worlds altogether: language in its world-creating, world-disclosing capacity.

Nor am I a drawing a distinction between words and experience (see also note 16 below). Words *are* the central experiences in the noosphere, but those verbal experiences cannot be represented without a common lifeworld. My further point in the text is that this also needs a developmental notion attached to it, because language is the home of being only if I pack my bags and move in. Shared lifeworlds mean a shared set not just of structural signifiers but also of *developmental* signifieds.

As for the meaning of "direct experience" and the possibility of such, see note 16 below.

14. Thus, I can *translate* a sign correctly (interpret it correctly or meaningfully) only if, among other things, I have previously *transformed* to the level of depth that discloses and supports the interior signified. I then exist in a system of relational exchange of *signifiers* that resonate at the same level of *signified* depth ("same-level relational exchange"). Otherwise it's all Greek to me, even though I can see perfectly all the physical marks of the written or spoken signifiers.

Just so, all holons are signs, each involved in translating reality (i.e., responding only to what it can register in the first place) according to its regime or codon set in a common worldspace of similar-depth holons. Outside of its worldspace it does not and cannot translate: it registers nothing. In other words, holons, as signs, only respond to, and register (translate), a narrow range of signifiers, via a regime that discloses a same-depth signified worldspace.

15. I have written extensively on this in *Eye to eye* [and *The Marriage of Sense and Soul*]. Serious readers will forgive the shortcuts I take in the text. Also, the three strands of knowledge accumulation are not to be confused with Habermas's three validity claims. The validity claims refer to the Big Three (subjective, objective, cultural; or I, it, we). I am maintaining that the three strands are operative in each of the Big Three validity claims (although they take quite different forms), and their operation helps assure us that the *referent* of knowledge is actually in the Kosmos (in a worldspace in the Kosmos), and not just in my imagination, or my misperception, or whatnot.

16. There is a great deal of semantic and philosophic confusion around the topic of "direct experience," usually centering on the question, does it even exist? Isn't all experience actually *mediated* by concepts, mental sets, cultural background values, and so forth? Since all knowledge/experience is situated and mediated, the best we can hope for is a hermeneutic study of individual pockets of "local knowledge," and any transcendental claims, of any sort, are largely unwarranted—or so the argument goes.

This has, among many other things, particularly caused turmoil in the comparative study of mystical experiences. Steven Katz and his colleagues, for example, in a series of articles and books, have made the strong claim that since all experience is mediated—Katz: "There are NO pure (i.e., unmediated) experiences"—then there can be no commonalities (or cross-cultural similarities) in mystical experiences. Since each culture and each belief system is different, and these different sets are partly constitutive of the mystical experience, then there is and can be no common mystical experience, and this also, it is said, undercuts the mystical claim to valid or universal knowledge.

This overall general approach has come loosely to be called "constructivist"—we don't receive knowledge of independent entities, we construct it based on various freeing practices; our basic present experiences are taken up and reworked in a type of neo-Kantian fashion, so that the final display in consciousness is an inseparable mixture of experience and mental-cultural molding.

But this approach suffers, ironically, from not being constructivist enough. To begin with, the dichotomy between experience and construction is a false dichotomy. It is not that there is experience on the one hand and contextual molding on the other. *Every* experience *is* a context; every experience, even simple sensory experience, is always already situated, is always already a context, is always already a holon. When Derrida says that "nothing is ever simply present," this is true of every holon. As Whitehead would have it, every holon is already a prehensive unification of its entire actual universe: nothing is ever simply present (this is also very similar to the Eastern notion of karma, or the past enfolded in the present). Everything is always already *a* context *in* a context.

And thus, every holon—and therefore every experience—is always already situated, mediated, contextual. It is not that "original experiences" arrive to be reworked by mental concepts; the original experiences are not original, but a contextual prehensive mediation of boundless contexts. That mind *further* contextualizes sensory contexts is neither new nor avoidable.

Everything (every holon) is a mediated context, but contexts touch *immediately*. It does not require "mystical pure consciousness" to be in immediate contact with the data of experience. When any point in the mediated chain is known (or experienced), that knowing or prehending is an *immediate* event in itself, an immediate "touching." The touching is not a touching of something merely present but rather is itself pure Presence (or prehension). If there were *only* a mere mediation forever, then nothing would or could ever be known or experienced; there would be nothing to stop the sliding chain from spinning contextually forever (there is no point that it *could* enter consciousness).

But in any moment of prehension/experience (and in any domain—sensory, mental, spiritual), there is *immediate* apprehension of what is given at the moment, and *that* immediate apprehension is the datum (which William James correctly defined as the given pure experience), and *that* experiential prehension is *pure* in the sense that when contexts touch, they touch without further mediation (even if they are always already situated in and as mediated contexts).

At the moment of touch, there is no mediation; if there is mediation, there is no touching. To say everything is *merely* mediated is simply a fancy modern twist on pure skepticism, which is profoundly self-contradictory (it says, "I have an unshakeable foundation belief that foundations are not possible," which simply allows the skeptic to trash everybody else's beliefs while conveniently leaving his own unexamined).

At the moment of touch, in any domain, there is no mediation, only prehension. This is why knowledge (and experience) of any sort is possible in the first place. That there is a "first place" means that mediation *has stopped* at some point (the point of touching). This is why when experience occurs in any domain (sensory, mental, spiritual), it is simply given, it is simply the case, it simply shows up, even though the experienced and the experiencer are forever situated and contextual. I find myself in *immediate* experience of *mediated* worlds. (And that Immediacy, that pure Presence, that touching, is, as we will see, one way to view Spirit: *immediateness* is Spirit's prehension *of* the world—and what is prehended is Spirit *in* the world contextually.)

In short, experience is immediate prehension of whatever mediated contexts are given, and that is why all experience is *both* pure (immediate) and contextual (capable of being refined and recontextualized *indefinitely*). As we will see, this is why Habermas maintains that all validity claims have both an *immanent* (culture-bound) and a *transcendent* (pure) component, and it is the transcendent component that allows intersubjective communication and learning to occur in the first place.

Thus, even Katz, as he ponders the "mediated nature of all experience" comes to a point where that notion itself, and his knowing of it, are not mediated. *At some point* in the mediation chain, the datum per se enters his consciousness directly, he directly touches the notion, and he says, "I know this" or "I believe this" or "I experience this." Whatever the space is that allows Katz to claim that all experience is mediated—that space is somehow free of mediation or he would never be able to make that statement. To say that all experience is mediated is to stand in a space that is not itself mediated. Put differently, if everything is only mediated, he would have no way to know it. The mediating chain would never stop long enough for him to touch it; it would have no point of *entry* to his awareness: if the self were merely situated, it would never know it. Even to be able to make the mistake of saying it is only situated demonstrates that it is not.

This undercuts Katz's whole line of attack on mysticism, because his attack applies to his own attack: it is self-defeating. That is, Katz claims that *all* experience is mediated and that this is true for *all* cultures, with no exceptions—and thus he is claiming to be in possession of a nonrelative truth that is true cross-culturally and universally, something that his formal thesis denies is possible for anybody. He is making universal/transcendental claims about reality which he denies to everybody else, thus hiding the real issue of how transcendental claims are in fact made, not just in "mystical" experiences but in *everyday ordinary communication.*

The question, then, is not whether culture mediates experience in a way that disallows cross-cultural similarities in mystical experience. The question is whether the unavoidably mediated *aspect* of all experience invalidates any similarities *in any experiences at all* (mysticism is completely secondary in this regard). In other words, are there any transcendental signifieds *at all*? Put differently, is there any chance of *identical signification* for *anybody*? Forget mysticism; can individuals from different cultures (or even the same culture) even talk about *anything*?

At this point in the argument, Derrida is often called on to support the notion that there are no transcendental signifieds at all (only sliding chains of signifiers and endless cultural mediation). But this is a misreading of Derrida. We already heard Culler explain that deconstruction does not disavow propositional truth but simply reminds us of its contextuality. Derrida himself points out that even if the transcendental signified is *situated*, "this does not prevent it [the transcendental signified] from functioning, and even from being indispensable within certain limits. For example, no translation would be possible without it" (*Positions*, p. 20). In other words, according to Derrida, the fact that we *can* translate languages to some significant degree means that there are genuine transcendental signifieds, and if this holds for ordinary experience, there is no reason it shouldn't hold for other experiences as well (from science to mysticism). We *can* translate languages because, even if all contexts are situated, a great number of contexts are *similarly situated* across cultures. "Context" does not automatically mean "relative" or "incommensurable." *It often means "common"*: hence the existence of real transcendental signifieds. Even Derrida concedes this elemental fact.

In that similarity—in the existence of those transcendental signifieds—lies the *possibility of communication* both *within* culture and *across* culture, even if mediated by language. This is why Habermas's validity claims are both *immanent* (contextual) and *transcendent* (common contextual), as McCarthy explains: "If communicative action is our paradigm, the decentered subject [the intersubjective subject] remains as a participant in social interaction mediated by language. On this account . . . language use is oriented to validity claims, and validity claims can in the end be redeemed only through intersubjective recognition. . . . The internal relation of meaning to validity means that communication is not only always 'immanent'—that is, situated, conditioned—but also always 'transcendent'—that is, geared to validity claims that are meant to hold *beyond any local context* and thus can be indefinitely criticized, defended, revised" (*Philosophical discourse*, p. xvi).

As Habermas puts it: "Validity claims have a Janus face [they are, of course, holons]. As claims, they *transcend* any local context; at the same time, they have to be raised here and now and be de facto recognized. The transcendent moment of *universal* validity bursts every provinciality asunder; the obligatory moment of accepted validity claims renders them carriers of a *context-bound* everyday practice" (his italics; *Philosophical discourse*, p. xvii).

This is why, even though we are linguistically situated, the validity claims place us in relation to *extralinguistic* aspects of reality (that are not in all ways *merely* situated). McCarthy: "The [validity] claims to truth, truthfulness, and rightness place the speaker's utterance in relation to extralinguistic orders of reality" (Introduction to Habermas, *Communication and the evolution of society*, p. xix). In other words, even everyday communication is always situated in aspects of reality that are not merely culture-bound (or solely provincial).

In my view, these extralinguistic orders (that is, orders that are not merely linguistic, even if linguistically situated) do indeed refer to the Big Three (I, we, it), as Habermas maintains, but *each* of those domains can be divided into prepersonal, personal, and transpersonal components. Or prelinguistic (the sensorimotor worldspace, or eye of flesh), paralinguistic (the noosphere, or eye of mind), and translinguistic (the theosphere, or eye of contemplation).

But even if we leave aside the transpersonal domain for the moment, Habermas's point is that each of these various orders *places constraints* on our situatedness, prevents the "construction" of the world from being *merely arbitrary,* and thus *allows learning to occur at all* (if there were "only" the text, there would be nothing to surprise the text, and learning would never occur).

Those not-merely-arbitrary worldspaces anchor the validity claims of any communicative act, and prevent contexts from spinning out of control endlessly . . . and meaninglessly. And if these extralinguistic or transcendent (or transcendental signified) factors *function even in common everyday communication and everyday experience* (as Habermas maintains and even Derrida concedes), then we transpersonalists don't have to defend mysticism as being somehow exceptional and in need of any special defense *in this regard.*

Thus, if Katz's claim were true, not only would it invalidate cross-cultural mystical claims, it would invalidate ordinary and everyday communication itself, at which point his position turns on itself and thoroughly dissolves its own credibility. Katz's position is an amalgam of neo-Kantian aphorisms, pressed into the service of a deconstructive atmosphere of self-contradictory (and self-congratulatory) rhetoric. It is shot through with aperspectival madness, the dominant form of intellectual confusion for the postmodern mind.

Beyond that, and looking to the transpersonal, what is most disturbing about Katz's whole approach is that it is a group of scholars trying to rationally think through transrational concerns. They are armchair contemplatives, so to speak, and they are trying to make sweeping pronouncements about spiritual referents without possessing the corresponding developmental signified. It's just talk-talk religion, not transformative spirituality. By their own accounts, they just want to *study texts*, they do not want to take up the injunctions and the paradigms and do the real science. This is like judging a bakeoff by reading the recipes.

I will return to Katz's argument in a note at the end of chapter 8, after we have looked at mystical development, and give a simpler critique (see note 58 for chap. 8; see also note 32 for chap. 13).

For an excellent discussion of the Katz controversy, see the superb volume edited and contributed to by Robert Forman, *The problem of pure consciousness*. Forman takes the basic approach of centering on one of the most common of mystical experiences, that of pure contentless consciousness (which I refer to as the causal unmanifest); and the various contributors to the volume present extensive evidence (empirical, phenomenological, and theoretical) that it is indeed a common, cross-cultural experience, and that the experience per se is *identical* wherever it shows up, simply because it is itself formless, and therefore constructed forms (and cultural differences) do not enter into the actual "experience" itself ("experience" being not quite the right word, as most contributors point out; the subject/object duality of experiencer/experienced is temporarily dissolved; further, "pure Consciousness" or "pure Emptiness" is not itself an experience but the ultimate openness in which experiences arise and pass).

I am in complete agreement with that general approach, but I would like to point out that the "identicalness" problem of experiences is, once again, *not confined to mysticism* and its "end-limit" of cessation or unmanifest Emptiness. In Habermas's theory of communicative action, for example, the speaker has to assume *identical signification* or else the conversation never gets started. Subsequent communication, driven by an omega point of mutual understanding, keeps refining, via validity claims, the identicality. Once again, there is nothing special about mysticism *in this regard*.

That different cultures give mystical experiences different interpretations and different tastes, as it were, is true enough (Katz bizarrely concedes that there is "the mystical reality" that is then culturally molded), but, in itself, this claim is trivial. That charge applies to *any* experience, from sunset to food, and if that charge doesn't invalidate sunsets (it does not), then it doesn't invalidate Spirit either. That part of the argument centers *merely* on the immanent aspect of validity claims, and conveniently ignores both the transcendent component and the identical signification mandatory for communication in the first place (except, of course, that Katz's position *implicitly* assumes transcendental/universal validity for *his* position that there is no transcendental/universal validity—the standard performative contradiction in the relativist's position).

And this means we don't have to defend similarity of types of mystical experience based solely on the strong version of formlessness. Mystical experiences, of whatever variety, simply face no types of problems that aren't also found in ordinary experiences and ordinary communication, and the difficulties in the latter simply cannot be selectively used to invalidate the former (without simultaneously invalidating the attack itself).

What is special about mysticism is the claim, as I construe it, that development can continue beyond the rational-ego (and centaur) and into higher/deeper domains of awareness. This approach is doubly immune from Katz's type of criticism. Even if the cynical (and deeply confused) version of Katz's constructivism were true, and no worldviews are similar, *types* of worldviews are (and could be). There are levels of mediation, levels of worldviews. And thus the approach is to abstract, at a very deep level, the deep structures of the various worldviews themselves (this approach is already at work in Piaget, Habermas, Gebser, etc.).

Common deep structures with culturally situated surface structures seem to me to steer a course between "no similarities at all" and "mostly or only a common core." At the same time, I am not suggesting that all the contemplative traditions possess all the stages in the overall map I am presenting; many aim at one stage and make it paradigmatic. Nonetheless, many of the traditions do possess all of these stages; and, as is the case with all developmental stages, these stages, when they emerge, emerge in an order than cannot be altered by social conditioning (just as images emerge before concepts and no amount of reinforcement, in any culture, can reverse that order).

As for the Abyss itself, or pure Emptiness, it is indeed pure identity, and that is why I often refer to the "theosphere" as being translinguistic: as consciousness approaches unmanifest absorption, all contexts—all holons—are temporarily suspended or temporarily dissolved (there is then only pure Immediacy or pure Presence, the *same* pure Presence that apprehends mediated contexts when they are present); but even that "experience," contentless itself, awakens to find itself situated in a cultural context that will partially provide interpretations, but will not totally determine the experience, or else the experience could never surprise anybody, and it surprises everybody.

Donald Rothberg's contribution to Forman's volume is significant. Rothberg's treatment of Katz is fair and fairly devastating. (The one point I would interpret differently is his conceptualization, following Brown, of the stages of meditation as a deconstruction of the previous structures. What is deconstructed is not the previous *basic* structures but the previous *exclusivity* structures; that is, the meditator does not permanently lose the capacity to perceive space, time, self, or individuality; rather, consciousness is no longer identified *exclusively* with those structures; again, preserves the structures, negates the partiality.)

17. They become more complicated in some cases (when we form a theory, for example, we run through the three strands twice, first to generate the data, then to generate and test how the theory—itself a mental datum—fits the other data); and although it is never a simple one-step-at-a-time process, the same three strands are involved.

Likewise, the same three strands anchor knowledge in both the Right- and Left-Hand paths. In one we check the data, in the other we check the data *and* the interpretations of the within of the data, the point being that the interpretations are themselves mental data that are then subjected to the same three strands. Same essential process, but with different notions about what we will allow as data (exteriors versus interiors).

In the text, I am centering mainly on the initial data apprehension (whether sensory, mental, or spiritual). In the monological approaches, the data is, so to speak, taken at face value. No attempt is made to understand it, only report it or describe it. In the dialogical approaches, the interior depths of the data are approached via interpretation, using the same three general strands (because the interpretation itself is another datum, subject to the same three strands). In the translogical approaches, the subjective perceiver and objective data are both pursued to their common or nondual source (guided by the same three strands).

18. Even the dog itself *exists* only in a *particular* worldspace; it does not exist in the worldspace of an atom, a molecule, a cell, or an autistic (primary narcissism) infant; the dog exists in the sensorimotor worldspace. And the as-if dog exists only in the rational worldspace.

19. Driven by evolutionary telos to mutual understanding and mutual transcendence.

20. Kuhn, *Structure of scientific revolutions*, 2nd ed., p. 206.

21. Walsh and Vaughan's *Paths beyond ego* remains by far the best introduction to, and survey of, the transpersonal field, and interested readers might start there.

22. See, for example, *Eye to eye, Transformations of consciousness, Grace and grit.*

CHAPTER 8. THE DEPTHS OF THE DIVINE

1. P. 99. All quotes, unless otherwise indicated, are from *Ralph Waldo Emerson, Selected prose and poetry*, ed. R. Cook.

Notice that Emerson handles Habermas's "identical signification" in a very direct way: it is not that we merely assume identical signification in order to get the conversation going; it is that on the deepest level we share a common Self or Nature, namely, God, and that is why the conversation can get going! Habermas's omega point of mutual understanding, while still true, is outcontextualized by Emerson's omega point of mutual identity (and in this Emerson is in a long line of descendants from Plotinus through Schelling to Emerson, as we will see). For Habermas, the "who" of Dasein is found in the *circling* of the intersubjective circle; for Emerson, the "Who" is simply God.

Thus Emerson refers to the Over-Soul as "that common heart of which all sincere conversation is the worship." Hölderlin: ". . . we calmly smiled, sensed our own God amidst intimate conversation, in one song of our souls."

2. P. 96.

3. Ibid.

4. Again, this does not mean that the Self is not situated; it means it is not merely situated. As Spirit manifests, it manifests in and as the four quadrants, or simply the Big Three—I, it, we. As Spirit/Consciousness evolves into the transpersonal domain, the Big Three appear as Buddha (the ultimate I), Dharma (the ultimate It), and Sangha (the ultimate We). Many mystics have an understandable tendency to emphasize the I-strand, where one's mind and Buddha Mind become, or are realized to be, one Self (which is no-self individually), and Emerson is no exception in this regard.

But by recognizing that all manifestation is in the *form* of the four quadrants, this tendency to a certain solipsism is carefully guarded against. At the same time, the timeless and eternal nature of Spirit can nonetheless be intuited in (or through) any of the four quadrants (I, it, or we), and Emerson, like so many mystics, is expressing that intuition here in the form of the I-quadrant, as the Over-self or Over-soul. But inseparable from that, I will maintain, is an "Over-it" and "Over-we," or higher truth and wider community.

5. Pp. 73, 81–82. My italics. In this and a few subsequent quotes I am combining excerpts because they are related by a common theme.

6. Pp. 95, 107, 52, 95.

7. The typical charge against cosmic consciousness—that it violates the physical boundaries of the individual organism—applies to every other sense of selfness as well. The sense of self-identity is *never* merely body-bound, except at fulcrum-2. At every other stage of development, the sense of central self is a set of feelings and preferences *within* an identity with the family, or the tribe, or the culture, or a cause, or a nation, or an ideal; it is inextricably bound up with possessions, friends, relationships, job, a massive set of values—all of which are beyond the boundaries of the body. The very sense of I-ness is *never* just the simple sensations this body is feeling, but always (past fulcrum-2) is inextricably bound up with identifications that violate the boundaries of the body. Every moral impulse I have violates the boundaries of the body; every anticipation of future action violates the boundaries of the body; every reflexive awareness I have of myself violates the boundaries of the body. The fact that cosmic consciousness also violates these boundaries is not even a vaguely justifiable criticism. (See the "Single-Boundary Fallacy" in note 3 for chap. 6, note 32 for chap. 13, and note 17 chap. 14.)

8. Emerson rarely uses the term *World Soul*, but it is implicit in his notion of the Over-Soul. I will often use the term, because it is an important reminder of the community-component (the we) of Spirit's manifestation. Like the term Over-Soul, the World Soul always refers to the psychic-level.

Because I am letting these individuals represent the various transpersonal structure/ stages, keep in mind that they will be presenting each stage with their own individual and cultural surface structures. Thus, Emerson tends to speak of the psychic level in terms of the "Over-Soul"—that does not mean that other persons or other traditions will use those terms (or even have that experience).

To be more specific: if for the moment we use the term *mysticism* in a very general sense to mean any form of awareness beyond the conventional space-time centered on the individual ego/bodymind, then the four general stages of transpersonal or mystical development (psychic, subtle, causal, and nondual) refer to mystical states that take, as a major part of their *referent*, elements in the gross/waking realm, in the subtle/ dream realm, in the causal/dreamless, and in the nondual (respectively).

Thus, the psychic level, which is the realm of initial transpersonal or mystical awareness, often involves a great number of seemingly unrelated phenomena, from various types of actual paranormal cognitions and events (e.g., "astral travel," out-of-the-body experiences) to numerous preliminary meditative states; kundalini awakening (especially of the first five chakras); reliving of birth and pre-birth states; temporary identification with plants, animals, humans, aspects of nature, or even all of nature (nature mysticism, cosmic consciousness)—to name a very few.

What all of those have in common is that the mystical experience moves beyond ordinary or conventional reality (the gross/waking realm), but still takes as *part of its referent* the gross/waking realm. Thus, no matter how extraordinary or "far-out," all mystical experiences of the psychic level are still *related* to gross/waking reality—the experiences still refer, in whole or part, to elements in the gross/waking realm. Thus, even though "cosmos" technically refers to the physiosphere, I will continue to use "cosmic consciousness" (and "nature mysticism") for the psychic level, as a reminder of its gross-realm orientation.

This is why, for example, Da Avabhasa places most kundalini experiences at this level. As he points out (in *The paradox of instruction*), the kundalini experiences are still somatically or *bodily felt*, they are still predominantly registering in the gross body (or still involve aspects of the gross body); they are subtler energies, to be sure, but energies still oriented to or related to gross determinants. The same holds for cosmic consciousness (which Avabhasa, Aurobindo, and Plotinus all situate at this psychic level); however different from kundalini experiences it might be, they both are mystical experiences related in large measure to gross reality (and the same is true for all forms of paranormal cognitions and activities). This is why, in general terms, all psychic-level phenomena can be referred to generically as nature (gross-related) mysticism, however much their surface structures vary dramatically.

Subtle-level mysticism (deity mysticism), on the other hand, has few, if any, referents in the gross realm. The interior luminosities, sounds, archetypal forms and patterns, extremely subtle bliss currents and cognitions (*shabd*, *nada*), have no major gross-realm referents. The reason is that these subtle archetypal luminous forms are the seed syllables or seed patterns *from which* gross-realm phenomena are derived (involutionally); and thus, at this point, there are no gross phenomena to be referred to in the first place. Pure subtle-level mysticism thus has few actual referents in the gross (natural) world.

This is also why Avabhasa refers to psychic-level mysticism as that of *yogis* and subtle-level mysticism as that of *saints*. Most yogic phenomena are gross-relating (control of bodily functions, leading to bodily awakening of kundalini currents), whereas saintly phenomena are interior halos of light and sound that have less relation to the gross body. More specifically, according to the *Paradox of instruction*—as only one example—the first five chakras are especially gross-oriented (or yogic); the sixth chakra, which contains numerous sublevels, is the beginning of the subtle (or saintly)—it has few gross-bodily referents—and the seventh chakra begins to transcend the psychic-subtle dimension altogether and open onto the radiant causal (which Avabhasa refers to as *sagely*).

Causal-level mysticism (or pure unmanifest absorption) has no gross or subtle referents; it has no referents at all, except its own self-existing Emptiness. And nondual mysticism is the identity of Emptiness and all Form, so its referents are *whatever* is arising at the moment (the nondual siddha, or spontaneously-so awakened one).

Thus, the four major types of mysticism (nature, deity, formless, nondual) have, as their deep structures, the four major levels of transpersonal development (psychic, subtle, causal, nondual), which are identified by their *predominant* referent (gross, subtle, causal, nondual). These are easily identifiable deep structures, no matter how much their surface structures vary. The most variation occurs in the psychic level, and there is a reason for this:

The psychic is on the border between the gross and the subtle states. As such, not only it is the home of all sorts of various preliminary and initial mystical phenomena (which can be expected to vary considerably from person to person), it is itself the broad transition state from gross to subtle. As such, it is mysticism, but mysticism with one foot still in the gross. It is gross-oriented mysticism (and that is what *all* of these wildly different phenomena have in common, from paranormal to kundalini to nature mysticism to cosmic consciousness: mysticism, yes, but gross-related, with one foot in each state, and that is why its actual surface structures can vary so dramatically but still issue from this same general realm, the psychic, the border between gross and subtle). Thus, the psychic is, and can be, the home of anything from initital meditation experiences to paranormal phenomena, from out-of-the-body experiences to kundalini awakenings, from a simple state of equanimity to full-blown cosmic consciousness: they are *all* the subtle realm breaking into the gross realm at the common border: the psychic.

Further, somebody passing through the psychic (or any other level, for that matter) would not be expected to experience *all* of the phenomena that are potentially disclosed at that level, just as somebody passing through the verbal dimensions will not automatically develop an understanding or experience of every language in existence; nor is it *necessary* to learn all languages before one can transcend the verbal. Nor is it necessary to *master* even a single language (one does not have to be Shakespeare to transcend the verbal). It is necessary only that the general level be objectified, and this can occur with or without conscious *mastery* (just as one can pass through the verbal without ever consciously learning and describing all the rules of grammar).

Thus, for example, in Zen, which aims directly for the causal/nondual, all sorts of psychic/subtle phenomena will arise in meditation (*everybody* has some sort of these experiences, however brief), but most are quickly dismissed as *makyo*, inferior apprehensions, and the student is urged to pass quickly through them (they are *objectified*, and thus generically transcended, and no more is required in that regard). Other traditions value these psychic and subtle phenomena, and cultivate them mightily, even to the point of mastery and explicit articulation (which is a wonderful contribution). But no stage can be merely and totally bypassed in full or integral adaptation, a point stressed by Aurobindo, Plotinus, and Avabhasa (to name a very few). *Peak experiences* from any of the higher realms are possible at virtually any point in development (since these higher levels are structural potentials present from birth), but full adaptation requires solid objectification of these levels, since they themselves actually embody a series of subjects (a series of selves), and all selves have to die (be made into objects) before the pure Witness stands forth and then dies itself in Emptiness. Any general stage/level *not objectified* will remain as a *hidden subject* (a hidden self-sense), obscuring no-self or pure nondual awareness.

This is why Aurobindo (and Plotinus and Avabhasa, etc.) is always saying things like: "The spiritual evolution obeys the logic of a successive unfolding; it can take a new decisive main step only when the previous main step has been sufficiently conquered: even if certain minor stages can be swallowed up or leaped over by a rapid and brusque ascension, the consciousness has to turn back to assure itself that the ground passed over is securely annexed to the new condition; a greater or concentrated speed [which is indeed possible] does not eliminate the steps themselves or the necessity of their successive surmounting" (*The life divine*, II, p. 26).

9. As with so much in development, this is a matter of degree. *Every* stage of development liberates the Witness from a previous identification: symbols witness images, concepts witness symbols, rules witness concepts, and so forth. By the time of the ego, and especially the centaur, this Witness as *observing self* emerges in consciousness (although present throughout)—we saw the emergence of the observing self with Broughton's research, for example—and at the psychic becomes a direct and powerful experience; this process gains in clarity and intensity all the way to the causal, as we will see (where the Witness finally dissolves in and as pure Emptiness).

Likewise, Emerson's own insights and awakenings often pass into the causal and the nondual, but it is a matter of degree, and his paradigmatic presentation is of the psychic-level Over-Soul.

10. By an "identity with all manifestation," I mean in this section all gross and gross-reflecting manifestation; nation-nature mysticism does not generally recognize the subtle or causal dimensions (see note 8 above). It is an identity with all of the

waking-state (gross) realm, and not with the subtle domain nor the deep-sleep (formless) domain.

11. P. 95.

12. Pp. 5–6.

13. Pp. 12–13, 97, 83, 84.

14. Emerson uses the term *Reason*, by which he means vision-logic (that is, he refers to the German Idealist's use of *Vernunft*); when this vision-logic is "stimulated to more earnest vision," as he puts it, then that is pure vision itself, which is the characteristic cognitive mode of the psychic level (as I suggested in *Transformations*: where centauric *vision-logic* adds up the integral, psychic-level *vision* simply and directly sees the integral); I have used the more common term *intuition* to cover direct psychic vision, and Emerson himself says "we denote this primary wisdom as Intuition."

15. Pp. 24, 36, 25.

16. Many ecophilosophers want to use *Nature*, with a capital N, to mean the same thing as God or Goddess or Spirit, except without the "otherworldly" connotation. For them, Nature is spirit-as-this-world, and we don't need any *transcendental* aspect to it. These ecophilosophers (whom I will discuss in great detail in volume 2) maintain that culture is simply another production of Nature—that all things are in some sense productions of Nature—and Nature is thus the *one ground* of *all* that is. *Nature* is just another word for Spirit, but a thoroughly immanent Spirit. (In chapter 12, I will further distinguish between Nature, or the Right-Hand empirical world, and NATURE, or the All. In this note, Nature means NATURE, or the entire Kosmos.)

But this introduces an irreconcilable problem: what is the relation between Nature and nature? That is, between the great Force that produces All, and the merely biospheric component of the All? These ecophilosophers slip back and forth between these two very different definitions: Nature (as the All) and nature (as the biosphere), precisely because they want to privilege the biosphere and make the laws of nature into the laws of Nature.

Thus, on the one hand, they have a brutal dualism: nature is good, culture is (or can be) very bad (because culture can hurt and despoil nature), and thus we have to get back to nature and heal the planet, etc. But if culture can so wildly *deviate* from nature, then culture is obviously *not* a product of nature, is *not* something nature itself is doing (unless we wish to define nature as suicidal, which is rather unbecoming in a god). Culture, then, is not simply nature.

But since nature is also supposed to be the *great force underlying All*, these theorists then switch to Nature with a big N, and this Nature is then said to underlie *both* cultural productions (which are simply a "second nature") *and* the natural world as well.

And the question remains: what then is the relation of nature and Nature? If Nature underlies both nature and culture, then Nature cannot simply be nature. Because they want to privilege the biosphere (or nature) as against culture, but because they also want the ecocentric notion to embrace *everything* (and thus be "spiritual"), they must *simultaneously* conceive Nature as being different from culture *and* as the all-encompassing great Force that includes culture as well. And they simply shift back and forth between these definitions, between nature and Nature, as suits their purposes. They refer to nature to show how culture is ruining it (because nature and culture are indeed different), and then they switch to Nature to impose the biocentric solution that must also include culture in its totalitarian umbrella.

If the biosphere is really God/dess, how could human beings possibly be destroying it? How could human beings actually be destroying God?

Because these particular ecophilosophers confuse fundamental with significant, they do not see that the ever-widening ontological circle goes from nature to culture to Spirit, or, if you wish, nature to culture to Nature, where Nature as Spirit is here correctly used to indicate "transcends and includes." Culture can indeed destroy segments of the biosphere, precisely because the biosphere is not Nature but merely nature. And culture, in transcending nature, can repress and dissociate it and indeed harm it (and since nature is an internal *component* of the cultural compound individual, then that is *purely suicidal*). But culture cannot destroy God or Nature or Spirit, any more than the tail can wag the dog. And further, nature and culture can be healed, never in nature, but only in Nature (Spirit).

Many of these ecophilosophers point to Spinoza to support their version of "God = Nature," and here they give a quite distorted reading. Spinoza maintained that Spirit is radically other to this world, and yet totally and completely embraces this world as a logical premise embraces all its consequents (his example). Spirit contains an *infinite* number of attributes, only two of which show up in "this world." As Karl Jaspers points out, according to Spinoza "it is a mark of the radical difference between God and the world that of God's infinitely many attributes only two are available to us." This was such a radical cleavage between God and this world that Hegel, who studied Spinoza carefully, called Spinoza's system "acosmism": the world is virtually nonexistent or illusory.

But Hegel goes a little too far; as Spinoza makes clear, in the two attributes that we *can* know—thought and extension—God is fully immanent, and thus, while a pitifully small portion of God is in this world, all of this world is in God (again, the higher transcends and includes the lower, so that all of the lower is in the higher but not all the higher is in the lower). This has led to the "popular" view that Spinoza actually said that this finite world itself is God. At no point whatsoever did Spinoza equate the sum total of finite nature with the whole of Spirit or God; this is utterly contrary to what he was trying to say.

This is why he himself differentiates carefully between Nature (as God or Spirit, *natura naturans*), which is infinitely transcendental, and nature (as this world, *natura naturata*), which is merely immanent. And holding up Spinoza as someone who equated Nature and nature is suspicious to say the least.

If by *Nature* (with a big *N*) these ecophilosophers mean a Reality with infinite attributes that infinitely transcends this world of nature, but totally pervades it and upholds it, then I agree entirely. For Spinoza, that ultimate Reality is approached and known in thought, not in sensory extension or nature—a nature sensory experience is for Spinoza an example of what he calls "mutilated, confused" knowledge, the lowest of his three levels of knowing.

I disagree with Spinoza on how that Reality is known (it is actually known transrationally, a "fourth" level beyond Spinoza's three, with which I otherwise agree), but ecophilosophers should realize what tree they're barking up. Why Spinoza feels that Spirit totally transcends the world but totally embraces it will become clearer, I believe, later in this chapter.

17. This was Hölderlin's view as well, as one of his major translators explains: "As the force of All-Unity, Nature is of course not limited to physical reality, but is present in transcendence ('Heaven') and in the historical being of 'nations' " (Unger, *Hölderlin's major poetry*, p. 109).

18. P. 32. My italics.

19. Or, more precisely, the Over-Soul is a new and higher holon embracing the physiosphere, biosphere, and noosphere as junior components in its own compound individuality.

20. Schopenhauer, *World as will and representation*.

21. This was also one of Hegel's criticisms of Kant, that in Kant's "autonomy" the external tyrant of heteronomy had been replaced by the internal tyrant of the categorical *imperative*, whereas authentic morality, issuing from the realization of self-positing Spirit, would be *spontaneously* expressed as one's own deepest nature.

Warwick Fox, in *Transpersonal ecology*, attempts to resurrect this Hegelian notion (following Naess), but he tries to fit it into a purely flatland ontology, with unpleasant results (which I will examine in volume 2). Although there is much to recommend Fox's approach, and certainly much to applaud in its clarity of expression and integrity of intent, it is crippled by a recognition of only the narrower/wider dimension, and not also the deeper/higher dimension—a legacy of the fundamental Enlightenment paradigm. And, typical of many ecophilosophers, Fox wants to *extend* this paradigm to a flatland identification with all flatland holons.

As for the supersession of omega points from the *mutual understanding* of the worldcentric perspective (of the ego-centaur) to the *mutual identity* of the Over-Soul, see note 1 above. Also, as we will see in the following section, the Over-Soul finds its own omega in the subtle and causal dimensions.

22. "The foundation of morality."

23. "The foundation of morality."

I would like to propose the thesis that the Basic Moral Intuition (BMI)—present at all stages of human growth—is "Protect and promote the greatest depth for the greatest span." This BMI represents (*is a direct result of*) the manifestation of Spirit in the four quadrants (or simply the Big Three)—the depth in I expanded to include others (we) in a corresponding objective state of affairs (it). That is, all individuals intuit

Spirit, and since Spirit *manifests as* the Big Three, then the basic spiritual intuition is felt in all three domains, and thus the basic spiritual intuition ("Honor and actualize Spirit") shows up as "Promote the greatest depth for the greatest span," or so I maintain.

My further claim is that if we take the BMI and apply it to the various worldviews (magic, mythic, rational, psychic, etc.) we can generate the typical moral stance of those stages, because each stage has the *same* BMI (because Spirit is one) but a *different* definition of self, others, and objects (because Spirit unfolds different depths in the course of its self-evolution).

In the egocentric stance, for example, depth is extended only (or primarily) to the self, and all others are merely (or usually) extensions of the self (its moral stance has a *span* of 1, and it promotes and protects the depth in that span only, namely, itself; this is also the typical hedonistic ethic).

In the sociocentric stance, depth is acknowledged to exist in others, but only those of one's particular culture (*span* extends to all believers in the mythology, and *their* depth is to be protected and promoted; this is typically the duty ethic). All other individuals, however, are some version of infidels, who possess no depth, no soul, and thus are not worthy of being protected and promoted (and, indeed, are often sacrifices to the glory of the culture's god).

In the worldcentric rational stance, depth is extended to all human beings; span now includes the human race as such. (And, as a side twist, with the flattening of the Kosmos to the objective natural world or cosmos, depth came to be interpreted merely as a mono-level happiness, in which case the BMI is unpacked as "Promote the greatest happiness for the greatest numbers," with no consideration of different types of happiness or levels of depth that should be extended in the first place. "Better an unhappy Socrates than a contented pig" cannot be handled in this flatland BMI.) Nonetheless, the rational worldcentric unpacking of the BMI extends its depth or potential depth to all human beings (if not in equality, then in equal opportunity, regardless of race, gender, creed), and may begin to extend it to nonhuman beings as well.

In the transpersonal domains, the BMI unfolds as Buddha (I), Dharma (It), and Sangha (We), and the ultimate Sangha is the community of all sentient beings as such. (But this doesn't prevent us from making relative distinctions as well. The BMI is to "protect and promote the greatest depth for the greatest span," and thus, if we had to choose between killing a mosquito or the Dalai Lama, we swat the mosquito. Even though both are perfect expressions of the same Self or Buddha Mind, the latter realizes more depth and thus is to be more protected.) There are variations on this theme in the different stages of the transpersonal domains, and I will deal with those elsewhere (see note 58 below).

In other words, I believe we all intuit Spirit to one degree or another, and thus we all possess the Basic Moral Intuition; but we unfold that intuition only at our present level of development. Since Spirit *always* manifests simultaneously in and as the four quadrants, then this spiritual intuition is (and would have to be) unpacked according to how each stage of development cognizes the four quadrants. Putting these horizontal and vertical factors together gives us a multidimensional grid that, as far as I can tell, can generate all of the major ethical stances that have been advanced.

Further, even within a given level of development, the Basic Moral Intuition is just that, an intuition, not an engraved tablet with written instructions. It is given form and shape by the surface structures of the culture in which it finds itself, and—precisely because it is an intuition—individuals within a given stage can *legitimately* disagree on precisely what depth is, what span should be included, and what measures should be taken to objectively implement it (I, we, it). (A higher stage would involve a more *authentic* unpacking of the BMI, but that stage would still face its own problems of legitimacy on its own level, and so on.)

To give a notorious example from the rational level: Should we or should we not extend the span of what we recognize as depth to include the fetus? Does "Protect and promote the greatest depth for the greatest span" mean the fetus has a real depth and thus should be protected? Or is the fetus basically just a potential depth (and thus accorded fewer rights than actual depth)? Or again, is the depth *already* present in the mother so much greater than the *potential* but *unrealized* depth of the fetus, that if having the baby would severely impair the mother's depth according to her, then the "greatest depth for the greatest span" would actually be *facilitated*, not hindered, by abortion? (The feminists uniformly take this stance, even if they don't articulate it in those terms.)

My point is that reasonable men and women can legitimately disagree on these issues, with each of them attempting to unpack the BMI as best they can. (Sociocentric or mythic-membership individuals, on the other hand, who believe in the mythogonic origin of life, feel the fetus is already a fully fledged person and thus has equal depth and needs equal protection; abortion is murder. Interestingly, this argument would extend only to unborn but potential *believers*; an unbeliever has no soul, or, without baptism, is destined for hell, and you are definitely allowed to murder infidels; what really irks them about abortion is not the loss of human life but the loss of a potential convert; they unreflexively say "all human life" because they automatically and deeply assume all human life belongs to their God, and theirs alone. A moral dilemma at *this* stage is: should I kill the killers of God's children? And, in fact, one physician has already been murdered in America on precisely these "grounds." After Dr. Gunn's murder, the fundamentalist newsletters indeed had a heated debate about his death: not whether he should have been killed—that was never questioned—but if it shouldn't have been done instead by stoning, this being the acceptable biblical death for such a sinner.)

I will return to the BMI later (see note 13 for chapter 14). My point for now—and this relates to Emerson and Schopenhauer—is that expressing the BMI in terms of the "one Self" or "common Over-Soul" is fine, as long as we remember that is only one aspect of the BMI (namely, the subjective strand) and only one component of a full ethical theory (which would have to involve all four quadrants). Otherwise we will very soon slide into solipsism and subjective idealism, which plays heavily into the hyper-agentic, hyper-masculine, disengaged and dangling subject of the fundamental Enlightenment paradigm.

24. Habermas does not recognize any stages (in any domain) higher than mature and differentiated/integrated, decentered, autonomous communicative reason (vision-logic). Communicative reason is the final omega, the end of history for Habermas. But the whole point of the transpersonal orientation is that that is an altogether premature closing of the history books.

Likewise, although I have made frequent use of Habermas's overall theory, we would of course differ on various aspects of that general framework. Habermas is particularly unclear, it seems to me, on the individual, subjective domain, that of "inner nature" (including the therapeutic and the aesthetic). On the one hand he states that the three domains are open to rationalization and cumulative learning; on the other, he states that the inner domain is only particular and not open to universal claims. Rothberg believes that Habermas would maintain that the form (or structure) of the inner domain evolves (and is universal in that manner), but the content is individualistic and particular (and not open to universality). If this is so (and it does seem to reflect Habermas's general position), it is not that different from my own claim that the deep structures of all domains evolve in a broadly universal form, but the surface structures are everywhere "particularistic" and dependent upon cultural and personal contingencies (the inner domain being no different, in this regard, than the external and the social domains; the deep structures of selfhood or subjectivity evolve in essentially the same framework, and along the same lines, as the it and the we; the "public" nature of the it and we is actually no more public than the I: the recognition of each domain is set in the same patterns of relational exchange). To say that the inner/subjective world has features that are idiosyncratic is true but trivial in this regard; what is not trivial is that the *types* of subjective space, and the *types* of data or inward apprehension, evolve in a broad pattern that is as cross-cultural as the unfolding of *types* of moral response and *types* of technical-objective potentials; to say the specific contents vary in one is to simply say they vary in all. Besides, privileging the it and the we over the I, when it comes to universal validity, ultimately undercuts the privileging itself.

While I agree with much of Habermas's critique of the philosophy of the subject—as we will see in detail in chapters 12 and 13—I believe he has tossed the baby with the bathwater in this particular issue. When he says that "nothing can be learned in an objectivating attitude about inner nature qua subjectivity," I know what he means—the subject ceases to be a subject precisely at the moment it is objectified; but the whole point about inner development is that what is embedded-subjectivity at one stage becomes generally object of awareness at the next, and the higher or contemplative stages of development are exactly ones where much is learned about inner nature precisely by objectifying it in intense introspective modes, an objectifying that is a freeing or detachment from the subjective-embeddedness or identification: much is learned about inner nature qua subjectivity precisely by objectifying it and robbing it of its power to subjectify the world—and further, the claim of the contemplative traditions is that general maps of this transsubjectifying process of development can be reconstructed, maps that possess a general or broadly universal character, and these maps, constructed with the warrant of the three strands of valid knowledge accumulation, are every bit as "public" as any other accumulation. (In chapters 12 and 13, when I criticize the merely objectifying mode of the fundamental Enlightenment paradigm—the philosophy of the disengaged self or subject—it is a criticism not of objectifying but of only objectifying: of the failure to see that the subject still exists in patterns of relational exchange with similarly depthed subjects, which finally anchors not only the we but the I.)

Rothberg discusses many of these topics in an excellent overview, "Rationality and religion in Habermas's recent work." Rothberg's conclusion that a "reconstructed" Habermasian framework is nonetheless broadly compatible with the phenomenology of contemplative religion is a conclusion I support. These topics are dealt with in detail in volume 3; in the meantime, critics might forgive my rather generalized use of Habermas's framework (likewise, I have based most of my comments in this volume on Habermas's work up to, but not including, his recently released magnum, *The theory of communicative action*, of which two volumes are now available. My discussion of these is reserved for later publications).

Accordingly, when Charles Taylor complains that Habermas's framework allows any true spiritual/substantive claims to "slip through the cracks," I believe this is true only if the Habermasian framework is not extended into the transpersonal strands of each of the three domains (each of the four quadrants). At the same time, my major criticism of Taylor is that he is inordinately fond of magico-mythic syncretisms. Time and again, when he comes to give examples of the ontic logos he feels that modernity "lost," he reverts to syncretic/mythic "wholes." A little more "trans," and a little less "pre," would, I believe, allow him to recognize the value of a "reconstructed" or extended (transpersonal) Habermasian framework.

But Habermas and Taylor are good examples of the impact of an ambivalent modernity on presentday theorists. Although both theorists are acutely appreciative of the good news/bad news nature of modernity, Habermas exuberantly celebrates "No more myths!," and Taylor sadly laments the "No more Ascent!"

25. P. 99. All quotes, unless otherwise indicated, are from *Interior castle*, trans. E. Peers.

26. Pp. 85, 86.

27. This cessation, which is the *single* primary transformative state in all of Teresa's stages, is a glimmer of *nirvikalpa samadhi*, or formless absorption. This cessation or suspension (both her terms) is hinted at in the Prayer of Quiet but comes to the fore at this stage and continues to be the transformative event in the sixth stage as well. This is why she says "the experiences [of cessation/absorption] of this [5th] mansion and of the next are almost identical" (pp. 106–107).

She therefore describes this state of suspension in the fourth, the fifth, the sixth, and even the seventh stage. Here are several quotes: "In this state of suspension the faculties are so completely absorbed that we might describe them as dead, and the senses so as well . . ." (p. 150). "Here we are all asleep, and fast asleep, to the things of the world, and to ourselves (in fact, for the short time that the condition lasts, the soul is without consciousness and has no power to think, even though it may desire to do so) . . ." (p. 97). "This is a delectable death; if it still breathes, it does so without realizing it; even if it has any consciousness, neither hands nor feet can move . . ." (p. 98). "For as long as the soul is in this state, it can neither see nor hear nor understand . . ." (p. 101). "Sometimes it seems doubtful if there is any breath in the body. This lasts only for a short time because, when this profound suspension lifts a little, the body seems to come partly to itself again, and draws breath, though only to die once more, and, in doing so, to give fuller life to the soul . . ." (p. 155). (Note the similarity to the Transcendental Meditation phenomenon of transcendence and breath suspension repeated in slow cycles.)

But as for the soul being "asleep" or "not conscious" in this state of cessation, Teresa finally points out that it is actually that the soul "is fully awake, while asleep as regards all attachment" (p. 155).

This suspension, absorption, cessation is referred to in Eastern texts as "conditional nirvikalpa samadhi," or formless awareness not yet fully established. Each of Teresa's stages are anchored in a taste of this cessation, and that taste triggers the development through the psychic and subtle stages with their various visions, raptures, voices, illuminations, and so forth (*savikalpa samadhi*).

This cessation/absorption is, for Teresa, pure union with God (and the Hindu texts would agree; however, this cessation is indeed "conditional" because there remain traces of dualism in the perception; if those traces were uprooted, God and soul would both give way to pure causal Godhead, or *jnana samadhi*, which we will examine in the next section with Eckhart and Ramana).

28. P. 101.
29. P. 106.
30. P. 155.
31. Pp. 129, 131. Compare the Buddhist "pseudo-*duhkha*"; see *Transformations*.
32. P. 199.
33. Pp. 134, 160. But this does not, she points out, mean that they cannot be communicated, and indeed confirmed, to and with those who have also had similar experiences. "Nor do I think I shall succeed in describing them in such a way as to be understood, except by those who have experienced it; for these are influences so delicate and subtle that they proceed from the very depth of the heart and I know no comparison that I can make which will fit the case" (p. 134)—that is, no signifier will work unless there is a developmental signified in the listener.

Hence these states are not merely arbitrary in a relative-constructivist sense: the true and the false still obtain: "But I fancy that even now you will not be satisfied, for you will think that you may be mistaken, and that these interior matters are difficult to investigate. In reality, what has been said will be sufficient for anyone who has experienced, for there is a great difference between the false and the true" (p. 100). Habermas is always wanting a simple "Yes or No," and Teresa would be more than glad to give it to him.

34. Pp. 149, 152.

35. The high-subtle, with intimations of low-causal. Teresa says that the lesser visions and raptures and illuminations (of the psychic and low subtle) are the result of the union of *aspects* of the soul with God, or the union of some faculties of the soul with God; but this final union is of the "the whole soul and God."

36. P. 213.

37. Pp. 213–14.

38. See also *The Atman project* and *Transformations of consciousness.*

39. P. 87.

40. *Ascent of Mount Carmel,* chap. 5.

41. Pp. 216, 210.

42. *Life divine,* pp. 704–705.

43. P. 169.

44. Pp. 218, 217. Unless otherwise indicated, all quotes are from Fox's new translations, *Breakthrough.*

45. P. 215.

46. Pp. 104, 178, 242.

47. Pp. 190, 177, 140, 128, 256.

48. Pp. 277, 294, 181.

49. Pp. 298, 309, 301.

50. P. 102. All quotes, unless otherwise indicated, are from *Talks with Ramana Maharshi.*

51. P. 32.

52. P. 13.

53. Pp. 110, 121.

54. Pp. 53, 93.

55. Pp. 182, 210, 82.

56. Pp. 132, 316.

57. Pp. 317, 316.

58. The Nondual traditions make a series of very fine distinctions here, which at least should be briefly mentioned. And we can return to Katz and offer a simpler critique of his position.

The manifest All as a totality, the consummate Holon, is *saguna Brahman,* final-God with form (the low causal). This manifest All-Unity therefore evolves as All-Form, the summit "application" of the twenty tenets. Pure unmanifest (or high causal) Spirit (*nirguna Brahman*) does not evolve or involve or in any way enter or leave the manifest stream, but "underlies" it as a mirror "underlies" equally all its objects, whatever the state of those objects; it is both the Source of all other levels and their ultimate Goal (but it is not a summation or addition of manifestation, as is saguna, because it is utterly unmanifest).

Godhead as nondual Suchness (as opposed to unmanifest Source and Goal) is neither manifest nor unmanifest; it is as fully present in the lowest holon as in the highest holon, and in any manifest or unmanifest state, as the isness or suchness of any phenomenon.

Roughly: Whitehead's Primordial Nature of God refers to archetypal patterns in the high subtle prior to gross involution (cf. "concretion"); Whitehead believes these patterns to be "eternal objects," but many traditions, such as Mahayana Buddhism (of Asanga and Vasubandhu), see these archetypal patterns (*vasanas* in the *alaya-vijnana*, high subtle to low causal), as being cosmic memory-habits, very much like Sheldrake's view. Whitehead's Consequent Nature of God is similar to saguna Brahman (high-subtle to low-causal), the ever-evolving consummate Holon.

But primordial and consequent nature do not cover, or even refer to, the unmanifest Godhead (high causal) or Godhead as Suchness (ultimate or nondual): Whitehead seems unfamiliar with those domains.

Of course, these types of broad similarities are the types of comparisons Katz and company do not think are theoretically warranted in any fashion. (We dealt in detail with Katz's position in note 16 for chap. 7, and I will here give a related but simpler critique; also, see note 32 for chapter 13 for a discussion of Kant and experience).

Katz's general pluralistic position is that mystical experience is mediated by cultural/concepts, which are everywhere different, and thus no mystical experiences are comparable. We saw that if this were true in the strong version that he believes, it would apply not only to mystical experiences but to everyday ordinary experiences as well, at which point simple communication itself (and translation between languages) would be impossible; and that, on the contrary, theorists from Habermas to Derrida maintain that transcendental signifieds anchor signification in ways that prevent interpretive reality from being *merely* constructed, since they are significantly anchored in extralinguistic factors.

The mystic's claim is that, in just the same way, mystical validity claims are anchored in extralinguistic realities that, however much they are molded by cultural factors, are not merely the product of shifting cultural and provincial-only fashions (the referents of the transcendental signifiers exist in a worldspace that is disclosed to those with the appropriate developmental signifieds, even if these are always already culturally situated. Thus, cultural context does not prevent natural science from making and redeeming universal claims, nor likewise does it prevent mystical science from making and redeeming not-merely-culture-bound claims).

And indeed, Katz implicitly (and sometimes explicitly) concedes the *existence* of these transcendental referents (which he calls "the mystic reality"). Katz: "Thus, for example, the nature of the Christian mystic's pre-mystical consciousness informs the mystical consciousness such that he experiences *the mystic reality* in terms of Jesus, the Trinity, or a personal God, etc., rather than in terms of the non-personal . . . Buddhist doctrine of nirvana" (my italics; "Language, epistemology and mysticism").

But how does Katz know that "the mystic reality" even exists, since supposedly nobody has access to it in anything other than culturally limited and nonuniversal terms? What allows him to make universal/transcendental claims that he denies to everybody else?

Fact is, in the back of his mind he recognizes quite obvious cross-cultural commonalities—he recognizes "the mystic reality"—which he *then* tries to erase by centering merely on the local cultural contexts, even as he continues to *assume the existence* of the mystical reality which *alone* can anchor his analysis (even though by his formal stance he should be able to make no universal statements about it at all. The standard performative contradiction).

The recent advance in transpersonal studies is that not only does this mystical reality exist and possess certain not-merely-cultural transcendental referents (such that cross-cultural commonalities can indeed be recognized, just as Katz in fact does recognize them), but the "mystical reality" itself unfolds in a developmental and holarchical fashion, and likewise cross-cultural commonalities can be found in the unfolding as well. Since Katz *already* recognizes the existence of "the mystical reality," no further tools are needed to recognize "stages in the mystical reality." That is, Katz has already taken the difficult, major, and big first step: *the mystical reality is there*. The transpersonal advance has simply refined that step: there are stages in its unfolding, *immanent* stages anchored in *transcendental referents* that (1) prevent "the mystical consciousness" from being merely provincial, (2) allow *learning* to occur in the unfolding (which could *not* occur if the unfolding were merely textual with nothing extralinguistic to ground it and correct it), (3) anchor mystical validity claims and make them something more profound than literary chit-chat and idiosyncratic interpretations, (4) usher one into worldspaces that are deeper disclosures of the Kosmic terrain, and (5) altogether make Spirit a reality and not simply a poem.

Take the sequence:

The self experiences an overpowering interior luminosity; self and luminosity merge; self and luminosity disappear into cessation; cessation gives way to nondual unity with all manifestation. Those are four very distinct *types* of mystical experience (they are, in fact, typical of the psychic, subtle, causal, and nondual domains). Those *types* of experiences can most definitely be found cross-culturally. Nobody is denying that a Buddhist will interpret the luminosity as the Sambhogakaya, the Christian will interpret it perhaps as an angel or as Christ himself, a Jungian will interpret it as an archetypal emergence, and so on. This is true but utterly trivial with regard to the ontological status (and impact) of the referent itself: it applies to *all* experiences (do you and I even see that patch of red color in precisely the same way? Who can say? But does this invalidate redness altogether, or mean that we never have anything in common and thus cannot even communicate our experiences?). To use this argument to undermine the mystical claim is simply to undermine any transcendental signifieds at all, which is to say: to render any communicative exchange theoretically impossible.

Just so, significant aspects of "the mystical consciousness" are anchored in referents ("the mystical reality") that, even if always culturally situated, are never *merely* culturally situated, which differentiates them from (1) personal hallucinations, (2) provincial cultural fictions, (3) idiosyncratic quirks, (4) mythic dogmatisms, (5) bogus power claims (ideology).

Likewise, precisely because "the mystical reality" is there, anchoring a transcendental signifier, this (1) allows learning to occur vis-à-vis the mystical reality, (2) ensures that learning will unfold, like all learning, in a developmental arc, (3) allows the stages of that arc to be identified, (4) allows the traditions themselves to spot when a trainee is making a mistake, (5) allows a rational reconstructive science of the mystical unfolding in abstract terms (deep structures), and (6) frankly acknowledges that Spirit *is there* prior to humans and their cultural tinkerings, and is not simply a product of men's and women's capacity to make symbols: Do we create God? or is that just slightly backwards?

Finally, in this regard, I mentioned in note 23 for chap. 4 that different languages are often used with (and seem best to metaphorically capture) the four major levels of transpersonal awareness. (I mentioned this in connection with Foucault's tracing of different discursive structures). Picking up the story with the rational-ego, I would tentatively (and very briefly) outline these languages as follows.

Rational-ego—the languages of representation and reflection. This is the typically monological representational paradigm that defined (and defines) the formal operational mind in its reflection on (mostly) exteriors. These languages describe the world as if it were simply pregiven to the equally pregiven (and disengaged) subject. Various depths are recognized, but these, too, tend to be described in simple monological terms (empiric-analytic).

Centaur/vision-logic—the languages of depth and development. Vision-logic tends more easily to recognize that beneath the "obvious" surfaces appearing to "pregiven" reflection, there lurk depths that can be disclosed, both in their exterior forms (e.g., neostructuralism) and in interior interpretations (e.g., hermeneutics). This is part of the move that Foucault traced from the languages of representation to the languages of depth (e. g., psychoanalysis, linguistics). Likewise, depth is generally understood to have *developed*—whereas surfaces simply appear, depths tend to unfold or develop or evolve: they are not simply "pregiven." The languages of the centaur thus tend to be dialectical, dialogical, network-oriented, developmental (evolutionary in the broadest sense)—in short, the languages of depth and development. (As I indicated, this volume is intentionally written in these languages.)

Psychic—the languages of vision and vibration. Vision-logic gives way to direct vision, and the developmental view is supplemented with a vibratory view, where vibration is used to convey not so much a physicalistic nature as a *quality* of *intensity* of awareness. These languages are most common in kundalini yoga and the early stages of tantric unfolding, but can be found to some degree in virtually all beginning stages of intense meditation. Since the psychic-level, as we have seen, takes as part of its referent aspects of the gross domain, vibration (felt bodily vibration) seems best to capture the subtle energies and consciousness that begin to transcend the gross domain but can still be intensely felt in that domain. But for whatever reasons, the languages of vision and vibration seem to be used quite often to describe this realm (and its subtle architecture of *chakras, nadis, tigles,* etc).

Subtle—the languages of luminosity and archetype. I mean "archetype" in the non-Jungian and more traditional sense (subtle Kosmic forms or patterns of manifestation, upon which all other patterns are based in involution, as we will see in chapter 9). Vision gives way to intense luminosities (sometimes audible illuminations and bliss-luminosities), and vibration is supplemented with a grasp of the subtle forms *from which* the gross domains radiate or vibrate. The "saintly" domains of interior illuminations and archetypal deity-forms—extraordinary, awe-inducing, supernatural, out-blazing.

Causal—the languages of emptiness and dream. Luminosity gives way to pure Emptiness, and the entire world of form (archetypal or otherwise) is seen as a dream (also expressed as a drama or play, *lila*). Thus, whenever we hear philosopher-sages speak of the Abyss, the Void, the Urgrund, the Unborn, the Uncreate (etc.), they almost always also refer to manifestation as the great dream, the great illusion, the dancing play, the great shimmering image of no final substance. We are in the languages of emptiness and dream.

Nondual—the languages of the extraordinary ordinary. Emptiness is Form, Form is Emptiness, and therefore the final realization is *not other in the slightest* to anything presently arising, and therefore can be directly pointed to with, for example, this from a Zen Master:

> *The spring breeze in a tree*
> *Has two different faces:*
> *A southward branch looks warm,*
> *A northward branch looks cool.*

That is a perfect pointing to the absolute truth: how obvious, how ordinary. This glass of water is cool, the sun is bright! Hear the rain on the rooftop? Who is not enlightened?

Where is one's Original Face? In the universe of One Taste, where else can we point for the absolute truth if not right here? Where *else* could it be? The languages of *just this*.

In all of these languages, notice that, as I have presented them, the two "components" (such as depth and development, or luminosity and form) refer, in a very general way, to the interior and exterior aspects of the particular level (until these two become finally not-two or nondual). The interior components refer to the subject-like side: vision-logic (or depth) to vision to luminosity to Emptiness (pure Witness); and the exterior components to the objective-like or exterior side: development to vibration to archetype to dream—until final Emptiness and final Dream/Form are not-two: the summer breeze is cool, the birds are singing loudly.

And notice that, in each of these languages, *words work just fine*. The signifiers are *perfectly adequate. If* you have the developmental signified, the referent is fairly obvious. If you say to a subtle-level practitioner, "I saw a light a thousand times brighter than the sun," that practitioner will know pretty much what you mean, as "pretty much," for example, as if you said, "Last night I had an intense orgasm." That is, in both cases, the words work equally well (or equally poorly), depending upon our own lived experience as a point of actual reference. I might not know *precisely* what you mean by "intense orgasm," but that applies to *all* signifiers, and in no way exclusively disarms mystical signifiers. As I said, that applies to "dogginess" as well as "Buddha-nature," but this epidemic linguistic slipperiness does not mean dogs and Buddhas are simply linguistic fairy tales.

The analysis of the types of languages of the transpersonal domains is, I believe, another of the central projects of postmodernity. Provided, of course, that it is not done as a merely academic endeavor, where a bunch of disengaged dunderheads attempt to analyze the signifiers without having a clue as to the actual referents.

59. Pp. 67–68, 138, 96, 191, 249.

Let us note the transformations that *worldcentric* has undergone. We have seen overall development move from physiocentric to biocentric to egocentric to ethnocentric to worldcentric—and it is with *rationality* that the various worldcentric conceptions first begin to emerge. And what all worldcentric perspectives have in common is that a certain *freedom* is to be extended (or made available) to all human beings, regardless of race, color, creed, or gender. *All stages at and beyond rationality are fundamentally worldcentric in nature.*

But the *types* of worldcentric freedoms that are offered unfold in a developmental fashion, and these *types of freedom* (and types of morality) are part of the defining characteristics of the higher and transpersonal stages of development.

To pick up the story with rationality and its worldcentric stances, we already saw Habermas, for example, point out that rationality first demands a legal freedom, then a moral freedom, then a political freedom for all members of a world society (note the three *types* of progressive freedom). And that takes us to the global, centauric vision-logic and its moral stance.

Continuing into the transpersonal—into Kohlberg's stage 7, or universal/spiritual morality (or "post-postconventional"), which actually contains several stage/levels—we still see a *worldcentric* orientation, but further freedoms are being offered beyond the legal or political: namely, *spiritual freedoms*. The same worldcentric *span*, but disclosing even further *depths*.

These progressive depths, we have seen, are psychic, subtle, causal, and nondual: and each of these offers a progressively deeper and wider *type* of spiritual freedom—a freedom that results from an identification with a worldspace that, so to speak, contains more room. They are *freedoms from* the lesser and more confining spaces, by finding *freedom in* the deeper and wider.

Each of these freedoms is thus often simultaneously described as a *freedom from* and a *freedom in*: a freedom from the previous, lesser domain, and a freedom in the new and "roomier" domain (as we will see in a moment). The *moral stance* of each of these freedoms thus relates to the *type* of new freedom that is therefore offered to all beings, worldcentrically (the Basic Moral Intuition inspires us to extend this new freedom, this new depth, to as many sentient beings as possible—greatest depth for greatest span—and this marks the *new types* of moral imperatives that issue from these new depths of freedom).

These increasing spiritual freedoms, and their corresponding moral stances, have been referred to, for example, as yogic, saintly, sagely, and nondual. These are spiritual freedoms *in reference to* the gross, subtle, causal, and nondual worldspaces, and thus in each case, these spiritual freedoms are described as *freedom from* the limitations of the lesser and *freedom in* the expanse of the greater: freedom from the gross in the subtle; freedom from the subtle in the causal; and freedom from the causal in (and as) the nondual. The lesser freedoms are not lost (just as legal and political freedoms are not denied); they are simply subsumed in a deeper space that embraces their concerns from a greater perspective.

Thus, to give a few condensed examples: the psychic/subtle disposition is a *freedom from* the ordinary gross-level fluctuations of pleasure/pain and sensual-mental desire and frustration (via yogic harnessing of awareness away from sensory immersion); it is a *freedom from* gross limitations and *freedom in* the deeper space of psychic and subtle awareness that moves prior to the entire gross domain of fluctuating sensory-mental desires and aversions.

Likewise, the causal disposition is a *freedom from* the limitations of the subtle domain and manifestation in general (samsara); a *freedom from* the limitations of an awareness identified with the mechanism of suffering and separation, and a *freedom in* the vast expanse of boundless, unmanifest, formless Emptiness that is consciousness as such. Never entering manifestation, causal freedom is never engulfed by the mechanics of madness and the mathematics of pain: the world of duality and its inherent lacerations.

The Nondual disposition is a *freedom from* the extremely subtle limitations of the causal unmanifest (namely, the causal tension or rarefied fear of manifestation: the tension around the heart of nirvikalpa samadhi that is the final Phobos of sahaj: the final fear of dissolving the boundary between emptiness and form and thus awakening as all Form, endlessly)—a *freedom from* the limitations of the causal and a *freedom in* (and as) the entire Kosmos of Emptiness and Form, spontaneously so.

These *types* of new spiritual freedom are thus also frequently described using the *various languages* as I briefly outlined them. Thus, vision-logic is a freedom from the limitations of mere representation; subtle-level nada illuminations are free of gross-bodily-felt vibrations; the causal is a freedom from the illusory dream of all manifestation; the nondual is a freedom from "stone Buddha" emptiness (to give the briefest examples).

Likewise, these four types of spiritual freedom (psychic, subtle, causal, nondual) are often correlated with Paths of Renunciation (yogic/psychic), Paths of Purification (subtle to causal), Paths of Transformation (causal to nondual), and Paths of Spontaneous Self-Liberation (nondual as such)—each with a different moral disposition that reflects the Basic Moral Intuition as unfolded and interpreted at that new depth.

I will leave the details of those correlations for later publications. All I wish to emphasize here is that the *worldcentric stance*—which *begins* with rationality and marks *all stages* at and beyond rationality, and which extends freedom to (at least) the span of all human beings: that stance of worldcentric *span* itself increasingly discloses more *depths*, or deeper and deeper *types* of freedoms that are to be extended *universally* via a new moral imperative (and then, not only to humans but all sentient beings as well).

This is yet another justification for the fact that reason is the great gateway to the beyond. Some individuals (and some cultures) can indeed "tap into" transpersonal domains, but they funnel the intuitions into prerational structures, into egocentric or sociocentric/ethnocentric dispositions, which merely serves to reinforce the egocentricity or ethnocentricity: they, and they alone, now have God on their side, at which point they begin to suspect that there is no other God. Short of *worldcentric rationality*, the intuition *cannot* be seen to operate in *all* human beings (and eventually, all sentient beings), and thus, short of worldcentric rationality, all spiritual intuitions, no matter how temporarily authentic, become hijacked and frozen in structures that prevent their authentic and compassionate manifestation.

Freeing these otherwise authentic intuitions from their crippling confinement to prerational structures and less-than-worldcentric embrace is part of the postmodern project.

60. For an interesting discussion of this theme, see Peter Russell, *The white hole in time*.

Various theorists have pointed to an undeniable acceleration of the pace of evolution, where major transformations seem to be occurring exponentially. This is certainly true from one point of view, but does not affect the argument that I will present in the text.

61. We will investigate the "source" of such omegas in the next two chapters, though we can already see at this point that the mystics would call that force Eros and that Goal God.

62. Fukuyama, *The end of history and the last man*, p. xiii.

The economic reasons will be discussed in volume 2, after we have looked at the techno-economic base of each of the major worldviews. Since every interior has an exterior, each major worldview is correlated with (but not reducible to) a major force of material production: magic (foraging), magic-mythic (horticulture), mythic (agrarian), rational (industrial), vision-logic (informational).

In volume 2, the correlations and characteristics of these various modes of production will be examined at length, and, most important, the status of the sexes will be examined in each period, with the goal of deciphering the roles of the sexes both yesterday . . . and tomorrow.

The nature of the global transformation now in progress cannot adequately be discussed without an understanding of the role of "the base" (the techno-economic infrastructure) in relation to various worldviews, and that, as I said, is a major topic of volume 2.

But this part of Fukuyama's argument (the economic) is the Right-Hand component, the material component, and he understandably finds it related to the conclusions of natural (RH) science: "Modern natural science [is] a regulator or mechanism to explain the directionality and coherence of History. Modern natural science is a useful starting point because it is the only important social activity that by common consensus is both cumulative and directional" (*End of history*, p. xiv).

The "struggle for recognition" is the Left-Hand component of his argument—the theme that *mutual recognition*, or the free exchange of *mutual self-esteem* among all peoples, is an omega point that pulls history and communication forward, toward its free emergence, which is then the end of history in that special sense.

63. *End of history*, p. xviii.

64. Ibid., p. xii.

65. I would like to mention several factors that New Age theorists tend to overlook in the "Greening of America" and "Age of Aquarius" theories. Even if we *collectively* evolve to, say, the psychic Over-Soul—which would mean that the center of social gravity revolved around political, educational, and cultural institutions organized toward the acknowledgment and realization of the Over-Soul in each and every being, much as we now organize around the recognition of rational egoic self-esteem—even if that were the case, three important factors should be kept in mind by the Aquarian utopianists:

(1) The average-expectable level of psychocultural development in any given society acts as a *pacer of development* up to that level but does not *guarantee* that development in all individuals will so proceed. Very few people even in "developed" countries reach a firm base in worldcentric, postconventional awareness (one study found only 4 percent of the American population at the higher postconventional stage—and that is the stage that *precedes* and is necessary for the Over-soul descent!). The same problems will beset any and all Over-Soul societies.

Conversely, no matter what the collective average-expectable level of a given society, individuals are always capable of evolving and developing beyond it to any of the higher structures of consciousness, since these are structural potentials of the human holon, although (a) they can only be *exercised* in microcommunities of the similarly depthed (sanghas), and (b) the higher the average-expectable level of a given society, then the easier (but not necessarily more likely) higher development will be.

(2) Even if society collectively evolves to the average-expectable level of the Over-Soul, every single person born in that society will nevertheless still *start development at square one*, as a single-celled zygote: and have to begin the arduous holoarchic

climb, physiocentric to biocentric to egocentric to sociocentric to worldcentric to theocentric. The pace of this climb can be accelerated, but the fundamental stages cannot be bypassed (nobody can form concepts without having images, and so on).

And at *every* stage in development, *things can go wrong*. The more stages, the more nightmares of possible developmental miscarriages.

(3) As for the higher stages themselves, they bring new and glorious potentials . . . and new and exquisitely wretched possible pathologies. Read again Teresa's account of the early subtle level (similar accounts are found in Hindu, Buddhist, and Sufi texts). Won't it be fun when society as whole is going through *that* stage?

The dialectic of progress covers even the face of God, and an idealized "Paradise" exists nowhere in the manifest Kosmos—for reasons we will further see.

66. The totality of all manifestation at any time—the All—subsists in the low causal, as the sum total of the consequent and primordial nature of Spirit (in roughly Whitehead's sense), and this Totality is the *manifest* omega pull on each individual and finite thing: as such, it is ever-receding: each new moment has a new total horizon that can never be reached or fulfilled, because the moment of fulfillment itself creates a new whole of which the previous whole is now a part: cascading whole/parts all the way up, holons endlessly self-transcending and thus never finally self-fulfilled: rushing forward ceaselessly in time attempting to find the timeless.

And the *final* Omega, the *ultimate* and *unmanifest* Omega, the causal Formless, is the magnet *on the other side* of the horizon, which never itself enters the world of Form as a singularity or as a totality (or any other phenomenal event), and thus is never found at all in any version of manifestation, even though all manifestation will rest nowhere short of this infinite Emptiness.

Thus, from any angle, there is no ultimate Omega to be found in the world of Form. There is no Perfection in the manifest world. Were the world of Form to find Perfection and utter fulfillment, there would be nothing else for it to do and nowhere else for it to go: nothing further to want, to desire, to seek, to find: the entire world would cease its search, stop its drive, end its very movement: would become without motion, time, or space: would become the Formless. But the Formless is already there, on the other side of the horizon, which is to say, the Formless is already there as the deepest depth of this and every moment.

This Deepest Depth is the desire of all Form, which cannot itself be reached in the world of Form, but rather is the Emptiness of each and every Form: when all Forms are seen to be always already Formless, then dawns the Nondual empty Ground that is the Suchness and the Thusness of each and every display. The entire world of Form is always already Perfectly Empty, always already in the ever-present Condition of all conditions, always already the ultimate Omega that is not the goal of each and every thing but the Suchness of each and every thing: *just this*. The search is always already over, and Forms continue their eternal play as a gesture of the Divine, not seeking Spirit but expressing Spirit in their every move and motion.

CHAPTER 9. THE WAY UP IS THE WAY DOWN

1. The ironic point is that Lovejoy displays no comprehension of the unifying Heart that integrates these two movements (and about which, strictly speaking, only a "thunderous silence" can be maintained), and so he is baffled by the actual nature of the ensuing titanic battle between the Descenders and the Ascenders, and, most lamentably, passes over those systems—such as that of Plotinus—that perfectly integrate the two. But, like a good reporter, he gives us all the necessary information to flesh out the picture for ourselves.

I should also mention that Lovejoy is not an advocate of the ideas whose history he is tracing; as for the Great Chain and its three main tenets, Lovejoy himself rejects two of them (as I will explain); so I am not quoting from an authority predisposed to support my position. Lovejoy, we would say, is a hostile witness, but no matter.

2. P. 41. All quotes, unless otherwise indicated, are from *The great chain of being*.

As for the pure Ascending One being timeless, unchanging, eternal, etc., there are several reasons to suppose that this was a direct spiritual intuition whose historical origin (in the West) we might trace to Parmenides. What I am about to suggest is rather speculative, but I believe future research will tend to bear it out. The facts are fairly straightforward; their interpretation is tricky.

"One day, around 460 B.C., the great philosopher Parmenides came to town accompanied by his pupil Zeno. Parmenides impressed Socrates with the idea of Permanence. Reality, Parmenides argued, is Unchanging. Zeno supported his mentor's position by reducing to absurdity any assertion that motion and change really do exist" (Cavalier, *Plato*, p. 31).

That is a popular account of the story Plato gives in the dialogue entitled *Parmenides*. Zeno's "demonstrations" have usually been taken as an example of a certain type of philosophical argument (dialectical refutation), and I do not doubt that they were that. But if they are compared with, for example, similar dialectical arguments put forth by the Buddhist genius Nagarjuna, it appears Zeno (and Parmenides) might have been attempting something else as well: namely, a direct pointing to reality freed of all differentiating conceptualizations.

Reading the remaining fragments of Parmenides, one gets the strongest impression that he had directly glimpsed the causal, formless "One," against which all manifest objects are fleeting and ultimately unreal shadows. He thus makes the standard distinction (as does Nagarjuna) between the Way of Truth and the Way of Appearance. To the world of appearance belongs all differentiation, all generation and destruction, all motion and change, whereas the Truth is as it is, perfectly self-existing, and not open to any differentiation or distinctions of any sort. As many scholars have noted, the Way of Truth is seeing the world *sub specie aeternitatis*, and not according to the mere beliefs of mortals or the Way of Appearance.

If we take his assertion that motion does not exist as being literally true, then Parmenides has to be seen as being rather confused. But if these statements were in fact part of the "pointing out instructions" for recognizing primordial awareness (free of differentiating conceptualization), then they take on rather profound meaning.

From Dzogchen Buddhism to Vedanta Hinduism, for example, we often find statements such as "that which moves is not Real." This is not to deny relative motion; it is rather an attempt to directly point to primordial awareness which is prior to motion or rest, which doesn't enter the stream of time as a particular object (moving or resting), but rather is the immediateness—the opening or clearing—in which all objects arise and fall.

This primordial awareness (itself neither at rest nor in motion) can in fact be rather easily pointed out to somebody using these types of "pointing out instructions" (as evidenced, for example, in Dzogchen training [see *The eye of spirit*, chap. 12, and *One taste* for examples of pointing out instructions]). Reading Parmenides, I get the distinct impression that this was in fact a large part of what he was attempting. In numerous places it seems simply unmistakable. Parmenides thoroughly and continuously denies *any* description of the world that presupposes that *difference is real*: he is directly pointing to the One, which can be found in one's own direct awareness *prior* to differentiation.

Thus, Parmenides's Absolute wasn't *literally* a round sphere, as many scholars have supposed, but simply the completeness and fullness of Being in itself, prior to differentiation and prior to the Way of Appearance. And the dialectic was in part a way to directly point to reality prior to its differentiation and illusory appearances, in which mortals become lost because they mistake these appearances for the undivided Truth.

If this is so, then to Parmenides goes the honor of being the first influential Westerner (as far as I can tell) to penetrate to the causal One, however briefly (although perhaps this might also be traced to Parmenides's own teacher, whom tradition names as the Pythagorean Ameinias).

Some scholars believe the meeting with Socrates actually occurred; others believe its invention was a liberty taken by Plato. But the dates match up fairly well: Parmenides would have been about sixty-five, Zeno about forty, and Socrates "very young." I am inclined to the view that the meeting occurred; but in either case the influence of Parmenides on subsequent Greek thought is undeniable. Moreover, the scholar T. Murti, whose *Central philosophy of Buddhism* is generally regarded as the finest treatment of Nagarjuna in English, points to Plato's *Parmenides* as the first real dialectic in the West that is similar to Nagarjuna's. Whether Plato fictionalized the meeting or not, that he chose Parmenides to present the dialectic shows that the meeting—if not "historical" then certainly "philosophical"—had profound meaning for him. And it makes all the more credible Plato's assertions that what he was really up to could not be put into written words, but had to be pointed out directly, from teacher to pupil, by a sudden illumination.

3. P. 42. My italics.

4. P. 45. My italics.

5. P. 45. My italics.

6. P. 53. My italics.

7. P. 49.

8. P. 48. My italics. I believe that this integration was grounded in a direct taste of causal/nondual awareness (however brief or initial it might have been; remember Teresa's point: one taste can be transformative). There can be no doubt that his Neoplatonic descendants thought so, and no doubt that they themselves inculcated this practice (as we will see with Plotinus). But at the very least, it is quite obvious that the overall Platonic theory involved this integration of Ascent and Descent, as Lovejoy makes quite clear.

9. This negation-and-affirmation (negate and preserve!) is found in all nondual contemplative schools. In Zen, to give only one example, it is said, "If all things return to the One, to what does the One return?" The answer: "The Many" (though the student has to demonstrate this understanding and not merely verbalize it; repeating the signifier without the developmental signified earns only a whack from the Master). So also the famous: "Before I studied Zen, mountains were mountains; when studying Zen, mountains were no longer mountains; when I finished studying Zen, mountains were once again mountains." The dividing line is the realization of Emptiness, or pure formless awareness (mountains were not mountains), in which all things then perfectly and freely arise as Emptiness just as they are (mountains were once again mountains, self-liberating). But one has first to realize the great death of Emptiness, or one remains merely lost in the mountains (shadows).

10. Hamilton (trans.), *Phaedrus*, p. 52. My italics.

11. *Phaedrus*, pp. 54–56.

12. Later in this chapter, we will see that Thanatos integrated is not the "death instinct" but Agape; Agape divorced from Eros becomes Thanatos, the death instinct.

In *Eye to eye* I pointed out that the life and death "instincts" or drives or forces have actually been used by theoreticians in two very different ways, which we might call vertical and horizontal. "Vertical life" is Ascent or Eros, as I am using it here (a move to higher and wider integrations), and "vertical death" is Thanatos (a move to lower and shallower occasions, thus less unified, resulting ultimately in insentient holons).

"Horizontal life," on the other hand, refers to the *preservation* of the elements of any given level, and "horizontal death" refers to the *negation* or surrendering of those elements. Thus, vertical Life (Eros) accepts the death or negation of a previous level while preserving its elements: it negates and preserves: a greater Life involves the death of the lesser life.

Thus, horizontal death is part of greater Life, whereas vertical Death is simply dissolution. This, of course, is a semantic nightmare, but failure to distinguish these different types of "life and death" have led to numerous theoretical difficulties (see *Eye to eye*, chap. 7, for an extensive discussion of these issues).

13. I.20–21. All quotes, unless otherwise indicated, are from Inge, *The philosophy of Plotinus*, vols. 1 & 2, indicated as I and II.

14. I.120. These are Inge's words paraphrasing Porphyry.

15. Again, *experience* is the wrong word; the Nondual is not an experience, but the vast openness in which all experiences come and go; the "apprehension" of the nondual carries with it an exhausting of any interest in merely experiential endeavors. Nonetheless, "direct experience" has the advantage of conveying the fact that this realization is not simply theoretical or abstract or whatnot (which is why I will continue to use "direct experience" on occasion).

Porphyry reports that he had four major breakthroughs or "satoris," if we may call them that; Plotinus tells us only that he himself had the realization "often," as in "often the case" or "typical." Karl Jaspers comments, "What, according to Porphyry, would seem to have been a rare, anomalous experience, is, in the statement of Plotinus, the natural reality."

Also noteworthy is that, whereas the mystical experiences of the psychic and subtle level are often quite dramatic, ecstatic, and extraordinary (and they are indeed *experiences*), the causal/nondual realization is usually "ordinary," "nothing special," simply because it is a realization of the divine in every single ordinary act. Thus when Jaspers, also known for his extensive work on psychopathology, reports the following, it is very telling: "Ecstatic states and mystical experiences play an important part in the history of all cultures. They are a field of psychological observation in which certain basic forms of experience always recur [what we call the deep structures of the psychic and subtle]. Plotinus's transcending of thought seems to differ from such experience. His accounts contain the barest minimum of psychological phenomena and no psychopathological indications whatever. One is impressed by the simplicity of his communication from the depths" (*Great philosophers*, p. 55).

16. This is the Upper-Left quadrant, including the higher or transpersonal stages discussed in the last chapter. This is not to imply that Plotinus or Aurobindo were necessarily ultimate Realizers in a permanent or perfected sense, but that they are superb representatives of a full-spectrum approach to human growth and development based on their own experiential disclosures of the higher domains.

17. II.134. My italics.

18. II.ix, 137.

19. II.58–59.

20. II.139. My italics.

21. He refers to it as "the within that is not within"—i.e., the within that is without, or subject-object nonduality.

22. II.xi. Jaspers emphasizes: "Plotinus's writings are an unexcelled record of this fundamental experience. The experience he pursues is not one that is enjoyed as an event in time; it pervades all moments of [time and] existence and is the source of all meaning. He actually experienced as a perfect whole what in this immanent world we can know only in the duality of loving and loved, beholding and beheld, that is, he experienced the goal that gives our imperfect yearning its direction" (p. 54).

I would amend that slightly to say that an experience that is not in time is not really an experience in the typical sense of the word, but rather the ground or opening in which all experiences arise (since experiences have a beginning and an end in time). This nondual realization is the immediateness of all experience and is not itself any particular experience. This, of course, could be said for any nondual realizer, from Buddha to Christ to Krishna to al-Hallaj, as their own teachings make quite clear. But it is not just the "highest goal" that Plotinus "experiences" or realizes and asks us to experimentally reproduce in the laboratory of our own awareness; it is every developmental stage leading to it, and then the "final" grasp of the Ground that is found *in each and every stage*, high or low: the hierarchy, having served its purpose as a ladder, having been long ago abandoned.

23. In other words, these higher stages are as real as any other developmental stages (linguistic, cognitive, ego, moral, etc.). They are reconstructions of the characteristics repeatedly displayed by those who have developed competence at the particular stage; they are structured wholes with defining patterns (deep structures) that govern the socioculturally constituted surface structures; and these deep structures stably emerge in an order than cannot be altered by conditioning. (See *Transformations of consciousness*.)

This is similar to Tart's notion of a "state-specific science," except that "state" is a very general notion; preop, conop, and formop, for example, all exist in the "waking state," and yet they are dramatically different worlds. It is a structure-specific science, along with a state-specific science, that is required, and one that is additionally reconstructive and not simply comparative.

The notion of "states" is an important conceptual tool—states do indeed exist—but they cover only discrete possibilities (hence, "discrete state of consciousness" has replaced "altered state of consciousness"); but it is when discrete states are converted into enduring traits (structures) that development occurs. See note 17 for chap. 14. [See *Integral psychology*.]

24 Aristotle had, in *De Anima*, first outlined the holonic notion of a "higher grade": each higher grade or level possesses the essentials of the lower (but not all the incidentals of the lower) plus essential and differentiating characteristics of its own that are not found in the lower, even though it shares a common boundary with the lower.

25. I.221.

26. This is the whole topic of involution and evolution, a topic that we will be following in all three volumes of this series, each time adding more perspectives to it.

For the moment, we should note the following: in Plotinus's view, in evolution or Reflux, if we represent the lowest level as A, the next level up is A + B, the next is A + B + C, and so on. But in involution or efflux, if we start with the highest and represent it as A, the next level down is A − B, the next is A − B − C, and so forth, since each involutionary efflux is a subtraction or a stepping-down from the ground of its predecessor. This is why Plotinus says that efflux (involution) is to be "understood in minuses." What is left over after all subtractions—namely, matter—is thus also that which is most *fundamental* (most basic) for the whole sequence, and that which is still present, even in the lowest realm, I have summarized with the twenty tenets (which are therefore operative in all domains as well, since they are what even the least possess: the weakest noodle science).

For Plotinus, in efflux or ontological essence, destroy any higher dimension and you destroy all lower, which is the *opposite* of evolution, where manifestation, because it is being built back up to Unity, rests upon each previous lower manifestation. Plotinus emphasizes both: that the higher is in no way dependent upon the lower for its being or essence, but does depend upon it for its manifestation. (Here Plotinus and Hegel would part company; contemplative evidence supports Plotinus, as does Aurobindo's view.)

For a good introduction to this topic (without the twenty tenets), see Schumacher's *Guide for the perplexed*.

To such questions as, Where was mind before it emerged? Plotinus would answer, Where was two plus two equals four? Everywhere and nowhere.

This phenomenon—namely, the presence of the higher pre-ontological worldspaces as *present structural potentials* in the human bodymind—is behind a vast number of theories attempting to account for and describe the involutional or Descending currents in the Kosmos and in the human being. Mystical experiences (of a psychic, subtle, causal, or nondual nature) are so common to humankind, and share at least a broad similarity wherever they occur (despite the inevitable cultural colorings given to them)—and when they do occur, virtually all of them carry the overwhelming conviction that one is not stumbling onto something *sui generis*, made up, constructed at the moment, or fabricated by the individual human mind; rather, one has the overwhelming apprehension that one is recognizing something once known but long forgotten, that one is being ushered into a worldspace that in some extremely fundamental sense existed prior to its recognition: it was *there* all along, just not seen.

And one of the many amazing things about mystical experiences is that, however much culture does indeed mold these apprehensions, more often than not the apprehensions do not confirm a person in his or her cultural beliefs, but totally explode any cultural beliefs imaginable: the person is totally undone, like Paul on the road to Damascus, or totally surprised, or completely taken aback. Die-hard materialists can have these experiences as easily as purebred idealists, and both are completely stunned into awestruck silence: the depths of the Mystery are disclosing themselves, and a muted mind must only bend in reverential awe.

The involutionary theories—from Plotinus to Hegel, from Asanga to Aurobindo, from Schelling to Shankara, from Abhinavagupta to the *Lankavatara Sutra*—are all attempts to take into account that the depths of the higher structural potentials are already present but not seen. This is not so different from mathematics (and, indeed mathematics is usually included in the involutionary theories): granted that math is only formulated by the human mind, do we really think that mathematics applies only to the human mind? Or that its patterns weren't operating prior to the emergence of the human brain?

The involutionary theories would rather say that the formative tendencies and patterns of the Kosmos that *produced* the human brain are being, to some degree, *recognized by* that brain: in this special sense, they are dis-coveries, un-coveries, re-cognitions, anamnesias, recollections, apprehensions of patterns and worldspaces present as potentials but only now being actualized or apprehended in individual cases.

This in no way denies that the surface structures of these worldspaces have to be worked out in cultural, historical, social, and intentional settings. Nor does it deny or denigrate the fact that human beings must, in large measure, work out their non-determined fate with diligence. It only says that the *apprehension* is in some profound sense *embedded* in the structure and the dynamic process of the Kosmos as well—it is not merely a subjective, personal, idiosyncratic fantasy—and this is why the mystical experiences always carry the sense of recognition and rediscovery, however initially surprising.

Many of the involutionary theories are admittedly rather crude by today's analytic standards. Plato's recollection of the Forms was always molded on a type of Pythagorean mathematics that we would now view as rather limited; Hegel's Logic (precisely the attempt to articulate the potential patterns present prior to manifestation but not apart from manifestation, patterns that would then unfold in evolution or development)—is terribly dated and culturally situated; Shankara's archetypes are still loaded with mythic and culturally bound Forms; Plotinus's Efflux is limited by the particular knowledge of the world disclosed at that time; Asanga and Vasubandhu's alaya-vijnana is perhaps more sophisticated in that the supra-individual Forms are the sum total of Kosmic memory-habits, but suffers from not being able to articulate the origination of such memories (the Hua-yen philosophy of causation by mutual interpenetration would arise to address just that lack).

Nonetheless, all of those are sincere attempts to outline involution, and all of them absolutely agree on the central and crucial issue: the presence of the higher world-spaces as *potentials* given to us now, but not yet realized. And in that deep form, the mystics of the world are in virtually unanimous and unyielding agreement, and this on the basis of their experiential evidence disclosed and discussed in a community of intersubjective interpreters. It is the only interpretation that makes sense of the mystical experience: a Kosmic depth disclosed, not an individual subjective fantasy conjured up.

27. The concepts "potential" and "actual" go back particularly to Aristotle, where Matter is pure potential and Form is actuality; the lower is in the higher potentially, the higher is in the lower actually; or the higher is the actual of which the lower is potential—the lower the level, the more it is "infected" with mere potentiality; the higher the level, the more actual, less potential. Plotinus generally follows suit. Modern usage would keep the meaning but often reverse the semantics, saying that the higher is an unrealized potential in (or for) the lower that needs to be realized or actualized in its own case; it comes to the same thing, and the meaning is as expressed in the paragraph.

28. Inge, I.248. Plotinus has a remarkable theory of the self, which is still altogether viable. Aside from the notion that we have "two souls" (the timeless witness and the temporal self), the self (or Soul) for Plotinus has two different meanings. One, it is a particular level or dimension of existence—the World Soul (or psychic level). But two, the Soul is what we moderns would call the "self-system," the actual navigator of the great holarchy—not a particular stage but the "traveler" of each stage; not a rung in the ladder but the climber of the ladder. As such, the Soul or self-system can span the entire spectrum of development (except the higher limits, where it is transcended or becomes the Absolute). Inge: "It [the Soul] touches every grade in the hierarchies of value and existence, from the super-essential Absolute to the infra-essential Matter. It has its own centre, a life proper to itself; but it can expand infinitely in every direction. No limit has been set to its possible expansion" (I.203).

Development, then, is a case of the Soul growing and expanding, taking more and more of the external world into itself, as Plotinus puts it. And what causes the Soul to be at a particular level of development? According to Plotinus, "All souls are potentially all things. Each of them is characterized by the faculty which it chiefly exercises. The souls, thus contemplating different objects, are and become that which they contemplate." In other words, the level at which the self places its attention is the level at which it remains: you become what you contemplate.

Therefore, contemplate the One. This begins, Plotinus says, when "loving-kindness burns like a fire" and the Soul immerses itself in a contemplation of Spirit, where, as Plotinus puts it, "it will see God and itself and the All; it will not at first see itself as the All, but being unable to find a stopping-place, to fix its own limits and determine where it ceases to be itself, it will give up the attempt to distinguish itself from universal Being, and will arrive at the All without change of place, abiding there, where the All has its home"—there being, of course, here (I.203).

Finally, for Plotinus, these dimensions are not just lying around waiting to be perceived in a passive fashion by the Soul. Plotinus's theory of perception is one of "making and matching"—perception is active, not passive. This theory, via Schelling, would lead to much of Gestalt psychology and autopoietic theories of cognition.

29. Taylor, *Sources of the self*, p. 250.

30. Likewise, Grace is Agape, and is given freely to all; it shines on all as the omega pull of the Source-Goal, urging Eros to return, to reflux, to that Source. (This doctrine of grace is also extensively found in the East, even—I would say especially—in Buddhism, where it is behind all forms of guru yoga.)

In Japanese Buddhism, a distinction is made between self-power (Eros) and other-power (grace or Agape), as represented respectively in Zen and Shin, but both schools agree that the distinction is ultimately based on the subject-object dualism, that there is neither self nor other, neither eros nor agape—again, Eros and Agape united only in the nondual Heart.

31. As we saw, because Freud didn't follow Ascent to its true Goal (he stops at rationality), he could not see Descent in its fullness (as Agape). When "plugged into

its source," Descent is not death but Goodness, the creative Effulgence of the Divine manifest in and as all things, including "dead" matter.

Even though Freud doesn't see the Great Circle of Ascent and Descent, he of course makes use of what small segment of it he does understand: in psychoanalytic therapy, one engages in "regression in service of ego"—that is, one descends back into a lower level that was alienated and split off by fear and anxiety (by Phobos), reintegrates that level by embracing it with love and acceptance (Agape), and thus releases that lower level from being a regressive pull, releases it from being in the grips of Thanatos, and allows it to rejoin the ongoing march of a higher identity and a wider love of others (Eros).

But Freud, fractured note to Plato, cannot see all this in its larger context. Descent, for Freud, is merely destruction, which is true only when it is disconnected from Ascent; he therefore cannot see that Descent is the creative movement of emanation, of superabundance, of spiritual overflowing that reaches even into matter and glorifies all that it finds and all that it creates—and manifests then not as death but as Goodness.

32. I.198, 205.

33. I.254.

34. This is why even Plato and Plotinus, *situated* in an average-mode mythic worldview, would dramatically underrate, although not ignore, the power and importance of empiricism; not even Aristotle would grasp the importance of empirical *measurement*. Plotinus (and Plato) were situated in an average-mode mythological background worldview (mythic-rational), against which they had to fight (while delicately and unavoidably embracing aspects of it). They spoke from the center of a mythic worldview, even as their own substantive Reason transcended it, and even as their own contemplation transcended Reason; but mythically situated they could not avoid. The mythic was simply the *average* level of consciousness, so of course some individuals could move beyond it, but of course they still had to speak through it to some degree.

Rationality, as a basic organizing principle of average consciousness, was indeed deeper and wider than mythology, and that was its positive and evolutionary advance, which allowed the Enlightenment *philosophes* to wildly deconstruct so much of the predecessor worldview: they no doubt had a great deal of truth on their side. But *differentiating* the Big Three is one thing; subsequently *collapsing* the Big Three to the Big One (flatland holistic systems-language, it-language, the Right-Hand path, the monological gaze): that was something else altogether.

Plotinus, and Plato, were of course using rationality (and beyond), but, precisely because they were not situated in the monological climate of early modernity (lucky in that regard!), the rationality was fuller and deeper and truer to itself; it was, as Taylor points out, substantive and dialectical and dialogical, not *merely* instrumental and procedural and monological.

But with the shift in *average-mode* worldview from mythic-rational to rational (for that *was* the Enlightenment), then reason, captured first by its positivistic powers, saw the entire Kosmos in terms of its Right-Hand or exterior or empirical components, and thus the multidimensional Kosmos collapsed into the flatland holism of interlocking orders, with everything in the cosmos interlocking so tightly ("they were wont to discourse at length on the Great Universal System") that no Left-Hand component (whether soul or consciousness or introspection or interpretation) had any place at all. All things being merely strands in the web meant that all things were merely instrumental, and thus there was only a technological, monological, instrumental mode of reason.

We will return to this in more specific detail in chapters 12 and 13.

CHAPTER 10. THIS-WORLDLY, OTHERWORLDLY

1. The "One," strictly speaking, refers to *causal-level absorption*, whether that causal level is viewed as Source (of involution) or Summit (of evolution). But I will sometimes use *One*, as does Plotinus, to also mean the Absolute as Nondual (ultimate) Ground. The same is true for terms such as *Godhead* and *Spirit*—they can mean causal oneness or ultimate nondualism (this double usage is found in Ramana Maharshi and many others). I will do this only when the context makes clear, I believe, which meaning is intended. This is always difficult, because there are only so many words to go around; but the realizations and the meanings are indeed distinct.

2. II.62.

3. II.62. If by "One" we mean the infinite, by "All" we mean the sum total of finite manifestation, and by "Each" we mean each particular manifestation, the Nondual "stance" is: One-in-Each, Each-is-One, Each-in-All, All-in-Each, One-in-All. The pantheistic stance as popularly advanced is: One-is-All, Each-in-All. The holographic "paradigm" is: Each-in-All, All-in-Each; as an overall worldview, this is magical syncretism (which is why it is often hooked up with foraging worldviews).

When the Nondual traditions speak of the One-in-the-Many and the Many-in-the-One, they mean the One-in-Each-and-All and Each-and-All-is-the-One.

The realization of One-in-Many and Many-in-One, is, of course, common and definitive for all Nondual schools, whether of the East or the West. D. T. Suzuki, for example, maintained that the essence of Zen realization is contained in the *Avatamsaka Sutra*, where the interpenetration of the Absolute (*li*) and the relative (*shih*)— One-in-Each (*shih li wu ai*)—allows the complete interpenetration of the Each with Each and All (*shih shih wu ai*)—precisely Plotinus's point.

At the same time, the "final" philosophical position of Zen, and all of Mahayana, is expressed in the *Vimalakirti-Nirdesha Sutra*, which is a discourse on Nonduality. Each of the assembled sages gives his or her definition of Nonduality: it is the not-twoness of nirvana and samsara, it is the not-twoness of enlightenment and passions, it is the not-twoness of the many and the one, and so forth. The "correct" answer is finally given by Vimalakirti, who responds with "a thunderous silence."

4. That is to say, to each of those propositions corresponds a direct, repeatable, experiential disclosure, *as interpreted* in a community of those who have mastered the paradigm and displayed competence in the injunctions and exemplars.

5. Lovejoy, *Great chain*, p. 55.

6. Inge points out that not only are efflux and reflux nondual in Plotinus, but even Matter and Form—that intractable Greek dualism—"together are one illuminated reality." Inge points out that this nonduality can be clearly traced back to Plato—it is certain Plotinus thinks so—and thus Inge (among others) concludes flat-out that "Platonism is not dualistic."

7. Early Buddhism was a pure causal-level Ascending endeavor: flee the Many, hold only to the Unmanifest; samsara, the world of illusion, was to be altogether transcended in nirvana (and the pure cessation of *nirodh*), and the two (samsara and nirvana) are diametrically opposed; it was, in all essential respects, a pure Gnosticism.

Nagarjuna's Mahayana (Madhyamika) revolution, on the other hand, was Nondual (advaya) to the core, seeing that nirvana and samsara are "not two," which also gave rise to tantric or Vajrayana Buddhism, where even the lowest defilements were seen to be perfect expressions of primordial wakefulness (rigpa). This is Plotinian through and through. Interestingly, Plotinus and Nagarjuna lived around the same century (second–third century CE).

Thus, when Nagarjuna begins his attacks on the schools that he feels are the most distorted and partial, he does not so much attack Hinduism, as might be expected from a Buddhist. He mostly attacks early Buddhism (Theravadin and Abhidharma) with a vengeance; his attacks are unrelenting and powerful and altogether convincing. This is so similar to Plotinus's attack on the Gnostics that one is tempted is say that the World Soul was, at that particular time, definitely trying to win through to a new level of spiritual realization (namely, Nondual), and tapped these two extraordinary individuals on the shoulders to carry out the task.

But whatever the cause, these two souls wrought historically unparalleled spiritual revolutions in the West and the East. It is only a slight exaggeration to say that virtually every Nondual tradition in both East and West traces its lineage, in whole or part, to Plotinus or Nagarjuna.

In the West, Christianity was lifted out of its mere mythic involvements, and into contemplative endeavors, by the efforts of such as Augustine and Saint Dionysius, both directly and deeply influenced by Plotinus; from them it would influence virtually every subsequent Christian mystic, including Saint Teresa, Saint John, Dame Julian, Cusanus, Bruno, Blake, Boehme, and Eckhart. Plotinus would further be a main influence on Novalis and Schelling, and thus the entire Romantic-Idealistic movements (and from there to Schopenhauer, Nietzsche, Emerson, William James, Royce). Neoplatonism was likewise a major influence on Jewish mysticism (Kabbalah and Hasidism), and on Islamic mysticism (especially Sufism). None of those movements added substantially to the overall revolution wrought by Plotinus, which is exactly why Benn could maintain that "no other thinker has ever accomplished a revolution so immediate, so comprehensive, and of such prolonged duration."

Except Nagarjuna. All schools of Tibetan Buddhism, of Chinese Ch'an and Tien Tai, of Japanese Zen and Shin and Kegon, trace their lineages directly to Nagarjuna (in whole or part). But not only the extraordinary flowering of Mahayana and Vajrayana rests on Nagarjuna's shoulders: his dialectic was a major influence on Shankara, Vedanta's greatest philosopher-sage (Ramana Maharshi being one of Shankara's many descendants), and Shankara's Nondual (Advaita) Vedanta revolutionized all of subsequent Indian philosophy/religion (so similar was it in many respects to Nagarjuna's Madhyamaka that Buddhism simply died out in India, being almost, as it were, reabsorbed in essentials back into Hinduism).

Thus, around the world, East and West, North and South—it is only a slight exaggeration to say that all Nondual roads lead to these two most extraordinary souls, world heroes in the truest sense.

(For an extensive discussion of Nagarjuna, see note 1 for chap. 14).

8. Tillich, *A history of Christian thought*, p. 374. Both Tillich and I are referring to the mythic components of Christianity, not to its psychic and subtle-level direct realizations, many of which approached a pure nondual understanding (as we saw with Teresa and Eckhart). Rather, it is the translation downward into "mythic dissociation" (precisely as defined by Campbell) that I am here discussing, because this mythic dissociation, particularly in regard to the Incarnation (Descent) and the Ascension (Ascent), came to thoroughly define essential Christianity, divorce it from other religions, and set an entire tone for the culture it would subsequently define. God created this world (which indeed manifests Goodness to that degree), but our final destiny can be found nowhere in this world (the classical Ascending ideal of mythic dissociation).

9. *History of Christian thought*, p. 395.

10. They were a source of what Peter Berger calls *nihilation*, which, as I view it, is any threat to the cultural translation process of generating meaning and its correlative form of social integration and stability. And, as Berger points out, nihilation needs to be countered with therapia, or a dislodging of the source of nihilation. (Nihilation is the threatened negation of a given level of translation; therapia is the preservation or restoration of that level. Transformation or transcendence, on the other hand, is the negation *and* preservation of that level: its nihilation in a greater therapia or Eros that embraces and preserves—Agape—its own accomplishments.)

But since cultures generally must preserve their own average-level mode of translation (as the source of social stability and cultural cohesion) and ward off all negating threats, then all threats are simply and generically viewed as nihilation, and massive measures of therapia (including coercion to the cultural paradigm) are brought into play (cf. Foucault). In rational societies, we simply label someone "outlaw" and lock them up, or "mad" or "insane" and then drug them out of our misery (see *A sociable God* for a distinction between pre-law and trans-law, two very different forms of outlaw, but both of which are nonetheless simply viewed as nihilation to the average-mode cultural paradigm, and both of which are strenuously "marginalized"). But in mythic societies, the nihilating offense is much more consequential, because an insult to the state is also an insult to God, and thus anything from excommunication to burning at the stake will be required for cultural therapia. Because the Big Three aren't differentiated in mythic cultures, a threat to any one is a threat to all three, and must be treated accordingly.

11. We see this already happening with the Apostolic Fathers, particularly Clement of Rome and Ignatius of Antioch. The original "preachers" of early Christianity, apart from the apostles, were the traveling "pneumatics," those in whom "spirit was alive" as a direct experiential opening. But by the time of the Apostolic Fathers, this "pneumaticism" was on the wane and was, in fact, viewed as dangerous, because it led to "disorder." The Apostolic Fathers therefore asked, Why do we even need them? Everything that needed to be said had already been said, they felt—first in the Old Testament, then in the still-being-drawn-up New Testament, and codified in the Canon and the Apostles' Creed, a series of *beliefs* without any reference to *actual experience*. The Church became the *ekklesia* (ecclesiastic) assembly of Christ, and the only governor of the ekklesia was the local bishop, who possessed "right belief," and not the pneumatic or the prophet, who might possess spirit but couldn't be "controlled." The Church was no longer defined as the community of saints, but as the assembly of bishops.

With Tertullian the relationship would become almost purely *legal*, a matter of proper sacramental rites (whether the person understood them or not), and thus, for example, baptism could save an infant. From here it was a short step to Cyprian's binding of the Spirit to the *legal office of the Church*. One could become a priest by *ordination* and not by *awakening*. A priest was no longer holy (sanctus) if he was personally awakened or sanctified, but if he simply held the legal office.

Likewise, rituals became "objectively holy" and "binding" even if the giver of them was *not* subjectively holy, and even if the receiver of them did not have the faintest idea what was going on—one could be "saved" by ritual sacrament alone and not by direct experience and evolution: "He who does not have the Church as Mother cannot have God as Father" (Cyprian). For various complex reasons, Augustine sided with this view, and the Church became structurally a pure mythic-rational legalism.

Finally, since salvation now belonged to the lawyers, grace became the forgiveness of sin (Saint Ambrose) conceived as breaking the law—"You are forgiven"; gone is the conception of Grace as Deification or discovery of true Self (Saint Clement: "He who knows himself knows God").

Much of this is unavoidable in any organized realization; but it obviously can go too far, and this "too far" is part of the "translation downward" which I am discussing.

12. *History of Christian thought.*

13. Justin Martyr, the most important of the Apologists, was pivotal in this regard, and is a perfect example of how the realization of a universal Nous or Mind would be translated downward into an exclusive mythic property. He had three main points:

(1) "Those who live according to the Logos are Christians." This is fine; the Logos is the Nous or universal Mind in each of us, and if a Christian is defined as one who lives in Logos, then to the extent that any of us live in Logos, we can all be called Christians. This is very similar to D. T. Suzuki saying that all true religion has satori (realization of universal Mind or no-mind) as its basis, and therefore all true religion has Zen as a basis. There is nothing in Justin's position so far that is divisive or merely mythic.

(2) "What anybody has said about the truth belongs to us, the Christians." This is still fine, as Justin has defined these terms. As Saint Ambrose would later put it, "Whatsoever is true, by whomsoever it is said, is from the Holy Spirit." All Truth issues forth from Spirit, and if a Christian is one who is actually living in this Truth, then all Truth belongs to Christians, as defined. Again, this is indistinguishable from Suzuki's "All truth is Zen."

(3) "If the Logos in its fullness had not appeared in Jesus, no salvation would be possible for anybody." And there Justin steps over the line, or down the line, and into mythic possessiveness. It follows only from any such line of "Christian" reasoning that the Logos needed to descend into *some* particular embodiment or incarnation. If it is understood that Buddha or Krishna or Lao Tzu was also a complete embodiment (Descent or Incarnation) and realization (Ascent or Ascension) of Spirit, that Jesus is non-unique, then we have the notion of avatar or world teacher, which is fine. But the only way to claim that just one of them is "the absolute and unique incarnation of God" is to ground that claim in mythological dogma that is separated out and protected from reason, from evidence, and from direct spiritual realization, and given a privileged status that is merely sociocentric, not pneumatocentric, in origin.

And the sad thing about this proprietary stance is that (1) it separates and divides Christians from all other humans and other world citizens; (2) it divides Christ from all Christians, because Christ is, in the final analysis, unique; and (3) it ultimately divides God from this world altogether, since the Incarnation and the Ascension, for this earth, both happened only once, no matter how much the first two genuinely universal points of Justin's ever-repeated argument are stressed.

Christian apologists are razor-sharp in applying these distinctions, and so you have to put it to them very bluntly: "Is Jesus Christ unique? Is, to give just one example, Gautama Buddha also an absolute Incarnation and Ascension of the full Logos?" If they say yes, then they are following genuine or catholic faith, and the only true conclusion of their reasoning; if they say no, they are still stuck in their local volcano god and his "only begotten." It will do no good, at this point, to claim that Buddhists are participating in the same Logos that was uniquely manifested in Jesus. Many modern Christian mystics, incidentally, are perfectly happy to answer yes.

The Trinitarian Doctrine (three Divine Forms, One Substance) is almost identical to the Buddhist doctrine of the Trikaya (three Bodies of Buddha), as worked out in *The Lotus Sutra*, the *Lankavatara Sutra*, etc., and was designed to answer the same question: If Buddha-nature (*tathagatagarbha*) is timeless and eternal, how did it come to appear in the historical person Gautama Shakyamuni, and what is his relation to it? The answer was also essentially the same: the formless Ground or Truth Body (Emptiness, the Dharmakaya; Father) appears, in the world of form, as the Nirmana-kaya or Form Body, the historical manifestation (Gautama; the Son), mediated by the Transformation Body (the Sambhogakaya; Holy Spirit)—all three of which are different but equal aspects of the Svabhavikakaya (Godhead). But the Buddhists drew the only correct conclusion: there are therefore potentially an infinite number of equal Sons and Daughters of Dharmakaya; not only is Jesus accepted as a perfect Nirmana-kaya manifestation, it is expected.

14. See *Up from Eden.*

15. If any other individual fully Ascended, it would destroy the meaning and integration of society on the whole—politically, morally, and religiously. This is, needless to say, a strong disincentive to Self-realization.

16. As a direct experience, not merely as a verbal claim. It will do no good, for example, to claim that Jehovah is a creation-centered God or Alpha Source (which is certainly the orthodox claim); many magico-mythic productions are "nondualistic"- or "holistic"-sounding, as we have seen: the gods and goddesses can be embedded in nature—it rains when they cry, and so forth. Or "God created both heaven and earth"—the pure Alpha Source, it is said.

But the test is always: is this merely a metaphysical claim, or is it grounded in direct spiritual practice and direct, repeatable, reproducible experience? Anybody can say "All is One." The question is, what is the structure of consciousness that actually authors that statement? Magic? Mythic? Mental? Psychic? Subtle? Causal?—for all can and will claim ultimacy, and all can sound, on occasion, very "nondual."

Anybody familiar with my work knows my profound indebtedness to the many great Christian (and Jewish and Islamic) mystics, who rose above magic, mythic, and rational endeavors to ground a realization in psychic, subtle, and even causal endeavors. My point, rather, is that the closer they got to that actual Goal, the more they were looked upon with grave suspicion and much worse, for reasons I have tried to suggest.

17. In Lovejoy, *Great chain*, p. 68. My italics.

18. Tillich's summary of Augustine; *History of Christian thought*, p. 112.

19. The main difference between the Eastern and Western Vedanta is the degree of penetration. The Witness-stance can be pointed out to most people fairly easily, and they can begin to grasp and feel its implications. This is, we might say, a "beginning glimpse."

The next stage is to meditate on this understanding (in the Western sense of "thinking hard about it"), which can lead, as it did for Descartes, to an actual absorption in the "I AM" state for several hours at a time. This is, so to speak, the second stage of prolonged absorption, and it can so revolutionize one's awareness and thoughts that, as Descartes put it, when he emerged from that state his whole philosophy had been completely formed.

But according to Vedanta (and Zen and Vajrayana, etc.) those states have to be refined and polished and brought to a constant realization, and not merely a "peak" experience or a "plateau" experience, because it is only in the breakthrough (the dissolution) of the Witness that the Nondual is recognized. The theories that come out of temporary peak experiences are almost always, to use an apt metaphor, half-baked. We see this half-bakedness in Augustine and virtually all of his "descendants" (with some happy exceptions), but at least it is all a step in the right direction, and infinitely preferable to the impossible positivistic attempt to ground certainty in the exteriorness of things.

Likewise, from Augustine on we would see two radically different approaches to trying to demonstrate (or "prove") the existence of God—and, no surprise, they are the Right- and the Left-Hand paths. The Left-Hand path (Augustine's path) starts with the interior, with *immediate consciousness*, and attempts to plumb the depths of that immediateness, of that Basic Wakefulness, and finds enlightenment or Spirit in the direction of the absolute *interior* of consciousness, which itself then issues in the nondual awareness transcending *both* interior and exterior altogether (the move from the interior to the *nondual* superior). This is why virtually all Christian mystical schools would trace their lineage to Augustine (and Dionysius).

The Right-Hand path starts (and ends) with *exteriors*, and attempts to deduce, logically, the Author of the exterior pattern (argument from design, ontological argument, first-cause argument, etc.). This is preeminently the path of Saint Thomas, and this is why *the* archbattle within the Church has always been between the Augustinians and the Thomists (in their many forms)—between the interior-mystical-consciousness approaches and the exterior, objectivisitic, legalistic, dogma-scriptural approaches (an exterior approach that continues right down to today, with the "modern physics proves the Tao" arguments, which are all Right-Hand arguments).

Augustine would reply, first, that those Right-Hand "proofs" can't be done (there is no convincing exterior proof) and, second, even if there were, it would be worthless, because exterior "proof" alone does not result in inner *transformation* of consciousness, and it is this inner transformation that alone discloses Spirit. Not an objective belief, but a transformation of will—this alone discloses God, according to Augustine. In this, Augustine is brilliant, original, and altogether compelling.

The difficulty is that this half-baked Witness would play directly into the hands of the disengaged egoic-rationality of the Enlightenment paradigm. In the Mahayana and tantric traditions, the pure Self or empty Witness transcends all, embraces all, and thus is *situated* and *embodied* in the entire manifest realm (the Dharmakaya or Emptiness has Rupakaya or Form Body). The disengaged ego of the Enlightenment was fueled in part, no doubt, by an intuition of the Witness (as we will see), but it lacked the fully baked Rupakaya, its embodiment or embeddedness in the entire world of Form, and these merely dissociated and disengaged aspects contributed to the "dehumanized humanism" of the Enlightenment.

But when Husserl explains that the world could end and it wouldn't affect the pure

Self, or describes the splitting of the witnessing self from empirical self (e.g., in section 15 of *Cartesian meditations*), or when Fichte describes the pure Observing Self as being *infinite* and supraindividual Spirit—this is Western Vedanta at its finest. It just needs its Rupakaya.

As for the *general similarity* of this Western Vedanta with the Eastern—Basic Wakefulness is impossible to doubt because doubt itself presupposes it—here is only one example from Dzogchen ("the Great Perfection") Buddhism, generally regarded as the highest of the Buddha's teachings (this is from Paltrul's "Self-Liberated Mind"):

"At times it happens that some meditators say that it is difficult to recognize the nature of the mind [which means, in Dzogchen, the absolute reality of pure Emptiness or primordial Purity]. Some male or female practitioners believe it to be impossible to recognize the nature of mind. They become depressed with tears streaming down their cheeks. There is no reason at all to become sad. It is not at all impossible to recognize. Rest directly in that which thinks that it is impossible to recognize the nature of the mind, and that is exactly it."

And as for the Witness itself passing into nonduality beyond subject and object, beyond inside and outside (but embracing both): "There are some meditators who don't let their mind rest in itself [basic immediateness], as they should. Instead they let it watch outwardly or search inwardly. You will neither see nor find the mind by watching outwardly or searching inwardly. There is no reason whatsoever to watch outwardly or search inwardly. Let go directly into this mind that is watching outwardly or searching inwardly, and that is exactly it."

20. *History of Christian thought*, p. 103.
21. *Sources of the self*, p. 131.
22. Taylor's words, *Sources of the self*, p. 139.
23. *History of Christian thought*, p. 114.
24. Ibid., p. 121.
25. *Great chain of being*, p. 84.
26. In fact, the existence of evil was taken as a type of proof for the existence of Goodness. Evil, it was thought, was not an actual substance or a real existent; it was simply a shadow; but all shadows are cast by Light, the Light of descending Goodness, in this case. The point was much like Vedanta: manifestation is dualistic, both good and evil come into being as correlative polarities, and you can't have one without the other. This was the basis of the famous "optimism" of the seventeenth/eighteenth century, and it meant almost the opposite of what it would mean for us today, where optimism means you can get rid of evil; then it meant, you can never get rid of evil, and therefore this is the best world that could actually be constructed (that is, this is the *best* of the actually possible worlds). What the Descenders lacked was the other half of the equation, the nondual view: both good and evil can be transcended.

Parts of this and the next paragraph are a direct paraphrasing of Lovejoy. For an extended discussion of these "two strategies," see Da Avabhasa, *The dawn horse testament*, chap. 18. Da uses *alpha* and *omega* in an opposite semantic sense to mine; the actual meanings are similar.

27. P. 84. There were, of course, many happy exceptions who saw, with Eckhart, Bruno, Cusanus, the way of the coincidentia oppositorum—saw that the way up and the way down are to be united in every moment of perception. They did not always suffer happy fates for this realization.

28. *Great chain of being*, p. 51.

29. Ibid., p. 65.

30. Ibid., p. 58.

31. The Descenders would exclusively claim that there are no gaps anywhere; the Ascenders would emphasize "leaps are everywhere"—the former embracing flatland, the latter attempting to leap out of it.

Most present-day ecophilosophers, because they are not radically nondual in their approach, are forced to argue the continuity thesis, which earns for them the charge from critics of being eco-fascists, which bewilders the ecophilosophers, who imagine that a flatland holism is actually a liberating notion.

Lovejoy demonstrates, as I mentioned in chapter 1, that the Great Chain theorists all subscribed to three interrelated notions: plenitude, continuity, and gradation. And, he points out, there was, and still is, a tension between the second and the third (namely, "no gaps" and "leaps everywhere"). This is the theoretical tension inherent in emergence theories, an unavoidable tension still present in today's biological, social, and psychological developmental theories. For reasons that this tension cannot be resolved theoretically, see chapter 14, "IOU."

32. *Divine ignorance*, III.1.

33. Schumacher, *Guide for the perplexed*. Schumacher is something of a hero to ecologists and environmentalists, and rightly so, I think; but they are utterly perplexed by his complete and enthusiastic embrace of hierarchy, precisely because they confuse hierarchy with pathological hierarchy and thus fail to understand that the only way to construct a genuine holism is via a genuine holarchy—flatland holism is not holism at all, but a heapism and a reductionism that produces not deep ecology but span ecology.

34. *Great chain of being*, p. 120.

35. Ibid., p. 119.

36. Ibid., p. 207, 186.

CHAPTER 11. BRAVE NEW WORLD

1. Taylor, *Sources of the self*, p. x.

2. My general claim is that most of the aspects of modernity that many theorists have seen as positive and beneficial can in fact be resolved into variations on "No more myths" (since that also includes a transformation from conop to formop and thus a differentiation of the Big Three, the specific "dignity" of modernity); and likewise, most of the aspects of modernity that have been harshly criticized by theorists can be resolved into variations on "No more Ascent!" (since that also includes the

loss of a grounding in a source and a context that provided substantive integration and meaning, and set the rational-ego and its instrumental-purposive reason afloat in a world that no longer cared).

3. Their relational exchange was not merely with society at large, which would not comprehend their endeavors; their relational exchanges occurred in the micro-communities or sanghas or lodges or academies of the like-minded, like-spirited, like-depthed; just as, in our culture today, many people fall below, and a few above, the typical expectable level of consciousness, morality, cognition, and so forth.

4. As we saw, referents exist only in worldspaces (a given referent exists only in a given worldspace). And worldspaces are *not* simply arbitrary interpretations of a pregiven monological cosmos; rather, worldspaces are embedded in the Kosmos, and are constrained accurately by the depth that they *can* register (or make and match).

5. Quoted in Taylor, *Hegel*, p. 4.

6. See Feuerstein's *Structures of consciousness* for a superb elucidation of Gebser's work, which only recently became available in English translation.

7. Tillich, p. 333. One of the reasons for all of this, I am suggesting, other than the emergence of rationality itself, is that since the Ascendent One is no longer admitted, it becomes the Descendent or manifest All or Whole in which now resides Providence or Harmony; it is no longer the relation of Each and All to the One but merely of Each to the All—it emphasized, as I said, the descending "systems theory" of Plenitude, which is correct as far as it goes, but is only "half" the story.

In the terms of Kegon (or Hua-yen) philosophy of Chinese Buddhism (which is often misappropriated as a systems theory), there are four main principles of reality. The first is *shih*, or individual phenomena. The second is *li*, or absolute noumenon. The third is *shih li wu ai* (between absolute and phenomena there is no obstruction). The fourth is *shih shih wu ai* (between phenomenon and phenomenon there is no obstruction—that is, all phenomena interpenetrate). Systems theory is the last tenet, with no understanding of *li*, the absolute.

8. Lovejoy, *Great chain*, p. 201.

9. Tillich, *History of Christian thought*, p. 328.

10. All Chafetz quotes in this section are from *Sex and advantage*, chap. 1. All italics are mine.

11. Thus, the shift from the hoe to the plow—itself driven by the vastly superior capacity of the plow to produce foodstuffs—meant a massive shift from a female-participatory to a predominantly male productive work force, and thus the public/productive sphere shifted away from that 50–50 balance toward male specialization and monopolization—driven, as Chafetz indicates, by efficiency and expediency.

The same shift from "matrifocal" or "bifocal" modes of production toward patriarchal modes would occur in societies that were based on the mounted horse; an astonishing 97 percent of such societies were patrifocal/patriarchal (Lenski). This general shift from matrifocal/bifocal to patriarchal—the development of the plow and horse—began around 3000 BCE, and accelerated around 1000 BCE, especially with the invention of iron (swords and plows). The evidence (presented in vol. 2) shows quite clearly that ideology had virtually nothing to do with this transformation; different ideologies were woven around the same basic transformation in the infrastructure; and in all cases, ideology followed the base (e.g., male deities replaced female deities only after the horse and/or plow was introduced, not before).

Much has been made of this shift from matrifocal or bifocal modes ("equalitarian") to patriarchal modes by feminists who tend to see women as powerless victims, despite the fact that such oppression theories have consistently failed to match available data. As Habermas points out, these "imposition theories" are now regarded as empirically refuted (the evidence for this is presented in vol. 2).

The great advantage of the more recent feminist theories (from Chafetz to Nielsen, not to mention the popular rise of "power feminists"), other than matching the evidence, is that they refuse to paint women as subjugated sheep, and see them instead as equal co-creators, under the *given* circumstances, of the various forms of social interaction: it is overwhelmingly empowering to the female, and converts her from "defined by an Other" to "co-definer with an Other"—at each and every stage of human development.

Most first- and second-wave feminists were thrown off track by the fact that women in other times and places would actually *choose* values that did *not* match their own liberal-Enlightenment heritage, and the choosing of those values was therefore attributed to an *outside force* (and not a deliberate choice co-created by the female in the face of difficult circumstances); this *postulating* of an outside force inadvertently but automatically defined woman as primarily molded by an Other. This malevolent Other was simply assumed to be Generic Male, and Oppression Studies 101 was set in motion, with the bizarre goal of empowering women by first defining them as powerless.

Volume 2 is a look at the third-wave feminists who are quietly but dramatically rewriting the history books, telling the tale not of women *reacting* nobly in the face of oppression, but *pro-actively* making the very best decisions possible in the gruesome conditions known as life in this biosphere.

12. In volume 2, the correlations between the emergence of egoic-rationality and industrialization are traced. Egoic instrumental rationality was intimately related with instrumental machinery/industry (Upper-Left and Lower-Right, respectively). Thus, when Chafetz names *industrialization* as the single greatest factor in the liberation of the female, she is describing the Lower-Right aspect of this shift (and I agree). In the LR, industrialization made individual physical strength less and less important, which again was a factor freeing the female from the subservience inherent in the biosphere. This also explains why historically, where previous women's movements had briefly flourished (wherever *reason* first emerged, such as in Greece or India), they nonetheless could not be *sustained*. The material base (LR) had not the industry to sustain the differentiation.

This applies to the other liberation movements as well. Rationality might create a mental space for universal perspectivism and universal liberation; but industrialization allowed it to manifest. As Amory Lovins has pointed out, industrialization in Europe gave every man, woman, and child the equivalent power of several slaves (so that today, Lovins calculates, every person in the world has the *average* equivalent power of *fifty* slaves). And that made it all the more likely, and more possible, that actual slavery could and would be outlawed.

These "Marxist" (or Right-Hand) considerations are part of volume 2, which sets a counterweight to the more "idealistic" (or Left-Hand) presentation of this volume. My overall point, of course, is that an all-quadrant view would take both into equal account. The fact that I am not covering the material base in this volume certainly does not mean I am neglecting it or devaluing it; in many ways, as we will see, it was (and is) often the crucial dimension (particularly for the average-mode consciousness).

Finally, the role of industrialization in the liberation movements puts the *ecofeminists* in a particularly delicate position. Their existence as a movement depends upon industrialization, the same industrialization that they must aggressively condemn as leading to the despoliation of Gaia. They have to bite the hand that feeds them, so to speak. Or, we might say, they are individuals who are not on speaking terms with their parents. I will explore this tricky situation in volume 2, and point out why the structures of the centaur are the only ones integrative enough to address this situation.

13. Eisler, *Chalice and the blade*, p. 165.

14. Equality of men and women in the noosphere—equal access to the public domain of the noosphere and equal rights in that domain—does not mean that a rigid 50–50 parity must be maintained in all areas. Many feminists believe, for example, that women as mothers need special legal protections (paid maternity leave, for example); and the topic of women in frontline combat will probably always be debated with no clearcut winner (although most countries that have tried women in frontline combat positions—the Soviets in World War II and Israel in the Six-Day War—almost immediately abandoned the practice; apparently women are more averse to blowing people's brains out at close range).

Likewise, some feminist researchers believe that, given the unavoidable aspects of childbearing, a "parity" in the public/productive domain would be around 60–40 male/female.

Parity, in other words, can still take into account some of the biological constants and differences, without letting them dominate the scene or unfairly prejudice any situation; but a rigid 50–50 parity across the board is probably a bad yardstick. Nonetheless, whatever we decide about parity, these issues can be consciously and in good will debated, and in all cases that beats leaving the decision up to a hoe.

15. Eisler, *Chalice*, p. 151.

16. See, for example, Dale Spender, *Feminist theorists: Three centuries of key women thinkers* (New York: Pantheon, 1983), and Mary Beard, *Woman as a force in history* (New York: Macmillan, 1946).

As for the debate about whether women gained ground or lost it in the Renaissance (compared to the feudal period), Eisler is correct, I believe, when she states that "in *neither* period do we find any fundamental alteration of women's subservience" (p. 227). The subservience, of course, was not primarily to men, but of both men and women to the biosphere.

17. Eisler, *Chalice*, p. 159.

18. Which is why this type of specific "aesthetic dimension" is not the same as the Upper-Left quadrant per se; aesthetic-sensory cognition is simply a small part of the early development of the Upper Left; and when Kant uses it in this fashion, it fails miserably in its assigned task of integrating the Big Three. Higher aesthetic development, which Kant hints at with his notion of the sublime, would indeed refer to *aspects* of higher UL developments, but that is not what Kant has in mind for his integrating link.

Also, I do not simply equate Beauty and Art. Beauty is the transparency of any phenomenon to the One; Art is anything with a frame around it.

The frame is sometimes an actual frame (as around a painting), or an arch (around a stage), or even air (around a sculpture), but a frame that always says: look at me. Anything within that frame is Art, whether it be the Mona Lisa or a tomato soup can. The Mona Lisa is Art and Beauty; the tomato soup can is Art.

(All of these issues, and especially the nature of the Upper-Left quadrant, are discusses in greater detail in volume 3.)

19. If we define modernity loosely as consisting of any or all of the following factors—(1) the differentiation of the Big Three (Weber); (2) the rise of the philosophy of the subject (Habermas); (3) the rise of instrumental rationality (Heidegger); (4) the transparency of language (pre-Saussure); (5) the widespread turn within (Taylor); (6) belief in univalent progress—then postmodernity may be loosely defined as any attempt to develop beyond (or at any rate to respond to) those factors. Foucault, Habermas, Taylor, Derrida, Lyotard are thus, in very different ways, all postmodern, as I use the term. All are especially united in the "death of the subject," the "death of instrumental rationality," and the "death of univalent progress," though their proposed solutions often differ dramatically. The thesis I defend in volume 3 is that all of them are accessing vision-logic in various ways (integral-aperspectival), and thus all of them are particularly "postmodern." As for the "postmodernists" who simply and solely trash reason and the Enlightenment, the world regression, not integration, comes most to mind.

20. Paul Tillich's claim, and I concur.

21. Tillich, *History*, p.33.

22. Inge, *Philosophy of Plotinus*, I.97.

23. Ibid., II.196, I.35. Gutsy Aristotle was more direct and "attributes no scientific or philosophical value to mythology" (II.196).

24. II.24.

25. II.17.

26. I.103.

27. Lovejoy, *Great chain*, p. 107.

28. Ibid., pp. 99, 111.

29. Ibid., p. 109.

30. Ibid., p. 116.

31. They were both also the first to clearly express the theory of *relativity* inherent in decentering rationality. Bruno maintained that our perception of the world is relative to the position in space and time from which we view it, and consequently there are as many world-perceptions as there are positions; this monadology had a profound effect on Spinoza and Leibniz. Cusanus pointed out (in 1440!) that "it is evident that this earth really moves, though it does not seem to do so, for we apprehend motion only by means of a contrast with some fixed point. If a man on a boat in a stream were unable to see the banks and did not know that the stream was flowing, how would he comprehend that the boat was moving? Thus it is that, whether a man is on the earth or the sun or some other star, it will always seem to him that the position he occupies is the motionless centre and that all other things are in motion." Rational perspectivism and relativity at its best.

32. Lovejoy, *Great chain*, p. 262.

33. Ibid., p. 232.

34. Ibid.

35. Ibid., p. 211.

36. Ibid., p. 184.

37. Ibid., p. 186.

38. Ibid., p. 188.

Chapter 12. The Collapse of the Kosmos

1. In any of the structures of consciousness development, one can be aware of the Ascending and Descending currents up to and down from the particular level of one's present state. Thus, for example, if one has evolved to the mythic level, one has conscious access to the Ascending current from matter up to that level (i.e., matter to sensation to perception to impulse to image to symbol to concept to rule), and to the Descending current from that level down to matter (from rule to concept to symbol to image, etc., to matter). This doesn't mean that one necessarily can verbalize those ascending and descending currents, or consciously articulate them and theoretically objectify them; but rather it means that that is the degree and extent to which one is consciously "plugged in" to the Great Circle of efflux and reflux. At any level of development, then, one can be conscious of both Ascent and Descent up to and down from that level, and one can therefore emphasize Ascent, or Descent, or both ("up to" and "down from" that level).

The two currents experientially join, and yield their secret marriage, only at the formless or causal level; prior to that point one has access only to the arcs up to and down from the present level of development. I will return to the meaning of this when

I discuss Tantra (in volume 2), or the conscious union of the ascending and descending currents in the conjugal Heart.

2. In Lovejoy, *Great chain*, p. 191.

3. Habermas, *Philosophical discourse*, p. 112.

4. Ibid., p. 19.

5. Ibid., pp. 83–84.

Each of the Big Three was already integrated *in its own domain* (that was the differentiation/integration that constituted the transformation upward from mythic-rational), but a further differentiation/integration would be necessary to integrate the three domains themselves into a higher synthesis (the insoluble problems of one level being defused only at the next; the differentiating component of the new and required differentiation/integration would be—and still is—the differentiation of consciousness from a preoccupation with the individual domains to the exclusion of the others). I will argue that the transformation from formop (and the understanding) to vision-logic was the central platform of the Idealist movement, and that, in various ways, vision-logic (dialectical, dialogical, integral-aperspectival, interpenetration of opposites, intersubjective, feeling/vision) remains the cognitive goal, and aperspectival foundation, of the moments of truth of the postmodern theorists (this is one of the major themes of volumes 2 and 3).

6. Donald Rothberg also gives a superb short summary of Habermas's overall view of the result of this reduction: "On such a basis, there has occurred what Habermas calls 'the colonization of the lifeworld': the emergence of models and practices of social engineering and technical approaches to practical life and subjectivity; the increasing control by 'experts' of political and social relations and subjectivity analyzed by Foucault (Rabinow); and the decline of public spheres of open discussion and debate, that is, the decline of the core mechanisms ensuring democratic procedures (Habermas). In this way, the potentials of a vital public realm of informed, practical reason and action, associated with much of the promise of democratic movements, has also been only partly developed. Similarly, the potentials of modern subjectivity have also been severely constrained through extensive control, ideological influence, and commodification." ("The crisis of modernity and the emergence of socially engaged spirituality.")

7. Lewis Mumford. The "disqualification" was not due primarily to the quantification or mathematicalization of the world; it was due primarily to the objectification of the world, to the focusing merely on exterior, Right-Hand surfaces, which, precisely because they are marked by extension, can be best treated as quantities and thus easily mathematized.

8. Technically, the interior and exterior holarchies of depth and span were collapsed into only the exterior holarchies of depth and span. Holarchies were still recognized, as I said (the Upper-Right and Lower-Right domains are both holarchical, and this was not denied; and these holarchies, like all holarchies, possessed both a depth and a span). However, the only holarchies acknowledged were holarchies of exteriors,

of surfaces that could be seen empirically with the senses or their instrumental extensions, and thus any values (any Left-Hand correlates) were *automatically excluded.* Since the interior dimensions were denied, there were no holarchies of qualitative distinctions, of value, of consciousness, of goods or hypergoods; no transformations of consciousness required to get at the truth, and no truth other than what could be analytically formulated around empirical, value-free data.

Thus, even though the Right-Hand path acknowledges holarchies with depth and span, the *depths* of these holarchies all have the same value, namely, nothing (that is, they are "value-free" surfaces). And thus, from the view of the interior and Left-Hand path, all of the "empirical depth" of exterior holarchies (e.g., molecules have more depth than atoms) are no more than variations on the same span (namely, empirical or functional, and thus based solely on physical *size*: molecules are bigger than atoms). Value does not, and cannot, enter.

This collapse is also captured in a shorthand phrase I have been using: vertical and horizontal were collapsed into horizontal alone (although technically, the correct statement is that vertical and horizontal of the interior and exterior were collapsed into vertical and horizontal of the exterior alone). In this shorthand, vertical means interior, and horizontal means exterior (just as, in the same shorthand, depth means interior and span means exterior).

9. Taylor, *Hegel*, p. 4.

10. Dreyfus and Rabinow, *Michel Foucault*, p. 236.

11. Even if the empirical-sensorimotor world were "pregiven" in some fundamental ways—and this is the assumption made by most practicing scientists—nonetheless the *restriction* of *knowledge* to that empirical world is the reductionism that most concerns me (and that reductionism *is* the reflection paradigm).

The intrinsic features (Wilfrid Sellars) of the sensorimotor world are indeed, in my opinion, already laid down by evolution prior to the emergence of rational reflection on that world; the practicing scientist is more or less correct in that regard, I believe—fundamental aspects of sensorimotor holons are "pregiven" in that rudimentary sense. (Sensorimotor holons mutually *enact each other* and are not absolutely pregiven in that sense; but, relatively speaking, the human mind is handed that enaction after the fact and begins *its* work with holons that are in many important respects then preexisting, even though mind will then *enact its own world* with those holons as only a component of the overall gestalt; just as, for example, in the Piagetian system operational cognition works with, but not only with, the holons already delivered to it by sensorimotor cognition.) Empirical scientific paradigms govern the selection, disclosure, and elaboration, not the in toto creation, of sensorimotor holons (which is why science, as Kuhn pointed out, can indeed make progress).

But the aspects of those holons that can disclose, and the interpretation of those holons, and the capacity to reflect upon them in the first place, and the types of conceptual elaborations brought to bear upon the sensorimotor surfaces—all of this is not pregiven, but rather develops in an intersubjective space of dialogical disclosures.

But the fact that sensorimotor holons are, in at least some important respects, pre-existing in the sensorimotor worldspace (a diamond will indeed cut a piece of glass, no matter what words or what theories we use for "diamond," "cut," and "glass")—was extended (by the Enlightenment) to the notion that *all knowledge* is itself merely a recovery, via unproblematic representation, of altogether preexisting situations. Thus, whereas "diamonds" and "glass" exist in the sensorimotor world, *knowledge* of the one cutting the other—and a theory of why it does so—exists only in the rational, not empirical, worldspace, and thus the *overall* referents and gestalts of scientific knowledge are not simply lying around in the sensory world waiting for anybody to spot them. Further, the subject of the knowing itself develops in an intersubjective space that alone discloses aspects of these overall referents.

Moreover, whereas fundamental aspects of sensorimotor holons were laid down by evolution prior to reflection on them by rationality (atoms were there, and are there, before reason thinks about them)—whereas, that is, sensorimotor worldspace holons exist or subsist prior to mental formulation, cultural holons do not. When mind reflects on nature, much of the nature is preexisting or pre-mental (the sensorimotor components); but when mind interacts with mind, the result is culture, and culture is not simply preexisting, but is created with the interaction itself. Mind might *discover* various facts in nature, but mind *creates* culture. And in that creation of culture, a reflection paradigm is woefully inadequate, because preexisting holons are not being discovered, but rather newly created holons are being produced.

And while we can indeed "reflect on" and "represent" these created and produced cultural holons (such as language), we cannot understand them *merely* via objectifying and empirical knowledge, but must *also* enter into their meaning via interpretation and hermeneutical disclosures.

The depth of an atom is relatively minor and can, for most practical purposes, be ignored (as in monological inquiry)—mutual understanding is not really possible and thus not really sought (even though, strictly speaking, the atom still contains depth, and thus still enacts *its* world with a type of interpretation, or a bringing forth of the types of distinctions it can respond to: the atom's agency is an opening or clearing which mutually enacts its worldspace with similarly depthed holons). As I said, because this depth is relatively minor, it can be (and is) ignored for most practical purposes of human inquiry.

But as depth increases, less and less can we rely on mere empiricist reflection, on the monological mode; interpretation becomes more and more a significant portion of the knowledge quest, and interpretive depth can then be ignored only by inflicting a great deal of violence on the subject of inquiry.

Just so, intersubjective cultural understanding is not merely monological surface perception (or empirical experience), because *mutual understanding* requires *interpretation* of meaning, intention, and depth, none of which simply sit on the surfaces. The crucial act of knowing in this case is not monological *reflection* on a surface but *interpretation* of a depth; not perceiving patches but interpreting their inner intentions and meanings. No amount of "reflection paradigm" will disclose that which lies beneath the surface being reflected. When you and I talk, interpretation is the paradigm, depth the medium, intention the hidden variable, and mutual understanding the goal. And *none* of that makes it into the reflection paradigm.

But the restriction of all knowledge to monological reflection on preexisting occasions gave the impression that (1) particular cultural truths were likewise fixed and unchanging, (2) intersubjective processes of knowledge creation had no bearing whatsoever on disclosure of sensorimotor truths, (3) monological and representational knowledge was the only knowledge worth pursuing, (4) subjective and intersubjective processes could be reduced to empirical referents, (5) interpretation of depth could be subsumed in empirical reflection on surfaces. In short, all translogical and all dialogical truth could be reduced to the mute monological gaze.

It was especially the ignoring of the role of intersubjective processes in the disclosure of objective knowledge, coupled with the reduction of all knowledge to merely objective "recovery" from an *altogether* pregiven world—this was the reflection paradigm. Subjective-I and intersubjective-we processes were reduced to objective-it forms that were thought to be equally recoverable in simple "representation," and thus all hermeneutic interpretations, all depth disclosures, all cultural meanings, all deeper and higher subjective transformations—all were dispensed with in the simple monological gaze at a world of surfaces altogether "pregiven." (See also note 12 below.)

12. The interesting point is that empiric-analytic science actually requires, in its practitioners, the arduous transformation from sensorimotor to preop to conop to formop. But formop is now, in today's world, the average-expectable level for educated men and women. Whereas, in previous eras, winning a rational perspective in the midst of a mythological worldview was a difficult and dangerous task, now, with modernity, the rational structure is so taken for granted that its worldview is simply *assumed* to be *obvious* and pregiven for any educated person.

Foucault makes this point with regard to Descartes. Whereas most previous philosophers had to describe an inner discipline and transformation required in order to make truth—even *rational* truth—*accessible*, Descartes simply *starts* with the "obvious" rational worldview; he doesn't have to defend it or explain how to get to it.

Thus, when I say the rational Right-Hand path doesn't require any transformation, what I mean is that it doesn't require any further transformation beyond what is now the average-expectable level of development (formop). Past evolution has *already* won this arduous transformation, and so rational science could simply rest on those laurels, so to speak. Nothing higher or deeper—no exceptional inner transformation or *ascesis*—is required in order to have access to "all the truth that is fit to know."

Further, it is this entire series of previous interior transformations—sensorimotor to preop to conop to formop—that "rational common sense" *took entirely for granted*, and thus it *failed* to see that its "common and obvious" world supposedly open to simple "empirical viewing" was in fact a *particular* and generated *worldspace*, many of whose referents were *not* in fact lying around waiting to be perceived by all and sundry, but were *constituted* in an arduous series of Left-Hand transformations, all of which were taken for granted, all of which were *ignored*—an ignorance that absolutely crippled its epistemology and reduced it to "I see the rose," thus overlooking the crucial fact that the secret of the seeing was in how the "I" was generated in the first place, and not in how the simple sensation of the rose jumps to that taken-for-granted I. The intersubjectively created worldspace, which itself *allows* the disclosure of individuated subjects and objects, was ditched in favor a mindless staring at the *end result* (mistaken as a pregiven), a mindless staring at the monological objects thus disclosed.

13. The role of industrialization—the instrumentalization of the economic base—obviously played a central and often dominant role in the instrumentalization of reason: the positive feedback loop that began to spin out of control precisely in its attempt to gain further control.

As I have often mentioned, the role of the techno-economic base in the stages of human evolution (and the construction of various worldviews) is the topic of volume 2. All I can do in this volume is point out that not only am I not ignoring these factors, I believe they have often been the major (but not sole) determining factors in many of these developments. But this definitely requires a book-length treatment, and such is volume 2.

14. Taylor, *Sources of the self*, p. 232.

15. Ibid.

16. Taylor's summary, ibid., p. 243.

17. Ibid., p. 264.

18. Quoted ibid., p. 265.

19. Hedonic pleasure/pain is almost always the most basic (and often the sole) *inner* motivation acknowledged by the Right-Hand path, since that is *as far into the interior dimension* as they are willing to venture, and since it demands *relatively little interpretation*. Since physical sensations are indeed very close to the surface (i.e., very shallow, very low on the holarchy), and since the Right-Hand theorists have to find *something* to *push* their action terms in the first place, physical pleasure/pain works perfectly for them. It gives them much pleasure, I presume.

As I indicated in a previous note (note 23 for chap. 8), the utilitarian ethic was the unpacking of the Basic Moral Intuition (preserve and promote the greatest depth for the greatest span) in terms of *monological* reason, where depth is interpreted as flatland mono-happiness (or pleasure) and span becomes all rational beings (and sometimes all sentient beings, since they share the same lowest common denominator): thus, the greatest mono-happiness for the greater numbers. It is worldcentric, but worldcentric flatland, and recognizes no serious gradations of depth or types of happiness. (This is pursued in volume 3.)

Although in the text I am strongly criticizing the flatland aspects of the utilitarians, the fact that they were one of the first clear worldcentric-rational ethics should not be overlooked; and, from this perspective, they deserve high praise indeed. It is the collapse of all motivations into flatland terms and hedonic happiness that I am, of course, criticizing.

20. Freud, of course, would treat his own "discovery" of the pleasure principle as a monumental breakthrough in psychology. It was actually just a dreary uncovering of the simplest push in the purely Descended world; not exactly something, I would think, to write home about.

21. For a wonderful discussion of these themes, and a lament of their loss, see Alasdair MacIntyre's *After virtue*, 2nd ed.

22. Even though the *affirmation* of this particular ethic itself steps outside the interlocking order: this ethic *devalues* other ethics—that is, it is a hierarchy of value that denies hierarchies of value, as we saw in chapter 1; theories that deny qualitative distinctions themselves embody a hidden qualitative distinction.

Taylor again perfectly spots this contradiction as it appears in a Bentham or a Helvétius: "These thinkers recognized only one good: pleasure. The whole point was to do away with the distinction between moral and non-moral goods and make all human desires equally worthy of consideration. But in the actual content of its tenets, as officially defined, none of this can be said; and most of it makes no sense." That is, they valued this theory with a passionate commitment over rival values, when all values were supposed to be monovalent: they can't even justify their own stance (*Sources of the self*, p. 332).

23. Ibid., pp. 282, 264–65.

24. The rejection of higher/deeper motivations was also intensely driven by "No more myths!" being confused with "No more Ascent!": all higher aspirations (of any sort) were thought to be nothing but the ways in which religions devalued this world and were thus willing to impose domination on their subjects in the name of a glorious afterlife (thus confusing all holoarchic transcendence-and-inclusion with mere mythic dissociation).

It is true that some utilitarians attempted to reintroduce various types or degrees of "happiness" or the "good"; Taylor finds them all unconvincing (and it is certain Kant and Hegel were completely unimpressed).

25. Taylor, *Sources of the self*, pp. 328–29; my italics.

26. Flatland holism is always dualistic to the core, not only because it denies Ascent, but because it splits relatively autonomous agency from the networks of communion on its own level (as we will see in detail).

27. Taylor, *Hegel*, pp. 22–23.

28. Foucault, "The subject and power," in Dreyfus and Rabinow, *Michel Foucault*, pp. 208–209.

29. Taylor spots the same two poles of the Enlightenment paradigm: "The attempts . . . of the radical Enlightenment, of a Helvétius, a Holbach, a Condorcet, a Bentham were founded on this notion of objectivity, and the age of Enlight-

enment was evolving an anthropology which was an amalgam, not entirely consistent, of two things: the notion of self-defining subjectivity correlative to the new objectivity; and the view of man as part of nature, hence fully under the jurisdiction of this objectivity. These two aspects did not always sit well together." To put it mildly. *Hegel*, p. 10.

30. Nor is this simply my reading of this extraordinary story. To give only one example now, Charles Taylor's rather extraordinary *Sources of the self* concludes that the modern (and postmodern) worldview is marked by three main characteristics: (1) a sense of inwardness (i.e., the Ego); (2) the voice of nature (i.e., the Eco), and (3) the affirmation of ordinary life (by which Taylor specifically means a denial of a transcendental Source or Ground or ontic Logos, an ordinary life-world that denies higher qualitative distinctions; in other words, a purely Descended world).

31. Habermas, *Philosophical discourse*, p. 107.

I would say rather, rationality preserves the concrete operational capacities that are the foundation of myth but negates the exclusivity of conop that produces myth. Basic structures are preserved, exclusivity structures (and their violence) are negated.

We saw, in chapter 6, that part of the violence inherent in myth is due to the syncretic fusion of various holons that in fact possess their own dignity and deserve their own individuality. Habermas reaches the same conclusion from a slightly different angle: "Myth owes the totalizing power with which it integrates all superficially perceived phenomena into a [syncretic] network of correspondences, similarities, and contrasts"—such as "Seven orifices mean only seven planets can exist," a stance that "integrates" its world by doing violence to the Kosmos. "For example, language is not yet [differentiated] from reality to such an extent that the conventional sign is completely separate from its semantic content and its referents; the linguistic world view remains interwoven with the order of the world." This is the "holistic-sounding" syncretism so attractive to the New Age "paradigm," which ignores the other (and rather ugly) features that necessarily follow in its wake: "Mythic traditions [therefore] cannot be revised without danger to the order of things and to the identity of the tribe set within it. Only when contexts of meaning and reality, when internal and external relationships have been unmixed [differentiated], only when science, morality, and art are each specialized in one validity claim, when each follows its own respective logic and is cleansed of all cosmological, theological, and cultural [egocentric and ethocentric] dross—only then can the suspicion arise that the autonomy of the validity claimed by a theory (whether empirical or normative) is an illusion because secret interests and power claims have crept into its pores" (*Philosophical discourse*, pp. 115–16.)

32. Ibid., p. 116.

33. Further, the Ego camp would argue, we cannot say that acting for the good of the whole web can even offer a coherent guideline for my moral actions. If the whole is really *more* than the sum of its parts, and if I am basically a *part* of this greater whole, then I can *never* know what the whole actually has in mind, and so I can never genuinely or honestly claim to be acting for the good of the whole (who knows what the whole is actually up to? It might be using me in ways that I can never even begin to imagine, and so I can never use "the good of the whole" to guide my individual actions, since the good of the whole is beyond my knowledge).

On the other hand, if I claim to become "one with the whole," then I am claiming something that no other part has access to; I am claiming, in fact, that I am *not* merely a strand in the web (which is supposed to be my official position, and the position that is supposed to be the foundation of my ethics, a foundation that I have just negated). Thus, either way, the "strand-in-the-web" argument fails as a basis for moral action.

34. Taylor, *Hegel*, p. 32.

35. Kant is, in other words, fighting to introduce a true measure of genuine Ascent into the flatland ethos in general (and the flatland instrumental-utilitarian-holists in particular), and since freedom (and wisdom) is *always* found in the Ascending current of Eros, the moral freedom Kant offered was absolutely exhilarating to the entire era. In terms of the overall spectrum of consciousness, Kant was pointing out that happiness does not reside in merely bodily desire (which is egocentric), but rather resides in a transformation—an Ascent!—from egocentric and sociocentric (heteronomy) to worldcentric care, a worldcentric stance that, since it is generated by my own rational will (by my own rational capacity for universal perspectivism), is thus truly free, truly self-determined, truly autonomous (I would say, relatively more autonomous than its predecessors; even greater freedoms lie beyond the egoic-rational, as Schelling would soon announce).

But it was Kant's noble, and in some important ways quite successful, attempt to introduce a real Ascent into the monochrome and flatland world that would drive Fichte's culminating insight into the pure Freedom of the pure Self or *infinite* supra-individual Subject, and this Fichtean move of pure Ascent (and the inherent problems of a pure Ascent divorced form Descent) would trigger the attempted integration with the Descending current (itself often represented by Spinoza; we will return to this in detail in chapter 13). This problem of integration had already showed up in Kant's rational Ego divorced from the Eco, and this is the problem the text in now introducing (alongside the many positive accomplishments of the Ego in general).

36. Thus also the "political atomism" that coexisted uneasily with the "systems harmony" (see fig. 12-6): the holistic world left no room for the subjective

self, and this atomistic self then found in its own hyper-*agency* none of the *communions* that would automatically bind it, and find it, in community, and so consequently the disengaged self had recourse to a variety of *social contract* theories, as if selves could and would only come together as a business venture. The Idealists would point out (using vision-logic) that agency is *always* agency-in-communion (whether it realizes it or not)—a self is always a self-in-relationship. We will be returning to this topic, and the twin notions of rights (agency) and responsibilities (communion), throughout this volume.

37. Foucault, *Discipline and punish*, p. 200. Incidentally, the Age of *Man* is correctly named in the masculine; it was indeed a hyperemphasis on the agency that tends to be more characteristic of the male (according to radical feminists, and it's not hard to agree with them). Only hyper-agentic males could come up with a theory of community based on a *contract*. This is explored in volume 2.

38. Habermas, *Philosophical discourse*, p. 341.

39. Ibid., p. 246.

40. Ibid., p. 245. Taylor concurs with both Foucault and Habermas, and refers to the "anthropology of the Enlightenment" as "the 'objectification' of human nature" (*Hegel*, p. 13).

41. This doesn't mean that the objective study of human and social behavior—the Right-Hand path—is without importance (on the contrary!). Rather, the point is that when human subjects (and intersubjectivity—"subjects in communication") are reduced merely to their objective components, reduced merely to empirical aspects which then claim to be the "only" knowledge—just then do these endeavors become pseudo-science, because that *reduction itself* cannot be sustained as a validity claim, or as genuine knowledge. It can only be sustained as a power drive (according to both Foucault and Habermas).

I would like to add that that power drive is Thanatos. It is the reduction of the higher to the lower. Although, as I will argue in the text, the central negative motivation of the Ego camps was Phobos, or the fear of the lower, in this particular objectifying stance their own Thanatos was at work. The new "dehumanizing humanism" *was* dehumanizing precisely because the fingerprints of Thanatos were all over the corpse.

When I read Foucault in this light, I am struck by the integrity and the dignity of the man and his absolute outrage at the hand of Thanatos descending on ordinary citizens in the name of a higher "benevolence" and "enlightenment." Foucault, I think , will always be remembered as the great chronicler of the Thanatos underbelly of the Enlightenment paradigm, its monological reduction of "subjects in communication" to "objects of information," claiming to be benevolent, ironically enslaving.

42. Habermas, *Philosophical discourse*, p. 265.

43. "That is to say," as Habermas summarizes it, "the self-relating subjectivity

purchases self-consciousness only at the price of *objectivating* internal and external nature. Because the subject has to relate itself constantly to objects both internally and externally in its knowing and acting [the reflection and production paradigms], it renders itself at once opaque and dependent in the very acts that are supposed to secure self-knowledge and autonomy. This limitation, built into the structure of the relation-to-self [retroflected instrumentality], remains unconscious in the process of becoming conscious. From this springs the tendency toward self-glorification and illusionment, that is, toward absolutizing a given level of reflection and emancipation" (ibid., p. 55).

44. Habermas's words, *Philosophical discourse*, pp. 32–33.

45. Ibid., pp. 55–56.

46. This is Habermas's summary of Horkheimer and Adorno. Habermas agrees only with regard to monological, not dialogical, reason, which is also my view. (Ibid., p. 110.)

Moreover, there is a convincing reading of Foucault that places him as well within this general stance. That is, as Foucault himself made quite clear, it was not reason in toto that he was attacking, but reason in its *objectifying, monological, instrumental*, and *representational* modes (and the retroflection of those modes in subjectifying/subjugating ways).

Dreyfus and Rabinow are certainly of this opinion. On *instrumental-rationality*, Foucault demonstrated that "human needs were no longer conceived of as ends in themselves or as subjects of a philosophic discourse. . . . They were now seen instrumentally and empirically, as means for the increase of . . . power" (p. 141). On the merely *objectifying-rationality*: "Foucault's object of study is the objectifying practices . . . as they are embodied in a specific technology" (p. 144). On *representational-rationality*: "The theory of representation, linked with the social contract view and with the imperative of efficiency and utility, produced [quoting Foucault] 'a sort of general recipe for the exercise of power over men: the mind as a surface of inscription for power, with semiology as its tool; the submission of bodies through the control of ideas' " (p. 149). And we already heard Foucault himself on *monological-rationality*: men and women became "objects of information and never subjects in communication."

Thus, as Dreyfus and Rabinow conclude, it is those generally monological-modes of reason applied to humans that are especially shot through with thinly or thickly disguised *power* (which, I would add, is always the power of Thanatos, the *reduction* of intersubjective communication and mutual understanding to objectifying/subjectifying/subjugating modes of power-over). This is why Dreyfus and Rabinow conclude that in Foucault's project, "trying to show that the relations of truth and power have for good reasons been mistakenly held to be opposed is still a matter of applying a *new and modified form of reason against* a more highly complex version of power." A *new reason* against power. This, they say, "should be seen as an advance, not a refutation of the Weberian project. Foucault is eminently reasonable" (pp. 132–33).

This, too, why Foucault identified himself with the broad lineage of Kant, and why he went out of his way to identify his points of agreement with Habermas. Foucault: "There is the problem raised by Habermas: if one abandons the work of Kant or Weber, for example, one runs the risk of lapsing into irrationality. I am completely in agreement with this." The problem was not solved by the abandonment of reason, but a finer attunement to its dangers and abuses: "How can we exist as rational beings, fortunately committed to practicing a rationality that is unfortunately crisscrossed by intrinsic dangers? *What* is this Reason that we use? What are its limits, and what are its dangers?" ("Space, Power, Knowledge").

Thus, the notion that Foucault saw *all* knowledge and reason equally and thoroughly shot through with power/domination is altogether incorrect. Nonetheless, this is a very common misconception, held mostly by certain American academics (unflatteringly but not altogether incorrectly referred to as "tenured radicals"), who want to deconstruct all forms of accepted knowledge.

Foucault's more balanced view—"What is this Reason that we use? What are its limits, and what are its dangers?"—has been the topic of this "Ego-negative" section in the narrative, where we explored some of the more common charges against the *limits* and the *dangers* of reason, and especially reason captured by its monological/instrumental mode (and we will return to these limitations and dangers in chapter 13). These were especially dangers *inherent* in monological-rationality (we will explore similar dangers and limitations *inherent* in dialogical rationality in chapter 13).

But at this point I would like to mention possibly the worst danger of rationality in general, a danger not inherent or intrinsic in reason, but a horrible misuse to which it can and all too often is (and has been) put. Namely, the structures and products of rationality can be used by a self (or society) that nonetheless is itself at the archaic, the magic, or the mythic mode of motivation.

It has long been known, for example, that cognitive development is *necessary* but *not sufficient* for moral development. Thus, to use Piaget and Kohlberg as examples, a person who has reached formop cognitively, can nonetheless remain even at moral stage 1. Formop rationality *can* support a fully postconventional moral response (worldcentric mutual recognition), but the fact that a person has access to formop does *not* guarantee that she will live up to its standards: moral development often lags behind cognitive development (whereas the reverse is not the case: someone at a postconventional moral response is always using formop in that response: conop and preop *will not support* postconventional responses). Thus, a highly developed moral response always has a highly developed cognitive structure, but not vice versa: highly developed cognitive structures do not in themselves ensure a moral response from that higher level (the reason, put simplistically, is that it is one thing to be able to merely think from a higher level, but quite another to actually *inhabit* that level with one's whole being, and thus *respond* morally from that higher level).

This is the difference between self-stages and basic structures (in this case, moral-stages and basic cognitive capacity, the latter of which is necessary but not sufficient for the former). And that means that a self (or society) with access to formop can use those powerful structures to implement preconventional (magical) or conventional (mythic) agendas—the higher reason is used for quite base purposes, and there is the nightmare of the explosive power of rationality (and all its technical know-how) put in the hands of egocentric magic and ethnocentric myth.

What was the Holocaust but the use of extremely sophisticated technical-rationality in the service of an ethnocentric mythology? Rationality itself, in its forms of intersubjective communication and mutual recognition, does not seek domination but mutual understanding. Rationality seeking *less than that* is rationality shackled and in servitude to egocentric power or ethnocentric domination, forms of massive power which those structures could *not themselves create*, but structures that can all too readily exploit the technical powers of rationality in service of a moral response *less* than worldcentric—less than universal pluralism, less than mutual understanding in a discourse of tolerance and recognition: power to my ego, my tribe, my culture—using the extremely sophisticated tools of a hijacked rationality.

These disasters—the Holocaust will always stand as the unimaginable example—have often been blamed on reason per se, but that is a grave misinterpretation. These disasters are the depth-capacity of reason in servitude to the shallow moral designs of egocentric and ethnocentric magic and myth. It is quite true that those disasters would not and *could not* have happened *without* rationality (which has lead confused theorists to lay the blame *simply* at rationality's doorstep); but those disasters were *not implemented by* rationality, but rather by a moral response that could not live up to rationality's worldcentric perspectivism, and therefore simply hijacked rationality and its know-how down to their own self-promoting, other-destroying level, in service, in servitude, in shackles to a master race, a master creed, a master ethnic that seeks not world understanding but world domination, mythic-imperialism of the fatherland, of the motherland, of the chosen peoples, one way or another, and in all cases: precisely the *failure of reason*, not its true colors, and precisely the triumph of a will shackled to mythic-imperialism.

Just so, the ecological crisis of modernity could not have happened without rationality's technological power, a power that can always overrun the biosphere in dissociative ways (a danger that I have been emphasizing throughout this account). But the major *moral motivations* behind the eco-crisis are not due to rationality, but rationality (and its technical know-how) in service of ethnocentric dominance or tribal power ploys. To pollute a common atmosphere knowing it will kill your own people is not rational in any sense of the word; it is in all ways a *failure of reason* applied to reason's own capacities, and driven precisely by sociocentric and egocentric magic and mythic uncaring—there is the *drive* behind "modernity's" eco-crisis, even if the *means* are provided by a hijacked rationality. Not only is tribalism not the answer to the eco-crisis, it is precisely the primary motivational cause: reason hijacked to my tribe (or my group or my nation), and completely disregarding the global commons.

Where previously and historically tribalism could only deplete small portions of the eco-system before it was forced to move on, extended tribalism can now, with the tools of a higher rationality, despoil the entire biosphere. The *means* are rational; the *motives* are egocentric and sociocentric: the *motives* precisely are *not* up to reason.

47. Quoted in Taylor, *Sources of the self*, p. 354.

48. Ibid., p. 24.

49. Taylor, *Hegel*, p. 22.

50. Ibid., p. 25.

51. Ibid., p. 23.

52. Ibid., pp. 26, 25.

53. That culture (and the rational-ego) can indeed repress and dissociate natural/libidinal impulses is true enough, and those alienated impulses need to be recontacted, freed from the cultural repressions, and reintegrated into the psyche (regression in service of the ego). But when culture is seen as only or primarily a repressive force, then the cure is regression, period, and this is the self-defeating and self-contradictory stance that many Eco-Romantics, then and now, embraced.

Thus, for our salvation, we were to retreat, not prior to the *dissociation* (everybody agreed that was necessary), but prior to the *differentiation* that had allowed the dissociation: not simply prior to the disease, but prior to the depth itself!—which amounted to: cure the disease by becoming more shallow.

These specifically regressive Eco-movements were (and are) extremely vocal and condemnatory in their rhetoric, and they never tire of leveling *ad hominem* attacks at their opponents, who are never portrayed as sincere but mistaken, but rather as monstrous idiots. Since *ad hominem* seems to be the mode of argument here, my own observation is that these particular critics seem to gravitate to the past phylogenetic structure that corresponds with the ontogenetic structure in themselves that is immediately prior to their failed personal integration.

54. Taylor, *Hegel*, p. 26.

692 | Notes for Pages 475–477

55. *ReVision* 15, no. 4, p. 184.

56. Habermas, *Philosophical discourse*, p. 338.

57. In the ecomasculinist camp, theorists of the God-forsaken world write books with titles like *In the absence of the sacred*, where "sacred" is defined as the archaic and magic structure. That book completely ignores the difference between differentiation and dissociation, and all of its arguments rest upon that failure. Its author's Zoroastrian worldview is then used for a critique of modernity, but "modernity" for Mander means anything at or beyond the hoe.

In the ecofeminist camp, the prevailing view is that the Great Goddess was honored the most in horticultural (and a few early maritime) societies (a view spearheaded by Gimbutas, Eisler, Spretnak), but then the Goddess was destroyed, more or less literally, by the coming of the Sun Gods and the warrior patriarchy. In volume 2, I try to demonstrate that this confuses the mythological Great Mother figure with the actual Great Goddess.

The Great Mother figure was indeed largely embedded in the horticultural structure (much as the Great Father figure was embedded in the agrarian structure; see *Up from Eden*). This also shows quite clearly that the Great Mother corpus was *not* rooted in *farming* per se and the seasonal currents of nature, but rather was rooted in the hand-held hoe used by the *female* work force; when the plow was invented—still a purely *farming* culture—the deity figures almost universally switched to Great Father images, because the plow was operated solely by *males*, and the mythic-heavenly figures *in both cases* reflected who was most responsible for subsistence (there is considerable data for this, which I present in volume 2).

And in agrarian societies—still *farming* societies—female subordination and patriarchal rule was at its peak!, and matched only by horse/herding cultures. The agrarian/farming cultures were very much in tune with nature, for sure: very much in tune with nature's disregard for rights based on anything other than power and might.

But the Great Goddess is best conceived, not as any particular structure or epoch or nature, horticultural or otherwise, but as the overall movement of Efflux, of Creative Descent and Superabundance, of Agape and Goodness and Compassion. She is not Gaia, not nature, not planting mythology (although She embraces all of those in Her divinely Creative Matrix). She even embraces the patriarchy and was every bit as much behind that structure as any other. (Likewise, true God is not the Great Father figure, but the overall movement of Reflux and Ascent and Eros; all of which is discussed at great length in volume 2).

By confusing the Great Goddess with the horticultural Great Mother mythic corpus—which was indeed *superseded* with the coming of the patriarchy—it appears that male culture destroyed the Goddess (a handful of males on a small planet destroyed the Creative Efflux of the entire Kosmos?), and that what is required is a recovery and resurrection of the Great Mother mythos.

The same Zoroastrian dualism, the same assumption that the Goddess *could* be banished, when all that was banished was a poorly differentiated mythos that many ecofeminists have reinterpreted to fit their ideology. When the Great Goddess is instead seen to be the entire movement of Creative Descent and Efflux (at each and every epoch), then the cure is no longer regression to horticultural mythology, but progression to Goddess embodiment in the forms of today's integrations (Agape).

The Great Goddess is not a victim, nor could She be banished without destroying the entire manifest universe. But by claiming that She was banished, and by purporting to know how and why it happened, these ecofeminists are then in possession of a certain type of power, the power to tell the world what it must do in order to recapture an ethos of which these theorists are now the primary possessors. The Goddess is shackled to a horticultural planting mythology that ensures that She could never be integrated with modernity, and sees Her only as our Lady of the Eternal Victim. The very framing of the Goddess in those terms denies Her ever-present creative attributes and buries them in rhetoric.

That which one can deviate from is not the true Tao. Not only is modernity not devoid of the Goddess, her Goodness and Agape and Compassion are written all over it, with its radically new and emergent stance of worldcentric pluralism, universal benevolence, and multicultural tolerance, something that no horticultural society could even conceive, let alone implement. Her Grace grows stronger and more obvious with every gain in the liberation movements, and that we ourselves have not always lived up to her new Grace that was modernity's Enlightenment, shows only that we are still surly children not on speaking terms with our own divine parents.

CHAPTER 13. THE DOMINANCE OF THE DESCENDERS

1. Lovejoy, *Great chain*, p. 292.
2. Ibid., p. 313.
3. Ibid., p. 312.
4. Ibid., pp. 293–94.
5. Ibid., p. 313.
6. Taylor, *Sources of the self*, p. 285.
7. Ibid., pp. 297–98.
8. Ibid., p. 299.
9. Ibid., p. 301.
10. Ibid.
11. The following quotes are simply repeats from those given in chapter 8, and the references can be found there.

12. This is not magical indissociation (although it often verged on that), because the mononature that is now spirit is nonetheless a differentiated nature.

13. Foucault, *The history of sexuality* and *Discipline and punish*.

14. Dreyfus and Rabinow, *Foucalt*, p. 140.

15. Lovejoy, *Great chain*, p. 256.

16. Ibid., p. 281.

17. Schelling, *System of transcendental idealism*.

18. Fichte's *Wissenschaftslehre* had three basic postulates, according to Fichte. The first is that "the ego simply posits in an original way its own being." The second is "A non-ego is simply opposed to the ego." The third, "I posit in the ego a divisible non-ego as opposed to a divisible ego."

The first postulate refers to the Pure Ego or supra-individual (transpersonal) pure Self or pure Witness (the I-I). It is that which, according to Fichte, is the pure activity of perceiving but can never itself be perceived as an object (only intuited from within). This pure activity or pure openness is the condition and the ground of any and all objects, and thus "out of" this Pure Self (which Fichte maintains is infinite and supra-individual), comes all finite objects—and this is the second postulate.

Since the finite non-ego (or objectivity in general) now exists, it is perceived by a finite and individual ego—and that is the third postulate.

Thus, the absolute and infinite Self posits within itself the finite ego and finite nature ("divisible"). This is a very clear form of the "Western Vedanta" that I have mentioned on several occasions. And Fichte derives all of this, not so much as a matter of abstract philosophy (although it was certainly that), but as a direct inquiry into consciousness: he bases it on the interior, not the exterior (the latter he calls "dogmatism" leading to "determinism and mechanism"). He used to say to his students: "Be aware of the wall. Now be aware of who or what is aware of the wall. Now be aware of who is aware of that. . . ." In other words, pushing back further and further to that pure Witness which is the ground of all existence but cannot itself be seen as an object. That pure Activity (or pure Witness), according to Fichte, is the ground and necessary condition for the manifestation of any objects at all, and as such, there is precisely nothing individual and finite about it: it is *infinite, absolute, supra-individual*. (Fichte is sometimes accused of meaning the finite ego is the ground, but this is entirely incorrect, as Fichte himself often pointed out.)

The entire world is thus, according to Fichte, the product of the pure Self's "productive imagination" (thus doing away with Kant's thing-in-itself). This is all virtually indistinguishable from Shankara and the notions of pure Atman, its identity with pure Brahman, and the production of the world through Brahman-Atman's creative *maya*.

The main (and crucial) difference between these forms of "Western Vedanta" and the Eastern approaches is that the latter maintain that the intuition of pure Witness has to be sustained through rigorous yogic concentration and intense discriminating awareness (prajna), or it is "snapped up" too soon by the finite self or ego.

In Zen, for example, inquiry into "Who am I?" (e.g., "Who chants the name of

Buddha?" or "What is my Original Face?" or simply "*Mu*") can take six or seven years, on average, before a truly profound breakthrough can occur; Fichte, apparently, intuited only glimmers of this profound openness, which is why I also refer to most Western Vedanta as "half-baked."

And in this half-baked form, it would play into the hands of the finite, hyperagentic, disengaged subject of the Enlightenment paradigm. Nature, or the total non-ego, was not finally integrated in pure Self; rather, nature was simply the *instrumental* background against which the Ego had to struggle *morally* to free itself, and this process of extrication, of disenchantment, was *unending*. As such, when Schelling claimed that Fichte did not truly integrate the Ego and the Eco, Schelling was quite right, I believe. But this shouldn't stop us from appreciating the important and profound steps that Fichte took in the right direction, however much he stopped short of a true breakthrough and true integration. (I would especially mention, for example, that Fichte's presentation of "a pragmatic history of consciousness" not only was the precursor of Hegel's developmental phenomenology of consciousness, but is the grandfather of virtually all schools of developmental psychology in general.)

As for Fichte's final position, it comes to the same thing to say that Fichte ended up with virtually an unending Ascending path that divorced itself from any true Descent (see next note), and that nature remained, as it were, the necessary enemy.

19. It was an attempt at pure Ascent, pure transcendence, pure Reflux, radical detachment. Both Kant, and to a much greater extent Fichte, were trying to reintroduce an Ascending and antileveling current into the flatland ontology, and this is why the impact of their philosophies was always described in terms like "exhilarating"— freedom is *always* in the Ascending current, but the agonizing dilemma remained for them both: how to integrate this with the Descending and manifest world, for without that integration Ascent is always, we have seen, bought at the price of *repression*: Eros degenerates into Phobos.

20. Males, generically having trouble with relationships, are often drawn to the impersonal it-language of Spinoza's infinite Substance: no messy interpersonal dialogical understanding here. Many proponents of deep ecology have explicitly called on Spinoza to support a depersonalized spiritual substance and a denatured nature; it plays into the hyperagency of the masculine mode.

21. It was a pure Descent, a pure immanence, total Efflux, radical immersion. The Eco camps were working *within* the great interlocking order, the manifest system, and attempting to reintroduce Spirit into a world (and nature) virtually abandoned by the pure Ego of Kant and Fichte, attempting to reweave the repressed fragments of a disengaging Subject. And indeed, the compassionate, caring, inclusive Agape is *always* found in the Descending current—but again the dilemma: how to integrate that with some sort of true Ascent, for without Ascent, Agape always degenerates, we have seen, into Thanatos: pure *regression* to predifferentiated states, which is not healing but heaping (i.e., wounding).

22. In the East, the same dilemma showed up in the desire to unite the Pure As-

cended Self or Consciousness (Purusha) with the Pure Descended Substance of the manifest world (Prakriti), or again, pure Emptiness with pure Form; or yet again, nirvana and samsara. The early yogic schools had taken the Gnostic or purely Ascending solution: the dissolution of all Form in pure unmanifest Emptiness or absorption, the extinction of samsara in nirvana.

The first decisive and far-reaching step to pure Nonduality was taken by Nagarjuna, in a move quite similar to Plotinus (transcend and include, nirvana and samara are not-two), a Nondual breakthrough that was subsequently applied to Hinduism by Shankara, to Kashmir Shaivism by Abhinavagupta, to Vajrayana by Padmasambhava—to name a very few. (For an extended discussion of this theme, see note 1 for chap. 14.)

The second step, a thoroughgoing evolutionary Nonduality, was made most explicit in the East by Aurobindo, and in the West by the gentleman the text is about to introduce. This Eastern approach is explored in depth in volume 3.

23. Taylor, *Hegel*, p. 36.

24. Taylor, *Sources*, p. 382.

25. From various letters, in Kierkegaard, *The concept of irony* and *Notes of Schelling's Berlin Lectures*. As Schelling began deviating into mythology, Kierkegaard lost his high opinion of him, as is well known (and it is certainly not hard to share Kierkegaard's opinion in that case).

26. Copleston's wording, vol. 7.1, p. 133. The good books by (or about) Schelling are very hard to come by, which is extremely unfortunate. The one widely available account is Copleston's *History of philosophy*. This contains an excellent short summary of Schelling's work, and since it is also virtually the only one readily available, I have tried to draw most of my quotations from it.

27. Fichte, Schelling, and Hegel each give their own version of the Great Holarchy, which, however we slice it, is in most essentials Plotinian.

28. Copleston, vol. 7.1, p. 141.

29. Schelling's usage; Hegel would suggest a similar scheme but with different terminology and, of course, a bit of an argument as to the precise nature of the unifying Absolute, revolving around cows.

30. Copleston, vol. 7.1, p. 163.

31. Copleston's summary, vol. 7.1, p. 139.

32. Schelling's brilliant synthesis and integration of the Ego and the Eco stand in sharp contrast to most of today's approaches to the environment, which still typically pursue either the Egoic-rational calculative adventure ("reform environmentalism"), or fall into various forms of regressive Eco-Romantic "reenchantment."

Indeed, the Eco-Romantic advocates of the "reenchantment of the world" are still very much with us. Most of the "new paradigm" schools of thought betray a profound allegiance to the flatland interlocking order (the fundamental Enlightenment paradigm), and then within that flatland they wobble between some version of rationalistic systems theory (mostly the ecomasculinists) or nature as disclosed in sentiments (mostly the ecofeminists). To the extent that either attempts to escape the flat-land interlocking order at all, they do so by *regression* to agrarian alchemy, magical animism, astrology, horticultural planting mythology, or foraging human-nature in-dissociation.

In these approaches, there are the standard Eco-Romantic elements: (1) a confusion of differentiation and dissociation leading to pre/trans fallacies; (2) a leveling of those very differentiations that allow their quest in the first place; (3) the view that history and evolution are marked by a heinous Crime that removed us from the Paradise now Lost; (4) a concomitant eulogizing of the archaic indissociation state (both phyloge-netically and ontogenetically) as being "nondual" and "spiritual/mystical" or gener-ally "holistic" (instead of seeing that it is predifferentiated, not transdifferentiated); (5) a "critique" of culture (and modernity) based upon the recapture of the archaic Paradise Lost, now conceived as the Promised Land; (6) a regressive slide into divine egoism, in either the form of monological systems theory preserving the disengaged ego in its empirical flattening (Habermas), or in the form of biocentric feeling and sentimental emoting as the connecting link to the "divine" (Taylor); (7) the need for a "new paradigm" that recaptures the Paradise Lost in a "new" epistemology of syn-cretism and indissociation, based, again, on either systems theory or somatic-vital sentiment. To give some prominent examples:

Morris Berman, The Reenchantment of the World *and* Coming to Our Senses

Morris Berman begins his particular quest for the new paradigm by enthusiastically embracing Gregory Bateson's version of cybernetic reality (the systems theory ap-proach to flatland). In *The reenchantment of the world*, Berman begins with all the standard critiques of the "disenchanted" modern world. Failing to grasp specifically any of the dignities of modernity, he of course finds the "disenchantment" not in the *dissociation* of the Big Three, but in the *differentiation* of the Big Three; and there then follows the standard and glowing (and disingenuous) eulogizing of medieval alchemy-animism, astrology, and magico-mythic anything.

As an Eco-Romantic, and fully caught in that version of the pre/trans fallacy, Ber-man is locked into elevationism. There is "divisive" consciousness (everything mod-ern, scientific, rational) and there is "participatory consciousness" (*everything else*: he mentions alchemy, magic, psychosis, dreams, mysticism, shamanism, mythology, and anything premodern). Since modern rationality/science is pretty bad, everything else must be pretty good (or various degrees of pretty good). And so of course he must eulogize alchemy, astrology, primitive participation, etc.

Thus, as examples of medieval "unified consciousness" (his term), where everything is "interlinked" with everything else, Berman gives the example of sixteenth-century Oswald Croll, who offers the "fact" that "walnuts prevent head ailments because the meat of the nut resembles the brain in appearance." To which we might ask, as a test of truthfulness and sincerity, if Dr. Berman has a headache, does he take two walnuts?

This "walnut/brain" is, as we have seen, the Piagetian confusion of metaphor as literal (preop), where two holons with similar agency are thought to have causal power: syncretism romantically reread as "holism." Berman gives other extensive examples of this "unified consciousness," especially drawn from magic, alchemy, and astrology. Here is the Renaissance "magician" Agrippa von Nettesheim, in his *De Occulta Philosophia* of 1533: "All Stars have their peculiar natures, properties, and conditions, the Seals and Characters whereof they produce, through their rays; whence every natural thing receives from its Star shining upon it, some particular Seal, or character, stamped upon it; which Seal or character contains in it a peculiar Virtue. Every thing, therefore, hath its character pressed upon it by its star for some particular effect, especially by that star which doth principally govern it" (p. 75).

That "astro-logic" is a good example of the Left Hand of the magico-mythic worldview, *syncretically* interwoven with its Right-Hand components (predifferentiated). Berman comments that "given this system of knowledge, modern distinctions between inner and outer, psychic and organic (or physical), do not exist. If you wish to promote love, says Agrippa, eat pigeons; to obtain courage, lions' hearts."

This type of indissociated astro-logic must be eulogized because, compared to modern science, this syncretism appears "holistic" and "unified." But the "crime" of the Enlightenment (the "bad news") was not the loss of that undifferentiated syncretism (as Romantics imagine); the crime was that, with the final differentiation of that syncretism—the differentiation of the Big Three—the differentiation was *not yet* brought into a new and corresponding *integration*: in the newly emerged egoic-rational worldspace, the newly emerged *interior* of the new worldspace was not integrated with the newly disclosed type of exterior world (I and we and it fell into dissociation). Thus, merely comparing the *dissociated* components of modernity with the *integrated* components of medievalism, makes it appear, in this flatland comparison (a comparison only of differentiation/integration and not also of the different degrees of depth)— makes it appear that modernity is *only* a brutal mistake.

But Berman realizes that a simple return to magico-mythic animism is not possible, and maybe not even a good idea (although he's not sure exactly why). So he shifts course a bit and (in another standard elevationist ploy) says that what we actually need, and need desperately, is to *recapture* animism in a "mature form." The alchemical/magical/syncretic worldview he thus starts referring to as "*naive* animism," and our salvation lies rather in a *postmodern* animism (what he also calls—when he tries to distance himself from naive syncretism—a "nonanimistic participation." But then, what happened to animism?). And for this "unity" he finds the best candidate to be Gregory Bateson and a version of . . . no surprise: cybernetic systems theory. Envisioned, of course, as an enterprise of recovery of archaic modes that, well, aren't really archaic anymore.

"Our culture, with its heavy emphasis on the digital, could restore such a complementary relationship only by recovering what it once knew about archaic modes of thought," and this will lead to "the sort of society that might be congruent with the holistic or cybernetic vision," which "is a mature type of alchemical . . . reasoning adapted to the modern age" (pp. 252, 255, 270; Berman uses "dialectical reasoning" to cover dreams, psychoses, holism, alchemy, etc.; this is not vision-logic but paleo-logic). This Batesonian cybernetic holism, says Berman, is by far the best candidate for *world salvation* in the postmodern era.

But then, finally, Berman thinks about all of that and, in the closing pages of his book, effectively nullifies virtually everything he has just labored to endorse (strangely, he does the same thing in his next book, as we will see). In the course of "just mentioning" some of the problems of Batesonian cybernetics, Berman gives (intentionally or not) a devastating critique of flatland holism in general, pointing out (quite rightly, and for the all reasons we have seen, especially concerning the extraordinary power drives of monological systems theory) that "holism, in short, could become the *agent of tyranny.*" And the holistic worship of nature goes hand in hand with this tyranny, since "the *celebration of nature* versus artifice is a central tenet of *fascist ideology*" (pp. 291, 292; my italics): what others have frankly called eco-fascism.

He thus offers, for salvation, basically nothing. His pre/trans fallacy has left him holding the ontological bag: "We thus confront a choice that must be made and yet cannot be made: the awakening of an entire civilization to its repressed archaic knowledge." What we adults really need, in other words, is to embrace a mature form of kindergarten. The "choice that *must* but *can't* be made" is simply the choice based on his pre/trans confusions (*can't* is to *must* as *pre* is to *trans*, and "fusing" the latter pair means a fusion of the former, which leaves him paralyzed).

Still under sway of this pre/trans fallacy, and still thoroughly confusing differentiation and dissociation, Berman is sure the answer must lie *prior* to the differentiation of mind and body: that is to say, our salvation must be primarily bodily, sensory, and *somatic* in nature, and thus what we really need, in this sensory-bound, sex-drenched, hedonistic, pleasure-seeking, body-gratifying culture of ours is . . . just a little more body.

And so, in his next book, *Coming to our senses*, Berman turns to the somatic body in earnest and thoroughly abandons any pretense to cybernetic systems theory (indeed, the Batesonian cybernetic flatland that was postmodernity's best chance of salvation now gets nary a mention; in between these two books, Berman wrote several very severe criticisms of systems theory, often arriving, I believe, at several right conclusions for several wrong reasons). He switches from the systems theory of "holism" to the "nature as it makes me feel" version, and throws his allegiance to the body, to its feelings, to its vital-impulsive force and pure emoting, to sensory delight—and thus adopts, in the process, a purely and classically Descended worldview, a worldview that sees all Ascent, of any variety, as primarily an attempt (often evil) to overcome a state of affairs (the subject/object "split") that simply never should have happened in the first place.

Thus, echoing the two-millennia-old mantra of Descenders everywhere, Berman's conclusion is that it is time to do away with all Ascent entirely, and this strikes him as a wildly new and original idea: "There is another alternative to recycling the ascent structure, and that is to finally abandon it once and for all. There can be no healing of our culture and ourselves without taking this option. Nothing less is at stake than the chance to be finally, fully human" (p. 307).

And thus, in line with every Descended hedonist who ever found ascent too demanding, Berman believes that "this is the crucial point—that true enlightenment is to really know, really feel . . . your somatic nature" (and never mind that Zen defines Enlightenment as "body and mind dropped"!). "The shift away from ascent, and toward bodily presence in the world," he says, *there* is our salvation (pp. 310–11).

Thus, if we can really emote in a genuine way, all of our existential problems will melt away. That my feelings cannot even begin to take the role of other, that my feelings, as feelings, are self-reverberating, that they circle egocentrically in their own orbit, that in themselves they never ascend into the intersubjective circle where love alone truly shines and compassion alone can flourish—that promoting a "feel good" ethic as the primary moral impulse lands me squarely in divine egoism and existential hell: of all of that, nary a note from Berman. He wants merely "direct engagement with nature and bodily functions." So self-absorbed is this stance that the selfcentric nature of the stance doesn't even dawn on him.

THE SINGLE BOUNDARY FALLACY

In the modern and postmodern era, the retro-Romantic *psychological* project has always centered on the first fulcrums of present-day human development, during the first three years of life, and the horrendous "split" between self and other that occurs at that time (these theorists don't usually distinguish fulcrum-1 and fulcrum-2, and so they mean the "first" differentiation of the self and other, the subject and object, from the primal/archaic state of undifferentiated and indissociated participation, which primarily occurs, we have seen, in the first two-to-three years of life). This *necessary differentiation* (F-1 and F-2) is completely confused with dissociation and is therefore interpreted as a primal loss, a *primal alienation*, that forever divides the self from others, from itself, and from nature. And most subsequent human desire, drive, motivation, and cultural endeavors are then seen as a doomed series of twisted attempts to regain this Paradise Lost.

And indeed, Berman takes, as his point of departure, just these theorists. "For the 'French school' of philosophy and child psychology—Henri Wallon, Maurice Merleau-Ponty, Jacques Lacan—this moment [of self-other differentiation], which marks the birth of your identity as a being in the world, also marks the birth of your alienation from the world" (p. 36). Not that this differentiation *can* go too far, or develop poorly, into pathological dissociation and alienation, but that it *simply is itself* the Great Crime. (This position was taken most forcefully in America by Norman O. Brown, whom Berman frequently quotes with much approval.)

The great problem with all of these theories is that, working within the modern flatland paradigm, they only recognize *one basic boundary*: the horizontal boundary between self and other (or subject and object), and since all alienation does indeed require some sort of separation of self and other, then this primary differentiation (F-1 and F-2) must simply be equated, not with the *possibility* of alienation, but with *alienation itself*.

(For further discussion of the Single-Boundary Fallacy, see note 3 for chap. 6 and note 17 for chap. 14.)

What all of this overlooks is the vertical dimension of *different types* of boundaries altogether: there is the differentiation of *physical* self and physical other (F-1), of *emotional* self and emotional other (F-2), of *conceptual* self and conceptual other (F-3), of *cultural* role-self and role-other (F-4), and so on throughout the vertical spectrum, with each self marked by an important functional boundary.

And only *one* of those boundaries is actually physical, and actually follows the skin-boundary of the physical body (F-1). *All* of the other boundaries (emotional to mental to spiritual) do not follow the skin-boundary at all (they merely refer to it and are grounded in it). My mental self, for example, includes all sorts of identifications with family, values, causes, groups, nations, etc., none of which exist inside my skin-boundary.

Rather, an entity is *inside* these various boundaries (emotional, mental, etc.), not when it is inside the physical body, but when *it follows the rules* (the code, the regime, the agency) of that particular self-sense or holon (much as you are "in" a chess game if you are following the rules of the game, no matter where you are physically sitting: you can even phone in your moves; and you are "out" of the game, not if you move in physical space, but if you break the rules).

That is, an entity becomes "inside" the structure of *any* holon when it is following the rules of the deep structure of that holon (when it is assimilated into the code, pattern, or regime of the holon), and this is true for physical boundaries, emotional boundaries, mental boundaries, etc. When physical food is first swallowed and enters the stomach, it is dissolved and digested, and the needed nutrients enter the bloodstream, from there to be *assimilated* into the structures of the physical body (they become truly part of the physical body, truly "in" the body), and unneeded food continues to pass through the digestive tube and out of the body: it never becomes part of the body, it is never assimilated into the patterns of the body itself: it remains outside the body's game, the body's rules and patterns, regimes and codes.

Likewise, another person becomes part of your *emotional* makeup when that person is in your emotional space, not in your stomach, and that emotional space floats beyond the skin, only occasionally checking in with the physical boundaries. The person (as image and emotional representation) becomes a *part* of your emotional patterns (your overall affective mood states and structures), and thus is *in* that emotional space as surely as food is in your stomach, but these two boundaries are not at all the same simple physical boundary (and that is what I mean when I say that an entity is *in* a structure or a holon when it is following the patterns or rules of the deep structure of that holon: I can emotionally identify with people, causes, groups, nations, no matter where they are actually located: they become part of me when I take them into my emotional identity, my emotional space, and they then operate within that space, following its particular rules, or the patterns of my affective mood states, which themselves might accommodate, or shift deep structure, in order to assimilate the new items).

The mental space likewise floats outside the emotional (the cognitive-mental space can actually take the role of other, "see through their eyes," assimilate/accommodate new ideas, etc.); and the spiritual space floats outside the mental and into the Kosmos at large.

Each of those boundaries is crucial. Indeed, the *differentiation* of an earlier boundary is the *prerequisite* for the *integration* of the self at the next-higher boundary. Thus, the stable emotional-self rests on the stable physical-self boundary; the stable mental-self rests on a stable emotional-object constancy (stable emotional boundary), and so on. The previous differentiated self is integrated into the newly emerging self, which must likewise stably differentiate itself from the others in its newly emerging environment. The boundaries of the previous self-structure are preserved in their function and their capacity, but negated in their exclusivity or partialness. Boundaries are not simply lost or evaporated or erased; they are transcended and included.

And this continues until all the self/other boundaries have been thoroughly *preserved* (included) and utterly *negated* (transcended) in Spirit. Zen masters, for example, can perfectly well tell the difference between their physical body and a physical chair (or physical other): the Zen adept is *not* healing the split between chair and body; he or she is transcending and *including* all self/other boundaries in the Empty clearing that allows all boundaries to arise, and cherishes all boundaries precisely as they are. The Zen adept is perfectly aware of the difference and the boundary between chair and body, between ego and other, between up and down, between meow and bark. Satori is not a "meltdown"; it is a negating and preserving of the entire manifest domain as a transparent shimmering of the empty Divine.

But if all of those important (and necessary) boundaries are simply collapses into one, single, arche-boundary, which is merely called THE "self/other" boundary, then massive theoretical confusion results. Even in psychoanalysis, for example, there is still the lingering notion that all mental identification involves temporary regression to the oral stage, because there is only *one* boundary recognized, and thus taking something across the mental boundary must be *actually* regressing to the physical oral boundary (any introjection must follow the same pathway as food, for there is only "the" boundary).

Likewise, with the more avant-garde theorists (who dabble in Zen and mysticism): since higher states involve somehow a "nonduality," then it appears that all the crucial and developmental action centers around this "*single*" self/other boundary, and since the first important boundaries (physical and emotional) are indeed differentiated during the first three years of life, this seems to be THE fulcrum driving all subsequent human development: it seems that the loss of the archaic predifferentiated state is irrevocable alienation, whereas it is simply the first stage of waking up.

The existential terrors that *come with* the first boundaries are not created by those boundaries, but simply disclosed by them: it's called recognizing samsara. The consciousness of those terrors can indeed be blotted out by regressing prior to the differentiation; but the terrors are actually overcome, not by going prior to the boundaries, but by transcending them.

But Berman follows the "single boundary" flatland paradigm, and thus he refers to this "single" differentiation as "the basic fault" (following Balint), in the sense of "basic default" and "basic mistake," which, like a geological fault line, he says, lies ready to toss up earthquakes under stress. But Berman does not believe that this basic differentiation/default is universal, because, he says, premodern foraging tribes either did not possess it at all or possessed it only a little (no argument from me there). But beginning with the Great Crime of the Neolithic farming cultures (and here we go), which differentiated the tame and the wild, subsequent history has been one long slide into more and more misery and mistake, all driven by humanity's attempts to overcome THE basic default that separates self and other. And so, he blithely assures us, "Most of our history has been a kind of unnecessary artifact" (p. 311).

As an example of what he considers a "mistake," he gives especially the Neolithic distinction between "tame" and "wild": "Once Wild and Tame were established as categories of life, so that Self and Other necessarily became antagonistically related, the 'problem' of what to do with immediate forms surfaced as a problem" (p. 76). Note that he maintains that distinguishing tame and wild *necessarily* alienates subject and object (these types of "necessities" are very common when every differentiation is read as a dissociation, because every branch of the oak must then be seen as a different type of violation of the acorn).

Likewise, the Great Crime according to the (ecomasculinist) Romantics (who embrace foraging) is *always* horticulture (tame vs. wild), all of which overlooks the foraging-culture distinction of the Raw and the Cooked—its own attempt to alter nature to suit its own ends. But the retro-Romantic position *demands* an earlier state of human existence that did *not* attempt to "alter nature," because the existence of this state is demanded by their prejudged critique of modernity (i.e., a "paradise" *must have* existed in the past because differentiation itself is not embraced by Romantics, so the predifferentiated past *must* be the Lost Paradise—an ideology that then rides roughshod over massive amounts of unpleasant evidence to arrive at glowing interpretations of the noble savage, Philip Slater and Gilbert Brim being merely two of the most recent reenchanters).

And thus, true to form, Berman maintains that "Neolithic civilization opened up an opposition between the Self and the World. Within the matrix of Neolithic society, and the categories of Self/Other, or Tame/Wild, the schismogenic cycle is quite inevitable. Western consciousness is deeply influenced by this dualism"—even though Eastern consciousness, which equally developed the Wild/Tame distinction, is somehow "nondualistic" (which undercuts his entire argument at the crucial fulcrum, because this is not actually a nasty dualism but a growth differentiation).

In all of this, there is no serious and consistent distinguishing between differentiation and dissociation, and there is no distinguishing whatsoever between all the vastly different types of necessary and functional boundaries: there is simply the basic gap, the basic fault, the basic split. There is no predifferentiated to differentiated to postdifferentiated integration and union (like every other process in nature); rather, there is simply subject/object fusion versus subject/object alienation: there is simply the nice acorn versus the vicious violence done to it by the oak—the oak being the alienated, fractured, distorted acorn.

Thus, for Berman, everything *past* fulcrum-2 (that is, everything past the biocentric bodyego, the somatic-vital body) is *fundamentally* nothing but various ways to tragically and pathetically attempt to bridge "the gap"—bridge the "split" that foragers did not have and that all of subsequent history has pathetically attempted to overcome. Everything past berry picking is past or beyond the immediate, sensory-driven body, and thus fundamentally should simply not have happened ("a kind of unnecessary artifact"), because everything beyond the immediate sensory body is . . . bad.

This means, as well, that for Berman the only *ontological* reality is that of the felt-body, the vital-emotional body, prior to its "split" into mind and body (prior, that is, to fulcrum-3). Since the vital body (the "somatic reality") is the only fundamental reality, then all of history has to be coded in this vital body, and has to be seen as primarily the vicissitudes of the body and its primary feelings. Everything else is . . . the pathetic attempt to fill the gap, the basic default. And thus a real history would have to be the history of the body, period. "History is finally an activity of the flesh" and can only be disclosed by "techniques of somatic analysis. The truth can *only* be seen from a somatic point of view" (his italics; pp. 136–37).

This is a tall order, even for a dedicated reductionistic (even the Marxists abandoned this one), and so in order to make "the body" the *primary* mover of all history, Berman has to come up with some very clever definitions of "body" in order to cover all the bases. And this he does, implicitly giving "the body" five entirely different definitions, which gives him more than enough rope to hang history.

The five different uses: body as somatically felt, body as experience in general, body as interiorness, body as predifferentiated consciousness, and body as physiological (including the brain). But the first two are the most important, and I will focus on those.

1. By "body" Berman primarily means the somatic, physical, sensory, felt-body. Hence the title: *Coming to our senses.* Generally, what we mean by "the flesh" or "the senses" or "vital feelings."

2. But he also uses "body" to mean *any* "intense experience," and anything that is *intensely experienced* is therefore *somatic* in Berman's universe (in this he even includes intense mystical states where the felt body *disappears altogether*, as he readily concedes; but since this experience is intense, it *must* be body, according to this definition, whether or not that is phenomenologically completely wrong).

The real problem with "body = experience" is the fundamental problem with empiricism in general.

THE MEANING OF EMPIRICISM

The strength of empiricism has always been its insistence on grounding knowledge in experience; the disaster of empiricism is that it has never been able to define experience adequately—in fact, it *excludes* much experiential evidence and settles, by default, on "sensory experience," presumably (and wrongly) because it (sensory experience) alone is supposed to be public. Thus empiricism tends to switch from experience as general prehension (well and good) to experience as sensory prehension (reductionistic).

There is sensory experience, there is mental experience, and there is spiritual experience (to stick with the simplified holarchy of body, mind, spirit; the same analysis applies to all of the levels of the spectrum of consciousness, which is a spectrum of experience).

But empiricists usually take *mental experience* to be simply an *abstraction* from *sensory experience* (the reflection paradigm), with all the self-contradictions and inadequacies involved in that fundamental Enlightenment paradigm (and which we have discussed at length). Mathematics, for example, is not simply an abstraction from sensory experience. Mathematics, as it is performed, is an incredibly rich, thick, deliciously dense *mental experience*: I *directly perceive*, with the mind's eye, the display of symbols, ideas, and concepts that parade by in my awareness, and that are every bit as real, *in themselves*, as the *experience* of clouds or rain or rocks. After all, both symbols and rocks are, at the actual moment of their experience, given to me in immediate awareness. (To say that rocks are "real" and symbols "aren't" is a *mental deduction* away from the direct experience in immediate awareness.)

Mental experiences (such as images, symbols, and concepts) often get a "bad reputation" as being "dry and abstract," but only because, unlike sensory experiences, mental experiences, besides being direct and immediate in themselves, *can also* go one step further and *represent* sensory experiences. The word *rock* (the *mental experience* "r-o-c-k") can stand for the *sensory experience* of rock. And if I get caught up in having mental experiences substitute for sensory experience, then yes, those mental experiences become "mere abstractions" in the bad sense: I have a sign but I am avoiding the referent. But that is an *abuse* of mental experiences, not a *definition* of mental experiences, nor is it anything even remotely close to an *explanation* of mental experiences, most of which don't even have purely empirical/sensory referents in the first place: I can simply and empirically point to a rock, but I can't simply and empirically point to envy, pride, honor, or jealousy, which is why empiricist theories—sensory/reflection theories—of language fall apart when it comes to the dialogical phases of intersubjective linguistics, as we have repeatedly seen. Most mental experiences and signs point, *not* to sensory referents, but to *other* mental experiences or referents; they are not abstractions from sense data, but condensations of rich mental experiences that circulate in the hermeneutic circle. They possess sensory components, but cannot be *reduced* to sensory components precisely because they do not primarily *derive* from sensory components: they *transcend* and *include* sense data—and therefore do not merely *point to*, or *abstract from*, sensory impressions.

Thus, as hermeneuticists have pointed out, language does not just *represent* the sensory world, it *creates* worlds not found simply in the senses, it *discloses* worlds, *presents* worlds, and doesn't just represent them. Language is the stuff of the noosphere, which doesn't primarily point to the biosphere or physiosphere, but primarily transcends the naturic sphere and points to other mental and linguistic experiences—whether magical, mythical, rational, or existential. Some of these mental/linguistic experiences point to premental holons, some point to transmental holons, and many simply point to other mental holons—a point we will return to in a moment—but in no way are they merely abstractions from sensory experiences.

(That the noosphere is thus "constructed" and not merely "reflecting" or "representative" does not mean, as we have seen, that it is therefore merely arbitrary, as radical relativists maintain; it is still grounded in validity claims that allow cultural *learning* to occur, because mental holons follow their own law-governed patterns and do not simply float happily and freely in random space. But the true moment of relativism, and a truth I endorse, is that, precisely because the noosphere does *not* merely reflect the biosphere, cultures cannot be judged *merely* on how well they *mirror* the physical world according to empirical science: if knowledge and culture are mistaken as merely the "mirror of nature," and if true knowledge of nature is given only by modern empirical science, then all cultures have to be judged merely against modern science, and other cultures are then reduced to *only* being "near misses" or "prototypes" of Newton. Judgments *can* be made across cultures, but the *merely* empiricist judgment is as inadequate there as elsewhere.)

It is in dealing with the world-creating, world-disclosing, world-presenting capacity of language that the merely "reflective" empirical/behaviorist theories of language come up short. And, whatever one might otherwise think of Chomsky's theories of language, it is almost universally agreed that he dismembered these theories in his widely circulated critique of Skinner.

Now, in addition to direct *sensory experiences* and direct *mental experiences* (both of which, even if they are partially mediated, finally and always present themselves immediately), there are also direct *spiritual experiences*. A subtle-level illumination, for example, is presented to my awareness in the same direct, immediate, given fashion as the experience of a rock or the experience of a mental image: they simply show up and I prehend them (quite apart from whatever mediating chains deliver them to the display; when we discover these mediating chains, they, too, enter awareness, when they do enter, immediately).

This is why, even if empiricism is always and lamentably tending toward "sensory empiricism," many mystics speak of "mystical empiricism," meaning *direct mystical experience*, using "experience" in the wider and truer sense of "immediate awareness" and not just "immediate sensory awareness" (which is why so many mystics insist on calling their endeavors experiential, experimental, and scientific in that sense).

And here, too, mental experience can get into trouble, because it can use a mental symbol, such as the *mental experience* of the word "G-o-d," to stand for the *spiritual experience* of direct illumination (for example), and so here again it is caught in "mere abstractions": it is using mental experiences to try to cover experiences that *aren't* in themselves mental. These "representations" then become "mere metaphysics," and since the time of Kant, we all know that is a very bad idea: it won't hold water, which is to say, it hasn't any experiential grounding: this type of "mere metaphysics" is simply *empty categories* devoid of true knowledge, which is to say, devoid of true experience.

However, since Kant doesn't acknowledge *spiritual experience*, he *therefore* thinks metaphysics per se is dead, which is the point at which Schopenhauer, among others, leveled a devastating criticism of Kant (and the point where Katz's neo-Kantian argument also collapses). Kant demonstrated that mental symbols without experiential grounding are *empty*: but the real conclusion of his argument is that *all future metaphysics must be experiential*—that is to say: experimental, grounded in direct awareness and experience, coupled with validity claims that can be *redeemed* in the experiment of contemplation, and grounded in the three strands of all true knowledge accumulation: injunction/paradigm, apprehension, and confirmation/rejection.

Virtually every thinker from Kant onward (and following his pioneering lead) has announced "the death of metaphysics" and the "death of philosophy"—from Nietzsche to Heidegger, from Ayer to Wittgenstein, from Derrida to Foucault, from Adorno to Lyotard. And in the sense of the "death of empty categories," I agree entirely. But the real prolegomenon to any future metaphysics is, not that the endeavor is altogether dead, but that the real metaphysics can now, finally, get under way: actual contemplative development (grounded in genuine spiritual experience) is the future of metaphysics.

Thus, Kant's attempt to "abolish knowledge [representational metaphysics] in order to make room for faith" (in God), should be completed à la Nagarjuna: abolishing mere symbols and concepts (abstract or representational metaphysics) in order to make room, not for faith in God, but for direct experience of God.

The point, then, is that we want to take the best of empiricism in general: genuine knowledge must be anchored in validity claims of evidence and experiential grounding; and we then add what should have been obvious all along: there is sensory experience, mental experience, and spiritual experience (holarchically interwoven, so that, for example, mental experience provided by culture mediates and colors, but does not create in toto, sensory experience and, should it occur, spiritual experience). We then add one final point: don't confuse these *types* of experience, and don't use the categories of one to cover the others (category error).

Thus, the "death of metaphysics" correctly means the death of using mental experiences (symbols) to stand for spiritual experiences, and the real birth of genuine metaphysics means: discover those spiritual experiences directly (and communally shared in a sangha of intersubjective discourse of checks and balances, and thus thoroughly grounded in validity claims).

WHICH BODY?

To return to the main topic: I mention all of this because Berman, by making the *somatic-sensory body* mean *experience in general,* can sneakily cover the entire spectrum of consciousness and claim that *all of it is really* the somatic-body. He simply switches definitions whenever necessary: from somatic experience to all experience and back again.

And he actually does this, by coming up with what he calls "the five bodies." Based on a (rather confused) interpretation of the Great Chain given by Robert Masters, Berman maintains that there are actually *five* bodies available to humans: (1) the physical body; (2) the vital-biotic body, which he also calls the lived-body and, loosely, he says this is also "the mind"; (3) the deeper mind, astral body, or true Self, which he also calls "soul"; (4) the psychic or paranormal body; and (5) ultimate spirit or the World Soul body.

This, as I said, is a slightly confused reading of what we have been summarizing as matter, body, mind, soul, and spirit; but either of these versions of the Great Chain will suffice at this point. Berman's main contention is that all of these levels of consciousness are really *bodily* events; he even claims that they are (potentially) *universal* because they are rooted in deep physiological processes of the somatic reality (the five bodies possess what I have called universal deep structures; in his own way, he is saying that the Great Chain is not a cultural artifact but is rooted in universal deep structures).

If Berman stopped there, that would be fine, more or less. The traditions (such as Vedanta) themselves use "body" in the same extended sense of "experience," and since all of the levels of the spectrum of consciousness are *experiential realities,* they are often referred to as being "bodies" or "sheaths" (the gross body, the subtle body, the causal body; or the Three Bodies of Buddha: gross, subtle, and very subtle—same general notion).

But notice, for the traditions, that these "bodies" are not variations on the gross *somatic* body: the somatic body itself is simply one of numerous bodies, most of which are completely nonempirical, nonsensory, nonvital, nonbiocentric (I would say, vis-à-vis the four quadrants, that *all* of these other bodies have Upper-Right-quadrant *correlates,* but *none* of them can be reduced to those correlates, and *none* of them are merely Left-Hand gross-vital entities).

But Berman wants it both ways: since the *only* body that is really real is the gross somatic body, all of these other bodies must really be . . . distorted ways to fill the primary gap, the basic fault. Once the somatic body differentiates from the world (F-2)—which Berman does not really see as a necessary differentiation on the way to higher integration, but primarily as the "basic split, the basic alienation, the basic default"—once that "basic split" occurs, then all of these other bodies are really twisted and distorted ways to bridge the gap, to come to terms with the Original Sin (the "single" boundary), and thus all of these other bodies—that is, the entire spectrum of Ascending developments!—are simply and finally "unnecessary artifacts."

The entire dynamic of ascent and development is a contorted attempt to fill in the original gap, the gap created because we weren't "mirrored enough" by Mommy at age two.

Everything past fulcrum-2 is thus fundamentally unnecessary; and with this absurdity, Berman tosses the last ten thousand years of human development into the garbage can. Both Hitler and Jesus, he tells us, are driven by the same basic energy. Page after page he rattles on royally about the colossal stupidity of the entire human race, a stupidity that touched every single cultural artifact past the picking of berries.

Occasionally Berman pauses to grudgingly admire some cultural production (certain modern artists, alchemists, the mystical Cathars of the Albigensian heresy), and gives a certain backhanded compliment to their accomplishments, along the lines of: (1) the basic default (F-2) isn't universal, because foragers don't have it (or much of it); (2) but the post-berry-picking world is universally stuck with it; (3) but it can be used *creatively* or *destructively* (and here he gives the lip-service to the creative use of something that, after all, never should have happened in the first place); (4) and then he ends with a call, nevertheless, to *abandon ascent altogether* and sink back into the senses, emoting in genuine fashion, and stopping, not to smell the roses, but only to smell the roses.

In the course of his historical analysis, he inadvertently points out the *true* drive of much of history, and the true drive that has always been resisted and condemned as *heresy*, and this heresy is *not* the experience of the simple somatic body (as Berman painfully attempts to maintain), but rather, *actual experiential Ascent*.

Berman outlines "four heresies" that have marked the West: the Greek (or Gnostic, by which he usually means Plato/Plotinus, but which also includes the actual *mystical* foundations of Jewish and Christian spirituality); the Albigensian heresy (the Cathars of southern France, who began actual mystical practices and experiences, against which the Inquisition was originally instituted); the occult sciences (of the later Middle Ages, the early scientists who were also mystics, magicians, and/or occultists of various flavors); and Nazism/fascism (which is, sort of, twisted ascent).

Interestingly, almost all of the movements that he likes (from certain early mystical scientists to the Cathars) were involved in *actual Ascent* (which he calls "five body practice," which means roughly, spanning the spectrum), and this was viewed as *heresy* by the Church (for all the reasons we investigated). And all the movements he especially condemns (Nazism) were involved in frustrated, twisted, distorted, or falsely promised ascent.

But never mind: it's time to do away with Ascent altogether. Not time to differentiate between true Ascent (as in the Cathar movement, which Berman grudgingly acknowledges was one of the finest cultures ever devised in the West, post-foraging-wise) versus frustrated or twisted Ascent (the Nazis, even though most of the Nazi ideology, as Berman himself inadvertently makes clear, was of the purely Descended, biocentric, emotion-drenched, nature-appealing, eco-fascist variety).

No, it is time to simply do away with Ascent altogether. What we need is a "restoration of the body," a "coming to our senses." As I earlier pointed out, according to Berman our salvation must be primarily bodily, sensory, and somatic in nature, and thus what we really need, in this sensory-bound, pleasure-seeking, body-gratifying culture of ours is . . . just a little more body.

Back to the body, back to the senses, back to the purely Descended world. And if it seems to a critic such as myself that our culture is already choking on sensory gratification, a Berman will always respond that we are not *really* feeling our body, but providing substitutes for it. The standard Descended solution: our problem is that we just aren't *close enough* to the sensory Shadows.

Back to the body, back to the senses, back to preconventional, egocentric, nature as it makes me feel. A restoration of the body: a centering in divine egoism, as long as it feels really good. Berman uses as an example of the "pure love" that he advocates as our salvation: a type of emotion displayed by a two-year-old child. "I believe it is this sort of consciousness that can properly be called Love of Life." Perhaps; but not Love of Other. The two-year-old, however "spontaneous" and "open," does not take the role of the other, enter into genuine intersubjective care where love is first seeing the other, putting oneself into the shoes of the other, and then honoring the other, often at one's own expense, precisely because it promotes the greatest depth for the greatest span. The two-year-old, locked into its own feelings, and not troubled by that nasty Basic Fault that lies right around the corner: there is the retro-Romantic Eden, this egocentric wonderfulness standing down in the face of Kosmic compassion.

Berman thinks about all this, and, true to form, in the last two pages of his book he basically erases his entire argument. "I am left with one lingering doubt, namely, that I have perhaps overvalued the body as a vehicle for cultural integrity. For returning to the body, in and of itself, as Paul Ryan once pointed out to me, is a monistic solution—it can only give you monads."

Preconventional, egocentric, body-bound monads: coming to our senses.

Theodore Roszak, The Voice of the Earth

By far the most sophisticated reenchanter is Theodore Roszak, and in his highly readable *The voice of the earth*, Roszak sets out to create an "ecopsychology" which would heal the "split" between human and nature. This is a useful and interesting book, precisely because it shows what happens when the transpersonal dimensions are left out, and the prepersonal and personal are forced to do double duty: a strain that finally breaks them both.

Like all reenchanters, Roszak traces the rise of the personal out of the prepersonal (what he also calls the protopersonal), and then, instead of *further* development into the *transpersonal*, we are given merely the task of *uniting* the prepersonal and the personal. Instead of integrating Earth, Human, and Heaven (body, mind, and spirit, or prepersonal, personal, and transpersonal, with the understanding that "Heaven" is right here, right now, on *this* Earth, when we live in *transpersonal* awareness)— instead of recommending that overall integration, a balanced Earth/body and Human/

mind is simply confused with Heaven/spirit itself. And since children and primal people did not differentiate clearly between Earth and Human, they must have been living in Heaven, and evolution beyond that Eden is a Tragic Mistake. Ecopsychology will reverse this Horrific Crime, and we will all once again be ushered into Heaven.

The specific difficulty with these types of approaches is that, as we have seen, the real problem (and the real pathology) is not the differentiation of Earth and Human, but the dissociation of Earth and Human. Roszak rather thoroughly confuses the two, which throws him into profound pre/trans fallacies, and this shows up in his presentation as a considerable wobbling and ambivalence (and often blatant self-contradiction) about the "Edenic" state itself. On the one hand he senses (and states) that the primal state was thoroughly parochial (egocentric/ethnocentric to the core), yet on the other hand it *must* be the repository of universal ecological wisdom, so he must figure out a way to eulogize the primal undifferentiated state and distance himself from it at the same time.

Unlike many reenchanters, Roszak is simply too fine a thinker (not to mention talented writer) to try to gloss over or oppress the obvious facts of evolution itself, starting with its thoroughly *hierarchical* nature. The universe "reveals itself as an emerging hierarchy of interlocking systems (a scale of forms)" (p. 133). "All theories of deep systems are *hierarchical*. They map the universe as a *pyramid* of systems ranked from lesser to greater, lower to higher, simpler to more complex" (p. 173).

Likewise, he clearly grasps *emergence, novelty* or creativity, and the fact that each level in the holarchy is an actual *transcendence* or *liberation* involving increased relative *autonomy*. "Emergence is what happens when we move across the boundary that defines hierarchical levels; we encounter genuine novelty" (p. 172). "Where we cross [a boundary] between higher and lower system . . . this astonishing process by which each structural level is not only elevated above, but in a significant sense is *liberated* from the governance of the next level down. Each level both depends upon and yet transcends the level below" (p. 175).

Transcends and includes: which is precisely why differentiation, which allows transcendence, can go too far into dissociation, which denies the dependence upon the lower and thus alienates it (the real pathology). Roszak doesn't seem to grasp this elemental difference, but he does, nonetheless, approvingly quote L. L. Whyte in this regard: "When not pathological the human person is, like all viable organisms, a differentiated hierarchy, a superbly coordinated system of hierarchies. *Hierarchical structure is the basic feature common to matter and mind* [his italics]. When we are ill there is a failure of coordination at one or more levels in these hierarchies, and the clarification of the relation of body to mind and psychosomatic illness requires a hierarchical approach. Guilt, hypocrisy, heartbreak, and so on, are lesions in the hierarchy" (p. 181).

Pathology, in other words, is due not to differentiated hierarchy, but to lesions (dissociations) in the natural hierarchies. And it is these *lesions*, not the required differentiations, that are the problem: but this crucial distinction is thoroughly missing in Roszak's presentation, and thus he proceeds to attempt to cure the human race with ecopsychology when he has not yet correctly diagnosed the actual disease, and, as usually happens with such approaches, the cure is more frightful than the illness.

In the meantime, his frank grasp of the basic truths of evolution leads him to an extremely gutsy admission for an eco-philosopher: human beings are indeed, he says, squarely on top of the pyramid of evolution. "Each level is shaped by the needs of the next level up. This is what the universe has been doing in all the long while since the atom and the galaxy rose into existence. It has been reaching toward finer orders of complexity, toward realms so subtle and complex that they can be fabricated only out of the delicate dynamics of the human imagination. And what stands at the crest of the hierarchy holds a crowning position. It embodies the full potentiality of all that has gone before, realizing it, expressing it. It occupies the *frontier* of the cosmos. Evolution progresses as a hierarchy of dynamic systems, all of which culminate in the human world. Human centrality emerges at the culminating 'now' of a hierarchically evolving universe" (pp. 185, 191, 209).

This "human centrality" is behind Roszak's endorsement of the Anthropic Principle. And, he says, "The Chain of Being becomes the temporal progression of complex systems emerging and maturing since the Big Bang" (p. 209).

There is obviously much truth to this pyramid of development as Roszak presents it; the evidence is simply too overwhelming for him to ignore it. But at precisely the point that all genuine Great Chain theorists would *continue* the development beyond the typical human state and into the transhuman, transpersonal, spiritual domains, Roszak is forced to stop. Since he will not acknowledge any genuinely transpersonal development, humans in their present state (the average egoic-mental state of consciousness and its desires) are left as the *crown* of evolution.

And precisely because he lacks a transpersonal dimension, and because present-day humans (the personal domains) apparently lack anything that is genuinely spiritual, Roszak *must* situate his spirituality in . . . the prepersonal domain. That's all he has left in his pyramid. It is not, as in the genuine Great Holarchy theorists, that the prepersonal and the personal (Earth and Human, body and mind) have to be taken up, integrated, and healed in the transpersonal domain (Heaven and spirit): rather, our salvation now lies merely in getting back beneath the original differentiation of Earth and Human: getting back prior to the Great Crime.

But this presents grave theoretical difficulties for Roszak, because, after all, it implies that evolution—this drive toward what Roszak calls higher, greater, more complex forms, with each successor being a greater "crowning achievement" toward which all lower developments were aiming—implies that all of that, somehow, was operative for fifteen billion years, finally producing humans, whose own evolution . . . then proceeded to run backward. The primitives had it right, and everything else has been downhill, at which point, in all such accounts, the laws of evolution so eulogized up to that point are simply suspended and replaced with the author's favored ideology about how the world should have unfolded.

At this point Roszak's presentation simply begins to contradict itself on almost every important point. Primal or tribal consciousness, he says, is indeed parochial and ethnocentric to the core: "Every group has its own rituals, its own words of power" (p. 76). There is nothing *global* or *universal*, he says, in the tribal rituals, and tribal healers deny that there are common, worldcentric powers at work: the powers work only for their particular tribe: "There are tribal healers who have denied the validity of any attempt [to reach any peoples beyond the particular tribe]" (p. 76). So much for a global humanity. Thus, Roszak says, tribal structures and "therapies" are "stubbornly parochial."

Even though there is nothing in the primal/tribal structure itself that is worldcentric, Roszak *must* find some trace of his Heavenly Paradise here, because he has *nowhere else* to situate it. And so, flagrantly contradicting himself, he is forced to say things like: "Science permeates the lives of people everywhere in the modern world. It is the closest our species has come to a universal culture since the days of the hunters and gatherers" (p. 95). The foragers had a *universal* culture? When few of them even spoke the same language? And when he just denied any universality to them?

Likewise, the primal/tribal consciousness per se *must* have been a source of "ecological wisdom." And yet his own evidence shows, as he puts it, that in many instances "tribal societies have abused and even ruined their habitat. In prehistoric times, the tribal and nomadic people of the Mediterranean basin overcut and overgrazed the land so severely that the scars of the resulting erosion can still be seen. Their sacramental sense of nature did not offset their ignorance of the long-range damage they were doing to their habitat" (p. 226).

Likewise, other primal societies have, "in their ignorance, blighted portions of their habitat sufficiently to endanger their own survival. River valleys have been devastated, forests denuded, the topsoil worn away; but the damage was limited and temporary" (p. 69).

The correct conclusion, in other words, is that the primal/tribal structure per se—*in itself*—did *not* necessarily possess ecological wisdom, it simply *lacked the means* to inflict its *ignorance* on larger portions of the global commons.

The main difference between tribal and modern eco-devastation is not presence or absence of wisdom, but presence of more dangerous means, where the *same* ignorance can now be played out on a devastating scale. The *cure* is not to reactivate the tribal form of ecological ignorance (take away our means), nor to continue the modern form of that ignorance (the free market will save us), but rather, to evolve and develop into the centauric vision-logic that, *for the first time in history*, can integrate a biosphere and noosphere now irreversibly differentiated.

And one of the leads in this area is indeed the useful and true portions of systems theory and ecological science that, for the *first time in history* (as Roszak freely concedes) can *demonstrate* that the ignorance of ecology will kill us all. Neither indissociation (tribal) nor continuing dissociation (industrialization) will save Gaia: both have amply demonstrated their inherent failings in that regard.

And that is why, as Roszak constantly points out, "Nothing [referring to science] now stands a better chance of uniting us as one human family" (p. 95). But, he also stresses, science, as a rather dry and abstract endeavor will not move the hearts and emotions of millions, and so—he is still looking for *somewhere* to situate his Heaven, since he will not allow a transpersonal release—and so what is needed is science integrated with . . . primitive animism. ("If any part of an animist sensibility is to be reclaimed, the project will have to integrate with modern science": p. 94).

And here Roszak's pre/trans fallacy (with all its self-contradictions) comes to the fore. Roszak wants to take, from primals and from infants, the parts of animism that he *likes* and then *jettison* the rest (which he concedes are merely "superstitious"). And so, in his pick-and-choose paradigm, he announces that "I take the liberty of passing over practices one finds throughout tribal culture (such as hexes, curses, the evil eye, and human sacrifice) that are authentically 'superstitious' . . ." (p. 77). Those practices are not superstitious; they are integral to the cultural meaning and the functional fit at those stages of development; in every way it's a package deal.

Having confused differentiation and dissociation, and thus likewise confusing anything prerational with transrational release, Roszak leads us on a eulogizing chase through all of the standard elevationist ploys. He fails to see that development involves a *lessening* of narcissism (in the bad sense), and thus he must eulogize anything *egocentric* as a way out of ethnocentric, totally overlooking preconventional versus postconventional responses, and thus he must champion: narcissism in general, animism, typhonic bodyego impulse, biocentric immersion. Since he has no developed conception of postconventional society, anything preconventional steps in to announce our salvation. "If this be narcissism," he says, "make the most of it."

In the same way, he subscribes to "the" single boundary between self and other, a boundary that primals and infants do not possess, and the existence of which constitutes the Primal Crime. The "id" therefore really represents a union with the "entire world" prior to differentiation, and we must in some sense get back to that primal undifferentiated state to heal the planet. . . .

Lacking a Heaven principle, a genuinely transpersonal state, Roszak must find his salvation, not in the overall integration of Earth, Human, and Heaven (body, mind, and spirit), but only in the de-differentiation of Earth and Human—which drives most of his own self-contradictions.

Here is what I believe is very true and very useful about Roszak's presentation, and how I believe it can be reconstructed:

The differentiation of Earth and Human, body and mind, biosphere and noosphere, begins (as we have seen) with fulcrum-3 and is more or less completed with fulcrum-5. At each of those fulcrums, the differentiation process can go too far into pathological dissociation, and that indeed is part of the pathology of modernity (the "bad news"). And the cure does involve, as Roszak indicates, a "recontacting" of the alienated dimensions, that is, a recontacting of the biosphere. When Roszak says the core of the id is actually the biosphere, I agree entirely (and I made just that point in chapter 6).

Every neurosis, I suggested, is a miniature ecological crisis (repression of biosphere by noosphere), and the worldwide ecological crisis is in fact a worldwide psychoneurosis.

On all of that, I agree with Roszak completely. But because he confuses differentiation and dissociation, and because he does not recognize any "crowning" achievement in evolution higher than the human mind (no transpersonal, transmental Spirit), then he *must* look for his salvation—and ours—in a "mature" regression to infantile animism/narcissism (somehow integrated with modern science). He confuses prerational nature/id with transrational spirit.

And, indeed, Roszak concedes that the id/Gaia that we are to recontact is actually a *lower* structure, driven merely by survival instincts. It is "a wisdom like that of the body in its stubborn will to pursue the tasks that physical survival demands. In the classic metaphysical use of the word, this is what 'soul' meant: the principle of bodily life that only God could create, but which functioned at some *lower level* than the demands of mind or spirit" (p. 159).

So here we have body/id/Gaia—which we are supposed to recontact—being *lower* than mind or spirit. But of spirit we hear . . . precisely nothing. Our salvation is simply *recontacting the lower* and living with that.

Since the biosphere/Gaia has indeed been dissociated, I agree that part of the cure is "derepression of the shadow"—that is, *recontacting* the lower structure that has been alienated and distorted. But *recontacting* the *lower* is not at all the same as *discovering* the *higher*: and it is in the higher that the true healing, and the true integration, can occur: it is *only* in the higher (fulcrum-5 and beyond) that global consciousness, global awareness, global conceptions, and global solutions can even be entertained in the first place. And in this, Roszak leads us exactly in the wrong direction, under the banner of his pre/trans confusions.

To use my terms, Roszak traces the "split" as it occurs from fulcrum-3 to fulcrum-5 (from the bodyego to rationality); and since he recognizes no higher stages, the "cure" must be recontacting the biosphere prior to fulcrum-3 and resurrecting the magico-animistic primal state. But it is only with *worldcentric* rationality (which starts at fulcrum-5) that *global* solutions to the dissociations can be conceived and implemented. This global orientation intensifies at fulcrum-6 (vision-logic) and reaches a major culmination at fulcrum-7 (the Over-Soul that is the World Soul). But Roszak has already confused the World Soul with magical-animism, and so he has no reason to lead us forward into the higher and integrative stages, and so instead he simply looks for ways to mix infantile animism/narcissism with modern science.

But since the primal structure (phylo and onto) contained neither *inherent* ecological wisdom (as he clearly demonstrates) *nor* a worldcentric/global orientation (as he also demonstrates), what exactly is it that we are supposed to recapture? The special form of primal ignorance?

Reintegrating the lower is mandatory, as I fully agreed; but reintegrating the lower is not finding the higher. Yet in our aggressively Descended world, this is precisely the type of solution that is so attractive. No higher, transpersonal growth is required:

simply fall back into something we allegedly knew yesterday. No nasty Ascent here: let the id be our guide to paradise, and let biocentric immersion lead the way to the glorious spirit for all.

Instead of climbing *up* and *off* the pyramid in transpersonal release—and actually having done with this anthropocentric madness—let us instead climb right back down it, right back into egocentric, biocentric, narcissistic, self-crowning glory, there to heal the planet. Because Roszak (and many ecophilosophers in general) aggressively denies any truly transpersonal sphere, the prepersonal domain is forced to serve that spiritual function.

(See note 17 for chap. 14 for further "reenchanters" and the inherent problems thereof.)

The Devil for Eco-Romanticism

There seems to be a general formula running through most of the reenchanting ecophilosophers, and it is some version of: rationality is the Devil, Gaia is the God/dess.

One finds variations on this standard formula not only in Berman and Roszak, but in most of the "new paradigm" theorists, virtually all of the ecofeminists, and virtually all of the ecomasculinists (deep ecologists).

And I believe it has a fairly simple explanation. In *Up from Eden* I suggested that the God of one stage of evolution and development tends to become the Devil of the next. What at one stage is worshiped and identified with becomes at the next stage precisely that which must be transcended, fought, and differentiated from: the God of one stage becomes the Devil of the next, becomes a reminder of what we once were and should not slide back into, that slide being "sin," devolution, regression, retreat.

Thus, the "Pan" God of the pagan religions (half human, half animal) became the actual personification of the Devil for monotheism. And at the next stage, when Reason became the God of the Enlightenment, the God of monotheistic and mythic religions became the Devil oppressive of free thought and full human potential—the God of mythic religions now had to be fought, overcome, transcended: it became the Devil of the Enlightenment.

Likewise, since the postmodern world is beginning to move into postrational modes, the time is ripe for Reason to start looking like the great Devil. That reason ideally should be transcended *and included* doesn't really matter: what tends to hit people first is the difficult task of actually differentiating from a structure which generations revered as the Divine itself—the God of Reason—and in that difficult differentiation, fine distinctions about "transcend and include" are not often translated into mass consciousness: reason starts to the look like the great Devil.

But if reason has become the great Devil, where is the new God? If the World Soul (the structure actually awaiting the postrational reception) is simply *confused with Gaia*, then Gaia becomes the new God/dess, the new Spirit, and we have the standard eco-formula: analytic patriarchal rationality is the Devil, Gaia is the God/dess. Which then translates into a thousand derivative forms: reason is analytic and divisive, Gaia is holistic and wholeness; reason is nasty patriarchy, wholeness is sweet matriarchy; reason is authoritarian, Gaia/wholeness is freedom; and so on through a series of dualisms meant to announce the New Age.

In other words, I believe that in many cases a good and true intuition of the World Soul is being filtered through what is in fact the old structure of the Eco-pole of the fundamental Enlightenment paradigm, the flatland embrace of empirical mononature (or the flatland system of Gaia) as God (or Spirit or Goddess).

This is why most of these theories do not escape the flatland paradigm—which is why they so easily confuse great span with great depth and thus *can* embrace Gaia as being "more whole," instead of seeing that it is simply bigger but more shallow: confusing bigger with better *is* the flatland paradigm.

And all of this plays directly into the standard *regressive* (and "reenchanting") trend of many of the Eco-approaches: analytic reason is simply condemned, for the most part (a certain lip-service aside) and not also included. These approaches are therefore constantly tending to go prerational, as we have repeatedly seen (e.g., Berman and Roszak). And since Gaia per se (the biosphere per se) is regressive as well, we thus have the standard regressive slide into an egocentric/geocentric stance that marks so many of these approaches.

The ways in which otherwise good and true spiritual intuitions get "unpacked" in less than graceful ways—ways that actually abort the original intuition—are the topic of the next chapter.

33. P. Rieff, Euphorion, quoted in Lovejoy, p. 298. More than one commentator has pointed out the similarity of Hegel's system with that of Proclus, Plotinus's most brilliant successor.

34. Tillich, *History of Christian thought*, pp. 488–89.

35. *System of transcendental idealism* was published in 1800; *Origin of species* in 1859.

36. Copleston, vol.7, p. 141.

37. Lovejoy, p. 320.

38. Ibid., pp. 262, 198.

39. There are numerous excellent discussions of the untenability of natural selection as the primary mechanism of evolution. One of the best, for several reasons, is that given by Varela, Thompson, and Rosch in *The embodied mind*. For a discussion of their enactive paradigm, see notes 49 and 52 for chap. 2, notes 13 and 43 for chap. 4, and note 1 for chap. 14. See also Stuart Kauffman's brilliant *The origins of order*.

CHAPTER 14. THE UNPACKING OF GOD

1. Nor, I believe, is Spirit's literal interpretation as "no-self" very helpful either.

The Buddhist Doctrine of No-Self

As is well known, Theravada Buddhism (and the general Abhidharma doctrine, the oldest of the Buddhist schools, unkindly referred to by later schools as Hinayana, or

"Lesser Vehicle") has, as one of its central teachings, the doctrine of *anatta* (or *anatman* in Sanskrit), which literally means "no-self." The mental stream is said to be composed of five aggregates (*skandhas*), none of which is or has a self, but together these aggregates give rise to the illusion of a separate, grasping, desiring self. A more careful meditative analysis of the mind stream, it is said, will reveal that the skandhas, although real enough in themselves, do not constitute a real and enduring self (but are instead simply discrete and momentary elements of experience), and thus this liberating discovery is simultaneously a release from the pain (*duhkha*) of defending an entity that isn't even there.

This no-self or *anatman* doctrine was particularly leveled against the Brahmanical and Samkhya doctrines of a permanent, unchanging, absolute Subject of experience (Purusha, Atman). And against the Atman tradition, the Anatman doctrine was wielded with much polemical force: in place of substance, flux; in place of self, no-self; in place of unitary, pluralistic; in place of cohesive, discrete.

Thus, the Abhidharma accepted the reality of momentary, pluralistic, discrete, and atomistic elements (*dharmas*) and accepted their causal effects as real (codependent origination). This causal (karmic) linkage allows the discrete momentary states to give rise to the illusion of an enduring, cohesive self, much as whirling a flashlight at night gives the illusion of a real circle. This cohesively felt but illusory self is driven by fear, grasping, and ignorance (particularly the ignorance of the reality of momentary states).

However true aspects of that doctrine might be, it nonetheless generated considerable controversy in subsequent Buddhist development. Virtually all ensuing schools of Buddhism (the Mahayana, or "Great Vehicle," and the Vajrayana, or Tantric Vehicle) accepted the Abhidharma as a starting point, but none of them remained with it: none of them accepted it as the final word, so to speak. In fact, the Abhidharma doctrine as a complete and adequate system was aggressively attacked, *both* in its capacity to cover relative or *phenomenal* reality *and* in its ability to indicate *absolute* reality. Some schools, in fact, would claim that the Abhidharma, taken in and by itself, fundamentally misunderstood *both* illusion *and* reality, both relative and absolute truth: wrong on all counts, as it were.

Thus, Nagarjuna launched a devastating attack on the ultimate reality of the *skandhas* and *dharmas* themselves (they have apparent reality only), and the general Yogachara and Vajrayana tradition lambasted the *skandha* system because it only dealt with "coarse"-level reality (the gross realm, or Nirmanakaya) and did not cover the "subtle" and the "very subtle" realms (what we have called the subtle and the causal, and which Vajrayana identified with the Sambhogakaya and Dharmakaya). Higher stages of meditative development, Vajrayana claimed, would disclose these subtle and very subtle dimensions of consciousness (particularly in the *anuttarayogatantra* tradition, which we will discuss in a moment), and the Abhidharma system, they claimed, was completely inadequate in this regard: it covered only the Nirmanakaya (gross-form-oriented or sensorimotor-oriented consciousness).

One would think that as a Buddhist, Nagarjuna would busy himself with the standard attacks on the Samkhya and Atman traditions; in fact, his main target is the Abhidharma of early Buddhism. "Nagarjuna himself applied the dialectic [critical analysis] against the Abhidharmika system—the doctrine of Elements [*dharmas*]. The *Madhyamika Karikas* [one of Nagarjuna's major works] are a sustained attempt to evolve the Śunyata [Emptiness] doctrine out of a *criticism* of the realistic and dogmatic interpretations of early Buddhism. His criticism of the Samkhya and other systems of the Atma tradition is rare and implicit" (Murti, p. 165).

T. R. V. Murti's *The Central Philosophy of Buddhism*, although not without its difficulties and occasional inaccuracies, is nonetheless a classic in the field, and it still manages to pinpoint several crucial factors in this area (many scholars, as David Loy points out, feel that "Murti's study remains perhaps the best work in English on the subject": p. 59). Without endorsing all of Murti's conclusions, I would nevertheless like to draw on those aspects of his presentation that remain valid. And all of this will eventually bring us to the genuine problems that have arisen in interpreting Spirit in terms of a "no-self" doctrine.

Because above all, for Nagarjuna, absolute reality (Emptiness) is radically Nondual (*advaya*)—in itself it is *neither* self nor no-self, neither *atman* nor *anatman*, neither permanent nor momentary/flux. His dialectical analysis is designed to show that all such categories, being profoundly dualistic, make sense only in terms of each other and are thus nothing in themselves (the Emptiness of all views and all phenomena). This dialectical analysis applies to all things, all thoughts, all categories: they are all mutually dependent upon each other and thus are nothing in themselves. They therefore have a *relative* or phenomenal reality, but not *absolute* or *unconditioned* reality (which is Emptiness disclosed in nondual *prajna*, which is not a reality apart from the relative world of Form, but is itself the Emptiness or Suchness of all Forms).

"In the *Prajnaparamita* and the literature of the Madhyamika [the school founded by Nagarjuna], the one basic idea that is reiterated *ad nauseum* is that there is no change, no origination, no cessation, no coming in or going out; the real is neither one, nor many; neither atman nor anatman; it is as it is always.

"Origination, decay, etc. [codependent origination of Abhidharma], are imagined by the uninformed; they are speculations indulged in by the ignorant. The real is utterly devoid (Śunya [Empty]) of these and other conceptual constructions; it is transcendent to thought and can be realized only in non-dual knowledge—Prajna or Intuition, which is the Absolute itself.

"We are also expressly warned not to consider Śunyata as another theory, the Dharmata [the Real] as other than the phenomenal world. The Absolute in one sense *transcends* phenomena as is devoid of empiricality, and in a vital sense is *immanent* or identical with it as their reality.

"The butt of the criticism is the dogmatic speculations (the reality of skandha, dhatu, ayatana, etc.) of the earlier Buddhism; the skandhas are *not impermanent*; but they are Śunyata, lacking a nature of their own. Pratitya-Samutpada is not the *tempo-

ral [and causal] *sequence* of entities [as in the Abhidharma] but their *essential dependence*"—that is, each entity depends on the others and thus is *nothing in itself*: it is *not ultimately real*—it belongs to *appearance only*. Thus, the entire doctrine of no-self and dependent origination "receives a deeper interpretation as *appearance*; and by a relentless dialectic it is shown that nothing escapes this predicament" (p. 86, my italics).

Thus, what for the Abhidharma were marks of the real (impermanence, no-self, and flux) are all shown by Nagarjuna to be *marks of the unreal*: they are dualistic notions dependent upon their opposites, and thus are nothing in themselves. And dependent origination—the notion of temporal causal linkage between phenomena—is shown by Nagarjuna to mean that no phenomenon is independent, unconditioned, self-generating; rather, they are all *mutually dependent*, which looks like a nice systems theory, and indeed it is; but Nagarjuna then adds the crucial point: *mutual dependence is a mark of the unreal*, since the dependent entities are nothing in themselves—they have apparent or relative reality only, not *unconditioned* or absolute reality: "phenomena are unreal or śunya because they are dependent; mutual dependence is a mark of the unreal" (p. 106).

Nor can we escape this by taking dependent origination itself as absolutely real, because, says Nagarjuna, the same argument can be applied to the *series itself*: if all things dependently originate, does origination itself originate? If so, it is relative and *unreal*; if not, it contradicts itself, and is thus *unreal*: either way, unreal.

Thus, for Nagarjuna, all views, of whatever nature, are smashed by the dialectic. "Relativity or mutual dependence is a mark of the unreal. No phenomenon, no object of knowledge, escapes this universal relativity. Tattva or the Real is something in itself, self-evident and self-existing. Only the Absolute as the unconditioned is real, and for that very reason it cannot be conceived as existence or non-existence, atma or anatma" (p. 139).

And, according to Nagarjuna (and perfectly à la Augustine) we *can* criticize all partial views because we have, in our own awareness, access to prajna, or nondual Perfection of Wisdom, against which imperfections stand out like a sore thumb. (For a further discussion of Nagarjuna's dialectic, see note 4 below.)

Absolute reality, then, is neither self nor no-self, neither substance nor flux, neither permanent nor impermanent, nor any other combination of such dualistic notions, but rather is the nondual Emptiness of all phenomena, all views, all stances. As the *Ratnakuta Sutra* (one of the early Mahayana texts) puts it: "That everything is permanent is one extreme; that everything is transitory is another. . . . that 'atman is' is one end, that 'atman is not' is another; but the Middle Way between the atma and anatma views is the Inexpressible." Is Emptiness.

Thus, "for the Madhyamika, the Real is neither one nor many, neither permanent nor momentary, neither subject nor object . . . , neither atma nor anatma. These are relative to each other and are equally unreal. Nagarjuna says: 'If the apprehension of the impermanent as permanent is illusion, why is the apprehension of the indeterminate as impermanent not illusion as well?' " (p. 239).

The question then arises, if no-self and impermanence do *not* actually apply to the Absolute (which is neither self nor no-self, nor both nor neither, etc.), do the notions of impermanence and no-self apply to the *phenomenal* stream itself (as maintained by the Abdhidharma)? If no-self is not *absolutely* true, is it at least *relatively* true? If it does not apply to noumenon, does it at least apply to phenomena?

And here the doctrine fares no better. Nagarjuna (and several of his followers) demonstrate that on the *phenomenal* level, the states (skandhas) *cannot exist without a self*, and the *self cannot exist without the states*: they are mutually dependent. Likewise, substance does not exist without modes or flux, and vice versa. Neither the states in themselves (without a self), nor the momentary modes, can even offer an *adequate explanation of phenomena*!

"If we confine ourselves to the phenomenal point of view, if we propose merely to give a transcription of what obtains in everyday experience, we must accept, besides the states or moments, the activity and the agent. From the noumenal point of view of the Unconditioned truth, the moments too are as unreal as the activity which the earlier Buddhism rightly rejects. The correct Madhyamika standpoint is that modes by themselves cannot offer an adequate explanation of phenomena. Substance too must be accepted" (p. 187).

Likewise for the self as cohesive agent in the phenomenal stream: "Nagarjuna explicitly says that there can be no act without an agent or vice versa; he calls them ignorant of the true meaning of the Buddha's teaching who take the reality of the atman only or of the states as separate from it; if there is no atman apart from the states, there are no states too apart from the atman. In fact, the entire Madhyamika position is developed by a trenchant criticism of the one-sided modal view [selfless flux] of the Abhidharmika system, by being alive to *the other side of the picture equally exhibited* in the empirical sphere [my italics].

"In the same strain Chandrakirti [a principal successor of Nagarjuna] complains that the Abhidharmikas have not given an adequate picture of the empirical even. 'If it is sought to depict the empirically real then besides momentary states, the activity and the agent too must be admitted.' Chandrakirti shows, in a sustained criticism of the view of mere attributes or states without any underlying self in which they inhere, that this does violence to common modes of thought and language; it fails as a correct picture of the empirical; nor [as we have already seen] can it be taken as true of the unconditioned real"—*wrong on both counts*, as it were (pp. 249–50).

These criticisms pointed up "the inadequacy of a stream of elements to account for the basic facts of experience, memory, moral responsibility, spiritual life, etc.: the states (skandhas) cannot completely substitute the atman; a permanent synthetic unity must be accepted" (as a relative truth; p. 81). "On the modal view [*anatma*, no-self], there are the different momentary states only; there is no principle of unity. But mental life is inexplicable without the unity of the self" (p. 205).

But again, on the phenomenal plane, this does not mean that the self-view alone covers the scene: "The self of the [early] Brahmanical systems is a bare colorless unity bereft of difference and change, which alone impart significance to it." Rather, according to Nagarjuna, "the self has no meaning apart from the states . . . , and there are no states apart from the self. The two are mutually dependent"—and therefore, both are phenomenally real but ultimately unreal, ultimately Empty (pp. 204–6, 249).

In short, *both* the cohesive self and the momentary states are relatively real, but *both* are ultimately Empty: the absolute is neither self nor no-self (nor both nor neither), neither momentary nor permanent (nor both nor neither), but is rather the "Thatness" disclosed by nondual Prajna or primordial awareness, a "Thatness" which, being radically unqualifiable, cannot be captured in any concepts whatsoever.

(We would say: the *signifier* "Emptiness" can only be understood in those who possess the developmental *signified* that is Prajna, whereupon the actual *referent* "Emptiness" is *directly perceived* as the Emptiness of all forms whatsoever, and not the privileging of one form or concept, such as no-self, over its opposite, such as self. But the *referent* of the signifier "Emptiness" becomes *obvious* only upon the awakening of Prajna, and Prajna is not an idea or a theory but an *injunction* or a paradigm: it begins its *practice* by categorically rejecting every conceivable category of thought to embrace the Real, or Dharmata. This categorical rejection—the dialectic—creates an opening or clearing in awareness in which the primordial and unobstructed nature of the Real can shine forth nondualistically as the Suchness of all phenomena. And all of *those* words are nevertheless still *signifiers* whose actual *referents* are disclosed *only* in that "opening" or "clearing," and prior to the discovery of that opening or Emptiness, they are all equally off the mark. Even the phrase "Emptiness is free of thought concepts" would itself be denied: that's just more words. Where is the actual referent? Where is your Original Face right now?)

Scholars are still arguing about the exact influence of Nagarjuna on subsequent Eastern thought, but most agree that the Nondual Madhyamika was a profound revolution that, to one degree or another, influenced virtually all succeeding schools of Asian thought (either directly or indirectly). Because Emptiness was not a realm *apart* from or divorced from other realms, but rather was the reality or Suchness or Emptiness of *all* realms, then nirvana was not, indeed could not be, sought apart from the phenomenal world. This, indeed, was revolutionary.

Thus the almost purely Ascending and Gnostic bent of the Samkhya, Yoga, and Abhidharma systems (nirvana as the extinction or utter cessation of samsara) gave way to a variety of Nondual systems (Yogachara, Vedanta, Tantra), all of which maintained, in their various ways, that "Emptiness is not other than Form, Form is not other than Emptiness"—the union and integration of the Ascending and Descending Paths, of Wisdom and Compassion, of Eros and Agape, of Ascent to the One (formless nirvana) perfectly and fully embracing the world of the Many (samsara and Form).

In all of these developments, a Path of Ascending (and Ascetic) *Purification* and *Renunciation* (where defilements are exterminated in cessation) gave way to *Paths of Transformation* (where the defilements are seen to be the *seeds* of *corresponding wisdoms*, since nirvana and samsara are ultimately not-two: the Ascending wisdom has a Descending compassion, and thus the defilements—indeed, the entire manifest do-mains—are seen to be *expressions* of the Absolute, not *detractions* from it: the integra-tion of Ascending and Descending, Emptiness and Form, Wisdom and Compassion). And these in turn often gave way to *Paths of Self-Liberation*, where the defilements are seen to be *already* self-liberated just as they are, and just as they arise, since their basic nature is *always already* primordial Purity (pure Emptiness in pure Presence: the radical and *already spontaneously accomplished* union of Emptiness and awareness/clarity/form: the *already accomplished* union of Ascent and Descent: Emptiness and Awareness, Emptiness and Clarity, and Emptiness and Form).

And all of this, in a sense, was opened up by Nagarjuna. Already in the Madhya-mika we find the twin principles of Prajna (Wisdom) and Karuna (Compassion), the former seeing that all Forms are Empty, the latter seeing that Emptiness manifests as all Form, and thus each and every Form is to be treated with care and compassion and reverence. (Cf. the whole discussion of this theme of ascending Wisdom—the Good—and descending Compassion—Goodness—in chapters 9 and 10.) "Śunyata [Emptiness] and Karuna [Compassion] are the two principal features of the Bodhicitta [awakened mind]. Śunyata is Prajna, [Wisdom or] intuition, and is identical with the Absolute. Karuna is the active principle of compassion that gives concrete expression to Emptiness in phenomena . . . , a free phenomenalizing act of grace and compassion [Descent/involution or manifestation itself as an act of Agape/Compassion]. If the first [Prajna] is transcendent and looks to the Absolute [Eros], the second [Karuna] is fully immanent and looks down toward phenomena [Agape; thus together: Ascent and Descent, evolution and involution]. The first is the universal reality of which no deter-minations can be predicated [Emptiness disclosed in and as Prajna]; it is beyond the duality of good and evil, love and hatred, virtue and vice; the second is goodness, love and pure act. Buddha and the Bodhisattva are thus amphibious beings with one foot in the Absolute and the other in phenomena [nondual]. They are virtuous and *good* [Eros] and the source of all *goodness* in the world [Agape]. With its phenomenalizing aspect, Karuna, the formless Absolute (Śunya) manifests itself as the concrete world. But the forms neither exhaust nor do they bring down the Absolute. It is through these forms again that individuals ascend and find their consummation with the uni-versal Real" (pp. 264, 109).

This profoundly Nondual conception had its equally profound impact. "Śunyata is the pivotal concept of Buddhism. The entire Buddhist philosophy turned on this. The earlier realistic phase of Buddhism, with its rejection of substance and uncritical erec-tion of a theory of elements, was clearly a preparation for the fully critical and self-conscious dialectic of Nagarjuna. Not only is the Yogachara idealism [Empty Con-sciousness] based on the explicit acceptance of Śunyata, but the critical and absolutist [Nondual] trend in the Vedanta tradition [of Gaudapada and Shankara and Ramana Maharshi] is also traceable to this" (p. 59).

And likewise the rise of Tantra. Although general tantric practices have a long history, nonetheless "it is Śunyata that provided the metaphysical basis for the rise of

Tantra"—largely because, as I indicated, nirvana was no longer conceived as being away from samsara (and body and sex and flesh and the senses in general) but rather was to be found within them. "It is the Śunya of the Madhyamika that made Tantricism possible. It may thus be said to have initiated a new phase in Buddhist philosophy and religion; this had its due influence on the corresponding phase on the Brahmanical side." As Bhattacharya put it, "There is hardly a Tantra in Hindu literature which is not tinged with Buddhistic ideas; it is no exaggeration to say that some Tantras of Hindus are entirely Buddhist in origin. It is thus amply proved that the Buddhist Tantras greatly influenced the Hindu Tantric literature" (all quotes p. 109).

These are a few of the reasons that I earlier likened Nagarjuna's influence in the East to that of Plotinus in the West: they brought the Nondual revolution.

Perhaps the major difference between the Madhyamika and the subsequent Nondual traditions which it sparked is that Madhyamika (as Prasangika) remained the pure *via negativa* school: no concepts at all (including the concept of no concepts) could be applied to the Real, which is "what" is disclosed in Prajna, not in concepts or ideas (I'll return to this in a moment: it was primarily a statement about signifiers and referents and corresponding injunctions/paradigms). Emptiness was not a conceptual view, but the *Emptiness of all views*, which itself is *not another view*. As Nagarjuna trenchantly put it, "Emptiness of *all* views is prescribed by the Buddhas as the way of liberation. *Incurable indeed* are they who take Emptiness itself as a view. It is as if one were to ask, when told that there is nothing to give, to be given that nothing" (p. 163).

Thus ultimately Emptiness takes no sides in a conceptual argument; Emptiness is not a view that can dislodge other views; it cannot be brought in to support one view as opposed to another: it is the Emptiness of all views, period. The relative merits (or relative truth) of various views are to be decided on their own terms.

The subsequent Nondual schools would, in various ways, relax the purity of this intense *via negativa* and, also in various ways, would look for phenomenal *metaphors* for Emptiness. Advaita Vedanta and Yogachara/Vijnanavada (Consciousness-only) schools "both reach advaitism [Nondualism]—the advaita of Pure Being (Brahman) and the advayata of Pure Consciousness (vijnapti-matrata) by rejecting appearance through dialectical methods—through negation. Their Absolutes partake of the form of the Madhyamika Śunyata in being transcendent to thought and being accessible only to non-empirical Intuition [prajna, jnana]. They also have recourse to the two truths [absolute and relative]" (p. 59).

Thus, "In the Madhyamika, Vijnanavada [Consciousness-only], and Vedanta systems, the Absolute is non-conceptual and non-empirical [perceived with the eye of contemplation, and not merely with the eye of mind or the eye of flesh]. It is realized in a transcendent non-dual experience, variously called by them prajna-paramita, lokottara-jnana, and aparoksanubhuti respectively. All emphasize the inapplicability of empirical determinations to the Absolute, and employ the language of negation. They are all agreed on the formal aspect of the Absolute [i.e., its strict *unqualifiability* with merely phenomenal categories]."

However, "The Vedanta and Vijnanavada identify the absolute with something that is experienced in some form even empirically—the Vedanta with Pure Being which is Atman and the Vijnanavada with Consciousness. Taking these as real, they try to remove the wrong ascriptions which make the absolute appear as a limited empirical thing. When, however, the atman or vijnana is absolute, *it is a misuse of words to continue to call it by such terms*; for there is no *other* from which it could be distinguished. They are also [at this point] reduced to [and agree with] the Madhyamika position of the Absolute as utterly inexpressible. Words can only be used *metaphorically* to *indicate* it" (p. 236, my italics).

But in that *indication* lay the direction that subsequent Nondual schools would follow (even though always resting, as it were, in final Emptiness). "The Vedanta and Vijnanavada [and Tantra], owing to their [metaphorical] identification of the real with Atman or Vijnana [or Tantra: great bliss] are seemingly more able to *provide a bridge* between the world of appearance and the Absolute. The transition seems easier. The Madhyamika by his insistence on the sheer transcendence of the absolute and his refusal to identify it with anything met with in his experience is too abrupt and harsh. But in principle, however, there is no difference in the form of the Absolutes in all these systems. Śunyata represents the form of all Nondualism" (p. 237).

As for the "bridge" between Absolute and phenomena: "Both Vedanta and Vijnanavada analyze illusion [samsara] and show that the illusory appears on a real ground *but for which* illusion itself would not be possible [again, cf. Augustine]. The world-illusion is thus a super-imposition on Brahman or Vijnana. It is not true to say that the Madhyamika conceives illusion to occur without any underlying ground. Tattva [Suchness] as Dharmata [realm of the Real] is accepted by the Madhyamika as the underlying ground of phenomena. But it is not [actually] shown by him to be immanent in experience, how Dharmata activates and illumines empirical things. Not that the Madhyamika takes the Absolute and the world of phenomena as two different sets of entities; but the Absolute is nowhere *explicitly* shown to be *in* things constituting their very reality. The relation between the two is not made abundantly clear. This may be said to constitute a drawback in the Madhyamika conception of the Absolute" (p. 237).

And this "drawback" is what Vedanta, Yogachara, and Tantra set about to redress. In their various conceptions, the pure Self constitutes a bridge to the Absolute (Vedanta), pure Consciousness freed of duality constitutes a bridge (Yogachara), and the emotions and desires of ordinary awareness constitute a bridge (Tantra). Thus, in Vedanta, "Brahman is no doubt devoid of determinations [is pure Emptiness]; it cannot be made an object of thought as a particular thing is. But it is self-evident [nondual intuition of Brahman is self-evident, self-certifying immediateness] and because of this anything becomes evident; it implicitly, invariably and unconditionally *illumines all things*. In a slightly different manner Vijnanavada shows that the object is dependent on consciousness, and not vice versa." Likewise, Tantra would show that even ordinary emotions and desires and confusions, if entered into with open awareness, would disclose the wisdoms at their base. Bridges each and all, whereas for "Madhyamika the Absolute is also held to be immanent, but epistemologically it is not shown to be such"—the drawback (p. 237).

Before we move on to the next and closely related topic (namely, the subtler stages of consciousness revealed beyond the five gross skandhas of Hinayana), let me finish with Murti by noting his assessment of the relation of the Consciousness-only schools with aspects of Western Idealism: "The parallel for Vijnanavada in the West is the system of Fichte or Hegel, both of whom conceive the Pure Ego (Fichte) or Reason (Hegel) as self-legislative, as containing and creating both the categories and the objects on which the categories function. The difference between Hegel and Vijnanavada is that the Hegelian absolute is thought or Reason [vision-logic]; the Vijnanavada absolute is above reason and is non-dual" (p. 317). I believe that the assessment I give later in this chapter is more specific, but the general agreement with Murti will be obvious.

The tantric Buddhism of Tibet (Vajrayana) would (particularly in the oldest, or Nyingma, tradition) divide the overall Buddhist teachings into nine *yanas* (vehicles, levels, stages): the first two were Hinayana, the third was Mahayana, and the remaining six were Tantric, divided into the three lower or outer Tantras (Kriya, Charya, Yoga), and the three higher or inner Tantras (Mahayoga, Anuyoga, and Atiyoga).

I will here briefly focus on the two highest *yanas*, the Anuyoga (or "Highest Yoga Tantra") and the Atiyoga, or Dzogchen, the "Great Perfection."

The "bridges" that tantric Anuyoga would disclose between the ordinary, gross-oriented mind and the enlightened mind of Clear Light Emptiness centered on the subtler states of consciousness and their associated "energies" or "winds." Consciousness was said to be divided into "gross," "subtle," and "very subtle" (causal) dimensions, with each dimension possessing a "body" or "medium" or "energy," commonly called a "wind" (so that we have a gross bodymind, a subtle bodymind, and a very subtle bodymind). Meditation consisted in a *developmental unfolding* of gross mind to subtle mind to very subtle (causal) mind, accomplished by a meditative manipulation of the winds or energies or bodies that supported each mind.

Thus, the existence of the five skandhas is fully acknowledged, as in the Abhidharma, but the skandhas are, as it were, just the beginning of the story. The five skandhas are generally listed as (1) physical form, (2) sensation, (3) perception/impulse, (4) emotion/image ("dispositions"), and (5) symbolic/conceptual consciousness. (Note that these are also exactly the first five basic structures of consciousness in the Upper-Left quadrant as I presented them in the text; see, for example, figs. 4-1, 5-1, and 9-1.)

For the higher Tantras, the last skandha—or mental consciousness in general—is then *further divided* into three general domains: the gross, the subtle, and the very subtle. The *gross* realm is the sensorimotor realm, and the gross mind is the mind tied to, or reflective of, the sensorimotor world (and supported by the gross or *vital wind*). The *subtle* domain is the mind freed from all gross-conceptions (as revealed in higher meditative states, certain dream states, and so on), supported by the subtle wind (and channels and drops—all part of the "subtle anatomy" of consciousness). The *very subtle* domain is the mind of Clear Light Emptiness, supported by the "eternal inde-structible drop" and very subtle wind in the center of the Heart.

These three domains (gross, subtle, and very subtle or causal) are also said to corre-spond to the Nirmanakaya, the Sambhogakaya, and the Dharmakaya; and to *waking*, *dream*, and *deep sleep* states (at which point the similarities with the Vedanta become quite striking; see Cozort 1986 and Gyatso 1982).

Geshe Kelsang Gyatso explains: "The five physical sense consciousnesses—those of the eye, ear, nose, tongue, and body—are necessarily gross levels of mind. The sixth consciousness—mental consciousness itself—has three divisions of gross, subtle, and very subtle. All eighty indicative conceptions [waking-state concepts and skandhas] belong to the gross level of mind. These are the 'used minds' and include the different thoughts we remember, what we think, and so forth. They are 'used minds' because they realize, hold and cognize their objects [all in the gross realm]; they think and meditate and we—the person—use them. In addition, the first four signs of meditation [its early stages relating to the five elements] are all gross minds because their mount-ing winds are gross" (p. 139).

As the gross mind (and wind) subsides with further meditation, the *subtle dimen-sions* (of which there are three) begin to unfold in developmental sequence.

Using my terminology, these three subtle minds (beyond the gross-oriented ego/centaur) begin with the psychic level (and the psychic anatomy of subtle winds and channels that begins to disclose at that level) and move into the subtle realm itself (experienced as various *interior illuminations*, as we will see), *culminating* in perfect *cessation* or pure unmanifest causal absorption (experienced as "black near-attain-ment").

Thus, Gyatso: "Beginning with the fifth sign [of advanced meditation, which is called *white luminosity appearance*] the subtle minds are experienced. They manifest from the beginning of the mind of white appearance to the mind of red increase [which are both subtle-level illuminations] to the end of the mind of black near-attain-ment [causal cessation]. Each successive mind is subtler than the last. Each is classified as subtle because during its arraisal there are *no gross dualistic conceptual thoughts*" (p. 139). (Compare this with note 8 for chapter 8, where I defined the subtle level as having no gross-referents in cognition; there is still, however, a subtle dualism between subject and object, and this is also pointed out by Anuyoga.)

This subtle-level development culminates in "black near-attainment," "black" be-cause all objective awareness ceases, and "near-attainment" because it is close to pure Nondual Emptiness/Awareness (or the Clear Light). As I would put it, black near-attainment is the transition between causal and nondual. "Finally, after the mind of black near-attainment has ceased, the mind of clear light arises. This is called the very subtle mind [causal/nondual] because there is no subtler mind than this" (p. 139).

It is this very subtle (causal/nondual) mind that is the Empty Ground of Enlightenment, and thus it is also called the "Root Mind" or the "Foundation Mind," and it is associated with the very subtle wind that is the "eternal indestructible drop or empty essence" in the heart. Gyatso: "The very subtle wind and the very subtle mind that is mounted upon it reside in the indestructible drop in the center of the heart" (p. 74). "At that time, the very subtle primordial wind and mind of clear light become manifest. It is called 'the eternal indestructible drop' because the continuum of the very subtle wind within it is never broken" (Cozort, pp. 76, 72). Gyatso: "Without utilizing this very subtle mind and wind there is absolutely no possibility of reaching the perfect enlightenment of Buddhahood" (p. 140).

Gyatso continues: "This very subtle mind residing within the heart is referred to both as the root mind and the resident mind. This latter name is employed to differentiate it from the gross and subtle levels of mind, which are temporary. While the gross body and mind are temporary bases upon which the 'I' is imputed [mistaken for the Real], the *primary* and *continuously residing* bases of imputation are the very subtle mind and its mounted wind" (his italics; pp. 137, 195.)

Gyatso then drives to the major point: "The very subtle body of the continuously residing continuum never dies [is timeless]. In the same way, the continuously residing body—the very subtle wind—is never separate from the *continuously residing self*. You have never been separated from it in the past nor will you ever be separate from it in the future" (p. 195, my italics).

Note the use of "continuously residing self"; we will return to this terminology shortly; for the moment simply notice that, as a *metaphor* (and it remains *merely* a metaphor until one has awakened to the direct realization for oneself), it is perfectly appropriate to use "continuously residing self" to indicate this state.

This very subtle mind (and continuously residing self)—that is, the pure Empty Awareness as Clear Light—is the *root cause* and *support* of *all* lesser states and domains (hence "causal"). "The distinguishing factor of secret mantra [Vajrayana] is its assertion that the deluded mind of self-grasping depends upon its gross mounted wind. This gross wind developed from a subtle one which in turn developed from the very subtle wind mounted by the all empty mind of clear light." And there is a precise description of Involution or Efflux (causal to subtle to gross) (p. 194).

And therefore, after one has developed (or evolved) to an awakening of the very subtle consciousness (nondual Emptiness/Awareness), one then consciously *reverses* this path of *Ascent* and deliberately re-creates the *complete path of Descent*: one reanimates from the causal to the subtle to the gross, *reentering* on the "way down" *all* of the meditative signs and all of the actual domains that one first met on the "way up," thus uniting and integrating the Path of Ascent with the Path of Descent (the real secret of Tantra), balanced and upheld by the indestructible empty essence that is the Heart.

Thus Gyatso: "When you arise from the appearance of Clear Light the first thing you will experience is the mind of black near-attainment of reverse order. Then comes the mind of red increase, the mind of white appearance, the eighty gross conceptual minds [and senses] and so forth in an order that is the reverse of the sequence in which the winds originally dissolved. *The mind of clear light is the foundation of all other minds.* When the gross and subtle minds and winds dissolve into the indestructible drop at the heart [pure Ascent], you perceive only the clear light and it is from this clear light that all other minds—each one more gross than the one it follows—are generated [pure Descent]" (p. 76, my italics).

As for referring to the Real (Emptiness) as a continuously residing self (or True Self, or pure Consciousness, etc.): since Nagarjuna had already demonstrated that the Real is neither self nor no-self, but that in the *phenomenal* realm, there is no self without the states and no states without the self, then the metaphor of a True Self could in fact serve as a much better bridge:

Not that a phenomenal self gives way to no-self (for pure Emptiness is neither self nor no-self); and *not* that a phenomenal no-self gives way to pure Emptiness (there is no phenomenal no-self); but rather, a phenomenal self gives way to pure Emptiness (that strictly speaking is neither self nor no-self nor both nor neither).

And since in the *phenomenal* realm the self is necessary and useful (as Nagarjuna and Chandrakirti pointed out), then as a *bridging* metaphor, it was more adequate to speak of the phenomenal self (relatively real but ultimately illusory or phenomenal-only) giving way to a True Self (that was no-phenomenal-self, and that strictly speaking was neither self nor no-self but pure Emptiness, free of all conceptual elaborations).

Thus Emptiness as True Self, Emptiness as pure Consciousness, Emptiness as Rigpa (pure knowing Presence), Emptiness as primordial Wisdom (*prajna, jnana, yeshe*), Emptiness as primordial Purity, even Emptiness as Absolute Subjectivity: all of these bridging notions began to spring up in the Mahayana and Vajrayana to supplement (or even replace) the notion of no-self, which, strictly speaking, was wrong both phenomenally and noumenally.

And indeed, starting with the *Nirvana Sutra*, the absolute was often metaphorically categorized as "Mahatman," the "Great Self" or "True Self," which was no-phenomenal-self: the selfless Self, so to speak (still metaphorical). And down to today (to give just a few examples), Zen master Shibayama would find that the ultimate state could best be metaphorically indicated as "Absolute Subjectivity." As he puts it, "The Master does not refer to the subjectivity that stands over against objectivity. It is 'Absolute Subjectivity', which transcends both subjectivity and objectivity and freely creates and uses them. It is 'Fundamental Subjectivity,' which can never be objectified or conceptualized and is complete in itself, with the full significance of existence in itself" (*Zen comments on the Mumonkan*, p. 92).

Likewise, Shibayama uses "True Self" to mean no-separate-self: "Thus Mumon says, 'There is nowhere to hide the True Self.' When the world is I-myself, there is no self. When there is no self, the whole world is nothing but I-myself, and this is the true no-mind in Zen. The ancient Masters were never tired of pointing out that the 'True Man of no title' [True Self that is no self] is the Master, or Absolute Subjectivity, and this is one's original True Self. To realize one's antecedent determinant is to awaken to the 'the True Self that is prior to the birth of one's parents.' It is to be born anew as a person of Absolute Subjectivity" (pp. 173, 123, 338). As Suzuki Roshi would put it, small mind finds itself in Big Mind.

Likewise, by the time we reach the highest vehicle in the Vajrayana (which is Maha-Ati, or Dzogchen, "The Great Perfection"), the absolute is most often described in terms of permanence, singularity, freedom, unbroken continuity, ongoing pure Presence. Thus the great Dzogchen teacher, Chagdud Tulku Rinpoche, who is one of a handful of Tibetan masters giving the entire Dzogchen teachings, A to Z, including *rushan, togyal,* and *trekchod,* explains: "From the day we were born until the day we die, our life experience is an ever-changing relative truth that we hold to be very real. It is not, however, absolutely real or permanent. This is very important to understand. When you wake up from your dream of life, all of your experiences, which seemed true, were not really true in the absolute sense.

"The criterion we can use for understanding truth is permanence. If something is permanent, it is true. If it is impermanent, it is not true, because it is going to disappear. To wake us up so we can see the illusory quality of our relative reality and understand our foundational absolute nature is the goal of the Buddha Dharma. A completely awakened state is enlightenment, the unwavering recognition of the absolute nature of our being. Absolute nature pervades everything and is separate from nothing, but we have gone so far into mind's dualistic delusion, that we have lost sight of what is absolute. Seeing separateness where there is none, we suffer in our experience of relative truth, [quenched only by] the unchanging, deathless absolute, the unchanging bliss of enlightenment" (*Life in relation to death*, pp. 10, 11, 14).

Likewise the Dzogchen master Namkhai Norbu, describing the self-knowing absolute or primordial state of pure Presence (*yeshe/rigpa*), points out that the Primordial State (or pure Presence, pure Emptiness) is not a *particular experience*, whether of pleasure or clarity or "voidness" itself, but rather is that which cognizes all experiences (or the pure Presence of any experience): "There is a great deal of difference between a sensation of pleasure and one of voidness, but the inherent nature of both the two experiences is *one and the same*. When we are in a state of voidness [no-thoughts], there is [nonetheless] a presence that continues all the time, a presence which is just the same in an experience of pleasurable sensation [or any other experience]. This Presence is unique and beyond the mind. It is a non-dual state which is the basis of all the infinite forms of manifestation."

Thus, he continues, "All that appears to us as a dimension of objects ['out there'] is not, in fact, really something concrete at all, but is an aspect of our own primordial state appearing to us. Different experiences can arise for us, but the presence *never changes*" (*Dzogchen*, pp. 52, 53, 50).

Different meditation practices engineer different states and different experiences, but pure Presence itself is unwavering, and thus the highest approach in Dzogchen is "Buddhahood without meditation": not the creation but rather the *direct recognition* of an already perfectly present and freely given primordial Purity, of the pure Emptiness of this and every state, embracing equally all forms: embracing a self, embracing a no-self, embracing whatever arises.

But in no case is primordial Emptiness a particular state versus another state, or a particular concept versus another concept, or a particular view versus a different view: it is the pure Presence in which any and all forms arise. It certainly is not "no-self" as opposed to "self." It is rather the opening or clearing in which, right now, all manifestation arises in your awareness, remains a bit, and fades: the unwavering clearing itself never enters the stream of time, but cognizes each and all with perfect Presence, primordial Purity, fierce Compassion, unflinching Embrace.

This unwavering Presence is not entered. There is no stepping into it or falling out of it. The Buddhas never entered this state, nor do ordinary people lack it (the Buddhas never entered it because nobody ever fell out of it). It is absolutely *not an experience*—not an experience of momentary states, not an experience of self, not an experience of no-self, not an experience of relaxing, not an experience of surrendering: it is the Empty opening or clearing in which *all* of those experiences come and go, an opening or clearing that, were it not always already perfectly Present, no experiences could arise in the first place.

This pure Presence is not a change of state, not an altered state, not a different state, not a state of peace or calm or bliss (or anger or fear or envy). It is the simple, pure, immediate, present awareness in which all of those states come and go, the opening or clearing in which they arise, remain, pass; arise, remain, pass. . . .

And yet there is something that does not arise or remain or pass—the simple opening, the immediateness of awareness, the simple feeling of Being, of which all particular states and particular experiences are simply ripples, wrinkles, gestures, folds: the clouds that come and go in the sky . . . and you are the sky. You are not behind your eyes staring out at the clouds that pass; you are the sky in which, and through which, the clouds float, endlessly, ceaselessly, spontaneously, freely, with no obstruction, no barriers, no contractions, no glitches: no moving parts in one's true nature, nothing to break down. In spring it rains; in winter it snows. Remarkable, this empty clearing.

You do not *become* this opening or clearing; you do not become the sky. You are not always the sky, nor are you already the sky: you are *always already* the sky: it is always already spontaneously accomplished: and that is why the clouds can come and go in the first place. The sunlight freely plays on the water. Remarkable! Birds are already singing in the woods. Amazing! The ocean already washes on the shore, freely wetting the pebbles and shells. What is not accomplished? Hear that distant bell ringing? Who is not enlightened?

And yet, and yet: how best to refer to this always already Emptiness? What words could a fish use to refer to water? How could you point out water to a fish? Drenched in it, never apart from it, upheld by it—what are we to do? Splash water in its face? What if its original face *is* water?

Twisting in this linguistic dilemma, the Mahayana/Vajrayana often uses equations such as: Big Mind = no-mind, True Self = no-self, Original Face = no-face, Permanent Vajra = impermanence of all objects, and so on. But it is quickly added that both "Self" and "no-self" (or "permanent" and "impermanent") are mere words, that is, mere *signifiers*. And thus in order for either of them to actually be true (absolute truth), one has to *recognize* the pure Emptiness, the pure opening or clearing in which all words, all things, all processes, spontaneously arise and fall. "Emptiness" or "True Self" or "no-mind" or "no-self" are all, finally, *signifiers* whose *referent* discloses itself only upon following the *injunctions* (paradigms) of contemplative competence which unfolds the appropriate developmental *signified* resulting in the direct recognition of always-already pure Emptiness, the recognition of one's "True Nature," which is, so to speak, all-pervading water: the *referent* is the universe of One Taste. But the Zen masters don't want you to *explain* Emptiness or nonduality to them; they want you to directly *show* them Emptiness, which might be as simple as yawning or snapping your fingers: Just this!

Thus, Shibayama, after repeatedly referring to reality as "Absolute Subjectivity," quickly adds: "It is Absolute Subjectivity, one's original True Self. But however precisely you may describe 'it,' you will miss it if you ever try to describe it at all; you have to grasp it yourself if you want to really know it. To call it by these names [Absolute Subjectivity, no-mind, True Self, no-self] is already a mistake, a step toward objectification and conceptualization. Master Eisai therefore remarked, 'It is ever unnameable' " (pp. 309, 93).

And Chagdud adds: "As the Buddha taught, in absolute truth, nothing really comes, and nothing really goes. Nothing is born, nothing ceases. Neither something nor nothing, [pure Presence] is neither one nor many. Absolute truth is beyond all of these ordinary concepts. Words can't name it"—unless one has recognized it, and then words work just fine: the cherry blossoms are in bloom, the spring air is cool.

Now I mention all of this because no-self has been mistaken (particularly by several Western theorists) as an *accurate account of the phenomenal mind stream* (which is incorrect: self and states are both relatively real), and also as a *description* of the *ultimate* Reality (also incorrect: the Ultimate is *shunya* of self and no-self), and this has caused an enormous amount of confusion, a confusion that has not helped the spread of Dharma in the West.

When "no-self" is *literally* applied to the Absolute, or the primordial state, or pure Emptiness, then that misunderstanding locks Emptiness into an extremely dualistic notion. Whether the experience of self or the experience of no-self arises, *both* are equally manifestations of the Primordial State, self-cognizing Emptiness and spontaneous luminosity. "Emptiness" is not a conceptual doctrine which one uses to advance the theoretical position of no-self as against the position of cohesive self—and yet it is being used, by some Western interpreters, to do just that.

Likewise, the no-self doctrine is being used by many Westerners to describe the lived, phenomenal mind stream: the phenomenal stream is supposed to be without a cohesive self. And here this confused and untenable notion runs into massive amounts of psychiatric and psychological evidence about what happens when the phenomenal mind lacks a cohesive self: the result is borderline neurosis and psychosis. If the self-system, in its formative phases (particularly fulcrum-2), fails to clearly differentiate self and object representations, the self boundary is constantly open to emotional flooding, on the one side, and to a sense of pervasive emptiness or hollowness or depression, on the other. A cohesive self fails to emerge and consolidate, and thus the mind stream remains a series of often fragmented "no-coherent-self" states, which is not Buddhist heaven but psychological hell.

And thus, when orthodox psychological researchers hear that the mind is "really without cohesive self," they are simply flabbergasted. You mean, people *want* to be borderline? They honestly cannot imagine what these "Buddhists" mean, because the self is being denied reality *precisely on the plane that it is indeed real*—and the loss of this cohesive self is not Enlightenment but some of the most painful psychological disturbances known.

These particular Western Buddhist interpreters are, with very good intentions, trying to integrate East and West, but in using the inadequate "no-self" doctrine, they invariably can point to only a few Western theorists. There is the obligatory quoting of Hume's paragraph from *Treatise* (which actually demonstrated nothing but the lame fact that if I look only at objects, I will never find a subject); and then some passing reference to Jacques Lacan (who himself committed the Single Boundary fallacy and is no friend of Emptiness; see note 3 for chap. 6 and note 32 for chap. 13). And recently attempts have been made to connect "no-self" with poststructuralism (ditto Hume) and "selfless" cognitive science (about which, more later).

But the rest of the traditions, both East and West (the vast majority of which do not deny a phenomenal cohesive self) are sealed out of their "integration," which has unfortunately led to a type of Buddhist arrogance: their phenomenal "no-self" doctrine (which indeed is unique to Hinayana) is felt to be the only correct doctrine, and everybody else is profoundly wrong (they go absolutely apoplectic if you mention Aurobindo or Ramana Maharshi).

This is why the work of such theorists as Jack Engler is so important. Engler, trained in both Buddhist *vipassana* (mindfulness) meditation and Western psychotherapy (and who works as a practicing therapist), made a much more useful theoretical bridge: "You have to be somebody before you can be nobody." That is, it is necessary to form a stable, cohesive self before one can transcend (or deconstruct) that self in pure Emptiness. Condemning the ego for not being Emptiness is like condemning an acorn for not being an oak—and, as we have seen, it is profoundly inadequate both phenomenally and numenally. (Engler's approach was consonant with my own work on the pre/trans fallacy—pre-ego to ego to trans-ego, and one has to form a stable ego before one can stably go trans-ego—and with Daniel Brown's work on the higher stages of Buddhist meditation, and thus we became the three main coauthors of *Transformations of Consciousness*; Engler's quote is on p. 49.)

But Engler's analysis contains another very important insight. Based on his clinical experience, he began to suspect that some individuals who were lacking a fully cohesive self were in fact drawn to this notion of phenomenal "no-self" because it seemed to speak to their condition, and certainly seemed to rationalize it. Since many researchers feel that a general borderline malaise is actually increasing in America (and not simply being reported more often [Masterson]), the appeal of this "no-self" notion is perhaps understandable. The Hinayana "no-self" speaks especially to individuals with an already precarious sense of self-cohesiveness.

Further, given the fact that the pure and exclusive "no-self" doctrine of the Abhidharma is part and parcel of an exclusively Ascending and Gnostic path, culminating in complete withdrawal from (and condemnation of) samsara and the manifest realm, this would directly appeal to those who were already having trouble befriending samsara in the first place. Instead of embracing the entire manifest realm with love and Agape, one simply opts out entirely (with Eros gone Phobos). As one Theravadin teacher put it, when asked about the suffering of others, "I don't care, I'm getting off."

Engler and I are not suggesting that any theorist who subscribes to the literal "no-self" doctrine is personally borderline. We are suggesting that, in addition to its own insuperable difficulties on both phenomenal and noumenal planes, this literal no-self doctrine is all too easily confused with a *borderline worldview*; and, indeed, this does especially appeal to individuals who are already having difficulties forming a cohesive self.

This phenomenal "no-self" doctrine: what difficulties it has caused in the attempt to integrate East and West. Perhaps it is time to throw our synthesizing net a little wider, a little more generously. If there is indeed Buddha-nature, it would not leave itself without a witness in other cultures, other places, other times: and in the ears of sensitive men and women, it seems to have whispered the equivalent of "Mahatman" much more often than "anatman."

Francisco Varela's Enactive Paraligm

Finally, to return to *The embodied mind*, by Varela, Thompson, and Rosch (see previous comments: notes 49 and 52 for chap. 2 and notes 13 and 43 for chap. 4). As I indicated, in my opinion I find their overall approach enormously rewarding and significant. Not only is their *enactive paradigm* one of a very few attempts to integrate the Left- and Right-Hand paths in cognitive science, it is the only major paradigm (using the word in its technically correct sense as injunction) that takes a meditative/mindfulness approach to the Upper-Left dimensions in an attempt to integrate them with Right-Hand cognitive science, an approach that alone can embody a scientific (i.e., repeatable, communal, injunctive) disclosure of the domains of awareness.

My major criticisms, all within this prior admiration, relate to several topics mentioned above:

(1) Varela et al. attempt to build most of their bridges between direct, lived experi-

ence (Left Hand) and objectivistic cognitive science (Right Hand) using the Abhi-dharma doctrine of "selfless minds" and the theory of discrete, selfless states and aggregates. But this Hinayana approach, as I tried to indicate, denies the cohesive self on precisely the plane where it is real and pragmatically mandatory. The authors are constantly emphasizing that, for most people, the cohesive self is unavoidable, habit-ual, inescapable; it is only with mindfulness training that we can actually interrupt the formation of a self out of the momentary states: presumably a two-year-old child cannot perform mindfulness training, and thus the self-sense is inevitable in human development. But an *inevitable* illusion is not *merely* an illusion: it is serving *some* purpose (it is, in fact, preventing borderline fragmentation, among many other things).

This type of atomistic analysis (à la Abhidharma) thus fails to account for the extremely important developmental period where a cohesive self is crucial. And thus:

(2) It tends to play into the borderline worldview, as I described above, and thus throws us into the retro-Romantic notion that at some point in development a horri-ble mistake occurred, and we have *undo the mistake* by digging backward, instead of *evolving forward* to higher integrations that *overcome a partialness.*

Each stage of self-development is indeed a type of deconstruction of the previous self-sense, not by plopping straight into pure Emptiness, but by growing a deeper and wider self that undoes the lesser self, until all selves are undone in Emptiness (which is neither self nor no-self). The mind stream, after all, *does not deconstruct itself*: it is deconstructed by a finer employment of attention and will, which, by any other name, means a self strong enough to finally die in pure Emptiness, not a self weak enough to regress to fragmented "no-self" discrete atomistic experiences.

(3) Modern cognitive science, as we have seen (note 43 for chap. 4), postulates something like a computational mind (or even a society of mind) that in itself is *com-pletely lacking consciousness.* That is, cognitive science believes in the existence of *totally* unconscious intentional computations (or whatnot) that perform all the requi-site functions of cognition, and your presence is not required. What is *really* happen-ing in your "mind" is a series of objective functions that possess no consciousness at all; consciousness remains either a puzzling epiphenomena, or is pronounced to be "not good for anything."

The authors refer to this as the discovery of the "selfless mind," and they therefore wish to integrate this scientific "discovery" with the Abhidharma "no-self" analysis of phenomenal experience, thus building a bridge between experience and theory. The "selfless minds" of cognitive science are supposed to be the correlate of the "selfless minds" of Abhidharma analysis. The practice of *mindfulness* (which discloses "no-self") can thus be brought together with the theoretical discovery of "selfless minds," they maintain.

But cognitive science has not discovered selfless minds; it has discovered mindless minds. There is in cognitive science no more room for mindfulness/awareness factors than there is room for a conscious self: *all* of those are banned from discussion, and banned from any sort of real existence, by the very mode of investigation itself. As I suggested earlier, far from being a major discovery, the "mindless minds" notion is a *fait accompli* of the objectivistic, representational paradigm that guides the theoretical, objectivist notion in the first place. Varela et al. are building a bridge between an *inadequate* (if not downright wrong) skandhas theory and the *reductionistic* portion of cognitive science.

(4) Because of their Abhidharma no-self atomistic analysis of experience, the authors likewise are constantly "juxtaposing" this with Emptiness *and* with the "selfless minds" of cognitive science. Both of these reductionistic approaches (Abhidharma and cognitive science) are supposed to be "evocative" of Emptiness. But, as I tried to argue, to bring Emptiness to bear on a particular view—as a crowbar to dislodge a particular view (instead of all views)—is to reduce Emptiness to one notion among others. It comes perilously close to "incurable view" Emptiness.

Thus, with reference to the above four points, it's hard to avoid the impression that the authors are taking the inadequate (or even wrong) aspects of all four and trying to *integrate those inadequacies*: a truncated/atomistic Hinayana psychology with the reductionistic aspects of cognitive science, flirting with an incurable view of Emptiness, and tending toward an embrace of the borderline worldview.

The authors are very much aware of some of these difficulties, and after working hard to establish these points, they begin to carefully qualify or even gingerly withdraw from them. Of cognitive science's "discovery" of "no-self" (which is really just a reductionism inherent in the approach), they finally state: "In cognitive science and in experimental psychology, the fragmentation of the self occurs because the field is trying to be scientifically objective. Precisely because the self is taken as an object, like any other external object in the world, as an object of scientific scrutiny—precisely for that reason—it disappears from view" (p. 230). Exactly. In *this* regard, cognitive science has not discovered an important *fact* of existence that needs to be integrated (as they earlier maintained), it has rather *distorted* the subjective domain in a way that needs to be abandoned (just as the Abhidharma distortion needs to be).

Likewise, of Hume's self-vanishing act, which they had earlier called "brilliant," they now point out, with reference to objectifying: "Nowhere is sleight of hand between inner and outer [subjective and objective] more evident than in the work of David Hume" (p. 230).

Of Emptiness and its "juxtaposition" with "no-self" analysis and the "selfless minds" of cognitive science, which they previously suggested, the authors now state: "This is a crucially important point. There is a powerful reason why some Madhyamika schools only refute the arguments of others and refuse to make assertions. Any conceptual position can become a ground, which vitiates the force of the Madhyamika (this is also why pragmatism is not the same as the middle way of Madhyamika). We would be doing a great disservice to everyone concerned—mindfulness/awareness practitioners, scientists, scholars, and any other interested persons—were we to lead anyone to believe that making assertions about enactive cognitive science was the same thing as allowing one's mind to be experientially processed by the Madhyamika" (p. 228).

And as for the Abhidharma theory of Elements (skandhas, states, dharmas), the authors are certainly aware of its fate in subsequent Buddhist schools: "The basic element analysis received certain kinds of devastating criticism from philosophers such as Nagarjuna" (p. 117). And, of course, Zen's approach was characteristically right to the point: "They were traditionally burned in Zen" (p. 121).

The authors then state that what they are actually trying to do is use these analyses (selfless minds, no-self phenomenal stream, enaction, etc.) simply to help *evoke* the true groundlessness of pure Emptiness. "Just as the Madhyamika dialectic, a provisional and conventional activity of the relative world, points beyond itself, so we might hope that our concept of enaction could, at least for some cognitive scientists and perhaps even for the more general milieu of scientific thought, point beyond itself to a truer understanding of groundlessness" (p. 228).

Personally, I find the enactive paradigm itself very appealing and a genuine advancement of knowledge, but not for that reason, because the "groundlessness" that they point to is, as I suggested, often atomistic, reductionistic, borderline, fragmented. It builds bridges precisely to the aspects of various theories that ought to be rejected, not integrated.

Because of this bias to a "no-self" analysis of the phenomenal mind stream, the authors likewise (and ironically) fail to appreciate aspects of the enactive paradigm that actually go a considerable way to solving some of the problems that their "no-self" analysis prematurely and unfortunately dissolves or disrupts. To mention a few:

The authors, early on in their attempt to build a bridge between the atomistic aspects of Abhidharma and the reductionistic aspects of cognitive science, point to Minsky's notion of the mind as a "society" (the society of mind), the idea being that we can't find a self here either: an *individual* mind is really a *society* of agents. And while there is some truth to that (nobody believes the self is the monolithic monster dominating awareness that many Enlightenment writers seemed to imagine), nonetheless there is a profound and significant difference between individuals and societies (although both are holons within holons within holons): individuals have a locus of self-awareness. That is, a compound individual is a society that further serves as a locus of self-sense. And whether that self-sense is "real" or "illusory" doesn't change the fact that individuals have it, societies don't.

Moreover, this self-sense develops. It develops, as we have seen, through nine (or so) major fulcrums of differentiation/integration (before fully recognizing Emptiness). This Left-Hand developmental process is not simply a matter of *monological* and sensorimotor cognition (although grounded in it). It is also, and especially, a matter of *dialogical* exchange and *intersubjective* mutual understanding (on the way to *translogical* Emptiness). And at that intersubjective and dialogical point, sensorimotor phenomenology begins to fail us miserably (after serving its grounding function).

And this indeed is one of the primary reasons that phenomenology as a discipline so often gave way to various forms of structuralism and neostructuralism (and even poststructuralism): the intersubjective functions of cognition display patterns that are not obvious, and not available, to simple phenomenological apprehension (that is, they follow patterns that are not available to or disclosed by the senses or immediate lived experience). The phenomenologists were simply no match for such items as linguistic intersubjectivity and the patterns that it displayed, patterns that could not be recovered in phenomenology.

Foucault recalls, in listening to a lecture by Merleau-Ponty (the theorist that Varela et al. point to as most resembling their own approach): "So the problem of language appeared and it was clear that phenomenology was no match for structural analysis in accounting for the effects of meaning that could be produced by a structure of the linguistic type. And quite naturally, with the phenomenological spouse finding herself disqualified by her inability to address language, structuralism became the new bride" (Miller, *Passion*, p. 52). This in turn would lead to Piaget, and Habermas, and more adequate accounts of the development of intersubjective competence. (See note 28 for chap. 4 for an extended discussion of the inadequacies of phenomenology.)

I am not suggesting an abandonment of phenomenology (nor unadulterated embrace of genetic structuralism). Rather, phenomenology (especially of the Merleau-Ponty variety) forms a type of foundational edifice quite similar, I believe, to that proposed by Varela et al. But its monological apprehensions need to be supplemented with dialogical recognitions, and this begins to take us away from the too-heavily sensorimotor anchoring in which Varela and his colleagues seem a bit mired.

This sensorimotor-heaviness shows up in the authors' presentation, I believe, as a constant confusion of organism/environment with inner/outer, which in fact are two very different types of interaction and enaction (the former is interaction between Upper Right and Lower Right—or objective organism and objective environment—the latter is interaction between Left and Right—or interior and exterior). They quote Lewontin and Oyama in support of the contention (which I accept) that the organism and the environment co-generate each other. But the argument of both of those authors (and most similar micro/macro arguments) are couched in purely outer (or objective) terms: they are talking about the interaction and mutual codetermination of Upper Right and Lower Right: that is, the mutual determination of the *organism* (its genetic material, its physiological structure, its nervous system, and so on) and its ecological *environment*, which are all *exteriors*.

But that is most definitely not the same as the interaction between the interior lived experience (Left Hand) with the outer material forms (Right Hand), whether those exterior forms are organismic or environmental. To equate the two is, once again, to buy into the flatland (Enlightenment) paradigm, even if we are trying to redress that paradigm by having the Upper Right (organism) and Lower Right (environment) mutually codetermine each other. Important as that insight is, it still buys the fundamental Enlightenment paradigm: it reduces the Left-Hand experience to the Right-Hand description, monological in result (thus Lewontin uses subject and object interchangeably with organism and environment: precisely the problem).

This has the effect, in the authors' presentation, of confusing sensorimotor as organic structure (Upper Right) with sensorimotor as immediate lived experience (lower portions of Upper Left), and this keeps their enactive paradigm heavily grounded in a biologistic bias. Thus, to the question "What is cognition?" the authors answer that it is "Enaction: A history of structural coupling that brings forth a world." Fair enough. But then, to the question "How does it work?" they answer, "Through a network consisting of multiple levels of interconnected, sensorimotor subnetworks." And that doesn't simply ground awareness in sensorimotor networks, it reduces the significant aspects of awareness to its more fundamental foundations: reduces dialogical understanding to monological apprehension (p. 206).

I do not doubt that basic sensorimotor cognition and the early mental categorization process has many of the features outlined by the enactive paradigm. But even so, that covers only the first three fulcrums of a nine-fulcrum awareness.

Thus, the authors point, as example, to the work of Mark Johnson (*The body in the mind*); and, as far as it goes, I am in complete agreement with their assessment. "Humans, [Johnson] argues, have very general cognitive structures called *kinesthetic image schemas*. These schemas originate in bodily experience, can be defined in terms of certain structural elements, have a basic logic, and can be metaphorically projected to give structure to a wide variety of cognitive domains." These are simple sensorimotor/physical schemas referring to such elementary distinctions as inside, outside, boundary, and so on (all anchored in the preconceptual *sensorimotor worldspace*).

"On the basis of a detailed study of these kinds of examples, Johnson argues that image schemas emerge from certain basic forms of sensorimotor activities and interactions and so provide a *preconceptual structure* to our experience"—very similar to Piaget's role of sensorimotor schemas in subsequent cognition. Johnson "argues that since our conceptual understanding is shaped by experience, we also have image-schematic concepts. These concepts have a basic logic, which imparts structure to the [higher] cognitive domains into which they are imaginatively projected. Finally, these projections are not arbitrary but are accomplished through metaphorical and metonymical mapping procedures that are themselves motivated by the structures of bodily experience" (pp. 175–76).

All of which I heartily endorse, as far as it goes. As I would put it, precisely because the mind transcends but includes the sensorimotor body, the foundational structures of the mind (and higher conceptual/formal thought) do indeed rest upon preconceptual bodily and sensorimotor foundations (the "include" part). But those foundations do *not* fully explain or account for the functions, patterns or capacities of the higher structures themselves (the "transcend" part). Johnson's book is thus quite aptly named: the mind is not in the body, the *body is in the mind*. And for just that reason, significant aspects of mental cognition and awareness are not simply, or even especially, sensorimotor networks.

(Likewise, spirit is not in mind, mind is in spirit; and thus mental interpretations color, but do not create, spiritual realities).

The authors' sensorimotor/biologistic tilt further tempts us to overlook the equally important types of *structural coupling* that are occurring in the *Left-Hand dimensions themselves*. That is, not only is the sensorimotor organism interacting with its environment (the Upper Right interacting with the Lower Right, all of which can be investigated and described monologically), but also the inner subject is interacting with its *intersubjective* pool of *mutual understandings* (i.e., the Upper Left is interacting with the Lower Left), and this mutual intersubjectivity cannot be *described* or *investigated* monologically: it is not simply the interaction or enaction of the organism with its environment: it is the enactment of the understanding of one person by another, with *mutually shared* experience the coinage and *mutual understanding* the goal. All of those have empirical and sensorimotor correlates; none of them reside in those correlates.

And this is where Varela's notion of *structural coupling* gains an added usefulness, even if not fully acknowledged by the author: it is not just the structural coupling of the objective organism and its enacted environment, but also of the subjective experience of that organism with the intersubjective experience of its enacted cultural worldspace (where, again, it creates and is created by its forms of intersubjectivity, and not just its forms of interobjectivity).

Thus, where Varela et al. attack the representational paradigm for its insistence that there is a pregiven world that the pregiven organism recovers in perception (I agree with that attack), nonetheless the alternative they propose is still monological: the monological (sensorimotor-specified) organism *enacts* a monological world (the perception of color being their favorite example). All of which is true enough, and all of which is foundational (or the sensorimotor starting point), but all of which totally overlooks the further nonreducible developments in the intersubjective sphere, which cannot be recovered in an enactive monological paradigm, but must include an enactive dialogical paradigm as well.

Thus, the authors (correctly, I believe) switch from a monological pregiven world to a monological enactive world (which is correct as far as it goes); but fail to move on to the crucial components of a dialogically enacted cultural world, which does *not* consist in the enactment of monological *perception* of *surfaces* (a patch of color), but the intersubjective *interpretation* of *depths* not given merely by the surfaces. The authors thus escape the crude mirror of nature paradigm (monological *and* pregiven), but only by attacking the pregiven part, not the monological part: the paradigm shifts from the monological *mirror* of a sensorimotor world to the monological *enactment* of a sensorimotor world. They grasp correctly that the subject enacts an object, but they fail to address the fact that, further, subjects mutually enact subjects via intersubjectivity, which is not the perception of a surface patch but the interpretation of dialogical depth (itself contingent upon the developmental history of structural couplings in the cultural worldspace, and not just structural couplings in the natural/environmental/sensorimotor worldspace).

And *further*, it is only *through* the dialogical enaction that one *gets to* translogical

Emptiness, and not through a regressive dissolution of dialogical intersubjectivity into atomistic monological states and reductionistic mindless cognitive mechanism, the path the authors often stray into.

(A more intensely nonreductionistic approach would likewise allow a more forceful recognition of telos/emergence in development itself. Like all drift, natural drift occurs only where there is some sort of current; in this case, the current of the Kosmos, or Eros. Enaction of the presently given worldspace is not just based on *past* structural coupling. Rather, based also on the emergent pressure of the future, the present world is enacted.)

That said, let me finish by repeating the strengths of the enactive paradigm, which first of all actually *is* a real paradigm, because it embodies a specific program of injunctive research (namely, the open-ended investigation of direct lived experience with a view to its integration with theoretical models). We might see it this way:

In Vajrayana (which the authors endorse), it is typical to embrace all three Buddhist vehicles in something of a developmental arc. One begins with Hinayana mindfulness, which is the monological investigation of immediate awareness. One then moves to Mahayana, and the specific approach of exchanging self and other (*tonglen*), that is, investigating the *dialogical* and intersubjective circle of awareness. One then moves to Vajrayana, or the final dissolution of exclusively monological apprehension and dialogical exchange, and rests in the nondual luminous Emptiness that playfully manifests all worlds (translogical).

Given the very crude state of modern cognitive science, the authors, in a sense, had no choice but to give a Hinayana-level account, and in many ways I think this can serve as a beginning foundation for the yet-higher developments of cognitive understanding, to which, no doubt, the authors of this volume will contribute in no small measure. (For a critique of phenomenology, see note 17 below.)

2. Nowhere are the inadequacies of taking monological sciences and trying to make them into a complete "new paradigm" more obvious than in the "new physics and mysticism" writers and theorists, whose names are legend. When reductionists have a spiritual experience (not generally dealt with in physics textbooks), it usually acts as an incitement to commit philosophy, and the result is not for the faint-hearted.

However wonderfully well intentioned, most of these theories—which play on the theme that the "new physics" (quantum and relativistic) supports/suggests/proves a mystically unified worldview—are crippled by trying to simply *extend* a flatland monological paradigm into dialogical and translogical domains (more of the flatland "bigger portions of bad food" approach).

They generally take certain mathematical formalisms (especially the Schroedinger wave equation and its collapse upon measurement) and give them a very wide interpretation (despite the fact that physicists themselves are sharply divided over how to interpret the formalisms), and they then wed this very loose and generous interpretation with their often equally loose interpretation of mystical spirituality, and the result is something like, the new physics supports or even proves a mystical worldview.

This admixture is promptly called "the new paradigm." Danah Zohar: "The idea of a 'quantum society' stems from a conviction that a whole new paradigm is emerging from our description of quantum reality and that this paradigm can be extended to change radically our perception of ourselves and the social world we want to live in. A wider appreciation of quantum reality can give us the conceptual foundations we need to bring about a positive revolution in society" (*The quantum society*).

From formalisms describing the lowest, shallowest, least conscious, least-depthed holons in existence, "*extended* to a paradigm" that is supposed to cover dialogical, intersubjective, cultural exchange based on mutual understanding and mutual recognition: this is more than a quantum leap, it is a leap of faith. Quantum formalisms cannot even account for the fundamentals of biology and autopoiesis, let alone economics, psychology, literature, poetry, morals, and ethics, to name a vital few. But physicalists are so used to thinking that "most fundamental" means "most significant" that they believe all higher branches of knowledge must be grounded in least-depthed holons or not be grounded at all. Thus the constant tendency to "extend physics" directly to any and all domains.

Thus Fred Alan Wolf: "I believe that quantum physics, the most powerful and rigorous science devised to date, will provide the basis for the formation of a new psychology—a *true humanistic psychology*."

This is simply painful, and it is a notion that is by no means confined to Dr. Wolf. A true humanistic psychology is built upon processes of intersubjective understanding and mutual recognition, about which quantum physics has not a single thing to say. Surely we do not want to base our new humanism on a power-driven monologue with rocks.

This extension of the monological hegemony and aggressive, even violent, reductionism suffers at both ends of the reduction (and everywhere in between). Not only is reality automatically assumed to revolve *most significantly* around those holons that actually possess the *least* significance, but mysticism itself—there are at least four quite different varieties of mysticism, as we have seen—is homogenized into some version of the unified field or dynamic web or quantum vacuum (or some such creative reading of the mathematical formalisms), and the two homogenized messes ("quantum" and "mystical") are crudely pressed together and presented as covering all the bases.

Besides distorting both ends of the spectrum of existence (*physis* and *theos*), it guts everything in between. Fred Alan Wolf's drug-induced mystical experience led him to this realization: "I was on this quest [using the psychedelic vine ayahuasca] trying to understand shamanism from the point of view of physics. What I discovered is that it can't be understood from the point of physics alone." Only a dedicated physical reductionistic could make that statement—forget shamanism: not even biology can be "understood from the point of physics alone," nor can art or economics or psychology or sociology. . . . That physics doesn't cover shamanism should be the least of his worries, but it perplexes Dr. Wolf no limit that physics doesn't apply here. And so he decides that "We need a new breed of physicists, who study living processes rather than the qualities of matter." We have them, and they're called biologists.

As we have seen repeatedly throughout this volume, the *shallower* one goes in search of significance, the more *egocentric* the belief system, because egocentrism is always the correspondingly shallowest point in the human holon. Lacking any real depth/height, one finds only *one's own interior* surveying all of reality, and all of a sudden, everything starts relating directly to *you*, and the word magic of egocentric association starts to actually look profound, and profoundly desirable.

"When my intent is clear," explains Wolf, "pathways appear as if by magic, taking me from one place to the next. Certain connections get made; key phone calls come in." Perhaps that is some actual psychic connection; a therapist might suggest it is narcissism and delusions of reference (is the world really that accommodating to your ego?). Who can tell? For Wolf, it is all based on the collapse of the quantum wave packet, so that, as his interviewer summarizes his position, "If you intend an ashtray, then the cloud of possibilities qwiffs it into being."

Fred puts this qwiffing to good use. In Lima, Peru, before his ayahuasca experience, he is shown a movie, *The Winds of Ayahuasca*, with what most people would probably find to be a rather typical and unimaginative plot of an American academic trying the drug, getting his paradigm blown, and, of course, falling in love with a beautiful native girl at the same time (take my paradigm, please). Sure enough, Fred takes the drug, has his mind blown, and bags a Peruvian babe, all in a no doubt memorable month.

"It was as if, astonishingly, inexplicably, the film had been written as a kind of life-script training entirely for him." Wolf reports breathlessly: "That was the realm of myth grabbing me by the collar and saying, Wake up, Fred! We're going to show you why you came down here to Peru." Many people would have settled for drugs and sex, but apparently there's more. "Here's this movie," he continues, "made years before you arrive in Lima, yet in it you see your life laid out before you. *Was the whole movie made for me?*"

He quickly decides in the affirmative, and so then must ask, "If so, who was scripting it?" That is, who was scripting this movie made just for him, and long before he arrived? He decides that "the Australian Aborigines came up with the best answer. The Great Spirit, they tell us, dreams us and all of material reality into existence." Let's assume that is indeed the case; but the Great Spirit then scripts everything that is happening, and no movie, and no event, was scripted wholly or especially for Fred Alan Wolf. That *particular* movie, in fact, was scripted by a bunch of not very imaginative filmmakers, and I would worry mightily, not celebrate, if it "captured my life perfectly."

Perhaps it really was genuinely psychic, and perhaps it was narcissistic reference. But my point is that in either case, the monological maps will tilt our interpretations, no matter how genuine, in the direction of preconventional egoism, because there is nothing in the maps to pull one up to dialogical endeavors, and thus no way to escape genuinely into the translogical. One is left merely with: I have a new paradigm, and it is your job to change your beliefs to accord with mine. There is no other interior development either recommended or mentioned or even hinted at (the embrace of the new monological paradigm is simply supposed to rather automatically usher one into homogenized mystical mush; drugs, presumably, are optional).

As I said, always the monological views play into the hands of divine egoism and its power drives, which is a point that we have seen emphasized by theorists from Foucault to Habermas. Without a paradigm of mutual dialogical recognition and care, there is no way to pull anyone out of divine egoism and into worldcentric compassion, and from there into the Over-Soul that is the World Soul, on the way to the mystery of the Deep altogether, and so instead we watch a life scripted by divine egos, for divine egos, about divine egos, and this is meant to be the basis of a glorious new paradigm. And this, it seems to me, is exactly the essence of so many "new physics and mysticism" notions—purely monological to the core.

No doubt. Dr. Wolf is attempting to stretch the knowledge quest by moving outside the realm of physics, and for this attempt we can all applaud. But in my opinion, if we are not much more careful in how we interpret this expanded knowledge quest—if we do not draw on all four quadrants equally, and if instead we simply orient ourselves around basically the lowest holons in the Upper-Right quadrant—we are constantly in danger of tilting our interpretations in prerational, not transrational, directions.

(Richard Leviton, "Through the Shaman's Doorway: Dreaming the Universe with Fred Alan Wolf," *Yoga Journal*, July-August 1992.)

3. This lack of cultural coherency in America has accelerated in the last few decades or so, with the result that virtually every group or class or type of citizen in the country is attempting to become its own nation, its own special group certified by special legal rights. The easiest and fastest way to secure special rights is to compete for the coveted status of victim, for this allows the victim group, in effect, to claim war reparations. The manufacture of rights has gone into overdrive, with the various classes lining up in court to outclaim each other as to who has been the most victimized, and thus who deserves the most war reparations. This has resulted in a series of massive and bizarre IOUs, with everybody trying to extract payment from everybody else, and nobody willing to pay.

At the beginning of this free-for-all for rights, the one great pool of war reparations was the white male, a pool from which everybody eagerly dipped. But this apparently unlimited fund of reparations has been totally splintered and divided up (and dried up), because white males now include males that themselves are vying for special rights as victims: drug abusers, emotionally disturbed, obese, short, hearing-impaired, handicapped, satanically ritually abused. There is no single and simple fund of bad guys left. Everybody, it seems, has victimized everybody else, and thus everybody is demanding special rights to protect them from everybody else.

There are no rights without responsibilities, no agency without communion. It is physically, theoretically, and existentially impossible to have one without the other. But in the splintering culture of America, each group wants agency (rights) with no communion (responsibility). I happen to be in favor of many of these rights; but unless this country finds a more common union and overriding principle of coherency—unless these rights are plugged into correlative responsibilities and communions—then the manufacture of rights alone simply *contributes* to the further *fragmentation* of the country, which alone can secure and protect the new rights in the first place. The IOUs are coming due. . . .

4. We can now finish our account with deconstruction, and set it in the context of Nagarjuna's Emptiness (a topic I will return to in volume 3 in much greater detail; the following may be taken as a preliminary summary).

We have seen that deconstruction is based on the twin principles, as Culler put it, that meaning is context-dependent and contexts are boundless—precisely another approach to the IOU principle (the contexts are whole but are never completed, and if they are, they self-contradict: complete or coherent, never both). A literary text, to the extent it attempts to be a "whole work," a "work of art," "a self-contained whole," will therefore issue a hidden IOU, an implicit IOU—hidden or implicit because language tends to take its own nature as given and unproblematic and thus unexamined (which we'll return to in a moment).

But first, to give a counterexample, mathematics makes its own transfinite IOUs *explicit*—the sets must expand indefinitely or mathematics self-destructs—and mathematicians do this *explicitly* because, so to speak, their job is to scrutinize mathematical language in its very syntax, its very structure, and so eventually they were forced to make explicit these IOUs. Mathematics had not always done this; in fact, the IOUs were very much implicit, hidden, unconscious until the Tarski/Gödel revolution, which, so to speak, psychoanalyzed math and dug up, from the depths of its foundations, the dirty little secret: mathematics is not, and never can be, coherent and complete.

When Tarski and Gödel wrenched this secret from the depths of mathematics, it first caused an intense commotion, almost panic, because the very foundations of mathematics—long taken to be the one, true, self-sufficient, self-complete, and self-certain form of human knowledge—was now shown to be inherently wobbling between self-contradiction, on the one hand, and incompleteness (un-self-certainty) on the other. But nowadays it's all taken rather in stride: let the sets expand forever, and let us get on about our business as usual. As I would put it, *all* holons issue an IOU to the Kosmos, and so what else is new?

What else was new was that essentially the same drama began to unfold in the study of everyday language itself, and literary language in particular. Language speakers and even language theorists had long been operating on hidden IOUs, and on the naive assumption that language is simply a transparent window through which one merely gazes undistortedly at a pregiven world (this was, for example, Foucault's point about the Classical Age: language was unproblematic). With the "linguistic turn" (dating roughly to Saussure), where language began to look at itself, the thick tangle and convoluted web of systematically generated meaning became a massive problem: how in fact did language represent a world, when inherently any word can stand for any thing? The relations *between signs themselves*, and not so much the simpler relation between signs and things, became the focus of intense investigation.

We saw this turn take two broad directions, following either the Left-Hand path of hermeneutics (meaning is generated by background and often nondiscursive social practices, which can only be grasped from *within* the context itself) or the Right-Hand path of structuralism (meaning is generated by structures or supra-individual patterns of rule-governed linguistic behavior that can only be grasped from the outside in an exterior mode of investigation). Hermeneutics focused on the *signifieds* (LH), which could only be grasped from within by empathic participation, and structuralism focused on the *signifiers* (RH), which can best be approached in a distancing stance of exterior study.

This Right-Hand structuralism soon gave way to the neo- and poststructuralists (Foucault to Barthes to Lacan to Derrida), which emerged with the insight that the signifiers themselves were largely "free-floating," anchored not in any pregiven empirical world, but rather anchored all too often in politics, power, or prejudice. That is, meaning was not simply lying around on the surface ready to be unambiguously perceived like a patch of color (as in the inadequate empirical theories), but rather was a complex self-referential system that was profoundly context-dependent, and . . . contexts are boundless.

And *that* gave the poststructural deconstructionists the tool to aggressively (often wildly) deconstruct meanings that were previously felt to be engraved in concrete and set in the unvarying ways that "things always are": now meaning, and the world of meaning, was disclosed not as a perception, but as a construction, and to the extent this construction was mistaken for eternal verities, it could be textually deconstructed just to point out how fleeting its "given truths" were, how context-dependent they were, and how contexts shifted endlessly, and how in principle they could never be mastered. . . .

The deconstructionists (particularly starting at Yale, home of a trio of professors— Derrida himself, the brilliant Paul de Man, and J. Miller—who were dubbed "the boa-deconstructors") began by examining *literary texts* (Derrida's examination of Rousseau's double use of the term *nature* still being a classic). As I earlier indicated, a literary text, to the extent it attempts to be a "whole work," a "work of art," "a self-contained whole," will therefore issue one or more hidden IOUs, implicit IOUs— hidden or implicit because language tends to take its own nature as given and transparent. Although the deconstructionists did not at first articulate the principles of their own procedure (they did not articulate the IOU principle, even though they were using it), they nonetheless wielded it with brutal efficiency. That is:

A literary text, like all holons, issues some sort of IOU: it will contradict itself at certain crucial points; it will use certain key terms but give them opposite meanings;

it will assume truths that it cannot itself generate; it will double back on itself self-reflexively and thus erase its own message; it will create a hierarchy of meaning only to have the text itself upset the hierarchy; it will create binary oppositions that pretend to be separate but actually undo each other; it will marginalize some notion as being not central yet define the central notion in terms of the marginal—the list is virtually endless, and the deconstructionists zeroed in on just these IOUs, just these hidden contradictions and inconsistencies and self-betrayals that the text itself would *inherently* issue. And that was exactly the agenda of the deconstructionists: find the points in the text where the text refers to itself, eventually contradicts itself, and thus self-deconstructs (just as mathematics does when it tries to become complete and coherent).

And just as the initial discovery of the IOU principle in mathematics caused an uproar verging on panic, so the deconstructionists were the terrorists of the linguistic world, sending panic echoing down the halls of establishment academia. *Every* text, without exception, could be deconstructed (because every holon issues an IOU; in the literary text, simply look for the meaning the text is trying to convey, show that that meaning is dependent upon a context beyond the text and thus totally out of control of the text itself, and thus the text becomes "undone," becomes not a self-contained and coherent artistic statement, but merely the unconscious mouthpiece of contexts beyond its control, contexts that swamp its own proclaimed message and set it afloat on a sea of signifiers of which it is only dimly aware: far from carrying an artistic and noble impact, it is a duped lackey of contexts that use and abuse it, contexts it is apparently too ignorant to grasp).

The deconstructionists, having (implicitly) stumbled on the IOU principle in literary holons, eventually let this truth spin all out of control, and in ways simply not warranted by that truth itself—ways that therefore turned finally on the terrorists themselves and deconstructed the deconstructionists. We have seen repeatedly throughout this book that sliding contexts do not mean the contexts are *merely* arbitrary and therefore imposed *merely* by power or prejudice. Contexts slide, but they often slide in relatively stable ways; meaning is context-bound, but many contexts are common or shared (and can be so disclosed in mutual understanding); contexts are boundless, but that doesn't prevent them from being fixed in relative but stable ways; that sociocultural reality is largely constructed does not mean it is constructed arbitrarily, for linguistic meaning is itself exposed to extralinguistic restraints (in both I, we, and it domains), and that allows *meaning* to be tied to *validity claims* that can be redeemed in the intersubjective circle, validity claims of truth, truthfulness, cultural justness, and functional fit. As I summarized it in chapter 5, the deconstructionists (and extreme relativists) went from saying that no perspective is finally privileged to saying that all perspectives are simply equal—at which point this stance turns on itself, undercuts and deconstructs its own claim to validity, and collapses under its own weight.

Just as the explicit disclosure of the IOU principle in mathematics moved it from closed and self-sufficient truths to stable, open, relative truths (with a period of panic), so the literary world began to slowly make a similar transition (also through much initial panic). But it soon became obvious that the end result of deconstructionist literary critique was *always the same*: the text blows up. Deconstruction became a generic, all-purpose solvent; it told us something about all literary holons and about language in general (all texts issue an IOU), but about specific texts it had precisely nothing to say. As literary criticism, it was virtually worthless (even Foucault referred to Derrida as a literary terrorist).

But here was a "literary criticism" made to order for the tenured radicals of the sixties: haven't the wits to build a building? No problem, just blow one up instead. [See Wilber, *Boomeritis* (forthcoming)]. Thousands of Ph.D. dissertations in deconstructionist themes were issued by American universities, with individuals like Stanley Fish leading the way. By 1979, Derrida was the most frequently cited authority in papers submitted to the journal of the Modern Language Association.

Deconstruction as a movement never caught on in Germany or France or England (or anywhere else, for that matter), and thus, at the height of the tear-down frenzy, Jacques Derrida exclaimed, "America is deconstruction!" It never caught on elsewhere because, apart from its undeniable contribution vis-à-vis the IOU principle in literature, it had nothing else of substance to say: it could only deconstruct what somebody else had the creativity to construct: it was basically parasitic. And it ended, as has often been pointed out, in pure nihilism, at which point it dissolved its own stance, and certainly its own credibility.

The end of any serious deconstructionist movement came with the Paul de Man debacle. On the morning of December 1, 1987, the *New York Times* reported that a young Belgian scholar named Ortwin de Graef had recently discovered incontrovertible evidence that in the 1940s Paul de Man, the most gifted and brilliant of the American deconstructionists, had been a Nazi sympathizer (writing articles and reviews for Belgian newspapers supporting the Nazi cause—something he had gone to great pains to hide) and had authored such articles as "The Jews in Contemporary Literature," where he had stated, among other things, that "a solution to the Jewish question . . . would not involve deplorable consequences for the West. It would lose, all told, a few personalties of mediocre value. . . ."

The deconstructionists immediately leapt to de Man's defense with all the conceptual tools at their slippery disposal, and furiously attempted to demonstrate that inherently sliding contexts meant that de Man's pro-Nazi actions were actually the *opposite* of a Nazi sympathizer, that his helping the Nazis was really not helping the Nazis, that his anti-Jewish stance was really pro-Jewish, that the criminal was really the victim, and other variations on the theme that "yes" really means "no" because contexts are boundless. . . . (Curiously, the deconstructionists did not likewise think that "less salary" was the same as "more salary," or that "no tenure" was really "tenure." Amazing how language straightens right up when you're sincere.)

Put bluntly, the deconstructionist defense came down to: since there is no way to anchor truth and right, there can be no false and wrong either, so Paul de Man's crime was really its binary opposite, innocence—and another nasty value hierarchy had once again been deconstructed and inverted, an inversion that converted a vicious act into a merely linguistic predicament, and excused an ontological demon as being merely an embarrassing literary faux pas, about as grievous as, say, a dangling participle.

Aperspectival madness—the inevitable endgame of confusing sliding contexts with merely arbitrary contexts, and the inevitable endgame of disconnecting meaning from validity claims, claims that seize contexts and hold them still long enough to establish clearly enough just who did, and who did not, help the Nazis.

As I indicate in the narrative, the best way to understand the real import of the partial truths that the deconstructionists stumbled on is to set them in the much deeper context of Nagarjuna's more radical (and, paradoxically, much more sane) deconstruction, where relative truths are conceded a relative reality (mathematics works just fine in the finite realm, as long as it admits its IOUs and does not pretend to the Absolute); but, demonstrates Nagarjuna, if any relative truth is pushed to the absolute, it contradicts and *deconstructs itself* (which is what happened to deconstruction when it thought it had become absolute). See note 1 above.

For Nagarjuna, the deconstruction of relative truths leaves not nihilism but Emptiness: it clears away the conceptual rubble in the mind's eye and thus allows the space of nondual intuition to disclose itself, and thus it follows to the limit the whole point of the IOU game: if you don't want to be a complete self-contradiction, then you must come to rest in infinite Emptiness, which alone redeems all IOUs, and which alone sets the soul free on the ocean of infinite Mystery.

5. Taylor, *Hegel*, p. 39. My italics.

6. Ibid., p. 44. My italics. I have changed "men" to "men and women."

7. Quoted in Weiss, *Hegel*, p. 15. As I put it earlier (in chap. 8), higher development is important, not because it actually moves closer to Ground, but because it makes it harder to deny the ever-present Ground.

8. As for their use of vision-logic: Kant had carefully differentiated the Big Three (theory, morals, judgment), expressed theoretically in empiric-analytic inquiry and pragmatically in categorical morality ("act in such a way that the law governing your behavior might be universal"—i.e., postconventional), and however much he expressed the hope that aesthetic judgment might bridge the gap between theory and practice, his own efforts at such, although very suggestive to a whole generation, were not altogether satisfactory.

The first thing Fichte did was attempt to break the hold of formal rationality by demolishing Kant's "thing-in-itself." If, as Kant had said, the thing-in-itself can never in any way be known, then, said Fichte, it doesn't exist. In its place he substituted the power of the productive imagination of the infinite and supra-individual Self, which in essence meant: it is not that there is some forever unknowable thing-in-itself, quite different from consciousness, that impinges on consciousness and "causes" perception; it is rather that there is only one dynamic Life process that knows itself in various degrees and from various angles. The world-knows-itself in various ways, and in order to do so it first cuts itself up, so to speak, into finite subject and finite object—but both subject and object issue from the same ground, so their apparent incompatibility is never ultimate, and the ground can be recovered in pure nondual perception.

But the ground of which Fichte speaks is almost always the Over-Soul, the psychic-level Witness (his pure Ego); he speaks constantly of it as "ever-receding," which is very typical of psychic-level Witness intuitions (the causal Witness *never* recedes). He expresses this psychic vision in centauric vision-logic, which shows up most notably in his moral position, which demands "an organization which transcends the individual State, namely, a federation of all States," and which meant a worldwide community of world citizens. Schelling was of similar opinion.

As for *existential meaning*, which, as we have seen, so often accompanies centauric vision-logic: "In these romantic groups there was an ironical transcending, a going beyond, the given forms of social existence. This [paradoxically but typically] results in a feeling of emptiness with respect to the meaning of life. You see now that the central problem of the twentieth century, namely, the question of the meaning of life, came out very strongly in the second period of Romanticism" (Tillich, p. 385).

Hegel's Reason (*Vernunft*)—as opposed to the understanding (*Verstand*) of formal operations—is one variety of pure vision-logic: dialectical, uniting opposites, members "passing into the other" at every point, and one member implying/containing all the others (network-logic). This "interpenetration of opposites" was one of the famous summaries of vision-logic (Engels).

As for their glimpse of higher or transpersonal domains: as I just mentioned, Fichte seems to have had a very strong psychic-level stance; his pure Ego is almost pure psychic-level Witness (or Over-Soul), which Fichte always approaches by the injunction: "Observe the observing self," which of course you can't, but which for precisely that reason leads you to an "ever-receding" intuition of the Seer as that which cannot be seen but through which all that is seen moves. Fichte's method opens one, I believe, into psychic-level Witnessing, but not stably into subtle or causal Witnessing; it was nevertheless a giant step forward from formal operational, and a true blossoming of Western Vedanta.

Hegel's Absolute is very similar to the low-causal saguna Brahman—God's "consequent Nature," the ultimate (but ever-in-process) Holon that embraces all manifest holons (and is, in that special sense, dependent upon them), and his Categories are very similar to the archetypes of the high-subtle ("primordial Nature"), although many traditions, such as the Yogachara of Asanga and Vasubandhu, see the archetypes as habit-memories more than eternal Forms, though all agree that the archetypes are formative processes prior to but not other to any particular manifestation. Hegel had no experience, and no conception, it seems, of the pure causal unmanifest, and his joke on Schelling ("that dark night in which all cows look alike") actually works against him here. His "infinite-in-the-finite" is thus "low-causal in the finite," and not "ultimate-in-the-finite."

This is also shown by reference to two aspects of the high-causal, both of which

Hegel denied, and both of which therefore situate his infinity elsewhere (i.e., in the low- and not high-causal). One, the unmanifest or high-causal is in no way dependent upon, or affected by, any manifestation—manifestation itself does not add to or subtract from it in any way whatsoever (the whole world, said Eckhart, adds nothing to Godhead; this is also Plotinus's view, and Ramana's). Two, we cannot say that creation *necessarily* arises from the causal; neither, however, can we say it does not—those are two alternative views of Spirit, but Spirit is not an alternative (is not dualistic). When the sages speak of the "superabundance of the One" and its "overflowing in and as the Many," that is a poetic concession to the verbal world; it is most definitely not the basis of a principle of sufficient reason. Whenever "spontaneous superabundance" is reduced to the principle of sufficient reason (and we see this in thinkers as diverse as Spinoza, Leibniz, aspects of Saint Thomas, Hegel, etc.), it is obvious that they are trying to "think their way" to or from Spirit, and that will never work.

Turn now to the low-causal (final-God, *saguna* Brahman): as the (ever-receding) Holon of all manifest holons, it participates in the experience of every holon and is, in a sense, the "result," at any stage, of the sum total of all holons' experience at that point in manifestation, although it is not simply the summation (but the organizing process as well). This is, of course, very close to Whitehead's consequent Nature and Hegel's self-actualizing Spirit, in which all is preserved (along with its "primordial nature" of formative processes in the high-subtle, which are prior to but not other to any manifestation).

Schelling's brief but unmistakable experiences of the high-causal show up in this doctrine of the pure undifferentiated Abyss (*Urgrund*)—pure Emptiness—which is indeed a dark night in which all cows do in fact look alike. Like Hegel, however, he seems to have no clear experience or conception of the ultimate stages; his "infinite-in-the-finite" is "high-causal-in-the-finite."

All of these experiences (or intuitions) were largely, as I said, translated downward into terms of vision-logic, and this left no injunctive proof for their existence—the particularly fatal flaw of Idealism. The Idealists undoubtedly glimpsed significant portions of the transpersonal domain, and at least made them a theoretical cornerstone of their systems. But glimpsing is not living—Idealism had no contemplative path to assure the reproducible results of its random intuitions. Further, since vision-logic was the actual level at which they were most stably adapted, they had a tendency to confuse Spirit with vision-logic itself; and, in fact, this is exactly what Hegel does ("The real is rational [Reason = vision-logic or dialectical-network], the rational is real").

In this sense, the Idealists were indeed "Idea-ists" or "Mind-ists," and not "Pneuma-ists" or "Spirit-ists." In Yogachara terms, the Idealists identified Reality with *manas* (or, at best, with tainted *alaya-vijnana*) and did not clearly break through to pure Alaya (or Amala), or causal/nondual Thusness.

This led Hegel to the belief that all the essentials of the manifest world could be *deduced* by dialectical reasoning, and this led to his confusing of logical deduction with historical procession.

It didn't take a Russell or a Moore to fundamentally undo that proposition. Many people were struck by its "forced" or "mechanical" nature. It wasn't so much that Hegel's general conclusions were wrong as that his methodology was entirely out of place for what are, after all, transrational domains. But even within his own system, Hegel could not deliver the "reconciliation" (what we might call "ultimate integration") that the system promised. As I would word it, the dialectic might get you to vision-logic, and point tantalizingly to higher (and even causal) occasions—no mean feat—but it could not actually bring you to those higher levels, and thus it could not finally or in any ultimate fashion integrate the Ascending and Descending currents in the human being (although, we might say, it would be a step in the right direction).

Schelling was aware of this inadequacy and devoted his remaining years to developing existential ("positive") philosophy, which, I would maintain, was altogether appropriate, since the centauric-existential level was and always had been the Idealists' home base. Centauric vision-logic can integrate Ascent and Descent up to and down from that level, but not beyond to their common source and actual "reconciliation." As Schelling's erstwhile friend Schiller would put it, "In reality there will always remain a preponderance of one of these elements [Ascent and Descent] over the other; and the highest point to which experience can attain [for all they knew] consists in an oscillation between these two principles"—between, that is, fleeing the shadows or embracing the shadows, with no discovery of that nondual awareness that unites transcendence and embrace in every act of choiceless awareness.

9. As I will argue in volume 3, the great quest of an integral postmodernity is for centauric vision-logic, or the aperspectival-integral bodymind as a collective center of social gravity—a development that, I will argue, awaited the evolution of the computer and information technology; the industrial base could support only an instrumental rationality.

In short, as I construe it: the quest of modernity was for the Ego in rationality; the quest of postmodernity is for the Centaur in aperspectival vision-logic (and I will relate them to the industrial and informational base in volume 2). [See also *The Marriage of Sense and Soul and Integral Psychology*.]

As indicated in note 5 for chapter 11, we can loosely define modernity as consisting of any or all of the following factors: (1) the differentiation of the Big Three (Kant); (2) the rise of the philosophy of the ego-subject (Habermas); (3) the rise of instrumental rationality (Heidegger) and its base of industrialization; (4) the transparency ("nonproblematic" nature) of language (pre-Saussure); (5) the widespread turn within (Taylor); (6) the belief in univalent progress—all of which I have summarized as "the Ego in rationality." Likewise, postmodernity may then be loosely defined as any attempt to develop beyond (or at any rate to respond to) those factors (focusing particularly on the "centaur in vision-logic").

In many ways, the first factor (the differentiation of the Big Three) remains the crucial item, and the *integration* of the Big Three, now that they have been irrevocably differentiated, remains the crucial problem, in the development from modernity to postmodernity. In a global world increasingly united only by the language of natural science (and similar monological it-languages, whether financial, informational, economic, or ecological), the recalcitrant dilemma is the integration of the increasingly isolated individual (the I) with meaningful forms of community (the we) against the backdrop of a universalist natural science (the it). And in societies where the collective we has managed to evolve beyond tribal magic and beyond mythic-imperialism (i.e., where the Big Three have actually been differentiated), the integration of the individual and the culture is still painfully problematic.

Nowhere is this more obvious (and more exacerbated) than in America, and astute observers of American life—starting most notably with Alexis de Tocqueville—have returned time and again to the difficulties and dilemmas of integrating the newly emerged Ego (and its "individualism") with the larger currents of social life. Tocqueville, in fact, helped give currency to a new word: " 'Individualism' is a word recently coined," he wrote, "to express a new idea. Individualism is a calm and considered feeling which disposes each citizen to isolate himself from the mass of his fellows and withdraw into the circle of family and friends; with this little society formed to his taste, he gladly leaves the greater society to look after itself." There is the disengaged and autonomous Ego of the radical Enlightenment, taken to its social limit, exacerbated in an America of puritan-to-cowboy self-reliance.

Robert Bellah and The Good Society

This "new individualism" has been penetratingly examined by Robert Bellah and his colleagues (Richard Madsen, William Sullivan, Ann Swindler, Steven Tipton), most notably in *Habits of the heart* and *The good society*. In *Habits*, Bellah et al. identify four main cultural currents that mold the American character: biblical, republican, utilitarian, and expressive (represented by, for example, Winthrop, Jefferson, Franklin, and Whitman). While all four strands have a strong emphasis on the individual, the first two (biblical and republican) are forms of what the authors call "social realism," the belief that society is at least as real and as important as the individual, with meaning and reward found in relationships and community; whereas the last two (utilitarian and expressive) are forms of "ontological individualism," the belief that society is a "second-order, derived or artificial construct" and the individual is the primary reality. In a sense, the first two stress the importance of communion, the last two stress individual agency.

Bellah and company would like to see the two strands of social realism (or communion-emphasis) somehow "updated" and "reappropriated" as a counterblast to rampant individualism (and hyperagency), but just how to accomplish this is exactly the problem, and their analysis, although often profound, is not altogether compelling. The difficulty, as I see it, is that *both* of the communion-strands derive from agrarian societies (and their intense mythic-membership modes and plow-patriarchy), and thus these "social glues" are often shot through with sexism, racism, elitism, dominator hierarchies, etc. (all of which Bellah et al. concede).

The last two strands—the highly individualistic strands—derive, of course, from modernity and the Enlightenment: utilitarian individualism (stressing calculating rationality) stems from the utilitarian English Enlightenment (Locke to Bentham), and the expressive strand (stressing expression of feelings and "feel good" ethos) from the Eco-Romantic rebellion. According to Bellah, both of these (utilitarian and expressive) have often ironically fused to produce the extremely agentic, hyperindividualistic, almost "atomistic" individualism of America (which I think is very true).

As my own analysis suggested, this "hyper-Ego" (whether utilitarian rational Ego or expressive divine egoism) went hand in hand with a collapse of the Kosmos into monological, flatland terms *devoid* of any (explicit) *qualitative distinctions*; and this was intimately related with the *lack of integration* of the Big Three once they had been irrevocably differentiated. Bellah: "If popular culture makes a virtue of lacking all qualitative distinctions, and if the intellectual culture, divided as it is, hesitates to say anything about the larger issues of existence, how does our culture hold together at all?" (p. 281).

The answer is, "tenuously": "The culture of separation offers two forms of integration—or should we say pseudo-integration?—that turn out, not surprisingly, to be derived from utilitarian and expressive individualism. One is the dream of personal success. . . . The second is the portrayal of vivid personal feeling. . . . But a strange sort of integration it is, for the world into which we are integrated is defined only by the spasmodic transition between striving and relaxing and is *without qualitative distinctions* of time and space, good and evil, meaning and meaninglessness" (p. 281; my italics).

According to Bellah et al., a world lacking all qualitative distinctions is therefore construed, not according to what is *worth* pursuing, but simply in terms of *what works* (as we put it, "What does it mean?" is reduced to "What does it do?": *efficient means* replace any criteria of worthy *ends*). Bellah: "For most of us, it is easier to think about how to get what we want than to know what exactly we should want" (p. 21).

This is exemplified in the two dominant "figures" of the culture of separation: the manager (utilitarian) and the therapist (expressive). "Like the manager, the therapist takes the functional organization of industrial society for granted, as the unproblematical context of life. The goal of living is to achieve some combination of occupation and 'lifestyle' that is economically possible and psychically tolerable, that 'works.' The therapist, like the manager, takes the ends *as they are given*; the focus is upon the effectiveness of the means" (p. 47; my italics; see note 42 for chap. 4 for "adaptation" as covert acceptance of "unproblematic" context).

"Between them, the manager and therapist largely define the outlines of twentieth-century American culture." They both focus *on the individual*, "presumed able to choose the roles he will play and the commitments he will make, not on the basis of higher truths but according to the criterion of life-effectiveness as the individual judges it" (p. 47).

And particularly as the individual *feels* it. Since all "values" are held to be equivalent (there are no explicit qualitative distinctions), then whatever makes me *feel* good must *be* good, as long as it doesn't overtly harm someone else. "Moral standards," says Bellah, "give way to aesthetic tastes" (p. 60).

Bellah then drives to the heart of the matter: "If the self is defined by its ability to choose its own values, on what grounds are those choices themselves based? For many, there is simply no objectifiable criterion for choosing one value or course of action over another. One's own idiosyncratic preferences are their own justification. The right act is simply the one that yields the agent the most exciting challenge or the most good feeling about himself" (pp. 75–76).

"Now if selves are defined by their preferences, but those preferences are arbitrary, then each self constitutes *its own moral universe*, and there is finally no way to reconcile conflicting claims about what is good in itself" (p. 76; my italics).

And here we have the all-too-familiar slide of the expressivist (Eco-Romantic) current into divine egoism and biocentric feeling—"the culture of Narcissus" (Lasch). Small wonder that large segments of American culture would be drawn to everything from sensory/material drenching in consumerism, sexually titillating anything, egocentric/biocentric tribalism dressed as a new paradigm, body-bound expressive "therapies," "do your own thing" idiosyncratic "values," fluctuating "meanings" that follow the ecocentric ebb and flow of passing feelings, and "relationships" whose primary aim is to "feel good," relationships that form—and dissolve—just as fast as the fleeting feelings that drive them.

Bellah et al. go on to point out that in this atmosphere of "feel good," all values eventually are reduced to economic values, because where feelings predominate, money alone can buy the free time to indulge them—the more money you have, the more free feelings you are allowed. "In the absence of any objectifiable criteria of right and wrong, good or evil, the self and its feelings become our only moral guide. What kind of world is inhabited by this self, perpetually in progress, yet without any fixed moral end? There each individual is entitled to his or her own 'bit of space' and is utterly free within its boundaries. But while everyone may be entitled to his or her own private space, only those who have enough money can, in fact, afford to do their own thing" (p. 76). Under these circumstances, "utility replaces duty; 'being good' becomes 'feeling good' "—and *that* depends primarily on bucks.

Thus the "strange and spasmodic integration" of the culture of separation: work as hard as you can (utilitarian) in order to retreat to the "freedom" of a private space of egoic feeling (expressivist) or shared feelings in a lifestyle enclave.

And thus, to return to my original point, what has failed miserably is still the *integration* of the Big Three (exacerbated in America, but evidenced aplenty in modernity at large). Particularly under the onslaught of divine (and regressive) egoism, alternating with a utilitarian-individualism [see *Boomeritis*], Bellah finds that "what has failed at every level—from the society of nations to the national society to the local community to the family—is *integration*: we have put our own good, as individuals, as groups, as a nation, ahead of the common good" (p. 285).

And he ends with a general conclusion that is certainly supported by the analysis offered in this book: "What we find hard to see is that it is the extreme fragmentation of the modern world that really threatens our individuation; that what is best in our separation and individuation, our sense of dignity and autonomy as persons, *requires a new integration* if it is to be sustained.

"The notion of a *transition to a new level of social integration* may also be resisted as absurdly utopian, as a project to create a perfect society. But the transformation of which we speak is both necessary and modest. Without it, indeed, there may be very little future to think about at all" (p. 286; my italics).

The difficulty that the authors (and postmodernity) are grappling with, as I see it, is constituted precisely in the tension between the two "communal" strands (biblical and republican) and the two hyperagency strands (utilitarian and expressivist). The agrarian-based biblical and republican strands—both variations on the duty ethic first formulated in mythic-membership cultures—seem to be the only available counter-blasts to the hyperindividualism and "unencumbered self" of modernity. And, indeed, this tension is reflected in the constant debate between liberal individualism (the unencumbered self) and various forms of communitarian theorists (of the saturated/situated self).

But, in my opinion, stripping the biblical and republican strands of their racism, sexism, dominator hierarchies, etc., is not "updated" enough to carry the burden of the integration of the Big Three, while the emerging forms of "integration" of the individualistic strands are indeed and all-too-often "pseudo-integrations." This dilemma—the dilemma of postmodernity—is a central theme of volumes 2 and 3.

The Evolution of Religion

As for Bellah's pioneering work on the evolution of religious systems, see *Beyond belief*. Bellah presents a standard version of some of the twenty tenets, and of course I agree with him on those points. For example, "Evolution at any system level I define as a process of increasing differentiation [tenet 12b] and complexity of organization [12a] that endows the organism, social system, or whatever the unit in question may be [any evolving holon] with greater capacity to adapt to its environment, so that it is in some sense more autonomous [12d] relative to its environment than were its less complex ancestors," although, of course, "simpler forms can prosper and survive alongside more complex forms" (p. 21).

Further, Bellah points out, "Since it has been clear for a long time that levels of social and cultural complexity are best understood in an evolutionary framework, it seems inevitable that religion too must be considered in such a framework" (p. 16).

Religious symbolism, says Bellah, "tends to change over time, at least in some instances, in the direction of more differentiated, comprehensive forms [tenet 12]. . . ." Bellah then traces *five major stages* of religious evolution, which he calls primitive, archaic, historic, early modern, and modern, and which correspond quite closely with our archaic/magic, mythic, mythic-rational, early rational/egoic, and rational/existential. A comparison of his stages with my descriptions will show the very strong and wide-ranging correlations and similarities.

Bellah concludes that "freedom has increased [tenet 12d] because at each successive stage the relation of humans to the conditions of their existence has been conceived as more complex, more open and more subject to change and development. This scheme of religious evolution has implied at almost every point a general theory of social evolution," which is why Habermas has also found Bellah's work inspirational (pp. 24, 44).

At the same time, Bellah is fully aware of what we have called the *dialectic* of progress; it is not all simply sweetness-and-light progression: "At every stage the increase in freedom is also an increase in the freedom to choose destruction" (p. xvi).

Beyond these large areas of agreement, my major criticism of this approach is that is based almost entirely on the analysis of social action systems according to the general cybernetic model (largely following Parsons): the standard Lower-Right systems theory that always tends to dissolve qualitative depth into functional action terms and behavioristic variables, and thus dissolve the distinction between authentic and legitimate (the capacity for vertical transformation and depth-disclosure is dissolved into the capacity for horizontal translation and functional fitness: the standard Right-Hand path that converts being into doing).

Bellah: "An action system may be defined as the symbolically controlled, systematically organized behavior of one or more biological organisms." Symbolic messages are thus eventually reduced to *representational* and monological recovery of internal or external messages: "Information in such a system consists largely of symbolic messages that *indicate* [represent] something about either the internal state or the external situation of the action system" (pp. 9, 10).

That is standard monological reductionism, and it leads naturally to relegating the entire Left-Hand dimensions to that which is simply "left over" after the systems analysis: "Thus, the set of symbolic patterns existing in a system will be partially determined by the nature of the external world with which that system has had to deal and by the nature of the laws governing the cognitive processes of the brain [all Right-Hand]. Within these limits there is a wide range of freedom, within which alternative symbolic patterns may operate with equal or nearly *equal effectiveness*"— and thus questions of the *worth* or authenticity of those "equal alternatives" get little or no attention within the constraints of functional fit and legitimacy (p. 10).

This reductionism leads naturally to a view of "symbolic action" as a *monological linkage* of meaning and motivation in a *behavioristic* system, and thus the role of religion is reduced to: "Religion is the most general mechanism for integrating meaning and motivation in action systems," and therefore any symbolic system that offers this "integration" is "religious," whether we would call it "true" or not (p. 12). The truth-value of all religious propositions is thus reduced to the functional capacity of the propositions to *appear* to be true and meaningful, which then provides the "integrating" link with motivation. But this is simply to reduce truth to the functional value of holding-something-as-true, which also happens to dissolve this theory's own claim to *be* true.

Thus, Bellah is perhaps too optimistic about the capacity of the "cybernetic model" to integrate Left- and Right-Hand paths. "It is precisely the stress on autonomy, learning capacity, decision, and control that gives the cybernetic model the ability, lacking in previous mechanistic and organic models, to assimilate the contributions of the humanistic disciplines—the *Geisteswissenschaften*—without abandoning an essentially scientific approach" (p. 10).

But it is precisely the *incapacity* of monological, cybernetic, representational systems theories to integrate dialogical, intersubjective, and interpretive occasions that has increasingly become obvious. The subjective space that builds those representational cybernetic models is *not itself* built only of representations but also of interpretive and intersubjective occasions, themselves *not* modeled in the theory that is supposed to explain them. The belief that such monological models *can* explain them *is* the fundamental Enlightenment paradigm in all its inadequate aspects, is everything bad about the disengaged and hovering Cartesian ego.

That religion seeks to integrate meaning as a Left-Hand dimension with motivation as a Right-Hand behavior is true enough, but the meaning component cannot be reduced to monological representation without completely destroying the sought integration.

I am not denying most of the specifics of Bellah's functional, Right-Hand analysis; I am denying it is the only, or even the most basic, analysis of religion. Bellah has refined many of these views, and he discusses some of these issues in the introduction to *Beyond belief*. And, as the previous section on *Habits of the heart* makes obvious, Bellah has become quite sensitive to the importance of qualitative distinctions and ontological values in providing the meaning component of the desired integration, none of which can be covered by cybernetic or systems theory.

10. As we put it earlier, by insisting that the noosphere is *part of* the biosphere, the actual integration is utterly blocked, because the noosphere is not part of the biosphere, the biosphere is part of the noosphere (and parts of the human being are parts of both). The Eco camp's "integration" is fabricated only by an ontological regression that confuses great span with great depth and thus jeopardizes both.

11. A small faction of ecofeminists have rebelled against a flatland bioequality, and I will examine their important contributions in volume 2 (Zimmerman, for one, has fruitfully drawn on their work, as has Cheney, all of which is carefully evaluated).

By and large, however, the ecofeminists have often substituted egoic-communion for egoic-agency, still caught as they are in the other pole, the Eco-communion pole, of the fundamental Enlightenment paradigm. And thus, when they attempt to escape the violence and leveling of the flatland view, they end up perpetuating it in subtler forms.

Charlene Spretnak, as only one example, in *States of Grace*, continues her several-books-long attack on anything agentic (which she ascribes mostly to the masculine mode) and emphasizes instead a thoroughly communion-based approach (more feminine, as it were). But then, when it comes to the spiritual orientation that will supplant and heal this agency-laden "patriarchal" worldview, she ends up almost incomprehensibly championing a broad form of *vipassana*/Theravadin Buddhism, the East's archetypal and merely Ascending path, which is radically dualistic and hyperagentic, a pure Goddess-denying, Descent-denying, Goodness-denying, Plenitude-denying path, a path that historically has denied and devalued the body, the earth (samsara), sex (and the ultimate sin-temptation, woman). Spretnak's stance is known as out of the frying pan and into the fire.

Sexual Universals

At the same time, I believe Spretnak and other feminists, who insist on the existence of universal native differences between the male and female value spheres, are quite right. But most of the attempts to delineate these "strong sexual universals" suffer by not taking a multidimensional view. In volume 2, based on rather extensive evidence, I suggest that a much more adequate view can be advanced by, at the very least, delineating the ways that translation and transformation operate in men and women. With this approach, we then have, *in addition* to such physical universals as strength/mobility, sexual profligacy, child birthing/lactation, and tactile differences (and other commonly advanced biological differences), the following: I advance the view that, on average, men tend to *translate* with an emphasis on agency, women tend to translate with an emphasis on communion. And men tend to *transform* with an emphasis on Eros (transcendence), women tend to transform with an emphasis on Agape (immanence).

These two "basic universals" (referring to the male/female native differences in ways of translation and transformation) help to explain an enormous amount of cross-cultural data and male/female differential values. In volume 2, I attempt to show the applicability of these universals by reformulating the various stages of present-day psychological, moral, cognitive, and object-relations development as they appear (sometimes quite differently) in men and women (both men and women translate with both agency and communion, but, as I suggested, males tend to place a higher emphasis on agency, women on communion; and likewise with transformation: men and women access both Eros and Agape, but men tend to emphasize the former, women the latter). I likewise trace these universals as they appear in historical development (during the six or so major stages of technological/worldview evolution).

The point, as I suggested in chapter 5, is to isolate a set of constants or universal differences, and then follow these universals as they manifest in different forms due to different evolving worldviews, modes of production, techno-economic infrastructures, and so on—because these various structures, at different stages of evolution, placed different values on the male and female spheres, emphasizing and prizing now one, now the other, occasionally both. But oppression has virtually no explanatory power or place in this scheme, which relieves us from the men-are-pigs, women-are-duped-sheep view.

On the historical status of men and women (down to today), most radical feminist and ecofeminist analyses suffer badly from taking a one-dimensional view (analyzing merely the differences in translation and not also transformation: agency versus communion is their only scale). Because they do not factor in the various stages of historical transcendence and transformation (Ascent), these feminist analyses are thus forced to make agency and communion cover the ground that actually belongs to Eros and Agape. They are thus forced to see agency (and "men") as "bad," communion (and "female") as "good." The bad agency then must have oppressed the good communion, and women are then necessarily defined as primarily molded by an Other. To say that men forced women to be molded is to define women as weaker. To say that women didn't object to this oppression because they were brainwashed is to define them as stupider. This is useless and self-defeating. When a more multidimensional analysis is applied, these categories of molded/weaker/stupider are unnecessary to get the game going.

And finally, these basic universals give us a series of strong clues as to the male and female faces of Spirit, and to the differences in male and female spirituality, spiritual practices, and spiritual goals (all of which are carefully explored).

My point with regard to most forms of ecofeminism can thus be put very simply: in denying any true transcendence and Ascent (Eros), most ecofeminists emphasize only immanence and Descent (Agape). In thus denying any true Ascent, they remain thoroughly wedded to flatland, and as such, they simply substitute the more feminine communion for the more masculine agency: they don't calculate flatland, they commune with flatland.

The same lack of vertical depth and transcendence, the same "nonhierarchical" heterarchy of shadows, the same failure to transform, and merely instead a different way to translate the same old flatland: mononature is not now God, mononature is the Goddess.

It's not so much that such approaches are altogether wrong as that they desperately need to be complemented and supplemented and rounded out. Because the dominant theme of Western culture, certainly from the time of the Ego-Enlightenment, has emphasized hyperagency (as we have amply seen), the general Eco-Romantic rebellion (of which ecofeminism is an offshoot) tended always to emphasize communion and thus to give a greater (sometimes total) emphasis to the feminine. But, as we also saw, that Eco-Romantic emphasis on feminine/communion was simply the other pole of the fundamental Enlightenment paradigm (communing with the great interlocking order of mononature): the other side of the flatland street.

What both the Ego (agency) and the Eco (communion) lacked—and what they both still lack—is the balance, not just of agency and communion, but also of Ascent and Descent, Eros and Agape, transcendence as well as immanence. What is not needed is simply more of the same dualistic view, now from the other side of the same failed coin: only Agape over Eros, only immanence over transcendence. This is just more of the fractured, schizoid notes to Plato. [For a more inclusive view, see *The eye of spirit*, "Integral Feminism," chap. 8.]

12. In volume 2, I will look at how agency/communion as rights/responsibilities plays out in the masculine/feminine modes, drawing, of course, on Gilligan's observation that male modes tend to the agentic/rights, and female to the communion/relational/responsibilities; this is a prelude to the masculine and feminine faces of Spirit, set in the context of the Great Holarchy of Being. But this is not just a theoretical venture; the actual status of men and women in the five or six major technological epochs are examined carefully for clues to their native orientation.

13. That less significant holons have fewer rights (and fewer responsibilities) than more significant holons does *not* mean they have *no* rights, and the recognition of their relative rights is part of the Liberation of Life (Birch and Cobb) that is inherent in the Eco-Noetic Self and its Community of all sentient beings.

As suggested in previous notes (see especially note 23 for chap. 8), the Basic Moral Intuition is "protect and promote the greatest depth for the greatest span." That is, when we intuit Spirit, we are actually intuiting it as it appears in all four quadrants (because Spirit manifests as all four quadrants—or, in short, as I and we and it). Thus, when I am intuiting Spirit clearly, I intuit its preciousness not only in myself, in my own depth, in my I-domain, but equally in the domain of all other beings, who share Spirit with me (and as me). And thus I wish to protect and promote that Spirit, not just in me, but in all beings possessing that Spirit, and I am moved, if I intuit Spirit clearly, to *implement* this Spiritual unfolding in as many beings as possible: I intuit Spirit not only as I, and not only as We, but also as a drive to implement that realization as an Objective State of Affairs (It) in the world.

Thus, precisely because Spirit actually manifests equally as all four quadrants (or as I, we, and it), then Spiritual intuition, when clearly apprehended, is apprehended as a desire to extend the depth of I to the span of We as an objective state of affairs (It): Buddha, Sangha, Dharma. Thus, protect and promote the greatest depth for the greatest span.

Precisely *how* to implement this Basic Moral Intuition (BMI)—the actual details of its application—that is not given in pure intuition. Rather, the details of application are part of the intersubjective and cultural and social project that all of us, in open communication free of domination, must discuss and decide. That is why the human moral response is a fine mixture of the Divine and the Human. That I clearly intuit Spirit in all I's and all We's and all Its does not mean that I automatically know all the details (scientific, philosophic, ethical, cultural, and otherwise) that must go into the final decision. In fact, I might have a fine intuition of Spirit but unpack it poorly owing to lack of relative understanding and relative knowledge.

Thus, for example, a Zen master might have a profound intuition of oneness with the Kosmos, with Gaia and all its inhabitants, and be dedicated to not harming a single living being, but that doesn't mean that the Zen master will automatically know that Styrofoam will kill life. The BMI is the infusion of the Divine into our awareness, not the working out of specific details.

I earlier gave as example one of the most agonizing of modern moral dilemmas,

762 | *Note for Page 546*

abortion. To the egocentric level of development (preconventional), abortion is in most cases completely acceptable, because this level of development recognizes depth only in itself (it extends depth only to a span of 1, namely, divine egoism; as Kohlberg's research showed, *morality* at this stage is "whatever I wish"). This level intuits Spirit (*all levels do*), but with its limited consciousness, the Basic Moral Intuition (protect and promote the greatest depth for the greatest span) shows up as: take care of number one, primarily, and then anybody who supports number one. (Infanticide is thus quite common in tribes organized around preconventional arbitration; Lenski reports that about one-third of all children at birth were "thrown away" in these societies; this is nonetheless perfectly *legitimate* at that level, because the ecological-carrying capacity of the tribe was quite limited—usually around forty people maximum—and infanticide was the most acceptable form of population control; the greatest depth for the greatest span was served by that practice).

With the expansion of awareness to the sociocentric/conventional level, span is extended to include one's culture, group, or nation. The BMI (greatest depth for greatest span) then shows up as "my country right or wrong" (depth is extended to the span of all people in the chosen group, but to few or no outsiders). In most *agrarian* mythic-membership cultures, abortion is considered a sin, because the fetus is a potential convert to that mythic God and thus a part of the chosen group, and thus infanticide partakes of mythic theocide, which might bring catastrophic retribution from the God (within the mythic circle, that is a very *legitimate* belief).

On the other hand, in the earlier *horticultural* mythic societies, infanticide as a *sacrifice* to the Goddess—this was considered acceptable, even mandatory, because the sacrifice of that particular depth *increases* the greatest depth for the greatest span (by ensuring a bountiful harvest, for example). *Within* that horticultural/mythic structure, that is also a *legitimate* unpacking of the BMI (and becomes "immoral" only when a greater collective depth is *actually* available, which *itself* would condemn this practice as unacceptable: the classic case of Abraham, first asked to sacrifice his son, then having the sacrifice called off: that is exactly the transition between horticultural/mythic and agrarian/mythic-rational).

With the expansion of awareness to the postconventional (rational-worldcentric) level, span is extended to include all humans universally and non-ethnocentrically (regardless of "chosen group" mythic status), and it is here that abortion becomes, for the first time, a profound *dilemma*: there are cases where it might be okay, and cases where it might not.

The difficulty, I believe, becomes precisely because the *form* of the BMI is protect and promote the greatest depth for the greatest span, and that *always* means that some depth can be sacrificed for a greater depth across greater span. The feminist argument in favor of legalization of abortion (or choice), though not put in exactly these terms, amounts to this: the depth of the mother (and consequently society) might be so profoundly harmed by having a child that the greatest depth for the greatest span is actually served by legalized abortion (the Soviet Union, for example, used this argument explicitly; over a third of adult women consequently had abortions).

This argument (pro-choice) is directly related to the issue of *reproductive rights*: in attempting to grow beyond patriarchy (or the once-appropriate, now-outmoded, predominantly male rights of agency in the noosphere), it is essential that women be given rights over their own reproductive capacity; since "rights" *always* means "wholeness-value" or *"autonomous value"* (as explained in the text), reproductive *rights* means that the woman will make the decision herself, *autonomously*, and not as *part* (or responsibility) of some *other* agency (such as the government/society). And, as always, if these rights are not recognized, then the *wholeness*, the *autonomy* announced by those rights, is *not* sustained: the woman is not her own wholeness, her own person, in this regard (i.e., if the rights are not recognized, that particular wholeness dissolves; the woman's rights revert to another agency, in this case, society, and women who have abortions are pronounced criminal).

Those are all legitimate arguments in favor of abortion/choice (which is usually limited to the first trimester, another example of the unpacking of the BMI: in the first trimester, the *depth* of the fetus is so relatively shallow that its sacrifice to a larger scheme is felt to be more acceptable; however, at the point that the fetus becomes "viable"—that is, at the moment it can or could sustain its own life—it is an autonomous whole, and we are then forced to recognize its *rights* as an independent *wholeness*; sacrificing it at that point is murder).

My point is that, whatever we decide on this (and other) issues, the BMI doesn't deliver hard and fast decisions, but a simple infusion of Spiritual Concern, which is unpacked according to the level of the depth of the person doing the intuiting, and is always of the form "protect and promote the greatest depth for the greatest span" (as best as that person can understand it).

Just so with animal rights and our understanding of them. That animals have relatively fewer rights (less depth) than humans does not mean that they have no rights. That is, they possess a certain depth, a certain wholeness, they are ends in themselves possessing *intrinsic value*. How much weight to give animal rights is open to *fair discussion*; my point is that, with the expansion of awareness beyond the egoic-rational (worldcentric human) to the centauric and even psychic (all life forms), then the recognition of rights other than the merely human is then the form taken by the more expanded (deeper and wider) Basic Moral Intuition.

This doesn't mean we can't kill animals if genuinely necessary; it means that it will upset anybody with this deeper intuition if we kill animals needlessly or heedlessly. And the greater the depth of the animal killed (ape versus worm, for example), the greater the intrinsic value that we have just destroyed. This killing will upset more highly evolved individuals, because the greatest depth for the greatest span is now starting to include nonhuman life as well.

Of course, this does not mean that all individuals upset by the killing of animals are operating from a centauric or psychic level—in fact, often the contrary: too many animal rights activists seem shot through with an ideological fury that betrays not moral sensitivity but divine egoism hooked to a convenient cause.

Further, many animal rights activists implicitly—and sometimes quite explicitly—value animal life above human life: if the analysis I have presented is accurate, then these individuals are in fact caught in what appears to be a regressive moral pathology.

That, in the best sense, we try to extend certain elementary rights to animals (especially more developed animals, all mammals and certainly all primates) simply means that we, *as a culture*, have *evolved* to the point where an expansion of awareness is *recognizing* that the BMI extends to nonhuman life, and when we extend these elementary rights to animals, it means that we have the evolved capacity to *recognize* the *wholeness* value, the intrinsic value, in those holons, and the *rights* protect the *wholeness* of that holon, that particular animal (however far down we decide to push those rights). If we don't recognize those rights, then that wholeness is not honored, and we will indiscriminately dissolve that wholeness (i.e., kill that animal heedlessly).

That is the intrinsic-value side of the argument for animal rights (it is wrong, in and by itself, to destroy depth); the extrinsic-value side is that our own lives depend upon the mutual flourishing of the greatest variety of animals (diversity of eco-system). We are harming ourselves, finally.

But the interesting thing about the Basic Moral Intuition is that it does not carry the command "Protect only depth." It carries the command "Protect all four quadrants as best you can," that is: protect not just depth but depth × span (and that always demands various sacrifices, for in a world of limits, not all can flourish paradisaically). This shows up even in animal population control, for example. We have long recognized that limiting animal populations (by, yes, killing some animals) allows the greater animal population to flourish more easily: we sacrifice particular depth for the greater depth × span. (This totally confuses the more rabid animal rights activists, who want to make all rights absolute, which will never work.)

And, in an era of limits, we are starting to face these same dilemmas with human beings. In health care, a miniature version of this shows up in our decisions about what diseases to cover in national health insurance. Which is better: spending $100,000 to cure one person with a rare disease, or letting that person die and spending the money to save a dozen people with "less expensive" diseases? In an era of limits, we are forced to choose (and probably choose the latter). Go one step further: what if that one person is the President, or Albert Einstein, or Mozart, and the twelve are criminals on death row? You see how the dilemma works? How do we unpack the intuition that we are to protect and promote the greatest depth for the greatest span? The intuition is given; the unpacking is our moral dilemma, always.

Thus, the Basic Moral Intuition is an infusion of Care for all four quadrants, but does not come with instructions engraved in stone tablets. What's worth more, one ape or a thousand frogs? Probably the ape. But my point is that *all* of our moral decisions (and dilemmas) are ultimately of that form, because they are all driven by the intuition "the greatest depth for the greatest span," and various *types* of depth and *degrees* of span are always necessarily being sacrificed for a greater depth × span, and how we choose those sacrifices is always our moral dilemma, the precise reflection of samsara's duality and its IOUs. (The recognition of *types* of depth—an entire spectrum of depth—is what separates this from the flatland utilitarians, who interpret all depth as the monohappiness of sensory pleasure, which prevents them from saying which is preferable, a happy pig or an unhappy Socrates. The utilitarian position, as I earlier argued, is how the BMI appears to the monological reason of flatland.)

Likewise, because the BMI says to protect depth × span, and not merely depth only, this places a powerful curb on the tendency to judge merely in terms of intrinsic value alone. A greater wholeness, with a greater depth, does indeed possess a greater intrinsic value, but that is only "half" the story, so to speak. To return to the one ape, one thousand frogs example: on intrinsic value alone, we would chose the ape. But both ape and frogs exist in networks of extrinsic communions as well (span), and if we discover that the frogs are part of a fragile ecosystem and their death would disrupt the entire system (since they are more fundamental than the ape), then we would choose to save the frogs, since that would preserve the greatest depth for the greatest span, including probably the lives of other apes (*demonstrating* that is the tricky issue, which is why, again, the BMI doesn't offer details, only Concern).

Likewise with delicate issues in human affairs, such as affirmative action. If it can be demonstrated that the depth of one group of people has been systematically oppressed by another group, then the greatest depth for the greatest span might indeed involve attempts to redress the imbalance. Demonstrating that is, again, the tricky issue.

I mention these examples not to come down in favor of one or the other, but only to show that the BMI does not simply center on depth alone (which would tend toward fascism), nor does it center on span alone (which would tend toward totalitarianism and the tyranny of the majority), but rather centers on both. The BMI delivers to us an intuition to implement the greatest depth across the greatest span, a spiritual Concern for all four quadrants, and the painful dilemmas of how to implement this intuition are the dilemmas of finite beings in finite circumstances attempting to honor an infinite Care.

All of these issues, and some of my suggested solutions (although "solutions" is too strong a word) are gone into in detail in volume 3. I wish here only to draw attention to the fact that all holons possess a degree of *depth*, with its correlative *rights*, existing in a *span* with correlative *responsibilities*, and that as our own awareness evolves to greater depth itself, it more adequately unpacks the Basic Moral Intuition, which infuses us with an awareness, and a drive, and a demand, to extend the greatest depth to the greatest span, as best we can under the ridiculous circumstances known as samsara.

14. And, as we have seen, since flatland eco-holism confuses great span with great depth, it gets the instrumentalization itself exactly backward. The biosphere does not depend upon the noosphere for its existence, but the noosphere does depend upon the biosphere for its existence (yet another way of saying that the noosphere is not part

of the biosphere, but rather the biosphere is part of the noosphere). Which means, the biosphere has instrumental value to the noosphere (and noospheric humans), not vice versa.

But my approach does not stop with instrumental value in any domain. Even though the biosphere is instrumental to the noosphere, the biosphere itself still possesses equal Ground-value (as the visible, sensible God/dess), and further, each biospheric holon possesses its own degree of instrinsic value (an ape is more intrinsically valuable than an atom, but the latter is more extrinsically valuable than the former). Thus, the entire Kosmos exists as a network of rights and responsibilities correlative with degrees of depth and consciousness. Nothing is ever merely instrumental to anything, and everything ultimately, finally, has perfectly equal Ground-value as a perfect manifestation of primordial Purity, radiant Emptiness.

15. In presenting its empirical "holistic" worldview, most Eco-camps take for granted, and thus overlook, the vast networks of intersubjective meaning and dialogical fabric that allow them to present and even comprehend a holistic web in the first place: they discuss none of the extensive dynamics of intersubjective communicative exchange that allows and upholds their objective web-of-life systems theories, and thus they have no actual recommendations as to how to reproduce that intersubjective agreement and mutual understanding in others or in the world at large—they can only insist that everybody agree with them and accept their systems view, ignoring how the intersubjective worldspace develops from egocentric to sociocentric to worldcentric comprehension.

This taking-for-granted (and thus ignoring) of the interior, intersubjective worldspace (the Left-Hand dimensions of the Kosmos), and consequently believing that all that is necessary is to produce a more "accurate" holistic map of the natural world: this is exactly the fundamental Enlightenment (and Cartesian) paradigm, the "proud culture of reflection," that "monster of arrested development," which contributed inexorably to the actual fracturing of the Kosmos, the collapse of the Kosmos, and the consequent despoliation of Gaia precisely in the *incapacity* to engage the interior processes of transformation upon which reflection itself depends in the first place, and upon which mutual agreement and mutual accord, which alone will save Gaia, must proceed.

16. *The Guru Papers*, a book by Kramer and Alstad, is a typical example of a purely Descended approach claiming to have the whole story. The authors mistake what certainly seems to be their ignorance of the Ascending current for a critique of all Ascending endeavors, and then interpret any *challenge* to their *own* stance as being "authoritarian." Their effort is admirable when confined to the half of reality they admit, but it is an effort that remains, in its partialness, steeped in the dominance of Descent.

Likewise, Goldsmith's *The Way: An Ecological World-view* is a splendid recital of the Eco-systems worldview, and I agree with a fair amount of its presentation, as far as it goes. But it is aggressively monological and purely Descended, and thus is replete with the *regressive* yearning for what it calls "the earliest period when people everywhere really knew how to live in harmony in with the natural world" (the naiveté of that stance is simply astonishing). Moreover, this monological view typically has no conception of the actual dynamics of the entire Left-Hand dimension of the Kosmos; in this regard it is quite dualistic and fragmented, and so of course it claims to be THE Way, as its title puts it, and states this claim very boldly: "These principles form an all-embracing and self-consistent model . . ." (apparently never having heard of Gödel or Tarski).

These two books are archetypal examples of the Descended path in today's world. Although these approaches lack any vertical *depth*, they make up for that in a type of fearless shallowness. In both books, the lack of any Ascending current, any genuinely transformative interior disclosures, allows the bearers of the worldview to go on about their business fundamentally *unchallenged*. There is no consciousness higher or deeper than *their* present state, only a different monological worldview that *others* should accept. Their salvation is not a matter of transcendence in their own case, but in altering the views of others. Hence the dominance inherent in Descent; hence the authoritarian stance claiming to combat authoritarianism; hence the flatland claiming to be the All.

17. The evidence for both of these claims is presented in volumes 2 and 3.

In this regard, a final word about Stanislav Grof and Jean Gebser, and a comment on the work of Richard Tarnas. Starting with Grof (these comments are in light of my previous discussion of Grof's work—note 3 for chapter 6—which the reader should have read first):

Stan Grof

Grof, in *Beyond the Brain* (1985), pinpoints two major areas of "disagreement" or at least "mismatch" between his psychological model and mine ("in spite of the otherwise far-reaching agreement"). The first involves the basic perinatal matrices. "In my opinion," he points out, "without a genuine appreciation of the paramount significance of [perinatal birth/death], the understanding of human nature is bound to be incomplete and unsatisfactory. The integration of these elements would give Wilber's model more logical consistency and greater pragmatic power. Without this, his model cannot account for important clinical data, and his description of the therapeutic implications of his model will remain the least convincing part of his work for clinicians used to dealing with the practical problems of psychopathology" (pp. 136–37).

My explanation of fulcrum-0 (note 3 for chap. 6) should satisfy, in a general fashion, most of those objections. I have, since 1982, considered fulcrum-0 an integral part of the overall spectrum model (due primarily to Stan's research and John Rowan's clarifications).

The second major "disagreement" mentioned by Grof and others is the "linearity" of my model. As Stan puts it, "My own observations suggest that, as conscious evolution proceeds from the centauric to the subtle realms and beyond, it does not follow a linear trajectory, but in a sense enfolds into itself. In this process, the individual returns to earlier stages of development, but evaluates them from the point of view of a mature adult. At the same time, he or she becomes consciously aware of certain aspects or qualities of these stages that were implicit, but unrecognized when confronted in the context of linear evolution" (p. 137).

My presentation throughout this book should make it clear that I am in general agreement with that statement, but perhaps even more radically than Grof. Not simply each stage beyond the ego/centaur, but *every* stage of development unfolds/enfolds its predecessors. Developmentalists from Loevinger to Kohlberg to Piaget have long pointed out that every stage involves various types of reworkings of previous patterns, taken up, enfolded, and reassimilated into the new structures. The "linear" nature of the various stages simply refers to broad patterns of *stable* adaptation (not merely temporary experiences), which unfold "linearly" in the way that an acorn unfolds "linearly" into an oak. Many temporary experiences can bounce all over the spectrum of potential experiences, but stable and continuous adaptation depends upon ongoing holarchical structuralization that must differentiate/integrate its predecessors: unfold, enfold (Grof's "enfolds into itself").

At the same time, because individuals have general access to gross/subtle/causal *states* (waking, dreaming, deep sleep), temporary experiences can, as I said, move all over the broad spectrum of possible experiences. And these temporary "peak" experiences tend to be emphasized in Grof's work, which is fine (and they often have considerable therapeutic effect, which is also just fine).

I agree with all of that, and I have all along (but beginning with *Up from Eden*, I pointed out that temporary experiences/states must be converted into stable patterns if actual growth is to occur; we will return to this point in a moment). Some critics, focusing on the "stages" portion of my model, think that I have some sort of monolithic, rigid, one-way street in mind, through which development moves, clunking and chunking a step at a time. But this type of narrow one-way street conception is *never* assumed by developmentalists of any variety; the various stages not only spiral, meander, wobble, and regress, they are themselves simply mileposts in the winding road of growth (and, as such, are extremely important, but not monolithic). This is largely taken for granted by developmentalists.

Moreover, in my model, there are the basic levels or waves of consciousness (matter to body to mind to soul to spirit), and the various developmental lines or streams (cognition, affects, morals, self-identity, spirituality/faith, defenses, etc.) that move relatively independently through those levels. Thus, for example, a person can be a high level of development cognitively, a medium level emotionally, and a low level spiritually. There is nothing linear about overall development.

Still, as I will emphasize in this note, even though overall development is not linear, and even though the self has access to all sorts of altered states and temporary peak experiences—none of which are linear—if those states are to become *enduring traits*, they must enter the stream of development and be converted from states to structures (or enduring capacities). In any developmental line that this occurs (cognitive, moral, affective, etc.), the development is "linear" as defined by developmentalists: namely, each stage builds upon certain competencies laid down by its predecessors (e.g., atoms to molecules to cells to organisms), and no amount of temporary peak experiences can turn atoms into cells without passing through molecules.

In short, while the model I have proposed makes ample room for *both* altered states and developmental structures, Stan's model accounts mostly for the former. This, in my opinion, is a major limitation in his model, since it means that it cannot easily account for development at all. Since permanent realization and stable access to higher domains is one of my primary interests, I have focused less on temporary states and more on enduring traits, but I fully acknowledge both.

Thus, from my perspective, the self-system can span the entire spectrum of potential experiences (which I will qualify in a moment). At the very least—and this is virtually indisputable—the self has constantly available to it the three great states of gross, subtle, and causal occasions (if nothing else, in waking, dream, and deep sleep states). And therefore, at any of its stages of stable growth and development, the self has access to temporary experiences ("influxes" or "infusions" or "transfusions") from these states.

These temporary infusions and altered states are not necessarily, in and of themselves, the stuff of stable adaptation and permanent structuralization (see below); but these general peak experiences can, and do, occur to the self at virtually any of its stages of growth and development (because it is immersed in or surrounded by the three states at every stage). This is why I have always agreed that children can indeed have a type of transpersonal influx (usually from the subtle state). And why "channeling" of subtle occasions can occur at virtually any stage of development. And why paranormal phenomena are not a particular stage of growth but an influx that may (or may not) occur at virtually any stage of growth. And why psychosis is sometimes invaded by mystical transfusions (indeed, *The Atman Project* was one of the first clear attempts to delineate the influx of transpersonal energies due to the breakdown of egoic translations in certain psychotic states).

But many theorists have so focused on these dramatic and temporary influxes or transfusions that the actual growth and development process has been severely misconstrued and even misunderstood. Temporary infusion (as important and profound as it sometimes is) becomes the be-all and end-all of consciousness occasions. This has led more than one researcher to sing the glories of psychosis as primarily a wonderful means for mystical growth (e.g., Laing) and the joys of infancy as a paradisaical and mystical state tragically fractured by growing up (e.g., Norman O. Brown, Michael Washburn).

Peak experiences can indeed glimpse various nonordinary states, but stable adaptation proceeds in a growthlike fashion (which is why neither the child nor the psychotic nor the channeler can actually "own" and stably adapt to the infusion; moreover, the latter two "transpersonal" states all rest on a dissociative base, which is why they occur outside the autonomy of the self-system; I will return to this shortly).

But most important, focusing on peak experiences has caused an enormous confusion about what is *actually available* in these experiences. Because, contrary to the peak theorists, not *all* of the spectrum of consciousness is lying around, fully formed but submerged, like a sunken treasure chest, waiting to be dug up. Consciousness *creates* as much as it *unearths*, and thus, as we will see, we have to be extremely careful in how we interpret these peak experiences—and most important, about the intersubjective space *in which* these experiences occur.

In the first place, that a transpersonal experience might involve the "reentering" or "reworking" or "reexperiencing" of a prepersonal occasion (archaic images, pleromatic indissociation, phylogenetic heritage, animal/plant identification, perinatal patterns, etc.) does not mean the transpersonal awareness *resides* in those archaic structures: it is the transpersonal awareness that is doing the entering, not the archaic modes—the question is always: what can those modes accomplish *on their own*, when they alone are manifest? Not a single prepersonal structure can itself, in itself, generate intrinsic transpersonal awareness, but it can become the object of transpersonal consciousness, and thus be "reentered" and "reworked," and it then becomes a type of *vehicle* of transpersonal awareness, but never its *source*. The pre/trans fallacy, in this instance, remains firmly in place.

Further, as I suggested in *A Sociable God*, we have to distinguish between peak experience, plateau experience, and stable adaptation; and as I suggested in "Paths beyond Ego in the Coming Decades" (in Walsh and Vaughan, *Paths beyond ego*), higher development is in part a conversion of *temporary states* into *enduring structures* of stable adaptation. That we have a type of (limited but fluid) access to transpersonal states should not detract from the fact that temporary experiences are not the stuff of enduring wisdom-adaptation (for an insightful reflection on these issues, see Reynolds 1994).

And this is why we have to be especially careful about the claim that all psychological domains are simply open to fluid access. As Reynolds (1994) summarizes the Grofian view: "The clinical data Grof has collected during therapy sessions of LSD and holotropic breathwork indicates a very fluid accessibility to *all* the domains of consciousness."

But, in fact, *all* the domains of consciousness are *not* available (holographically or otherwise) to a particular subject of consciousness. For example, an adult at moral stage 2 does *not* have access to the consciousness of moral stage 5. Moral stage 5 (with all its enormous richness and characteristics) is *not* available to moral stage 2; no amount of LSD, breathwork, or any other intense experiential drama will build the *intersubjective* structures necessary to take the role of other, reflect on those roles, criticize those roles, and move into *universal* compassion, through which alone stable transpersonal awareness can shine.

(There is a sense in which the deep structures of higher stages are indeed present as *potentials* in the individual—what I have called the "ground unconscious"—much as, for example, Chomsky believes that the brain simply comes with all languages embedded in it as potentials. But just as that does not mean that a two-month-old child can simply and fluidly access French and start speaking it, so an individual at moral stage 2 cannot simply access moral-stage-5 experiences.)

Thus, a moral-stage-2 person might experience a temporary dissolution of self/ other boundary (and might indeed do so after regressing to BPM), but that person will "contain" or "hold"—and therefore *manifest*—only a *preconventional* understanding of that "nondual" state: which is to say, that person will understand it virtually not at all. A person at moral stage 2, who cannot fluidly operate from the consensus role of the generalized other, will not stably adapt to universal mind.

POSTMODERNISM AND THE INTERSUBJECTIVE SPACE

Grof is aware of these general issues, but I feel he might be neglecting this crucial area. The tendency in his writing is to take this *intersubjective worldspace* for granted (as I will explain in a moment) and then simply *describe* the monological subject having monological experiences: the Cartesian subject has an experience, sees an archetypal image, experiences volcanic pleasure/pain, identifies with plants, has an out-of-the-body experience, relives birth, or perhaps dissolves altogether; and Grof has given a truly extraordinary catalog of these phenomenological experiences (which, for the most part, I accept as phenomenology).

But all of that phenomenology *takes for granted*, and thus neglects, the crucial point that the *subject* of those experiences is *formed* only through an *intersubjective* and *dialogical* process, and exists *only* by virtue of those (hidden) intersubjective structures, which themselves *never* come jumping into awareness with LSD or breathwork or yoga or any simply phenomenological awareness. The Grofian lists of transpersonal experiences are simply a phenomenology of experiences, and accurate and important as that is, it fails (or comes up short) at precisely the point that *all* phenomenology fails: it takes the subjective and intersubjective patterns *doing* the phenomenology for granted, and thus fails to disclose the processes and intersubjective structures *necessary* for the experiences *to be able* to unfold at all (see below).

For these experiences do not unfold in pregiven subjective space; that subjective space exists only in, and *by virtue of*, an intersubjective space formed by a developmental process of mutual understanding and mutual recognition, which is the actual space or "background" of the subjective experiences (i.e., it is the intersubjective space *in which* the experiences come and go), a background that *does not itself* enter awareness as an *experience*, and therefore *cannot* be found phenomenologically, and thus *can be found in none of Grof's cartographies* (important as they otherwise are).

(For a discussion of the background and structural unconscious—the hidden patterns of the intersubjective space—see note 28 for chap. 4. The discovery of the importance of the constructed intersubjective space *is* the essence of the genuinely post-Cartesian paradigms, from structuralism to hermeneutics to genealogy and archaeology and grammatology—a point we will return to below.)

By taking the intersubjective space for granted, then the isolated individual subject appears to be having a series of monological experiences, and Grof has given us the phenomenology of many of those experiences. But his analysis leaves intact the taken-for-granted nature of the intersubjective space (in which the particular subject and object arise). This approach is thus still caught in the fundamental Enlightenment paradigm, which it alters by *extending* what the subject *can* experience (the cartographies of nonordinary experiences), but it does not examine the intersubjective space that supports and allows the particular subject and the particular experiences to arise in the first place. The subjective space in which these experiences occur is not pregiven and unproblematic: it develops.

And *that* examination leads to the developmental and formative processes that are hidden to phenomenology, an examination that discloses the stages of the formation of the individual subject and thus discloses *what that subject can assimilate* (and stably adapt to) among the monological experiences that bombard it in dramatic experiential episodes (breathwork, LSD, etc.). That aspect of the examination leads not just to a cartography and phenomenology of ordinary and extraordinary *experiences*, but also to a developmental cartography of the intersubjective spaces *in which* the experiences arise and *can* arise (and therefore, into which they *can* be stably integrated). We will see several examples of this in the next section.

Thus, in my own approach, the study of the peak (and temporary) experiences of phenomenology (important but limited) are incorporated into a broader framework that includes the formative stages of structures of stable adaptation: the subject having these experiences did not simply parachute to earth fully formed, and thus the *types* of experiences available to it are *not* in all ways holographically pregiven and available to fluid access for *any* subject (a broad and general range of transpersonal experiences, yes; a stable adaptation and assimilation, no. I will return to this in a moment).

In short, the overcoming of the fundamental Enlightenment (Lockean to Cartesian) paradigm consists not merely in giving a new and *expanded* description of the world, but also in recognizing the intersubjective construction of the consciousness that is doing the describing, the intersubjective unfolding of the worldspace *in which* subjects and objects manifest (see note 15 above). As I think I have demonstrated throughout this book, most of the "new paradigm" theorists (who claim to be non-Cartesian, and most of whom Grof quotes with approval) are still Cartesian to the core: the subject of consciousness is simply giving a new, expanded, process-oriented, nonatomistic, nonmechanistic description of the world, without any understanding that the subject doing the describing is itself created by a dialogical process that does not enter awareness as an immediate object and thus *never* enters their "new paradigm" descriptions. (They fail to recognize the various types of the "structural unconscious," the "background hermeneutic unconscious," and the "functional unconscious" described in note 28 for chap. 4, all of which constitute the actual breakthroughs in the genuinely post-Cartesian paradigms from Heidegger to Foucault to Habermas.)

It is precisely these formative processes of the intersubjective worldspace that create an opening or clearing in which different types of subjects and objects can stably emerge: *these developmental processes define the primary arc of consciousness evolution and adaptation*. And *none* of those processes make it into the "new paradigms," and none of them show up on Grof's extended cartographies of phenomenological experiences.

Peak experiences can indeed bounce all over the spectrum of possibilities, in a specific sense: the average adult has access to the entire phylogenetic (archetypal) image forms, the various perinatal matrices, the psychodynamic unconscious, as well as the overall *general states* of gross/subtle/causal (waking/dreaming/sleep), and thus in every twenty-four-hour period the average adult sweeps through the general spectrum of consciousness (gross to subtle to causal and back again: ascent and descent, evolution and involution, the Great Circle once each day, not to mention miniature versions throughout). But those states (and the temporary experiences they generate) can be converted into stable, conscious adaptations only through a process of holarchical structuralization: converting states into traits.

No doubt a peak or plateau experience can profoundly alter people's lives, and totally shake up their accepted worldviews, and shatter old beliefs. But then, where do they carry it? What is done with it? Where does fit into their structure? How can they *manifest* it? Are they ready for stable adaptation and conscious responsibility for the *burdens* of compassion and wisdom that transpersonal structures *demand*? Or does the "insight" come and go as another sensational experience, and fade away into the night, a vaguely fond memory of what might have been?

To give only one example, again using moral stages (since these indicate how people *actually orient* themselves in interpersonal space, as opposed to their own "private" experiences): Kohlberg's moral stages have now been tested in numerous widely different cultures (first-, second-, and third-world), and *no exceptions to his basic stages have yet been found* (there are minor variations, but all are within the broad outlines). As far as we can tell, they are indeed universal deep structures (Gilligan altered the surface, not the deep, forms). [See *Integral psychology* for references.]

Thus, a person at moral stage 2 who has a profound satori-like experience (say, a temporary dissolution of self in the causal unmanifest, which is indeed ever-available), might have his belief system and worldview profoundly shaken, and this might indeed lead him to *transform* his actual moral orientation in the world, which means he would move to . . . moral stage 3. *There is nowhere else for him to go.* He will not simply jump to permanent stage-6 (or higher) adaptation (among other things, stage 6 operates upon various structures in stage 5 and stage 4, and these are not manifest, are not present: it would be like building cells with no molecules).

By centering on the fact that temporary satori-like experiences are indeed available to virtually anybody ("fluid access"), we tend to miss the crucial fact that, once the experiential fireworks have subsided, the arc of consciousness returns to its structural patterns, which themselves evolve and develop in intersubjective formative processes. I am not saying experiential disclosures aren't important (just the opposite); but I am saying they address the phenomenal realm and tend to take the subject of the experiences as unproblematic and monological, as a preexisting center of awareness against which we will throw experiences until it surrenders, instead of seeing that the intersubjective worldspace allows the arising of both the subject and object in stable awareness: altering the objective experiences without addressing the intersubjective developmental space in which the objects and subjects arise will thus be of limited lasting effect (that is the whole point of the genuinely post-Cartesian paradigms).

Some Examples

Let me give a few examples of the intersubjective space and why typical phenomenology does not disclose it (and why it took postmodernism to get beyond Cartesian phenomenology). Say you are watching a card game of poker. The cards are the actual phenomena; you can see them directly; and you could do a complete phenomenology by describing each of them very carefully. But the cards are being used according to various *rules* (the rules of poker), and those rules are not written on the cards. Those rules appear nowhere on the phenomena, and thus even a complete and exhaustive description (and phenomenology) of all of the cards would not tell you what the rules are.

In order to discover the rules—the actual *structure* (or grammar or syntax) of the game of poker—you have to perform some sort of structuralism: you have to investigate the *collective behavior* of the cards, and study their *patterns of relationships*. You have to discover which card goes where and when; you have to spot which combinations of cards "win"; you have to uncover the specific ways and patterns that the cards are used. None of this is written on any of the cards. In short, you have to study the *intercard relationships* (the intersubjective relationships). No amount of studying just the subjects (or just the objects: we are using the cards to mean either individual subjects or objects)—no amount of phenomenology—will show you the intersubjective patterns. *And yet every card is obeying those intersubjective patterns.*

The postmodern, post-Cartesian approaches are all united by a belief that background intersubjective (and interobjective) patterns govern much of the phenomenology of individual subjects and objects, a background that cannot be spotted by a mere phenomenology or a description of those individual subjects and objects. (This is why we heard Foucault, in note 1 above, explain why phenomenology could not even handle the existence of linguistic meaning: subjective phenomenology cannot spot intersubjective structures.)

Thus, we do not overcome the Cartesian paradigm by giving a more expanded phenomenology, important as that might be (and this is what Grof has contributed). For even if we have superconscious subjects and mega-cards undergoing altered states, those cards are still following patterns that can be found on none of them. The genuine post-Cartesian approaches therefore focus on the intersubjective and interobjective patterns (described at length in note 28 for chap. 4) that create a space or clearing in which the individual subjects and objects arise. And they investigate the influence of this background on individual states of consciousness (a background that explicitly shows up in none of the individual states of consciousness).

What does a "space" or "clearing" mean in this regard? Originating with Heidegger, but taking on numerous other (yet related) meanings, the idea is essentially this: reality is not a pregiven monological entity lying around for all to see; rather, various social practices and cultural contexts create an opening or clearing in which various types of subjects and objects can appear. For example, as I would put it, the magic worldspace creates a clearing in which animistic objects can appear; the mythic worldspace creates a space in which a caring God can appear; the rational worldspace creates a clearing in which worldcentric compassion can appear; the psychic worldspace creates a space in which the World Soul can appear; the causal worldspace creates a clearing in which the Abyss can be recognized. None of those are simply lying around out there and hitting the eyeballs of everybody. There is no single "pregiven world" (*the* essential insight of postmodernism).

This is why postmodernism is generally united in recognizing the fact that it is the intersubjective space (with its correlate, the interobjective realm) that *allows* individual subjects and objects to be seen and recognized in the first place. This is why the postmodern poststructuralists trace part of their lineage to Saussure, who demonstrated that language is an intersubjective structure; and to Nietzsche, who emphasized interpretation over empiricism. This is also why the great postmodern thinkers have all given their version of the background intersubjective space that cannot be grasped by mere descriptive phenomenology (contra Husserl). Heidegger uses hermeneutics or interpretive ontology (as we saw: surfaces can be seen—phenomenology— but depths must be interpreted); Foucault uses archaeology and genealogy to dig beneath and behind the phenomena (even if he insists there are only surfaces, he is looking at interobjective surfaces and their rules of behavior); Derrida uses grammatology to spot the play of signifiers and their endless contexts (none of which can be found in a phenomenology of signifiers). It is this postmodern turn—which, in its constructive form, demands that all four quadrants be taken into account—that I have explicitly attempted to incorporate, along with the important (but by itself inadequate) phenomenology of consciousness (which, by itself, remains firmly rooted in the Cartesian space).

When it comes to individual psychology, a typical example is the one we have been using: moral structures and their development. People generally feel that they are free to think whatever they like, and within limits that is true. But a person at moral-stage 2 will not, under any circumstances, have a moral-stage-5 thought. As I said, stage-5 is built of components from all of the earlier stages, just as letters go into words, go into sentences, go into paragraphs, and no one has ever seen a paragraph without sentences. In other words, the intersubjective moral structures *create a space in which individual thoughts will arise*, and thoughts that do not fit that space will simply not arise. This is what I mean when I say that the intersubjective structures create an opening or clearing in which various types of subjects and objects *can* arise (and that is perhaps the root insight of the various postmodern movements, including even Habermas).

Two points about this example. One, no altered state or peak experience will produce a moral-stage-5 thought in a moral-stage-2 person. Two, no phenomenology of any subjective experience, at any stage, will disclose the structures of that stage. Phenomenology has to be supplemented with structuralism (or any background analysis as described in note 28 for chap. 4). That is what virtually all of the "expanded cartographies of consciousness" miss; and that is why, however transpersonal they are, they are also still Cartesian. An "all-level, all-quadrant" approach is meant specifically to address this limitation.

Past Actuals and Future Potentials

This further points up the profound difference between experiences that, on the one hand, are essentially a process of *past recovery* (or reliving), and those that are essentially a process of *future discovery* or developmental building. To take them in that order:

The average adult already has, for example, all of the structures in place that involve the birth trauma. Those perinatal structures have *already* been laid down; they are simply unconscious, or unrecovered. All that is required, so to speak, is a lowering of the repression or amnesia threshold (however conceived). The same is true of the Freudian or personal dynamic unconscious and the phylogenetic or archaic heritage: they are *already there* in actual structural form (but are simply "forgotten" or "submerged").

Experiential *recovery* or *reliving* of those already formed structures can thus, without too much fanfare, be therapeutic in and of itself (I fully agree with Grof on that issue). Of course, these experiences will still have to be worked through, and they will still have to fit into the ongoing march of structuralization that defines all growth and evolution (tenet 12c). And they will still have to be assimilated into the subjective space of the self-sense. Thus, to stick with moral stages as an example: a person at moral stage 5 might experience plant identification as an indication of worldcentric One Life; moral stage 1 will experience it as an extension of its own narcissistic self and arrive at an ontological theory of the Wonder of Being Me.

But those previous, *already evolved* structures (phylogenetic and ontogenetic) can indeed be *recovered* by any techniques that sufficiently lift the repression/amnesia barrier (LSD, yoga, breathwork, therapy, intense stress, etc.), and thus, working through those experiences does not, *in itself*, demand a fundamental and massive transformation of the self-sense. The reliving and working-through can indeed dissolve various pathologies created by the repression or submergence of the past structures. And this recovery (apart from dissolving certain pathologies) might also lead to a transformation that notches the self up from, say, moral stage 2 to stage 3, or stage 3 to stage 4, as we saw.

But primarily this is a reworking (often therapeutic) of structures already present but submerged: the *primary* threat is in lifting the pain (or amnesia) barrier that constituted the repression; the threat is not that the self will have to transform upward by fundamentally surrendering its present subjective space: it simply has to work, into that space, some material that could just as well *already* have fit into that space if it weren't for the repression. And even if (by, for example, plant identification) the self changes its objective worldview, a change of objective worldview is not necessarily a change in the subjective space that holds that worldview. Saints as well as egotists can champion "One Life."

But that *recovery* of past structures *already laid down* (by phylogenetic or ontogenetic development) but subsequently repressed or submerged (or at any rate forgotten)—that is quite different from building future structures as they unfold, for the first time, in manifest consciousness. Here the intense experiential disclosures do not have to be worked into a subjective structure already present, but rather have to be part of a subjective and intersubjective process of building a structure not yet in existence: experiences have to be part, not of structural uncovering, but of structural building.

In short, there are very important differences between digging up *past actuals* and bringing down *future potentials*. Past actuals (structures *already* laid down in evolution but subsequently submerged) include the phylogenetic heritage and the ontogenetic heritage (from material/plant/animal identification to perinatal matrices to Freudian psychodynamic unconscious to collective archaic unconscious). Although it might take a transpersonal awareness to access some of these more remote structures, the point is that they are *already* formed, already evolved, already laid down, and so of course we have a type of "fluid access" to them (once the amnesia barrier is penetrated by whatever means).

But future potentials are available only as *potentials*: the structures themselves have *not yet* been built in the particular evolution of the particular individual (just as the higher moral stages are only future potentials to the lower stages). And thus, at most, the average individual has access to random, *temporary experiences* (across the broad range of states, gross and subtle and causal)—experiences which, if they are to enter consciousness as permanent, stable, responsible traits, must be built into the ongoing stream of subjective and intersubjective structuralization. A person at moral stage 2 cannot simply unearth and instantly have fluid access to all of the characteristics of moral stage 5—these future potential structures must be built and created *intersubjectively*; they are *not* fully formed, buried treasure chests lying around and waiting to pop to the *subjective* surface under hyperventilation.

The same is true of linguistic and grammatical structures, cognitive development, Loevinger's self-development, Habermas's communicative competence, psychoanalytic object relations—the list is virtually endless, and none of those involve fluid access (and unearthing) of an already formed occasion, but rather interpersonal and intersubjective building and creating. All of that tends to get overlooked in the excitement of digging up a past actual (e.g., plant identification) or of experiencing an infusion from subtle or causal states (e.g., satori-like experiences).

Temporary peak experiences from higher domains (subtle or causal) can indeed *accelerate transformation* toward those domains, but cannot, for example, simply jump a person at moral stage 2 to moral stage 5 or beyond. There is no evidence that suggests that can happen, unless you *ignore* the intersubjective space, at which point it *appears* that a simple transpersonal experience has created a transpersonal subject, bypassing the unavoidable stages of subjective and intersubjective growth.

I previously gave the example of a moral-stage-1 individual who has a temporary peak experience of phylogenetic identity with all plants/animals: the experience is eventually and necessarily interpreted in terms of the self-structure at moral stage 1: namely, narcissistic and preconventional. We see this all too often in "weekend enlightenment seminars," where the person dips into the transpersonal well and emerges as a "transformed" person, namely, as an insufferable egotist, around whom the world now revolves because the person is "empowered." This is not the dissolution of the Cartesian ego, but its hyperinflation to cosmic proportions: a temporary transfusion of higher domains has empowered a monster, as Charles Taylor might put it.

(These experiences may indeed—and often do—alleviate certain pathologies by virtue of putting the person in touch with *past actuals* that were previously submerged and then liberatingly unearthed in a dramatic recovery experience. And these experiences may, *at best*, notch the person up a stage or two on the interpersonal scale. But just as often, they otherwise simply *reinforce*, via a jolting transfusion, the person's already present level of adaptation.)

If the dynamics of these two very different processes (past actuals and future potentials) are confused and conflated, and thus both are equally modeled on the "already fully accessible" paradigm (which is true only of the submerged unconscious of past actuals and certain temporary transpersonal experiences and states), then it appears that the engineering of subtler objective experiences is the primary task, instead of also centering on the subjective and intersubjective processes of growth and development which alone can stably accommodate the new experiential disclosures.

It is in the growth of subjective structures, themselves set in intersubjective modes of mutual understanding and mutual recognition, that the Western paths of orthodox developmental psychology meet up with, and run into, the contemplative paths of deeper/higher growth and evolution (mystical traditions East and West). This is an area that I believe Grof has neglected. My respect for Stan's work is well known. I am simply suggesting that he might be a little more sensitive to *where* these experiences will fit in the ongoing developmental process of subjective (and intersubjective) space, into which and through which these experiences come and go.

At the same time, this would open Grof's model to the vast amount of research (clinical, philosophical, experimental, dialogical, and therapeutic) on the formation of these intersubjective structures of competence (Piaget, Austin, Searle, Selman, Loevinger, Kohlberg, G. H. Mead, Heidegger, Gadamer, Habermas, etc.), research that, after all, covers an enormous amount of ground, which no comprehensive model can afford to ignore—and research that is at the heart of the *genuinely* post-Cartesian revolution. [See *The eye of spirit*, chap. 7, for an extended discussion of Grof's work.]

The Passion of the Western Mind

Richard Tarnas's *The Passion of the Western Mind* is a fine narrative, thoughtful and insightful. But in the epilogue, Tarnas's presentation points up the extremely difficult and subtle issue of actually coming to terms with modernity, both what it accomplished and what it demolished.

The general idea, as Tarnas presents it, is that the modern mind has emerged from an earlier, premodern *participation mystique* (which he also calls "undifferentiated unitary consciousness"), through a painful process of alienation and separation and dualism (the Ego of modernity), and is now on the brink of a dialectical synthesis and reunion of subject and object, mind and nature, male and female, inner and outer, knower and known.

This scheme finds support, Tarnas believes, in the discoveries of depth psychology (Freud, Jung, Hillman, Grof), since these approaches have undermined the absolute autonomy of the disengaged and rational Ego (as we have seen, depth psychology was indeed one of numerous approaches that began to decenter the Ego).

At this point, Tarnas turns to a Grofian explanation based on the Basic Perinatal Matrices (or fulcrum-0, which moves from a prior undifferentiated unity of fetus and womb, through a volcanically painful differentiation process, to a post-uterine resolution and integration). Tarnas wants to see *all of Western history playing out these three major subphases of the perinatal birth trauma.*

Although Tarnas is careful to point out that this "birth drama" of the Western mind cannot simply be reduced to the biological birth trauma, there is a constant tendency to do exactly that. "The dynamics of this archetypal development [of the collective Western mind] appear to be essentially *identical* to the dynamics of the perinatal process." (All quotes are from pp. 430–44.)

And this means that all of the various fulcrums of differentiation/integration will be collapsed into one single, overarching perinatal Fulcrum. Fulcrum-0 becomes *the* fulcrum of the entire Western mind. And this means Tarnas wants to see all of Western history as having one long "metatrajectory" with three stages. The Western collective psyche, he says, started out in an undifferentiated oneness with nature and the whole universe ("connection with the whole," although what the "whole" means is not really specified; we will return to this in a moment). Then the collective psyche began a painful process of separation, differentiation, and alienation (there is no distinction here between differentiation and dissociation; there is simply alienation). And that dualistic alienation brings us up to the present, up to the modern and postmodern mind (on the brink of tomorrow's reunion with that which was historically *lost* in the middle phase of alienation).

Thus, "the fundamental subject-object dichotomy that has governed and defined modern consciousness—that has *constituted* modern consciousness, that has been generally assumed to be absolute, taken for granted as the basis for any 'realistic' perspective and experience of the world—appears to be rooted in a specific archetypal condition associated with the unresolved trauma of human birth."

With this allegedly *unresolved birth trauma* (the collective Western mind stuck, as it were, in the second subphase of fulcrum-0), "an original consciousness of undifferentiated organismic unity with the mother, a *participation mystique* with nature, has been outgrown, disrupted, and lost."

Note the curious juxtaposition of "outgrown" with "disrupted and lost." When we normally describe something as "outgrown" (e.g., I have outgrown thumbsucking) the "outgrowing" is usually a good and desirable thing, whereas something "disrupted and lost" is usually bad and alienating. But confusing these two is part of what seems to be Tarnas's confusion of differentiation and dissociation. Modernity's necessary differentiation of the subject from an unreflexive immersion in magico-mythic syncretism must therefore actually be an unresolved birth trauma that is *primarily* alienating.

Tarnas is clearly aware of the necessary movement of modernity's "separation," but because he doesn't distinguish differentiation and dissociation, he is most taken with modernity in its alienating modes. He constantly refers to "separation" and "dichotomy" and "dualism" as being the same thing. "Here [in the unresolved birth trauma], on both the individual and the collective levels, can be seen the source of the profound dualism of the modern mind: between man and nature, between mind and matter, between self and other, between experience and reality—that pervading sense of a separate ego irrevocably divided from the encompassing world."

All differentiations are thus converted merely to dissociations. At the same time, Tarnas collapses all fulcrums into fulcrum-0. That is, a further confusion, as I see it, is that the *dynamic* of fulcrums *in general*—a dynamic that always moves from fusion to differentiation to integration (that is the form of all nine fulcrums that we examined—this general dynamic of all fulcrums is simply treated as being *basically perinatal* (simply because the perinatal fulcrum also follows the same general dynamic, and is the first to do so). And thus, as soon as any intense differentiation/integration process occurs (as soon as any fulcrum is spotted), it is called a perinatal process. Consequently, all of the other and significant fulcrums tend to get collapsed into the most fundamental fulcrum (F-0).

This leaves Tarnas with only three major stages of growth and development (and only *one* birth process) through which he must squeeze all of Western history, instead of seeing that the death/rebirth (or differentiation/integration) process has collectively occurred at least five or six major times, each of which was a profound, world-creating, world-birthing transformation (e.g., archaic to magic to mythic to rational to centauric). To say that in each of these cases development basically repeated the perinatal drama on a different level is simply to privilege the most fundamental fulcrum and try to make it explain the more significant fulcrums that followed upon it (hence the enduring danger of reductionism that is vigorously denied and nevertheless constantly embraced as explanatory).

Likewise, the collapse of all fulcrums into the perinatal fulcrum leaves Tarnas with the Single Boundary fallacy, which, as I suggested (note 3 for chap. 6 and note 32 for chap. 13) tends to mark all retrogressive (and reductionistic) theories of development: since there is only one major boundary in consciousness (or one major fulcrum of differentiation/integration, instead of the nine or so boundaries or fulcrums that more adequately describe development), then the creation of that "single" boundary tends to be seen as a primal alienation, a primal loss, instead of a necessary and appropriate process of growth and differentiation (thus the juxtaposition of "outgrown" with "disrupted," as if the oak were simply an alienating disruption of the acorn).

With his Single Boundary fallacy, Tarnas can then reduce *all* differentiations (and all dissociations) to the perinatal boundary and the unresolved birth trauma that is somehow inhabiting the entire Western mind.

That, as I said, is to reduce the boundary phenomenon of *modernity* to a perinatal birth boundary. But the *defining* boundary of modernity (modernity's Ego/world boundary) is a fulcrum-5 boundary: it is five complete boundaries down the road from the birth trauma. (All of the previous boundaries are enfolded in the compound individual of modernity, and they unfold in its own ontogeny, but they do not *define* the *emergent* component *added* by modernity.) That modernity's *specific dissociations* might indeed have reactivated a similar perinatal birth matrix (F-0 subphase 2) is certainly possible, but that completely *reverses the causal order* of Tarnas's scheme. Tarnas is invoking, as cause, that which is actually effect; and further, his "cause" is deeply misread as perinatal (based on the Single Boundary fallacy).

Thus, modernity's *specific* (fulcrum-5) boundaries actually involved the standard dynamic of growth from a previous relatively undifferentiated state (in this case, mythic-syncretism) to differentiated (the Big Three) to integrated (still in progress), an ordeal that can go pathological as fusion or failure to differentiate (on the one hand) or dissociation and hyperdifferentiation (on the other). But specific fulcrum-5 problems cannot be reduced to fulcrum-0 disasters: that wipes out the major transformations of fifty thousand years that prepared the way for modernity.

If there are F-0 pathologies, then these can indeed infect subsequent development. But an F-0 birth trauma could not be the major cause of modernity because it implies premodern societies had no similar birth traumas at all (since they certainly had no modernity). Rather, the dissociations in the F-5 boundaries of modernity could *anchor* themselves in the "existential" subphase of F-0 (subphase 2) in a way that would *not* happen in premodern times because there were no collective F-5 boundaries in premodernity to do the anchoring, even though they went through essentially the same perinatal birth process.

Either way, the perinatal matrices are not the cause of modernity. Rather, modernity's *own problems*, generated primarily in fulcrums 5 and 6, might indeed have reactivated a fulcrum-0 subphase and *caused* it to be *selected* as the bodily grounding of its *own* self-generated dilemmas: being spat out of the womb has hit every human being from day one; modernity hit only a recent few.

Thus, as a general point of agreement, I would say that modernity was indeed caught in subphase 2 (the differentiation subphase) of a developmental fulcrum, but it is fulcrum-5, not fulcrum-0. It is fulcrum-5 subphase 2 (and not BPM II/III, or fulcrum-0 subphase 2) that is Hegel's "monster of arrested development," or aborted birth in this case. As I repeatedly pointed out in the text, modernity's own differentiations fell into dissociation by lacking the higher, integrative, and resolving stance of vision-logic (which would fulfill fulcrum-5 and act as the base of fulcrum-6, thus integrating the "monster"; we will return to this actual alienation in a moment).

Since the biological birth matrix, like all fulcrums, also has a subphase 2 (painful differentiation), there is a broad sense in referring to modernity's "birth pains." And that, I suppose, is the general way Tarnas (and Grof) use the BPMs. But after vigorously denying any biological reductionism, they often tend to immediately do so anyway.

This certainly seems like an attempt to reduce a very complex and multidimensional situation to a simplistic and highly "ordered" state of affairs, as if complexity must somehow be defused in a simple 1-2-3 move, instead of seeing that, as I said, the "1-2-3" move (a fulcrum of development) has repeated itself in numerous profound (and profoundly different) transformations across vast stretches of historical time.

There then follows the standard retro-Romantic pre/trans fallacy: the "original unity with the womb/world," which has been historically "disrupted and lost," will now be *resurrected* in a *mature* form, and Tarnas makes exactly that claim.

An altogether fatal problem with these types of views is that the alleged "original union with the whole world" is *not* a "union," for one thing, and it does *not* involve the "whole world," for another. Even if development were just this single and simple 1–2–3 move (a single historical fulcrum as a "metatrajectory), it is *still* inadequate on both counts, in my opinion.

The original "union" state (whether conceived as the actual womb or as the premodern and prehistorical *participation mystique*) is *not* a union because it is simply an undifferentiation. That would be like saying the acorn unifies all the branches and all the leaves of the oak, whereas they are not actually differentiated in the first place: it is not a *union* of these later elements, because these elements do not yet exist. (We can say it "unifies" them *in potential*, I suppose, but the point again is that the particular union is indeed a potential, not an *actual*, and so, in the future, there is no *past actual* to resurrect. Thus, in either case, *there is no actual union to recontact*. The only thing the acorn actually unifies is its own cellular components, and the oak continues to do this or it would simply die, it would not limp into the future "alienated"; the "problem" the oak faces is how to unify its leaves and roots and branches, and this the acorn does not even remotely do. Just so, neither the womb nor the premodern state was a union of the Big Three, because those were not yet differentiated entities that could be united; they either did not exist or were largely undifferentiated.)

Likewise, this primal "union" is *not* one with "the whole world" because the leaves and branches are not yet even part of the world; the only "whole world" that is actually present is minuscule, relatively speaking. Likewise, for humans: the fetus in the womb is not one with the *whole world* of intersubjective morals, art, logic, poetry, history, economics, etc., from which it will supposedly "separate" and "alienate" itself upon birth; rather, it is one with a biophysical matrix that it will necessarily differentiate. (And the premodern matrix did not unify the Big Three, but simply failed to differentiate them in the first place; this is a fusion, not an integration or an actual union.)

From any conceivable angle, there is *no actual union* to resurrect; there is only a predifferentiation that is about to be outgrown. It is surely unacceptable to call this diminutive and undifferentiated state a "union with the whole world."

Incidentally, for a typical psychological approach to development based on this confusion, see Washburn's *The Ego and the Dynamic Ground* (1988), which stands as a classic attempt to define oaks in terms of recaptured acorns; and likewise Nelson's attempt, in *Healing the Split* (1994), to integrate a developmental model with Washburn's regressive orientation. Both Washburn's and Nelson's attempts are based on variations of the Single Boundary fallacy, which usually brings numerous pre/trans confusions in its wake. The "original union" of the infant state is said to be an "unconscious union with the ground or whole," which is supposed to be the essential *spiritual state* that will be *resurrected* in a conscious, mature form, once the ego regresses to this state and reunites with it. The infant is said to be "more open" to this spiritual state than the adult.

But note that, as all parties would acknowledge, the infant *cannot take the role of other* and therefore cannot possess anything resembling genuine love and compassion and interpersonal care and concern. Genuine love and compassion involves at least the capacity to put oneself in another's shoes and consciously act in their behalf, often against one's own inclinations, and this the autistic infantile self cannot do—it doesn't even recognize an other, let alone take its role.

In other words, this "original union" cannot step outside of its own egocentric orbit, and it thoroughly lacks intersubjective love and authentic compassion. Why would that egocentric lack of love and compassion be "spiritual," and why would we want to recapture that? If Spirit, among other things, is Love, how could the infantile self be "more open" to spirit when it cannot evidence love or compassion in any authentic sense? To counter by saying that the infantile self is spiritual because subject and object are still one, is precisely to fall into the Single Boundary fallacy, which is exactly the confusion that drives these theories.

Some adults, under regressive-type situations, do indeed recover childhood memories, but they *inhabit* those memories from within a now broadly rational and *perspectival* cognition. That is, they might "remember" taking the *role of other* at age two or three, and report *vividly* what this entailed. But clinical and empirical evidence repeatedly and consistently shows that actual children at that age *do not* and cannot take the role of other (the Three Mountain experiment, for example).

Memory traces of childhood, some of which are broadly accurate (and many of which are not), are, in the adult, reactivated and reinhabited by an *irrevocably pers-pectival* consciousness, and then those memories are "read" by that highly evolved consciousness *as if* that were the way they were originally experienced. Once one has learned to take the role of other, that capacity is not lost (except under the severest pathologies), and so no matter how much the adult "regresses," the recovered "mem-ories" are *recovered in an intersubjective and perspectival space*, which then instantly and automatically interprets them *from that space*, and then reads them back into childhood as if they were accurate and indelible photographs of the original occasion, even if children at that age show no evidence of being able to do any of this.

Certain types of memory traces are in some cases actually there and broadly accu-rate, perhaps even back into the womb (as some researchers claim), and we may suppose that they leave their imprints, like bruises, on the psyche. But when adults "recover" these imprints, they reanimate them with a consciousness that brings to life aspects of these imprints that were simply not registered at the time. The imprints, assuming they are there, are run through a highly evolved intersubjective space that can then pull out implications and perspectives that were not present when the im-prints were laid down, and this "pulling out" is then mistakenly ascribed in all ways to a *recapturing*, a recapturing that goes quite beyond past actuals and into a *reconsti-tuting* of past actuals created by highly sophisticated structures of present conscious-ness. All of this imbues these lower structures with capacities that massive amounts of empirical evidence deny they possess.

No doubt, as we said, infants and young children can have an infusion of transper-sonal states (usually subtle), but this is an infusion *into* an intersubjective space that is fundamentally *preconventional* and *egocentric*, not *postconventional* and *compas-sionate*, and thus they cannot be carried in any evolved spiritual awareness. [See *Inte-gral psychology* for an extensive discussion of this issue.]

This infantile "spiritual union" seems especially attractive (and "divine") when the Ego itself is painted in the very worst possible light: following the Romantic agenda, the Ego is defined as a dissociation and not as a transformative and transcending differentiation. The mature Ego is simply called "divisive," "abstract," "analytic," and so forth, overlooking the fact that it is the mature Ego that eventually builds a universal and worldcentric consciousness, the *first* consciousness in all of development through which a truly postconventional and universal Love and Compassion can begin to flow, a worldcentric perspective that unites more, integrates more, contains more union and more wholeness than the infantile self could even begin to imagine (literally).

As a *generalization*, then, it is the Ego, not the infantile self, that is "more open" to the spiritual state (and the trans-egoic states are *even more* open, and this "even more" does not include a return to the infantile lack of love and lack of compassion in order to "recapture" this "spiritual ground" that was "lost"—a notion that can only be sustained by the Single Boundary fallacy). [See *The eye of spirit*, chap. 6, for an extensive critique of Washburn's model.]

Sometimes these theorists (and Tarnas is one) insist that by "whole world" they actually mean the "universal matrix" or the Ground of Being. True, the undifferentiated womb state (however conceived) is indeed one with the Ground of Being, but then so is absolutely everything else; thus, there is no reason whatsoever to *privilege* this state in that regard. Further, in terms of *consciously realizing* this union with Ground, this state is the *farthest removed*, since it has the most number of stages of growth lying ahead of it before it can *consciously* awaken to and as the empty Ground.

If (as an entirely separate issue) some pathology occurs in the differentiation process (it either remains stuck in fusion or goes too far into dissociation), then healing that pathology does indeed often mean recontacting the alienated aspects (the *past actuals*), but these past actuals are aspects of those lower levels themselves that were once present and then alienated: they are *not* the higher level that was never present but will emerge to integrate the overall system. Again, the *defining* aspect of the higher integration is not *recontacting* a past actual, but rather involves the *emergence* of something largely novel (never historically present and thus never historically lost).

All of these factors get collapsed by Tarnas into the simple 1–2–3 scheme (around the Single Boundary): original unity with the "whole world" or "universal matrix," then historical alienation from "the" world, then reunion with "the" world: "Yet full experience of this double bind, of this dialectic between the primordial unity [the 'whole world'] on the one hand and the birth labor and subject-object dichotomy on the other, unexpectedly brings forth a third condition: a redemptive reunification of the individuated self with the universal matrix."

This seems to be a confusion between the process of *involution* itself and specific events in *evolution*. Involution, or the superabundant Efflux of the One into the Many, is simply confused with biological birth. The "universal matrix" (the One) thus becomes the actual womb, and the biological birth process becomes the separation of the individual (and the Many) from the One, which produces the "dualistic world," which is then somehow reunited with the One after the biological birth trauma is "resolved" by the Mother embracing the infant (which becomes the "nondualistic" state).

This reductionism is sweeping: "A redemptive reunification of the individuated self with the universal matrix. Thus the child is born and embraced by the mother, the liberated hero ascends from the underworld to return home after his far-flung odyssey. The individual and the universal are reconciled. The suffering, alienation, and death [birth trauma] are now comprehended as necessary for birth, for the creation of the self. A situation that was fundamentally unintelligible is now recognized as a necessary element in a larger context of profound intelligibility. The dialectic is fulfilled, the alienation redeemed. The rupture from Being is healed. The world is rediscovered in its primordial enchantment. The autonomous individual self has been forged and is now reunited with the ground of its being."

There is indeed a type of "recovering" or "reuniting" with the One (the whole process of Ascent back to the One, as we have seen; and we will return to this process in a moment). But the first thing to notice about that "reunion" is that the "oneness" or the "whole" that we are to "recapture" *never existed anywhere in previous evolution*. The "primordial enchantment" that was the historical "womb" of collective modernity was *not* a sweet holistic integration with the entire Kosmos and the One: it was an often brutal indissociation that demanded seven planets for seven bodily orifices and was, on more than one occasion, willing to burn anybody who disagreed. There is no way—and little reason—to "reunite" with that, any more than we would want to reunite with "mature" thumbsucking.

The *alienation* that marked modernity was not an alienation from the previous mythic and allegedly "unified" structure (which was often violently syncretic), but rather, after the Big Three had been irrevocably differentiated, there was as yet no obvious way to reconcile and integrate them: nothing was *lost*, something was simply *not yet* found, and this let the Big Three fall into fragments. *Within* the entirely new worldspace of modernity, a struggle to unite the *newly emergent* components of that worldspace was as yet unfulfilled. The fulfillment *itself* would not consist in any sort of reuniting with anything lost, but rather involved a newly emergent integration, which would bring a new synthesis entirely unprecedented.

That modernity *also* involved some repression of *previous* structures is in itself nothing new: myth repressed magic wherever it could. Modernity had its own spectacular repressions, but those repressions are not what *defined* modernity—which was the differentiation of the Big Three—and thus merely reversing its repressions of any *previous* structures would not itself constitute the *emergence* of the *higher* structure (e.g., vision-logic), which could integrate the defining domains of modernity itself. To say that modernity was defined *primarily* by its repressions *is* the standard retro-Romantic stance, the pre/trans fallacy that confuses differentiation with dissociation and thus sees salvation as a recapturing of something lost with the Crime, instead of moving forward toward something not-yet-born at the time.

Thus, modernity did not *alienate* a nature that was *integrated* in premodernity. A differentiated nature was *never* present in premodernity to be integrated in the first place. Modernity differentiated nature and then failed to integrate *that*: it did *not* fail at something premodernity had succeeded in. Rather, it failed in the second part of a task that no previous period had even attempted. It did not lose something that must be regained: it has not yet found that which will integrate the new differentiations.

As with most subscribers to the pre/trans fallacy, Tarnas himself realizes that this new integration that will "heal modernity" cannot *actually* be a reunion, and so, after laboring to demonstrate that it *is* a reunion, he quickly but gingerly begins to distance himself from that view. Actually, he says, the "recovery of the whole" will really be "a *new* and *profoundly different level* from that of the primordial unconscious unity." It involves, he says, something "*fundamentally new* in human history" (my italics).

Profoundly different, fundamentally new. This is no more a reunion than an oak's discovering that it is one with the whole forest is a reunion with its acornness. But this does not stop Tarnas from unfortunately lapsing back into a *recovery* of *past actuals*: the new state will be "a reconciliation with lost unity, a triumphant and healing reunion" (with something once possessed in previous history or prehistory).

A more adequate view involves, I believe, the recognition that the dynamic of the stages of *evolutionary growth* does indeed involve a series of deaths and rebirths: a series of undifferentiation to differentiation to integration; each new integration is then, in its own way, undifferentiated with respect to further growth, and so it becomes the new ground for a new round of fusion/differentiation/integration (each is a new *fulcrum* of growth). The birth trauma and perinatal process is the *most fundamental*, and *least significant*, of these many fulcrums winding their way back to God.

Most *fundamental*, because it is the *first* fulcrum in individual evolution. It is the foundation, the first manifestation of growth back to the Self; it sets the basic perinatal matrices and patterns that will ground the manifestation of consciousness (and thus I have no doubt that in intense experiential disclosures, as with LSD or breathwork, it might be the bodily form reassumed in the disclosure, as Grof repeatedly maintains). But in itself it is also the *least* significant, because the separation/individuation from the biophysical perinatal matrix explains nothing that is *specifically* cultural, mental, social, artistic, economic, political, scientific, etc., which are all worlds that emerge after the natal drama is finished. Yes, pathologies at fulcrum-0 can influence, but do not cause, higher fulcrums. Each higher fulcrum is an *emergent* that brings new capacitates, new desires, new cognitions, new motivations, and new pathologies that cannot be reduced to, or explained by, the birth fulcrum.

If the higher boundaries and fulcrums are all collapsed to the most fundamental (fulcrum-0), and the Single Boundary fallacy is embraced, then all subsequent movements of fusion/separation/integration must be seen as a replay of the first fulcrum: reductionism at its worst, it seems to me.

The passion of the Western mind (and the Eastern mind as well) is not to recover what was prior in evolution, but rather what is prior in involution—to recover what is prior to manifestation itself, prior to the whole movement of involution, prior to time and duality and wombs and tombs and birth and death altogether: the self-existing, self-luminous, self-revealing, self-unfolding Spirit that empties itself into and as the entire Kosmos. To recapture *that* is the passion of the Western mind.

And the evolution of that Western mind was not an alienating move away from the Answer, but a series of profound (and heroic and painful) steps heading toward that Answer, the Chaotic Attractor as a future potential that was the Source and the Goal of the entire display, a future potential that became *actual* in any individual who stepped off the display and into the Abyss, from which the individual was reborn, not from the womb, but from Emptiness itself—and *as* the entire Kosmos.

Jean Gebser

A final word about Jean Gebser's work, since he has written at length about the coming transformation of the integral-aperspectival mind, now haltingly in progress.

Gebser believed that the integral-aperspectival structure at its fullest was a transparency to *the* whole, the Itself, the fully Spiritual, an unobstructed "access" to the total horizon, *das Ganze* (the whole). But this should not surprise us; we have seen the ever-present tendency to mistake one's present wholeness for Wholeness itself, whereas it is merely a phase in an ever-greater unfolding. And when Gebser suggests that Jesus and Eckhart were embodying the integral structure, we know that he is short of the mark (beyond the integral are the psychic, the subtle, the causal, and the nondual). Georg Feuerstein, in *Structures of Consciousness*, points out, with reference to a continuous satori (which would indeed mark the ultimate or nondual stage), that "it would appear Gebser was only marginally aware of this and that he might even have misunderstood it," and that there is "a depth of spiritual realization in Eckhart that is absent in Gebser." [Feuerstein has subsequently agreed that beyond Gebser's integral structure lie at least psychic, causal, and nondual. See his "Jean Gebser's Structures of Consciousness and Ken Wilber's Spectrum Model," in Crittenden et al., *Kindred Visions*.]

As for Gebser's confusion regarding the archaic structure and the Origin, or *Ursprung*. Of the archaic structure, Gebser says it is "closest to and presumably originally identical with origin." In fact, what the archaic structure was closest to was not the Origin but the great apes and hominids. Feuerstein chastises me for accusing Gebser of a form of the pre/trans fallacy, and of course in any sort of blatant fashion Gebser does not do so; but it is rampant in his confusing of prerational archaic with transrational causal, prerational magic with transrational psychic, and prerational mythic with transrational subtle. The causal Witness, for example, is not "close to the archaic" at all; it lies in precisely the opposite direction. Nor are subtle archetypes located in the mythic structure: all dimensions are based on the transrational Forms lying next to the causal, not next to the magical.

In the terms of Da Avabhasa, the integral-aperspectival structure is a stage 3 to 4 structure. There lie beyond it stages 4 (psychic), 5 (subtle), 6 (causal), and 7 (nondual). Are we really to imagine that *any* of the examples Gebser gives of the integral structure—say, Stravinsky, Cézanne, Frank Lloyd Wright—were anywhere near the seventh stage?

Feuerstein tries to support Gebser's claim that the archaic consciousness is "closest to and presumably originally identical with origin" by saying that "it is closest to the ever-present origin solely in terms of the simplicity of its internal configuration." But that will not do, first, because Origin is neither simple nor complex, but acategorical: it does not lie closer to simplicity than to complexity, but embraces both equally in transparency. *That* cannot be used as a yardstick, and even if it were, then why not push it back further and recognize that, given the even greater simplicity of the ape mind, the ape was even closer to the Origin? That, again, is pushing it precisely in the wrong direction. But there is no reason to privilege archaic humans in that regard: worms were even closer to *that* origin.

Feuerstein then points out that what Gebser must really mean, in part, is that the archaic is close to the Origin in *potential*, but then Feuerstein himself immediately points out that that is true of *every* structure, since Origin is equally the ground of each. And, I would add, since each structure is *in potentia* identically close to Origin, the only other measure is *actually* (or self-actualizingly) close to Origin, and by that only acceptable measure, the archaic is, of course, the farthest from the Origin.

I am a great fan of Georg Feuerstein's work, and have been honored to write a foreword for one of his many superb books, and this is probably the only thing we have ever disagreed on sharply, in a friendly sort of way. [In the above-mentioned article, Feuerstein, while not completely agreeing with my view of the archaic/origin, does acknowledge that Gebser's position on this issue is something of a "cop-out" that "contains unresolved difficulties."]

In the meantime, both Feuerstein and I definitely agree that beyond the rational-ego is indeed the integral-aperspectival mind; that it is holonic in nature, dialectical, dialogical, with a bodymind integration, linguistically transparent, and opening onto a more truly conscious spiritual orientation. We both see evidence of that mind collectively emerging, and that is one of the topics of the next volume.

References

Abend, S.; M. Porder; and M. Willick. 1983. *Borderline patients: Psychoanalytic perspectives.* New York: International Univ. Press.

Abhinavagupta. 1988. *A trident of wisdom.* Trans. J. Singh. Albany: SUNY Press.

Abraham, R.; and C. Shaw. 1985. *Dynamics.* 3 vols. Santa Cruz: Aerial.

Abrams, M. 1971. *Natural supernaturalism.* New York: Norton.

Abrose, E. 1990. *The mirror of creation.* Edinburgh: Scottish Academic Press.

Achterberg, J. 1985. *Imagery in healing.* Boston: Shambhala.

———. 1990. *Woman as healer.* Boston: Shambhala.

Adair, R. 1987. *The great design.* New York: Oxford Univ. Press.

Adler, G. 1979. *Dynamics of the self.* London: Coventure.

Adorno, T. 1973. *Negative dialectics.* New York: Continuum

Albert, M., et al. 1986. *Liberating theory.* Boston: South End Press.

Alexander, F. 1931. Buddhist training as an artificial catatonia. *Psychoanalytic Review* 18: 129–45.

Alexander, S. 1950. *Space, time and deity.* New York: Humanities Press.

Alkon, D. 1992. *Memory's voice.* New York: HarperCollins.

Allaby, M. 1989. *Green facts.* London: Hamlyn.

Allison, J. 1968. Adaptive regression and intense religious experiences. *J. Nervous Mental Disease* 145: 452–63.

Allport, G. 1955. *Becoming.* New Haven: Yale Univ. Press.

Almaas, A. 1986. *Essence.* York Beach, Maine: Samuel Weiser.

———. 1988. *The pearl beyond price.* Berkeley: Diamond Books.

Anderson, L. 1991. *Sisters of the earth.* New York: Vintage.

Anderson, S., and P. Hopkins. 1991. *The feminine face of God.* New York: Bantam.

Anthony, D.; B. Ecker; and K. Wilber, eds. 1987. *Spiritual choices.* New York: Paragon.

Apel, K. 1980. *Towards a transformation of philosophy.* London: Routledge & Kegan Paul.

Aquinas, T. 1969. *Summa theologiae.* 2 vols. New York: Doubleday/Anchor.

Arieti, S. 1955. *Interpretation of schizophrenia.* New York: Brunner.

———. 1976. *The intrapsychic self.* New York: Basic Books.

Aristotle. 1984. *The complete works of Aristotle.* Ed. J. Barnes. Princeton: Princeton Univ. Press.

Armstrong, A., ed. 1953. *Plotinus.* London: George Allen & Unwin.

Artaud, A. 1965. *Artaud anthology.* San Francisco: City Lights Books.

Assagioli, R. 1965. *Psychosynthesis.* New York: Viking.

Astavakra. 1982. *The song of the self supreme.* Clearlake, Calif.: Dawn Horse Press.

Augros, R., and G. Stanciu. 1988. *The new biology.* Boston: Shambhala.

Aurobindo. n.d. *The life divine* and *The synthesis of yoga.* Pondicherry: Centenary Library, XVIII–XXI.

Avalon, A. 1974 (1931). *The serpent power.* New York: Dover.

Ayala, F., and J. Valentine. 1979. *Evolving.* New York: Benjamin-Cummings.

Baars, B. 1988. *A cognitive theory of consciousness.* Cambridge: Cambridge Univ. Press.

Bachelard, G. 1987. *On poetic imagination and reverie.* Dallas: Spring.

Badiner, A. 1990. *Dharma Gaia.* Berkeley: Parallax.

Bailey, R. 1993. *Eco-scam.* New York: St. Martin's Press.

Baldwin, J. 1975 (1906–1915). *Thought and things.* New York: Arno Press.

Bandura, A. 1971. *Social learning theory.* New York: General Learning Press.

———. 1977. Self-efficacy: Toward a unifying theory of behavioral change. *Psychological Review* 34: 191–215.

Barbour, I. 1990. *Religion in an age of science.* San Francisco: Harper.

Barbour, I., ed. 1973. *Western man and environmental ethics.* Reading, Mass.: Addison-Wesley.

Baring, A., and J. Cashford. 1991. *The myth of the Goddess.* New York: Viking Arkana.

Barnstone, W., ed. 1984. *The other bible.* San Francisco: Harper.

Barrow, J., and F. Tipler. 1988. *The anthropic cosmological principle.* Oxford: Oxford Univ. Press.

Barthes, R. 1972. *Critical essays.* Evanston, Ill.: Northwestern Univ. Press.

———. 1975. *S/Z.* London: Cape.

———. 1976. *The pleasure of the text.* London: Cape.

———. 1982. *A Barthes reader.* Ed. S. Sontag. New York: Hill & Wang.

Bataille, G. 1985 (1927–1939). *Visions of excess.* Ed. A. Stoekl. Minneapolis: Univ. of Minnesota Press.

Bateson, G. 1979. *Mind and nature.* New York: Dutton.

Bateson, M. 1990. *Composing a life.* New York: Plume.

Baynes, K.; J. Bohman; and T. McCarthy. 1987. *After philosophy.* Cambridge: MIT Press.

Beall, A., and R. Sternberg, eds. 1993. *The psychology of gender.* New York: Guilford Press.

Beard, M. 1946. *Woman as a force in history.* New York: Macmillan.

Beck, A.; A. Rush; B. Shaw; and G. Emery. 1979. *Cognitive therapy of depression.* New York: Guilford Press.

Becker, E. 1973. *The denial of death.* New York: Free Press.

Belenky, M., et al. 1986. *Women's ways of knowing.* New York: Basic.

Bellah, R. 1970. *Beyond belief.* New York: Harper & Row.

Bellah, R.; R. Madsen; W. Sullivan; A. Swindler; and S. Tipton. 1993. *The good society.* New York: Vintage.

———. 1985. *Habits of the heart.* Berkeley: Univ. of California Press.

Benoit, H. 1955. *The supreme doctrine.* New York: Viking.

Bentham, J. 1990. *The principles of morals and legislation.* Buffalo, N.Y.: Prometheus Books.

Benton, T. 1984. *The rise and fall of structural Marxism.* New York: St. Martin's Press.

Benvenuto, B., and R. Kennedy. 1986. *The works of Jacques Lacan.* London: Free Association Books.

Berger, P. 1977. *Facing up to modernity.* New York: Basic Books.

———. 1979. *The heretical imperative.* New York: Doubleday.

Berger, P., and H. Kellner. 1981. *Sociology reinterpreted.* Garden City, N.Y.: Doubleday.

Berger, P., and T. Luckmann. 1966. *The social construction of reality.* Garden City, N.Y.: Doubleday.

Bergson, H. 1944. *Creative evolution.* New York: Random House.

Berkeley, G. 1982 (1710). *A treatise concerning the principles of human knowledge.* Cambridge: Hackett.

Berman, M. 1981. *The reenchantment of the world.* Ithaca, N.Y.: Cornell Univ. Press.

———. 1989. *Coming to our senses.* New York: Simon & Schuster.

Berne, E. 1972. *What do you say after you say hello?* New York: Bantam.

Bernstein, R. 1983. *Beyond objectivism and relativism.* Philadelphia: Univ. of Penn. Press.

———, ed. 1985. *Habermas and modernity.* Cambridge: MIT Press.

Berry, T. 1988. *The dream of the earth.* San Francisco: Sierra Club.

Berry, W. 1977. *The unsettling of America.* San Francisco: Sierra Club.

Bertalanffy, L. von. 1968. *General system theory.* New York: Braziller.

Binswanger, L. 1956. Existential analysis and psychotherapy. In F. Fromm-Reichmann and J. Moreno, eds., *Progress in psychotherapy.* New York: Grune & Stratton.

Birch, C., and J. Cobb. 1990. *The liberation of life.* Denton, Tex.: Environmental Ethics Books.

Bird, D. 1974. *Born female.* New York: David McKay.

Blanck, G., and R. Blanck. 1974. *Ego psychology: Theory and practice.* New York: Columbia Univ. Press.

———. 1979. *Ego psychology II: Psychoanalytic developmental psychology.* New York: Columbia Univ. Press.

———. 1986. *Beyond ego psychology.* New York: Columbia Univ. Press.

Blofeld, J. 1970. *The tantric mysticism of Tibet.* New York: Dutton.

Blos, P. 1962. *On adolescence: A psychoanalytic interpretation.* New York: Free Press.

———. 1967. The second individuation process of adolescence. *Psychoanalytic Study of the Child* 22: 162-86.

Blumberg, R. 1984. A general theory of gender stratification. In Collins 1984: 23-101.

Bohm, D. 1973. *Wholeness and the implicate order.* London: Routledge.

Bonifaci, C. 1978. *The soul of the world.* Lanham, Md.: Univ. Press of America.

Bookchin, M. 1990. *Remaking society.* Boston: South End Press.

———. 1991. *The ecology of freedom.* 2nd ed. Montreal: Black Rose.

Boorstein, S. 1983. The use of bibliotherapy and mindfulness meditation in a psychiatric setting. *J. Transpersonal Psych.* 15 2: 173-9.

———, ed. 1980. *Transpersonal psychotherapy.* Palo Alto, Calif.: Science and Behavior Books.

Bosanquet, B. 1921. *The meeting of extremes in contemporary philosophy.* New York: Macmillan.

Boss, M. 1963. *Psychoanalysis and daseinanalysis.* New York: Basic Books.

Boucher, S. 1993. *Turning the wheel.* Boston: Beacon.

Boudon, R. 1971. *The uses of structuralism.* London: Heinemann.

Boulding, E. 1976. *The underside of history.* Boulder, Colo.: Westview.

Bourdieu, P. 1989. *The logic of practice.* Oxford: Basil Blackwell.

Bowie, M. 1991. *Lacan.* Cambridge: Harvard Univ. Press.

Bowlby, J. 1969, 1973. *Attachment and loss.* 2 vols. New York: Basic Books.

Bradley, F. 1902. *Appearance and reality.* New York: Macmillan.

———. 1922. *The principles of logic.* Oxford: Clarendon Press.

Bragdon, E. 1990. *The call of spiritual emergency.* San Francisco: Harper.

Brainerd, C. 1978. The stage question in cognitive-developmental theory. *Behavioral and Brain Sciences* 2: 173-213.

Branden, N. 1971. *The psychology of self-esteem.* New York: Bantam.

Brandon, R., and R. Burian, eds. 1984. *Genes, organisms, and populations.* Cambridge: MIT Press.

Brandt, A. 1980. Self-confrontations. *Psychology Today*, Oct.

Brennan, A. 1988. *Thinking about nature.* London: Routledge & Kegan Paul.

Brentano, F. 1973. *Psychology from an empirical standpoint.* London: Routledge & Kegan Paul.

Bridenthal, R., and C. Koonz, eds. 1977. *Becoming visible.* Boston: Houghton Mifflin.

Brim, G. 1992. *Ambition.* New York: Basic Books.

Brittan, A., and M. Maynard. 1984. *Sexism, racism and oppression.* New York: Blackwell.

Broad, C. 1925. *The mind and its place in nature.* London: Routledge & Kegan Paul.

Broughton, J. 1975. The development of natural epistemology in adolescence and early adulthood. Doctoral dissertation, Harvard.

Brown, D. P. 1977. A model for the levels of concentrative meditation. *International J. Clinical and Experimental Hypnosis* 25: 236-73.

———. 1981. Mahamudra meditation: Stages and contemporary cognitive psychology. Doctoral dissertation, University of Chicago.

Brown, D. P., and J. Engler. 1980. The stages of mindfulness meditation: A validation study. *J. Transpersonal Psychology* 12 2: 143-92.

Brown, D. P.; S. Twemlow; J. Engler; M. Maliszewski; and J. Stauthamer. 1978. The profile of meditation experience POME, Form II, Psychological Test Copyright, Washington, D.C.

Brown, L., and C. Gilligan. 1992. *Meeting at the crossroads.* New York: Ballantine.

Brown, N. O. 1959. *Life against death.* Middletown, Conn.: Wesleyan Univ. Press.

Bruner, J. 1983. *In search of mind.* New York: Harper & Row.

Buddhaghosa, B. 1976. *The path of purification.* 2 vols. Boulder, Colo.: Shambhala.

Bulkeley, K. 1994. *The wilderness of dreams.* Albany: SUNY Press.

Burke, E. 1990. *Reflections on the revolution in France.* Buffalo, N.Y.: Prometheus Books.

Cabezon, J., ed. 1992. *Buddhism, sexuality and gender.* Albany: SUNY Press.

Callenbach, E. 1977. *Ectopia.* New York: Bantam.

Callicott, J. 1986. The metaphysical implications of ecology. *Environmental Ethics* 8: 301-16.

———. 1986. The search for an environmental ethic. In Regan 1986: 381-424.

Campbell, Jeremy. 1982. *Grammatical man.* New York: Simon & Schuster.

Campbell, Joseph. 1959–1968. *The masks of God.* Vols. 1-4. New York: Viking.

Campbell, Joseph. 1968. *The hero with a thousand faces.* New York: World.

Campenhausen, H. von. 1955. *The fathers of the Greek church.* New York: Pantheon.

Caponigri, A. 1963. *Philosophy from the Renaissance to the romantic age.* Notre Dame, Ind.: Univ. of Notre Dame Press.

Capra, F. 1982. *The turning point.* New York: Simon & Schuster.

Capra, F., and D. Steindl-Rast. 1991. *Belonging to the universe.* San Francisco: Harper.

Caputo, J. 1986. *The mystical elements in Heidegger's thought.* New York: Fordham Univ. Press.

Cassirer, E. 1951. *The philosophy of the Enlightenment.* Boston: Beacon.

Casti, J. 1989. *Paradigms lost.* New York: Avon.

———. 1990. *Searching for certainty.* New York: William Morrow.

Chafetz, J. 1984. *Sex and advantage.* Totowa, N.J.: Rowman & Alanheld.

Chagdud Tulku. 1987. *Life in relation to death.* Cottage Grove, Ore.: Padma.

———. 1991. *Mirror of freedom.* Various volumes. Junction City, Calif.: Padma.

Chang, G. 1971. *The Buddhist teaching of totality.* University Park: Pennsylvania State Univ. Press.

———. 1974. *Teachings of Tibetan yoga.* Secaucus, N.J.: Citadel.

Chase, S., ed. 1991. *Defending the earth.* Boston: South End Press.

Chaudhuri, H. 1981. *Integral yoga.* Wheaton, Ill.: Quest.

Chirban, J. 1981. *Human growth and faith.* Washington, D.C.: University Press of America.

Chittick, W. 1992. *Faith and practice of Islam.* Albany: SUNY Press.

Chodorow, N. 1978. *The reproduction of mothering.* Berkeley: Univ. of Calif. Press.

———. 1989. *Feminism and psychoanalytic theory.* New Haven: Yale Univ. Press.

Chomsky, N. 1980. *Rules and representations.* New York: Columbia Univ. Press.

———. 1984. *Modular approaches to the study of mind.* San Diego: San Diego Univ. Press.

———. 1986. *Knowledge of language.* New York: Praeger.

Chopel, G. 1992. *Tibetan arts of love.* Ithaca, N.Y.: Snow Lion.

Chopra, D. 1992. *Unconditional life.* New York: Bantam.

Churchland, P. 1984. *Matter and consciousness.* Cambridge: MIT Press.

———. 1986. *Neuorphilosophy.* Cambridge: MIT Press.

Churchman, C. W. 1979. *The systems approach and its enemies.* New York: Basic Books.

Clifford, T. 1984. *Tibetan Buddhist medicine and psychiatry.* York Beach, Maine: Samuel Weiser.

Collins, A. 1987. *The nature of mental things.* Nortre Dame, Ind.: Univ. of Notre Dame Press.

Collins, R., ed. 1984. *Sociological theory.* San Francisco: Jossey-Bass.

Commoner, B. 1990. *Making peace with the planet.* New York: Pantheon.

Commons, M.; F. Richards; and C. Armon. *Beyond formal operations.* New York: Praeger.

Connell, R. 1987. *Gender and power.* Stanford: Stanford Univ. Press.

Copleston, F. 1960–1977. *A history of philosophy.* 9 vols. New York: Image.

———. 1982. *Religion and the one.* New York: Crossroad.

Corballis, M. 1991. *The lopsided ape.* New York: Oxford Univ. Press.

Cornell, D.; M. Rosenfeld; and D. Carlson. 1992. *Deconstruction and the possibility of justice.* London: Routledge.

Cornell, J., ed. 1988. *Bumps, voids, and bubbles in time.* Cambridge: Cambridge Univ. Press.

Cott, N., and E. Pleck, eds. 1979. *A heritage of her own.* New York: Simon & Schuster.

Cousins, M., and A. Hussein. 1984. *Michel Foucault.* London: Macmillan.

Coveney, P., and R. Highfield. 1990. *The arrow of time.* New York: Fawcett Columbine.

Cowan, P. 1978. *Piaget with feeling.* New York: Holt.

Coward, H. 1990. *Derrida and Indian philosophy.* Albany: SUNY Press.

Cozort, D. 1986. *Highest tantra yoga.* Ithaca, N.Y.: Snow Lion.

Crittenden, J. 1992. *Beyond individualism.* Oxford: Oxford Univ. Press.

Culler, J. 1982. *On deconstruction.* Ithaca, N.Y.: Cornell Univ. Press.

Da Avabhasa. 1977. *The paradox of instruction.* San Francisco: Dawn Horse.

———. 1978. *The enlightenment of the whole body.* San Francisco: Dawn Horse.

———. 1982. *The liberator.* Clearlake, Calif.: Dawn Horse.

———. 1989a. *The basket of tolerance.* Clearlake, Calif.: Dawn Horse.

———. 1989b. *The Da upanishad.* Clearlake, Calif.: The Dawn Horse.

———. 1991. *The dawn horse testament.* Clearlake, Calif.: Dawn Horse.

———. 1992. *The method of the siddhas.* Clearlake, Calif.: Dawn Horse.

Daly, M. 1973. *Beyond God the father.* Boston: Beacon.

———. 1978. *Gyn/ecology.* Boston: Beacon.

Dargyay, E. 1978. *The rise of esoteric Buddhism in Tibet.* New York: Weiser.

Darwin, C. 1872. *Origin of species* and *The descent of man.* New York: Modern Library.

Dattatreya. 1980. *Tripura rahasya.* Tiruvannamalai, India: Sri Ramanasramam.

———. n.d.. *Avadhuta gita.* Mylapore, Madras: Sri Ramakrishna Math.

Davies, P. 1991. *The cosmic blueprint.* New York: Simon & Schuster.

———. 1992. *The mind of god.* New York: Simon & Schuster.

Davies, P., and J. Gribbin. 1991. *The matter myth.* London: Viking.

Dawkins, R. 1976. *The selfish gene.* New York: Oxford Univ. Press.

Deatherage, O. 1975. The clinical use of "mindfulness" meditation techniques in short-term psychotherapy. *J. Transpersonal Psychology* 7 2: 133-43.

de Beauvoir, S. 1952. *The second sex.* New York: Bantam.

Deikman, A. 1982. *The observing self.* Boston: Beacon.

Delaney, G., ed.. 1993. *New directions in dream interpretation.* Albany: SUNY Press.

Deleuze, G., and F. Guattari. 1983. *Anti-Oedipus*. Minneapolis: Univ. of Minnesota Press.

Delphy, C. 1984. *Close to home*. Amherst: Univ. of Mass. Press.

Dennett, D. 1978. *Brainstorms*. Cambridge: MIT Press.

————. 1984. *Elbow room*. Cambridge: MIT Press.

————. 1991. *Consciousness explained*. Boston: Little, Brown.

Derrida, J. 1976. *Of grammatology*. Baltimore: Johns Hopkins.

————. 1978. *Writing and difference*. Chicago: Univ. of Chicago Press.

————. 1981. *Positions*. Chicago: Univ. Chicago Press.

————. 1982. *Margins of philosophy*. Chicago: Univ. of Chicago Press.

Descartes, R. 1911. *The philosophical works of Descartes*. 2 vols. Cambridge: Cambridge Univ. Press.

Deutsche, E. 1969. *Advaita Vedanta*. Honolulu: East-West Center.

Devall, B., and G. Sessions, eds. 1985. *Deep ecology*. Layton, Utah: Gibbs Smith.

Dew, P. 1987. *Logics of disintegration*. London: Verso.

Dewey, J. 1981. *The philosophy of John Dewey*. Ed. J. McDermott. Chicago: Univ. of Chicago Press.

Diamond, I., and G. Orenstein. 1990. *Reweaving the world*. San Francisco: Sierra Club.

Diamond, M. 1974. *Contemporary philosophy and religious thought*. New York: McGraw-Hill.

Dinnerstein, D. 1976. *The mermaid and the minotaur*. New York: Harper & Row.

Dionysius the Areopagite. 1965. *The mystical theology and celestial hierarchy*. Surrey: Shrine of Wisdom.

Dogen. 1993. *Rational Zen*. Trans. T. Cleary. Boston: Shambhala.

Doore, G., ed. 1988. *Shaman's path*. Boston: Shambhala.

Doresse, J. 1986. *The secret books of the Egyptian Gnostics*. Rochester, Vt: Inner Traditions.

Dossey, L. 1993. *Healing words*. San Francisco: Harper.

Douglas, M. 1966. *Purity and danger*. London: Routledge.

————. 1978. *Implicit meanings*. London: Routledge.

————. 1982. *In the active voice*. London: Routledge.

Dozier, R. 1992. *Codes of evolution*. New York: Crown.

Dretske, F. 1988. *Explaining behavior*. Cambridge: MIT Press.

Dreyfus, H. 1979. *What computers can't do*. New York: Harper & Row.

————. 1982. *Husserl: Intentionality and cognitive science*. Cambridge: MIT Press.

Dreyfus, H., and P. Rabinow. 1983. *Michel Foucault: Beyond structuralism and hermeneutics*. Chicago: Univ. Chicago Press.

Dudjom Lingpa. 1994. *Buddhahood without meditation*. Junction City, Calif.: Padma.

Dudjom, J. 1991. *The Nyingma school of Tibetan Buddhism*. 2 vols. Boston: Wisdom.

Durkheim, E. 1990. *Ethics and the sociology of morals*. Buffalo, N.Y.: Prometheus Books.

Dyer, D. 1991. *Cross-currents of Jungian thought*. Boston: Shambhala.

Eccles, J. 1984. *The human mystery*. London: Routledge.

Eckersley, R. 1992. *Environmentalism and political theory*. Albany: SUNY Press.

Eckhart, Meister. 1941. *Meister Eckhart*. Trans. R. Blakney. New York: Harper.

———. 1980. *Breakthrough*. Trans. M. Fox. New York: Image.

Eco, U. 1976. *A theory of semiotics*. Bloomington: Indiana Univ. Press.

———. 1984. *Semiotics and the philosophy of language*. Bloomington: Indiana Univ. Press.

Edinger, E. F. 1992. *Ego and archetype*. Boston: Shambhala.

Edwards, P., ed. 1967. *The encylopedia of philosophy*. New York: Collier Macmillan.

Ehrlich, P., and A. Ehrlich. 1991. *Healing the planet*. New York: Addison-Wesley.

Eichenbaum, L., and S. Orbach. 1983. *Understanding women*. New York: Basic Books.

Eisler, R. 1987. *The chalice and the blade*. San Francisco: Harper.

Eliade, M. 1964. *Shamanism*. Princeton: Princeton Univ. Press.

———. 1969. *Yoga: Immortality and freedom*. Princeton: Princeton Univ. Press.

Elior, R. 1993. *The paradoxical ascent to God*. Albany: SUNY Press.

Ellenberger, H. 1970. *The discovery of the unconscious*. New York: Basic Books.

Emerson, R. 1969 (1909-1914). *Ralph Waldo Emerson: Selected prose and poetry*. Ed. R. Cook. San Francisco: Rinehart.

Engels, F. 1942 (1884). *The origin of the family, private property, and the state*. New York: International Publishers.

Engler, J. 1983. Buddhist satipatthana-vipassana meditation and an object relations model of therapeutic developmental change: A clinical case study. Doctoral dissertation, University of Chicago.

———. 1984. Therapeutic aims in psychotherapy and meditation: Developmental stages in the representation of self. *J. Transpersonal Psychology* 16 1: 25-61.

Erdelyi, M. 1985. *Psychoanalysis: Freud's cognitive psychology*. New York: Freeman.

Erdmann, E., and D. Stover. 1991. *Beyond a world divided*. Boston: Shambhala.

Eribon, D. 1991. *Michel Foucault*. Cambridge: Harvard Univ. Press.

Erikson, E. H. 1950, 1963. *Childhood and society*. New York: Norton.

————. 1959. *Identity and the life cycle*. New York: International Univ. Press.

Evans, D. 1993. *Spirituality and human nature*. Albany: SUNY Press.

Evans-Wentz, W. 1971. *Tibetan yoga and secret doctrines*. London: Oxford Univ. Press.

Fairbairn, W. 1952. *Psychoanalytic studies of the personality*. New York: Basic Books.

————. 1954. *An object relations theory of the personality*. New York: Basic Books.

Fausto-Sterling, A. 1985. *Myths of gender*. New York: Basic Books.

Feigl, H. 1967. *The 'mental' and the 'physical.'* Minneapolis: Univ. of Minnesota Press.

Fenichel, O. 1945. *The psychoanalytic theory of neurosis*. New York: Norton.

Ferrucci, P. 1982. *What we may be*. Los Angeles: Tarcher.

Feuerbach, L. 1990. *The essence of Christianity*. Buffalo, N.Y.: Prometheus Books.

Feuerstein, G. 1987. *Structures of consciousness*. Lower Lake, Calif.: Integral.

————. 1989. *Yoga*. Los Angeles: Tarcher.

————, ed. 1989. *Enlightened sexuality*. Freedom, Calif.: Crossing.

Feuerstein, G., and T. L. Feuerstein, eds. 1993. *Voices on the threshold of tomorrow*. Wheaton, Ill.: Quest.

Feyerabend, P. 1978. *Against method*. London: Verso.

Fields, R. 1991. *The code of the warrior*. New York: Harper.

Fingarette, H. 1958. The ego and mystic selflessness. *Psychoanalytic Review* 45: 5-40.

Fischer, R. 1971. A cartography of the ecstatic and meditative states: The experimental and experiential features of a perception-hallucination continuum. *Science* 174: 897-904.

Flanagan, O. 1984. *The science of the mind*. Cambridge: MIT Press.

Flavell, J. 1963. *The developmental psychology of Jean Piaget*. Princeton, N.J.: Van Nostrand.

————. 1970. Concept development. In P. Mussen (ed.), *Carmichel's manual of child psychology*. Vol. 1. New York: Wiley.

Fodor, J. 1975. *The language of thought*. Cambridge: Harvard Univ. Press.

————. 1983. *The modularity of mind*. Cambridge: MIT Press.

Fodor, J., and E. Lepore. 1992. *Holism*. Oxford: Blackwell.

Forman, R., ed. 1990. *The problem of pure consciousness*. New York: Oxford Univ. Press.

Foss, L. and K. Rothenberg. 1987. *The second medical revolution: From biomedicine to infomedicine*. Boston: Shambhala.

Foucault, M. 1965. *Madness and civilization*. New York: Random House.

————. 1970. *The order of things*. New York: Random House.

————. 1972. *The archaeology of knowledge*. New York: Random House.

————. 1975. *The birth of the clinic*. New York: Random House.

————. 1978. *The history of sexuality*. Vol. 1. New York: Random House.

————. 1979. *Discipline and punish*. New York: Vintage.

————. 1980. *Power/knowledge*. New York: Pantheon.

Fowler, J. 1981. *Stages of faith: The psychology of human development and the quest for meaning*. San Francisco: Harper & Row.

Fox, M. 1979. *A spirituality named compassion*. San Francisco: Harper.

————. 1991. *Creation spirituality*. San Francisco: Harper.

Fox, W. 1990. *Toward a transpersonal ecology*. Boston: Shambhala.

Frager, R., and J. Fadiman, 1984. *Personality and personal growth*. New York: Harper & Row.

Frank, J. 1961. *Persuasion and healing: A comparative study of psychotherapy*. Baltimore: Johns Hopkins.

Frankfurt, H. 1988. *The importance of what we care about*. Cambridge: Cambridge Univ. Press.

Frankl, V. 1963. *Man's search for meaning*. Boston: Beacon.

————. 1969. *The will to meaning*. Cleveland: New American Library.

French, M. 1985. *Beyond power: On women, men, and morals*. New York: Ballantine.

Freud, A. 1946. *The ego and the mechanisms of defense*. New York: International Univ. Press.

————. 1963. The concept of developmental lines. In *The psychoanalytic study of the child*, vol. 8, pp. 245-65. New York: International Univ. Press.

————. 1965. *Normality and pathology in childhood*. New York: International Univ. Press.

Freud, S. 1937. Analysis terminable and interminable. *Standard Edition* (SE). 23, pp. 209-53. London: Hogarth Press.

————. 1959 (1926). Inhibitions, symptoms and anxiety. SE 20. London: Hogarth Press.

————. 1961 (1923). *The ego and the id*. se 19. London: Hogarth Press.

————. 1961 (1930). *Civilization and its discontents*. New York: Norton.

————. 1964 (1940). *An outline of psychoanalysis*. SE 23. London: Hogarth Press.

————. 1971. *A general introduction to psychoanalysis*. New York: Pocket Books.

————. 1974. *New introductory lectures in psychoanalysis*. New York: Norton.

Frey-Rohn, L. 1974. *From Freud to Jung*. New York: Delta.

Friedan, B. 1963. *The feminine mystique*. New York: Dell.

Friedl, E. 1984. *Women and men*. Prospect Heights, Ill.: Waveland Press.

Fromm, E.; D. T. Suzuki; and R. DeMartino, 1970. *Zen Buddhism and psychoanalysis*. New York: Harper & Row.

Fukuyama, F. 1992. *The end of history and the last man*. New York: Avon.

Futuyma, D. 1986. *Evolutionary biology*. 2nd ed. Sunderland, Mass.: Sinauer.

Gadamer, H. 1976. *Philosophical hermeneutics*. Berkeley: Univ. of Calif. Press.

———. 1992. *Truth and method*. 2nd ed. New York: Crossroad.

Gadon, E. 1989. *The once and future goddess*. New York: Harper.

Galileo, G. 1990. *Dialogues concerning two new sciences*. Buffalo, N.Y.: Prometheus Books.

Gallup, G. 1982. *Adventures in immortality*. New York: McGraw-Hill.

Gard, R. 1962. *Buddhism*. New York: Braziller.

Gardner, H. 1972. *The quest for mind*. New York: Vintage.

———. 1991. *The unschooled mind*. New York: Basic Books.

Gardner, M. 1985. *The mind's new science*. New York: Basic Books.

Gazzaniga, M. 1985. *The social brain*. New York: Basic Books.

———. 1992. *Nature's mind*. New York: Basic Books.

Gebser, J. 1985. *The ever-present origin*. Athens: Ohio Univ. Press.

Gedo, J. 1979. *Beyond interpretation: Toward a revised theory for psychoanalysis*. New York: International Univ. Press.

———. 1981. *Advances in clinical psychoanalysis*. New York: International Univ. Press.

Geertz, C. 1973. *The interpretation of cultures*. New York: Harper & Row.

Gergen, K. 1991. *The saturated self*. New York: Basic Books.

Gibson, J. 1979. *The ecological approach to visual perception*. Boston: Houghton Mifflin.

Giddens, A. 1971. *Capitalism and modern social theory*. Cambridge: Cambridge Univ. Press.

———. 1977. *Studies in social and political theory*. London: Hutchinson.

Gilligan, C. 1982. *In a different voice*. Cambridge: Harvard Univ. Press.

Gilson, E. 1941. *God and philosophy*. New Haven, Conn.: Yale Univ. Press.

Gimbutas, M. 1991. *The civilization of the Goddess*. San Francisco: Harper Collins.

Gleick, J. 1987. *Chaos*. New York: Viking.

Globus, G. 1990. Heidegger and cognitive science. *Philosophy Today* (Spring), 20-30.

Goldberg, A., ed. 1980. *Advances in self psychology*. New York: International Univ. Press.

Goldsmith, E. 1993. *The way: An ecological world-view*. Boston: Shambhala.

Goldstein, J. 1983. *The experience of insight*. Boston: Shambhala.

Goleman, D. 1988. *The meditative mind*. Los Angeles: Tarcher.

Goleman, D., and M. Epstein. 1980. Meditation and well-being: An Eastern model of psychological health. *ReVision* 3: 73-85.

Goodwin, B.; N. Holder; and C. Wyles, eds. 1983. *Development and evolution.* Cambridge: Cambridge Univ. Press.

Gore, A. 1992. *Earth in the balance.* New York: Houghton Mifflin.

Gossman, L. 1990. *Between history and literature.* Cambridge: Harvard Univ. Press.

Gottesman, I., and M. Schields. 1972. *Schizophrenia and genetics: A twin study vantage point.* New York: Academic Press.

Gottlieb, R., ed. 1990. New York: Crossroad.

Gould, S. 1977. *Ontogeny and phylogeny.* Cambridge: Harvard Univ. Press.

Green, E.; A. Green; and D. Walters. 1970. Voluntary control of internal states: Psychological and physiological. *J. Transpersonal Psychology* 2: 1-26.

Greenson, R. 1967. *The technique and practice of psychoanalysis.* New York: International Univ. Press.

Greenstein, G. 1988. *The symbiotic universe.* New York: Morrow.

Greist, J.; J. Jefferson; and R. Spitzer, eds.. 1982. *Treatment of mental disorders.* New York: Oxford Univ. Press.

Griffin, D. 1989. *God and religion in the postmodern world.* Albany: SUNY Press.

———, ed. 1988. *The reenchantment of science.* Albany: SUNY Press.

Griffin, D., and R. Falk, eds. 1993. *Postmodern politics for a planet in crisis.* Albany: SUNY Press.

Griffin, D., et al. 1989. *Varieties of postmodern theology.* Albany: SUNY Press.

Griffin, S. 1978. *Woman and nature.* New York: Harper.

Groddeck, G. 1949. *The book of the It.* New York: Vintage.

Grof, C., and S. Grof. 1990. *The stormy search for the self.* Los Angeles: Tarcher.

Grof, S. 1975. *Realms of the human unconscious.* New York: Viking.

———. 1985. *Beyond the brain.* Albany: SUNY Press.

———. 1988. *The adventure of self-discovery.* Albany: SUNY Press.

Grof, S., with H. Bennett. 1992. *The holotrophic mind.* San Francisco: Harper.

Gross, R. 1993. *Buddhism after patriarchy.* Albany: SUNY Press.

Group for the Advancement of Psychiatry (GAP). 1976. *Mysticism: Spiritual quest or psychic disorder?* New York: GAP (publication 97).

Guénon, R. 1945. *Man and his becoming according to Vedanta.* London: Luzac.

Guenther, H. V. 1989. *From reductionism to creativity: rDzogs-chen and the new sciences of mind.* Boston: Shambhala.

Guntrip, H. 1969. *Schizoid phenomena, object relations and the self.* New York: International Univ. Press.

———. 1971. *Psychoanalytic theory, therapy, and the self.* New York: Basic Books.

Gupta, B., ed. 1987. *Sexual archetypes, East and West.* New York: Paragon House.

Gyatso, K. 1986. *Progressive stages of meditation on emptiness.* Oxford: Longchen Foundation.

———. 1982. *Clear light of bliss.* London: Wisdom.

Habermas, J. 1971. *Knowledge and human interests.* Boston: Beacon.

———. 1973. *Theory and practice.* Boston: Beacon.

———. 1975. *Legitimation crisis.* Boston: Beacon.

———. 1979. *Communication and the evolution of society.* Trans. T. McCarthy. Boston: Beacon Press.

———. 1984-1985. *The theory of communicative action.* 2 vols. Trans. T. McCarthy. Boston: Beacon.

———. 1990. *The philosophical discourse of modernity.* Trans. F. Lawrence. Cambridge: MIT Press.

Haley, J., and L. Hoffman, eds. 1968. *Techniques of family therapy.* New York: Basic Books.

Hall, E. 1959. *The silent language.* New York: Doubleday.

Halliwell, J. 1992. *Quantum cosmology.* Cambridge: Cambridge Univ. Press.

Hampshire, S. 1975. *Freedom of the individual.* Princeton: Princeton Univ. Press.

Hanly, C. and J. Masson. A critical examination of the new narcissism. *International J. Psychoanalysis* 57: 49-65.

Hargrove, E. 1989. *Foundations of environmental ethics.* New Jersey: Prentice Hall.

Harland, R. 1987. *Superstructuralism.* London: Methuen.

Harman, W. 1988. *Global mind change.* New York: Warner.

Harner, M. 1980. *The way of the shaman.* New York: Harper.

Harris, E. 1991. *Cosmos and anthropos.* Atlantic Highlands, N.J.: Humanities Press International.

———, ed. 1992. *Cosmos and theos.* Atlantic Highlands, N.J.: Humanities Press International.

Harris, K. 1978. *Carlyle and Emerson.* Cambridge: Harvard Univ. Press.

Harris, M. 1974. *Cows, pigs, wars and witches.* New York: Random House.

———. 1977. *Cannibals and kings.* New York: Random House.

Harris, R. 1993. *The linguistics wars.* New York: Oxford Univ. Press.

Hartmann, H. 1958 (1939). *Ego psychology and the problem of adaptation.* New York: International Univ. Press.

Hartmann, N. 1932. *Ethics.* New York: Macmillan.

Hartshorne, C. 1984. *Omnipotence and other theological mistakes.* Albany: SUNY Press.

Hartsock, N. 1985. *Money, sex, and power.* Boston: Northeastern Univ. Press.

Hastings, A. 1990. *Tongues of men and angels*. New York: Holt, Rinehart & Winston.

Haught, J. 1984. *The cosmic adventure*. Ramsey: Paulist Press.

Hawkes, T. 1977. *Structuralism and semiotics*. Berkeley: Univ. California Press.

Hawking, S. 1988. *A brief history of time*. New York: Bantam.

Hayward, J. 1987. *Shifting worlds, changing minds*. Boston: Shambhala.

Hayward, J., and F. Varela, 1992. *Gentle bridges: Conversations with the Dalai Lama on the sciences of mind*. Boston: Shambhala.

Heard, G. 1963. *The five ages of man*. New York: Julian.

Hebb, D. O. 1949. *The organization of behavior: A neuropsychological theory*. New York: Wiley & Sons.

Hegel, G. 1949. *The phenomenology of mind*. Trans. J. Baille. New York: Humanities Press.

———. 1971. *Hegel's philosophy of mind*. Oxford: Clarendon.

———. 1974. *Hegel: The essential writings*. Ed. F. Weiss. New York: Harper Torchbooks.

———. 1977. *Phenomenology of spirit*. Trans. A. Miller. Analysis by J. Findlay. Oxford: Oxford Univ. Press.

———. 1993. *Hegel's science of logic*. New Jersey: Humanities Press International.

Heidegger, M. 1959. *Introduction to metaphysics*. New Haven: Yale Univ. Press.

———. 1962. *Being and time*. New York: Harper & Row.

———. 1968. *What is called thinking?* New York: Harper & Row.

Heidegger, M. 1977. *Basic writings*. Ed. D. Krell. New York: Harper & Row.

Hill, G. 1992. *Masculine and feminine*. Boston: Shambhala.

Hillman, J. 1972. *The myth of analysis*. New York: Harper & Row.

———. 1975. *Re-visioning psychology*. New York: Harper & Row.

———. 1989. *A blue fire*. Ed. Thomas Moore. New York: Harper & Row.

Hinding, A. 1986. *Feminism*. Minnesota: Greenhaven.

Hixon, L. 1989. *Coming home*. Los Angeles: Tarcher.

Hobbes, T. 1990. *The Leviathan*. Buffalo, N.Y.: Prometheus Books.

Hockett, C. 1987. *Refurbishing our foundations*. Philadelphia: John Benjamins.

Hoeller, K. 1982. *Merleau-Ponty and psychology*. New Jersey: Humanities Press.

Hoffman, E. 1981. *The way of splendor*. Boston: Shambhala.

Hofstadter, D. 1989. *Gödel, Escher, Bach*. New York: Vintage.

Holton, G. 1988. *Thematic origins of scientific thought*. Cambridge: Harvard Univ. Press.

Horkheimer, M., and T. Adorno. 1972 (1944). *Dialectic of enlightenment*. New York: Continuum.

Horner, A. 1979. *Object relations and the developing ego in therapy*. New York: Aronson.

Horney, K. 1950. *Neurosis and human growth.* New York: Norton.

———. 1967. *Feminine psychology.* New York: Norton.

Horowitz, M. 1988. *Introduction to psychodynamics.* New York: Basic.

Houston, J. 1980. *Life force.* New York: Delta.

Hoy, D., ed. 1991. *Foucault: A critical reader.* Cambridge, Mass.: Basil Blackwell.

Hume, D. 1964. *A treatise of human nature.* Oxford: Clarendon.

———. 1990. *An enquiry concerning human understanding.* Buffalo, N.Y.: Prometheus Books.

Hume, R., trans. 1974. *The thirteen principal Upanishads.* London: Oxford.

Humphrey, N. 1983. *Consciousness regained.* New York: Oxford Univ. Press.

Husserl, E. 1962 (1931). *Ideas.* New York: Collier.

———. 1965. *Phenomenology and the crisis of philosohphy.* New York: Harper & Row.

———. 1970. *The crisis of European sciences and transcendental phenomenology.* Evanston, Ill.: Northwestern Univ. Press.

———. 1991 (1950). *Cartesian meditations.* Boston: Kluwer.

Huxley, A. 1944. *The perennial philosophy.* New York: Harper & Row.

Huxley, J. 1990. *Evolutionary humanism.* Buffalo, N.Y.: Prometheus Books.

Inada, K. 1970. *Nagarjuna.* Tokyo: Hokusiedo Press.

Inge, W.R. 1968 (1929). *The philosophy of Plotinus.* vols. 1 & 2. Westport, Conn.: Greenwood.

Jackendoff, R. 1987. *Consciousness and the computational mind.* Cambridge: MIT Press.

Jacobi, J. 1942. *The psychology of C. G. Jung.* London: Routledge.

Jacobson, E. 1964. *The self and the object world.* New York: IUP.

Jaggar, A. 1988. *Feminist politics and human nature.* New Jersey: Rowman & Littlefield.

Jahn, R., and B. Dunn. 1989. *Margins of reality.* San Diego: Harcourt Brace Jovanovich.

Jaki, S. 1978. *The road of science and the ways to God.* Chicago: Univ. of Chicago Press.

Jakobson, R. 1980. *The framework of language.* Michigan Studies in the Humanities.

———. 1990. *On language.* Cambridge: Harvard Univ. Press.

James, W. 1950 (1890). *Principles of psychology.* 2 vols. New York: Dover.

———. 1961 (1901). *The varieties of religious experience.* New York: Colliers.

Jameson, F. 1972. *The prison-house of language.* Princeton: Princeton Univ. Press.

———. 1981. *The political unconscious.* Ithaca, N.Y.: Cornell Univ. Press.

Jamgon Kongtrul the Third. 1992. *Cloudless sky.* Boston: Shambhala.

Janeway, E. 1980. *Powers of the weak.* New York: Knopf.

Jantsch, E. 1980. *The self-organizing universe*. New York: Pergamon.

Jantsch, E., and C. Waddington, eds. 1976. *Evolution and consciousness*. Reading, Mass.: Addison-Wesley.

Jantzen, G. 1984. *God's world, God's body*. Philadelphia: Westminster Press.

Jaspers, K. 1966. *The great philosophers*. New York: Harcourt.

Jee, L. 1988. *Kashmir Shaivism*. Albany: SUNY Press.

John of the Cross. 1979. *The collected works of St. John of the Cross*. Trans. K. Kavanaugh and O. Rodriguez. Washington, D.C.: ICS Publications.

Johnson, C. 1993. *System and writing in the philosophy of Jacques Derrida*. Cambridge: Cambridge Univ. Press.

Johnson, M. 1987. *The body in the mind*. Chicago: Univ. of Chicago Press.

Jonas, H. 1958. *The gnostic religion*. Boston: Beacon.

Jordan, J., et al. 1991. *Women's growth in connection*. New York: Guilford.

Jung, C. G. 1957. *The undiscovered self*. New York: Mentor.

———. 1961. *Analytical psychology: Its theory and practice*. New York: Vintage.

———. 1964. *Man and his symbols*. New York: Dell.

———. 1971. *The portable Jung*. Ed. J. Campbell. New York: Viking.

Kakar, S. 1982. *Shamans, mystics, and doctors*. New York: Knopf.

Kalupahana, D. 1986. *Nagarjuna*. Albany: SUNY Press.

Kalweit, H. 1992. *Shamans, healers, and medicine men*. Boston: Shambhala.

Kant, I. 1949. *Kant's Critique of Practical Reason and other writings in moral philosophy*. Trans. L. Beck. Chicago: Univ. of Chicago Press.

———. 1951. *Critique of judgement*. New York: Hafner.

———. 1990. *Critique of pure reason*. Buffalo, N.Y.: Prometheus Books.

———. 1993. *Prolegomena*. Chicago: Open Court.

Kaplan, R. D. 1994. The coming anarchy. *Atlantic*, Feb.

Kapleau, P. 1965. *The three pillars of Zen*. Boston: Beacon.

Katz, J. 1972. *Semantic theory*. New York: Harper & Row.

———. 1990. *The metaphysics of meaning*. Cambridge: MIT Press.

Katz, S., ed. 1978. *Mysticism and philosophical analysis*. Oxford: Oxford Univ. Press.

———, ed. 1983. *Mysticism and religious traditions*. Oxford: Oxford Univ. Press.

Kaufman, G. 1985. *Theology for a nuclear age*. Oxford: Manchester Univ. Press.

Kaufmann, W. 1974. *Nietzsche*. Princeton: Princeton Univ. Press.

Keegan, J. 1993. *A history of warfare*. New York: Knopf.

Keeton, W., and J. Gould. 1986. *Biological Science*. 4th ed. New York: Norton.

Kelley, G. 1955. *The psychology of personal constructs*. 2 vols. New York: Norton.

Kelly, J. 1982. Early feminist theory and the *Querelles des femmes*, 1400-1789. *Signs* 8 (Autumn): 4-28.

Kernberg, O. 1975. *Borderline conditions and pathological narcissism.* New York: Aronson.

———. 1976. *Object relations theory and clinical psychoanalysis.* New York: Aronson.

Kessler, S., and W. McKenna. 1978. *Gender.* Chicago: Univ. of Chicago Press.

Khan, Inayat. 1977. *The soul: Whence and whither.* New York: Sufi Order.

Khyentse, D. 1992. *The heart treasure of the enlightened ones.* Boston: Shambhala.

Kierkegaard, S. 1953. *Fear and trembling* and *The sickness unto death.* New York: Anchor.

———. 1957. *The concept of dread.* Princeton: Princeton Univ. Press.

———. 1992. *The concept of irony* and *Notes of Schelling's Berlin lectures.* Trans. Hong and Hong. Princeton: Princeton Univ. Press.

Kimball, R. 1991. *Tenured radicals.* New York: Harper.

King, A., and B. Schneider. 1991. *The first global revolution.* New York: Pantheon.

Klein, A. 1986. *Knowledge and liberation.* Ithaca, N.Y.: Snow Lion.

Klein, M. 1932. *The psychoanalysis of children.* London: Hogarth.

Koestenbaum, P. 1976. *Is there an answer to death?* New York: Prentice-Hall.

Koestler, A. 1964. *The act of creation.* New York: Dell.

———. 1976. *The ghost in the machine.* New York: Random House.

Kohlberg, L. 1981. *Essays on moral development.* Vol. 1. San Francisco: Harper.

Kohut, H. 1971. *The analysis of the self.* New York: IUP.

———. 1977. *The restoration of the self.* New York: IUP.

Kojeve, A. 1969. *Introduction to the reading of Hegel.* Ithaca, N.Y.: Cornell Univ. Press.

Kramer, J., and D. Alstad. 1993. *The guru papers.* Berkeley: North Atlantic Books.

Krishna, G. 1972. *The secret of yoga.* London: Turnstone.

Kroy, M. 1982. *The rationality of mysticism.* Privately published.

Kuhn, T. 1970. *The structure of scientific revolutions.* Chicago: Univ. of Chicago Press.

Kunsang, E., trans. and ed. 1986. *The flight of the garuda.* Kathmandu: Rangjung Yeshe Publications.

Kurzweil, E. 1980. *The age of structuralism.* New York: Columbia Univ. Press.

LaBerge, S. 1985. *Lucid dreaming.* Los Angeles: Tarcher.

Lacan, J. 1968. *The language of the self.* Baltimore: Johns Hopkins.

Lacan, J. 1982. *Feminine sexuality.* New York: Norton.

Laing, R. D. 1969. *The divided self.* New York: Pantheon.

Lake, M. 1991. *Native healer.* Wheaton, Ill.: Quest.

Lakoff, G. 1987. *Women, fire and dangerous things*. Chicago: Univ. of Chicago Press.

Lakoff, G., and M. Johnson. 1980. *Metaphors we live by*. Chicago: Univ. of Chicago Press.

Lakoff, R. 1975. *Language and woman's place*. New York: Octagon Books.

Lancaster, B. 1991. *Mind, brain, and human potential*. Rockport, Mass.: Element.

Lasch, C. 1979. *The culture of narcissism*. New York: Norton.

Laszlo, E. 1972. *Introduction to systems philosophy*. New York: Harper & Row.

————. 1987. *Evolution: The grand synthesis*. Boston: Shambhala.

————. 1994. *The choice: Evolution or extinction?* Los Angeles: Tarcher.

Laughlin, C., J. McManus, and E. d'Aquili. 1992. *Brain, symbol and experience*. New York: Columbia Univ. Press.

Lawson, H. 1985. *Reflexivity*. La Salle, Ill.: Open Court.

Layzer, D. 1990. *Cosmogenesis*. New York: Oxford Univ. Press.

Leach, E. 1974. *Claude Lévi-Strauss*. Chicago: Univ. of Chicago Press.

Lecky, P. 1961. *Self-consistency*. Hamden, Conn.: Shoe String Press.

Lehman, D. 1991. *Signs of the times*. New York: Poseidon.

Leibniz, G. 1990. *Discourse on method* and *Monadology*. Buffalo, N.Y.: Prometheus Books.

Lenski. G. 1970. *Human societies*. New York: McGraw-Hill.

Leopold, A. 1981 (1949). *A sand county almanac*. Oxford: Oxford Univ. Press.

Lerman, H. 1986. *A mote in Freud's eye*. New York: Springer.

Lerner, G. 1979. *The majority finds its past: Placing women in history*. New York: Oxford Univ. Press.

————. 1986. *The creation of patriarchy*. New York: Oxford Univ. Press.

Leslie, J., ed. 1989. *Physical cosmology and philosophy*. New York: Macmillan.

Lévi-Strauss, C. 1966. *The savage mind*. London: Weidenfeld.

————. 1972. *Structural anthropology*. New York: Penguin Books.

Levin, M. 1979. *Metaphysics and the mind-body problem*. Oxford: Oxford Univ. Press.

Levinson, D., et al. 1978. *The season's of a man's life*. New York: Knopf.

Levy, H., and A. Ishihara. 1989. *The tao of sex*. Lower Lake, Calif.: Integral.

Lewontin, R. 1983. The organism as the subject and object of evolution. *Scientia* 118: 63-82.

Liebes, Y. 1993. *Studies in the Zohar*. Albany: SUNY Press.

Lightman, A. 1991. *Ancient light*. Cambrdige: Harvard Univ. Press.

Locke, J. 1990. *A letter concerning toleration*. Buffalo, N.Y.: Prometheus Books.

————. 1990. *Second treatise on civil government*. Buffalo, N.Y.: Prometheus Books.

Loevinger, J. 1977. *Ego development*. San Francisco: Jossey-Bass.

Loewald, H. 1978. *Psychoanalysis and the history of the individual.* New Haven: Yale Univ. Press.

Longchenpa. 1977. *Kindly bent to ease us.* 3 vols. Trans. H. V. Guenther. Emeryville, Calif.: Dharma Press.

Lovejoy, A. 1964 (1936). *The great chain of being.* Cambridge: Harvard Univ. Press.

Lovelock, J. 1988. *The ages of Gaia.* New York: Norton.

Lowe, V. 1966. *Understanding Whitehead.* Baltimore: Johns Hopkins Press.

Lowen, A. 1967. *The betrayal of the body.* New York: Macmillan.

Loy, D. 1987. The cloture of deconstruction. *International Phil. Quarterly* 27: 1, 60-80.

————. 1989. *Non-Duality.* New Haven: Yale Univ. Press.

Luk, C. 1962. *Ch'an and Zen teaching.* 3 vols. London: Rider.

Lukacs, G. 1971. *History and class consciousness.* Cambridge: MIT Press.

Lukoff, D.; F. Lu; and R. Turner. 1992. Toward a more culturally sensitive DSM-IV. *Journal of Nervous and Mental Diseases* 80: 673-82.

Lycan, W. 1987. *Consciousness.* Cambridge: MIT Press.

Lyons, J. 1991. *Chomsky.* Cambridge: Cambridge Univ. Press.

Lyons, W. 1986. *The disappearance of introspection.* Cambridge: MIT Press.

Lyotard, J. 1984. *The postmodern condition.* Minneapolis: Univ. of Minnesota Press.

Lyotard, J., and J. Thebaud. 1986. *Just gaming.* Manchester: Manchester Univ. Press.

Maccoby, E., and C. Jacklin. 1974. *The psychology of sex differences.* Stanford: Stanford Univ. Press.

Machiavelli, N. 1990. *The prince.* Buffalo, N.Y.: Prometheus Books.

MacIntyre, A. 1984. *After virtue.* 2nd ed. Notre Dame, Ind.: Univ. of Notre Dame Press.

————. 1990. *Three rival versions of moral enquiry.* Nortre Dame, Ind.: Univ. of Notre Dame Press.

Macy, J. 1991. *World as lover, world as self.* Berkeley: Parallax.

Madison, G., ed. 1993. *Working through Derrida.* Evanston, Ill.: Northwestern Univ. Press.

Mahasi Sayadaw. 1972. *Practical insight meditation.* Santa Cruz, Calif.: Unity Press.

Mahler, M. 1968. *On human symbiosis and the vicissitudes of individuation.* New York: IUP.

Mahler, M.; F.Pine; and A. Bergman. 1975. *The psychological birth of the human infant.* New York: Basic Books.

Mahoney, M. 1991. *Human change processes: The scientific foundations of psychotherapy.* New York: Basic Books.

Maliszewski, M.; S. Twemlow; D. Brown; and J. Engler. 1981. A phenomenological typology of intensive meditation. *ReVision* 4.

Mander, J. 1991. *In the absence of the sacred.* San Francisco: Sierra Club.

Manes, C. 1990. *Green rage.* Boston: Little, Brown.

Manjusrimitra. 1987. *Primordial experience.* Boston: Shambhala.

Marcel, A., and E. Bisiach, eds. 1988. *Consciousness in contemporary science.* Oxford: Oxford Univ. Press.

Marcuse, H. 1955. *Eros and civilization.* Boston: Beacon Press.

Margenau, H. 1987. *The miracle of existence.* Boston: Shambhala.

Marin, P. 1975. The new narcissism. *Harper's.* Oct.

Marsh, J.; J. Caputo; and M. Westphal. 1992. *Modernity and its discontents.* New York: Fordham Univ. Press.

Martin, M., and B. Voorhies. 1975. *Female of the species.* New York: Columbia Univ. Press.

Marx, K. 1974. *Political writing.* 3 vols. New York: Random House.

———. 1977. *Capital.* 3 vols. New York: Random House.

Maslow, A. 1968. *Toward a psychology of being.* New York: Van Nostrand Reinhold.

———. 1970. *Religions, values, and peak experiences.* New York: Viking.

Maslow, A. 1971. *The farther reaches of human nature.* New York: Viking.

Masterson, J. 1981. *The narcissistic and borderline disorders.* New York: Bruner/Mazel.

———. 1988. *The search for the real self.* New York: Free Press.

Maturana, H., and F. Varela. 1992. *The tree of knowledge.* Rev. ed. Boston: Shambhala.

May, R. 1969. *Love and will.* New York: Norton.

———. 1977. *The meaning of anxiety.* Rev. ed. New York: Norton.

———, ed. 1969. *Existential psychology.* New York: Random House.

May, R.; E. Angel; and H. Ellenberger, eds. 1958. *Existence.* New York: Basic.

Mayr, E. 1982. *The growth of biological thought.* Cambridge: Harvard Univ. Press.

McCarthy, T. 1978. *The critical theory of Jurgen Habermas.* Cambridge: MIT Press.

McCawley, J. 1988. *Thirty million theories of grammar.* Chicago: Univ. of Chicago Press.

McCulloh, W. 1965. *Embodiments of mind.* Cambridge: MIT Press.

McFague, S. 1987. *Models of God.* Philadelphia: Fortress Press.

McGaa, E. 1990. *Mother earth spirituality.* San Francisco: Harper.

McKibben, B. 1989. *The end of nature.* New York: Random House.

McMullin, E., ed. 1985. *Evolution and creation.* Notre Dame, Ind.: Univ. of Notre Dame Press.

Mead, G. H. 1934. *Mind, self, and society.* Chicago: Univ. Chicago Press.

Merchant, C. 1983. *The death of nature.* San Francisco: Harper.

———. 1989. *Ecological revolutions.* Chapel Hill: Univ. of North Carolina Press.

———. 1992. *Radical ecology.* London: Routledge.

Merleau-Ponty, M. 1962. *Phenomenology of perception.* London: Routledge & Kegan Paul.

———. 1963. *The structure of behavior.* Boston: Beacon.

Merquior, J. 1985. *Foucault.* London: Fontana.

———. 1986. *From Prague to Paris.* London: Verso.

Metzner, R. 1986. *Opening to inner light.* Los Angeles: Tarcher.

———. 1993. The split between spirit and nature in European consciousness. *ReVision* 15 (4): 177-84.

Meyendorff, J. 1975. *Byzantine theology.* New York: Fordham Univ. Press.

Meyer, J. 1975. *Death and neurosis.* New York: IUP.

Midgley, M. 1978. *Beast and man.* Ithaca, N.Y.: Cornell Univ. Press.

Milbrath, L. 1989. *Envisioning a sustainable society.* Albany: SUNY Press.

Mill, J. S. 1990. *On liberty.* Buffalo, N.Y.: Prometheus Books.

———. 1990a. *The subjection of women.* Buffalo, N.Y.: Prometheus Books.

———. 1990b. *Utilitarianism.* Buffalo, N.Y.: Prometheus Books.

Miller, J. 1976. *Toward a new psychology of women.* Boston: Beacon.

———. 1993. *The passion of Michel Foucault.* New York: Simon & Schuster.

Minsky, M. 1985. *The society of mind.* New York: Touchstone.

Minsky, M., and S. Papert. 1987. *Perceptrons.* Cambridge: MIT Press.

Mitchell, N., ed. 1985. *Nobel prize conversations.* San Francisco: Saybrook.

Moltmann, J. 1985. *God in creation.* London: SCM Press.

Money, J., and A. Ehrhardt. 1972. *Man and woman, boy and girl.* Baltimore: Johns Hopkins.

Monod, J. 1971. *Chance and necessity.* New York: Knopf.

Monroe, R. 1971. *Journeys out of the body.* New York: Doubleday.

Moody, R. 1988. *The light beyond.* New York: Bantam.

Mookerjee, A. 1982. *Kundalini.* New York: Destiny Books.

Morris, C. 1971. *Writings on the general theory of signs.* The Hague: Mouton.

Muller-Ortega, P. 1989. *The triadic heart of Shiva.* Albany: SUNY Press.

Mumford, L. 1966. *The myth of the machine.* New York: Harcourt.

Murphy, M. 1992. *The future of the body.* Los Angeles: Tarcher.

Murphy, M., and S. Donovan. 1989. *The physical and psychological effects of meditation.* San Rafael, Calif.: Esalen.

Murphy, N. 1990. *Theology in the age of scientific reasoning.* Ithaca, N.Y.: Cornell Univ. Press.

Murti, T. 1970. *The central philosophy of Buddhism.* London: Allen & Unwin.

Naess, A. 1989. *Ecology, community, and lifestyle*. Cambridge: Cambridge Univ. Press.

Nagao, G. 1989. *The foundational standpoint of Madhyamika philosophy*. Albany: SUNY Press.

———. 1991. *Madhyamika and Yogacara*. Albany: SUNY Press.

Nagel, T. 1986. *The view from nowhere*. New York: Oxford Univ. Press.

Narada. 1975. *A manual of Abhidhamma*. Kandy: Buddhist Publication Society.

Nash, R. 1989. *The rights of nature*. Madison: Univ. of Wisconsin Press.

Needham, J. 1956. *Science and civilization in China*. Vol. 2. Cambridge: Cambridge Univ. Press.

Needleman, J. 1980. *Lost Christianity*. Garden City, N.Y.: Doubleday.

Neisser, U. 1976. *Cognition and reality*. Ithaca, N.Y.: Cornell Univ. Press.

Nelson, J. 1994. *Healing the split*. Albany: SUNY Press.

Nerval, G. de 1993. *Aurelia* and *Sylvie*. Santa Maria: Asylum Arts.

Neufeldt, R. 1986. *Karma and rebirth*. Albany: SUNY Press.

Neumann, E. 1954. *The orgins and history of consciousness*. Princeton: Princeton Univ. Press.

Newmeyer, F. 1986. *Linguistic theory in America*. New York: Academic Press.

Nicholson, S., and B. Rosen. 1992. *Gaia's hidden life*. Wheaton, Ill.: Quest.

Nielsen, J. 1990. *Sex and gender in society*. 2nd ed. Prospect Heights, Ill.: Waveland Press.

Nietzsche, F. 1965. *The portable Nietzsche*. Ed. W. Kaufmann. New York: Viking.

———. 1968. *Basic writings of Nietzsche*. Trans. and ed. W. Kaufmann. New York: Modern Library.

Nisbet, R. 1980. *History of the idea of progress*. New York: Basic Books.

Nishitani, K. 1982. *Religion and nothingness*. Berkeley: Univ. of California Press.

Norbu, N. 1989. *Dzogchen: The self-perfected state*. London: Arkana Penguin.

———. 1989. *Self-liberation through seeing with naked awareness*. Barrytown, N.Y.: Station Hill.

———. 1992. *Dream yoga and the practice of natural light*. Ithaca, N.Y.: Snow Lion.

Norris, C. 1987. *Derrida*. London: Fontana.

Novak, P. 1993. Tao how? Asian religions and the problem of environmental degradation. *ReVision*, 16 (2): 77-82.

Nyanaponika. 1973. *The heart of Buddhist meditation*. New York: Weiser.

Nye, A. 1988. *Feminist theory and the philosophies of man*. London: Routledge.

Nyima, C. 1989. *The union of Mahamudra and Dzogchen*. Kathmandu: Rangjung Yeshe Publications.

Nyima, C. 1991. *The bardo guidebook*. Kathmandu: Rangjung Yeshe Publications.

O'Brien, T., ed. 1988. *The spiral path*. St. Paul, Minn.: Yes International.

Ochs, C. 1983. *Women and spirituality*. New Jersey: Rowman & Allanheld.

Odajnyk, V. 1993. *Gathering the light*. Boston: Shambhala.

Ornstein, R. 1991. *The evolution of consciousness*. New York: Prentice Hall.

Ortner, S., and H. Whitehead, eds. 1981. *Sexual meanings*. Cambridge: Cambridge Univ. Press.

Otto, R. 1969. *The idea of the holy*. New York: Oxford Univ. Press.

Oyama, S. 1985. *The ontogeny of information*. Cambridge: Cambridge Univ. Press.

Paehlke, R. 1989. *Environmentalism and the future of progressive politics*. New Haven: Yale Univ. Press.

Paine, T. 1990. *Rights of man*. Buffalo, N.Y.: Prometheus Books.

Palmer, H. 1988. *The enneagram*. San Francisco: Harper.

Parfit, D. 1984. *Reasons and persons*. Oxford: Oxford Univ. Press.

Parkin, S. 1989. *Green parties*. London: Heretic Books.

Parsons, T. 1951. *The social system*. New York: Free Press.

Parsons, T. 1966. *Societies*. Engelwood Cliffs, N.J.: Prentice-Hall.

Passmore, J. 1980. *Man's responsibility for nature*. London: Duckworth.

Peacocke, A. 1986. *God and the new biology*. London: J. M. Dent & Sons.

Peacocke, A., ed. 1981. *The sciences and theology in the twentieth century*. Stocksfield: Oriel/Routledge.

Pearce, J. C. 1992. *Evolution's end*. San Francisco: Harper.

Peat, F. 1991. *The philosopher's stone*. New York: Bantam.

Peirce, C. 1931-1958. *Collected papers*. 8 vols. Cambridge: Harvard Univ. Press.

Peirce, C. 1955. *Philosophical writings of Peirce*. Ed. J. Buchler. New York: Dover.

Penrose, R. 1990. *The emperor's new mind*. New York: Oxford Univ. Press.

Perls, F.; R. Hefferli; and P. Goodman. 1951. *Gestalt therapy*. New York: Delta.

Peters, T., ed. 1989. *Cosmos as creation*. Nashville: Abingdon Press.

Phillips, A. 1988. *Winnicott*. Cambridge: Harvard Univ. Press.

Piaget, J. 1977. *The essential Piaget*. Eds. H. Gruber and J. Voneche. New York: Basic Books.

Plant, J., ed. 1989. *Healing the wounds*. Philadelphia: New Society.

Plaskow, J., and C. Christ, eds. 1989. *Weaving the visions*. San Francisco: Harper.

Plato. 1973. *Phaedrus and letters VII & VIII*. Trans. W. Hamilton. New York: Penguin.

Polkinghorne, D. 1983. *Methodology for the human sciences*. Albany: SUNY Press.

Polkinghorne, J. 1989. *Science and providence*. Boston: Shambhala.

Popper, K. 1974. *Objective knowledge*. Oxford: Clarendon.

Popper, K., and J. Eccles. 1983. *The self and its brain*. London: Routledge.

Porritt, J., and D. Winner. 1988. *The coming of the greens*. London: Fontana.

Post, S. C., ed. 1972. *Moral values and the superego concept in psychoanalysis*. New York: International Univ. Press.

Postal, P. 1974. *On raising*. Cambridge: MIT Press.

Priest, S. 1991. *Theories of the mind*. Boston: Houghton Mifflin.

Prigogine, I. 1980. *From being to becoming*. San Francisco: Freeman.

Prigogine, I., and I. Stengers. 1984. *Order out of chaos*. New York: Bantam.

Putnam, H. 1987. *The faces of realism*. LaSalle, Ill.: Open Court.

———. 1988. *Representation and reality*. Cambridge: MIT Press.

Pylyshyn, Z. 1984. *Computation and cognition*. Cambridge: MIT Press.

Quine, W. 1985. *The time of my life*. Cambridge: MIT Press.

Rabinow, P., ed. 1984. *The Foucault reader*. New York: Pantheon.

Radha, S. 1992. *From the mating dance to the cosmic dance*. Palo Alto: Timeless Books.

Ram Dass and P. Gorman. 1985. *How can I help?* New York: Knopf.

Ramana Maharshi. 1984. *Talks with Sri Ramana Maharshi*. Tiruvannamalai: Sri Ramanasramam.

———. 1972. *The collected works*. London: Rider.

Rangdrol, T. 1989. *Lamp of Mahamudra*. Trans. Erik Pema Kunsang. Boston: Shambhala.

———. 1990. *The circle of the sun*. Kathmandu: Rangjung Yeshe Publications.

Rangjung Dorje. 1992. *Song of Karmapa*. Kathmandu: Rangjung Yeshe Publications.

Regan, T., ed. 1986. *Matters of life and death*. New York: Random House.

Reich, W. 1971. *The function of the orgasm*. New York: World Publishing.

Reiter, R., ed. 1975. *Toward an anthropology of women*. New York: Monthly Review Press.

Reynolds, B. 1994. A consideration of the transpersonal models of Stanislav Grof and Ken Wilber. Unpublished manuscript.

Ricoeur, P. 1978. *The philosophy of Paul Ricoeur*. Boston: Beacon.

———. 1981. *Hermeneutics and the human sciences*. Cambridge: Cambridge Univ. Press.

Rifkin, J. 1984. *Algeny*. New York: Penguin.

———. 1991. *Biosphere politics*. San Francisco: Harper.

Ring, K. 1980. *Life at death*. New York: Coward McCann & Geoghegan.

———. 1984. *Heading toward omega*. New York: Morrow.

Rizzuto, A. 1979. *The birth of the living God*. Chicago: Univ. Chicago Press.

Roberts, B. 1989. *What is self?* Austin: Goens.

Robinson, H. 1982. *Matter and sense*. Cambridge: Cambridge Univ. Press.

Rodgers-Rose, L. F., ed. 1980. *The black woman*. Beverly Hills: Sage.

Rogers, C. 1961. *On becoming a person*. Boston: Houghton Mifflin.

Rolston, H. 1987. *Science and religion.* New York: Random House.

———. 1988. *Environmental ethics.* Philadelphia: Temple Univ. Press.

Roos, P. 1985. *Gender and work.* Albany: SUNY Press.

Rorty, R. 1979. *Philosophy and the mirror of nature.* Princeton: Princeton Univ. Press.

———. 1982. *Consequences of pragmatism.* Minneapolis: Univ. of Minnesota Press.

Rosaldo, M., and L. Lamphere, eds. 1974. *Woman, culture, and society.* Stanford: Stanford Univ. Press.

Rosen, H. 1985. *Piagetian dimensions of clinical relevance.* New York: Columbia Univ. Press.

Rosenthal, D., ed. 1991. *The nature of mind.* New York: Oxford Univ. Press.

Rosewater, L., and L. Walker, eds. 1985. *Handbook of feminist therapy.* New York: Springer.

Rossi, A. 1977. A biosocial perspective on parenting. *Daedalus,* 106 (2): 1-31.

Rossman, N. 1991. *Consciousness.* Albany: SUNY Press.

Roszak, T. 1992. *The voice of the earth.* New York: Touchstone.

Rothberg, D. 1986. Philosophical foundations of transpersonal psychology. *Journal of Transpersonal Psych.* 18: 1-34.

Rothberg, D. 1986. Rationality and religion in Habermas' recent work. *Philosophy and Social Criticism* 11: 221-43.

———. 1990. Contemporary epistemology and the study of mysticism. In R. Forman (ed.), *The problem of pure consciousness.* New York: Oxford Univ. Press.

———. 1992. Buddhist nonviolence. *Journal of Humanistic Psychology* 32 (4): 41-75.

———. 1993. The crisis of modernity and the emergence of socially engaged spirituality. *ReVision* 15 (3): 105-15.

Rothenberg, P., ed. 1988. *Racism and sexism.* New York: St. Martin's Press.

Rothschild, J., ed. 1989. *Machina ex dea.* New York: Pergamon.

Rousseau, J. 1983. *The essential Rousseau.* New York: Meridian.

Rowan, J. 1990. *Subpersonalities.* London: Routledge.

———. 1993. *The transpersonal.* London: Routledge.

Rowbotham, S. 1974. *Women, resistance and revolution.* New York: Vintage.

Rubin, G. 1975. The traffic in women. In Reiter 1975: 157-210.

Ruether, R. 1983. *Sexism and God-talk.* Boston: Beacon.

Russell, P. 1992. *The white hole in time.* San Francisco: Harper.

Rutter, M., and M. Rutter. 1993. *Developing minds.* New York: Basic Books.

Sabom, M. 1982. *Recollections at death.* New York: Harper & Row.

Sadawii, N. 1980. *The hidden face of Eve: Women in the Arab world.* London: ZED Press.

Sade, M. de. 1965. *Justine, Philosophy in the bedroom, and other writings.* New York: Grove Weidenfeld.

Sadock, J. 1991. *Autolexical syntax.* Chicago: Univ. of Chicago Press.

Sahlins, M. 1972. *Stone age economics.* New York: Aldine.

Sanday, P. 1981. *Female power and male dominance.* Cambridge: Cambridge Univ. Press.

Sanday, P., and R. Goodenough, eds. 1989. *Beyond the second sex.* Philadelphia: Univ. of Penn. Press.

Sandel, M. 1982. *Liberalism and the limits of justice.* Cambridge: Cambridge Univ. Press.

Sandel, M., ed. 1984. *Liberalism and its critics.* New York: New York Univ. Press.

Sannella, L. 1976. *Kundalini: Psychosis or transcendence?* San Francisco: Dakin.

Sarup, M. 1989. *An introductory guide to post-structuralism and postmodernism.* Athens: Univ. of Georgia Press.

Saussure, F. 1966 (1915). *Course in general linguistics.* New York: McGraw-Hill.

Sayers, J. 1982. *Biological politics.* New York: Tavistock.

Scarce, R. 1990. *Eco-warriors.* Chicago: Noble.

Schafer, R. 1976. *A new language for psychoanalysis.* New Haven: Yale Univ. Press.

Scharfstein, B., ed. 1978. *Philosophy East/Philosophy West.* London: Basil Blackwell.

Schaya, L. 1973. *The universal meaning of the Kabbalah.* Baltimore: Penguin.

Scheffer, V. 1991. *The shaping of environmentalism in America.* Seattle: Univ. of Washington Press.

Schelling, F. 1978 (1800). *System of transcendental idealism.* Trans. P. Heath. Charlottesville: Univ. Press of Virginia.

Schmitt, C., and Q. Skinner, eds. 1988. *The Cambridge history of Renaissance philosophy.* Cambridge: Cambridge Univ. Press.

Schneider, S. 1989. *Global warming.* New York: Vintage.

Schneiderman, S. 1983. *Jacques Lacan.* Cambridge: Harvard Univ. Press.

Schopenhauer, A. 1969. *The world as will and representation.* 2 vols. New York: Dover.

Schumacher, E. F. 1977. *A guide for the perplexed.* New York: Harper & Row.

Schuon, F. 1975. *Logic and transcendence.* New York: Harper.

———. 1976. *The transcendent unity of religions.* New York: Harper & Row.

———. 1986. *The essential writings of Fritjof Schuon.* Ed. S. Nasr. Shaftesbury: Element.

Searle, J. 1969. *Speech acts.* Cambridge Univ. Press.

———. 1983. *Intentionality.* Cambridge: Cambridge Univ. Press.

———. 1984. *Minds, brains, and science.* Cambridge: Harvard Univ. Press.

———. 1992. *The rediscovery of the mind*. Cambridge: MIT Press.

Searles, H. 1960. *The nonhuman environment in normal development and in schizophrenia*. New York: IUP.

Segal, H. 1976. *Introduction to the work of Melanie Klein*. London: Hogarth.

Selman, R., and D. Byrne. 1974. A structural analysis of levels of role-taking in middle childhood. *Child Development* 45.

Shapiro, D., and R. Walsh, eds. 1984. *Meditation: Classic and contemporary perspectives*. New York: Aldine.

Sheldrake, R. 1981. *A new science of life*. Los Angeles: Tarcher.

———. 1989. *The presence of the past: Morphic resonance and the habits of nature*. New York: Viking.

———. 1990. *The rebirth of nature*. London: Century.

Shepherd, L. 1993. *Lifting the veil: The feminine face of science*. Boston: Shambhala.

Shibayama, Z. 1974. *Zen comments on the Mumonkan*. New York: Harper & Row.

Shoemaker, S., and R. Swinburne. 1984. *Personal identity*. London: Basil Blackwell.

Silburn, L. 1988. *Kundalini*. Albany: SUNY Press.

Singer, P. 1977. *Animal rights*. New York: Avon.

Singh, K. 1975. *Surat shabd yoga*. Berkeley: Images.

Sivard, R. 1985. *Women: A world survey*. Washington, D.C.: World Priorities.

Sjoo, M., and B. Mor. 1987. *The great cosmic mother*. San Francisco: Harper.

Skolimowski, H. 1981. *Eco-philosophy*. Salem, N.H.: Marion Boyars.

Slater, P. 1991. *A dream deferred*. Boston: Beacon.

Smith, A. 1990. *Wealth of nations*. Buffalo, N.Y.: Prometheus Books.

Smith, H. 1976. *Forgotten truth*. New York: Harper & Row.

———. 1989. *Beyond the postmodern mind*. Wheaton, Ill.: Quest.

———. 1991. *The world's religions*. San Francisco: Harper.

Smith, P., and O. Jones. 1986. *The philosophy of mind*. Cambridge: Cambridge Univ. Press.

Smuts, J. 1926. *Holism and evolution*. London: Macmillan.

Sogyal Rinpoche. 1990. *Dzogchen and Padmasambhava*. California: Rigpa Fellowship.

Sorokin, P. 1937-1941. *Social and cultural dynamics*. 4 vols. New York: American Book Co.

Spence, J., and R. Helmreich. 1978. *Masculinity and feminity*. Austin: Univ. of Texas Press.

Spengler, O. 1939. *The decline of the West*. New York: Knopf.

Sperry, R. 1983a. *Science and moral priority*. New York: Columbia Univ. Press.

———. 1983b. *Science and moral priority*. New York: Columbia Univ. Press.

Spinoza, B. 1985. *The collected works of Spinoza*. Princeton: Princeton Univ. Press.

Spiro, M., ed. 1965. *Context and meaning in cultural anthropology*. New York: Free Press.

Spitz, R. 1959. *A genetic field theory of ego formation*. New York: IUP.

———. 1965. *The first year of life*. New York: IUP.

Spragens, T. 1990. *Reason and democracy*. Durham: Duke Univ. Press.

Spretnak, C. 1991. *States of grace*. San Francisco: Harper.

———, ed.. 1982. *The politics of women's spirituality*. New York: Anchor.

Spretnak, C., and F. Capra. 1985. *Green politics*. London: Paladin.

Stace, W. 1987. *Mysticism and philosophy*. Los Angeles: Tarcher.

Stavenhagen, R. 1975. *Social classes in agrarian societies*. Garden City, N.Y.: Doubleday.

Steinberg, D., ed. 1992. *The erotic impulse*. Los Angeles: Tarcher.

Stephanou, E. 1976. *Charisma and gnosis in Orthodox thought*. Fort Wayne, Ind.: LMOR.

Stern, D. 1985. *The interpersonal world of the infant*. New York: Basic.

Stevens, A. 1983. *Archetypes*. New York: Quill.

Stich, S. 1983. *From folk psychology to cognitive science*. Cambridge: MIT Press.

Stone, C. 1987. *Earth and other ethics*. New York: Harper & Row.

Stone, M. 1976. *When God was a woman*. New York: Harvest.

———. 1980. *The borderline syndromes*. New York: McGraw-Hill.

Strauss, L. 1989. *Liberalism ancient and modern*. Ithaca, N.Y.: Cornell Univ. Press.

Stuard, S., ed. 1976. *Women in medieval society*. Philadelphia: Univ. of Penn. Press.

Sturrock, J. 1979. *Structuralism and since*. Oxford: Oxford Univ. Press.

Sullivan, H. 1953. *The interpersonal theory of psychiatry*. New York: Norton.

Suzuki, D. T. 1959. *Zen and Japanese culture*. Princeton: Princeton Univ. Press.

———. 1968. *Studies in the Lankavatara Sutra*. London: Routledge.

———. 1970. *Essays in Zen Buddhism*. 3 vols. London: Rider.

Swimme, B., and T. Berry. 1992. *The universe story*. San Francisco: Harper.

Szarmach, P., ed. 1984. *Introduction to the medieval mystics of Europe*. Albany: SUNY Press.

Taimni, I. 1975. *The science of yoga*. Wheaton, Ill.: Quest Books.

Takakusu, J. 1956. *The essentials of Buddhist philosophy*. Honolulu: Univ. Hawaii Press.

Tannahill, R. 1992. *Sex in history*. Scarborough House.

Tannen, D. 1990. *You just don't understand*. New York: Morrow.

Tarnas, R. 1991. *The passion of the Western mind*. New York: Harmony.

Tarnas, R. 1993. The Western mind at the threshold. *Quest.* 6 2 (Summer): 25-31.

Tart, C. T. 1975. *States of consciousness.* New York: Dutton.

———. 1986. *Waking up.* Boston: Shambhala.

———. 1989. *Open mind, discriminating mind.* San Francisco: Harper.

———, ed. 1992. *Transpersonal psychologies.* New York: Harper Collins.

Taylor, C. 1975. *Hegel.* Cambridge: Harvard Univ. Press.

———. 1985. *Philosophy and the human sciences—philosophical papers 2.* Cambridge: Cambridge Univ. Press.

———. 1989. *Sources of the self.* Cambridge: Harvard Univ. Press.

Taylor, P. 1986. *Respect for nature.* Princeton: Princeton Univ. Press.

Teilhard de Chardin, P. 1961. *The phenomenon of man.* New York: Harper Torchbooks.

———. 1964. *The future of man.* New York: Harper Torchbooks.

Teresa of Ávila. 1961. *Interior castle.* Trans. E. Peers. Garden City, N.Y.: Image.

Thompson, W. I. 1989. *Imaginary landscapes.* New York: St. Martin's Press.

———, ed. 1987. *Gaia: A way of knowing.* Hudson, N.Y.: Lindisfarne Press.

———, ed. 1991. *Gaia 2: Emergence.* Hudson, N.Y.: Lindisfarne Press.

Thondrup, T. 1989. *Buddha mind.* Ithaca, N.Y.: Snow Lion.

Thornton, M. 1989. *Folk psychology.* Toronto: Univ. of Toronto Press.

Thurman, R. 1984. *Tsong Khapa's speech of gold in the essence of true eloquence.* Princeton: Princeton Univ. Press.

Tillich, P. 1967. *A history of Christian thought.* New York: Simon & Schuster.

Tocqueville, A. de. 1969. *Democracy in America.* New York: Anchor.

Tolstoy, L. 1929. *My confession, my religion, the gospel in brief.* New York: Scribners.

Tong, R. 1989. *Feminist thought.* San Francisco: Westview.

Torrance, T. 1981. *Divine and contingent order.* Oxford: Oxford Univ. Press.

———. 1989. *The Christian frame of mind.* Colorado Springs: Helmers & Howard.

Toulmin, S. 1982. *The return to cosmology.* California: Univ. of California Press.

Toynbee, A. 1972. *A study of history.* Oxford: Oxford Univ. Press.

Trungpa, C. 1976. *The myth of freedom.* Boston: Shambhala.

———. 1981. *Glimpses of Abhidharma.* Boulder: Prajna Press.

———. 1988. *Shambhala: The sacred path of the warrior.* Boston: Shambhala.

———. 1991. *Crazy wisdom.* Boston: Shambhala.

Turkle, S. 1984. *The second self.* New York: Simon & Schuster.

Ulanov, A. 1971. *The feminine.* Evanston, Ill.: Northwestern Univ. Press.

Underhill, E. 1955. *Mysticism.* New York: Meridian.

Unger, R. 1975. *Hölderlin's major poetry: The dialectics of unity.* Bloomington: Indiana Univ. Press.

Unno, T., ed. 1989. *The religious philosophy of Nishitani Keiji*. Berkeley: Asian Humanities Press.

Urgyen, Tulku. 1988. *Vajra heart*. Kathmandu: Rangjung Yeshe Publications.

Vaillant, G. 1977. *Adaptation to life*. Boston: Little, Brown.

Varela, F. 1979. *Principles of biological autonomy*. New York: North Holland.

Varela, F.; E. Thompson; and E. Rosch. 1993. *The embodied mind*. Cambridge: MIT Press.

Varenne, J. 1976. *Yoga and the Hindu tradition*. Chicago: Univ. Chicago Press.

Vattimo, G. 1989. *The end of modernity*. Baltimore: Johns Hopkins.

Vaughan, F. 1979. *Awakening intuition*. New York: Doubleday.

———. 1986. *The inward arc*. Boston: Shambhala.

Venkatesananda, trans. 1981. *The supreme yoga*. Australia: Chiltern.

Verman, M. 1992. *The books of contemplation*. Albany: SUNY Press.

Vicinus, M., ed. 1972. *Suffer and be still: Women in the Victorian age*. Bloomington: Indiana Univ. Press.

Waldrop, M. 1992. *Complexity*. New York: Simon & Schuster.

Walsh, M., ed. 1987. *The psychology of women*. New Haven: Yale Univ. Press.

Walsh, R. 1977. Initial meditative experiences: I. *J. Transpersonal Psychology* 9: 151-192.

———. 1978. Initial meditative experiences: II. *J. Transpersonal Psychology* 10: 1-28.

———. 1980. Meditation. In R. Corsini (ed.), *A handbook of innovative psychotherapies*. New York: Wiley.

———. 1984. *Staying alive: The psychology of human survival*. Boston: Shambhala.

———. 1989. Can Western philosophers understand Asian philosophies? *Crosscurrents* 34: 281-99.

———. 1990. *The spirit of shamanism*. Los Angeles: Tarcher.

Walsh, R., and F. Vaughan, eds. 1993. *Paths beyond ego*. Los Angeles: Tarcher.

Warren, K. 1987. Feminism and ecology. *Environmental Ethics* 9: 3-20.

Washburn, M. 1988. *The ego and the dynamic ground*. Albany: SUNY Press.

Watts, A. 1968. *Myth and ritual in Christianity*. Boston: Beacon.

———. 1972. *The supreme identity*. New York: Vintage.

———. 1975. *Tao: The watercourse way*. New York: Pantheon.

Weber, M. 1963. *The sociology of religion*. Boston: Beacon.

Weil, A. 1972. *The natural mind*. Boston: Houghton Mifflin.

Weinberg, S. 1992. *Dreams of a final theory*. New York: Pantheon.

Welwood, J. 1990. *Journey of the heart*. New York: Harper Collins.

Werner, H. 1964 (1940). *Comparative psychology of mental development*. New York: IUP.

West, C. 1989. *The American evasion of philosophy*. Madison: Univ. of Wisconsin Press.

West, M., ed. 1987. *The psychology of meditation*. Oxford: Clarendon.

White, J. 1979. *Kundalini, evolution,* and *enlightenment*. New York: Anchor.

White, L. 1973. The historical roots of our ecologic crisis. In Barbour 1973: 18-30.

Whitehead, A. 1957 (1929). *Process and reality*. New York: Macmillan.

———. 1966. *Modes of thought*. New York: Macmillan.

———. 1967. *Adventures of ideas*. New York: Macmillan.

———. 1967. *Science and the modern world*. New York: Macmillan.

Whitmont, E. 1982. *Return of the goddess*. New York: Crossroad.

Whorf, B. 1956. *Language, thought and reality*. Cambridge: MIT Press.

Whyte, L. L. 1950. *The next development in man*. New York: Mentor.

———. 1954. *Accent on form*. New York: Harpers.

Wilber, K. 1977. *The spectrum of consciousness*. Wheaton, Ill.: Quest.

———. 1980. *The Atman project: A transpersonal view of human development*. Wheaton, Ill.: Quest.

———. 1981. *Up from Eden*. New York: Doubleday/Anchor.

———. 1986. *A sociable god*. Boston: Shambhala.

———. 1989. *Eye to eye*. Boston: Shambhala.

———. 1991. *Grace and grit*. Boston: Shambhala.

Wilber, K.; J. Engler; and D. P. Brown. 1986. *Transformations of consciousness: Conventional and contemplative perspectives on development*. Boston: Shambhala.

Willis, J., ed. 1987. *Feminine ground*. Ithaca, N.Y.: Snow Lion.

Winkelman, M. 1990. The evolution of consciousness. *Anthropology of Consciousness*, 1 (3/4): 24-32.

———. 1993. The evolution of consciousness? *Anthropology of Consciousness*, 4 (3): 3-10.

Winnicott, D. 1958. *Collected papers*. New York: Basic Books.

Winograd, T., and F. Flores 1986. *Understanding computers and cognition*. Norwood, N.J.: Ablex.

Wit, H. 1991. *Contemplative psychology*. Pittsburgh: Duquesne Univ. Press.

Wittgenstein, L. 1953. *Philosophical investigations*. London: Basil Blackwell.

Wollstonecraft, M. 1990. *A vindication of the rights of women*. Buffalo, N.Y.: Prometheus Books.

Wolman, B., and M. Ullman, eds. 1986. *Handbook of states of consciousness*. New York: Van Nostrand Reinhold.

Woolhouse, R. 1983. *Locke*. Minneapolis: Univ. of Minnesota Press.

Wright, E. 1984. *Psychoanalytic criticism*. London: Methuen.

Wulff, D. 1991. *Psychology of religion*. New York: Wiley.

Wuthnow, R.; J. Hunter; A. Bergesen; and E. Kurzwell. 1984. *Cultural analysis.* London: Routledge.

Yalom, I. 1980. *Existential psychotherapy.* New York: Basic.

Yogeshwarand Saraswati. 1972. *Science of the soul.* India: Yoga Niketan.

Young, D. 1991. *Origins of the sacred.* New York: St. Martin's Press.

Yuasa, Y. 1987. *The body.* Albany: SUNY Press.

Ywahoo, D. 1987. *Voices of our ancestors.* Boston: Shambhala.

Zadeh, L. 1987. *Fuzy sets and applications.* New York: Wiley.

Zaehner, R. 1957. *Mysticism, sacred and profane.* New York: Oxford Univ. Press.

Zammito, J. 1992. *The genesis of Kant's critique of judgement.* Chicago: Univ. of Chicago Press.

Zanardi, C., ed. 1990. *Essential papers on the psychology of women.* New York: New York Univ. Press.

Zimmerman, M. 1981. *Eclipse of the self.* Athens: Ohio Univ. Press.

Zimmerman, M. 1990. *Heidegger's confrontation with modernity.* Bloomington: Indiana Univ. Press.

Zoeteman, K. 1991. *Gaiasophy.* Hudson, N.Y.: Lindisfarne.

Zohar, D. 1990. *The quantum self.* New York: Morrow.

Zweig, C., ed. 1990. *To be a woman.* Los Angeles: Tarcher.

Credits

The author thanks the publishers who granted permission to reprint material copyrighted or controlled by them:

From *Evolution: The Grand Synthesis* by Ervin Laszlo, © 1987 by Ervin Laszlo. Reprinted by permission of John White.

Figures 24 and 28 from *The Self-Organizing Universe: Scientific and Human Implications of the Emerging Paradigm of Evolution* by Erich Jantsch. Reprinted by permission of Elsevier Science Ltd, Pergamon Imprint, Oxford, England.

From *Communication and the Evolution of Society* by Jürgen Habermas, © 1979 by Beacon Press. Reprinted by permission of Beacon Press.

From *The Great Chain of Being: A Study of the History of an Idea* by Arthur O. Lovejoy (Cambridge, Mass.: Harvard University Press), © 1936, 1964 by the President and Fellows of Harvard College. Reprinted by permission of Harvard University Press.

From *Sources of the Self: The Making of Modern Identity* by Charles Taylor (Cambridge, Mass.: Harvard University Press), © 1989 by Charles Taylor. Reprinted by permission of Harvard University Press.

Index

transpersonal development and, 215, 322, 522–23, 640–43*n*. 23
and unconscious truth, 576–84*n*. 28
See also Great Holarchy of Being
Fowler, James, 267, 623*n*. 6
Fox, Warwick, 638–40*n*. 21
Frameworks, 34–36
Freedom. *See* Autonomy/freedom
Freud, Sigmund, 576–84*n*. 28
  concept of "id," 500
  concept of "the I" (ego), 237
  depth psychology, 572–73*n*. 19
  fixation/obsession and, 503, 564–65*n*. 78
  inability to see nondual context, 663–64*n*. 31
  influence of Schelling on, 517
  on metonym and metaphor, 617–18*n*. 12
  on mythic archeytpes, 228–29
  "pleasure principle," 503, 684*n*. 20
  pre/trans fallacy in, 212
  primary processes of, 226
  as reflection of interaction of Eco and Ego, 504–505
  separation of Eros and Thanatos, 340–41
Frobenius, Leo, 190
Fukuyama, Francis, 320–21, 653–54*n*. 62
Fulcrums of development, 611–15*n*. 3, 618–19*n*. 23. *See also* Cognitive development
Functional fit, 145, 148, 151–53, 588*n*. 41, 600–604*n*. 26. *See also* Validation
Functionalism, 576–84*n*. 28, 600–604*n*. 26

Gaia, 93–94, 411
  confusion of with God (Great Goddess), 434, 525, 692–93*n*. 57, 710–16*n*. 32, 758–60*n*. 10
  See also Biosphere; Ecodevastation; Nature
Galileo, 17, 20, 411
Gardner, Howard, 25, 26, 264–65
Gebser, Jean, 223
  critique of, 779–89*n*. 17
  on integral-aperspectival mind (vision-logic), viii, 192–96, 403
  stages of consciousness, 125
Gender differences, 159–64, 183, 395–96, 597*n*. 9. *See also* Female(s); Male(s)
General System Theory, 15
Genetic communication, 101–102
Geocentric identity, 229. *See also* Cognitive development
Gestalt psychology, 663*n*. 28

Gilligan, Carol, 5, 27–28, 320, 761*n*. 12, 767–73*n*. 17
Gimbutas, Maria, 692–93*n*. 57
Global culture; global consciousness
  development process, 190–97, 542, 610–11*n*. 48, 621*n*. 39
  global economies, 183–84
  multiculturalism/transnationalism, 204–209, 267–68
  religion and, 608–9*n*. 44
  and tribal awareness, 600*n*. 25
  See also Worldspaces
Global syncretism (cognitive anthropology), 171
Gnostics, critiques of, 351–54, 666–67*n*. 7
God, 33
  as Alpha Source (Deist view), 424
  Aristotle's view of, 359–60
  Augustine's view of, 368–69
  as basis for human connectedness, 633–34*n*. 1
  dissociation from in Western culture, 360–61, 365–66, 405, 409, 476–80
  deification of Nature/Gaia, 420, 491–92, 495, 497–98, 638–40*n*. 16, 716–17*n*. 32, 758–60*n*. 10
  determining truth of, 576–84*n*. 28, 646–51*n*. 58, 670–72*n*. 19
  nondual views of, 513, 646–51*n*. 58
  Plato's view of, 335, 359–60
  Union with, 644–45*n*. 27, 646*n*. 35
  See also Nondual traditions; Spirit
Gödel, Kurt, 529, 530
Godhead, 665*n*. 1
  Eckhart's views on, 749–52*n*. 8
  nondual views of, 646–51*n*. 58
  Plotinus' views on, 342
  transcendance into, 309–17
  See also God; Transcendence
Goethe, Johann Wolfgang von, 417
Goldsmith, Edward, 57, 766–67*n*. 16
Good, the, 332–33, 337, 356–59, 367, 375, 439. *See also* Moral development
Grace, 302–3, 663*n*. 30, 692–93*n*. 57
Gradation, grade. *See* Depth
Graef, Ortwin de, 745–49*n*. 4
Grammatology (Derrida), 573–74*n*. 24
Great Chain of Being paradigm. See Great Holarchy of Being
Great Father myth, 692–93*n*. 57
Great Goddess, 692–93*n*. 57. *See* Ecofeminism; Gaia; God